MECHANICAL COMPONENTS HANDBOOK

Other McGraw-Hill Handbooks of Interest

Baumeister • Marks' Standard Handbook for Mechanical Engineers
Bovay • Handbook of Mechanical and Electrical Systems for Buildings
Brady and Clauser • Materials Handbook
Brater and King • Handbook of Hydraulics
Chopey and Hicks • Handbook of Chemical Engineering Calculations
Dudley • Gear Handbook
Fink and Beaty • Standard Handbook for Electrical Engineers
Harris • Shock and Vibration Handbook
Hicks • Standard Handbook of Engineering Calculations
Juran • Quality Control Handbook
Karassik, Krutzsch, Fraser, and Messina • Pump Handbook
Kurtz • Handbook of Engineering Economics
Maynard • Industrial Engineering Handbook
Optical Society of America • Handbook of Optics
Pachner • Handbook of Numerical Analysis Applications
Parmley • Standard Handbook of Fastening and Joining
Peckner and Bernstein • Handbook of Stainless Steels
Perry and Green • Perry's Chemical Engineers' Handbook
Raznjevic • Handbook of Thermodynamic Tables and Charts
Rohsenow and Hartnett • Handbook of Heat Transfer
Rothbart • Mechanical Design and Systems Handbook
Schwartz • Metals Joining Manual
Seidman and Mahrous • Handbook of Electric Power Calculations
Shand and McLellan • Glass Engineering Handbook
Smeaton • Motor Application and Control Handbook
Smeaton • Switchgear and Control Handbook
Transamerica Delaval, Inc. • Transamerica Delaval Engineering Handbook
Tuma • Engineering Mathematics Handbook
Tuma • Technology Mathematics Handbook
Tuma • Handbook of Physical Calculations

MECHANICAL COMPONENTS HANDBOOK

ROBERT O. PARMLEY, P.E.
Editor in Chief

Morgan & Parmley, Ltd., Ladysmith, Wisconsin
Consulting Engineer

McGraw-Hill Book Company

New York St. Louis San Francisco Auckland Bogotá Hamburg
Johannesburg London Madrid Mexico Montreal New Delhi
Panama Paris São Paulo Singapore Sydney Tokyo Toronto

"While McGraw-Hill, the Editor, and the contributors to this Handbook have used their best efforts to ensure that the information contained in the Handbook is accurate and comprehensive, neither McGraw-Hill, the Editor nor the contributors warrants the total accuracy or completeness thereof. Further, neither McGraw-Hill, the Editor, nor the contributors shall have any liability to any third party for any damages resulting from the use, application, or adaptation of the information contained in this Handbook, whether such damages be direct or indirect, or in the nature of lost profits or consequential damages. This Handbook is published with the understanding that McGraw-Hill is not engaged in providing engineering, design, or other professional services."

Library of Congress Cataloging in Publication Data
Main entry under title:

Mechanical components handbook.

Includes index.
1. Machine parts—Handbooks, manuals, etc.
2. Machinery—Design—Handbooks, manuals, etc.
I. Parmley, Robert O.
TJ243.M43 1985 621.8 84-17147
ISBN 0-07-048514-3

1234567890 DOC/DOC 898765

ISBN 0-07-048514-3

The editors for this book were Harold B. Crawford and George F. Watson, the designer was Mark E. Safran, and the production supervisor was Teresa F. Leaden. It was set in Century Schoolbook by Bi-Comp, Incorporated.

Printed and bound by R. R. Donnelley & Sons, Inc.

To Wayne and Laura

CONTENTS

CONSULTANTS AND CONTRIBUTORS

SPECIAL CONSULTANTS

Robert C. Brown, Jr., *Manager,* Engineering Services, American Gear Manufacturers Association, Arlington, Virginia

Leonard Kirsch, *President,* Kirsch Communications, Garden City, New York

Dan S. Kling, P.E., *President,* Cooper Engineering Co., Inc., Rice Lake, Wisconsin

CONTRIBUTORS

Harry S. Brenner, P.E., *President,* Almay Research and Testing Corporation, Los Angeles, California

Joseph F. Briggs, *Manager,* Engineering Services, Aeroquip Corporation, Jackson, Michigan

William R. Edgerton, *Chief Engineer,* Whitney Chain Operations, Power Transmission Division, Dresser Industries, Inc., Hartford, Connecticut

Dr. Leslie A. Horve, P.E., *Vice-President of Technology,* Chicago Rawhide Manufacturing Company, Elgin, Illinois

Dr. Edmund Killian, P.E., *Chief Engineer,* Waldes Kohinoor, Inc., Truarc Retaining Rings Division, Long Island City, New York

Robert O. Parmley, P.E., CMfgE, *President,* Morgan & Parmley, Ltd., Ladysmith, Wisconsin

Members of the Power Transmission Design Staff, Penton/IPC, Inc., Cleveland, Ohio

Members of the Staff, PT Components, Inc., Indianapolis, Indiana

David Roos, Gates Rubber Company, Denver, Colorado

James D. Shepherd, Product Applications, Gates Rubber Company, Denver, Colorado

Members of the Staff, Spring Manufacturers Institute, Oak Brook, Illinois

Members of the Staff, Sterling Instrument, Division of Designatronic, Inc., New Hyde Park, New York

PREFACE

This handbook germinated from the research conducted during the preparation of *Standard Handbook of Fastening and Joining* and took root shortly after its publication. Therefore, *Mechanical Components Handbook* should be considered a companion handbook and technically compatible with fastening and joining technology.

The major goal of *Mechanical Components Handbook* is to categorize, define, and discuss basic mechanical components used in current mechanical technology. Its secondary purpose is to stimulate the reader's creative processes and perhaps open the gateway to new designs. For this reason, a section entitled "Innovative Design," has been included at the end of the work to help spark the creative processes.

Throughout this handbook, the reader will discover applicable tables, data, standards, and appropriate illustrations. A large portion of this information was provided by various technical institutes, societies, organizations, leading manufacturers, consulting firms, and publishers. The sources for this data have been properly noted and credit given at the appropriate places. The cooperation received from these various sources and the technical contributors has been tremendous. Their eagerness to participate in this project and their constructive suggestions have enabled us to produce a practical handbook. Each section is devoted to an individual topic. Separate presentations have been compiled by well-qualified engineers and liberally illustrated.

As editor in chief, I would like to note that it is an impossible task to include all known mechanical components in a reasonable-size handbook. The present "explosion" of technical information has overwhelmed the engineers' desks and crowded their bookshelves. As previously noted, our prime objective is to include major or basic mechanical components and to stimulate innovative designs.

McGraw-Hill has given me a relatively free hand in developing the outline, arranging the contents, selecting contributors, obtaining technical data, and finalizing the manuscript. McGraw-Hill's guidance throughout this project was extremely helpful. A board of special consultants was selected to periodically monitor the progress and ensure that the basic pattern was maintained. Their analysis and technical feedback has strengthened the handbook's presentation and preserved its original concept. A list of these consultants is included.

Special thanks go to Frank Yeaple, former editor of *Design Engineering* magazine, for his encouragement and assistance; Robert Kelley, editor of *Assembly Engineering;* and Harry Brenner, president of Almay Research and Testing, for their periodic advice. Additional thanks go to a fine young drafter, Paul Stauffer, who prepared many technical illustrations, all of superb caliber. Last, but certainly not least, a sincere thank you to Lana and Ethne for their untiring efforts in typing correspondence and the final manuscript.

Robert O. Parmley, P.E.
Editor in Chief

MECHANICAL COMPONENTS HANDBOOK

1

Gears and Gearing

ROBERT O. PARMLEY, P.E.
President/Morgan & Parmley, Ltd./Ladysmith, Wisconsin

INTRODUCTION

The material in this section was originally prepared by Dr. George Michalec, Professor of Mechanical Engineering at Stevens Institute of Technology, for Sterling Instrument, Division of Designatronics, Inc., New Hyde Park, New York, and appeared in their Master Catalog series. The editor wishes to express his thanks to Sterling Instrument for granting permission for its use in this handbook.

1-1 General Scope

This section is a technical coverage of gear fundamentals. It is intended to be a broad coverage written in a manner that is easy for any person interested in knowing how gear systems function to follow and understand. Since gearing is a specialty component, not all designers and engineers are expected to either possess or have been exposed to all aspects of this knowledge.

For those whose first encounter with gear components this is, it is suggested that this section be read in the order presented to obtain a logical development of the subject. Subsequently, and for those already familiar with gears, this material can be used item by item in random access as a design reference.

BASIC GEOMETRY OF SPUR GEARS

The spur gear is used to illustrate the fundamentals of gearing both because it is the simplest, and hence most comprehensible, type and because it is the most widely used, particularly for instruments and control systems.

1-2 Basic Spur-Gear Geometry

The basic geometry and nomenclature of a meshed spur-gear pair is shown in Fig. 1-1. The essential features of a gear-mesh design are:

1. Center distance
2. The pitch-circle diameters (or pitch diameters)
3. Size of the teeth
4. Number of teeth
5. Pressure angle of the contacting involutes

Details and definitions of these items and their interdependence are covered in subsequent sections.

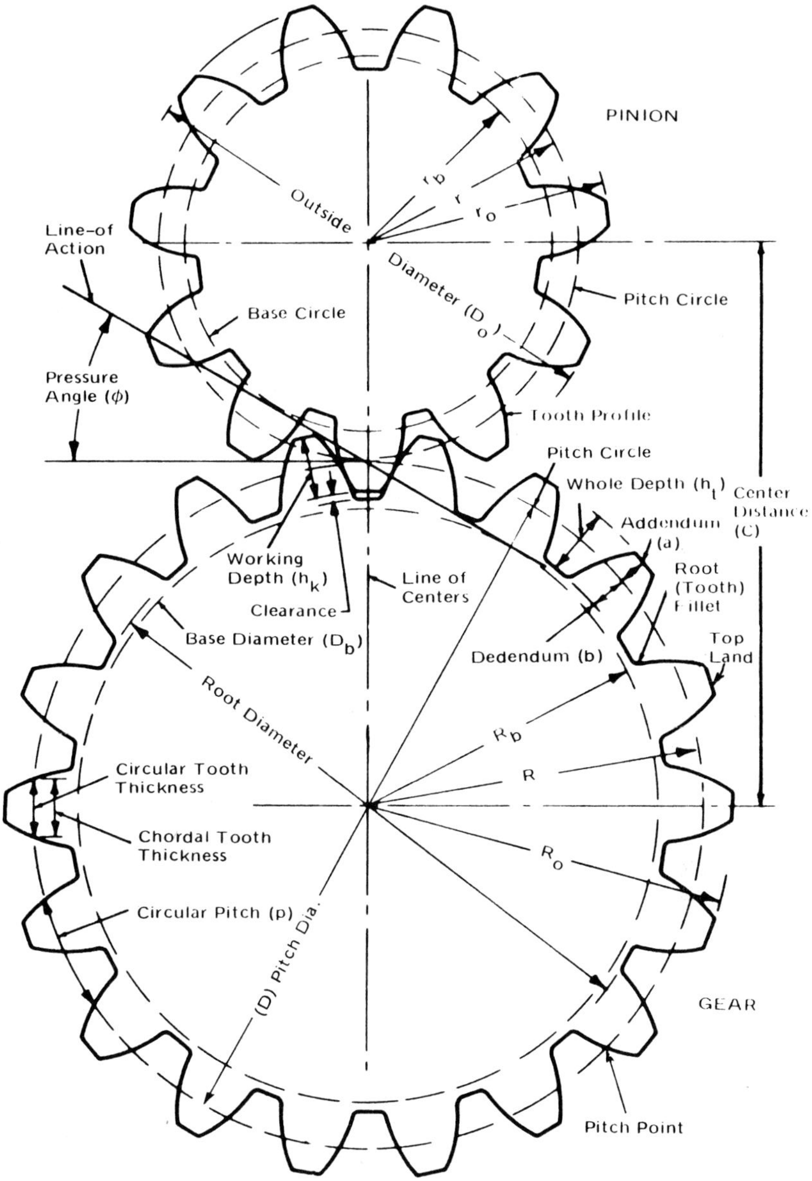

FIG. 1-1 Basic gear geometry.

1-3 The Law of Gearing

A primary requirement of gears is linearity of velocity or position transmission. Precision instruments require positioning fidelity. High-speed and/or high-power gear trains require transmission of constant angular rotation to avoid severe dynamic problems.

Linearity of motion is defined as "conjugate action" of the gear-tooth profiles. A geometric relationship can be derived from the form of the tooth profiles to provide conjugate action, which is summarized as the law of gearing:

> A common normal to the tooth profiles at their point of contact must, in all positions of the contacting teeth, pass through a fixed point on the line of centers called the *pitch point.*

Any two curves or profiles that engage each other and satisfy the law of gearing are conjugate curves.

1-4 The Involute Curve

An almost infinite number of curves can be developed to satisfy the law of gearing; many different curve forms have been tried in the past. Modern gearing (except for clock gears) is based on involute-curve teeth. The involute has three major advantages:

1. Conjugate action is independent of changes in center distance.
2. The basic rack tooth is straight sided, and therefore it is relatively simple and can be accurately made, and as a generating tool it imparts high accuracy to the cut gear tooth.
3. One cutter can generate all gear-tooth numbers of the same pitch.

The involute curve is most easily understood as the trace of a point at the end of a taut string that is unwound from a cylinder. See Fig. 1-2. The base cylinder, or circle as it is referred to in gear literature, fully describes the form of the involute; in a gear it is an inherent parameter, although it is invisible.

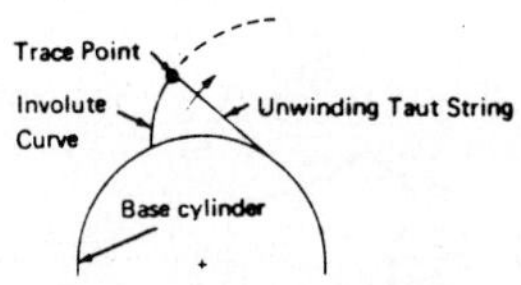

FIG. 1-2 Generation of an involute by a taut string.

The development and action of mated teeth can be evolved by imagining the taut string as being unwound from one base circle and wound on to another, as shown in Fig. 1-3*a*. Then, a single point on the string simultaneously traces an involute on each base circle; these are conjugate, since at all points of contact the common normal is the common tangent which passes through a fixed point on the line of centers. If a second winding and unwinding taut string is added around the base circles in the opposite direction, Fig. 1-3*b*, this generates oppositely

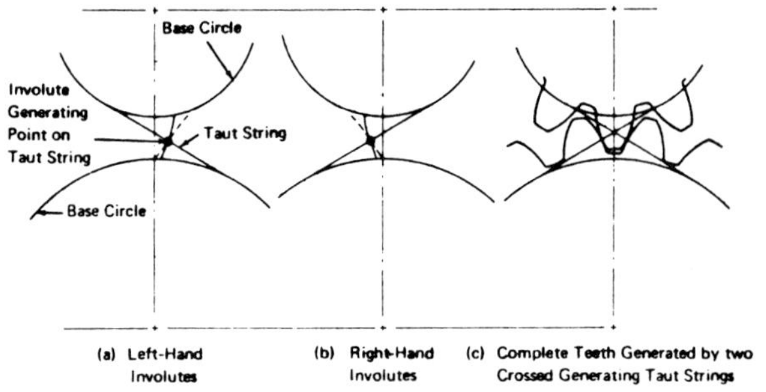

FIG. 1-3 Generation and action of gear teeth. (*a*) Left-hand involutes. (*b*) Right-hand involutes. (*c*) Complete teeth generated by two crossed taut generating strings.

curved involutes which can provide for reverse-direction drive. When the involute pairs are properly spaced, the result is the involute gear tooth, Fig. 1-3*c*.

1-5 Pitch Circles

Referring to Fig. 1-4, the tangent to the base circles is the line of contact, or line of action in gear vernacular. Where this line crosses the line of centers establishes the pitch point *P*. This in turn sets the size of the pitch circles, or, as they are commonly called, the pitch diameters. The ratio of the pitch diameters gives the velocity ratio:

$$\text{Velocity ratio} = Z = \frac{D_1}{D_2} \qquad (1\text{-}1)$$

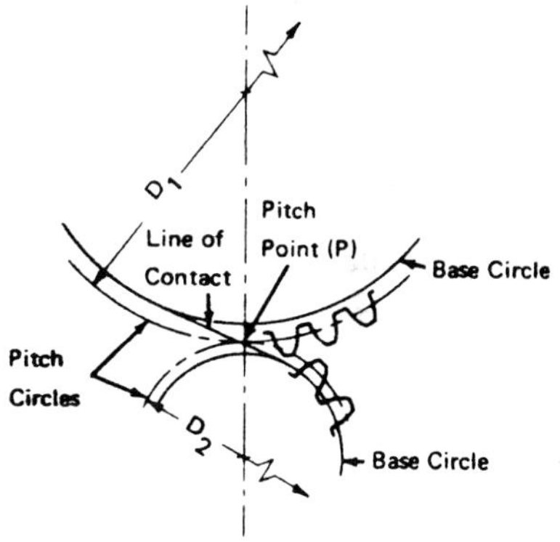

FIG. 1-4 Definition of pitch circle and pitch point.

1-6 Pitch

Essential to prescribing gear geometry is a measure of the size, or spacing, of the teeth along the pitch circle. This is termed "pitch," and there are two basic forms.

Circular Pitch This is a naturally conceived linear measure along the pitch circle of the tooth spacing. In Fig. 1-5, it is the linear distance (measured along the pitch-circle arc) between the same points of adjacent teeth. Its measure is the pitch-circle circumference divided by the number of teeth:

$$p_c = \text{circular pitch} = \frac{\text{pitch-circle circumference}}{\text{number of teeth}} = \frac{D\pi}{N} \qquad (1\text{-}2)$$

Diametral Pitch A more popularly used pitch measure, although geometrically much less evident, is one that gives the number of teeth per inch of pitch diameter. This is simply expressed as

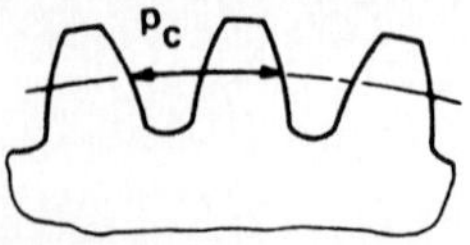

FIG. 1-5 Definition of circular pitch.

$$P_d = \text{diametral pitch} = \frac{N}{D} \tag{1-3}$$

Diametral pitch is used so commonly with fine-pitch gears that it is usually contracted to simply "pitch," with the fact that it is diametral implied.

Relation of Pitches From the geometry that defines the two pitches, it can be shown that they are related by the product expression

$$P_d \times p_c = \pi \tag{1-4}$$

This is a simple relationship to remember and permits an easy transformation from one measure to the other.

GEAR-TOOTH FORMS AND STANDARDS

Involute gear-tooth forms and standard tooth proportions are specified in terms of the basic rack, which has straight-sided teeth for involute systems. The American National Standards Institute has accepted certain standards prepared by technical committees of the American Gear Manufacturers Association (AGMA) under the approval committee development procedure and, after processing those standards for public review, has approved them as American National Standards. Although a large number of tooth proportions and pressure angle standards have been formulated, only a few are currently active and widely used. Symbols for the basic rack are given in Fig. 1-6 and pertinent standardized data pitch in Table 1-1.

Note that the data in Table 1-1 are based on diametral pitch equal to 1. To convert to another pitch, divide by diametral pitch.

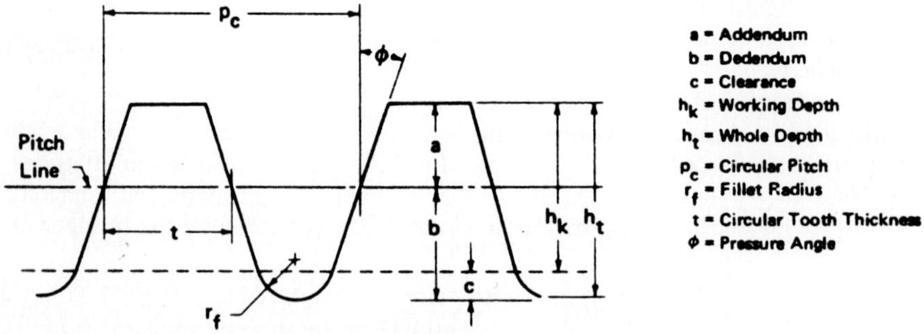

FIG. 1-6 Tooth proportions. (*From AGMA 201.02.*)

TABLE 1-1 Tooth Proportions of Basic Rack for Standard Involute Gear Systems

Tooth parameter	Symbol in rack (see Fig. 1-6)	14½° full-depth involute system	20° full-depth involute system	20° coarse-pitch involute spur gears	20° fine-pitch involute system
System sponsors	—	ANSI & AGMA	ANSI	AGMA	ANSI & AGMA
Pressure angle	ϕ	14½°	20°	20°	20°
Addendum	a	$1/P$	$1/P$	$1.000/P$	$1.000/P$
Dedundum	b	$1.157/P$	$1.157/P$	$1.250/P$	$1.200/P + 0.002$ in
Whole depth	h_t	$2.157/P$	$2.157/P$	$2.250/P$	$2.200/P + 0.002$ in
Working depth	h_k	$2/P$	$2/P$	$2.000/P$	$2.000/P$
Clearance	c	$0.157/P$	$0.157/P$	$0.250/P$	$0.200/P + 0.002$ in
Basic circular tooth thickness on pitch line	t	$1.5708/P$	$1.5708/P$	$\pi/2P$	$1.5708/P$
Fillet radius in basic rack	r_f	1⅓ × clearance	1½ × clearance	$0.300/P$	Not standardized
Diametral pitch range	—	Not specified	Not specified	19.99 and coarser	20 and finer
Governing standard:					
ANSI	—	B6.1	B6.1	—	B6.7
AGMA	—	201.02	—	201.02	207

TABLE 1-2 Preferred Diametral Pitches

Class	Pitch	Class	Pitch	Class	Pitch	Class	Pitch
	½		12		20		150
	1	Medium coarse	14		24	Ultra fine	180
Coarse	2		16	Fine	32		200
	4		18		48		
	6				64		
	8				72		
	10				80		
					96		
					120		
					128		

1-7 Preferred Pitches

Although there are no standards regarding pitch, a preference has developed among gear producers and users. This is given in Table 1-2. Adherence to these pitches is very common in the fine-pitch range, but less so among the coarse pitches.

TABLE 1-3*a* Basic Gear Data—AGMA: Nomenclature List

Number of teeth	$N = 30$ (external)	
Diametral pitch	$P_d = 6$	
Pressure angle	$\phi = 20°$	
Circular tooth thickness	$t = 0.2600$ in	
Pin diameter	$d = 0.2810$ in	
Pitch diameter	$D = \dfrac{N}{P_d}$	5.000 in
Circular pitch	$p_c = 0.5235988$ in	
Space width	$t_s = p - t$	0.2635988 in
From table of cosines	$\cos\phi = 0.9396926$	
From table of involute functions	$\text{inv}\,\phi = 0.0149044$	
Increase of involute of pressure angle	$\Delta\,\text{inv}\,\phi = \dfrac{(d/\cos\phi) - t_s}{D}$	0.0070870
Involute of pressure angle at center of pin	$\text{inv}\,\phi_m = \text{inv}\,\phi + \Delta\,\text{inv}\,\phi$	0.0219914
Pressure angle at center of pin	$\phi_m =$ (from table of involute functions)	22.658026°
Diameter at center of pins	$D_m = \dfrac{\cos\phi}{\cos\phi_m}$	5.0914 in
Diameter over pins	$D_M = D_m + d$	5.3724 in

Source: AGMA Standard 231.52.

TABLE 1-3*b* Basic Data—AGMA: Pin Measurement Table for Involute Spur Gears
(1 Diametral Pitch; 1.728-in Pin Diameter)

Number of teeth N	Dimension over pins D_M, in	Thickness factor K_m	Number of teeth N	Dimension over pins D_M, in	Thickness factor K_m	Number of teeth N	Dimension over pins D_M, in	Thickness factor K_m
10	12.3445	2.01	51	53.4053	2.50	91	93.4303	2.60
11	13.2332	2.05	52	54.4304	2.51	92	94.4441	2.60
12	14.3579	2.09	53	55.4074	2.51	93	95.4310	2.60
13	15.2639	2.12	54	56.4315	2.52	94	96.4445	2.60
14	16.3683	2.14	55	57.4093	2.52	95	97.4317	2.60
15	17.2871	2.17						
16	18.3768	2.19	56	58.4325	2.52	96	98.4449	2.61
17	19.3053	2.21	57	59.4111	2.53	97	99.4323	2.61
18	20.3840	2.23	58	60.4335	2.53	98	100.4453	2.61
19	21.3200	2.25	59	61.4128	2.53	99	101.4329	2.61
20	22.3900	2.26	60	62.4344	2.53	100	102.4456	2.61
21	23.3321	2.28	61	63.4144	2.54	101	103.4335	2.61
22	24.3952	2.29	62	64.4352	2.54	102	104.4460	2.61
23	25.3423	2.30	63	65.4159	2.54	103	105.4341	2.61
24	26.3997	2.32	64	66.4361	2.55	104	106.4463	2.62
25	27.3511	2.33	65	67.4173	2.55	105	107.4346	2.62
26	28.4036	2.34	66	68.4369	2.55	106	108.4466	2.62
27	29.3586	2.35	67	69.4186	2.55	107	109.4352	2.62
28	30.4071	2.36	68	70.4376	2.56	108	110.4469	2.62
29	31.3652	2.37	69	71.4198	2.56	109	111.4357	2.62
30	32.4102	2.38	70	72.4383	2.56	110	112.4472	2.62
31	33.3710	2.39	71	73.4210	2.56	111	113.4362	2.62
32	34.4130	2.40	72	74.4390	2.57	112	114.4475	2.62
33	35.3761	2.41	73	75.4221	2.57	113	115.4367	2.63

TABLE 1-3*b* Basic Data—AGMA: Pin Measurement Table for Involute Spur Gears (continued)
(1 Diametral Pitch; 1.728-in Pin Diameter)

Number of teeth N	Dimension over pins D_M, in	Thickness factor K_m	Number of teeth N	Dimension over pins D_M, in	Thickness factor K_m	Number of teeth N	Dimension over pins D_M, in	Thickness factor K_m
34	36.4155	2.41	74	76.4396	2.57	114	116.4478	2.63
35	37.3807	2.42	75	77.4232	2.57	115	117.4372	2.63
36	38.4178	2.43	76	78.4402	2.57	116	118.4481	2.63
37	39.3849	2.43	77	79.4242	2.58	117	119.4376	2.63
38	40.4198	2.44	78	80.4408	2.58	118	120.4484	2.63
39	41.3886	2.45	79	81.4252	2.58	119	121.4380	2.63
40	42.4217	2.45	80	82.4413	2.58	120	122.4486	2.63
41	43.3920	2.46	81	83.4262	2.58	121	123.4384	2.63
42	44.4234	2.46	82	84.4418	2.58	122	124.4489	2.63
43	45.3951	2.47	83	85.4271	2.59	123	125.4388	2.63
44	46.4250	2.47	84	86.4423	2.59	124	126.4491	2.64
45	47.3980	2.48	85	87.4279	2.59	125	127.4392	2.64
46	48.4265	2.48	86	88.4428	2.59	126	128.4493	2.64
47	49.4007	2.49	87	89.4287	2.59	127	129.4396	2.64
48	50.4279	2.49	88	90.4433	2.59	128	130.4496	2.64
49	51.4031	2.50	89	91.4295	2.60	129	131.4400	2.64
50	52.4292	2.50	90	92.4437	2.60	130	132.4498	2.64
131	133.4404	2.64	156	158.4521	2.66	181	183.4469	2.67
132	134.4500	2.64	157	159.4443	2.66	182	184.4538	2.67
133	135.4408	2.64	158	160.4523	2.66	183	185.4471	2.67
134	136.4502	2.64	159	161.4445	2.66	184	186.4539	2.67
135	137.4411	2.64	160	162.4524	2.66	185	187.4473	2.67
136	138.4504	2.64	161	163.4448	2.66	186	188.4540	2.67
137	139.4414	2.64	162	164.4526	2.66	187	189.4474	2.67

138	140.4506	2.65	163	165.4450	2.66	188	190.4541	2.67
139	141.4418	2.65	164	166.4527	2.66	189	191.4476	2.67
140	142.4508	2.65	165	167.4453	2.66	190	192.4542	2.67
141	143.4421	2.65	166	168.4528	2.66	191	193.4478	2.67
142	144.4510	2.65	167	169.4455	2.66	192	194.4543	2.67
143	145.4424	2.65	168	170.4529	2.66	193	195.4480	2.67
144	146.4512	2.65	169	171.4457	2.66	194	196.4544	2.67
145	147.4427	2.65	170	172.4531	2.66	195	197.4482	2.67
146	148.4513	2.65	171	173.4459	2.66	196	198.4546	2.67
147	149.4430	2.65	172	174.4532	2.66	197	199.4483	2.67
148	150.4515	2.65	173	175.4461	2.66	198	200.4547	2.67
149	151.4433	2.65	174	176.4533	2.67	199	201.4485	2.68
150	152.4516	2.65	175	177.4463	2.67	200	202.4548	2.68
151	153.4435	2.65	176	178.4535	2.67	201	203.4487	2.68
152	154.4518	2.65	177	179.4465	2.67	300	302.4579	2.70
153	155.4438	2.65	178	180.4536	2.67	301	303.4538	2.70
154	156.4520	2.66	179	181.4467	2.67	400	402.4596	2.71
155	157.4440	2.66	180	182.4537	2.67	401	403.4565	2.71
						500	502.4606	2.72
						501	503.4581	2.72
						∞	$(N + 2)$.4646	2.75

Source: AGMA Standard 231.52.

TABLE 1-3c Basic Gear Data—AGMA: Pin Diameters for Various Pitches

Diametral pitch P	For standard external gears $d = \frac{1.728 \text{ in}}{P}$	For standard internal gears $d = \frac{1.680 \text{ in}}{P}$	For long-addendum pinions $d = \frac{1.920 \text{ in}}{P}$
1	1.7280	1.680	1.920
1½	1.1520	1.120	1.280
2	0.864	0.840	0.960
2½	0.69120	0.6720	0.7680
3	0.5760	0.560	0.640
4	0.4320	0.420	0.480
5	0.34560	0.3360	0.3840
6	0.2880	0.280	0.320
7	0.24686	0.240	0.27428
8	0.2160	0.210	0.240
9	0.1920	0.18666	0.21333
10	0.17280	0.1680	0.1920
11	0.15709	0.15273	0.17454
12	0.1440	0.140	0.160
14	0.12343	0.120	0.13714
16	0.1080	0.1050	0.120
18	0.0960	0.09333	0.10667

Source: AGMA Standard 231.52.

1-8 Design Tables

In designing gear trains, it is helpful to have basic data available for each tooth number for the preferred pitches. Table 1-3 presents such data (a combination of data) for most of the preferred fine pitches.

1-9 AGMA Standards

In the United States most gear standards have been developed and sponsored by AGMA. They range from general and basic standards, such as those for tooth form, already mentioned, to specialized standards. The list is very long, and only a selected few, most pertinent to fine-pitch gearing, are listed in Table 1-4. These and additional standards can be procured from AGMA by contacting the headquarters office at 1901 North Fort Myer Drive, Suite 1000, Arlington, Virginia 22209 (Tel.: 703-525-1600).

INVOLUTOMETRY

This section includes basic calculations for gear systems for ready reference in design. More advanced calculations are available in the listed references.

TABLE 1-4 Selected List of AGMA Standards

General	AGMA 390	Gear Classification Manual
Spurs and helicals	AGMA 201	Tooth Proportions for Coarse-Pitch Involute Spur Gears
	AGMA 207	Tooth Proportions for Fine-Pitch Involute Spur Gears and Helical Gears
Nonspur	AGMA 208	System for Straight Bevel Gears
	AGMA 209	System for Spiral Bevel Gears
	AGMA 202	Zerol Bevel Gear System
	AGMA 330	Design Manual for Bevel Gears
	AGMA 203	Fine-Pitch On-Center Face Gears for 20° Involute Spur Pinions
	AGMA 374	Design for Fine-Pitch Worm Gearing

1-10 Nomenclature and Symbols

The terminology, nomenclature, and symbols used here are summarized in this section. While this list may not be totally consistent with gear literature currently published by AGMA and ANSI, this in no way invalidates the formulas presented. See Table 1-5.

1-11 Pitch Diameter and Center Distance

As already mentioned, the pitch diameters for a meshed gear pair are tangent to one another at a point on the line of centers called the "pitch point." See Fig. 1-4. The pitch point always divides the line of centers in proportion to the number of teeth in each gear.

$$\text{Center distance} = C = \frac{D_1 + D_2}{2} = \frac{N_1 + N_2}{2P_d} \tag{1-5}$$

The pitch-circle sizes are related as

$$\frac{D_1}{D_2} = \frac{R_1}{R_2} = \frac{N_1}{N_2} \tag{1-6}$$

1-12 Velocity Ratio

The gear ratio, or velocity ratio, can be obtained from several different parameters:

$$Z = \frac{D_1}{D_2} = \frac{N_1}{N_2} = \frac{\omega_1}{\omega_2} \tag{1-7}$$

TABLE 1-5 Nomenclature and Definitions

The symbols contained in this list shall be used throughout the following text and formulas, but do not necessarily reflect current AGMA or ANSI standards.

Symbol	Nomenclature and definition	Symbol	Nomenclature and definition
B	backlash, linear amount along pitch circle	Y	Lewis factor, diametral pitch
		Z	mesh velocity ratio
B_{LA}	backlash, linear amount along line of action	a	addendum
		b	dedendum
${}_aB$	backlash in minutes	c	clearance
C	center distance	d	pitch diameter, pinion
ΔC	change in center distance	d_w	pin diameter, for over-pins measurement
C_o	operating center distance		
C_{std}	standard center distance	e	eccentricity
D	pitch diameter	m_p	contact ratio
D_b	base circle diameter	n	number of teeth, pinion
D_o	outside diameter	n_w	number of threads in worm
F	face width	p_a	axial pitch
K	factor, general	p_b	base pitch
L	length, general; also lead of worm	p_c	circular pitch
		p_{cn}	normal circular pitch
M	measurement over pins	r	pitch radius, pinion
N	number of teeth, usually gear	r_b	base circle radius, pinion
N_c	critical number of teeth for no undercutting	r_o	outside radius, pinion
		t	tooth thickness, and for general use for tolerance
N_v	virtual number of teeth for helical gear		
		y_c	Lewis factor, circular pitch
P_d	diametral pitch	γ	pitch angle, bevel gear
P_{dn}	normal diametral pitch	θ	rotation angle, general
P_t	horsepower, transmitted	λ	lead angle, worm gearing
R	pitch radius, gear or general use	μ	mean value
		ν	gear-stage velocity ratio
R_b	base circle radius, gear	ϕ	pressure angle
R_o	outside radius, gear	ϕ_o	operating pressure angle
R_T	testing radius	ψ	helix angle
S	radial clearance	ω	angular velocity
T	tooth thickness, gear		
W_b	beam tooth strength		

1-13 Pressure Angle

The pressure angle is defined as the angle between the line of action (common tangent to the base circles in Figs. 1-3 and 1-4) and a perpendicular to the line of centers. See Fig. 1-7. It is obvious from the pictured geometry that the pressure angle varies (slightly) as the center distance of a gear pair is altered.

$$D_b = D \cos \phi \tag{1-8}$$

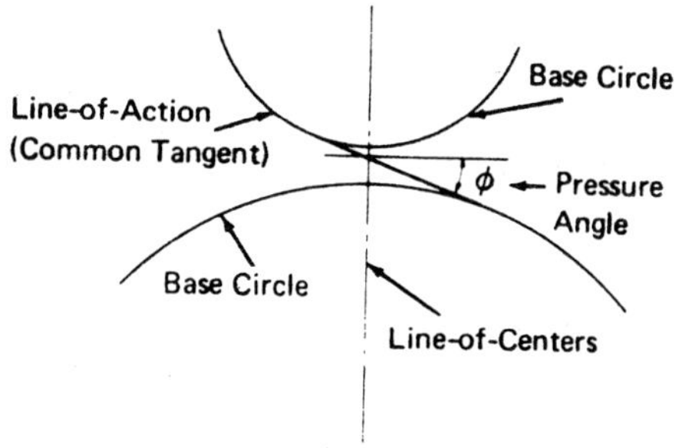

FIG. 1-7 Definition of pressure angle.

The larger the pressure angle, the smaller the base circle. Thus, 14½° pressure angle gears have base circles much nearer to the roots of the teeth. It is for this reason that 14½° gears encounter undercutting problems before 20° gears. This is further elaborated on in Sec. 1-17.

1-14 Tooth Thickness

This is measured along the pitch circle, but as a linear measure. For this reason it is specifically called circular tooth thickness. This is shown in Fig. 1-1. Tooth thickness is related to the pitch:

$$T = \frac{p_c}{2} = \frac{\pi}{2P_d} \tag{1-9}$$

The tooth thickness T_2 at any radial position is found from a known value T_1 at a known radius from the center R_1 and known pressure angle ϕ_1 at that point, using the equation

$$T_2 = T_1 \frac{R_2}{R_1} - 2R_2(\text{inv}\ \phi_2 - \text{inv}\ \phi_1) \tag{1-10}$$

To save time, tables giving the involute function have been computed and are available in the references. An abridged listing is given in Table 1-6.

1-15 Measurement over Pins

Often tooth thickness is measured indirectly by gaging over pins which are placed in diametrically opposed tooth spaces, or as close as possible to it for odd numbers of gear teeth. This is pictured in Fig. 1-8.

For a specified tooth thickness the over-pins measurement M is calculated from the knowns as follows:

For even numbers of teeth,

TABLE 1-6 Table of the Involute Function
Inv $\phi = \tan\phi - \phi$ for values of ϕ from 10 to 40°

	Minutes				
ϕ, degrees	0	12	24	36	48
10	0.00180	0.00191	0.00202	0.00214	0.00226
11	0.00239	0.00253	0.00267	0.00281	0.00296
12	0.00312	0.00328	0.00344	0.00362	0.00379
13	0.00398	0.00417	0.00436	0.00457	0.00476
14	0.00488	0.00520	0.00543	0.00566	0.00590
15	0.00615	0.00640	0.00667	0.00694	0.00721
16	0.00750	0.00779	0.00809	0.00839	0.00870
17	0.00902	0.00935	0.00969	0.01004	0.01039
18	0.01076	0.01113	0.01142	0.01191	0.01231
19	0.01272	0.01314	0.01357	0.01400	0.01444
20	0.01490	0.01537	0.01585	0.01634	0.01683
21	0.01734	0.01786	0.01840	0.01894	0.01949
22	0.02006	0.02063	0.02122	0.02182	0.02242
23	0.02304	0.02368	0.02433	0.02499	0.02566
24	0.02635	0.02705	0.02776	0.02849	0.02922
25	0.02998	0.03074	0.03152	0.03232	0.03313
26	0.03394	0.03478	0.03563	0.03650	0.03739
27	0.03829	0.03920	0.04013	0.04108	0.04204
28	0.04302	0.04402	0.04503	0.04606	0.04710
29	0.04816	0.04924	0.05034	0.05146	0.05260
30	0.05375	0.05492	0.05612	0.05733	0.05856
31	0.05981	0.06108	0.06237	0.06368	0.06502
32	0.06636	0.06773	0.06913	0.07055	0.07199
33	0.07345	0.07493	0.07644	0.07797	0.07952
34	0.08110	0.08270	0.08432	0.08597	0.08765
35	0.08934	0.09106	0.09281	0.09459	0.09639
36	0.09822	0.10008	0.10196	0.10388	0.10582
37	0.10778	0.10978	0.11180	0.11386	0.11594
38	0.11806	0.12020	0.12238	0.12459	0.12683
39	0.12911	0.13141	0.13375	0.13612	0.13853
40	0.14097	0.14344	0.14595	0.14850	0.15108

$$M = \frac{D\cos\phi}{\cos\phi_1} + d_w \tag{1-11}$$

For odd numbers of teeth,

$$M = \frac{D\cos\phi}{\cos\phi_1}\cos\frac{90^\circ}{N} + d_w \tag{1-12}$$

The value of ϕ_1 is obtained from

$$\text{inv } \phi_1 = \frac{T}{D} + \text{inv } \phi + \frac{d_w}{D \cos \phi} - \frac{\pi}{N} \tag{1-13}$$

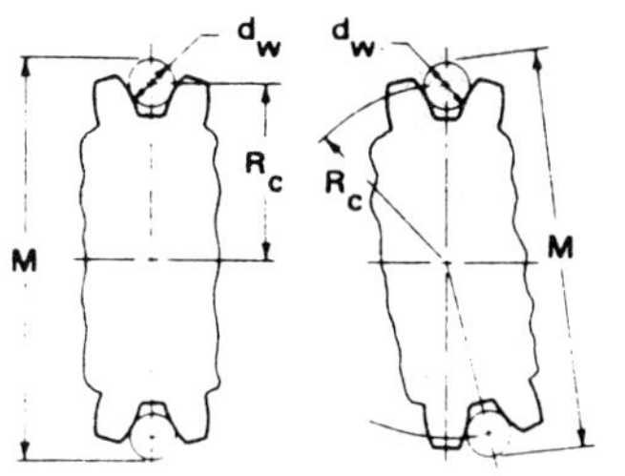

FIG. 1-8 Over-pins size measurement. (*a*) Even number of teeth. (*b*) Odd number of teeth.

Tabulated values of over-pins measurements for standard designed gears are given in Table 1-3. This provides a rapid means for calculating values of M, even for gears with slight departures from standard tooth thicknesses.

When tooth thickness is to be calculated from a known over-pins measurement M, then the equations can be manipulated to yield

$$T = D\left(\frac{\pi}{N} + \text{inv } \phi_c - \text{inv } \phi - \frac{d_w}{D \cos \phi}\right) \tag{1-14}$$

where

$$\cos \phi_c = \frac{D \cos \phi}{2R_c} \tag{1-15}$$

For even numbers of teeth,

$$R_c = \frac{M - d_w}{2} \tag{1-16}$$

For odd numbers of teeth,

$$R_c = \frac{M - d_w}{2 \cos 90°/N} \tag{1-17}$$

1-16 Contact Ratio

To assure smooth, continuous tooth action, as one pair of teeth passes out of action, a succeeding pair of teeth must have already started action. It is desired to have as much overlap as possible. A measure of this overlapping action is the contact ratio. This is a ratio of the length of action to the base pitch. Figure 1-9 shows the geometry. The length of action is determined from the intersection of the line of action and the gears' outside radii. The ratio of the length of action to the base pitch is calculated from

$$m_p = \frac{\sqrt{(R_o^2 - R_b^2)} + \sqrt{(r_o^2 - r_b^2)} - C \sin \phi}{p_c \cos \phi} \tag{1-18}$$

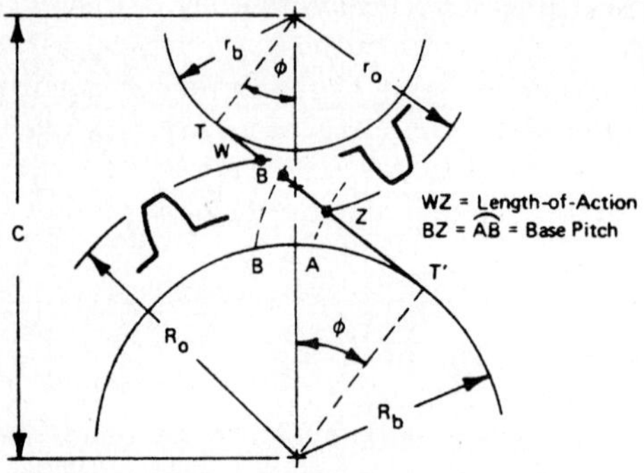

FIG. 1-9 Geometry of contact ratio.

It is good practice to maintain a contact ratio of 1.2 or greater. Under no circumstances should the ratio drop below 1.1, calculated for all tolerances at their worst-case values.

A contact ratio between 1 and 2 means that part of the time two pairs of teeth are in contact, and the rest of the time one pair are in contact. A ratio between 2 and 3 means that there are either 2 or 3 pairs in simultaneous contact at any given time. Such a high contact ratio generally is not obtained with external spur gears, but can be developed in the meshing of an internal and external spur-gear pair or in special design nonstandard external spur gears.

1-17 Undercutting

From Fig. 1-9 it can be seen that the maximum length of contact is limited to the length of the common tangent. Any tooth addendum that extends beyond the tangent points (T and T') not only is useless, but interferes with the root fillet area of the mating tooth. This results in the typical undercut tooth, shown in Fig. 1-10. The undercut not only weakens the tooth with a wasp-like waist, but also removes some of the useful involute adjacent to the base circle.

From the geometry of the limiting length of contact ($T - T'$, Fig. 1-9), it is

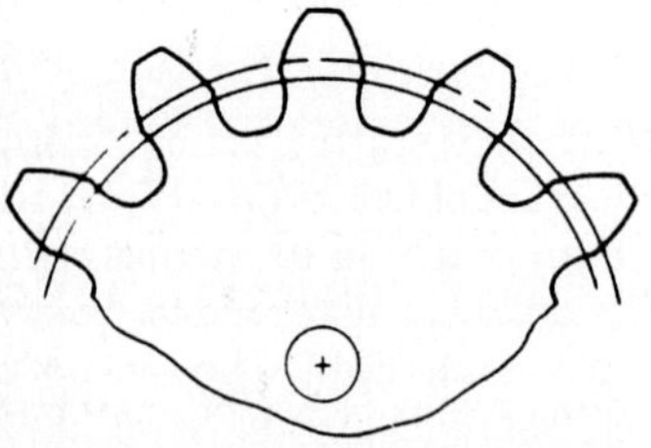

FIG. 1-10 Example of undercut standard design gear; 12 teeth, 20° pressure angle.

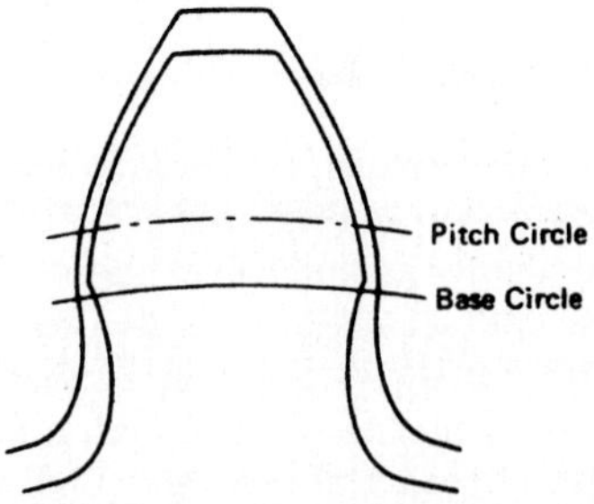

FIG. 1-11 Comparison of enlarged and undercut standard pinion; 13 teeth, 20° pressure angle, fine-pitch standard.

evident that interference is first encountered by the addendums of the gear teeth digging into the flanks of the mating pinion teeth. Since addendums are standardized by a fixed ratio ($1/P_d$), the interference condition becomes more severe as the number of teeth on the gear increases. The limit is when the gear becomes a rack. This is a realistic case, since the hob is a rack-type cutter. The result is that standard designed gears with tooth numbers below a critical value are automatically undercut in the generating process. The limiting number of teeth in a gear meshed with a rack is given by the expression

$$N_c = \frac{2}{\sin^2 \phi} \tag{1-19}$$

This indicates the minimum number of teeth if undercutting decreases with increasing pressure angle. For 14½° the value of N_c is 32, and for 20° it is 18. Thus, 20° pressure angle gears with low numbers of teeth have the advantage of much less undercutting, and therefore are both stronger and smoother acting.

1-18 Enlarged Pinions

Undercutting of pinion teeth is undesirable because of losses of strength, contact ratio, and smoothness of action. The severity of these faults depends upon how far below N_c the tooth number is. Undercutting for the first few numbers is small, and in many applications its adverse effects can be neglected.

For very small numbers of teeth, such as 10 or fewer, and for high-precision applications, undercutting should be avoided. This is done by pinion enlargement (or correction, as it is often called), wherein the pinion teeth, still generated with a standard cutter, are shifted radially outward to form a full involute tooth free of undercut. The tooth is enlarged both radially and circumferentially. Comparison of tooth form before and after enlargement is shown in Fig. 1-11.

The details of enlarged pinion design, mating gear design, and in general profile-shifted gears constitute a large and involved subject beyond the scope of this writing. References 1, 2, 3, and 4 offer additional information. For measurement and inspection of such gears, in particular, consult Ref. 3.

1-19 Backlash Calculations

Up to this point the discussion has implied that there is no backlash. If the gears are of standard tooth-proportion design and operate on standard center distance, they would function ideally with neither backlash nor jamming.

In practice, a small amount of backlash is desirable to provide for lubricant space and differential expansion between the gear components and the housing. The challenge is to have close control. Backlash is defined in Fig. 1-12*a* as the excess thickness of the tooth space over the thickness of the mating tooth. There are two basic ways in which backlash arises: Tooth thickness is below the no-backlash value, or the center distance of operation is greater than the no-backlash value.

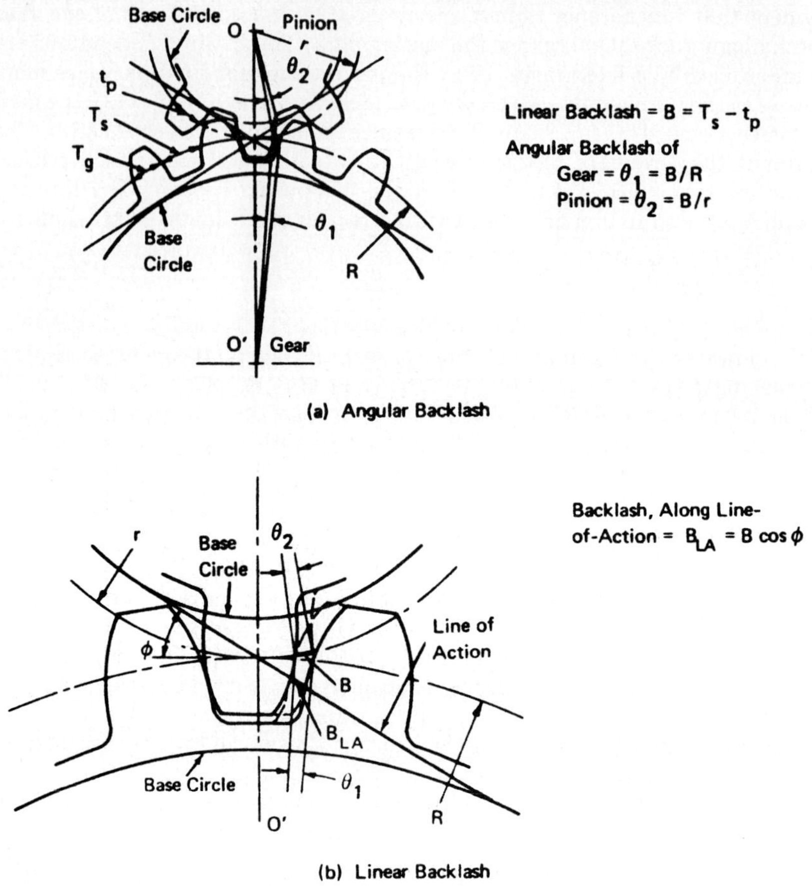

FIG. 1-12 Geometrical definition of backlash. (*a*) Angular backlash. (*b*) Linear backlash.

If the tooth thickness of either or both mating gears is less than the no-backlash value, the amount of backlash introduced in the mesh is simply this numerical difference.

$$B = T_{\text{std}} - T_{\text{act}} = \Delta T \tag{1-20}$$

where B = linear backlash measured along the pitch circle (Fig. 1-12*b*)
T_{std} = no-backlash tooth thickness on the operating-pitch circle, which is the standard tooth thickness for ideal gears
T_{act} = actual tooth thickness

When the center distance is increased by a relatively small amount ΔC, a backlash space develops between mated teeth, as in Fig. 1-13. The relationship between the center distance opening and the amount of linear backlash is

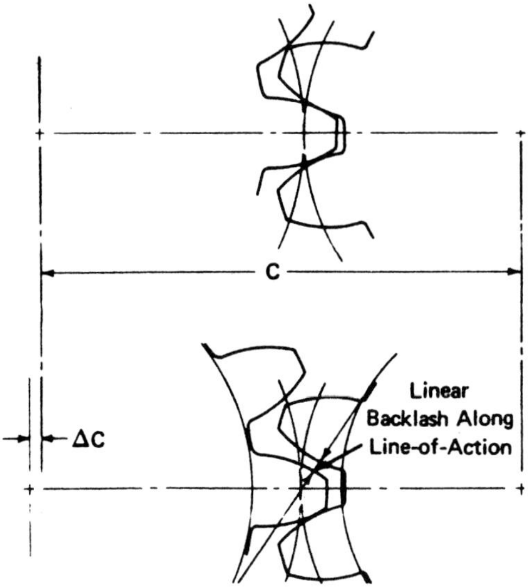

FIG. 1-13 Backlash caused by opening of center distance.

$$B_{LA} = 2(\Delta C) \sin \phi \tag{1-21}$$

This is a measure along the line of action; it is a useful equation when a feeler gage has been inserted between teeth to measure backlash.

The equivalent linear backlash measure along the pitch circle is given by

$$B = 2(\Delta C) \tan \phi \tag{1-22}$$

where ΔC = change in center distance
ϕ = pressure angle

Although this is an approximate relationship, it is adequately accurate for most uses. Its derivation, limitations, and correction factors are detailed in Ref. 3.

Note that backlash due to center-distance opening is dependent on the tangent function of the pressure angle. Thus, 20° gears have 41 percent more backlash than 14½° gears; this is one of the few advantages of the lower pressure angle.

Equation (1-22) is a more useful relationship, particularly for converting to angular backlash. Also, for fine-pitch gears the use of feeler gages for measurement is impractical, whereas an indicator at the pitch line gives a direct measure. The two linear backlashes are related by

$$B = \frac{B_{LA}}{\cos \phi} \tag{1-23}$$

The angular backlash at the gear shaft is usually the desired criterion of the gear application. As seen in Fig. 1-12*a*, this is related to the gear's pitch radius by

$$_aB = 3440 \frac{B}{R_1} \qquad \text{arcmin} \tag{1-24}$$

Obviously, angular backlash is inversely proportional to the gear radius. Also, since the two gears of a mesh usually have different pitch diameters, the linear backlash of the measure converts to different angular values for each gear. Thus, an angular backlash must be specified for a particular shaft or gear center.

1-20 Summary of Gear-Mesh Fundamentals

The basic geometric relationships of gears and meshed pairs given in the preceding sections are summarized in Table 1-7.

HELICAL GEAR

The helical gear differs from the spur gear in that its teeth are twisted along a helical path in the axial direction. It resembles the spur gear in the plane of rotation, but in the axial direction it is as if there were a series of staggered spur gears. See Fig. 1-14. This design brings forth a number of features that differ from those of the spur gear, with two of the most important being:

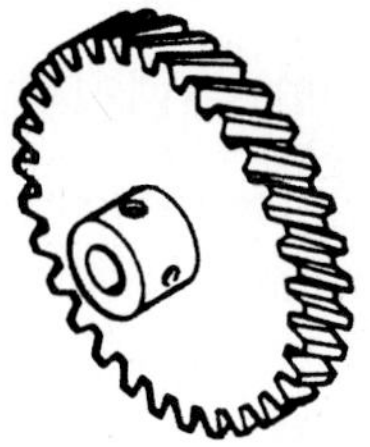

FIG. 1-14 Helical gear.

1. Tooth strength is improved because of the elongated helical wrap-around tooth base support.
2. Contact ratio is increased because of the axial overlap of teeth.

Helical gears are used in two forms:

1. Parallel shaft applications, which is the more common use
2. Crossed helicals (or spiral gears) for connecting skew shafts, usually at right angles

1-21 Generation of the Helical Tooth

The helical tooth form is involute in the plane of rotation and can be developed in a similar manner to the spur gear. However, unlike the spur gear, which can be essentially seen in two-dimensional representations, because axially

TABLE 1-7 Summary of Fundamentals

To obtain	From known	Formula
	Spur gears	
Pitch diameter	Number of teeth and pitch	$D = \frac{N}{P_d} = \frac{N - p_c}{\pi}$
Circular pitch	Diametral pitch or number of teeth and pitch diameter	$p_c = \frac{\pi}{P_d} = \frac{\pi D}{N}$
Diametral pitch	Circular pitch or number of teeth and pitch diameter	$P_d = \frac{\pi}{p_c} = \frac{N}{D}$
Number of teeth	Pitch and pitch diameter	$N = DP_d = \frac{\pi D}{p_c}$
Outside diameter	Pitch and pitch diameter or pitch and number of teeth	$D_o = D + \frac{2}{P_d} = \frac{N + 2}{P_d}$
Root diameter	Pitch diameter and dedendum	$D_R = D - 2b$
Base circle diameter	Pitch diameter and pressure angle	$D_b = D \cos \phi$
Base pitch	Circular pitch and pressure angle	$p_b = p_c \cos \phi$
Tooth thickness at standard pitch diameter	Circular pitch	$T_{\text{std}} = \frac{p_c}{2} = \frac{\pi D}{2N}$

TABLE 1-7 Summary of Fundamentals (continued)

To obtain	From known	Formula
Addendum	Diametral pitch	$A = \dfrac{1}{P_d}$
Center distance	Pitch diameters or number of teeth and pitch	$c = \dfrac{D_1 + D_2}{2} = \dfrac{N_1 + N_2}{2P_d} = \dfrac{p_c(N_1 + N_2)}{2\pi}$
Contact ratio	Outside radii, base radii, center distance, and pressure angle	$m_p = \dfrac{\sqrt{R_o^2 - R_b^2} + \sqrt{r_o^2 - r_b^2} - C \sin \phi}{p_c \cos \phi}$
Backlash (linear)	From change in center distance	$B = 2(\overline{\Delta C}) \tan \phi$
Backlash (linear)	From change in tooth thickness	$B = \Delta T$
Backlash, linear along line-of-action	Linear backlash along pitch circle	$B_{LA} = B \cos \phi$
Backlash, angular	Linear backlash	${}_aB = 6880 \dfrac{B}{D}$ (arcmin)
Minimum number of teeth for no undercutting	Pressure angle	$N = \dfrac{2}{\sin^2 \phi}$
	Helical gearing	
Normal circular pitch	Transverse circular pitch	$p_{cn} = p_c \cos \psi$
Normal diametral pitch	Transverse diametral pitch	$P_{dn} = \dfrac{P_d}{\cos \psi}$

Axial pitch	Circular pitches	$p_a = p_c \cot \psi = \dfrac{p_{cn}}{\sin \psi}$
Normal pressure angle	Transverse pressure angle	$\tan \phi_n = \tan \phi \cos \psi$
Pitch diameter	Number of teeth and pitch	$D = \dfrac{N}{P_d} = \dfrac{N}{P_{dn} \cos \psi}$
Center distance (parallel shafts)	Number of teeth and pitch	$C = \dfrac{N_1 + N_2}{2P_{dn} \cos \psi}$
Center distance (crossed shafts)	Number of teeth and pitch	$C = \dfrac{1}{2P_{dn}} \left(\dfrac{N_1}{\cos \psi_1} + \dfrac{N_2}{\cos \psi_2} \right)$
Shaft angle (crossed shafts)	Helix angles of two mated gears	$\theta = \psi_1 + \psi_2$
	Worm meshes	
Pitch diameter of worm	Number of teeth and pitch	$d_w = \dfrac{n_w p_{cn}}{\pi \sin \lambda}$
Pitch diameter of worm gear	Number of teeth and pitch	$D_g = \dfrac{N_g p_{cn}}{\pi \cos \lambda}$
Lead angle	Pitch, diameter, teeth	$\lambda = \tan^{-1} \dfrac{n_w}{P_d d_w} = \sin^{-1} \dfrac{n_w p_{cn}}{\pi d_w}$
Lead of worm	Number of teeth and pitch	$L = n_w p_c = \dfrac{n_w p_{cn}}{\cos \lambda}$
Normal circular pitch	Transverse pitch and lead angle	$p_{cn} = p_c \cos \lambda$

TABLE 1-7 Summary of Fundamentals (continued)

To obtain	From known	Formula
Center distance	Pitch diameters	$C = \frac{d_w + D_g}{2}$
Center distance	Pitch, lead angle, teeth	$C = \frac{p_{cn}}{2\pi}\left(\frac{N_g}{\cos\lambda} + \frac{n_w}{\sin\lambda}\right)$
Velocity ratio	Number of teeth	$Z = \frac{N}{n_w}$
	Bevel gearing	
Velocity ratio	Number of teeth	$Z = \frac{N_1}{N_2}$
Velocity ratio	Pitch diameters	$Z = \frac{D_1}{D_2}$
Velocity ratio	Pitch angles	$Z = \frac{\sin\gamma_1}{\sin\gamma_2}$
Shaft angle	Pitch angles	$\Sigma = \gamma_1 + \gamma_2$

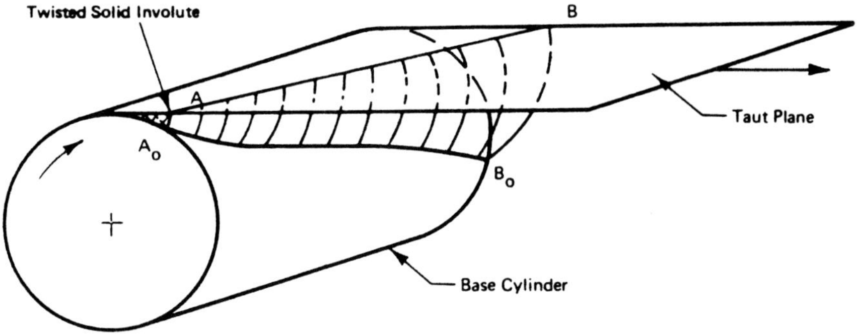

FIG. 1-15 Generation of the helical-tooth profile.

the gear features remain constant, the helical gear must be portrayed in three dimensions to show changing axial features.

Referring to Fig. 1-15, there is a base cylinder from which a taut plane is unwrapped, analogous to the unwinding taut string of the spur gear in Fig. 1-2. On the plane there is a straight line AB, which when wrapped on the base cylinder has a helical trace A_0B_0. As the taut plane is unwrapped, any point on line AB can be visualized as tracing an involute from the base cylinder. Thus, there is an infinite series of involutes generated by line AB, all alike, but phase displaced along a helix on the base cylinder.

Another concept analogous to spur-gear-tooth development is to imagine the taut plane being wound from one base cylinder to another as the base cylinders rotate in opposite directions. The result is the generation of a pair of conjugate helical involutes. If a reverse direction of rotation is assumed and a second tangent plane is arranged so that it crosses the first, a complete involute helicoid tooth is formed.

1-22 Fundamentals of Helical Teeth

In the plane of rotation, the helical-gear tooth is involute and all the relationships governing spur gears apply to the helical. However, the helical twist of the teeth introduces a helix angle. Since the helix angle must vary from the base of the tooth to the outside radius, the helix angle ψ is defined as the angle between the tangent to the helicoidal tooth, at the intersection of the pitch cylinder and the tooth profile, and an element of the pitch cylinder. See Fig. 1-16.

The direction of the helical twist is designated as either left or right. The direction is defined by the right-hand screw rule.

1-23 Helical-Gear Relationships

For helical gears there are two measurements of the usual pitches: One in the plane of rotation and the other in a plane normal to the tooth. In addition, there is an axial pitch. These are defined and related as follows.

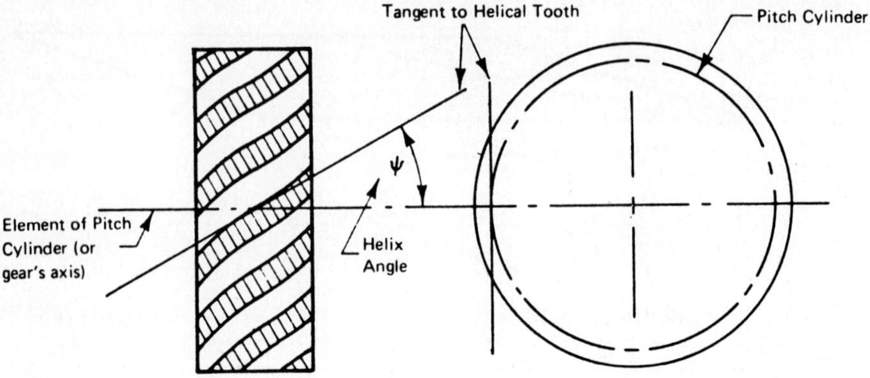

FIG. 1-16 Definition of helix angle.

Referring to Fig. 1-17, the two circular pitches are equated as:

$$p_{cn} = p_c \cos \psi = \text{normal circular pitch} \tag{1-25}$$

The normal circular pitch is less than the plane of rotation (or traverse) circular pitch by the cosine of the helix angle factor.

Consistent with this, the normal diametral pitch is a larger number than the transverse pitch, and they are related as

$$P_{dn} = \frac{P_d}{\cos \psi} = \text{normal diametral pitch} \tag{1-26}$$

The axial pitch of a helical gear is the distance between corresponding points of adjacent teeth measured parallel to the gear's axis. See Fig. 1-18. Axial pitch is related to the circular pitches by the expression

$$p_a = p_c \cos \psi = \frac{p_{cn}}{\sin \psi} = \text{axial pitch} \tag{1-27}$$

1-24 Equivalent Spur Gear

The true involute pitch and involute geometry of a helical gear are those in the plane of rotation. However, in the normal plane, looking at one tooth, there is

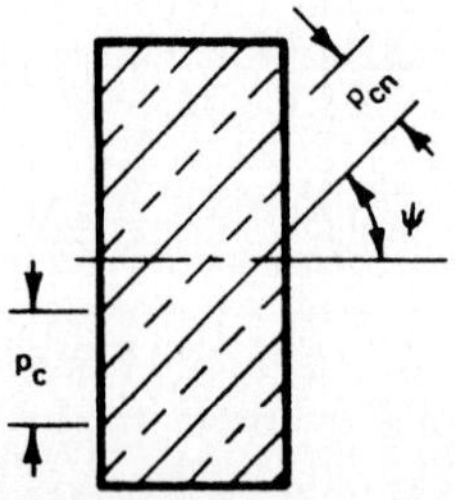

FIG. 1-17 Relationship of circular pitches.

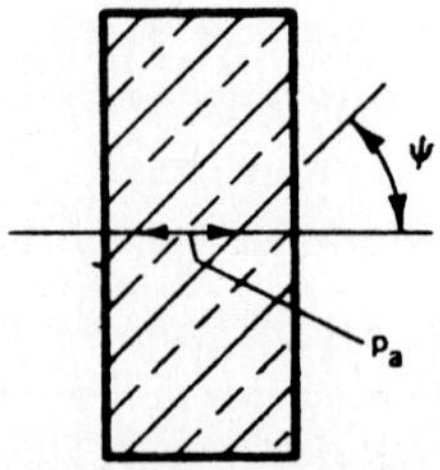

FIG. 1-18 Axial pitch of a helical gear.

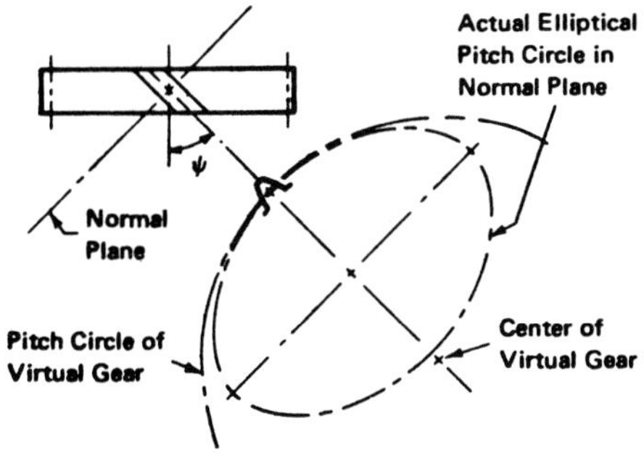

FIG. 1-19 Geometry of helical gear's virtual number of teeth.

a resemblance to an involute tooth of a pitch corresponding to the normal pitch. But, the shape of the tooth corresponds to that of a spur gear with a larger number of teeth, the exact value depending on the magnitude of the helix angle.

The geometric basis for deriving the number of teeth in this equivalent-tooth-form spur gear is given in Fig. 1-19. The result of the transposed geometry is an equivalent number of teeth, given as

$$N_v = \frac{N}{\cos^3 \psi} \tag{1-28}$$

This equivalent number is also called a virtual number because this spur gear is imaginary. The value of this number is its use in determining helical-tooth strength.

1-25 Pressure Angle

Although, strictly speaking, pressure angle exists only for a gear pair, a nominal pressure angle can be considered. For the helical gear there is a normal pressure angle as well as the usual pressure angle in the plane of rotation. Figure 1-20 shows the relationship, which is expressed as

$$\tan \phi = \frac{\tan \phi_n}{\cos \psi} \tag{1-29}$$

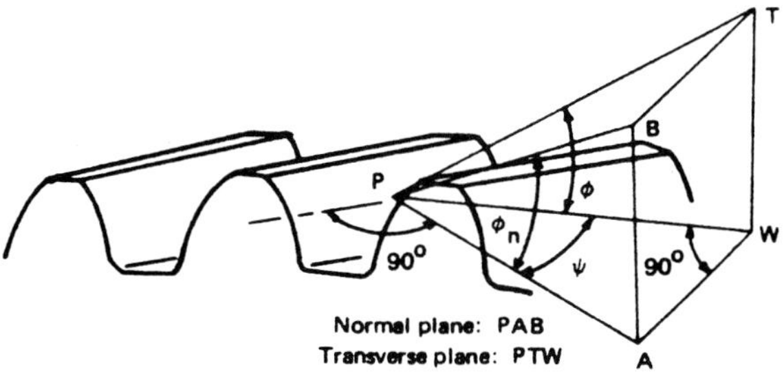

FIG. 1-20 Two-pressure-angle geometry.

1-26 Normal Plane Geometry

Because of the way in which gears are generated with a rack-type hob, a single tool can generate helical gears with all helix angles as well as spur gears. However, this means that the normal pitch is the common denominator, usually taken as a standard value. Since the true involute features are in the transverse plane, they will differ from the standard normal values and be odd valued. Hence, there is a real need to relate parameters in the two reference planes.

1-27 Helical-Tooth Proportions

Helical-tooth proportions follow the same standards as do those of spur gears. The values for the addendum, dedendum, whole depth, and clearance are the same regardless of whether they are measured in the plane of rotation or the normal plane. Pressure angle and pitch are usually specified as standard values in the normal plane, but there are times when they are specified in the transverse plane.

1-28 Parallel-Shaft Helical-Gear Meshes

The following information for the design of gear meshes is fundamental.

Helix Angle Both gears of a meshed pair must have the same helix angle. However, the helix directions must be opposite, i.e., a left-hand mates with a right-hand helix.

Pitch Diameter This is given by the same expression as for spur gears, but if the normal pitch is involved it is a function of the helix angle. The expressions are

$$D = \frac{N}{P_d} = \frac{N}{P_{dn} \cos \psi} \tag{1-30}$$

Center Distance Using Eq. (1-30), the center distance of a meshed pair is expressed as

$$C = \frac{N_1 + N_2}{2P_{dn} \cos \psi} \tag{1-31}$$

Note that for standard parameters in the normal plane, the center distance will not be a standard value, unlike with standard spur gears. Further, by manipulating the helix angle ψ, the center distance can be adjusted over a wide range of values.

Contact Ratio The contact ratio of helical gears is enhanced by the axial overlap of the teeth. Thus, the contact ratio is the sum of the transverse contact ratio, calculated in the same manner as for spur gears [Eq. (1-18)], plus a term involving the axial pitch.

$$m_{p,\text{total}} = m_{p,\text{trans}} + m_{p,\text{axial}} \tag{1-32}$$

where $m_{p,\text{trans}}$ = value from Eq. (1-18)

$$m_{p,\text{axial}} = \frac{F}{P_a} = \frac{F \tan \psi}{P_c}$$

F = face width of gear

Involute Interference Helical gears cut with standard normal pressure angles can have considerably higher pressure angles in the plane of rotation [Eq. (1-29)], depending on the helix angle. Therefore, referring to Eq. (1-19), the minimum number of teeth without undercutting can be significantly reduced, and helical gears with very low tooth numbers without undercutting are feasible.

1-29 Cross-Helical-Gear Meshes

These are also known as spiral and screw gears. They are used for interconnecting skew shafts, as in Fig. 1-21. They can be designed to connect shafts at any angle, but most applications are 90°.

Helix Angle and Hands The helix angles need not be the same. However, their sum must equal the shaft angle:

$$\psi_1 + \psi_2 = \theta \qquad (1\text{-}33)$$

where ψ_1, ψ_2 = respective helix angles of the two gears

θ = shaft angle (the acute angle between the two shafts when viewed in a direction paralleling a common perpendicular between the shafts)

Except for very small shaft angles, the helix hands are the same.

Pitch Because it is possible to have two different helix angles for the gear pair, the transverse pitches may not be the same. However, the normal pitches must always be identical.

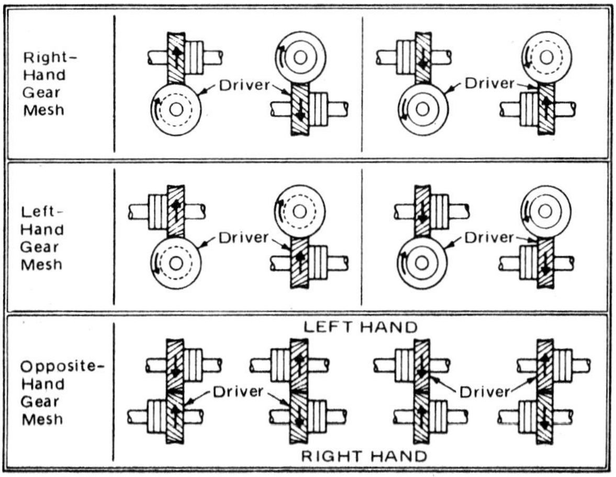

Note:
1) Helical gears of the same hand operate at right angles.
2) Helical gears of opposite hand operate on parallel shifts.
3) Bearing location indicates the direction of thrust.

FIG. 1-21 Gear-mesh combinations.

Center Distance The pitch diameter of a crossed-helical gear was given in Sec. 1-28, and the center distance is expressed as

$$C = \frac{1}{2P_{dn}}\left(\frac{N_1}{\cos\psi_1} + \frac{N_2}{\cos\psi_2}\right) \tag{1-34}$$

Again it is possible to adjust the center distance by manipulating the helix angle. However, the helix angles of both gears must be consistently altered in accordance with Eq. (1-33).

Velocity Ratio Unlike spur and parallel-shaft helical meshes, the velocity ratio (gear ratio) for cross-helical meshes cannot be determined from the ratio of pitch diameters, since these can be altered by juggling of helix angles. The speed ratio can be determined only from the number of teeth:

$$\text{Velocity ratio } Z = \frac{N_1}{N_2} \tag{1-35}$$

If pitch diameters are introduced, the relationship is

$$Z = \frac{D_1 \cos\psi_1}{D_2 \cos\psi_2} \tag{1-36}$$

1-30 Axial Thrust of Helical Gears

In both parallel-shaft and crossed-shaft applications, helical gears develop an axial thrust load. This is a useless force that loads gear teeth and bearings and must accordingly be considered in the housing and bearing design. In some special instrument designs this thrust load can be utilized to actuate face clutches, provide a friction drag, or for other special purpose. The magnitude of the thrust load depends on the helix angle and is given by the expression

$$W_T = W_t \tan\psi \tag{1-37}$$

where W_T = axial thrust load
W_t = transmitted load

The direction of the thrust load is related to the hand of the gear and the direction of rotation. This is depicted in Fig. 1-21.

RACKS

1-31 Description of Racks

Gear racks (Fig. 1-22) are important components in that they are a means of converting rotational motion into linear motion. In theory the rack is a gear with infinite pitch diameter, resulting in an involute profile that is essentially a straight line, and the tooth is of simple V form. Racks can be made in both

FIG. 1-22 Gear rack.

spur and helical types. A rack will mesh with all gears of the same pitch. Backlash is computed by the same formula used for gear pairs, Eq. (1-22). However, the pressure angle and the gear's pitch radius remain constant regardless of changes in the relative position of the gear and rack. Only the pitch line shifts accordingly as the gear center is altered. See Fig. 1-23.

INTERNAL GEARS

A special feature of spur and helical gears is their capability of being made in an internal form, in which an internal gear mates with a common external gear. This offers considerable versatility in the design of planetary gear trains and miscellaneous instrument packages.

1-32 Development of the Internal Gear

The gears considered so far can be imagined as equivalent pitch-circle friction disks which roll one upon the other with contact on their outsides. If, instead, one of the pitch circles rolls on the inside of the other, it forms the basis of internal gearing. In addition, the larger gear must have the material forming the teeth on the convex side of the involute profile, so that the internal gear is an inverse of the common external gear. See Fig. 1-24*a*.

The base circles, line of action, and development of the involute profiles and action are shown in Fig. 1-24*b*. As with spur gears, there is a taut generating string that winds and unwinds between the base circles. However, in this case

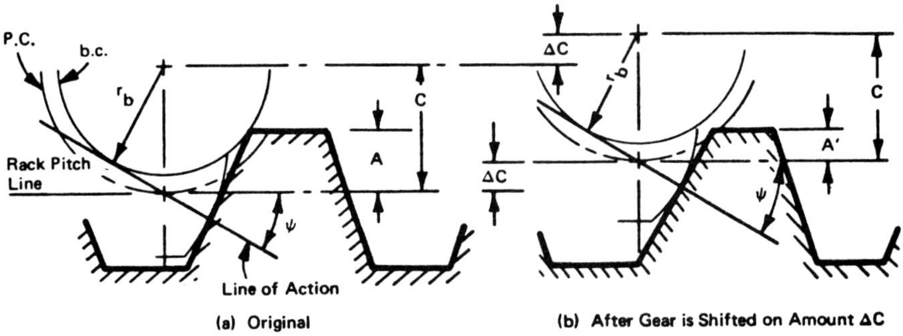

FIG. 1-23 Operational effects of changing the distance between a rack and a gear. (*a*) Original. (*b*) After gear is shifted an amount ΔC.

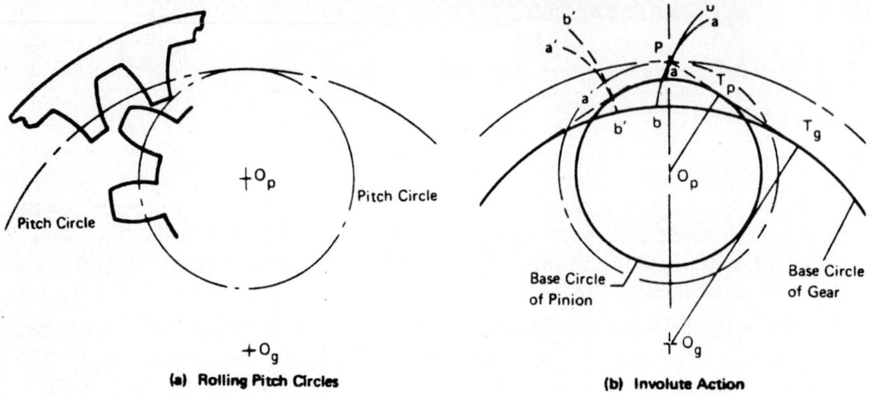

FIG. 1-24 Development of the internal gear. (*a*) Rolling pitch circles. (*b*) Base circles, line of action, and development of the involute profiles and action.

the string does not cross the line of centers, and actual contact and involute development occur on an extension of the common tangent. Otherwise, the action parallels that for external spur gears.

1-33 Tooth Parts of Internal Gear

Because the internal gear is reversed, the tooth parts are backward from the ordinary (external) gear. This is shown in Fig. 1-25. Tooth proportions and

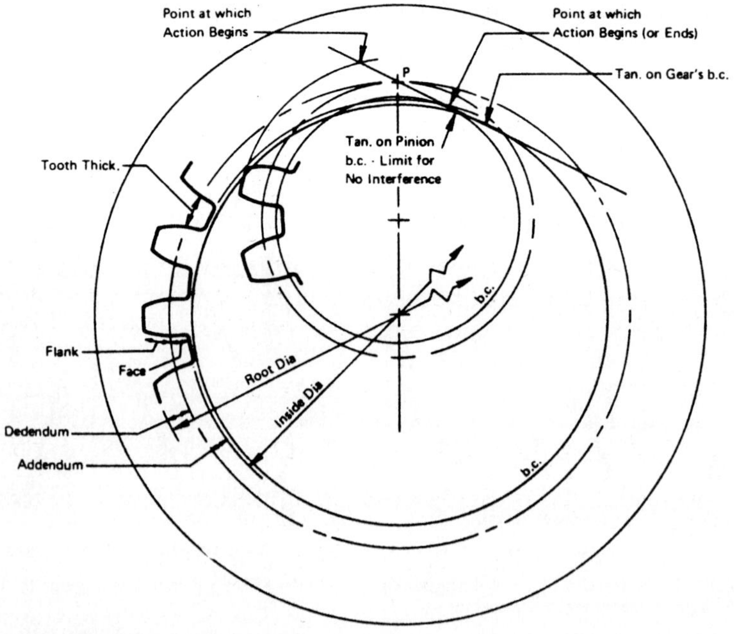

FIG. 1-25 Internal gear-tooth parts.

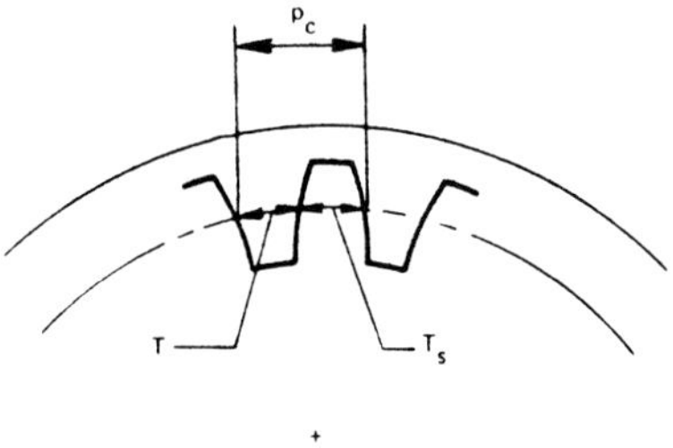

FIG. 1-26 Tooth and space thickness.

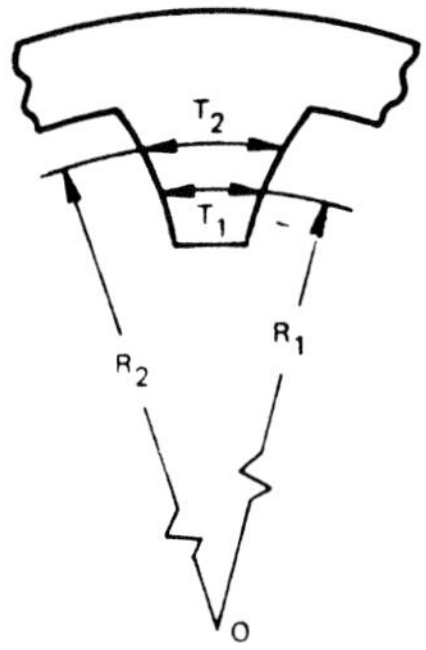

FIG. 1-27 Tooth thickness at various radii.

standards are the same as for external gears except that the addendum is reduced to avoid trimming the teeth in the fabrication process.

Tooth thicknesses of the internal gear can be calculated using Eqs. (1-9) and (1-20), but one must remember that the tooth and space thicknesses are reversed. See Fig. 1-26. Also, in using Eq. (1-10) to calculate tooth thicknesses at various radii, Fig. 1-27, it is the tooth space that is calculated; the internal gear-tooth thickness is obtained by a subtraction from the circular pitch at that radius. Thus, applying Eq. (1-10) to Fig. 1-27,

$$T_{s-1} = p_{c-1} - T_1 = \frac{2\pi R_1}{N} - T_1$$

$$T_{s-2} = T_{s-1}\frac{R_2}{R_1} - 2R_2(\text{inv}\,\phi_2 - \text{inv}\,\phi_1)$$

and

$$T_2 = p_{c-2} - T_{s-2} = \frac{2\pi R_2}{N} - T_{s-2}$$

1-34 Tooth Thickness Measurement

In a procedure similar to that used for external gears, tooth thickness can be measured indirectly by gaging pins, but this time between the pins, as shown in Fig. 1-28. Equations (1-11) through (1-13) are modified accordingly to yield:

For even number teeth,

$$M = 2\left(R_c - \frac{d_w}{2}\right) \tag{1-38}$$

For odd number teeth,

$$M = 2\left(R_c \cos\frac{90}{N} - \frac{d_w}{2}\right) \tag{1-39}$$

$$\text{inv}\,\phi_1 = \text{inv}\,\phi + \frac{\pi}{N} - \frac{T}{D} - \frac{d_w}{D\cos\phi}$$

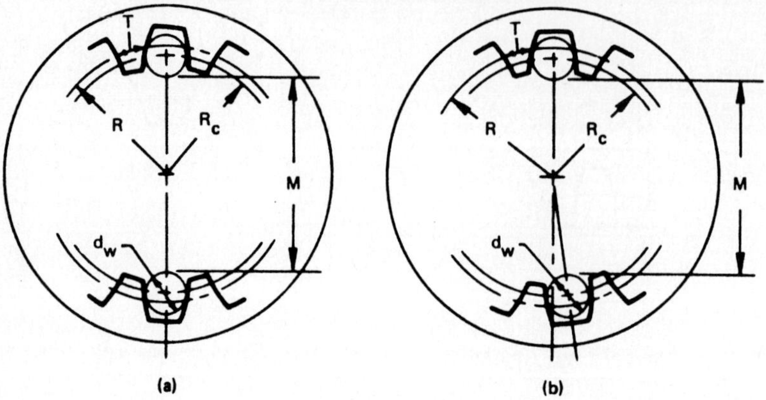

FIG. 1-28 Pin gaging. (*a*) Even number of teeth. (*b*) Odd number of teeth.

where

$$R_c = \frac{\cos \phi}{\cos \phi_1} R$$

1-35 Features of Internal Gears

General advantages:

1. Lead to compact design, since the center distance is less than for external gears.
2. A high contact ratio is possible.
3. Good surface endurance because a convex profile surface is working against a concave surface.

General disadvantages:

1. Housing and bearing support is involved because the external gear nests within the internal gear.
2. Low velocity ratios are unsuitable and in many cases impossible because of interference.
3. Fabrication is limited to the shaper generating process, and usually special tooling is required.

WORM MESH

The worm mesh is another type of gear used for interconnecting skew shafts, usually 90°; see Fig. 1-29. Worm meshes are characterized by high velocity ratios. Also, they offer the advantage of the higher load capacity of a line contact, in contrast to the point contact of the cross-helical mesh.

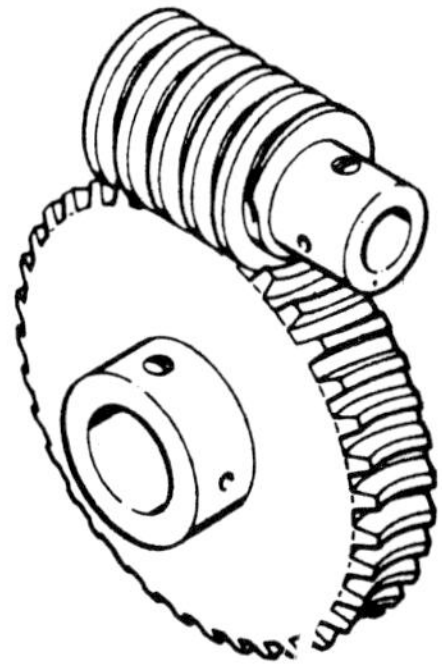

FIG. 1-29 Typical worm mesh.

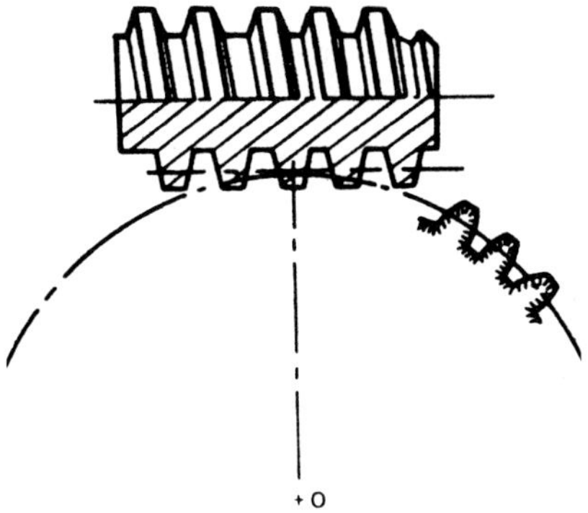

FIG. 1-30 Central section of a worm and worm gear.

1-36 Worm-Mesh Geometry

The worm is equivalent to a V-type screw thread, as is evident from Fig. 1-30. The mating worm-gear teeth have a helical lead. A central section of the mesh, taken through the worm's axis and perpendicular to the worm gear's axis, as in Fig. 1-30, reveals a rack-type tooth for the worm and a curved involute tooth form for the worm gear. However, the involute features are true only for the central section. Sections on either side of the worm axis reveal nonsymmetric and noninvolute tooth profiles. Thus, worm gearing is not a true involute mesh. Also, for conjugate action the center distance of the mesh must be an exact duplicate of that used in generating the worm gear.

To increase the length of action, the worm gear is made of a throated shape to wrap around the worm.

1-37 Worm-Tooth Proportions

Worm-tooth dimensions, such as addendum, dedendum, pressure angle, and others, follow the same standards as those for spur and helical gears. The standard values apply to the central section of the mesh. (See Fig. 1-31*a*.) A high pressure angle is favored, and in some applications values as high as 25° and 30° are used.

1-38 Number of Threads

The worm can be considered to resemble a helical gear with a high helix angle. For extremely high helixes, there is one continuous tooth or thread. For slightly smaller angles there can be two, three, or even more thread starts, designated n_w. Thus, a worm is characterized by the number of threads.

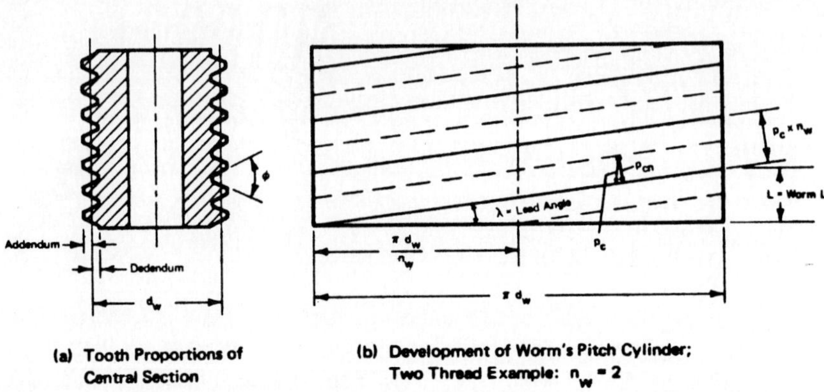

FIG. 1-31 Worm-tooth proportions and geometric relationships. (*a*) Tooth proportions of central section. (*b*) Development of worm's pitch cylinder; two-thread example: $n_w = 2$.

1-39 Worm and Worm-Gear Calculations

Referring to Fig. 1-31*b* and recalling the relationships established for normal and transverse pitches, the worm-mesh components can be defined in the following way.

Pitch Diameters, Lead and Lead Angle

$$\text{Pitch diameter of worm} = d_w = \frac{n_w p_{cn}}{\pi \sin \lambda} \qquad (1\text{-}40)$$

$$\text{Pitch diameter of worm gear} = D_g = \frac{N_g p_{cn}}{\pi \cos \lambda} \qquad (1\text{-}41)$$

where n_w = number of threads in worm

$$L = \text{lead of worm} = n_w p_c = \frac{n_w p_{cn}}{\cos \lambda}$$

$$\lambda = \text{lead angle} = \tan^{-1} \frac{n_w}{P_d d_w} = \sin^{-1} \frac{n_w p_{cn}}{\pi d_w}$$

$$p_{cn} = p_c \cos \lambda$$

Center Distance of Mesh

$$C = \frac{d_w + D_g}{2} = \frac{p_{cn}}{2}\left(\frac{N_g}{\cos \lambda} + \frac{n_w}{\sin \lambda}\right) \qquad (1\text{-}42)$$

1-40 Velocity Ratio

The gear ratio of a worm mesh cannot be calculated from the ratio of pitch diameters. It can be determined only from the ratio of tooth numbers:

$$\text{Velocity ratio} = Z = \frac{\text{no. teeth in worm gear}}{\text{no. threads in worm}} = \frac{N}{N_w} \tag{1-43}$$

BEVEL GEARING

For angular intersecting shaft centers, bevel gears offer a good means of transmitting motion and power. Most transmissions occur at right angles (Fig. 1-32), but the shaft angle can be any value.

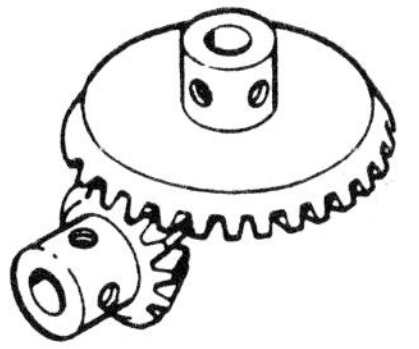

FIG. 1-32 Typical right-angle bevel gear.

1-41 Development and Geometry of Bevel Gears

Bevel gears have tapered elements because they are generated and operate (in theory) on the surface of a sphere. Pitch diameters of mated bevel gears belong to frusta of cones, as shown in Fig. 1-33. In the full development of the surface of a sphere, a pair of meshed bevel gears conjugately engage themselves and a crown gear, as shown in Fig. 1-34.

The crown gear, which is a bevel gear with the largest possible pitch angle (defined in Fig. 1-36), is analogous to the rack of spur gearing, and is the basic tool for generating bevel gears. However, for practical reasons the tooth form is not that of a spherical involute, and instead, the crown gear profile assumes a slightly simplified form. Although the deviation from a true spherical involute is minor, it results in a line of action that has a figure 8 trace in its extreme extension. See Fig. 1-35. This shape gives rise to the name "octoid" for the tooth form of modern bevel gears.

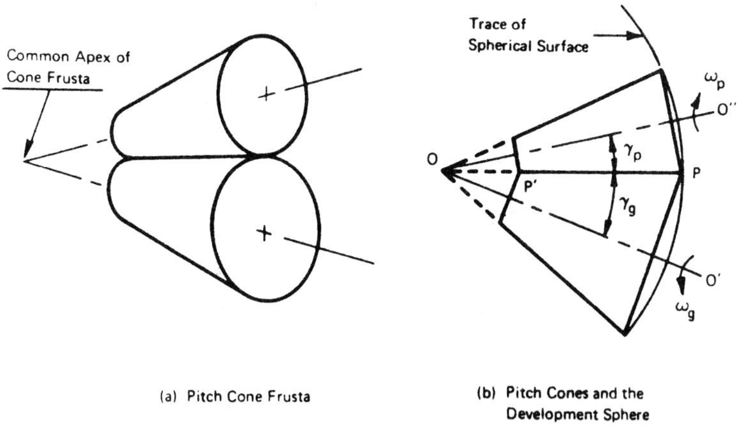

FIG. 1-33 Pitch cones of bevel gears. (*a*) Pitch cone frusta. (*b*) Pitch cones and development sphere.

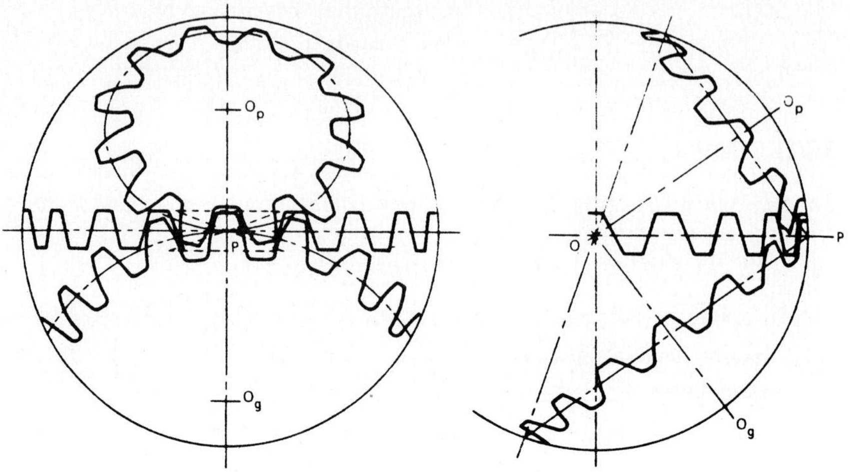

FIG. 1-34 Meshed bevel-gear pair with conjugate crown gear.

1-42 Bevel-Gear-Tooth Proportions

Bevel-gear teeth are proportioned in accordance with the standard tooth-proportion systems used for spur gears. However, the pressure angle of all standard-design bevel gears is limited to 20°. Small tooth number pinions are enlarged automatically when the design follows the Gleason system.

Since bevel-tooth elements are tapered, the outer end (heel) is where all the tooth dimensions and pitch diameter are designated. Since the narrow end of the teeth (toe) vanishes at the pitch apex (the center of the reference generating sphere), there is a practical limit to the length (face) of a bevel gear. The geometry and identification of bevel-gear parts are given in Fig. 1-36.

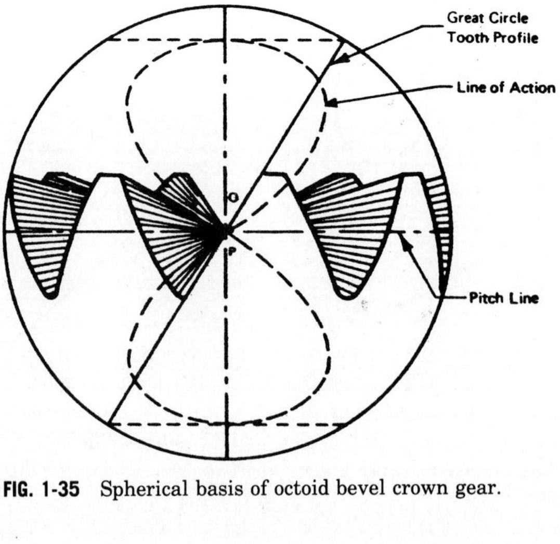

FIG. 1-35 Spherical basis of octoid bevel crown gear.

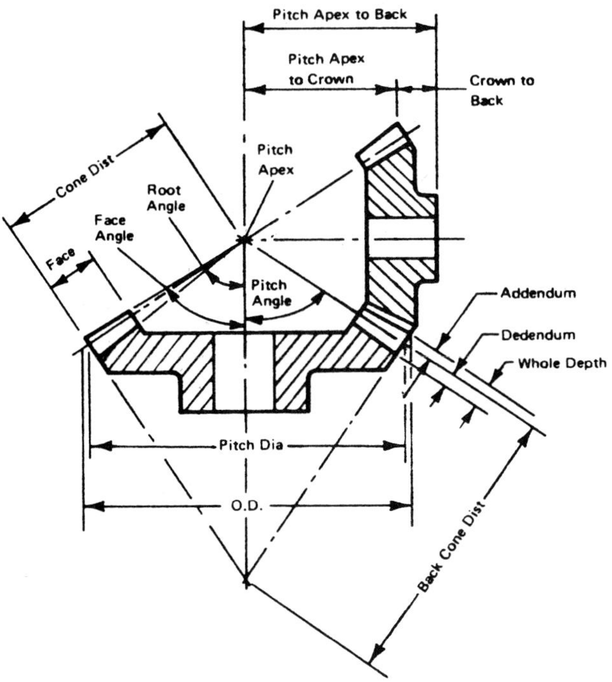

FIG. 1-36 Bevel-gear-pair design parameters.

1-43 Velocity Ratio

The velocity ratio can be derived from the ratio of several parameters:

$$\text{Velocity ratio} = Z = \frac{N_1}{N_2} = \frac{D_1}{D_2} = \frac{\sin \gamma_1}{\sin \gamma_2} \tag{1-44}$$

where γ = pitch angle (Fig. 1-36).

1-44 Forms of Bevel Teeth

In the simplest design the tooth elements are straight radial, converging at the cone apex. However, it is possible to have the teeth curve along a spiral apex, resulting in greater tooth overlap, analogous to the overlapping action of helical teeth. The result is a spiral bevel tooth. In addition, there are other possible variations. One is the zerol bevel, which is a curved tooth with elements that start and end on the same radial line.

Straight bevel gears come in two variations, depending on the fabrication equipment. All current Gleason straight bevel generators are of the Coniflex form, which gives the tooth surfaces an almost imperceptible convexity. Older machines produce true straight elements. Common forms of bevel-gear teeth are shown in Fig. 1-37.

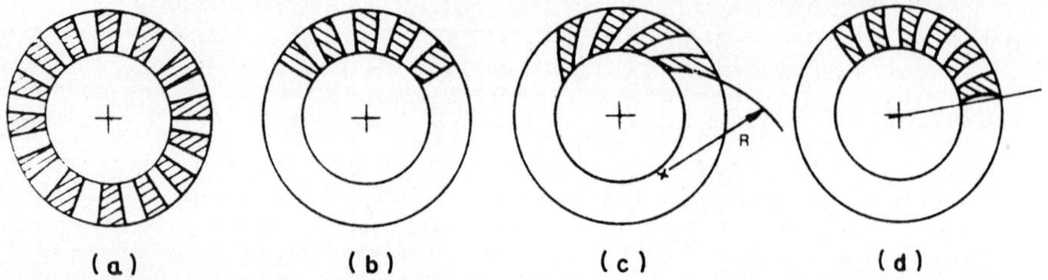

FIG. 1-37 Forms of bevel-gear teeth. (*a*) Straight teeth. (*b*) Coniflex teeth (exaggerated tooth curving). (*c*) Spiral teeth. (*d*) Zerol teeth.

CRITERIA OF GEAR QUALITY

In addition to the sizing of gear parameters, it is necessary to ensure that these specifications plus other controls result in the desired gear quality. This includes not only tolerances, but an understanding of what determines gear quality.

1-45 Basic Gear Formats

Specification of a gear requires a drawing that shows details of the gear body, the mounting design, face width, any special features, and the fundamental and essential gear data. These gear data can be efficiently and consistently specified on the gear drawing in a standardized block format. The format varies in accordance with gear type. A typical data block for standard fine-pitch spur gears is given in Fig. 1-38. Formats for coarse-pitch gears, helical gears, and other gear types are given in detail in the appendix of Ref. 3.

1-46 Tooth Thickness and Backlash

One of the most important criteria of gear quality is the specification and control of tooth thickness. The magnitude of tooth thickness, plus its tolerance, is a direct measure of backlash when the gear is assembled with its mate.

Although it is possible to set the tooth thickness and tolerance at any value within a wide range, convenient quality classes have been established by AGMA in *Gear Classification Manual 390.03.*

1-47 Position (or Transmission) Error

In many precision gear applications, the transmission of motion from shaft to shaft must have a high degree of linearity. This is known by several names: transmission linearity, angular transmission accuracy, and index accuracy. Theoretically, involute gears will function perfectly. However, in practice there are deviations from ideal motion transmission as a result of involute profile variations, spacing errors, pitch-line runout, and radial out-of-position. All these errors combined cause a net position error, which in the actual device results in position error.

RECOMMENDED MINIMUM FINE-PITCH SPUR AND HELICAL GEAR SPECIFICATIONS FOR GENERAL APPLICATIONS (See Note 1)

Arranged for Printing on a Standard Form Drawing or for Application to Drawing by Rubber Stamp
For proper use of capital or lower case lettering, see note #2

	SPUR GEAR DATA	
Basic Specification Data	Number of teeth (*N*)	
	Diametral pitch (*P*)	
	Pressure angle (ϕ)	
	Standard pitch diameter (*N/P*) (Ref.)	
	Tooth form (per AGMA 207.05) Max. calc. cir. thickness on std. pitch circle	
Mfg. and Inspection Data	Gear testing radius	
	AGMA quality number	
	Total composite tolerance	
	Tooth-to-tooth composite tolerance	
	Outside diameter	
	Master gear basic cir. tooth thickness at std. pitch circle (Ref.)	
	Master gear number of teeth (Ref.)	

	*HELICAL GEAR DATA	
BASIC SPECIFICATION DATA	NUMBER OF TEETH (*N*)	
	NORMAL DIAMETRAL PITCH (P_n)	
	NORMAL PRESSURE ANGLE (ϕ_n)	
	HELIX ANGLE — HAND ψ	
	STANDARD PITCH DIAMETER ($N/P_n \cos \psi$) (Ref.)	
	TOOTH FORM (PER AGMA 207.05)	
	MAX. CALC. NORMAL CIR. THICKNESS ON STD. PITCH CIRCLE	
MFG. AND INSPECTION DATA	GEAR TESTING RADIUS	
	AGMA QUALITY NUMBER	
	TOTAL COMPOSITE TOLERANCE	
	TOOTH-TO-TOOTH COMPOSITE TOLERANCE	
	LEAD	
	OUTSIDE DIAMETER	
	MASTER GEAR BASIC NORMAL CIR. TOOTH THICKNESS AT STD. PITCH CIRCLE (Ref.)	
	MASTER GEAR NUMBER OF TEETH (Ref.)	

*If desired, a combination format covering both spur and helical gears can be used by specifying the helix angle equal to zero degrees. This permits standardization on the helical drawing format for both spur and helical gears.

NOTE 1: For data on the determination of spur and helical tooth proportions, see Section 4 or Standard AGMA 207.05. For data on inspection, see Section 10. For data on quality number, see Section 6 or AGMA 390.03.

NOTE 2: The use of all upper case letters or both upper and lower case letters is optional. The spur gear format illustrates the proper use of both upper and lower case letters. The helical format illustrates the use of all upper case (capital) letters.

FIG. 1-38 Recommended minimum fine-pitch spur and helical-gear specifications. (*From AGMA Standards.*)

The single most important criterion of the position errors is the gear's total composite error (TCE). This is simply defined as the maximum variation in center distance as the gear is rolled, intimately meshed with a master gear, on a variable-center-distance fixture. The device has one center floating, and as the gears are rolled, any eccentricity, tooth-to-tooth variation, or profile deviation results in center-distance variation. This variation can be measured and plotted, as shown in Fig. 1-39. The TCE parameter encompasses the combination of runout and tooth-to-tooth errors, as indicated in Fig. 1-39. The latter,

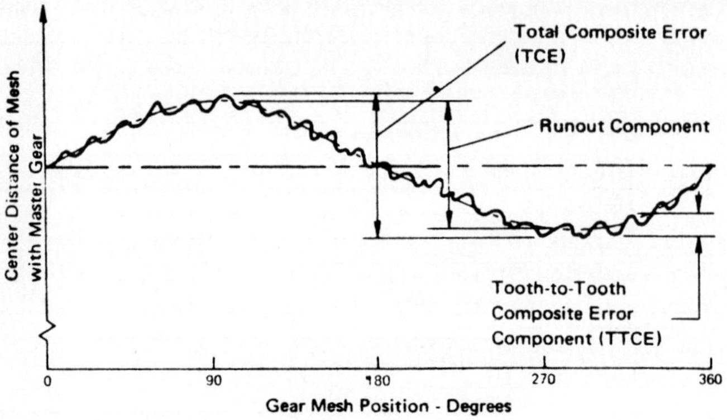

FIG. 1-39 Typical total composite error plot.

which is essentially the variation over a tooth cycle, is known as tooth-to-tooth composite error (TTCE). Both TCE and TTCE are controlled by specifying maximum values. Since TCE includes TTCE, it is necessary to specify both only when finer control of the TTCE is desired.

The relationship between TCE and transmission error is adequately approximated by the expression

$$E_T = \frac{E_{tc}}{2} \sin\theta \qquad (1\text{-}45)$$

where θ = mesh position of the gear.

This relationship indicates that position error fluctuates sinusoidally between maximum lead and lag values.

Equation (1-45) yields a linear position error measured in inches along the pitch circle. If an angular transmission error is desired, it is necessary to divide by the pitch radius of the gear. Thus

$$\begin{aligned} {}_aE_T &= \frac{E_{tc}}{2R}\sin\theta \qquad \text{rad} \\ &= \frac{3440}{D} E_{tc}\sin\theta \qquad \text{arcmin} \end{aligned} \qquad (1\text{-}46)$$

Equation (1-46) defines the error in a single gear. In practice, one is interested in the total error in a mesh arising from errors in both gears. Considering only the maximum error, to avoid the complexity of phase angle, the peak total mesh error is

$$\begin{aligned} \text{Maximum peak error} &= {}_aE_{T,\text{mesh}} = \\ &= \frac{(E_{tc})_1 + (E_{tc})_2}{R_{1,2}} 3440 \qquad \text{arcmin} \end{aligned} \qquad (1\text{-}47)$$

where subscripts 1 and 2 represent two meshed gears and R_1 and R_2 are the respective pitch radii yielding angular error values for the respective gear center of the particular pitch radius substituted.

1-48 AGMA Quality Classes

Using criteria that are indicators and measures of gear quality, AGMA has established convenient standards for a continuous spectrum of quality classes ranging from the crudest to the most precise gear. For coarse-pitch gears there are 13 classes, numbered 3 through 15. Fine-pitch gears are separately standardized because of the size effect of the smaller geometry and the much greater dependence on TCE and TTCE control parameters. There are 12 quality classes, numbered 5 through 16.

CALCULATION OF GEAR PERFORMANCE CRITERIA

Essential to proper application of gears is the derivation of values of functional criteria. Most important are backlash, transmission error, and total position error. In considering a meshed gear pair, their performance depends not only on specific gear parameters, but also on many installation and design features, such as bearings, shafting, and housing.

1-49 Backlash for a Mesh

A single gear cannot have backlash, only an allowance for backlash. Backlash occurs when two mating gears are assembled. The clearance between the teeth at mesh is backlash.

All backlash sources must be considered and summed to obtain a total backlash for the mesh. The sources are grouped into the following categories:

Design Backlash Allowance

1. Gear-size allowance—any specific reduction in tooth thickness (or testing radius) below nominal value
2. Center distance—any specific increase in center distance above nominal value

Major Tolerance Backlash Sources

1. Gear-size tolerance (tooth thickness or testing radius)
2. Center-distance tolerance

Gear Center Shift Due to Secondary Sources

1. Fixed bearing eccentricities:
   a. Ball-bearing outer race eccentricity
   b. Sleeve bearing's inside diameter and outside diameter runout

2. Radial clearances due to tolerances and allowances:
   a. Ball-bearing radial play
   b. Fit between shaft and bearing bore
   c. Fit between bearing outside diameter and housing bore

Backlash Sources Viable in Magnitude With gear rotation:

1. Total composite error
2. Clearance between gear bore and shaft
3. Runout at point of gear mounting
4. Ball bearing rotating race eccentricity

Miscellaneous Sources

1. Thermal dimensional changes
2. Deflections: teeth, gear body, shaft, and housing

A more complete and very detailed coverage of these backlash sources is given in Ref. 3.

Using the list of categories of backlash sources, those that contribute significantly can be evaluated and summed. Thus, the total backlash for a mesh is expressed as

$$B_{\text{mesh}} = \Sigma B_i \tag{1-48}$$

In using Eq. (1-48), it should be noted that all radial backlash sources, such as center-distance tolerance and radial shift due to eccentricities, must be converted to a backlash measure along the pitch circle, using Eq. (1-22), before they can be added to such sources as tooth thinning.

Also, note that the backlash sources can be separated into two categories: those constant with gear rotation, and those that vary in magnitude with gear rotation. The latter are the runout sources. Thus, backlash can be expressed as having two components:

$$B = B_c + B_v \tag{1-49}$$

where B_c = constant backlash
B_v = variable backlash

1-50 Transmission Error

The sources of transmission error originate from both the gears and the installation. Some of these sources also contribute to backlash. The complete list of usual sources is as follows:

Position Error in the Individual Gears

1. Total composite error (TCE)
   a. Single-cycle errors (pitch-line runout)
   b. High-frequency tooth-to-tooth composite errors (TTCE)

Installation Errors

1. Runout sources
   a. Clearance between gear bore and shaft
   b. Runout at point of gear mounting
   c. Ball-bearing rotating race eccentricity
   d. Miscellaneous runouts: component shaft and composite gear assembly
2. Miscellaneous error sources
   a. Shaft couplings
   b. Shaft and bearing creepage

These errors are transformed to angular position error in the same manner as TCE is converted by Eq. (1-46). Thus, the total transmission error for a mesh is the sum of all eccentricity error offsets:

$$_aE_{T,\text{mesh}} = \pm\frac{3440\Sigma E_i}{R} \qquad \text{arcmin} \tag{1-50}$$

where E_i = eccentricity (one-half runout value) of error contributors.

A more detailed explanation and analysis of transmission error can be found in Ref. 3. An example of backlash and transmission-error computation is given later in this text.

1-51 Integrated Position Error

Backlash and transmission error are independent functional discrepancies that arise from different causes and are not necessarily related in the gear's performance of its intended function. For example, in a servomotor gear train, backlash may be very important, whereas position error is immaterial. Alternatively, in a unidirection sensor positioning gear train, backlash is of little concern, but transmission error might be critical. However, in gear trains overall positional accuracy is often most important, and in such cases the backlash combines with the transmission error to yield an integrated position error (IPE). In essence the IPE is the total out of position. It is a sum of the backlash and transmission error. However, this is not simple to achieve, since many of the transmission-error sources are identical to the sources of variable backlash. Also, the transmission error varies between maximum lead and lag values. Details of the integration are beyond the scope of this book, but can be obtained in Ref. 3. The basic relationship for the peak value is

$$\text{Peak IPE} = E_I = \pm\left(E_T + \frac{B_c}{2}\right) \tag{1-51}$$

where B_c = backlash constant with rotation
E_T = transmission error (± peak value)

1-52 Control of Backlash

In the many cases in which it is necessary to minimize backlash, a proper control must be chosen. The direct approach of narrowing all allowances and tolerances on sources is effective. Accordingly, precision gear qualities are specified, particularly in regard to testing radius (tooth thickness) and TCE. However, there are practical limitations, since cost increases exponentially with precision. Some method of circumventing extremes of precision must be used. An alternative means of controlling backlash is to use adjustable centers or spring-load the gears, using one of several different designs. In this regard the spring-loaded scissor gear has particular merit, since all backlash is continually eliminated. However, it is limited to low-torque applications. Consult Ref. 3 for in-depth coverage of various types of backlash control and elimination schemes.

1-53 Control of Transmission Error

The methods available for controlling transmission error are much more limited than the means of controlling backlash. The most effective is direct control of error sources by specification of narrow tolerances. This means precision categories for TCE, TTCE, and installation components such as shafting and ball bearings.

In special cases, when the gear mesh is 1:1 ratio, it is possible to calibrate the gears and match the pitch-line runouts to provide error cancellation. However, besides being costly and not foolproof, this method is very limited, since it requires not only a 1:1 ratio, but also runout errors of like magnitudes in both gears.

GEAR-TRAIN DESIGN AND ANALYSIS

Optimum design of a gear train involves not only a total gear reduction, but choices regarding number of meshes, pitch diameter, total backlash, transmission error, and other such factors. There are many different considerations, and their relative importance will be determined by the application and the requirements. Several important design features and analyses are reviewed in the following sections.

1-54 Gear-Train Types and Gear Ratios

There are three basic types of gear trains, irrespective of the gear types involved: simple, compound, and planetary gear trains. The velocity ratio or gear ratio of the train is of primary importance. This is always arranged to be greater than 1.

Simple Gear Train A single mesh constitutes a simple gear train, which is defined as a gear train that has only one gear on each shaft. Two shafts are coupled by a meshed pair, as in Fig. 1-40. The gear ratio for such a train is merely the ratio of the two numbers of teeth (and for spurs, parallel-shaft helicals, and bevel meshes, also the ratio of pitch diameters).

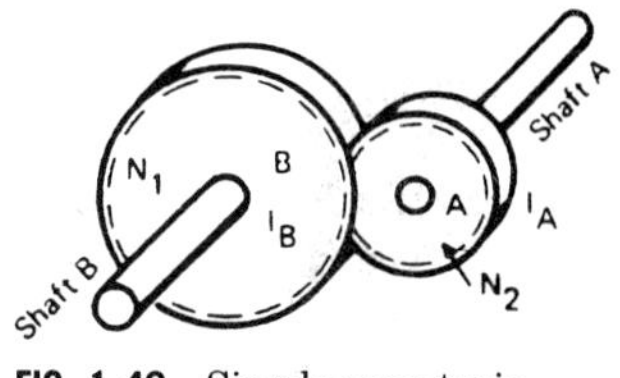

FIG. 1-40 Simple gear train.

$$\text{Gear ratio} = Z = \frac{N_1}{N_2} \tag{1-52}$$

An extension is a series of one or more gear meshes, as in Fig. 1-41. The overall gear ratio of such a train is also given by Eq. (1-52), where N_1 and N_2 define the numbers of teeth in the first and last gears, respectively. The in-between gears act as idlers and serve no purpose beyond changing the direction of rotation and extending the center distance between the first and last gears.

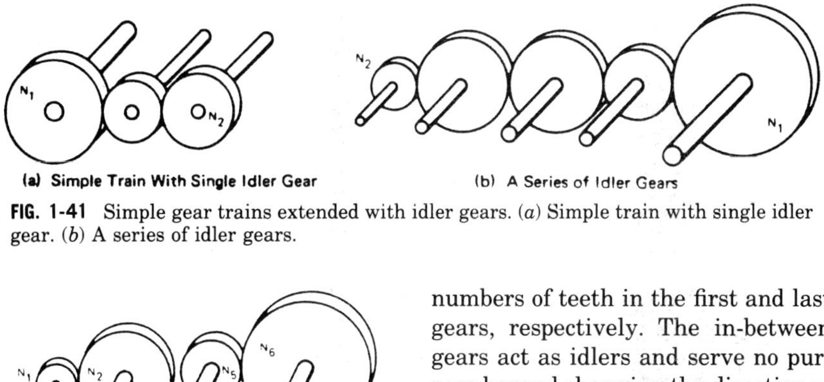

(a) Simple Train With Single Idler Gear (b) A Series of Idler Gears

FIG. 1-41 Simple gear trains extended with idler gears. (*a*) Simple train with single idler gear. (*b*) A series of idler gears.

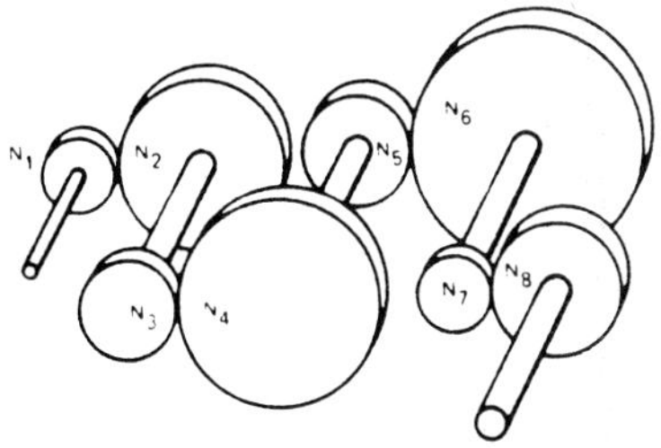

FIG. 1-42 A compound gear train.

Compound Gear Train This is the most prevalent type. It is defined as a gear train which has at least one shaft carrying two or more gears. A typical compound train is shown in Fig. 1-42. The feature of the compound train is the cumulative gear reduction. This is expressed as

$$\text{Gear-train ratio} = v = \frac{N_2}{N_1}\frac{N_4}{N_3}\frac{N_6}{N_5}\cdot\ .\ .\ = \Pi Z_i \tag{1-53}$$

where Z_i = individual mesh gear ratios

Planetary Gear Train This is a special type of train, also known as an epicyclic, in which one or more of the gear axes rotate. An example is shown in

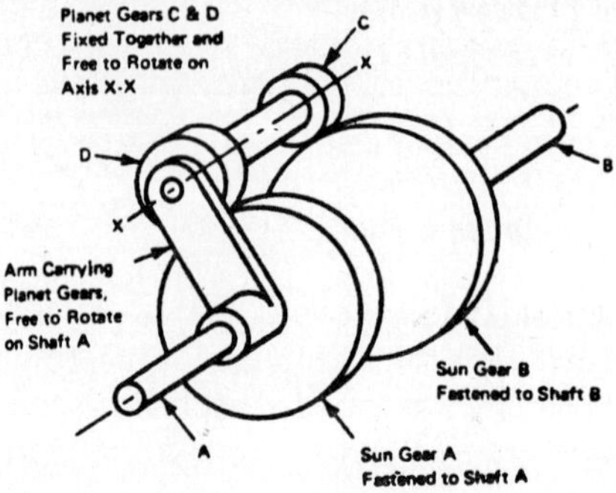

FIG. 1-43 Example of a planetary gear train.

Fig. 1-43. Determination of the gear ratio is much more involved than for the compound train. Several methods for calculating the gear ratio through a planetary train are available; these include relative velocity synthesis and tabulation and superposition of step motions. For details, consult Ref. 2.

Simplified Gearing Schematic Representing a gear train by a three-dimensional pictorial as in Fig. 1-42 is tedious and requires costly graphics. More often, trains are pictured in a two-dimensional stretch-out schematic, illustrated in Fig. 1-44. Numbers of teeth and pitch are designated as a fraction.

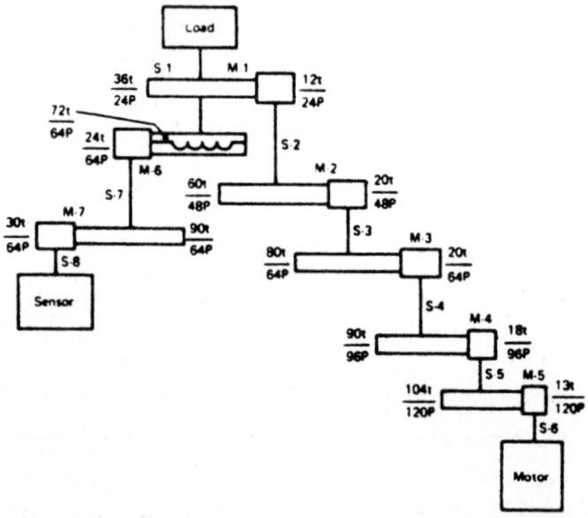

FIG. 1-44 Schematic representation of a gear train.

1-55 Gear-Train Backlash Computations

The first step in calculating the total backlash of a gear train is to determine the angular backlash of each mesh, using the procedures in Sec. 1-49. Then the total backlash for the train is obtained by summing the individual backlashes of each mesh, or stage, as it is frequently termed. However, the total backlash is referred to a specific shaft, such as the slowest shaft (usually the output). Each gear-stage backlash figure must be modified by the gear ratio factor from it to the reference measurement shaft. This is expressed as

$$ {}_aB_{\text{train}} = {}_aB_{m-1} + \frac{{}_aB_{m-2}}{v_1} + \frac{{}_aB_{m-3}}{v_2} + \cdots = \sum \frac{{}_aB_{m-i}}{v_i} \qquad (1\text{-}54) $$

where v_i = gear-stage ratio between a specific shaft in the ith mesh and the reference measurement shaft.

1-56 Gear-Train Transmission-Error Computation

In a procedure parallel to that for determining backlash, the total transmission error for a train is calculated by determining the angular transmission error for each mesh, using Eq. (1-50). Then all meshes are summed, taking into account the gear ratio of each stage. Thus,

$$ {}_aE_{T,\text{train}} = {}_aE_{T_{m-1}} + \frac{{}_aE_{T_{m-2}}}{v_1} + \frac{{}_aE_{T_{m-3}}}{v_2} + \cdots = \sum \frac{{}_aE_{T_{m-i}}}{v_i} \qquad (1\text{-}55) $$

1-57 Integrated Position Error for a Train

Calculation of this criterion follows from an extension of Eq. (1-51). Gear-train values of B_c and E_T are substituted into the equation to yield a peak IPE value.

1-58 Statistics and Probabilistic Design

The procedures outlined in the previous sections yield specific magnitudes of backlash, transmission error, and integrated position error for particular values of error sources. However, they neglect cancellation effects of the sources and errors that depend on mesh position and phasing. Thus, the procedures are satisfactory for calculating worst-case or maximum error values, such as when all contributing sources are at their extreme tolerance values.

Worst-case figures are of limited use, since they represent improbable magnitudes and combinations of error sources. Much more useful to designers are figures representing the expected distribution of the gear train's backlash, transmission error, and IPE. By utilizing known statistical procedures and data, it is possible to derive reliable probabilistic statistical error values. These are very useful in optimizing a design for functional performance, tolerances, and cost. Unfortunately, the subject is outside the scope of this text. Readers should refer to Chap. 5 of Ref. 3, which covers in detail the theory of probabilistic statistical design and analysis along with its application to gear trains.

1-59 Example of Backlash and Transmission-Error Computation

To illustrate error analysis, the following numerical example is presented. This is based on the gear-train schematic shown in Fig. 1-44 and pertinent design data given in Table 1-8. The layout of the gear train, showing orientation of the line of centers θ, is given in Fig. 1-45. The two meshes connecting the sensor to the load shaft have spring-loaded gears to eliminate all backlash.

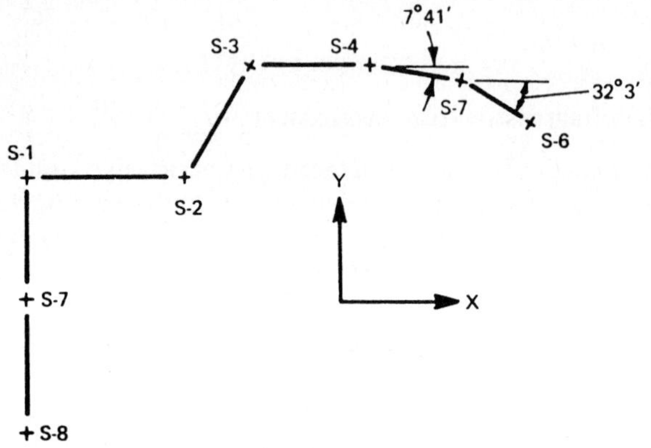

FIG. 1-45 Gear-train-centers layout.

The following quantities must be calculated: (1) the maximum backlash between the motor and the load shafts, measured at S-1; (2) the maximum transmission error between the load shaft and the sensor shaft S-8, measured at the load shaft S-1; and (3) the maximum integrated position error for item 2. Assume all gears are 20° pressure angle.

Maximum Backlash Calculation (S-1 to S-6) This is determined in accordance with Eq. (1-48), taking into account all the applicable backlash sources for each mesh.

With regard to center-distance variation, the orientation angle θ is significant, since it is specified by X-Y coordinates. Center-distance variation is related to the coordinate tolerance and angle θ by the expression*

$$t_e = t(\sin\theta + \cos\theta)$$

where t = coordinate tolerance ($\pm$ value)
t_e = effective center-distance tolerance ($\pm$ value)

Using this equation, a maximum center-distance opening can be determined for each mesh. In addition, any center-distance allowance must be applied.

* Derivations of this equation and cases of unequal coordinate tolerances are given on p. 88 of Ref. 3.

Summing all the backlash sources for each mesh, the results are:

Mesh 1

Max ΔC = center-distance allowance + center-distance tolerance + testing radius allowance + testing radius tolerance + TCE + bearing outer race runout + bearing inner race runout + bearing radial play + gear-to-shaft clearance + bearing to housing clearance + shaft runout

Substituting values from Table 1-8,

$$\text{Max } \Delta C = 0 + 2(0.0005) + 2(0.0005) + 2(0.0008) + 2(0.0005) + 2\left(\frac{0.0002}{2}\right) + 2\left(\frac{0.0002}{2}\right) + 2\left(\frac{0.0005}{2}\right) + \frac{0.0002 + 0.0002 + 0.0001}{2} + 2\left(\frac{0.0002 + 0.0003}{2}\right) + \frac{0.0002}{2} = 0.00635 \text{ in}$$

In this equation, many of the terms are doubled because the backlash source is applicable to both gears or shafts of the mesh. A number of the backlash sources are halved because the values stipulated in Table 1-8 are TIR (total indicated runout) values, and it is only one-half the runout that causes a maximum backlash.

The derived maximum increase in center distance is converted to a linear backlash using Eq. (1-22), and to an angle value using Eq. (1-24). Thus, for mesh 1,

$$B_{m-1} = (0.00635)(2 \tan 20°) = (0.00635)(2)(0.364) = 0.0046 \text{ in}$$

and the angular backlash is

$${}_aB_{m-1} = 3440\,\frac{B}{R} = 3440\,\frac{0.0046}{0.750} = 21.6 \text{ arcmin}$$

In a similar manner, meshes 2 through 5 are calculated, and the summary of the results is:

	Max ΔC	B	${}_aB$	Angular measure at shaft no.
Mesh 1	0.00635	0.0046	21.6	S-1
Mesh 2	0.00967	0.00704	38.7	S-2
Mesh 3	0.00970	0.00706	38.9	S-3
Mesh 4	0.00982	0.00715	52.4	S-4
Mesh 5	0.00101	0.00736	58.4	S-5

TABLE 1-8 Design Details of Pertinent Parameters for the Gear Train of Fig. 1-44

Parameter	Mesh 1	Mesh 2	Mesh 3	Mesh 4	Mesh 5	Meshes 6 and 7
1. Center distance:						
Allowance (per mesh)	0.0000	0.0002	0.0002	0.0005	0.0005	0.0002
Tolerance on *X*-*Y* coordinates	±0.0005	±0.0005	±0.0005	±0.0005	±0.0005	±0.0005
θ (Line-of-center angle with *X* axis)	0°	60°	0°	7°41′	32°3′	90°
2. Testing radius:						
Allowance (each gear)	0.0005	0.0010	0.0010	0.0010	0.0010	0.0005
Tolerance (each gear)	$\begin{pmatrix} \pm 0.0000 \\ -0.0008 \end{pmatrix}$	$\begin{pmatrix} \pm 0.0000 \\ -0.0010 \end{pmatrix}$	$\begin{pmatrix} \pm 0.0000 \\ -0.0012 \end{pmatrix}$	$\begin{pmatrix} \pm 0.0000 \\ -0.0012 \end{pmatrix}$	$\begin{pmatrix} \pm 0.0000 \\ -0.0012 \end{pmatrix}$	$\begin{pmatrix} \pm 0.0000 \\ -0.0012 \end{pmatrix}$
3. Total composite error (each gear)	0.0005	0.0010	0.0010	0.0010	0.0010	0.0005
4. Ball bearings:						
ABEC class	5	3	3	3	3	5
Fixed outer race runout	0.0002	0.0004	0.0004	0.0004	0.0004	0.0002
Inner race runout	0.0002	0.0002	0.0002	0.0002	0.0002	0.0002
Radial play	0.0001 to 0.0005	0.0001 to 0.0005	0.0001 to 0.0005	0.0001 to 0.0005	0.0001 to 0.0005	0.0001 to 0.0005

5. Gear bore to shaft clearance:						
Shaft diameter tolerance	$\begin{pmatrix}+0.000\\-0.0002\end{pmatrix}$	$\begin{pmatrix}+0.0000\\-0.0003\end{pmatrix}$	$\begin{pmatrix}+0.0000\\-0.0003\end{pmatrix}$	$\begin{pmatrix}+0.0000\\-0.0003\end{pmatrix}$	$\begin{pmatrix}+0.0000\\-0.0003\end{pmatrix}$	$\begin{pmatrix}+0.0000\\-0.0002\end{pmatrix}$
Bearing bore tolerance	$\begin{pmatrix}+0.0000\\-0.0002\end{pmatrix}$	$\begin{pmatrix}+0.0000\\-0.0002\end{pmatrix}$	$\begin{pmatrix}+0.0000\\-0.0002\end{pmatrix}$	$\begin{pmatrix}+0.0000\\-0.0002\end{pmatrix}$	$\begin{pmatrix}+0.0000\\-0.0002\end{pmatrix}$	$\begin{pmatrix}+0.0000\\-0.0002\end{pmatrix}$
Allowance	0.0001	0.0001	0.0001	0.0001	0.0001	0.0001
6. Bearing to housing fit:						
Bearing OD tolerance	$\begin{pmatrix}+0.0000\\-0.0002\end{pmatrix}$	$\begin{pmatrix}+0.0000\\-0.0003\end{pmatrix}$	$\begin{pmatrix}+0.0000\\-0.0003\end{pmatrix}$	$\begin{pmatrix}+0.0000\\-0.0003\end{pmatrix}$	$\begin{pmatrix}+0.0000\\-0.0003\end{pmatrix}$	$\begin{pmatrix}+0.0000\\-0.0002\end{pmatrix}$
Housing bore tolerance	$\begin{pmatrix}+0.0003\\-0.0000\end{pmatrix}$	$\begin{pmatrix}+0.0003\\-0.0000\end{pmatrix}$	$\begin{pmatrix}+0.0003\\-0.0000\end{pmatrix}$	$\begin{pmatrix}+0.0003\\-0.0000\end{pmatrix}$	$\begin{pmatrix}+0.0003\\-0.0000\end{pmatrix}$	$\begin{pmatrix}+0.0003\\-0.0000\end{pmatrix}$
Allowance	0	0	0	0	0	0
7. Shaft runout at gear mounting	0.0002	0.0003	0.0003	0.0003	0.0003	0.0002

NOTE: All pinions are integral with the shafts and there is no gear bore to shaft clearance.

The total backlash measured at shaft S-1 is given by Eq. (1-54) as

$$
\begin{aligned}
(\max)_a B_{\text{train}} &= 21.6 + \frac{38.7}{3} + \frac{38.9}{9} + \frac{52.4}{36} + \frac{58.4}{180} \\
&= 21.6 + 12.9 + 4.32 + 1.46 + 0.324 \\
&= 40.6 \text{ arcmin}
\end{aligned}
$$

The effect of gear ratio demagnifying backlash error is clearly seen in this example. The first mesh contributes 53 percent of the backlash, and successive meshes contribute progressively less. The fourth and fifth meshes contribute such a small portion that they can be neglected without seriously affecting the answer.

If the backlash at the motor is of interest, it will be the same backlash figure multiplied by the gear-train ratio between S-1 and S-6:

$$
{}_aB_{\text{train}} = \frac{40.6 \times 1440}{60} = 974° \qquad \text{at shaft S-6}
$$

Transmission Error Calculations (S-1 to S-8) Only meshes M-6 and M-7 are involved in calculation of this error. The sources of transmission error are TCE of the gears, runouts of the bearings and shafts, and effective runout of the shafts resulting from looseness between the gear bores and the mounting diameter of the shaft.

The error sources are summed for each mesh and converted to angular values using Eq. (1-50), as follows:

Mesh 6

$$
(\max)E_t = \pm \frac{1}{2}(\text{TCE} + \text{bearing inner race runout} + \text{gear bore to shaft clearance} + \text{shaft runout})
$$

In substituting values, the last two error sources do not apply to the pinion shafts, since they are integral. Therefore

$$
\begin{aligned}
(\max)E_T &= \pm \frac{1}{2}[2(0.0005) + 2(0.0002) + 0.0005 + 0.0002] \\
&= \pm 0.00105
\end{aligned}
$$

$$
(\max)_a E_T = \frac{E_T}{R} 3440 = \pm \frac{0.00105}{1.125/2} 3440 = \pm 6.42 \text{ arcmin}
$$

Mesh 7

This mesh is the same design as mesh 6, and therefore the same linear E_T results. Only the angular transmission errors differ as a result of the different pitch diameter of the gears on shafts S-7 and S-8. Thus,

$$(\max)_a E_T = \pm \frac{0.00105}{1.406/2} 3440 = \pm 5.14 \text{ arcmin}$$

Total transmission error from the sensor to the load shaft is calculated using Eq. (1-55) as

$$\begin{aligned} {}_aE_{\text{train}} &= \pm \left(6.42 + \frac{5.14}{3}\right) \\ &= \pm\ 8.13 \text{ arcmin} \qquad \text{measured at S-1} \end{aligned}$$

Integrated Position Error (S-1 to S-8) This is a combination of the transmission error and backlash meshes 6 and 7, from Eq. (1-51). Since mesh 6 is spring-loaded to eliminate all backlash, only the backlash of mesh 7 needs to be integrated. However, just the constant-backlash sources are to be considered. Thus, for mesh 7,

$$\begin{aligned} \Delta C = {} & \text{center-distance allowance} + \text{center-distance tolerance} \\ & + \text{testing radius allowance} + \text{testing radius tolerance} \\ & + \text{bearing outer race runout} \\ & + \text{bearing radial play} + \text{bearing-to-housing clearance} \end{aligned}$$

Substituting values from Table 1-8,

$$\begin{aligned} \Delta C &= 0.0002 + 2(0.0005) + 2(0.0005) - 2(0.0012) \\ &\quad + 2\left(\frac{0.0002 + 0.0005}{2}\right) + 2\left(\frac{0.0002 + 0.0003}{2}\right) \\ &= 0.0058 \text{ in} \end{aligned}$$

Converting this radial opening of the gear centers to linear backlash measure by Eq. (1-22),

$$B = (0.0058)(2 \tan 20^\circ) = (0.0058)(0.728) = 0.0042 \text{ in}$$

and the angular backlash is

$$B_c = \frac{0.0042}{1.406/2} 3440 = 20.5 \text{ arcmin}$$

From Eq. (1-51), the integrated position error from the load shaft S-1 to the sensor shaft S-8 measured at S-1 is

$$(\max)\text{IPE} = E_I = \pm \left(E_T = \frac{B_c}{2}\right)$$

Substituting the B_c value, modified by the gear ratio to S-1, and the transmission error previously found,

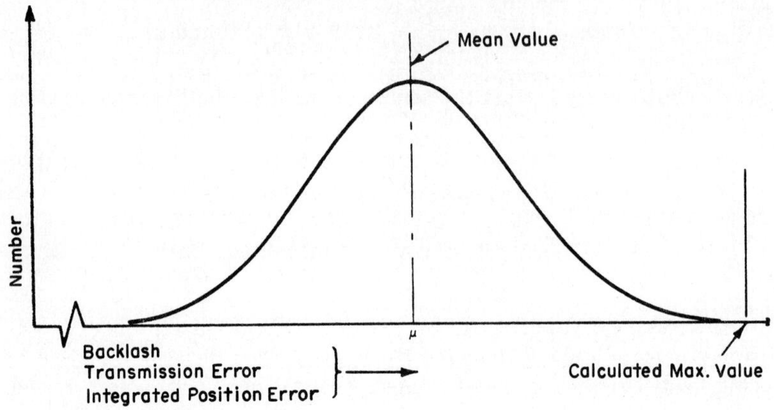

FIG. 1-46 Statistical distribution of gear-train backlash, transmission error, or integrated position error.

$$E_I = \pm\left(8.13 + \frac{1}{2}\frac{20.5}{3}\right) = \pm 11.5 \text{ arcmin} \qquad \text{measured at S-1}$$

These values are maximums and represent the worst possible results, with all error sources to be at their extreme tolerance values. In practice, a realistic distribution of parameter tolerances will result in a distribution of backlash, transmission error, and IPE far below the desired maximum values. This is typified by Fig. 1-46. Usually mean values will be only a fraction of the maximums. Details of how to derive statistical distributions of gear-train backlash, transmission error, and IPE are given in Ref. 3.

1-60 Gear-Train Inertia Calculation

A gear train's moment of inertia is important when the train is to respond in a minimum time, and also for vibration analysis. The inertia (abbreviation for moment of inertia) of a gear train is the sum of the inertias of all the gears, shafts, and elements tied to the shafts. However, in summing these inertias, the effect of gear ratio must be taken into account.

For the simple gear train of Fig. 1-40, it can be shown that the inertia of gear B, measured on shaft A, is inversely proportional to the square of the gear ratio; thus, measured on shaft A, the total inertia of both pairs is

$$I_{\text{pair}} = I_A + \frac{I_B}{Z^2} \qquad I = \text{mass moment of inertia} \tag{1-56}$$

Expanding this to an entire train, such as that of Fig. 1-42, the total inertia measured at the slowest speed shaft is

$$I_S = I_1 + \frac{I_2}{V_1^2} + \frac{I_3}{V_2^2} + \cdots \tag{1-57}$$

where I_s = total inertia of system
$I_1, I_2, I_3, \ldots$ = total inertia of the particular shaft assembly
$V_1, V_2, \ldots$ = stage gear ratio from the particular shaft to shaft 1

Note that Eq. (1-57) is set up for a speed-reducing train, in which one is interested in the reflected inertia of the low-speed stages as seen, for example, by the driving motor. Looking at the gear system from the slow-speed end, the stage inertias are multiplied by the gear-ratio square factors. This case, and all cases in general, can be handled by Eq. (1-57) if v_i is allowed to have values above and below 1, in accordance with the way in which the ith shaft relates to the total inertia measurement shaft.

To simplify calculations, a solid plain gear can be approximated to a disk with radius equal to the pitch radius and thickness equal to the face width. Thus the moment of inertia is given by

$$I_g = \frac{mr^2}{2} \tag{1-58}$$

where m = weight in mass units = w/g.

In calculating the inertia of each shaft, the significance of the shaft itself depends on the relative size of the gear. Pinion shafts may contribute significant inertia, whereas gears with significant diameter often make the shaft inertia negligible.

GEAR STRENGTH AND DURABILITY

Gear failure can occur as a result of tooth breakage or from surface failure in the form of fatigue and wear. The first is referred to as tooth strength and the latter as durability. Strength is determined, for both static and dynamic conditions, in terms of tooth-beam stresses. It follows well-established formulas and procedures that yield results in which one can have confidence. Durability rating is evaluated in terms of surface stresses, influenced not only by dynamics, but also by material combinations, lubrication, and a considerable number of empirically derived factors.

1-61 Bending Tooth Strength

Tooth loading produces stresses that can ultimately result in tooth breakage. This is not a prevalent type of failure because the mechanical properties of gear materials are well known, and the design equations are adequately accurate. Bending stresses are analyzed as follows:

In transmitting power the driving force acts along the line of action, and the tooth sees a moving force acting from the tip to the base, as shown in Fig. 1-47. The load can be resolved into a tangential force W_t causing bending and a normal force W_N causing compression. These are shown in Fig. 1-48 along with the net stresses.

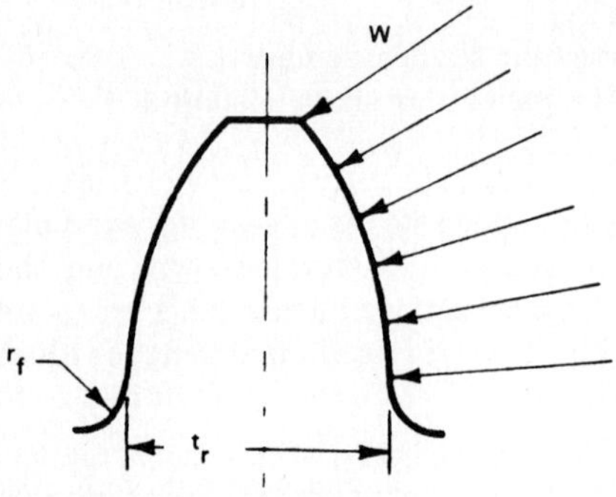

FIG. 1-47 Load action on tooth.

Lewis Formula An old and still useful tooth beam strength formula was developed by Wildred Lewis in 1893. He assumed that one tooth carried the load and, to make the condition even more extreme, placed the load at the top. Then he inscribed a constant-strength parabola cantilever in the tooth, as shown in Fig. 1-49. Details of the derivation of his formula are omitted, but the resulting stress/load expression is

$$W_t = \frac{SFY}{P_d} \qquad (1\text{-}59)$$

where W_t = transmitted load
S = maximum bending tooth stress, at the root outer fibers
F = face width of gear
Y = Lewis factor

The Lewis factor is dimensionless and independent of tooth size; it is only a function of shape. Lewis factors for standard teeth are given in Table 1-9.

A safe stress level depends on the material and the number of stress cycles to which the teeth are subjected. This can be evaluated from an *S-N* curve, modified Goodman diagram, Soderberg line, or equivalent data. Reference 4 contains helpful information on fatigue stress analysis.

Using the material's proper limiting stress value S_e in Eq. (1-59) results in a calculated tooth beam strength load W_b. For an acceptable design, $W_b \geqslant W_t$.

The transmitted load is calculated from horsepower being transmitted by

$$W_t = \frac{126{,}000P_t}{DN_r} \qquad (1\text{-}60)$$

where P_t = transmitted horsepower
N_r = gear speed, r/min
D = gear pitch diameter

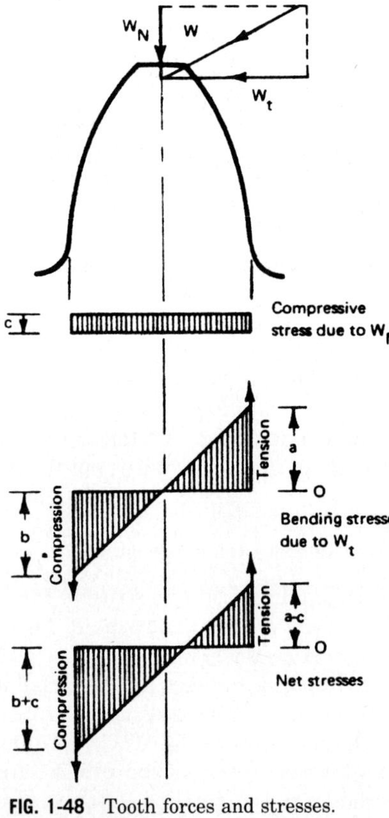

FIG. 1-48 Tooth forces and stresses.

The loading conditions assumed by the Lewis equation are very conservative, and a modification that results in a more realistic situation was made by Dudley (Ref. 2), giving

$$W_t = \frac{m_p SFY}{P_d} \tag{1-61}$$

where the contact ratio m_p takes into account the fact that when the load is at the tip, a second pair of teeth are in contact, sharing the load.

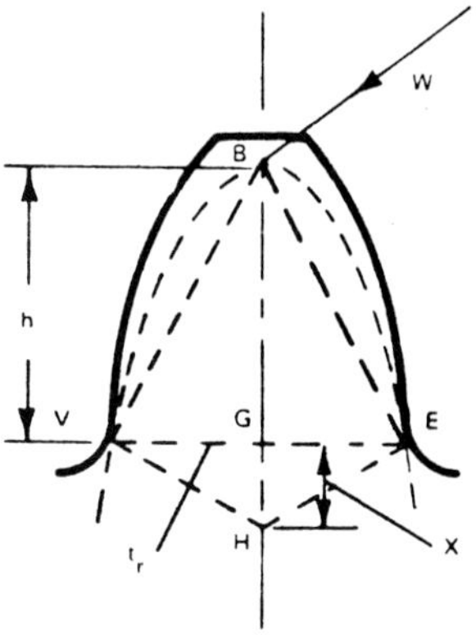

FIG. 1-49 Lewis formula tooth geometry.

1-62 Dynamic Strength

Equations (1-59) and (1-61) give adequate results for gear meshes that are in a static situation. When gears are in action, however, tooth loading is greater than the static value, due to dynamics. In gear systems, dynamic forces originate from a combination of the masses involved and the elasticity of the elements, plus a forcing function. Inaccuracies in gear-tooth profiles cause accelerations and decelerations during gear action which reflect as inertia forces; these can greatly exceed static tooth loadings. The severity of dynamic forces is a function of pitch-line velocity and tooth errors.

Predicting dynamic force precisely is very difficult, and to handle the situation various factors and formulas have been devised to increase the static tooth force to a value that handles the dynamic condition safely. A dynamic factor DF is used to modify static tooth strength, Eqs. (1-59) and (1-61), such that

$$W_d = W_t \cdot \mathrm{DF} \tag{1-62}$$

and, for acceptable design,

$$W_b \geqslant W_d$$

With the aid of empirical data, Buckingham established the increment of dynamic increase of the transmitted force as a function of profile errors, acceleration forces, elasticity properties, forces required to deform the teeth an amount equivalent to the errors, and pitch-line velocity. His simplified equation is

For spur gears,

$$W_d = W_t + \frac{0.05V(FC + W_t)}{0.05V + (FC + W_t)^{1/2}} \tag{1-63}$$

and for helical gears,

$$W_d = W_t + \frac{0.05V(FC \cos^2 \psi + W_t) \cos \psi}{0.05V + (FC \cos^2 \psi + W_t)^{1/2}} \tag{1-64}$$

where V = pitch-line velocity, f/min
F = active face width, in
C = deformation factor

TABLE 1-9 Lewis *Y* Factors

Number of teeth	14½° full-depth involute	20° full-depth involute
10	0.176	0.201
11	0.192	0.226
12	0.210	0.245
13	0.223	0.264
14	0.236	0.276
15	0.245	0.289
16	0.255	0.295
17	0.264	0.302
18	0.270	0.308
19	0.277	0.314
20	0.283	0.320
22	0.292	0.330
24	0.302	0.337
26	0.308	0.344
28	0.314	0.352
30	0.318	0.358
32	0.322	0.364
34	0.325	0.370
36	0.329	0.377
38	0.332	0.383
40	0.336	0.389
45	0.340	0.399
50	0.346	0.408
55	0.352	0.415
60	0.355	0.421
65	0.358	0.425
70	0.360	0.429
75	0.361	0.433
80	0.363	0.436
90	0.366	0.442
100	0.368	0.446
150	0.375	0.458
200	0.378	0.463
300	0.382	0.471
Rack	0.390	0.484

Values of C for common material combinations and a range of tooth action errors are presented in Table 1-10. These tooth errors can be equated to total composite and tooth-to-tooth composite errors.

1-63 Surface Durability

The Lewis formula and its modification to incorporate consideration of dynamic conditions is limited to beam stress analysis. In addition, there are stresses generated in the surface layers of teeth by direct crushing action of the forces. These stresses can exceed the limits of material structure and result in pitting, scoring, scuffing, seizing, and plastic deformation.

Pitting This is the removal of small bits of metal from the surface as a result of fatiguing, leaving small holes or pits. This is caused by heavy tooth load developing an excessive surface stress, a high local heat due to high rubbing speeds, or inadequate lubrication. Minute cracking of the surface develops, which spreads, and ultimately small bits break out of the tooth surface.

Scoring This is a heavy scratch pattern extending from the tooth root to the tip. It is as if a heavily loaded tooth pair had dragged foreign matter between the sliding teeth. It can be caused by lubricant failure, incompatible materials, and overload.

Scuffing This is a surface destruction that results from plastic material plus superimposed gouges and scratches caused by loose metallic particles acting as an abrasive between the teeth. Both scoring and scuffing are associated with welding (or seizing) and plastic deformation. It is often difficult to distinguish among the several failure types, as there is considerable intermingling.

There have been many attempts to derive expressions for calculating safe surface stress. The Buckingham durability equations, based upon Hertz con-

TABLE 1-10 Values of Deformation Factor C

Materials, pinion and gear	Tooth form	Error in action, in					
		0.0005	0.001	0.002	0.003	0.004	0.005
Cast iron and cast iron	14½°	400	800	1600	2400	3200	4000
Steel and cast iron	14½°	550	1100	2200	3300	4400	5500
Steel and steel	14½°	800	1600	3200	4800	6400	8000
Cast iron and cast iron	20° full depth	415	830	1660	2490	3320	4150
Steel and cast iron	20° full depth	570	1140	2280	3420	4560	5700
Steel and steel	20° full depth	830	1660	3320	4980	6640	8300
Cast iron and cast iron	20° stub tooth	430	860	1720	2580	3440	4300
Steel and cast iron	20° stub tooth	590	1180	2360	3540	4720	5900
Steel and steel	20° stub tooth	860	1720	3440	5160	6880	8600

tact stresses and the work of others, can be found in the references. All of the various design equations and procedures are closely related to specific empirical data and experience. Currently the AGMA equations are in wide use in the United States.

1-64 AGMA Strength and Durability Ratings

The AGMA rating formulas again combine analytics, approximations, and empirical data. A complete treatment of AGMA practices is too extensive for this writing, and only an introductory survey is offered. Complete details are available from AGMA literature and Chap. 11 of Ref. 4.

The AGMA formulas pertain to strength and surface durability, with dynamic and other conditions fully conditioned. The equations are as follows:

Tooth Strength (bending stress)

$$S_t = \frac{W_t K_o}{K_v} \frac{P_d}{F} \frac{K_s K_m}{I} \tag{1-65}$$

Surface Durability

$$S_c = C_P \left(\frac{W_t C_o}{C_v} \frac{C_s}{dF} \frac{C_f C_m}{I} \right)^{1/2} \tag{1-66}$$

which state that stress is a function of load, size, and stress distribution.

These calculated stresses must be less than the material's allowable stress values, which must take into account application conditions. These are related as follows:

Allowable bending stress:

$$S_t \leq S_{at} \frac{K_L}{K_r K_T} \tag{1-67}$$

Allowable surface durability stress:

$$S_c \leq S_{ac} \frac{C_L C_H}{C_R C_T} \tag{1-68}$$

The terms in Eqs. (1-67) and (1-68) are defined in Table 1-11.

This equation for tooth strength is a modified Lewis equation. The extent of departure and the better accommodation to actual performance depends on the chosen coefficients associated with each term.

Surface durability found using the equation is related to the long-in-use Hertz contact stress formula. Again, coefficients which better relate the theory to actual teeth should be used.

The coefficients in these equations are explained as follows:

TABLE 1-11 Definitions of Symbols in AGMA Rating Formulas

Term	Strength	Durability
Load:		
Transmitted load	W_t	W_t
Dynamic factor	K_v	C_v
Overload factor	K_a	C_a
Size:		
Pinion pitch diameter	—	d
Net face width	F	F
Transverse diametral pitch	P_d	—
Size factor	K_s	C_s
Stress Distribution:		
Load-distribution factor	K_m	C_m
Geometry factor	J	I
Surface condition factor	—	C_f
Stress:		
Calculated stress	s_t	s_c
Allowable stress	s_{at}	s_{ac}
Elastic coefficient	—	C_p
Hardness-ratio factor	—	C_H
Life factor	K_L	C_L
Temperature factor	K_T	C_T
Factor of safety	K_R	C_R

Load Distribution Factors C_m and K_m These factors cope with things that cause nonuniform load distribution across the gear width: profile errors, eccentricity of mounting, nonparallelism of shafts, and deflections and distortions. The effect of these errors is to cause a load concentration.

Overload Factors K_o and C_o These evaluate peak loads and other nonsmoothnesses of the driving and driven gear meshes.

Dynamic Factors K_v and C_v These relate to speed and gear errors which lead to dynamic loading. As pitch-line velocity increases, the dynamic load increment increases linearly. However, the dynamic effects of tooth errors are much more complex. Tooth-to-tooth errors, which come in a variety of forms, have a different dynamic effect than runout errors. Also, elastic tooth deflections cause apparent errors.

Life Factors K_L and C_L These factors are primarily intended to take into account the performance of gears whose life can be finite.

Factors of Safety K_R and C_R Although the factor of safety is old in engineering, in this usage it is related more rigorously to the degree of reliability sought.

Temperature Factors K_T and C_T These factors modify the design in accordance with adverse temperature effects upon lubricant performance. Usually this factor does not become significant until temperature exceeds 200°F.

Surface Factors C_f, C_H, and C_p The three durability factors, C_f, C_H, and C_p, for surface condition, hardness ratio, and elastic coefficient rate the ability of the gear's surface to resist wear.

Size Factors K_s and C_s These reflect the nonuniformity of material characteristics, such as hardness, and the dimension size of the gear. Within the size considerations are diameter, face width, tooth size, and ratio of case depth to tooth size.

Geometry Factors J and I These relate to the tooth proportions, primarily radii of curvature and load sharing. They are somewhat akin to the Lewis Y factors. For standard tooth proportions there are fixed values.

Allowable Stress S_{at} and S_{ac} This is the material's rated stress value as specified by the manufacturer or standards, or obtained from material testing. This value takes into account cyclic stressing and is the nominal endurance stress rating of the material.

Numerical Values of Factors Specific factor values are available from AGMA publications, and alternatively from self-made studies. Procedures for extracting factors are given in the AGMA literature. When either the conditions are such that a given factor is unimportant or there is insufficient information to establish it adequately, it is usually safe to equate the factor to 1. In most cases this results in a conservative or mid-value rating.

Evaluation of Equations This situation has outlined the AGMA procedures for determining strength and durability ratings. As an outline it cannot include detail, and therefore, to apply the procedures the reader should refer to the references.

GEAR MATERIALS

If gears are to achieve their intended performance, life, and reliability, gear material is very important. Often not all the requirements are compatible. High load capacity requires a tough, hard material which is difficult to machine, whereas high precision favors materials that are easy to machine, and therefore have lower mechanical property ratings. Light weight and small size favor light nonferrous materials, but high capacity dictates the opposite. Thus, trade-offs and compromise are required to achieve an optimum design.

Gear materials spread over a wide range extending from ferrous metals through the many nonferrous and lightweight metals, and down through the various plastics. The gear designer and user face a myriad of choices. The final selection should be based upon an understanding of materials properties and the requirements of the application.

1-65 Ferrous Metals

Despite the introduction of many new exotic metals and plastics with impressive characteristics in recent years, ferrous materials are still most widely used for gears because they offer high strength, response to heat treatment, and low cost. Cast iron and steel, carbon steels, and alloy steels are frequently used.

Cast Iron Cast iron is widely used for large gears where it is advantageous to reap machining savings by molding the gear blank. Cast steels offer similar advantage plus higher tensile and yield strengths, but cast iron has much better dynamic conditions, providing excellent internal properties.

Steel The steels are divided into two main divisions, plain carbon and alloy. The carbon steels offer low cost, reasonably easy machining, and ability to be hardened. A major disadvantage is no resistance to corrosion.

When other elements besides carbon are added to the iron, it is termed "alloy steel." These cover a wide range, from low-grade types to special alloys that offer exceptionally high strengths. Stainless steels are contained within this large grouping. Alloy steels offer a wide range of heat-treatment properties that make this gear material grouping the most versatile.

Stainless Steels Stainless steel can be divided into two groupings: the so-called 300 series true stainless steels, which resist nearly all corrosive conditions, and the 400 series, which, although not truly stainless, offers less corrosion resistance only in certain environments, such as certain acids and saltwater, and are otherwise considered stainless. The further bid distinction between the two series is that the 300 series generally are much more difficult to machine, nonmagnetic, and non-heat treatable, although somewhat responsive to cold working. The 400 series is magnetic, almost every alloy is heat treatable, and they have a much better index of machinability, corresponding to some of the carbon steels.

1-66 Nonferrous Metals

The commonly used nonferrous materials are the aluminum alloys and bronzes. Also, zinc die-cast alloys are used. Nonferrous metals generally or selectively offer good machinability, light weight, corrosion resistance, and nonmagnetism.

Aluminum As a gear material, aluminum has the special feature of light weight, with moderately good strength for the low weight. It is also corrosion resistant, and very easy to machine. A major disadvantage is its large coefficient of thermal expansion compared with the steels. There are many aluminum alloys offering different forming, machining, and casting features. Aluminum alloys respond to cold working and heat treatment.

Bronzes Bronzes have been used for a long time as gear materials; they offer good sliding properties when mated with steel gears. They are particularly good in worm meshes and crossed-helical meshes because of the large amount of sliding. Bronzes are extremely stable and offer excellent machinability. The material can be cast, but bar stock and forgings are superior. Their chief disadvantages are the high specific weight (highest of the gear materials) and relatively high cost.

There are many bronze alloys, but only a few are extensively used for gears.

1-67 Die-Cast Alloys

Many high-volume low-cost gears are produced by the die-cast process. Most are produced in alloys of aluminum and zinc, and a few in bronze and brass.

1-68 Sintered Powder Metal

This is a process of molding fine metal powder and alloying ingredients under high pressure, then firing to fuse the mass. It is a high-production means of producing relatively high-strength gears at low cost. Metals used for gears are iron-base mixtures, bronzes, and brasses. Powder metals are expensive, but, offsetting this, the scrap losses are very small.

1-69 Plastics

Plastic gears offer quiet operation, wear resistance, damping, light weight, noncorrosiveness, little or no lubrication, and low cost. On the debit side, they are difficult to machine to high precision and have large temperature-induced dimensional change and instability. Gears can be direct-finish molded with teeth, entirely machined from bar and plate stock, or cut from molded blanks.

Phenolic Laminates These have bases of paper, linen, or cotton cloth, with relative strengths in that order. They offer relatively good strength, and in cotton-canvas base are suitable for large gears and high loads.

Nylon Nylon has good wear resistance, even when operated without lubricant. A major disadvantage is its instability in the presence of moisture and humidity.

1-70 Applications and General Comments

Large gears and power applications use ferrous materials. The greater the load and durability requirements, the more essential are the high-alloy steels. Plain carbon steels are frequently used for low-quality commercial gears.

An exception in the ferrous group are the stainless steels. These are predominantly used in the small-gear, fine-pitch instrument fields because of their anticorrosion property. For fine-pitch precision applications, stainless steels are excellent. Although the 400 series is easier to machine and can have superior properties with heat treatment, stainless steel 303 has reasonable machinability and offers superior corrosion resistance. In addition, when used in conjunction with aluminum housings, its coefficient of thermal expansion is a much better match with aluminum than that of the 400 series.

The aluminum alloys, particularly 2024ST, are excellent instrument gear materials when used within their strength ratings. Aluminums have no value as a power-gear material and should not be used beyond low-load instrument-type applications.

Bronze is excellent for worm gears through the full range, from light loads to power applications. It is also useful for spur and helical meshes that have high velocity and/or significant loading.

Plastic materials are best suited for small gears of the instrument and light commercial product variety. Their quiet operation and minimal lubrication requirements make them particularly attractive for consumer products.

FINISH COATS

Thin finish coatings are often applied to metal gears for protection against the environment or for decorative purposes. The type of finish chosen is related to the material, corrosive conditions, and level of gear quality and precision.

1-71 Anodize

An excellent finish for aluminum gears is anodize. This is an artificially induced thin, but even and hard, coating of oxide. The thickness of the coating can be varied by process control, and it can become sufficient to be troublesome to the maintenance of close tolerances. Consequently, anodizing of precision aluminum gears is usually limited to anodizing of the gear blank before the teeth are cut.

Because the oxide film is somewhat porous, it can be impregnated with dyes, and a spectrum of colors plus black is possible. Anodized gears not only have an improved appearance, but offer significant protection against many corrosive atmospheres and salt spray.

1-72 Chromate Coatings

Applicable to aluminum, bronze, zinc, and magnesium, these are low-temperature dip-bath processes that produce a chemical chromate film that is extremely thin and does not alter the dimensions. However, the thin film has little wear resistance and offers corrosion protection only against nonabrasive environments. The color of the coating varies with the particular metal and alloy. Most often there is an iridescent color, which generated the common trademark Iridite. Dyes can be added to give a wide assortment of colors. Because there is no dimensional change, all gears, including precision gears, can be chromated after the teeth are cut.

1-73 Passivation

This is not a coating, in the strict sense, but a conditioning of the surface. It is particularly applicable to stainless steels. The process is essentially a low-strength nitric acid dip. It results in an invisible oxide film that develops the "stainless" property, removes "tramp iron," and reduces the metal's anodic potential in the galvanic series. Passivation causes no dimensional changes and does not discolor or otherwise alter the natural surface. If anything, it prevents random staining due to "free iron" particles left from machining. Stainless gears of all qualities can be passivated after complete machining, since dimensions and stability are unaffected.

1-74 Platings

The common electroplates, cadmium, chromium, nickel, and copper, are not suitable for gear surfaces, since they alter dimensions. Also, their susceptibility to localized buildup precludes their use on any precision part. Use of these platings should be limited to application of the coating before cutting of teeth and any other close dimensions.

1-75 Special Coatings

In recent years special extra-thin coatings have been developed; these are available under various commercial names. Some claim surface hardness, wear resistance, low coefficient of friction, anticorrosive qualities, and so on. There are many successful applications on record; however, each case should be investigated and tested.

1-76 Application of Coatings

It is advisable to finish-coat all gears that are to operate in a corrosive environment or must meet the requirements of military equipment specifications. In addition, appearance considerations may compel a protective finish.

Aluminum gears are best protected when anodized, in a natural color, without the coating on tooth surfaces. A chromate coating is adequate for many applications and is acceptable for many military specifications.

Passivation of stainless steels is required by both good practice and military equipment standards. Even for nonmilitary applications this is advisable to preclude discoloration from free iron particles and minimize galvanic action with other parts.

Bronze gears could be chromate-coated after cutting or cadmium-plated in the blank state, followed by chromating after teeth generation.

LUBRICATION

Lubrication serves several purposes, but its basic and most important function is to protect the sliding and rolling tooth surfaces from seizing, wearing, and other phenomena associated with surface failure by film separation. This is particularly pertinent to power gearing. In addition, lubrication aids all gearing by reducing friction and protecting against corrosion.

1-77 Lubrication of Power Gears

Power gear trains require sealed housings with a lubricant bath. Depending on the amount of power transmitted and the speed, it may be necessary to have a circulating system with lubricant cooling. Lubricant can be supplied as a liquid bath or as a fine spray. Lubrication of small, low-power gear trains can be accomplished with a grease pack in some cases. Many consumer home products are lubricated in this way.

1-78 Lubrication of Instrument Gears

Because of their much smaller size and capacity, generally lower speeds, and small or negligible power transmission, lubrication of instrument gears is very different from that of power gears. Often the lubricant's main purpose is to reduce friction.

Instrument gears that are relatively highly loaded and working near full capacity require lubrication systems as good as those of power gears. The difference is that in these extremely low powers the heat dissipation is no problem, and the unit can be packed and sealed with no concern for lubricant circulation, filtering, etc.

Lightly loaded gear trains can be of the open variety, in which a thin lubricant film is brushed on the teeth once at the outset, and reapplied only as maintenance and usage dictate. When lubricant is applied in this way, it is important that gear speeds not be so great that the lubricant is flung away by centrifugal force. Also, the lubricant should have a minimum "spreading" rating. For this reason greases are often favored.

Open-housing gear trains are subject to contamination, and it is advisable to guard against excessive exposure. Instruments whose outer enclosures must be removed often for maintenance of other items should be worked on in clean and controlled environments. Where prolonged or uncontrollable exposure occurs, temporary or permanent inner dust covers for the gear train are recommended. This is particularly advisable in hybrid electronic instrument boxes, where the danger of solder and other debris is high.

1-79 Oil Lubricants

Oils are the most common lubricants and come in various types. The compounding of oils provides a number of combinations and generates various properties. The most basic lubricant is petroleum, with which animal, vegetable, and synthetic oils and additives are combined to obtain specific properties.

Oils offer a wider range of operating speeds than greases. Also, they are easier to handle and are more effective because of their liquid nature.

1-80 Grease

Grease is a combination of liquid and solids, in which the latter serves as a reservoir for the liquid lubricant as well as imparting certain of their own properties. Grease has the advantage of remaining in place and not "spreading" as oils do, and has a much lower evaporation rate. Also, it can provide a lubricant film at heavy loads and at low speeds.

1-81 Solid Lubricants

In recent years a number of "dry film" lubricants have been developed. These have the following advantages: they have a wide temperature range; they do not disperse or evaporate, and hence they are good for space and other vacuum applications; and they are easier to use in open gearing, since they do not contaminate as rapidly as oils and grease. However, most solid films significantly, and some drastically, alter dimensions, and this cannot be tolerated in quality gearing. Also, these are one-shot lubricants that must last the life of the gears, with the film being continually eroded and worn away from the start of its use.

1-82 Typical Lubricants

The choice of lubricants is very wide. Military specifications govern most types and classes of lubricants, and many manufacturers' products qualify. Table 1-12 lists typical gear lubricants and their applications.

GEAR FABRICATION

The fabrication of a complete gear normally includes most or all of the following operations:

1. Blank fabrication
2. Generation of teeth
3. Refining of teeth
4. Heat treatment
5. Deburring and cleaning
6. Finish coating

Although it is not necessary to apply all six operations to every gear, items 1, 2, and 5 in themselves determine the quality level of a gear.

Blank fabrication involves all the general and special features of the gear body. Generation of teeth is unique to machine-cut or ground gears, as in other fabrication methods the teeth and body are formed simultaneously. The refining operation (shaving, grinding, or honing) is a special means of improving quality, particularly in high volume. Heat treatment is limited to gears requiring surface hardness and/or strength. Deburring and cleaning is essential for all gears, regardless of how they are made or of their quality. Finish coats are limited to certain materials and environments where corrosion protection or improved appearance is required.

TABLE 1-12 Viscosity Ranges for AGMA Lubricants

Rust- and oxidation-inhibited gear oils (AGMA lubricant no.)	Viscosity range,* mm^2/s (cSt) at 40°C	Equivalent ISO grade†	Extreme-pressure gear lubricants‡ (AGMA lubricant no.)	Viscosities of former AGMA system¶ (SSU at 100°F)
1	41.4–50.6	46		193–235
2	61.2–74.8	68	2 EP	284–347
3	90–110	100	3 EP	417–510
4	135–165	150	4 EP	626–765
5	198–242	220	5 EP	918–1122
6	288–352	320	6 EP	1335–1632
7 Comp§	414–506	460	7 EP	1919–2346
8 Comp§	612–748	680	8 EP	2837–3467
8A Comp§	900–1100	1000	8A EP	4171–5098

NOTE: Viscosity ranges for AGMA lubricant numbers will henceforth be identical to those of ASTM 2422.

* "Viscosity System for Industrial Fluid Lubricants," ASTM 2422. Also British Standards Institute, B.S. 4231.

† "Industrial Liquid Lubricants—ISO Viscosity Classification." International Standard, ISO 3448.

‡ Extreme-pressure lubricants should be used *only* when recommended by the gear-drive manufacturer.

¶ AGMA 250.03, May 1972 and AGMA 251.02, November 1974.

§ Oils marked Comp are compounded with 3 to 10 percent fatty or synthetic fatty oils.

Source: AGMA 250.04.

There are a wide variety of modern methods of producing gear teeth, listed as:

1. Machine cutting
2. Grinding
3. Casting
4. Molding
5. Forming (drawing, extruding, rolling)
6. Stamping

Each method offers particular quality, production quantity, cost, material, and application features.

1-83 Generation of Gear Teeth

Machining is the most important method of generating gear teeth. It is suitable for high-precision gears in both small and large quantities.

Rack Generation This is the basic method of producing involute teeth by having a rack cutter form conjugate tooth profiles on the blank as the rack and blank are given proper relative motion by the generating machine's drive mechanism. As the rack traverses the gear blank, it is reciprocated across the blank face. Cutting edges on the rack teeth generate mating conjugate teeth into the blank. The chief disadvantage of this method is that the rack has a limited length, necessitating periodic indexing. This limits both operating speed and accuracy.

Hob Generation This is the most widely used method of cutting teeth. It is similar to rack generation, but the rack is in the form of a worm. Referring to Fig. 1-30, the hob's central section is identical to that of the worm and mate. The differences are that the hob is axially gashed or fluted through the thread teeth in several places to form cutting edges, and the sides and top of thread teeth are relieved behind the gash surface to permit proper cutting action. This arrangement gives, in effect, an infinitely long rack, and cutting is both steady and continuous. To generate the full width of the gear, the hob is slowly traversed across the face of the gear as it rotates. Thus, the hob has a basic rotation motion and a unidirection traverse at right angles. Both movements are relatively simple to effect, resulting in a very accurate process.

A further advantage of hobbing is that the hob can be swiveled relative to the blank axis. This permits helical gears of all angles to be cut with the same tooling.

With regard to accuracy, hobbing is superior to the other cutting processes. Gears can be directly hobbed to ultra-precision tolerances without the need for any secondary refining processes.

Gear-Shaper Generation This process, unlike the other two, employs a gear-shaped cutter as the tool instead of a rack or equivalent. Also, as with the rack, a given gear is conjugate to all tooth numbers of that pitch. Thus, a gear made as a cutting tool can generate mating teeth in a blank when the two are rotated at proper speeds. The cutting tool can be imagined as a gear that axially traverses the blank with a reciprocating axial motion as it rotates. The teeth on the gear tool are properly relieved to form cutting edges on one face.

Although the shaping process is not suitable for direct cutting of ultra-precision gears and generally is not as highly rated as hobbing, it can produce precision-quality gears. Generally it is a more rapid process than hobbing.

Two outstanding uses of shaping are for generating shouldered gears and internal gears. Compound gears and shaft gears frequently are so compactly designed that a hob cutter interferes with adjacent material. In such cases, shaping generation can be used, since the shaper-cutter stroke requires very little runout space on one side of the gear. With regard to internal gears, the shaping process is the only basic way of generating them.

The shaping process can be used to generate helical gears. However, each helix angle requires special tooling. Therefore, shaping is not as convenient and is more expensive than hobbing for helical gears.

Top Generating This is a fabrication option utilizing cutters that finish-cut the outside diameter of the teeth simultaneously with the cutting of the tooth profiles.

It is done in both the hobbing and shaping processes, although it is more prevalent in hobbing and among the fine pitches. The main advantages of topping are:

1. Liberal tolerances can be applied to the outside diameter of the blank.
2. The deburring problem is reduced.

1-84 Gear Grinding

Although grinding is often associated with quantity fabrication of high-quality gears as a secondary refining operation, it is also a basic process for producing hardened gears. In addition, many high-precision fine-pitch gears have their teeth entirely ground from the blank.

Grinding teeth can be done by either form grinding or generating grinding. The latter is basically more accurate because the grinding wheel is straight-side dressed.

1-85 Plastic Gears

These can be produced by the normal hobbing and shaping machine-cutting processes. In addition, they can be produced by various molding techniques. Gears produced by the latter methods are not as accurate as cut gears as a result of shrinkage, mold variations, and flow inconsistencies.

Regardless of method, the fabrication of plastic gears compared with that of metal gears suffers as a result of temperature instability, material flow, and generally poorer cutting qualities. Attainable quality is less than for metals and varies with the particular plastic.

GEAR INSPECTION

Functional performance of a gear can be assured only by confirming the critical dimensions and parameters. With increasing gear precision, adequate and proper inspection becomes more paramount.

There are many aspects of gear inspection, and the subject is too vast to be completely covered here. However, two basic and important inspection criteria are TCE and tooth thickness.

1-86 Variable-Center-Distance Testers

Both TCE and tooth thickness can be measured by means of roll testing with a variable-center-distance fixture. There are many varieties, but essentially all consist of a fixture with two parallel shafts, one fixed and the other floating on smooth, low-friction ways. One shaft mounts the gear to be tested and the other an accurate, known-quality master gear. The pair is held in intimate contact by spring loading or an equivalent. As the test gear is rotated, tooth-to-tooth errors and runout are revealed as a variation in the center distance of the pair. This variation can be sensed, amplified, and displayed as a dial reading or recorded on a chart. See Fig. 1-39. Sensitivity is of the order of 50 to 100 millionths of an inch.

The unique feature of gear roll testing is that the inspection parallels the actual use of the gear. Thus, roll testing is a functional inspection.

Total Composite Error The TCE is clearly revealed in roll testing, and its components can be identified. Referring to Fig. 1-39, it is evident that the magnitude of runout and TTCE can be extracted, and from this the gear quality judged. Also, when parameters are out of tolerance, the fabricator can identify the source of difficulty and take appropriate corrective action.

Gear Size If the center-distance setting of the roll tester is carefully established, then the absolute readings are an indication of tooth thickness. Thus, in Fig. 1-39, the mean line of the trace is a measure of tooth thickness. The high and low readings indicate the extreme variations of tooth thickness at the nominal pitch radius. Changes in center distance are an indirect measure of tooth thickness and must be converted using Eq. (1-22).

Advantages and Limitations of Variable-Center-Distance Testers Functional test of a gear is a good feature, as it reveals characteristics that occur in the real application. Also, the method is rapid, and therefore suitable for inspection of production gears. The ability to obtain a hard copy record is also a distinct advantage.

Roll testing gives excellent results with regard to TCE and TTCE. Repeatability and absolute measure are usually good, of the order of 0.0001 in. On the other hand, size measurement is not as reliable as an absolute measure. This is due to the nature of the fixture and the integration of several error sources in the calibration process. A repeatability of 0.0002 in is considered good, and often it is greater than that.

1-87 Over-Pins Gaging

The equations relating tooth thickness and a measurement over cylindrical pins or rolls inserted between the teeth were given in Sec. 1-15. This is a widely used method of gaging gears during fabrication while they are still in the gear-generating machine, and for final inspection. Accuracy of the over-pins measurement is of the order of 0.0001 in.

A major disadvantage of over-pins gaging is the inability to perfectly correlate it with variable-center-distance measurements. This is because the method is insensitive to pitch-line runout. On the other hand, rolling a gear necessarily involves the TCE and its runout component. The best correlation is obtained by equating the over-pins measurement with the average center-distance value found in the roll test.

Beyond the correlation problem, over-pins measurements by themselves are inadequate because the undetected runout can be out of control, causing interference with its mate. It is necessary to have a runout control and inspection.

1-88 Other Inspection Equipment

In addition to the basic inspection methods and equipment previously described, there is additional special-purpose equipment: involute profile form checkers, tooth-spacing gages, runout checkers. Also, for high-precision gears there is equipment for inspecting the position error of individual gears and the transmission error of a gear train.

1-89 Inspection of Fine-Pitch Gears

Because of their small geometry, fine-pitch gears do not easily lend themselves to the kind of detailed tooth measurements suitable for large, coarse-pitch gears. Consequently, fine-pitch gears are inspected almost exclusively by functional testing on a variable-center-distance fixture.

Over-pins measurements are also used, but these generally are restricted to a reference measurement. This is primarily used in the fabrication process as a set-up dimension, and also used by inspection departments that are not equipped to roll-test gears.

1-90 Significance of Inspection

The inspection operation is essential to obtaining a quality product. In effect it is a policing operation that ensures conformance with dimension tolerances and other drawing specifications.

The effort, care, and cost of inspection are related to the quality level. Inspection of precision gears demands much more than inspection of low-quality gears. Equipment must be of the best grade, calibrated periodically, and used only by qualified personnel. Control of temperature environment is essential for measurements of the order of 0.0001 in. Also, cleanliness of equipment, gears, and working area is imperative.

REFERENCES

1. Buckingham, Earle: *Manual of Gear Design,* 3 vols., Industrial Press, New York, 1935.
2. Dudley, D. W.: *Gear Handbook,* McGraw-Hill, New York, 1962.
3. Michalec, G. W.: *Precision Gearing: Theory and Practice,* Wiley, New York, 1966.
4. Shigley, J. E.: *Mechanical Engineering Design,* McGraw-Hill, New York, 1963.
5. Steeds, W.: *Involute Gears,* Longmans, Green, London, 1948.

LITERATURE OF GENERAL INTEREST

AGMA Standards, 1901 North Ft. Myer Drive, Suite 1000, Arlington, VA 22209.
Woodbury, R. W.: *History of Gear Cutting Machines,* MIT Technology Press, Cambridge, Mass., 1958.
Dudley, D. W.: *The Evolution of Gear Art,* AGMA paper No. 990.14. (Published in book form by AGMA, January 1969.)

ACKNOWLEDGMENT

The editor wishes to express his sincere thanks to Robert C. Brown, Jr., Manager of Engineering Services, American Gear Manufacturers Association (AGMA), for his expert guidance on this section.

2
CHAINS AND SPROCKETS

WILLIAM R. EDGERTON
Chief Engineer, Whitney Chain Operations
Dresser Industries, Inc.
Hartford, Connecticut

2-1 Historical Background

The fundamental concept of creating a strong, yet flexible, chain structure by joining together a consecutive series of individual links is an idea that dates back to the earliest human utilization of metals. The use of iron for this purpose probably dates to the eighth century B.C.

The second step in the development process was the fashioning of wheels adapted to interact with the flexible chains, by the provision of teeth and pockets on the circumference of the wheels. These specially adapted wheels, known as "sprocket wheels," but usually referred to simply as "sprockets," were first developed by the military engineers of Greece some 22 centuries ago. From the writings of Philo of Byzantium, c. 200 B.C., we learn of chain and sprocket drives being used to transmit power from early water wheels, of a pair of chains fitted with buckets to lift water to higher elevations, and of a pair of reciprocating chain drives which acted as a tension linkage to feed and cock a repeating catapult.

The first two of these instances of chain and sprocket interaction probably used simple round-link chain, but the third involved a flat-link chain concept designed by Dionysius of Alexandria while working at the Arsenal at Rhodes. The design conceived by Dionysius employed what is now known as the inverted-tooth chain-sprocket engagement principle, a major advance over the cruder round-link design.

Despite its very early origins, rather little practical use was made of chain and sprocket interaction for the transmission of power or the conveyance of materials until the advent of the Industrial Revolution, which took place largely during the nineteenth century. The development of machinery to mechanize textile manufacture, agricultural harvesting, and metalworking manufacturing brought with it a need for the positive transmission of power and accurate timing of motions that only a chain-and-sprocket drive could provide.

The earliest sprocket chains manufactured in the United States employed cast components, usually of malleable iron, and many configurations of detachable link chain and pintle chain were produced in large quantities. As the need for higher strength and improved wear resistance became evident, chains employing heat-treated steel components were introduced. The use of rolled or drawn steel as a raw material required manufacturing machinery which provided greater dimensional accuracy than was possible in foundry practice, with the result that certain of the new types of sprocket chains came to be known as "precision chain." This developed somewhat earlier in Great Britain than in the United States, starting with the Slater chain, patented in England in 1864. The Slater design was further refined by Hans Renold with the development of precision roller chain, patented in England in 1880.

Chain manufacture in the United States continued to be principally concerned with cast and detachable link designs until the American introduction of the "Safety Bicycle" in 1888. Drop-forged steel versions of the cast detachable chains were first used, then precision steel block chain for bicycle driving, and progressively larger sizes were manufactured in the U.S. as the horseless carriage craze swept the country in the 1890s.

Precision inverted-tooth chain, popularly known as "silent chain," was introduced in the late 1890s, with many proprietary styles being developed during the early part of the twentieth century.

The first efforts toward standardization of roller chain were begun in the 1920s, resulting in the publication of the first chain standard, American Standard B29a, on July 22, 1930. Since that time, eighteen B29 standards have been developed, covering inverted-tooth chain; detachable chain; pintle and offset-sidebar chains; cast, forged, and combination chains; mill and drag chains; and many other styles.

2-2 Classification of Chain Types

There are eighteen American National Standards which relate to the various types of sprocket chains in general use. This family of standards is the result of over 50 years of standardization activity, which had its beginning in the work that led to the publication of *American Standard B29a—Roller Chain, Sprockets, and Cutters* in 1930. The chain types covered by the current standards are as follows:

ANSI B29.1	Precision roller chain
ANSI B29.2	Inverted-tooth (or silent) chain
ANSI B29.3	Double-pitch roller chain for power transmission
ANSI B29.4	Double-pitch roller chain for conveyor usage
ANSI B29.6	Steel detachable chain
ANSI B29.7	Malleable iron detachable chain
ANSI B29.8	Leaf chain
ANSI B29.10	Heavy-duty offset-sidebar roller chain
ANSI B29.11	Combination chain
ANSI B29.12	Steel-bushed rollerless chain
ANSI B29.14	Mill chain (H type)
ANSI B29.15	Heavy-duty roller-type conveyor chain
ANSI B29.16	Mill chain (welded type)

ANSI B29.17	Hinge-type flat-top conveyor chain
ANSI B29.18	Drag chain (welded type)
ANSI B29.19	Agricultural roller chain (A and CA types)
ANSI B29.21	Chains for water and sewage treatment plants
ANSI B29.22	Drop-forged rivetless chain

The basic size dimension for all types of chain is pitch—the center-to-center distance between two consecutive joints. This dimension ranges from 3⁄16 in (in the smallest inverted-tooth chain) to 30 in (the largest heavy-duty roller-type conveyor chain).

2-3 Sprockets and Sprocket Tooth Form

Chains and sprockets interact with each other to convert linear motion to rotary motion or vice versa, since the chain moves in an essentially straight line between sprockets and moves in a circular path while engaged with each sprocket. A number of tooth-form designs have evolved over the years, but the prerequisite of any tooth form is that it must provide:

1. Smooth engagement and disengagement with the moving chain
2. Distribution of the transmitted load over more than one tooth of the sprocket
3. Accommodation of changes in chain length as the chain elongates as a result of wear during its service life

The sprocket layout is based on the pitch circle, the diameter of which is such that the circle would pass through the center of each of the chain's joints when that joint is engaged with the sprocket. Since each chain link is rigid, the engaged chain forms a polygon whose sides are equal in length to the chain's pitch. The pitch circle of a sprocket, then, is a circle that passes through each corner, or vertex, of the pitch polygon. The calculation of the pitch diameter of a sprocket follows the basic rules of geometry as they apply to pitch and number of teeth. This relationship is simply

$$\text{Pitch diameter} = \frac{\text{pitch}}{\sin\,(180°/\text{number of teeth})} \tag{2-1}$$

The action of the moving chain as it engages with the rotating sprocket is one of consecutive engagement. Each link must articulate, or swing, through a specific angle to accommodate itself to the pitch polygon, and each link must be completely engaged, or seated, before the next in succession can begin its articulation.

As depicted in Fig. 2-1, with a sprocket tooth at top-dead-center position, link *AB* is completely seated; with rotation of one-half a link, so that the space between two teeth is at top-dead-center position, Fig. 2-2, link *BC* has articulated through just half the angle to engage with the sprocket. Continued rotation now brings the next tooth to top-dead-center position, completing the

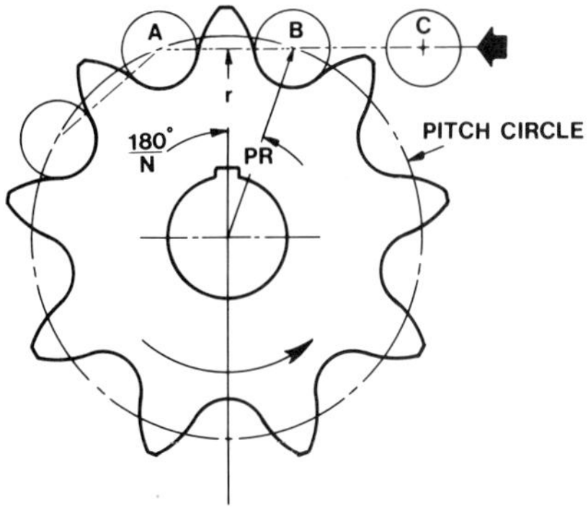

FIG. 2-1 Chordal action, tooth top dead center. *(Source: Whitney Chain Operations, Dresser Industries, Inc., Hartford, Conn.)*

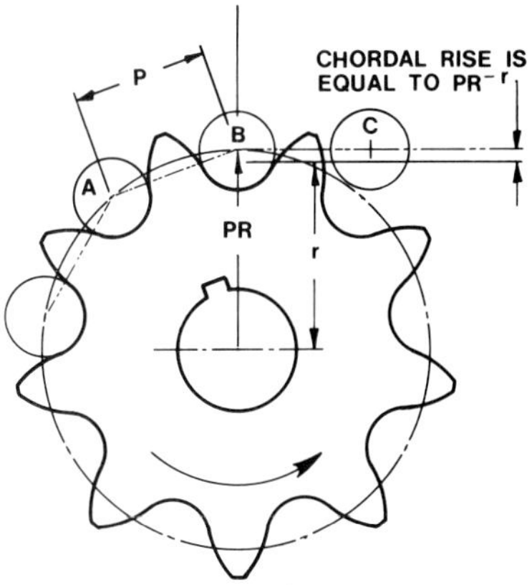

FIG. 2-2 Chordal action, space top dead center. *(Source: Whitney Chain Operations, Dresser Industries, Inc., Hartford, Conn.)*

articulation of link *BC* through the angle required (or 360°/number of teeth) to become completely seated. The cycle will repeat with the link that follows, and so on.

Since the center-to-center dimension of each engaged link forms a chord relative to the pitch circle of the sprocket, this angular swing of each link as each joint articulates is usually referred to as chordal action. It will be noted that the incoming straight chain strands position moves slightly up or down, depending on whether a chain joint or a tooth is at top-dead-center position. This result of each link's chordal action is referred to as chordal rise and fall. Because of chordal rise and fall, there are slight variations in the linear velocity of the incoming chain strand. The important conclusion to be drawn from this action is that as the number of teeth in a sprocket increases, the angle of articulation, and therefore the chordal action, decreases. For example, for 10 teeth, velocity variation is 5.15 percent; for 20 teeth , 1.25 percent; and at 30 teeth, variation is only 0.55 percent.

Sprocket tooth forms vary according to the type of chain with which they are to operate. The tooth form must be accurately machine-cut for roller chain, double-pitch roller chain, and silent chain. For engineering-class chain, the tooth form is often machine-cut, but it also can be produced by casting to shape in cast iron, or flame-cut from steel plate. For cast-link, combination, and welded-link chains, cast-to-shape and flame-cut sprockets are most commonly used.

Machine-cut sprockets can be produced by milling, hobbing, or shaping. Milling cutters and hobs which will produce the standard tooth form are readily available from tool manufacturers. Cutters for Fellows Gear Shapers are available from the Fellows Company. In addition, many small sprockets are now made by powder metallurgy, and some light-duty items are being manufactured by injection molding of plastic materials.

2-4 Chain Length and Center Distance

For the simplest of chain arrangements, the two-sprocket layout, chain length and center distance can be readily calculated. For these calculations, center distance is always considered in units of chain pitch to be consistent with length in links, or pitches, and sprocket size, identified by number of teeth, which actually is a count of the number of pitches of chain which would be engaged in complete wrap, or moved by one revolution, of the sprocket.

When both sprockets are the same size, for a 1:1 ratio, both chain strands are parallel, and chain length is simply twice the center distance plus the number of teeth in one sprocket. In most layouts, one sprocket has more teeth than the other, which means that the straight chain strands between the two sprockets will not be parallel. The various methods for calculating length or center distance make use of the number of teeth in the two sprockets to take the angle between the two strands into account. Using N for the larger sprocket and n for the smaller sprocket, the quantities $N + n$ and $N - n$ are used as follows (see Fig. 2-3):

For chain length, in pitches, when center distance, in pitches, is known:

$$\text{Length} = \frac{N + n}{2} + 2(\text{center distance}) + \frac{K}{\text{center distance}} \tag{2-2}$$

where K is a factor varying in value with the magnitude of $N - n$. The values are given in Table 2-1.

For quick approximation of chain length, the nomograms in Figs. 2-4 and 2-5 will provide satisfactory results.

Since chain length must be rounded off to an integral number of pitches, which in turn changes the center distance that will produce a taut fit, the

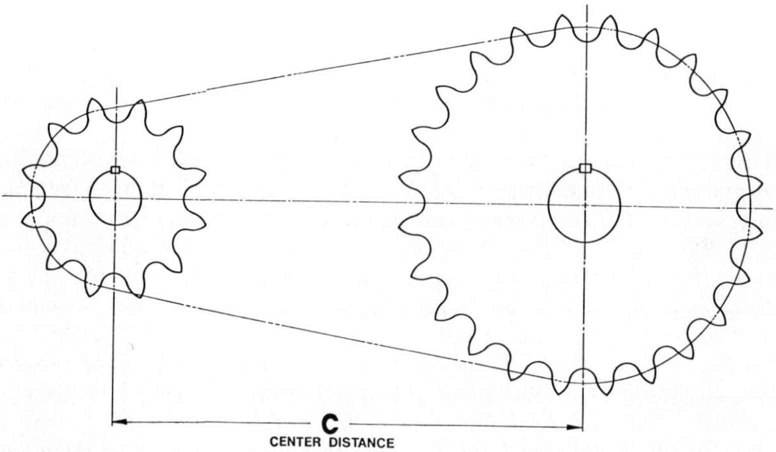

FIG. 2-3 Center distance. (*Source: Whitney Chain Operations, Dresser Industries, Inc., Hartford, Conn.*)

usual design procedure is to select an approximate center, calculate chain length for this distance, round off to the nearest integral number, preferably the nearest integral even number, and then calculate the final center distance from the integral number of pitches. The following formula provides an accurate determination of the center distance:

$$\text{Center distance} = \frac{1}{4}\left[L - \frac{N-n}{2} + \sqrt{\left(\frac{N-n}{2}\right)^2 - W(N-n)}\right] \qquad \text{(2-3)}$$

where W is a factor which varies in value from 0.20264 to 0.22302, depending on the magnitude of the expression $(L - n)/(N - n)$ from 12.000 down to 1.008. For $(L - n)/(N - n)$ greater than 12.000, use 0.20264, and for this quantity less than 1.008, use 0.22302. See Table 2-2.

The calculated center distance is the theoretically exact value for taut fit with sprockets at exact pitch diameter and chain at exact length. Since sprocket manufacturing tolerances are minus only and chain length tolerance is plus only, the calculated center distance will normally provide a satisfactory amount of slack in the chain's return strand.

In applications where shaft center distance is already established as a result of some other consideration, chain length must be calculated from this established distance, then rounded upward to the next integral number. If slack is excessive, an idler may be required to control chain action in the slack strand. A number of adjustable and spring-tensioned idler mountings are available for this purpose.

For multiple-sprocket layouts or serpentined arrangements, chain length is usually determined from an accurate drafter's layout of the design. The sprocket pitch circles are drawn, then the chain strands are drawn as straight lines tangent to the pitch circles. Starting at a chosen beginning point, chain pitches can be marked off on the straight runs with a scale, then stepped off around the pitch circles with an accurately set pair of dividers.

Whenever possible, drives should be set up with adjustable center distance; that is to say, at least one of the shafts should be mounted in such a way that it can be moved outward from the theoretically exact dimension. This will provide a way to adjust for chain-wear elongation as it occurs during the service life of the chain. The amount of movement to be provided should be equal to at least one pitch of chain. Beyond this point, the chain length can be shortened by the removal of links, and the centers adjusted inward to compensate.

2-5 Attachments

To adapt chains for various conveying and material-handling assignments, the normal link plates, or pins, can be replaced with specially shaped components that allow the chain to hold, carry, or suspend articles and containers, and to push, pull, or clamp as necessary. These specialized link plates and pins are known as "chain attachments."

For most of the principal classes of chain, there are several standardized attachment designs, and the dimensions of these components are listed in the

TABLE 2-1 Chain Length—Values of *K*

$N - n$	K	$N - n$	K	$N - n$	K	$N - n$	K	$N - n$	K	$N - n$	K
1	0.03	32	25.94	63	100.54	94	223.82	125	395.79	156	616.44
2	0.10	33	27.58	64	103.75	95	228.61	126	402.14	157	624.37
3	0.23	34	29.28	65	107.02	96	233.44	127	408.55	158	632.35
4	0.41	35	31.03	66	110.34	97	238.33	128	415.01	159	640.38
5	0.63	36	32.83	67	113.71	98	243.27	129	421.52	160	648.46
6	0.91	37	34.68	68	117.13	99	248.26	130	428.08	161	656.59
7	1.24	38	36.58	69	120.60	100	253.30	131	434.69	162	644.77
8	1.62	39	38.53	70	124.12	101	258.39	132	441.36	163	673.00
9	2.05	40	40.53	71	127.69	102	263.54	133	448.07	164	681.28
10	2.53	41	42.58	72	131.31	103	268.73	134	454.83	165	689.62
11	3.06	42	44.68	73	134.99	104	273.97	135	461.64	166	698.00
12	3.65	43	46.84	74	138.71	105	279.27	136	468.51	167	706.44
13	4.28	44	49.04	75	142.48	106	284.67	137	475.42	168	714.92
14	4.96	45	51.29	76	146.31	107	290.01	138	482.39	169	723.46
15	5.70	46	53.60	77	150.18	108	295.45	139	489.41	170	732.05

16	6.48	47	55.95	78	154.11	109	300.95	140	496.47	171	740.68
17	7.32	48	58.36	79	158.09	110	306.50	141	503.59	172	749.37
18	8.21	49	60.82	80	162.11	111	312.09	142	510.76	173	758.11
19	9.14	50	63.33	81	166.19	112	317.74	143	517.98	174	766.90
20	10.13	51	65.88	82	170.32	113	323.44	144	525.25	175	775.74
21	11.17	52	68.49	83	174.50	114	329.19	145	532.57	176	784.63
22	12.26	53	71.15	84	178.73	115	334.99	146	539.94	177	793.57
23	13.40	54	73.86	85	183.01	116	340.84	147	547.36	178	802.57
24	14.59	55	76.62	86	187.34	117	346.75	148	554.83	179	811.61
25	15.83	56	79.44	87	191.73	118	352.70	149	562.36	180	820.70
26	17.12	57	82.30	88	196.16	119	358.70	150	569.93	181	829.85
27	18.47	58	85.21	89	200.64	120	364.76	151	577.56	182	839.04
28	19.86	59	88.17	90	205.18	121	370.86	152	585.23	183	848.29
29	21.30	60	91.19	91	209.76	122	377.02	153	592.96	184	857.58
30	22.80	61	94.25	92	214.40	123	383.22	154	600.73	185	866.93
31	24.34	62	97.37	93	219.08	124	389.48	155	608.56		

Source: Whitney Chain Operations, Dresser Industries, Inc., Hartford, Conn.

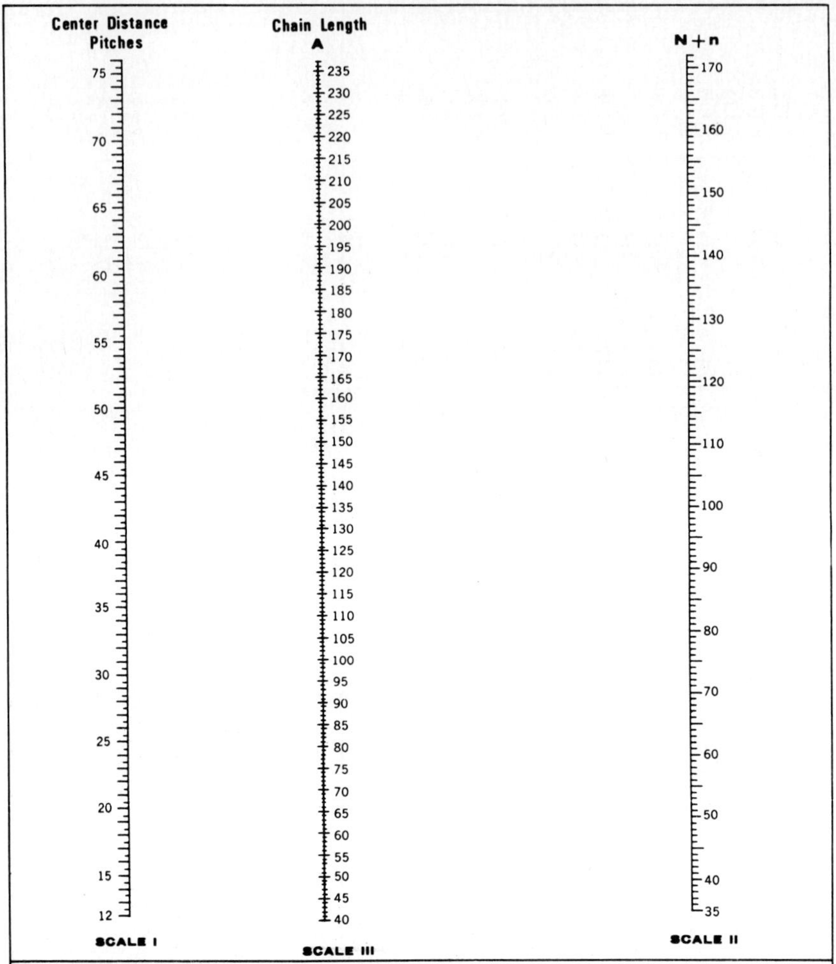

quickest way to determine chain length in pitches

Step 1 Lay a straight edge on the points representing the center distance in pitches, Scale I, and the *total number* (N+n) of teeth on the large (N) and small (n) sprocket Scale II. Read chain length "A" on Scale III.

Step 2 Lay a straight edge on the points representing the *difference in the number* (N—n) of teeth on the large (N) and small (n) sprockets, Scale IV, and the center distance in pitches, Scale V. Read chain length "B" on Scale VI.

Step 3 Add chain length "A" and "B" for total chain length in pitches.

Step 4 Answer should be rounded off to nearest even number of pitches.

FIG. 2-4 Chain length nomograph, length value *A*. (*Source: Whitney Chain Operations, Dresser Industries, Inc., Hartford, Conn.*)

individual American National Standards covering each class. In addition to these standardized attachments, thousands of nonstandard attachment link plates and pins have been developed through cooperative efforts of the chain

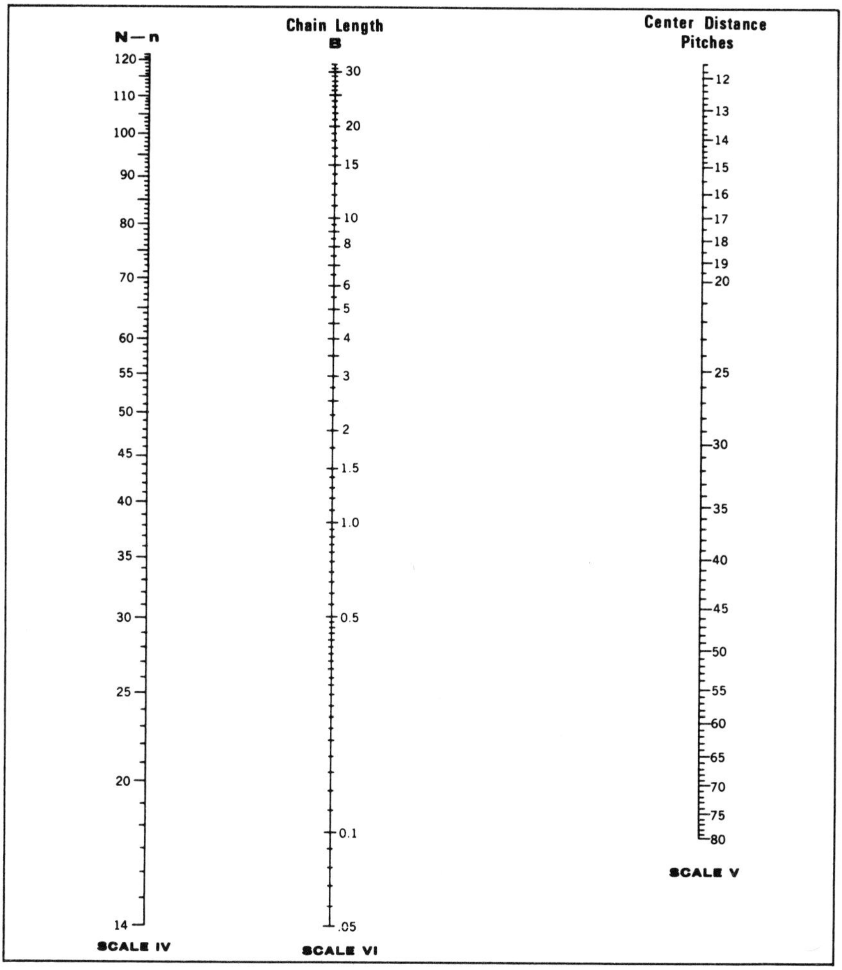

FIG. 2-5 Chain length nomograph, length value *B*. (*Source: Whitney Chain Operations, Dresser Industries, Inc., Hartford, Conn.*)

manufacturers and the designers of all kinds of special machinery.

For precision roller chain (ANSI B29.1), the standardized attachment link plates are provided with a lug, or tab, which can be supplied so that the lug projects directly outward from the link plate contour, or bent 90° so that it projects outward perpendicular to the basic plates. In addition, these lugs are provided with a pierced hole, so that a screw or rivet can be utilized to fasten whatever type of cradle, platform, cross piece, etc., the designer might wish to use.

Precision roller chain can also be supplied with standardized extended pins,

TABLE 2-2 Center Distance—Values of W

$\frac{L-n}{N-n}$	W	$\frac{L-n}{N-n}$	W	$\frac{L-n}{N-n}$	W	$\frac{L-n}{N-n}$	W	$\frac{L-n}{N-n}$	W
12.00	0.20264	2.70	0.20297	1.33	0.20574	1.130	0.20929	1.034	0.21648
11.00	0.20265	2.60	0.20301	1.32	0.20584	1.120	0.20967	1.032	0.21681
10.00	0.20266	2.50	0.20306	1.31	0.20595	1.110	0.21011	1.030	0.21715
9.00	0.20267	2.40	0.20311	1.30	0.20607	1.100	0.21061	1.028	0.21751
8.00	0.20268	2.30	0.20318	1.29	0.20619	1.090	0.21119	1.026	0.21789
7.00	0.20269	2.20	0.20325	1.28	0.20632	1.080	0.21184	1.024	0.21828
6.00	0.20271	2.10	0.20334	1.27	0.20645	1.070	0.21258	1.022	0.21869
5.00	0.20272	2.00	0.20345	1.26	0.20659	1.060	0.21343	1.020	0.21914
4.80	0.20272	1.90	0.20358	1.25	0.20674	1.058	0.21360	1.019	0.21937
4.60	0.20273	1.80	0.20374	1.24	0.20689	1.056	0.21378	1.018	0.21961
4.40	0.20273	1.70	0.20396	1.23	0.20705	1.054	0.21397	1.017	0.21987
4.20	0.20274	1.60	0.20424	1.22	0.20721	1.052	0.21416	1.016	0.22014
4.00	0.20276	1.50	0.20463	1.21	0.20738	1.050	0.21437	1.015	0.22042
3.80	0.20277	1.40	0.20517	1.20	0.20755	1.048	0.21458	1.014	0.22072
3.60	0.20279	1.39	0.20523	1.19	0.20774	1.046	0.21481	1.013	0.22104
3.40	0.20281	1.38	0.20530	1.18	0.20793	1.044	0.21505	1.012	0.22137
3.20	0.20284	1.37	0.20538	1.17	0.20815	1.042	0.21531	1.011	0.22173
3.00	0.20288	1.36	0.20546	1.16	0.20839	1.040	0.21558	1.010	0.22212
2.90	0.20291	1.35	0.20555	1.15	0.20865	1.038	0.21586	1.009	0.22255
2.80	0.20294	1.34	0.20564	1.14	0.20895	1.036	0.21616	1.008	0.22302

Source: Whitney Chain Operations, Dresser Industries, Inc., Hartford, Conn.

which project outward substantially beyond the pin link plate surface. Chains with extended pins are often used in pairs, with tubular cross flights fitting over the pin extensions, thus creating a connection between parallel runs of chain. (See Figs. 2-6 to 2-10.)

Engineering steel chains and cast malleable chains are also available with standardized lug-type attachment links and with larger projections, called wings, that adapt the chains to a variety of scraper and bucket devices.

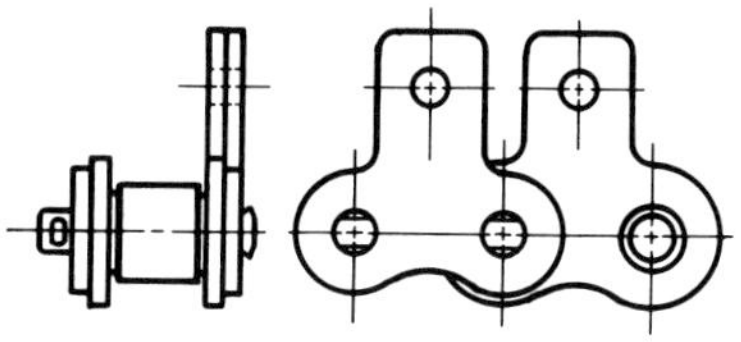

FIG. 2-6 Straight attachment lugs, one side. (*Source: Whitney Chain Operations, Dresser Industries, Inc., Hartford, Conn.*)

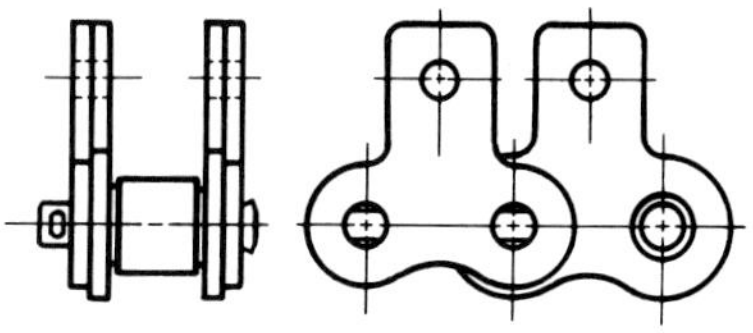

FIG. 2-7 Straight attachment lugs, both sides. (*Source: Whitney Chain Operations, Dresser Industries, Inc., Hartford, Conn.*)

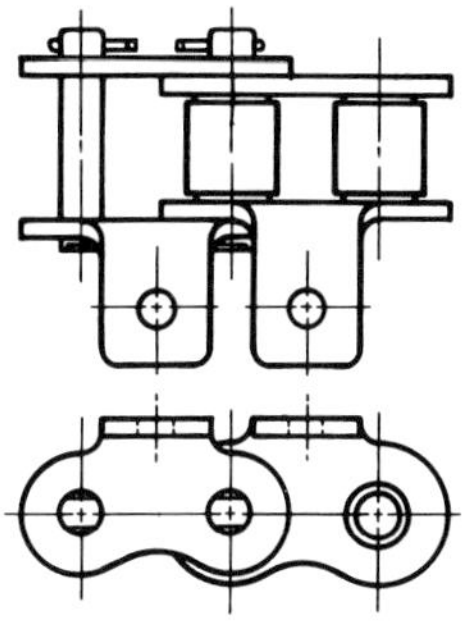

FIG. 2-8 Bent attachment lugs, one side. (*Source: Whitney Chain Operations, Dresser Industries, Inc., Hartford, Conn.*)

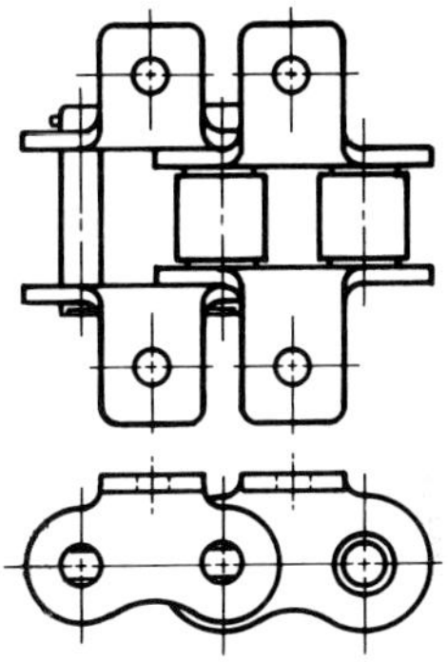

FIG. 2-9 Bent attachment lugs, both sides. (*Source: Whitney Chain Operations, Dresser Industries, Inc., Hartford, Conn.*)

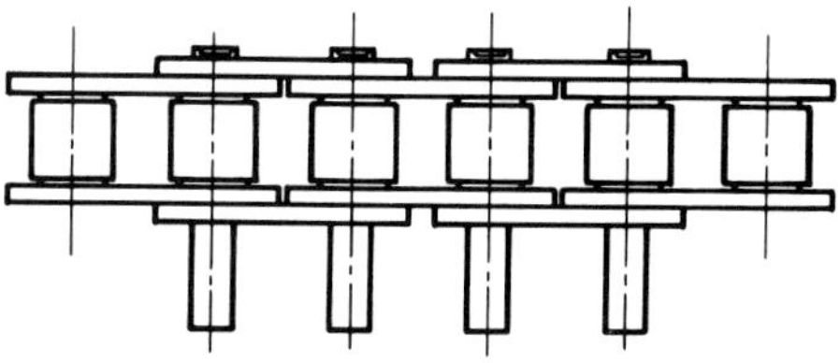

FIG. 2-10 Extended-pin attachments. (*Source: Whitney Chain Operations, Dresser Industries, Inc., Hartford, Conn.*)

2-6 Selection and Application

The load-carrying ability or power-transmitting capacity of the various chain types is usually stated in either rated horsepower capacity or rated working load at specific speeds, and usually is related to sprocket size. For transmission roller chains and for silent chain, the rated capacity is presented in the form of horsepower tables. A typical table, in this case no. 80, 1-in pitch roller chain of the ANSI B29.1 type, is shown as Table 2-12. In this example, sprocket sizes from 9 to 45 teeth are tabulated vertically, and speeds of rotation from 10 to 3400 r/min are tabulated horizontally.

The values listed are the horsepower capacity ratings for the many combinations of sprocket size and speed for single-strand chain; factors for calculating the rating of multiple-strand chain are provided below the table. Additionally, recommended lubrication practice is indicated by the use of shading to delineate lubrication type boundaries.

For chain types where rated capacity is presented in terms of allowable working load, this load value is subject to modification by tabulated speed factors which take sprocket size and chain speed into account. The speed factors for a number of engineering steel chains are shown as Table 2-3. In this example, sprocket sizes from 6 to 24 teeth are tabulated vertically, and linear speeds from 10 to 1000 ft/min are tabulated horizontally.

For all chain types, in order to select the proper pitch size and the optimum number of sprocket teeth, the designer must analyze the characteristics of the task to be performed. The type of input power to be used and the nature of the driven load are both considered in assigning a service factor to the application. This relationship is typically shown in tabular form, as seen in Table 2-4.

For applications which must operate in adverse atmospheric or environmental conditions, an additional factor is often used to take these effects into consideration.

Begin with the horsepower to be transmitted or the load to be carried; this value is then multiplied by the applicable service factors to produce "design horsepower," or "design working load," as the case may be. It is this design value that is compared with the capacity ratings of the available sizes of the chain type under consideration.

For some chain types where capacity ratings are set forth as horsepower tables, it is possible, by plotting the tabulated values on a common graph sheet, to provide a selection aid called a "quick selector chart." This chart is helpful in narrowing down the range of sizes to be checked. A quick selector chart for precision roller chain is reproduced as Fig. 2-11.

2-7 Precision Roller Chain—Base Pitch (ANSI B29.1)

This chain type is an alternating assembly of inner links, called "roller links," and outer links, called "pin links." The pins of each pin link are free to articu-

TABLE 2-3 Speed Factor—Values of S_F

Speed Factor Table for Steel Chains

No. of Teeth	Ft/min 10	25	50	75	100	125	150	175	200	225	250	275	300	400	500	600	700	800	900	1000
6	0.917	1.09	1.37	1.68	2.00	2.40	2.91	3.57	4.41	5.65	7.35	10.6	16.7	—	—	—	—	—	—	—
7	0.855	0.971	1.13	1.27	1.44	1.61	1.81	2.04	2.29	2.60	2.96	3.42	3.95	8.62	—	—	—	—	—	—
8	0.813	0.909	1.04	1.16	1.26	1.37	1.49	1.63	1.76	1.93	2.10	2.29	2.48	3.62	6.21	—	—	—	—	—
9	0.794	0.870	0.980	1.07	1.17	1.26	1.36	1.45	1.55	1.65	1.76	1.88	2.00	2.56	2.94	4.29	6.09	9.90	—	—
10	0.775	0.840	0.943	1.02	1.09	1.16	1.24	1.31	1.37	1.45	1.53	1.61	1.68	2.03	2.41	2.81	3.31	3.82	4.48	5.37
11	0.758	0.820	0.901	0.971	1.03	1.09	1.15	1.22	1.28	1.34	1.40	1.46	1.52	1.78	2.05	2.33	2.63	2.96	3.37	3.82
12	0.741	0.787	0.862	0.926	0.990	1.05	1.10	1.16	1.21	1.26	1.32	1.37	1.42	1.63	1.34	2.05	2.26	2.51	2.77	3.05
14	0.735	0.769	0.833	0.855	0.935	0.980	1.02	1.07	1.11	1.15	1.19	1.24	1.28	1.47	1.61	1.78	1.94	2.10	2.29	2.48
16	0.725	0.763	0.813	0.885	0.893	0.935	0.971	1.01	1.05	1.08	1.12	1.16	1.19	1.34	1.48	1.63	1.77	1.93	2.09	2.28
18	0.719	0.752	0.800	0.833	0.877	0.909	0.943	0.980	1.01	1.04	1.08	1.11	1.14	1.27	1.40	1.53	1.67	1.80	1.95	2.11
20	0.717	0.746	0.787	0.826	0.855	0.893	0.917	0.952	0.980	1.01	1.04	1.07	1.10	1.22	1.34	1.45	1.57	1.69	1.82	1.96
24	0.714	0.735	0.769	0.800	0.820	0.847	0.877	0.901	0.935	0.962	0.980	1.01	1.04	1.15	1.26	1.37	1.48	1.59	1.71	1.84

Source: Jeffrey Chain Division, Dresser Industries, Inc., Hartford, Conn.

TABLE 2-4 Service Factor

Type of driven load	Type of Input Power: Internal-combustion engine with hydraulic drive	Electric motor or turbine	Internal-combustion engine with mechanical drive
Smooth	1.0	1.0	1.2
Moderate shock	1.2	1.3	1.4
Heavy shock	1.4	1.5	1.7

Source: Whitney Chain Operations, Dresser Industries, Inc., Hartford, Conn.

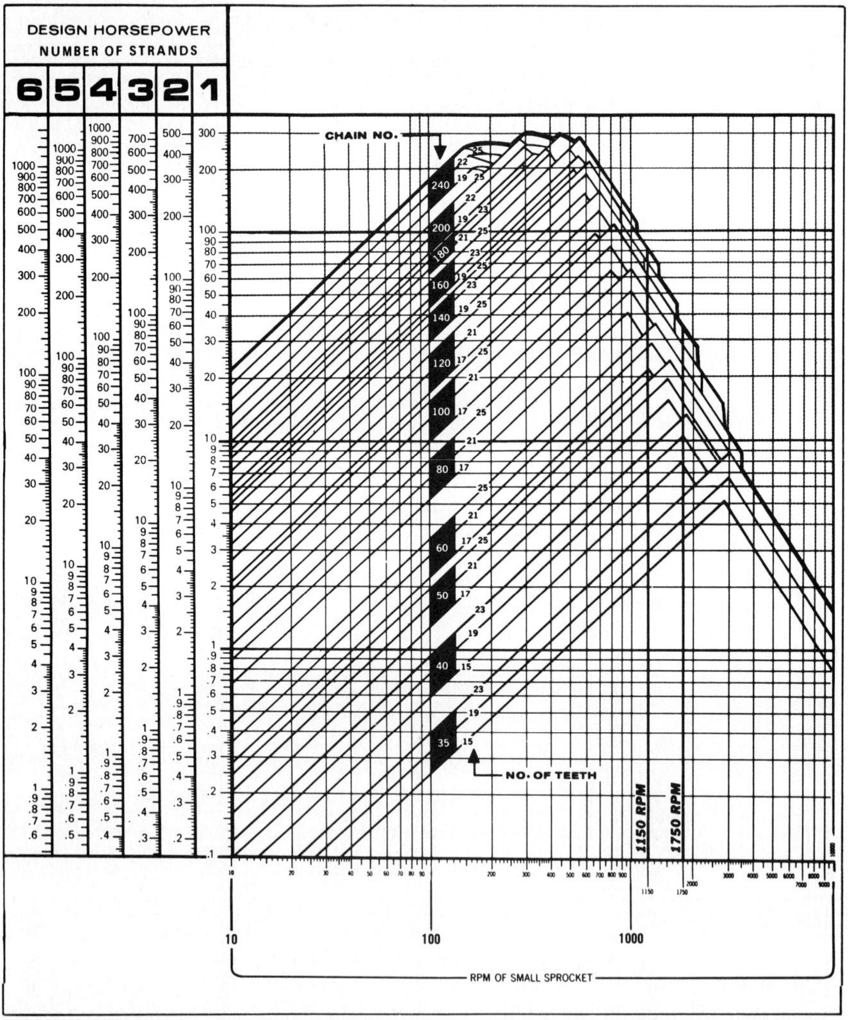

FIG. 2-11 Standard roller chain, quick selector chart. (*Source: Whitney Chain Operations, Dresser Industries, Inc., Hartford, Conn.*)

late within the bushings of adjacent roller links. Except in the two smallest pitch sizes in this group, the bushings are surrounded with free-turning rollers which contact the sprocket teeth. (In the smallest chains, the bushings' outer surfaces contact the teeth directly.) Thirteen pitch sizes are provided, ranging from 1/4 to 3 in. Additionally, nine of these pitch sizes are offered in a "heavy" series, in which the link plate thickness is increased to the gage normally used in the next larger pitch size. The heavy series size ranges run from the 3/4-in to 3-in pitch, inclusive. Also included in the standard is a lighter model in the 1/2-in pitch category, known commonly as "lightweight machinery chain." See Figs. 2-12 and 2-13.

FIG. 2-12 Single-strand roller chain. (*Source: Whitney Chain Operations, Dresser Industries, Inc., Hartford, Conn.*)

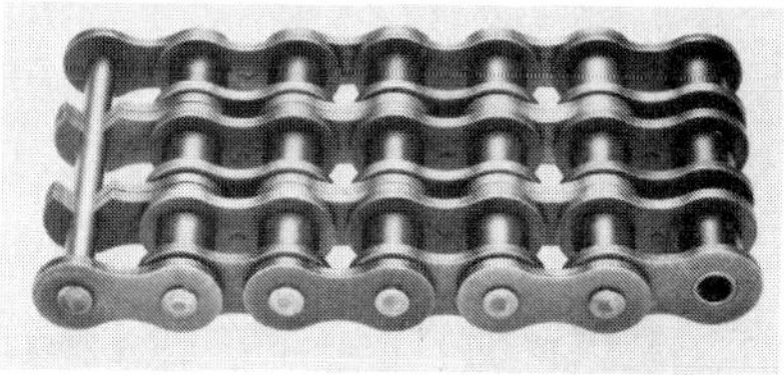

FIG. 2-13 Multiple-strand roller chain. (*Source: Whitney Chain Operations, Dresser Industries, Inc., Hartford, Conn.*)

Multiple-strand assemblies are available; these function similarly to a number of parallel single-strand drives, but are actually a unitized assembly, since longer pins pass completely through all strands in the assembly. Multiple-strand chains are commonly available in widths up to eight strands. See Tables 2-5 through 2-19.

2-8 Precision Roller Chains—Double Pitch (ANSI B29.3, .4)

Double-Pitch Power-Transmission Roller Chain (ANSI B29.3) The name of this type of roller chain, "double pitch," indicates that the dimensions of these chains are derived from the basic precision roller chains in ANSI B29.1. In the double-pitch design, the pins, rollers, and bushings of a specific base-pitch chain are assembled with link plates of twice the base-pitch dimension. The result is a lower-priced product suitable for use at low to medium speeds. The ANSI B29.3 chains are referred to as transmission chains because their link plate shape is the typically narrow waisted, or "figure 8," contour.

Six standard pitch sizes are available. The smallest is 1-in pitch, derived from base-pitch chain no. 40, and the largest is 3-in pitch, a derivative of base-pitch chain no. 120. Multiple-strand assemblies are not available as standard products. See Table 2-20.

Double-Pitch Conveyor Roller Chain (ANSI B29.4) This group of chains also derives many of its dimensions from the basic precision roller chains in ANSI B29.1. Since these chains are intended primarily for conveying applications, they have link plates with straight parallel sides rather than the figure 8 contour. Both pin link plates and roller link plates are of consistent

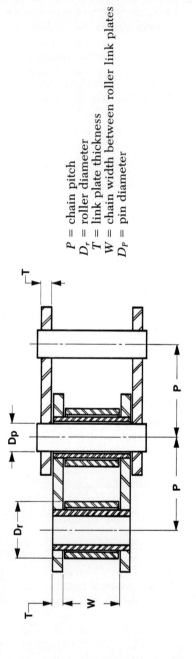

TABLE 2-5 Precision Roller Chain (ANSI B29.1)—General Dimensions

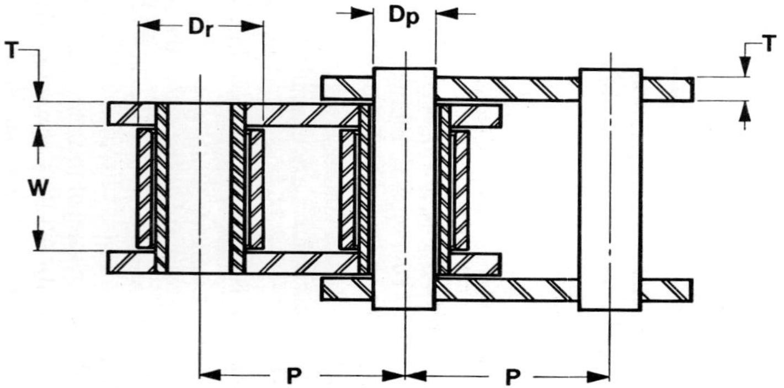

P = chain pitch
D_r = roller diameter
T = link plate thickness
W = chain width between roller link plates
D_P = pin diameter

(In Inches)

Standard chain no.	Pitch P	Maximum roller diameter D_r	Nominal width W	Nominal pin diameter D_p	Link plate thickness T		Measuring load, lb†	Length tolerance, in/ft	Minimum ultimate tensile strength, lb†
					Standard series	Heavy series			Standard & heavy series
25	0.250	0.130*	0.125	0.0905	0.030	—	18	0.031	780
35	0.375	0.200*	0.188	0.141	0.050	—	18	0.022	1,760
41	0.500	0.306	0.250	0.141	0.050	—	18	0.019	1,500
40	0.500	0.312	0.312	0.156	0.060	—	31	0.019	3,125
50	0.625	0.400	0.375	0.200	0.080	—	49	0.018	4,880
60	0.750	0.469	0.500	0.234	0.094	0.125	70	0.017	7,030
80	1.000	0.625	0.625	0.312	0.125	0.156	125	0.016	12,500
100	1.250	0.750	0.750	0.375	0.156	0.187	195	0.016	19,530
120	1.500	0.875	1.000	0.437	0.187	0.219	281	0.015	28,125
140	1.750	1.000	1.000	0.500	0.219	0.250	383	0.015	38,280
160	2.000	1.125	1.250	0.562	0.250	0.281	500	0.015	50,000
180	2.250	1.406	1.406	0.687	0.281	0.312	633	0.015	63,280
200	2.500	1.562	1.500	0.781	0.312	0.375	781	0.015	78,125
240	3.000	1.875	1.875	0.937	0.375	0.500	1000	0.015	112,500

* Bushing diameter, as these chains have no rollers.
† For single-strand chain.
Source: Extracted, with permission, from ANSI B29.1-1975, published by The American Society of Mechanical Engineers.

TABLE 2-6 Horsepower Capacity, No. 25 Roller Chain (ANSI B29.1)

¼-in Pitch, Single Strand

NO. OF TEETH SMALL SPKT.	REVOLUTIONS PER MINUTE / SMALL SPROCKET																							
	50	100	300	500	700	900	1200	1500	1800	2100	2500	3000	3500	4000	4500	5000	5500	6000	6500	7000	7500	8000	8500	9000	10000
9	0.02	0.04	0.12	0.18	0.25	0.31	0.41	0.50	0.58	0.67	0.79	0.93	1.06	1.02	0.86	0.73	0.63	0.56	0.49	0.44	0.40	0.36	0.33	0.30	0.26
10	0.03	0.05	0.13	0.21	0.28	0.35	0.45	0.56	0.65	0.75	0.88	1.04	1.19	1.20	1.00	0.86	0.74	0.65	0.58	0.52	0.47	0.42	0.39	0.35	0.30
11	0.03	0.05	0.14	0.23	0.31	0.39	0.50	0.62	0.73	0.83	0.98	1.15	1.32	1.38	1.16	0.99	0.86	0.75	0.67	0.60	0.54	0.49	0.45	0.41	0.35
12	0.03	0.06	0.16	0.25	0.34	0.43	0.55	0.68	0.80	0.92	1.07	1.26	1.45	1.57	1.32	1.12	0.97	0.86	0.76	0.68	0.61	0.56	0.51	0.47	0.40
13	0.04	0.06	0.17	0.27	0.37	0.47	0.60	0.74	0.87	1.00	1.17	1.38	1.58	1.77	1.49	1.27	1.10	0.96	0.86	0.77	0.69	0.63	0.57	0.53	0.45
14	0.04	0.07	0.19	0.30	0.40	0.50	0.65	0.80	0.94	1.08	1.27	1.49	1.71	1.93	1.66	1.42	1.23	1.08	0.96	0.86	0.77	0.70	0.64	0.59	0.50
15	0.04	0.07	0.20	0.32	0.43	0.54	0.70	0.86	1.01	1.17	1.36	1.61	1.85	2.08	1.84	1.57	1.36	1.20	1.06	0.95	0.86	0.78	0.71	0.65	0.56
16	0.04	0.08	0.22	0.34	0.47	0.58	0.76	0.92	1.09	1.25	1.46	1.72	1.98	2.23	2.03	1.73	1.50	1.32	1.17	1.05	0.94	0.86	0.78	0.72	0.61
17	0.05	0.08	0.23	0.37	0.50	0.62	0.81	0.99	1.16	1.33	1.56	1.84	2.11	2.38	2.22	1.90	1.64	1.44	1.28	1.14	1.03	0.94	0.86	0.79	0.67
18	0.05	0.09	0.25	0.39	0.53	0.66	0.86	1.05	1.24	1.42	1.66	1.96	2.25	2.53	2.42	2.07	1.79	1.57	1.39	1.25	1.12	1.02	0.93	0.86	0.73
19	0.05	0.09	0.26	0.41	0.56	0.70	0.91	1.11	1.31	1.50	1.76	2.07	2.38	2.69	2.62	2.24	1.94	1.70	1.51	1.35	1.22	1.11	1.01	0.93	0.79
20	0.06	0.10	0.28	0.44	0.59	0.74	0.96	1.17	1.38	1.59	1.86	2.19	2.52	2.84	2.83	2.42	2.10	1.84	1.63	1.46	1.32	1.20	1.09	1.00	0.86
21	0.06	0.11	0.29	0.46	0.62	0.78	1.01	1.24	1.46	1.68	1.96	2.31	2.66	2.99	3.05	2.60	2.26	1.98	1.76	1.57	1.42	1.29	1.17	1.08	0.92
22	0.06	0.11	0.31	0.48	0.66	0.82	1.07	1.30	1.53	1.76	2.06	2.43	2.79	3.15	3.27	2.79	2.42	2.12	1.88	1.69	1.52	1.38	1.26	1.16	0.99
23	0.06	0.12	0.32	0.51	0.69	0.86	1.12	1.37	1.61	1.85	2.16	2.55	2.93	3.30	3.50	2.98	2.59	2.27	2.01	1.80	1.62	1.47	1.35	1.24	1.06
24	0.07	0.13	0.34	0.53	0.72	0.90	1.17	1.43	1.69	1.94	2.27	2.67	3.07	3.46	3.73	3.18	2.76	2.42	2.15	1.92	1.73	1.57	1.44	1.32	1.12
25	0.07	0.13	0.35	0.56	0.75	0.94	1.22	1.50	1.76	2.02	2.37	2.79	3.21	3.61	3.96	3.38	2.93	2.57	2.28	2.04	1.84	1.67	1.53	1.40	1.20
26	0.07	0.14	0.37	0.58	0.79	0.98	1.28	1.56	1.84	2.11	2.47	2.91	3.34	3.77	4.19	3.59	3.11	2.73	2.42	2.17	1.95	1.77	1.62	1.49	1.27
28	0.08	0.15	0.40	0.63	0.85	1.07	1.38	1.69	1.99	2.29	2.68	3.15	3.62	4.09	4.54	4.01	3.47	3.05	2.70	2.42	2.18	1.98	1.81	1.66	1.42
30	0.08	0.16	0.43	0.68	0.92	1.15	1.49	1.82	2.15	2.46	2.88	3.40	3.90	4.40	4.89	4.45	3.85	3.38	3.00	2.68	2.42	2.20	2.01	1.84	1.57
32	0.09	0.17	0.46	0.73	0.98	1.23	1.60	1.95	2.30	2.64	3.09	3.64	4.18	4.72	5.25	4.90	4.25	3.73	3.30	2.96	2.67	2.42	2.21	2.03	1.73
35	0.10	0.19	0.51	0.80	1.08	1.36	1.76	2.15	2.53	2.91	3.41	4.01	4.61	5.20	5.78	5.60	4.86	4.26	3.78	3.38	3.05	2.77	2.53	2.32	1.98
40	0.12	0.22	0.58	0.92	1.25	1.57	2.03	2.48	2.93	3.36	3.93	4.64	5.32	6.00	6.68	6.85	5.93	5.21	4.62	4.13	3.73	3.38	3.09	2.83	2.42
45	0.13	0.25	0.66	1.05	1.42	1.78	2.31	2.82	3.32	3.82	4.47	5.26	6.05	6.82	7.58	8.17	7.08	6.21	5.51	4.93	4.45	4.04	3.69	3.38	2.89
	TYPE A				TYPE B									TYPE C											

The limiting r/min for each lubrication is read from the column to the left of the boundary line.
Type A, manual or drip lubrication; Type B, bath or disk lubrication; Type C, oil-stream lubrication.
Source: Whitney Chain Operations, Dresser Industries, Inc., Hartford, Conn.

MULTIPLE-STRAND FACTORS

NO. OF STRANDS	2	3	4	5	6
FACTOR	1.7	2.5	3.3	4.1	4.9

TABLE 2-7 Horsepower Capacity, No. 35 Roller Chain (ANSI B29.1)

¼-in Pitch, Single Strand

NO. OF TEETH SMALL SPKT.	50	100	300	500	700	900	1200	1500	1800	2100	2500	3000	3500	4000	4500	5000	5500	6000	6500	7000	7500	8000	8500	9000	10000
	REVOLUTIONS PER MINUTE/SMALL SPROCKET																								
9	0.08	0.15	0.39	0.62	0.84	1.06	1.37	1.68	1.98	2.27	2.65	2.17	1.73	1.41	1.18	1.01	0.88	0.77	0.68	0.61	0.55	0.50	0.46	0.42	0.36
10	0.09	0.16	0.44	0.70	0.95	1.19	1.54	1.88	2.21	2.54	2.97	2.55	2.02	1.65	1.39	1.18	1.03	0.90	0.80	0.71	0.64	0.58	0.53	0.49	0.42
11	0.10	0.18	0.49	0.77	1.05	1.31	1.70	2.08	2.45	2.82	3.30	2.94	2.33	1.91	1.60	1.37	1.18	1.04	0.92	0.82	0.74	0.67	0.62	0.57	0.48
12	0.11	0.20	0.54	0.85	1.15	1.44	1.87	2.29	2.70	3.10	3.62	3.35	2.66	2.17	1.82	1.56	1.35	1.18	1.05	0.94	0.85	0.77	0.70	0.64	0.55
13	0.12	0.22	0.59	0.93	1.26	1.57	2.04	2.49	2.94	3.38	3.95	3.77	3.00	2.45	2.05	1.75	1.52	1.33	1.18	1.06	0.95	0.87	0.79	0.73	0.62
14	0.13	0.24	0.63	1.01	1.36	1.71	2.21	2.70	3.18	3.66	4.28	4.22	3.35	2.74	2.30	1.96	1.70	1.49	1.32	1.18	1.07	0.97	0.88	0.81	0.69
15	0.14	0.25	0.68	1.08	1.47	1.84	2.38	2.91	3.43	3.94	4.61	4.68	3.71	3.04	2.55	2.17	1.88	1.65	1.47	1.31	1.18	1.07	0.98	0.90	0.77
16	0.15	0.27	0.7[illegible]	1.16	1.57	1.97	2.55	3.12	3.68	4.22	4.94	5.15	4.09	3.35	2.81	2.40	2.08	1.82	1.62	1.45	1.30	1.18	1.08	0.99	0.85
17	0.16	0.29	0.78	1.24	1.68	2.10	2.73	3.33	3.93	4.51	5.28	5.64	4.48	3.67	3.07	2.62	2.27	2.00	1.77	1.58	1.43	1.30	1.18	1.09	0.93
18	0.17	0.31	0.83	1.32	1.78	2.24	2.90	3.54	4.18	4.80	5.61	6.15	4.88	3.99	3.35	2.86	2.48	2.17	1.93	1.73	1.56	1.41	1.29	1.18	1.01
19	0.18	0.33	0.88	1.40	1.89	2.37	3.07	3.76	4.43	5.09	5.95	6.67	5.29	4.33	3.63	3.10	2.69	2.36	2.09	1.87	1.69	1.53	1.40	1.28	1.10
20	0.19	0.35	0.93	1.48	2.00	2.51	3.25	3.97	4.68	5.38	6.29	7.20	5.72	4.68	3.92	3.35	2.90	2.55	2.26	2.02	1.82	1.65	1.51	1.39	1.18
21	0.20	0.37	0.98	1.56	2.11	2.64	3.42	4.19	4.93	5.67	6.63	7.75	6.15	5.03	4.22	3.60	3.12	2.74	2.43	2.17	1.96	1.78	1.62	1.49	1.27
22	0.21	0.38	1.03	1.64	2.22	2.78	3.60	4.40	5.19	5.96	6.97	8.21	6.59	5.40	4.52	3.86	3.35	2.94	2.61	2.33	2.10	1.91	1.74	1.60	1.37
23	0.22	0.40	1.08	1.72	2.33	2.92	3.78	4.62	5.44	6.25	7.31	8.62	7.05	5.77	4.83	4.13	3.58	3.14	2.79	2.49	2.25	2.04	1.86	1.71	1.46
24	0.23	0.42	1.14	1.80	2.44	3.05	3.96	4.84	5.70	6.55	7.66	9.02	7.51	6.15	5.15	4.40	3.81	3.35	2.97	2.66	2.40	2.17	1.99	1.82	1.56
25	0.24	0.44	1.19	1.88	2.55	3.19	4.13	5.05	5.95	6.84	8.00	9.43	7.99	6.54	5.48	4.68	4.05	3.56	3.16	2.82	2.55	2.31	2.11	1.94	1.65
26	0.25	0.46	1.24	1.96	2.66	3.33	4.31	5.27	6.21	7.14	8.35	9.84	8.47	6.93	5.81	4.96	4.30	3.77	3.35	3.00	2.70	2.45	2.24	2.05	1.75
28	0.27	0.50	1.34	2.12	2.88	3.61	4.67	5.71	6.73	7.73	9.05	10.7	9.47	7.75	6.49	5.55	4.81	4.22	3.74	3.35	3.02	2.74	2.50	2.30	1.95
30	0.29	0.54	1.45	2.29	3.10	3.89	5.03	6.15	7.25	8.33	9.74	11.5	10.5	8.59	7.20	6.15	5.33	4.68	4.15	3.71	3.35	3.04	2.77	2.55	2.17
32	0.31	0.58	1.55	2.45	3.32	4.17	5.40	6.60	7.77	8.93	10.4	12.3	11.6	9.47	7.93	6.77	5.87	5.15	4.57	4.00	3.69	3.35	3.06	2.81	0
35	0.34	0.64	1.71	2.70	3.66	4.59	5.95	7.27	8.56	9.84	11.5	13.6	13.2	10.8	9.08	7.75	6.72	5.90	5.23	4.68	4.22	3.83	3.50	3.21	0
40	0.39	0.73	1.97	3.12	4.23	5.30	6.87	8.40	9.89	11.4	13.3	15.7	16.2	13.2	11.1	9.47	8.21	7.20	6.39	5.72	5.15	4.68	0		
45	0.45	0.83	2.24	3.55	4.80	6.02	7.80	9.53	11.2	12.9	15.1	17.8	19.3	15.8	13.2	11.3	9.79	8.59	7.62	6.82	0				

TYPE A — TYPE B — TYPE C

The limiting r/min for each lubrication is read from the column to the left of the boundary line shown. Type A, manual or drip lubrication; Type B, bath or disk lubrication; Type C, oil-stream lubrication.
Source: Whitney Chain Operations, Dresser Industries, Inc., Hartford, Conn.

MULTIPLE-STRAND FACTORS

NO. OF STRANDS	2	3	4	5	6
FACTOR	1.7	2.5	3.3	4.1	4.9

TABLE 2-8 Horsepower Capacity, No. 40 Roller Chain (ANSI B29.1)

NO. OF TEETH SMALL SPKT.	REVOLUTIONS PER MINUTE/SMALL SPROCKET																							
	10	25	50	100	200	300	400	500	700	900	1000	1200	1400	1600	1800	2100	2400	2700	3000	3500	4000	5000	6000	7000	8000
9	0.04	0.10	0.19	0.35	0.65	0.93	1.21	1.48	2.00	2.51	2.75	3.25	3.73	4.12	3.45	2.74	2.24	1.88	1.60	1.27	1.04	0.75	0.57	0.45	0.37
10	0.05	0.11	0.21	0.39	0.73	1.04	1.35	1.65	2.24	2.81	3.09	3.64	4.18	4.71	4.04	3.21	2.63	2.20	1.88	1.49	1.22	0.87	0.65	0.53	0.43
11	0.05	0.12	0.23	0.43	0.80	1.16	1.50	1.83	2.48	3.11	3.42	4.03	4.63	5.22	4.66	3.70	3.03	2.54	2.17	1.72	1.41	1.01	0.77	0.61	0.50
12	0.06	0.14	0.25	0.47	0.88	1.27	1.65	2.01	2.73	3.42	3.76	4.43	5.09	5.74	5.31	4.22	3.45	2.89	2.47	1.96	1.60	1.15	0.87	0.69	0.57
13	0.06	0.15	0.28	0.52	0.96	1.39	1.80	2.20	2.97	3.73	4.10	4.83	5.55	6.26	5.99	4.76	3.89	3.26	2.79	2.21	1.81	1.29	0.98	0.78	0.64
14	0.07	0.16	0.30	0.56	1.04	1.50	1.95	2.38	3.22	4.04	4.44	5.23	6.01	6.78	6.70	5.31	4.35	3.65	3.11	2.47	2.02	1.45	1.10	0.87	0.71
15	0.07	0.17	0.32	0.60	1.12	1.62	2.10	2.56	3.47	4.35	4.78	5.64	6.47	7.30	7.43	5.89	4.82	4.04	3.45	2.74	2.24	1.60	1.22	0.97	0.79
16	0.08	0.19	0.35	0.65	1.20	1.74	2.25	2.75	3.72	4.66	5.13	6.04	6.94	7.83	8.18	6.49	5.31	4.45	3.80	3.02	2.47	1.77	1.34	1.07	0.87
17	0.08	0.20	0.37	0.69	1.29	1.85	2.40	2.93	3.97	4.98	5.48	6.45	7.41	8.36	8.96	7.11	5.82	4.88	4.17	3.31	2.71	1.94	1.47	1.17	0.96
18	0.09	0.21	0.39	0.73	1.37	1.97	2.55	3.12	4.22	5.30	5.82	6.86	7.88	8.89	9.76	7.75	6.34	5.31	4.54	3.60	2.95	2.11	1.60	1.27	0
19	0.09	0.22	0.42	0.78	1.45	2.09	2.71	3.31	4.48	5.62	6.17	7.27	8.36	9.42	10.5	8.40	6.88	5.76	4.92	3.91	3.20	2.29	1.74	1.38	0
20	0.10	0.24	0.44	0.82	1.53	2.21	2.86	3.50	4.73	5.94	6.53	7.69	8.83	9.96	11.1	9.07	7.43	6.22	5.31	4.22	3.45	2.47	1.88	1.49	0
21	0.11	0.25	0.46	0.87	1.62	2.33	3.02	3.69	4.99	6.26	6.88	8.11	9.31	10.5	11.7	9.76	7.99	6.70	5.72	4.54	3.71	2.66	2.02	1.60	0
22	0.11	0.26	0.49	0.91	1.70	2.45	3.17	3.88	5.25	6.58	7.23	8.52	9.79	11.0	12.3	10.5	8.57	7.18	6.13	4.87	3.98	2.85	2.17	1.72	0
23	0.12	0.27	0.51	0.96	1.78	2.57	3.33	4.07	5.51	6.90	7.59	8.94	10.3	11.6	12.9	11.2	9.16	7.68	6.55	5.20	4.26	3.05	2.32	1.84	0
24	0.13	0.29	0.54	1.00	1.87	2.69	3.48	4.26	5.76	7.23	7.95	9.36	10.8	12.1	13.5	11.9	9.76	8.18	6.99	5.54	4.54	3.25	2.47	1.96	0
25	0.13	0.30	0.56	1.05	1.95	2.81	3.64	4.45	6.02	7.55	8.30	9.78	11.2	12.7	14.1	12.7	10.4	8.70	7.43	5.89	4.82	3.45	2.63	0	
26	0.14	0.31	0.58	1.09	2.04	2.93	3.80	4.64	6.28	7.88	8.66	10.2	11.7	13.2	14.7	13.5	11.0	9.23	7.88	6.25	5.12	3.66	2.79	0	
28	0.15	0.34	0.63	1.18	2.20	3.18	4.11	5.03	6.81	8.54	9.39	11.1	12.7	14.3	15.9	15.0	12.3	10.3	8.80	6.99	5.72	4.09	3.11	0	
30	0.16	0.37	0.68	1.27	2.38	3.42	4.43	5.42	7.33	9.20	10.1	11.9	13.7	15.4	17.2	16.7	13.6	11.4	9.76	7.75	6.34	4.54	3.45	0	
32	0.17	0.39	0.73	1.36	2.55	3.67	4.75	5.81	7.86	9.86	10.8	12.8	14.7	16.5	18.4	18.4	15.0	12.6	10.8	8.54	6.99	5.00	0		
35	0.19	0.43	0.81	1.50	2.81	4.04	5.24	6.40	8.66	10.9	11.9	14.1	16.2	18.2	20.3	21.0	17.2	14.4	12.3	9.76	7.99	5.72	0		
40	0.22	0.50	0.93	1.74	3.24	4.67	6.05	7.39	10.0	12.5	13.8	16.3	18.7	21.1	23.4	25.7	21.0	17.6	15.0	11.9	9.76	6.99	0		
45	0.25	0.57	1.06	1.97	3.68	5.30	6.87	8.40	11.4	14.2	15.7	18.5	21.2	23.9	26.6	30.5	25.1	21.0	17.9	14.2	11.7	0			
	TYPE A				TYPE B								TYPE C												

The limiting r/min for each lubrication is read from the column to the left of the boundary line shown.
Type A, manual or drip lubrication; Type B, bath or disk lubrication; Type C, oil-stream lubrication.
Source: Whitney Chain Operations, Dresser Industries, Inc., Hartford, Conn.

MULTIPLE-STRAND FACTORS

NO. OF STRANDS	2	3	4	5	6
FACTOR	1.7	2.5	3.3	4.1	4.9

TABLE 2-9 Horsepower Capacity, No. 41 Roller Chain (ANSI B29.1)

NO. OF TEETH SMALL SPKT.	REVOLUTIONS PER MINUTE/SMALL SPROCKET																								
	10	25	50	100	200	300	400	500	700	900	1000	1200	1400	1600	1800	2100	2400	2700	3000	3500	4000	5000	6000	7000	8000
9	0.02	0.05	0.10	0.19	0.36	0.51	0.66	0.81	1.10	1.38	1.52	1.27	1.01	0.82	0.69	0.55	0.45	0.38	0.32	0.25	0.21	0.15	0.11	0.09	0.07
10	0.03	0.06	0.11	0.21	0.40	0.57	0.74	0.91	1.23	1.54	1.70	1.49	1.18	0.96	0.81	0.64	0.53	0.44	0.38	0.30	0.24	0.17	0.13	0.11	0.08
11	0.03	0.07	0.13	0.24	0.44	0.64	0.82	1.01	1.37	1.71	1.88	1.71	1.36	1.11	0.93	0.74	0.61	0.51	0.43	0.34	0.28	0.20	0.15	0.12	0.10
12	0.03	0.07	0.14	0.26	0.49	0.70	0.91	1.11	1.50	1.88	2.07	1.95	1.55	1.27	1.06	0.84	0.69	0.58	0.49	0.39	.32	0.23	0.17	0.14	0.11
13	0.04	0.08	0.15	0.28	0.53	0.76	0.99	1.21	1.63	2.05	2.25	2.20	1.75	1.43	1.20	0.95	0.78	0.65	0.56	0.44	0.36	0.26	0.20	0.16	0.13
14	0.04	0.09	0.16	0.31	0.57	0.83	1.07	1.31	1.77	2.22	2.44	2.46	1.95	1.60	1.34	1.06	0.87	0.73	0.62	0.49	0.40	0.29	0.22	0.17	0.14
15	0.04	0.09	0.18	0.33	0.62	0.89	1.15	1.41	1.91	2.39	2.63	2.73	2.17	1.77	1.49	1.18	0.96	0.81	0.69	0.55	0.45	0.32	0.24	0.19	0.16
16	0.04	0.10	0.19	0.36	0.66	0.95	1.24	1.51	2.05	2.57	2.82	3.01	2.39	1.95	1.64	1.30	1.06	0.89	0.76	0.60	0.49	0.35	0.27	0.21	0.17
17	0.05	0.11	0.20	0.38	0.71	1.02	1.32	1.61	2.18	2.74	3.01	3.29	2.61	2.14	1.79	1.42	1.16	0.98	0.83	0.66	0.54	0.39	0.29	0.23	0.19
18	0.05	0.12	0.22	0.40	0.75	1.08	1.40	1.72	2.32	2.91	3.20	3.59	2.85	2.33	1.95	1.55	1.27	1.06	0.91	0.72	0.59	0.42	0.32	0.25	0
19	0.05	0.12	0.23	0.43	0.80	1.15	1.49	1.82	2.46	3.09	3.40	3.89	3.09	2.53	2.12	1.68	1.38	1.15	0.98	0.78	0.64	0.46	0.35	0.28	0
20	0.06	0.13	0.24	0.45	0.84	1.21	1.57	1.92	2.60	3.26	3.59	4.20	3.33	2.73	2.29	1.81	1.49	1.24	1.06	0.84	0.69	0.49	0.38	0.30	0
21	0.06	0.14	0.26	0.48	0.89	1.28	1.66	2.03	2.74	3.44	3.78	4.46	3.59	2.94	2.46	1.95	1.60	1.34	1.14	0.91	0.74	0.53	0.40	0.32	0
22	0.06	0.14	0.27	0.50	0.93	1.35	1.74	2.13	2.89	3.62	3.98	4.69	3.85	3.15	2.64	2.09	1.71	1.44	1.23	0.97	0.80	0.57	0.43	0.34	0
23	0.06	0.15	0.28	0.53	0.98	1.41	1.83	2.24	3.03	3.80	4.17	4.92	4.11	3.37	2.82	2.24	1.83	1.54	1.31	1.04	0.85	0.61	0.46	0.37	0
24	0.07	0.16	0.29	0.55	1.03	1.48	1.92	2.34	3.17	3.97	4.37	5.15	4.38	3.59	3.01	2.39	1.95	1.64	1.40	1.11	0.91	0.65	0.49	0.39	0
25	0.07	0.17	0.31	0.57	1.07	1.55	2.00	2.45	3.31	4.15	4.57	5.38	4.66	3.81	3.20	2.54	2.08	1.74	1.49	1.18	0.96	0.69	0.53	0	0
26	0.07	0.17	0.32	0.60	1.12	1.61	2.09	2.55	3.46	4.33	4.76	5.61	4.94	4.05	3.39	2.69	2.20	1.85	1.58	1.25	1.02	0.73	0.56	0	
28	0.08	0.19	0.35	0.65	1.21	1.75	2.26	2.77	3.74	4.69	5.16	6.08	5.52	4.52	3.79	3.01	2.46	2.06	1.76	1.40	1.14	0.82	0.62	0	
30	0.08	0.20	0.38	0.70	1.31	1.88	2.44	2.98	4.03	5.06	5.56	6.55	6.13	5.01	4.20	3.33	2.73	2.29	1.95	1.55	1.27	0.91	0.69	0	
32	0.09	0.22	0.40	0.75	1.40	2.02	2.61	3.20	4.33	5.42	5.96	7.03	6.75	5.52	4.63	3.67	3.01	2.52	2.15	1.71	1.40	1.00	0		
35	0.10	0.24	0.44	0.83	1.54	2.22	2.88	3.52	4.76	5.97	6.57	7.74	7.72	6.32	5.29	4.20	3.44	2.88	2.46	1.95	1.60	1.14	0		
40	0.12	0.27	0.51	0.96	1.78	2.57	3.33	4.07	5.50	6.90	7.59	8.94	9.43	7.72	6.47	5.13	4.20	3.52	3.01	2.39	1.95	1.40	0		
45	0.14	0.31	0.58	1.08	2.02	2.92	3.78	4.62	6.25	7.84	8.62	10.2	11.3	9.21	7.72	6.13	5.01	4.20	3.59	2.85	2.33	0			
	TYPE A				TYPE B								TYPE C												

The limiting r/min for each lubrication is read from the column to the left of the boundary line shown. Type A, manual or drip lubrication; Type B, bath or disk lubrication; Type C, oil-stream lubrication.
Source: Whitney Chain Operations, Dresser Industries, Inc., Hartford, Conn.

MULTIPLE-STRAND FACTORS

NO. OF STRANDS	2	3	4	5	6
FACTOR	1.7	2.5	3.3	4.1	4.9

TABLE 2-10 Horsepower Capacity, No. 50 Roller Chain (ANSI B29.1)

NO. OF TEETH SMALL SPKT.	REVOLUTIONS PER MINUTE/SMALL SPROCKET																								
	10	25	50	100	200	300	400	500	700	900	1000	1200	1400	1600	1800	2100	2400	2700	3000	3500	4000	4500	5000	5500	6000
9	0.09	0.19	0.36	0.67	1.26	1.81	2.35	2.87	3.89	4.88	5.36	6.32	6.02	4.92	4.13	3.27	2.68	2.25	1.92	1.52	1.25	1.04	0.89	0.77	0.68
10	0.10	0.22	0.41	0.76	1.41	2.03	2.63	3.22	4.36	5.46	6.01	7.08	7.05	5.77	4.83	3.84	3.14	2.63	2.25	1.78	1.46	1.22	1.04	0.90	0.79
11	0.11	0.24	0.45	0.84	1.56	2.25	2.92	3.57	4.83	6.06	6.66	7.85	8.13	6.65	5.58	4.42	3.62	3.04	2.59	2.06	1.68	1.41	1.20	1.04	0.92
12	0.12	0.26	0.49	0.92	1.72	2.47	3.21	3.92	5.31	6.65	7.31	8.62	9.26	7.58	6.35	5.04	4.13	3.46	2.95	2.34	1.92	1.61	1.37	1.19	1.04
13	0.13	0.29	0.54	1.00	1.87	2.70	3.50	4.27	5.78	7.25	7.97	9.40	10.4	8.55	7.16	5.69	4.65	3.90	3.33	2.64	2.16	1.81	1.55	1.34	0
14	0.14	0.31	0.58	1.09	2.03	2.92	3.79	4.63	6.27	7.86	8.64	10.2	11.7	9.55	8.01	6.35	5.20	4.36	3.72	2.95	2.42	2.03	1.73	1.50	0
15	0.15	0.34	0.63	1.17	2.19	3.15	4.08	4.99	6.75	8.47	9.31	11.0	12.6	10.6	8.88	7.05	5.77	4.83	4.13	3.27	2.68	2.25	1.92	1.66	0
16	0.16	0.36	0.67	1.26	2.34	3.38	4.37	5.35	7.24	9.08	9.98	11.8	13.5	11.7	9.78	7.76	6.35	5.32	4.55	3.61	2.95	2.47	2.11	1.83	0
17	0.17	0.39	0.72	1.34	2.50	3.61	4.67	5.71	7.73	9.69	10.7	12.6	14.4	12.8	10.7	8.50	6.96	5.83	4.98	3.95	3.23	2.71	2.31	2.01	0
18	0.18	0.41	0.76	1.43	2.66	3.83	4.97	6.07	8.22	10.3	11.3	13.4	15.3	13.9	11.7	9.26	7.58	6.35	5.42	4.30	3.52	2.95	2.52	0	
19	0.19	0.43	0.81	1.51	2.82	4.07	5.27	6.44	8.72	10.9	12.0	14.2	16.3	15.1	12.7	10.0	8.22	6.89	5.88	4.67	3.82	3.20	2.73	0	
20	0.20	0.46	0.86	1.60	2.98	4.30	5.57	6.80	9.21	11.5	12.7	15.0	17.2	16.3	13.7	10.8	8.88	7.44	6.35	5.04	4.13	3.46	2.95	0	
21	0.21	0.48	0.90	1.69	3.14	4.53	5.87	7.17	9.71	12.2	13.4	15.8	18.1	17.6	14.7	11.7	9.55	8.01	6.84	5.42	4.44	3.72	3.18	0	
22	0.22	0.51	0.95	1.77	3.31	4.76	6.17	7.54	10.2	12.8	14.1	16.6	19.1	18.8	15.8	12.5	10.2	8.59	7.33	5.82	4.76	3.99	3.41	0	
23	0.23	0.53	1.00	1.86	3.47	5.00	6.47	7.91	10.7	13.4	14.8	17.4	20.0	20.1	16.9	13.4	11.0	9.18	7.84	6.22	5.09	4.27	0		
24	0.25	0.56	1.04	1.95	3.63	5.23	6.78	8.29	11.2	14.1	15.5	18.2	20.9	21.4	18.0	14.3	11.7	9.78	8.35	6.63	5.42	4.55	0		
25	0.26	0.58	1.09	2.03	3.80	5.47	7.08	8.66	11.7	14.7	16.2	19.0	21.9	22.8	19.1	15.2	12.4	10.4	8.88	7.05	5.77	4.83	0		
26	0.27	0.61	1.14	2.12	3.96	5.70	7.39	9.03	12.2	15.3	16.9	19.9	22.8	24.2	20.3	16.1	13.2	11.0	9.42	7.47	6.12	5.13	0		
28	0.29	0.66	1.23	2.30	4.29	6.18	8.01	9.79	13.2	16.6	18.3	21.5	24.7	27.0	22.6	18.0	14.7	12.3	10.5	8.35	6.84	5.73	0		
30	0.31	0.71	1.33	2.48	4.62	6.66	8.63	10.5	14.3	17.9	19.7	23.2	26.6	30.0	25.1	19.9	16.3	13.7	11.7	9.26	7.58	0			
32	0.33	0.76	1.42	2.66	4.96	7.14	9.25	11.3	15.3	19.2	21.1	24.9	28.6	32.2	27.7	22.0	18.0	15.1	12.9	10.2	8.35	0			
35	0.37	0.84	1.57	2.93	5.46	7.86	10.2	12.5	16.9	21.1	23.2	27.4	31.5	35.5	31.6	25.1	20.6	17.2	14.7	11.7	9.55	0			
40	0.43	0.97	1.81	3.38	6.31	9.08	11.8	14.4	19.5	24.4	26.8	31.6	36.3	41.0	38.7	30.7	25.1	21.0	18.0	14.3	0				
45	0.48	1.10	2.06	3.84	7.16	10.3	13.4	16.3	22.1	27.7	30.5	35.9	41.3	46.5	46.1	36.6	30.0	25.1	21.4	0					
	TYPE A					TYPE B					TYPE C														

The limiting r/min for each lubrication is read from the column to the left of the boundary line shown.
Type A, manual or drip lubrication; Type B, bath or disk lubrication; Type C, oil-stream lubrication.
Source: Whitney Chain Operations, Dresser Industries, Inc., Hartford, Conn.

MULTIPLE-STRAND FACTORS

NO. OF STRANDS	2	3	4	5	6
FACTOR	1.7	2.5	3.3	4.1	4.9

TABLE 2-11 Horsepower Capacity, No. 60 Roller Chain (ANSI B29.1)
(Applicable to Regular and Heavy Series.)

NO. OF TEETH SMALL SPKT.	REVOLUTIONS PER MINUTE/SMALL SPROCKET																								
	10	25	50	100	150	200	300	400	500	600	700	800	900	1000	1100	1200	1400	1600	1800	2000	2500	3000	3500	4000	4500
9	0.15	0.33	0.62	1.16	1.67	2.16	3.12	4.04	4.94	5.82	6.68	7.54	8.38	9.21	9.99	8.77	6.96	5.70	4.77	4.08	2.92	2.22	1.76	1.44	1.21
10	0.16	0.37	0.70	1.30	1.87	2.43	3.49	4.53	5.53	6.52	7.49	8.44	9.39	10.3	11.2	10.3	8.15	6.67	5.59	4.77	3.42	2.60	2.06	1.69	1.41
11	0.18	0.41	0.77	1.44	2.07	2.69	3.87	5.02	6.13	7.23	8.30	9.36	10.4	11.4	12.5	11.9	9.41	7.70	6.45	5.51	3.94	3.00	2.38	1.95	1.63
12	0.20	0.45	0.85	1.58	2.26	2.95	4.25	5.51	6.74	7.94	9.12	10.3	11.4	12.6	13.7	13.5	10.7	8.77	7.35	6.28	4.49	3.42	2.71	2.22	1.86
13	0.22	0.50	0.92	1.73	2.49	3.22	4.64	6.01	7.34	8.65	9.94	11.2	12.5	13.7	14.9	15.2	12.1	9.89	8.29	7.08	5.06	3.85	3.06	2.50	0
14	0.24	0.54	1.00	1.87	2.69	3.49	5.02	6.51	7.96	9.37	10.8	12.1	13.5	14.8	16.2	17.0	13.5	11.1	9.26	7.91	5.66	4.31	3.42	2.80	0
15	0.25	0.58	1.08	2.01	2.90	3.76	5.41	7.01	8.57	10.1	11.6	13.1	14.5	16.0	17.4	18.8	15.0	12.3	10.3	8.77	6.28	4.77	3.79	3.10	0
16	0.27	0.62	1.16	2.16	3.11	4.03	5.80	7.52	9.19	10.8	12.4	14.0	15.6	17.1	18.7	20.2	16.5	13.5	11.3	9.66	6.91	5.26	4.17	3.42	0
17	0.29	0.66	1.24	2.31	3.32	4.30	6.20	8.03	9.81	11.6	13.3	15.0	16.7	18.3	19.9	21.6	18.1	14.8	12.4	10.6	7.57	5.76	4.57	3.74	0
18	0.31	0.70	1.31	2.45	3.53	4.58	6.59	8.54	10.4	12.3	14.1	15.9	17.7	19.5	21.2	22.9	19.7	16.1	13.5	11.5	8.25	6.28	4.98	4.08	0
19	0.33	0.75	1.39	2.60	3.74	4.85	6.99	9.05	11.1	13.0	15.0	16.9	18.8	20.6	22.5	24.3	21.4	17.5	14.6	12.5	8.95	6.81	5.40	4.42	0
20	0.35	0.79	1.47	2.75	3.96	5.13	7.38	9.57	11.7	13.8	15.8	17.9	19.8	21.8	23.8	25.7	23.1	18.9	15.8	13.5	9.66	7.35	5.83	0	
21	0.36	0.83	1.55	2.90	4.17	5.40	7.78	10.1	12.3	14.5	16.7	18.8	20.9	23.0	25.1	27.1	24.8	20.3	17.0	14.5	10.4	7.91	6.28	0	
22	0.38	0.87	1.63	3.05	4.39	5.68	8.19	10.6	13.0	15.3	17.5	19.8	22.0	24.2	26.4	28.5	26.6	21.8	18.2	15.6	11.1	8.48	6.73	0	
23	0.40	0.92	1.71	3.19	4.60	5.96	8.59	11.1	13.6	16.0	18.4	20.8	23.1	25.4	27.7	29.9	28.4	23.3	19.5	16.7	11.9	9.07	7.19	0	
24	0.42	0.96	1.79	3.35	4.82	6.24	8.99	11.6	14.2	16.8	19.3	21.7	24.2	26.6	29.0	31.3	30.3	24.8	20.8	17.8	12.7	9.66	7.67	0	
25	0.44	1.00	1.87	3.50	5.04	6.52	9.40	12.2	14.9	17.5	20.1	22.7	25.3	27.8	30.3	32.7	32.2	26.4	22.1	18.9	13.5	10.3	8.15	0	
26	0.46	1.05	1.95	3.65	5.25	6.81	9.80	12.7	15.5	18.3	21.0	23.7	26.4	29.0	31.6	34.1	34.2	28.0	23.4	20.0	14.3	10.9	8.65	0	
28	0.50	1.13	2.12	3.95	5.69	7.37	10.6	13.8	16.8	19.8	22.8	25.7	28.5	31.4	34.2	37.0	38.2	31.3	26.2	22.4	16.0	12.2	0		
30	0.54	1.22	2.28	4.26	6.13	7.94	11.4	14.8	18.1	21.4	24.5	27.7	30.8	33.8	36.8	39.8	42.4	34.7	29.1	24.8	17.8	13.5	0		
32	0.57	1.31	2.45	4.56	6.57	8.52	12.3	15.9	19.4	22.9	26.3	29.7	33.0	36.3	39.5	42.7	46.7	38.2	32.0	27.3	19.6	14.9	0		
35	0.63	1.44	2.69	5.03	7.24	9.38	13.5	17.5	21.4	25.2	29.0	32.7	36.3	39.9	43.5	47.1	53.4	43.7	36.6	31.3	22.4	17.0	0		
40	0.73	1.67	3.11	5.81	8.37	10.8	15.6	20.2	24.7	29.1	33.5	37.7	42.0	46.1	50.3	54.4	62.5	53.4	44.7	38.2	27.3	0			
45	0.83	1.89	3.53	6.60	9.50	12.3	17.7	23.0	28.1	33.1	38.0	42.9	47.7	52.4	57.1	61.7	70.9	63.7	53.4	45.6	32.6	0			
	TYPE A			TYPE B							TYPE C														

The limiting r/min for each lubrication type is read from the column to the left of the boundary line shown.
Type A, manual or drip lubrication; Type B, bath or disk lubrication; Type C, oil-stream lubrication.
Source: Whitney Chain Operations. Dresser Industries, Inc., Hartford, Conn.

MULTIPLE-STRAND FACTORS

NO. OF STRANDS	2	3	4	5	6
FACTOR	1.7	2.5	3.3	4.1	4.9

TABLE 2-12 Horsepower Capacity—Typical Tabulation (No. 80 Standard Roller Chain)
1-in Pitch, Single Strand (Applicable to Regular and Heavy Series)

NO. OF TEETH SMALL SPKT.	REVOLUTIONS PER MINUTE/SMALL SPROCKET																								
	10	25	50	100	150	200	300	400	500	600	700	800	900	1000	1100	1200	1400	1600	1800	2000	2200	2400	2700	3000	3400
9	0.34	0.78	1.45	2.71	3.90	5.05	7.28	9.43	11.5	13.6	15.6	17.6	17.0	14.5	12.6	11.0	8.76	7.17	6.01	5.13	4.45	3.90	3.27	2.79	2.32
10	0.38	0.87	1.63	3.03	4.37	5.66	8.16	10.6	12.9	15.2	17.5	19.7	19.9	17.0	14.7	12.9	10.3	8.40	7.04	6.01	5.21	4.57	3.83	3.27	2.71
11	0.42	0.97	1.80	3.36	4.84	6,28	9.04	11.7	14.3	16.9	19.4	21.9	23.0	19.6	17.0	14.9	11.8	9.69	8.12	6.93	6.01	5.27	4.42	3.77	1.70
12	0.47	1.06	1.98	3.69	5.32	6.89	9.93	12.9	15.7	18.5	21.3	24.0	26.2	22.3	19.4	17.0	13.5	11.0	9.25	7.90	6.85	6.01	5.04	4.30	0
13	0.51	1.16	2.16	4.03	5.80	7.52	10.8	14.0	17.1	20.2	23.2	26.2	29.1	25.2	21.8	19.2	15.2	12.5	10.4	8.91	7.72	6.78	5.68	4.85	0
14	0.55	1.25	2.34	4.36	6.29	8.14	11.7	15.2	18.6	21.9	25.1	28.4	31.5	28.2	24.4	21.4	17.0	13.9	11.7	9.96	8.63	7.57	6.35	5.42	0
15	0.59	1.35	2.52	4.70	6.77	8.77	12.6	16.4	20.0	23.6	27.1	30.6	34.0	31.2	27.1	23.8	18.9	15.4	12.9	11.0	9.57	8.40	7.04	6.01	0
16	0.63	1.45	2.70	5.04	7.26	9.41	13.5	17.6	21.5	25.3	29.0	32.8	36.4	34.4	29.8	26.2	20.8	17.0	14.2	12.2	10.5	9.25	7.76	6.62	0
17	0.68	1.55	2.88	5.38	7.75	10.0	14.5	18.7	22.9	27.0	31.0	35.0	38.9	37.7	32.7	28.7	22.7	18.6	15.6	13.3	11.5	10.1	8.49	7.25	0
18	0.72	1.64	3.07	5.72	8.25	10.7	15.4	19.9	24.4	28.7	33.0	37.2	41.4	41.1	35.6	31.2	24.8	20.3	17.0	14.5	12.6	11.0	9.25	7.90	0
19	0.76	1.74	3.25	6.07	8.74	11.3	16.3	21.1	25.8	30.4	35.0	39.4	43.8	44.5	38.6	33.9	26.9	22.0	18.4	15.7	13.6	12.0	10.0	8.57	0
20	0.81	1.84	3.44	6.41	9.24	12.0	17.2	22.3	27.3	32.2	37.0	41.7	46.3	48.1	41.7	36.6	29.0	23.8	19.9	17.0	14.7	12.9	10.8	0	
21	0.85	1.94	3.62	6.76	9.74	12.6	18.2	23.5	28.8	33.9	39.0	43.9	48.9	51.7	44.8	39.4	31.2	25.6	21.4	18.3	15.9	13.9	11.7	0	
22	0.90	2.04	3.81	7.11	10.2	13.3	19.1	24.8	30.3	35.7	41.0	46.2	51.4	55.5	48.1	42.2	33.5	27.4	23.0	19.6	17.0	14.9	12.5	0	
23	0.94	2.14	4.00	7.46	10.7	13.9	20.1	26.0	31.8	37.4	43.0	48.5	53.9	59.3	51.4	45.1	35.8	29.3	24.6	21.0	18.2	15.9	13.4	0	
24	0.98	2.24	4.19	7.81	11.3	14.6	21.0	27.2	33.2	39.2	45.0	50.8	56.4	62.0	54.8	48.1	38.2	31.2	26.2	22.3	19.4	17.0	14.2	0	
25	1.03	2.34	4.37	8.16	11.8	15.2	21.9	28.4	34.7	40.9	47.0	53.0	59.0	64.8	58.2	51.1	40.6	33.2	27.8	23.8	20.6	18.1	15.1	0	
26	1.07	2.45	4.56	8.52	12.3	15.9	22.9	29.7	36.2	42.7	49.1	55.3	61.5	67.6	61.8	54.2	43.0	35.2	29.5	25.2	21.8	19.2	16.1	0	
28	1.16	2.65	4.94	9.23	13.3	17.2	24.8	32.1	39.3	46.3	53.2	59.9	66.7	73.3	69.0	60.6	48.1	39.4	33.0	28.2	24.4	21.4	0		
30	1.25	2.85	5.33	9.94	14.3	18.5	26.7	34.6	42.3	49.9	57.3	64.6	71.8	78.9	76.6	67.2	53.3	43.6	36.6	31.2	27.1	23.8	0		
32	1.34	3.06	5.71	10.7	15.3	19.9	28.6	37.1	45.4	53.5	61.4	69.2	77.0	84.6	84.3	74.0	58.7	48.1	40.3	34.4	29.8	26.2	0		
35	1.48	3.37	6.29	11.7	16.9	21.9	31.6	40.9	50.0	58.9	67.6	76.3	84.8	93.3	96.5	84.7	67.2	55.0	46.1	39.4	34.1	0			
40	1.71	3.89	7.27	13.6	19.5	25.3	36.4	47.2	57.7	68.0	78.1	88.1	98.0	108	117	103	82.1	67.2	56.3	48.1	20.0	0			
45	1.94	4.42	8.25	15.4	22.2	28.7	41.4	53.6	65.6	77.2	88.7	100	111	122	133	123	98.0	80.2	67.2	54.1	0				
	TYPE A			TYPE B					TYPE C																

The limiting r/min for each lubrication type is read from the column to the left of the boundary line shown.
The manufacturer should be consulted regarding operation in the dark grey (galling range) area.
Type A, manual or drip lubrication; Type B, bath or disk lubrication; Type C, oil stream lubrication.
Source: Jeffrey Chain Division, Dresser Industries, Inc., Hartford, Conn.

MULTIPLE-STRAND FACTORS

NO. OF STRANDS	2	3	4	5	6
FACTOR	1.7	2.5	3.3	4.1	4.9

TABLE 2-13 Horsepower Capacity, No. 100 Standard Roller Chain (ANSI B29.1)

1¼-in Pitch, Single Strand (Applicable to Regular, Solid Roller, Heavy, and Super Strength Series)

NO. OF TEETH SMALL SPKT.	REVOLUTIONS PER MINUTE/SMALL SPROCKET																							
	10	25	50	100	150	200	300	400	500	600	700	800	900	1000	1100	1200	1300	1400	1600	1800	2000	2200	2400	2600	2700
9	0.65	1.49	2.78	5.19	7.47	9.68	13.9	18.1	22.1	26.0	29.6	24.2	20.3	17.4	15.0	13.2	11.7	10.5	8.57	7.19	6.13	5.32	4.67	4.14	0
10	0.73	1.67	3.11	5.81	8.37	10.8	15.6	20.2	24.7	29.2	33.5	28.4	23.8	20.3	17.6	15.5	13.7	12.3	10.0	8.42	7.19	6.23	5.47	4.85	0
11	0.81	1.85	3.45	6.44	9.28	12.0	17.3	22.4	27.4	32.3	37.1	32.8	27.5	23.4	20.3	17.8	15.8	14.2	11.6	9.71	8.29	7.19	6.31	1.29	0
12	0.89	2.03	3.79	7.08	10.2	13.2	19.0	24.6	30.1	35.5	40.8	37.3	31.3	26.7	23.2	20.3	18.0	16.1	13.2	11.1	9.45	8.19	7.19	0	
13	0.97	2.22	4.13	7.72	11.1	14.4	20.7	26.9	32.8	38.7	44.5	42.1	35.3	30.1	26.1	22.9	20.3	18.2	14.9	12.5	10.6	9.23	8.10	0	
14	1.05	2.40	4.48	8.36	12.0	15.6	22.5	29.1	35.6	41.9	48.2	47.0	39.4	33.7	29.2	25.6	22.7	20.3	16.6	13.9	11.9	10.3	9.05	0	
15	1.13	2.59	4.83	9.01	13.0	16.8	24.2	31.4	38.3	45.2	51.9	52.2	43.7	37.3	32.4	28.4	25.2	22.5	18.4	15.5	13.2	11.4	10.0	0	
16	1.22	2.77	5.17	9.66	13.9	18.0	26.0	33.6	41.1	48.4	55.6	57.5	48.2	41.1	35.7	31.3	27.7	24.8	20.3	17.0	14.5	12.6	11.1	0	
17	1.30	2.96	5.52	10.3	14.8	19.2	27.7	35.9	43.9	51.7	59.4	63.0	52.8	45.0	39.0	34.3	30.4	27.2	22.3	18.7	15.9	13.8	0.79	0	
18	1.38	3.15	5.88	11.0	15.8	20.5	29.5	38.2	46.7	55.0	63.2	68.6	57.5	49.1	42.5	37.3	33.1	29.6	24.2	20.3	17.4	15.0	0		
19	1.46	3.34	6.23	11.6	16.7	21.7	31.2	40.5	49.5	58.3	67.0	74.4	62.3	53.2	46.1	40.5	35.9	32.1	26.3	22.0	18.8	16.3	0		
20	1.55	3.53	6.58	12.3	17.7	22.9	33.0	42.8	52.3	61.6	70.8	79.8	67.3	57.5	49.8	43.7	38.8	34.7	28.4	23.8	20.3	17.6	0		
21	1.63	3.72	6.94	13.0	18.7	24.2	34.8	45.1	55.1	65.0	74.6	84.2	72.4	61.8	53.6	47.0	41.7	37.3	30.6	25.6	21.9	19.0	0		
22	1.71	3.91	7.30	13.6	19.6	25.4	36.6	47.4	58.0	68.3	78.5	88.5	77.7	66.3	57.5	50.4	44.7	40.0	32.8	27.5	23.4	20.3	0		
23	1.80	4.10	7.66	14.3	20.6	26.7	38.4	49.8	60.8	71.7	82.3	92.8	83.0	70.9	61.4	53.9	47.8	42.8	35.0	29.4	25.1	7.74	0		
24	1.88	4.30	8.02	15.0	21.5	27.9	40.2	52.1	63.7	75.0	86.2	97.2	88.5	75.6	65.5	57.5	51.0	45.6	37.3	31.3	26.7	0			
25	1.97	4.49	8.38	15.6	22.5	29.2	42.0	54.4	66.6	78.4	90.1	102	94.1	80.3	69.6	61.1	54.2	48.5	39.7	33.3	28.4	0			
26	2.05	4.68	8.74	16.3	23.5	30.4	43.8	56.8	69.4	81.8	94.0	106	99.8	85.2	73.8	64.8	57.5	51.4	42.1	35.3	30.1	0			
28	2.22	5.07	9.47	17.7	25.5	33.0	47.5	61.5	75.2	88.6	102	115	112	95.2	82.5	72.4	64.2	57.5	47.0	39.4	33.7	0			
30	2.40	5.47	10.2	19.0	27.4	35.5	51.2	66.3	81.0	95.5	110	124	124	106	91.5	80.3	71.2	63.7	52.2	43.7	10.0	0			
32	2.57	5.86	10.9	20.4	29.4	38.1	54.9	71.1	86.9	102	118	133	136	116	101	88.5	78.5	70.2	57.5	48.2	0				
35	2.83	6.46	12.0	22.5	32.4	42.0	60.4	78.3	95.7	113	130	146	156	133	115	101	89.8	80.3	65.8	55.1	0				
40	3.27	7.46	13.9	26.0	37.4	48.5	69.8	90.4	111	130	150	169	188	163	141	124	110	98.1	80.3	0					
45	3.71	8.47	15.8	29.5	42.5	55.0	79.3	103	126	148	170	192	213	194	168	148	131	117	45.3	0					
	TYPE A		TYPE B			TYPE C																			

The limiting r/min for each lubrication type is read from the column to the left of the boundary line shown. The manufacturer should be consulted regarding operation in the dark gray (galling range) area.

Source: Whitney Chain Operations, Dresser Industries, Inc., Hartford, Conn.

MULTIPLE-STRAND FACTORS

NO. OF STRANDS	2	3	4	5	6
FACTOR	1.7	2.5	3.3	4.1	4.9

TABLE 2-14 Horsepower Capacity, No. 120 Standard Roller Chain (ANSI B29.1)

1½-in Pitch, Single Strand (Applicable to Regular, Solid Roller, Heavy, and Super Strength Series)

NO. OF TEETH SMALL SPKT.	REVOLUTIONS PER MINUTE/SMALL SPROCKET																							
	10	25	50	100	150	200	300	400	500	600	700	800	900	1000	1100	1200	1300	1400	1500	1600	1700	1800	1900	2000	2100
9	1.10	2.52	4.69	8.76	12.6	16.3	23.5	30.5	37.3	43.2	34.3	28.1	23.5	20.1	17.4	15.3	13.5	12.1	10.9	9.92	9.06	8.31	7.67	7.10	6.60
10	1.24	2.82	5.26	9.81	14.1	18.3	26.4	34.2	41.8	49.2	40.1	32.9	27.5	23.5	20.4	17.9	15.9	14.2	12.8	11.6	10.6	9.74	8.98	8.31	7.73
11	1.37	3.12	5.83	10.9	15.7	20.3	29.2	37.9	46.3	54.6	46.3	37.9	31.8	27.1	23.5	20.6	18.3	16.4	14.8	13.4	12.2	11.2	10.4	9.59	0
12	1.50	3.43	6.40	11.9	17.2	22.3	32.1	41.6	50.9	59.9	52.8	43.2	36.2	30.9	26.8	23.5	20.9	18.7	16.8	15.3	13.9	12.8	11.8	10.9	0
13	1.64	3.74	6.98	13.0	18.8	24.3	35.0	45.4	55.5	65.3	59.5	48.7	40.8	34.9	30.2	26.5	23.5	21.0	19.0	17.2	15.7	14.4	13.3	12.3	0
14	1.78	4.05	7.56	14.1	20.3	26.3	37.9	49.1	60.1	70.8	66.5	54.4	45.6	39.0	33.8	29.6	26.3	23.5	21.2	19.2	17.6	16.1	14.9	8.94	0
15	1.91	4.37	8.15	15.2	21.9	28.4	40.9	53.0	64.7	76.3	73.8	60.4	50.6	43.2	37.4	32.9	29.1	26.1	23.5	21.3	19.5	17.9	16.5	0	
16	2.05	4.68	8.74	16.3	23.5	30.4	43.8	56.8	69.4	81.8	81.3	66.5	55.7	47.6	41.2	36.2	32.1	28.7	25.9	23.5	21.5	19.7	18.2	0	
17	2.19	5.00	9.33	17.4	25.1	32.5	46.8	60.6	74.1	87.3	89.0	72.8	61.0	52.1	45.2	39.6	35.2	31.5	28.4	25.8	23.5	21.6	19.9	0	
18	2.33	5.32	9.92	18.5	26.7	34.6	49.8	64.5	78.8	92.9	97.0	79.4	66.5	56.8	49.2	43.2	38.3	34.3	30.9	28.1	25.6	23.5	11.3	0	
19	2.47	5.64	10.5	19.6	28.3	36.6	52.8	68.4	83.6	98.5	105	86.1	72.1	61.6	53.4	46.8	41.5	37.2	33.5	30.4	27.8	25.5	0		
20	2.61	5.96	11.1	20.7	29.9	38.7	55.8	72.2	88.3	104	114	92.9	77.9	66.5	57.6	50.6	44.9	40.1	36.2	32.9	30.0	27.5	0		
21	2.75	6.28	11.7	21.9	31.5	40.8	58.8	76.2	93.1	110	122	100	83.8	71.6	62.0	54.4	48.3	43.2	39.0	35.4	32.3	29.6	0		
22	2.90	6.60	12.3	23.0	33.1	42.9	61.8	80.1	97.9	115	131	107	89.9	76.7	66.5	58.4	51.8	46.3	41.8	37.9	34.6	16.6	0		
23	3.04	6.93	12.9	24.1	34.8	45.0	64.9	84.0	103	121	139	115	96.1	82.0	71.1	62.4	55.3	49.5	44.6	40.5	37.0	0			
24	3.18	7.25	13.5	25.3	36.4	47.1	67.9	88.0	108	127	146	122	102	87.4	75.8	66.5	59.0	52.8	47.6	43.2	39.4	0			
25	3.32	7.58	14.1	26.4	38.0	49.3	71.0	91.9	112	132	152	130	109	92.9	80.6	70.7	62.7	56.1	50.6	45.9	41.3	0			
26	3.47	7.91	14.8	27.5	39.7	51.4	74.0	95.9	117	138	159	138	115	98.6	85.4	75.0	66.5	59.5	53.7	48.7	26.6	0			
28	3.76	8.57	16.0	29.8	43.0	55.7	80.2	104	127	150	172	154	129	110	95.5	83.8	74.3	66.5	60.0	54.4	0				
30	4.05	9.23	17.2	32.1	46.3	60.0	86.4	112	137	161	185	171	143	122	106	92.9	82.4	73.8	66.5	42.4	0				
32	4.34	9.90	18.5	34.5	49.6	64.3	92.6	120	147	173	199	188	158	135	117	102	90.8	81.3	73.3	0					
35	4.78	10.9	20.3	38.0	54.7	70.9	102	132	162	190	219	215	180	154	133	117	104	92.9	47.7	0					
40	5.52	12.6	23.5	43.9	63.2	81.8	118	153	187	220	253	263	220	188	163	143	127	59.5	0						
45	6.27	14.3	26.7	49.8	71.7	92.9	134	173	212	250	287	314	263	224	195	171	80.0	0							
TYPE A			TYPE B				TYPE C																		

The limiting r/min for each lubrication type is read from the column to the left of the boundary line shown. The manufacturer should be consulted regarding operation in the dark gray (galling range) area.

Source: Whitney Chain Operations, Dresser Industries, Inc., Hartford, Conn.

MULTIPLE-STRAND FACTORS

NO. OF STRANDS	2	3	4	5	6
FACTOR	1.7	2.5	3.3	4.1	4.9

TABLE 2-15 Horsepower Capacity, No. 140 Standard Roller Chain (ANSI B29.1)

1¾-in Pitch, Single Strand (Applicable to Regular, Solid Roller, Heavy, and Super Strength Series)

NO. OF TEETH SMALL SPKT.	REVOLUTIONS PER MINUTE/SMALL SPROCKET																								
	10	25	50	100	150	200	250	300	350	400	450	500	550	600	700	800	900	1000	1100	1200	1300	1400	1500	1600	1700
9	1.71	3.89	7.26	13.6	19.5	25.3	30.9	36.4	41.8	47.2	52.5	57.7	55.7	48.9	38.8	31.7	26.6	22.7	19.7	17.3	15.3	13.7	12.4	11.2	10.2
10	1.91	4.36	8.14	15.2	21.9	28.3	34.6	40.8	46.9	52.9	58.8	64.6	65.2	57.2	45.4	37.2	31.2	26.6	23.1	20.2	17.9	16.1	14.5	13.1	0
11	2.12	4.83	9.02	16.8	24.2	31.4	38.4	45.2	52.0	58.6	65.2	71.6	75.2	66.0	52.4	42.9	35.9	30.7	26.6	23.3	20.7	18.5	16.7	15.2	0
12	2.33	5.31	9.91	18.5	26.6	34.5	42.2	49.7	57.1	64.4	71.6	78.7	85.7	75.2	59.7	48.9	41.0	35.0	30.3	26.6	23.6	21.1	19.0	17.3	0
13	2.54	5.79	10.8	20.2	29.0	37.6	46.0	54.2	62.2	70.2	78.0	85.8	93.5	84.8	67.3	55.1	46.2	39.4	34.2	30.0	26.6	23.8	21.5	19.5	0
14	2.75	6.27	11.7	21.8	31.5	40.8	49.8	58.7	67.4	76.0	84.5	93.0	101	94.8	75.2	61.6	51.6	44.1	38.2	33.5	29.7	26.6	24.0	21.8	0
15	2.96	6.76	12.6	23.5	33.9	43.9	53.7	63.2	72.7	81.9	91.1	100	109	105	83.4	68.3	57.2	48.9	42.4	37.2	33.0	29.5	26.6	0	
16	3.18	7.24	13.5	25.2	36.3	47.1	57.5	67.8	77.9	87.8	97.7	107	117	116	91.9	75.2	63.1	53.8	46.7	41.0	36.3	32.5	29.3	0	
17	3.39	7.73	14.4	26.9	38.8	50.3	61.4	72.4	83.2	93.8	104	115	125	127	101	82.4	69.1	59.0	51.1	44.9	39.8	35.6	32.1	0	
18	3.61	8.23	15.4	28.6	41.3	53.5	65.3	77.0	88.5	99.8	111	122	133	138	110	89.8	75.2	64.2	55.7	48.9	43.3	38.8	35.0	0	
19	3.82	8.72	16.3	30.4	43.7	56.7	69.3	81.6	93.8	106	118	129	141	150	119	97.4	81.6	69.7	60.4	53.0	47.0	42.1	37.9	0	
20	4.04	9.22	17.2	32.1	46.2	59.9	73.2	86.3	99.1	112	124	137	149	161	128	105	88.1	75.2	65.2	57.2	50.8	45.4	0		
21	4.26	9.72	18.1	33.8	48.7	63.1	77.2	91.0	104	118	131	144	157	170	138	113	94.8	80.9	70.2	61.6	54.6	48.9	0		
22	4.48	10.2	19.1	35.6	51.3	66.4	81.2	95.6	110	124	138	151	165	178	148	121	102	86.8	75.2	66.0	58.6	52.4	0		
23	4.70	10.7	20.0	37.3	53.8	69.7	85.2	100	115	130	145	159	173	187	158	130	109	92.8	80.4	70.6	62.6	56.0	0		
24	4.92	11.2	20.9	39.1	56.3	72.9	89.2	105	121	136	151	166	181	196	169	138	116	98.9	85.7	75.2	66.7	59.7	0		
25	5.14	11.7	21.9	40.8	58.8	76.2	93.2	110	126	142	158	174	189	205	180	147	123	105	91.1	80.0	70.9	63.5	0		
26	5.37	12.2	22.8	42.6	61.4	79.5	97.2	115	132	148	165	181	198	214	190	156	131	112	96.7	84.8	75.2	0			
28	5.81	13.3	24.7	46.2	66.5	86.2	105	124	143	161	179	197	214	232	213	174	146	125	108	94.8	84.1	0			
30	6.26	14.3	26.7	49.7	71.6	92.8	113	134	154	173	193	212	231	249	236	193	162	138	120	105	93.2	0			
32	6.71	15.3	28.6	53.3	76.8	99.5	122	143	165	186	206	227	247	267	260	213	178	152	132	116	0				
35	7.40	16.9	31.5	58.7	84.6	110	134	158	181	205	227	250	272	295	297	243	204	174	151	130	0				
40	8.54	19.5	36.4	67.9	97.7	127	155	182	210	236	263	289	315	340	363	297	249	213	178	0					
45	9.70	22.1	41.3	77.1	111	144	176	207	238	268	298	328	357	387	434	355	297	237	92.7	0					
TYPE A		TYPE B										TYPE C													

The limiting r/min for each lubrication type is read from the column to the left of the boundary line shown. The manufacturer should be consulted regarding operation in the dark gray (galling range) area.
Source: Whitney Chain Operations, Dresser Industries, Inc., Hartford, Conn.

MULTIPLE-STRAND FACTORS

NO. OF STRANDS	2	3	4	5	6
FACTOR	1.7	2.5	3.3	4.1	4.9

TABLE 2-16 Horsepower Capacity, No. 160 Standard Roller Chain (ANSI B29.1)

2-in Pitch, Single Strand (Applicable to Regular, Solid Roller, Heavy, and Super Strength Series)

NO. OF TEETH SMALL SPKT.	REVOLUTIONS PER MINUTE/SMALL SPROCKET																								
	10	25	50	100	150	200	250	300	350	400	450	500	550	600	650	700	750	800	850	900	1000	1100	1200	1300	1400
9	2.48	5.65	10.5	19.7	28.3	36.7	44.8	52.8	60.7	68.5	76.1	71.5	62.0	54.4	48.2	43.2	38.9	35.3	32.2	29.6	25.3	21.9	19.2	17.0	0
10	2.77	6.33	11.8	22.0	31.7	41.1	50.3	59.2	68.0	76.7	85.3	83.7	72.6	63.7	56.5	50.5	45.6	41.4	37.8	34.7	29.6	25.7	22.5	20.0	0
11	3.07	7.01	13.1	24.4	35.2	45.6	55.7	65.6	75.4	85.0	94.5	96.6	83.7	73.5	65.2	58.3	52.6	47.7	43.6	40.0	34.1	29.6	26.0	23.0	0
12	3.38	7.70	14.4	26.8	38.6	50.1	61.2	72.1	82.8	93.4	104	110	95.4	83.7	74.2	66.4	59.9	54.4	49.6	45.6	38.9	33.7	29.6	26.3	0
13	3.68	8.40	15.7	29.2	42.1	54.6	66.7	78.6	90.3	102	113	124	108	94.4	83.7	74.9	67.5	61.3	56.0	51.4	43.9	38.0	33.4	29.6	0
14	3.99	9.10	17.0	31.7	45.6	59.1	72.3	85.2	97.8	110	123	135	120	105	93.6	83.7	75.5	68.5	62.6	57.4	49.0	42.5	37.3	33.1	0
15	4.30	9.80	18.3	34.1	49.2	63.7	77.9	91.7	105	119	132	145	133	117	104	92.8	83.7	76.0	69.4	63.7	54.4	47.1	41.4	0	
16	4.61	10.5	19.6	36.6	52.7	68.3	83.5	98.4	113	127	142	156	147	129	114	102	92.2	83.7	76.4	70.2	59.9	51.9	45.6	0	
17	4.92	11.2	20.9	39.1	56.3	72.9	89.1	105	121	136	151	166	161	141	125	112	101	91.7	83.7	76.8	65.6	56.9	49.9	0	
18	5.23	11.9	22.3	41.6	59.9	77.6	94.8	112	128	145	161	177	175	154	136	122	110	99.9	91.2	83.7	71.5	62.0	54.4	0	
19	5.55	12.7	23.6	44.1	63.5	82.2	101	118	136	153	171	188	190	167	148	132	119	108	98.9	90.8	77.5	67.2	59.0	0	
20	5.86	13.4	25.0	46.6	67.1	86.9	106	125	144	162	180	198	205	180	160	143	129	117	107	98.1	83.7	72.6	63.7	0	
21	6.18	14.1	26.3	49.1	70.7	91.6	112	132	152	171	190	209	221	194	172	154	139	126	115	105	90.1	78.1	68.5	0	
22	6.50	14.8	27.7	51.6	74.4	96.3	118	139	159	180	200	220	237	208	184	165	149	135	123	113	96.6	83.7	0		
23	6.82	15.6	29.0	54.2	78.0	101	124	146	167	189	210	231	251	222	197	176	159	144	132	121	103	89.5	0		
24	7.14	16.3	30.4	56.7	81.7	106	129	152	175	197	220	241	263	237	210	188	169	154	140	129	110	95.4	0		
25	7.46	17.0	31.8	59.3	85.4	111	135	159	183	206	229	252	275	252	223	200	180	164	149	137	117	101	0		
26	7.78	17.8	33.1	61.8	89.1	115	141	166	191	215	239	263	287	267	237	212	191	173	158	145	124	108	0		
28	8.43	19.2	35.9	67.0	96.5	125	153	180	207	233	259	285	311	298	265	237	214	194	177	162	139	120	0		
30	9.08	20.7	38.7	72.2	104	135	165	194	223	251	279	307	335	331	293	263	237	215	196	180	154	0			
32	9.74	22.2	41.5	77.4	111	144	176	208	239	269	300	329	359	365	323	289	261	237	216	198	169	0			
35	10.7	24.5	45.7	85.2	123	159	194	229	263	297	330	363	395	417	370	331	298	271	247	227	180	0			
40	12.4	28.3	52.8	98.5	142	184	225	265	304	343	381	419	457	494	452	404	365	331	302	257	0				
45	14.1	32.1	59.9	112	161	209	255	301	345	389	433	476	519	561	539	482	418	348	271	189	0				
TYPE A		TYPE B			TYPE C																				

The limiting r/min for each lubrication type is read from the column to the left of the boundary line shown. The manufacturer should be consulted regarding operation in the dark gray (galling range) area.

Source: Whitney Chain Operations, Dresser Industries, Inc., Hartford, Conn.

MULTIPLE-STRAND FACTORS

NO. OF STRANDS	2	3	4	5	6
FACTOR	1.7	2.5	3.3	4.1	4.9

TABLE 2-17 Horsepower Capacity, No. 180 Standard Roller Chain (ANSI B29.1)

2¼-in Pitch, Single Strand

NO. OF TEETH SMALL SPKT.	REVOLUTIONS PER MINUTE/SMALL SPROCKET																								
	10	25	50	100	150	200	250	300	350	400	450	500	550	600	650	700	750	800	850	900	950	1000	1050	1100	1150
9	3.42	7.80	14.5	27.1	39.1	50.7	61.9	73.0	83.8	94.5	92.0	78.5	68.1	59.7	53.0	47.4	42.8	38.8	35.4	32.5	30.0	27.8	25.8	24.1	0
10	3.83	8.74	16.3	30.4	43.8	56.8	69.4	81.8	93.9	106	108	92.0	79.7	70.0	62.1	55.5	50.1	45.5	41.5	38.1	35.1	32.5	30.2	28.2	0
11	4.24	9.68	18.1	33.7	48.6	62.9	76.9	90.6	104	117	124	106	92.0	80.7	71.6	64.1	57.8	52.4	47.9	43.9	40.5	37.5	34.9	32.5	0
12	4.66	10.6	19.8	37.0	53.4	69.1	84.5	99.6	114	129	142	121	105	92.0	81.6	73.0	65.8	59.7	54.6	50.1	46.2	42.8	39.7	37.1	0
13	5.08	11.6	21.6	40.4	58.2	75.4	92.1	109	125	141	156	136	118	104	92.0	82.3	74.2	67.4	61.5	56.5	52.1	48.2	44.8	0	
14	5.51	12.6	23.4	43.7	63.0	81.6	99.8	118	135	152	169	152	132	116	103	92.0	82.9	75.3	68.7	63.1	58.2	53.9	50.1	0	
15	5.93	13.5	25.3	47.1	67.9	88.0	108	127	146	164	182	169	146	129	114	102	92.0	83.5	76.2	70.0	64.5	59.7	55.5	0	
16	6.36	14.5	27.1	50.5	72.8	94.3	115	136	156	176	196	186	161	142	126	112	101	92.0	84.0	77.1	71.1	65.8	61.2	0	
17	6.79	15.5	28.9	54.0	77.7	101	123	145	167	188	209	204	177	155	138	123	111	101	92.0	84.4	77.9	72.1	0		
18	7.22	16.5	30.8	57.4	82.7	107	131	154	177	200	222	222	193	169	150	134	121	110	100	92.0	84.8	78.5	0		
19	7.66	17.5	32.6	60.8	87.6	114	139	164	188	212	236	241	209	183	163	145	131	119	109	99.8	92.0	85.2	0		
20	8.10	18.5	34.5	64.3	92.6	120	147	173	199	224	249	260	226	198	176	157	142	129	117	108	99.3	92.0	0		
21	8.53	19.5	36.3	67.8	97.6	126	155	182	209	236	262	280	243	213	189	169	152	138	126	116	107	99.0	0		
22	8.97	20.5	38.2	71.3	103	133	163	192	220	248	276	300	260	228	203	181	163	148	135	124	115				
23	9.41	21.5	40.1	74.8	108	140	171	201	231	260	290	318	278	244	216	194	175	159	145	133	123	0			
24	9.86	22.5	42.0	78.3	113	146	179	210	242	273	303	333	296	260	231	206	186	169	154	142	131	0			
25	10.3	23.5	43.9	81.8	118	153	187	220	253	285	317	348	315	277	245	220	198	180	164	151	139	0			
26	10.7	24.5	45.7	85.4	123	159	195	229	264	297	331	363	334	293	260	233	210	191	174	160	0				
28	11.6	26.6	49.6	92.5	133	173	211	249	286	322	358	394	374	328	291	260	235	213	194	178	0				
30	12.5	28.6	53.4	99.6	144	186	227	268	308	347	386	424	414	364	322	289	260	236	216	198	0				
32	13.4	30.7	57.2	107	154	199	244	287	330	372	414	455	456	401	355	318	287	260	238	0					
35	14.8	33.8	63.1	118	170	220	268	316	363	410	456	501	522	458	406	364	328	291	220	0					
40	17.1	39.0	72.9	136	196	254	310	365	420	473	526	579	575	524	465	398	324	244	0						
45	19.4	44.3	82.7	154	222	288	352	415	477	538	598	631	578	514	441	360	271	0							
	TYPE A	TYPE B			TYPE C																				

The limiting r/min for each lubrication type is read from the column to the left of the boundary line shown. The manufacturer should be consulted regarding operation in the dark gray (galling range) area.

Source: Whitney Chain Operations, Dresser Industries, Inc., Hartford, Conn.

MULTIPLE-STRAND FACTORS

NO. OF STRANDS	2	3	4	5	6
FACTOR	1.7	2.5	3.3	4.1	4.9

TABLE 2-18 Horsepower Capacity, No. 200 Standard Roller Chain (ANSI B29.1)

2½-in Pitch, Single Strand (Applicable to Regular, Solid Roller, Heavy, and Super Strength Series)

NO. OF TEETH SMALL SPKT.	REVOLUTIONS PER MINUTE/SMALL SPROCKET																							
	10	15	20	30	40	50	70	100	150	200	250	300	350	400	450	500	550	600	650	700	750	800	850	900	950
9	4.54	6.54	8.47	12.2	15.8	19.3	26.1	36.0	51.9	67.3	82.2	96.9	111	119	100	85.4	74.1	65.0	57.6	51.6	46.5	42.2	38.6	35.4	32.6
10	5.08	7.32	9.49	13.7	17.7	21.6	29.3	40.4	58.2	75.4	92.1	109	125	140	117	100	86.7	76.1	67.5	60.4	54.5	49.4	45.2	41.4	0
11	5.64	8.12	10.5	15.1	19.6	24.0	32.5	44.8	64.5	83.5	102	120	138	156	135	115	100	87.8	77.9	69.7	62.8	57.0	52.1	47.8	0
12	6.19	8.92	11.6	16.6	21.6	26.4	35.7	49.2	70.8	91.8	112	132	152	171	154	132	114	100	88.8	79.4	71.6	65.0	59.4	54.5	0
13	6.75	9.72	12.6	18.1	23.5	28.7	38.9	53.6	77.2	100	122	144	166	187	174	148	129	113	100	89.6	80.7	73.3	66.9	61.4	0
14	7.31	10.5	13.6	19.7	25.5	31.1	42.1	58.1	83.7	108	132	156	179	202	194	166	144	126	112	100	90.2	81.9	74.8	68.6	0
15	7.88	11.3	14.7	21.2	27.4	33.5	45.4	62.6	90.1	117	143	168	193	218	215	184	159	140	124	111	100	90.8	82.9	0	
16	8.45	12.2	15.8	22.7	29.4	36.0	48.7	67.1	96.6	125	153	180	207	234	237	203	176	154	137	122	110	100	91.4	0	
17	9.02	13.0	16.8	24.2	31.4	38.4	52.0	71.6	103	134	163	193	221	249	260	222	192	169	150	134	121	110	100	0	
18	9.59	13.8	17.9	25.8	33.4	40.8	55.3	76.2	110	142	174	205	235	265	283	242	209	184	163	146	132	119	109	0	
19	10.2	14.6	19.0	27.3	35.4	43.3	58.6	80.8	116	151	184	217	249	281	307	262	227	199	177	158	143	130	118	0	
20	10.7	15.5	20.1	28.9	37.4	45.8	61.9	85.4	123	159	195	229	264	297	331	283	245	215	191	171	154	140	0		
21	11.3	16.3	21.1	30.5	39.5	48.2	65.3	90.0	130	168	205	242	278	313	348	305	264	232	205	184	166	150	0		
22	11.9	17.2	22.2	32.0	41.5	50.7	68.7	94.6	136	177	216	254	292	330	366	327	283	248	220	197	178	161	0		
23	12.5	18.0	23.3	33.6	43.5	53.2	72.0	99.3	143	185	226	267	307	346	384	349	303	266	236	211	190	172	0		
24	13.1	18.9	24.4	35.2	45.6	55.7	75.4	104	150	194	237	279	321	362	402	372	323	283	251	225	203	184	0		
25	13.7	19.7	25.5	36.8	47.6	58.2	78.8	109	156	203	248	292	335	378	421	396	343	301	267	239	215	195	0		
26	14.3	20.6	26.6	38.4	49.7	60.7	82.2	113	163	212	259	305	350	395	439	420	364	319	283	253	228	0			
	TYPE A		TYPE B							TYPE C															

The limiting r/min for each lubrication type is read from the column to the left of the boundary line shown. The manufacturer should be consulted regarding operation in the dark gray (galling range) area.
Source: Whitney Chain Operations, Dresser Industries, Inc., Hartford, Conn.

MULTIPLE-STRAND FACTORS

NO. OF STRANDS	2	3	4	5	6
FACTOR	1.7	2.5	3.3	4.1	4.9

TABLE 2-19 Horsepower Capacity, No. 240 Standard Roller Chain (ANSI B29.1)

3-in Pitch, Single Strand

NO. OF TEETH SMALL SPKT.	REVOLUTIONS PER MINUTE/SMALL SPROCKET: 5	10	15	20	25	30	40	50	60	80	100	125	150	175	200	250	300	350	400	450	500	550	600	650	700
9	3.92	7.31	10.5	13.6	16.7	19.6	25.4	31.1	36.7	47.5	58.1	71.0	83.6	96.1	108	132	156	169	138	116	98.9	85.7	75.2	66.7	0
10	4.39	8.19	11.8	15.3	18.7	22.0	28.5	34.9	41.1	53.2	65.0	79.5	93.7	108	121	148	175	198	162	136	116	100	88.1	78.1	0
11	4.86	9.08	13.1	16.9	20.7	24.4	31.6	38.6	45.5	59.0	72.1	88.1	104	119	135	164	194	223	187	156	134	116	102	90.1	0
12	5.34	9.97	14.4	18.6	22.7	26.8	34.7	42.4	50.0	64.8	79.2	96.8	114	131	148	181	213	245	213	178	152	132	116	103	0
13	5.83	10.9	15.7	20.3	24.8	29.2	37.9	46.3	54.5	70.6	86.4	106	124	143	161	197	232	267	240	201	172	149	131	116	0
14	6.31	11.8	17.0	22.0	26.9	31.7	41.0	50.1	59.1	76.5	93.6	114	135	155	175	213	251	289	268	225	192	166	146	129	0
15	6.80	12.7	18.3	23.7	28.9	34.1	44.2	54.0	63.6	82.4	101	123	145	167	188	230	271	311	297	249	213	184	162	0	
16	7.29	13.6	19.6	25.4	31.0	36.6	47.4	57.9	68.2	88.4	108	132	156	179	202	247	290	334	328	274	234	203	178	0	
17	7.78	14.5	20.9	27.1	33.1	39.0	50.6	61.8	72.9	94.4	115	141	166	191	215	263	310	356	359	301	257	222	195	0	
18	8.28	15.4	22.3	28.8	35.2	41.5	53.8	65.8	77.5	100	123	150	177	203	229	280	330	379	377	328	280	242	213	0	
19	8.78	16.4	23.6	30.6	37.4	44.0	57.0	69.7	82.2	106	130	159	187	215	243	297	350	402	393	355	303	263	231	0	
20	9.28	17.3	24.9	32.3	39.5	46.5	60.3	73.7	86.8	112	138	168	198	228	257	314	370	423	407	383	328	284	249	0	
21	9.78	18.2	26.3	34.1	41.6	49.0	63.5	77.7	91.5	119	145	177	209	240	270	331	390	439	421	395	352	305	268	0	
22	10.3	19.2	27.6	35.8	43.8	51.6	66.8	81.7	96.2	125	152	186	220	252	284	348	410	454	435	406	369	324	271	0	
23	10.8	20.1	29.0	37.6	45.9	54.1	70.1	85.7	101	131	160	195	230	265	298	365	430	469	448	417	377	329	0		
24	11.3	21.1	30.4	39.3	48.1	56.7	73.4	89.7	106	137	167	205	241	277	312	382	450	483	460	427	385	334	0		
25	11.8	22.0	31.7	41.1	50.3	59.2	76.7	93.8	110	143	175	214	252	290	327	399	470	496	472	437	391	337	0		
26	12.3	23.0	33.1	42.9	52.4	61.8	80.0	97.8	115	149	183	223	263	302	341	416	491	509	483	446	397	340	0		
	TYPE A		TYPE B									TYPE C													

The limiting r/min for each lubrication type is read from the column to the left of the boundary line shown.
The manufacturer should be consulted regarding operation in the dark gray (galling range) area.
Source: Whitney Chain Operations, Dresser Industries, Inc., Hartford, Conn.

MULTIPLE-STRAND FACTORS

NO. OF STRANDS	2	3	4	5	6
FACTOR	1.7	2.5	3.3	4.1	4.9

TABLE 2-20 Double-Pitch Power-Transmission Roller Chain—ANSI B29.3—General Dimensions

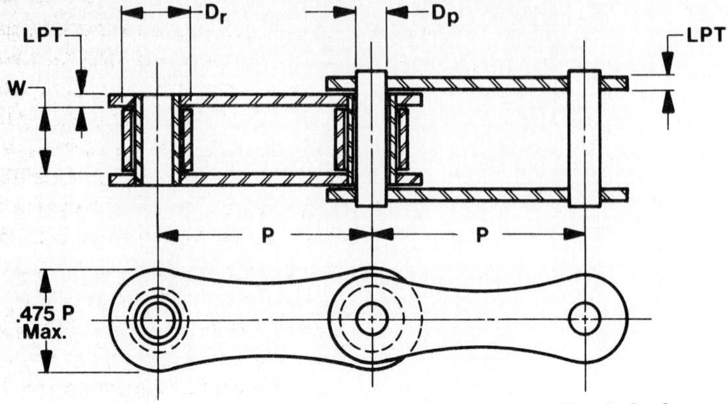

P = chain pitch
D_r = roller diameter
LPT = link plate thickness
W = chain width between roller link plates
D_p = pin diameter

(In Inches)

Standard chain no.	Pitch P	Roller diameter D_r	Chain width W (nom)	Pin diameter D_p	Link plate thickness LPT	Measuring load lbs
2040	1.000	0.312	0.312	0.156	0.060	31
2050	1.250	0.400	0.375	0.200	0.080	49
2060	1.500	0.469	0.500	0.234	0.094	70
2080	2.000	0.625	0.625	0.312	0.125	125
2100	2.500	0.750	0.750	0.375	0.156	195
2120	3.000	0.875	1.000	0.437	0.187	281

Source: Extracted, with permission, from ANSI B29.3-1977, published by The American Society of Mechanical Engineers.

height, so that the horizontal runs of the conveyor can be supported, by sliding contact, on rails or tracks.

Additionally, a second series of these chains is offered with larger-diameter rollers that project below the link plate edges to provide rolling support for horizontal conveyor runs. Rolling friction is typically half, or less, the level of sliding friction, so the large roller series are commonly used to minimize chain working load on long conveyor runs. See Fig. 2-14.

FIG. 2-14 Double-pitch conveyor roller chain, large roller series. (*Source: Whitney Chain Operations, Dresser Industries, Inc., Hartford, Conn.*)

Seven standard pitch sizes are

available, with each size offered with the choice of regular or larger rollers. Normally, the 1-in and 1¼-in pitch sizes are stocked with regular-thickness link plates only, while 1½-in through 4-in pitch sizes are stocked with heavy-series link plates, although regular-thickness plates are available. Multiple-strand assemblies are not available as standard products.

A wide range of conveying attachments are available for assembly into these chains. See Table 2-21.

TABLE 2-21 Double Pitch Conveyor Roller Chain—ANSI B29.4—General Dimensions

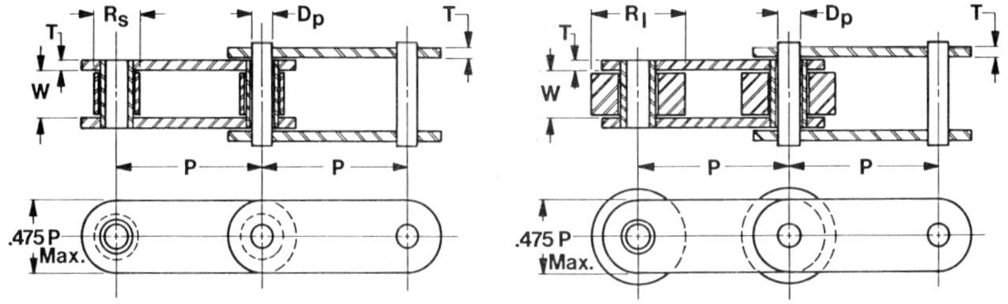

P = chain pitch

R_s = roller diameter, small-roller series

R_L = roller diameter, large-roller series

W = chain width between roller link plates

D_p = pin diameter

T = link plate thickness

(In Inches)

Chain pitch P	Chain number	Roller diameter R_s	Roller diameter R_L	Link plate thickness T	Pin diameter D_p	Width W_{nom}
1.000	C2040	0.312		0.060	0.156	0.312
	C2042		0.625			
1.250	C2050	0.400		0.080	0.200	0.375
	C2052		0.750			
1.500	C2060	0.469		0.094	0.234	0.500
	C2062		0.875			
	C2060H	0.469		0.125		
	C2062H		0.875			
2.000	C2080	0.625		0.125	0.312	0.625
	C2082		1.125			
	C2080H	0.625		0.156		
	C2082H		1.125			

TABLE 2-21 Double Pitch Conveyor Roller Chain—ANSI B29.4—General Dimensions (*Continued*)

(In Inches)

Chain pitch P	Chain number	Roller diameter R_s	Roller diameter R_L	Link plate thickness T	Pin diameter D_p	Width W_{nom}
2.500	C2100	0.750		0.156	0.375	0.750
	C2102		1.562			
	C2100H	0.750		0.187		
	C2102H		1.562			
3.000	C2120	0.875		0.187	0.437	1.000
	C2122		1.750			
	C2120H	0.875		0.219		
	C2122H		1.750			
4.000	C2160	1.125		0.250	0.562	1.250
	C2162		2.250			
	C2160H	1.125		0.281		
	C2162H		2.250			

Source: Extracted, with permission, from ANSI B29.4-1979, published by The American Society of Mechanical Engineers.

2-9 Inverted-Tooth (Silent) Chain (ANSI B29.2)

This chain type is more commonly known as "silent" chain, a name which was apparently applied by chain designers attempting to duplicate the quieter sliding contact that characterizes gear-tooth engagement. Silent chain is a jointed assembly of tooth-shaped link plates; the joint designs providing the articulating capability vary among the several manufacturers. Depending on the particular design, the joint may be composed of round cross-section pins. In addition to the toothed link plates, guide links are also provided to assure sideways stability of the chain relative to the sprocket face. See Fig. 2-15.

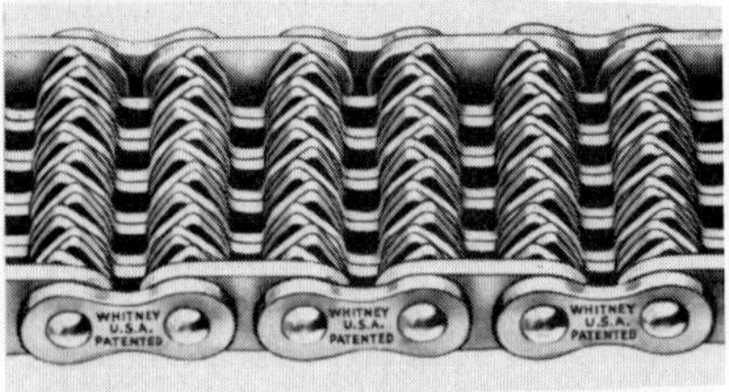

FIG. 2-15 Inverted tooth (silent) chain. (*Source: Whitney Chain Operations, Dresser Industries, Inc., Hartford, Conn.*)

Nine standard pitch sizes are available, ranging from a miniature 3⁄16-in model to 2-in pitch. Many widths are commonly provided, with the widest in each pitch size usually limited to 16 times the pitch dimension. See Tables 2-22 through 2-30.

TABLE 2-22 Horsepower Capacity per Inch of Width, 3/16-in Pitch Silent Chain (ANSI B29.2)

STANDARD WIDTHS 5/32, 7/32, 9/32, 11/32, 13/32, 15/32, 17/32, 19/32, 21/32, 23/32, 25/32, 27/32, 29/32, 31/32

No. teeth, small sprocket	Revolutions per minute, small sprocket											
	500	600	700	800	900	1200	1800	2000	3500	5000	7000	9000
15	0.28	0.33	0.38	0.43	0.47	0.60	0.80	0.90	1.33	1.66	1.94	1.96
17	0.33	0.39	0.44	0.50	0.55	0.70	0.96	1.05	1.60	2.00	2.40	2.52
19	0.37	0.43	0.50	0.55	0.61	0.80	1.10	1.20	1.80	2.30	2.76	2.92
21	0.41	0.48	0.55	0.62	0.68	0.87	1.22	1.33	2.03	2.58	3.12	3.35
23	0.45	0.53	0.60	0.68	0.75	0.96	1.35	1.47	2.25	2.88	3.50	3.78
25	0.49	0.58	0.66	0.74	0.82	1.05	1.47	1.60	2.45	3.13	3.80	4.10
27	0.53	0.62	0.71	0.80	0.88	1.15	1.58	1.72	2.63	3.35	4.06	4.37
29	0.57	0.67	0.76	0.86	0.95	1.21	1.70	1.85	2.83	3.61	4.40	4.72
31	0.60	0.72	0.81	0.91	1.01	1.30	1.81	1.97	3.02	3.84	4.66	5.00
33	0.64	0.75	0.86	0.97	1.07	1.37	1.90	2.08	3.17	4.02	4.85	—
35	0.68	0.80	0.92	1.03	1.14	1.45	2.03	2.21	3.41	4.27	5.16	—
37	0.71	0.84	0.96	1.08	1.19	1.52	2.11	2.30	3.48	4.39	5.24	—
40	0.77	0.91	1.04	1.16	1.29	1.64	2.28	2.50	3.77	4.76	—	—
45	0.86	1.02	1.15	1.30	1.43	1.83	2.53	2.75	4.15	5.21	—	—
50	0.95	1.12	1.27	1.37	1.58	2.00	2.78	3.02	4.52	5.65	—	—

Type I — Type II — Type III

* For best results, smaller sprocket should have at least 21 teeth.

Source: Whitney Chain Operations, Dresser Industries, Inc., Hartford, Conn.

TABLE 2-23 Horsepower Capacity per Inch of Width, 3/8-in Pitch Silent Chain (ANSI B29.2)

STANDARD WIDTHS 1/2, 3/4 1, 1 1/4, 1 1/2, 1 3/4, 2, 2 1/4, 2 1/2, 3, 4, 5, 6

No. teeth, small sprocket	Revolutions per minute, small sprocket												
	100	500	1000	1200	1500	1800	2000	2500	3000	3500	4000	5000	6000
*17	0.46	2.1	4.6	4.9	5.3	6.5	6.9	7.9	8.5	8.8	8.8	—	—
*19	0.53	2.5	4.8	5.4	6.5	7.4	7.9	9.1	9.9	10	11	9.8	—
21	0.58	2.8	5.1	6.0	7.3	8.3	9.0	10	11	12	12	12	10
23	0.63	3.0	5.6	6.6	8.0	9.3	10	12	13	14	14	14	12
25	0.69	3.3	6.1	7.3	8.8	10	11	13	14	15	15	15	14
27	0.74	3.5	6.8	7.9	9.5	11	12	14	15	16	18	18	16
29	0.80	3.8	7.3	8.5	10	12	13	15	16	18	19	19	18
31	0.85	4.1	7.8	9.1	11	13	14	16	18	19	20	20	19
33	0.90	4.4	8.3	9.8	12	14	15	18	19	21	21	21	20
35	0.96	4.6	8.8	10	13	15	16	19	20	23	23	23	21
37	1.0	4.9	9.1	11	14	15	16	20	21	24	24	24	—
40	1.1	5.3	10	12	15	16	18	21	24	25	26	26	—
45	1.3	6.0	11	13	16	19	20	24	26	28	29	—	—
50	1.4	6.6	13	15	18	20	23	26	29	30	—	—	—

Type I   Type II   Type III

* For best results, smaller sprocket should have at least 21 teeth.
Source: Whitney Chain Operations, Dresser Industries, Inc., Hartford, Conn.

TABLE 2-24 Horsepower Capacity per Inch of Width, 1/2-in Pitch Silent Chain (ANSI B29.2)

STANDARD WIDTHS 1/2, 3/4, 1, 1 1/4, 1 1/2, 1 3/4, 2, 2 1/4, 2 1/2, 2 3/4, 3, 3 1/2, 4, 5, 6, 8

No. teeth, small sprocket	Revolutions per minute, small sprocket										
	100	500	700	1000	1200	1800	2000	2500	3000	3500	4000
*17	0.83	3.8	5.0	6.3	7.5	10	11	11	11	11	—
*19	0.93	3.8	5.0	7.5	8.8	11	13	14	14	14	—
21	1.0	5.0	6.3	8.8	10	14	14	15	16	16	—
23	1.1	5.0	7.5	10	11	15	16	18	19	19	18
25	1.2	5.0	7.5	10	13	16	18	20	21	21	20
27	1.3	6.3	8.8	11	13	18	19	21	24	24	23
29	1.4	6.3	8.8	13	14	19	21	24	25	25	25
31	1.5	7.5	10	13	15	21	23	25	28	28	28
33	1.6	7.5	10	14	16	23	24	28	29	30	29
35	1.8	7.5	11	15	18	24	25	29	31	31	30
37	1.9	8.8	11	16	19	25	26	30	33	33	—
40	2.0	8.8	13	18	20	28	29	33	35	35	—
45	2.5	10	14	19	23	30	30	36	39	—	—
50	2.5	11	15	21	25	34	36	40	—	—	—

Type I Type II Type III

* For best results, smaller sprocket should have at least 21 teeth.

Source: Whitney Chain Operations, Dresser Industries, Inc., Hartford, Conn.

TABLE 2-25 Horsepower Capacity per Inch of Width, 5/8-in Pitch Silent Chain (ANSI B29.2)

STANDARD WIDTHS 1, 1¼, 1½, 1¾, 2, 2½, 3, 4, 5, 6, 7, 8, 10

No. teeth, small sprocket	Revolutions per minute, small sprocket									
	100	500	700	1000	1200	1800	2000	2500	3000	3500
*17	1.3	6.3	7.5	10	11	14	15	14	—	—
*19	1.4	6.3	8.8	13	14	16	18	18	—	—
21	1.6	7.5	10	13	15	19	20	20	20	—
23	1.8	7.5	11	15	16	21	23	24	23	—
25	1.9	8.8	11	16	19	24	25	26	26	24
27	2.0	10	13	18	20	26	28	29	29	26
29	2.1	10	14	19	21	28	30	31	31	29
31	2.4	11	15	20	23	30	31	34	34	31
33	2.5	11	16	21	25	33	34	36	36	34
35	2.6	13	16	23	26	34	36	39	39	35
37	2.8	13	18	24	28	36	39	43	41	—
40	3.0	14	19	26	30	39	41	44	—	—
45	3.4	16	21	29	34	44	46	—	—	—
50	3.8	18	24	33	38	48	40	—	—	—

Type I Type II Type III

* For best results, smaller sprocket should have at least 21 teeth.
Source: Whitney Chain Operations, Dresser Industries, Inc., Hartford, Conn.

TABLE 2-26 Horsepower Capacity per Inch of Width, ¾-in Pitch Silent Chain (ANSI B29.2)

STANDARD WIDTHS 1, 1¼, 1½, 2, 2½, 3, 3½, 4, 5, 6, 7, 8, 9, 10, 12

No. teeth, small sprocket	Revolutions per minute, small sprocket								
	100	500	700	1000	1200	1500	1800	2000	2500
*17	1.9	8.1	11	14	15	16	18	18	—
*19	2.0	9.3	13	15	18	20	21	21	—
21	2.3	10	14	18	20	23	24	25	24
23	2.5	11	15	20	23	25	28	28	28
25	2.8	13	16	21	25	29	31	31	30
27	2.9	14	18	24	28	31	34	35	35
29	3.1	15	20	26	30	34	36	38	38
31	3.4	15	21	28	31	36	40	41	41
33	3.6	16	23	30	34	39	43	44	44
35	3.8	18	24	31	36	41	45	46	46
37	4.0	19	25	34	39	44	48	49	49
40	4.4	20	28	36	41	48	51	53	53
45	4.9	23	30	40	46	53	56	58	—
50	5.4	25	34	45	51	58	61	—	—

Type I — Type II — Type III

* For best results, smaller sprocket should have at least 21 teeth.
Source: Whitney Chain Operations, Dresser Industries, Inc., Hartford, Conn.

TABLE 2-27 Horsepower Capacity per Inch of Width, 1-in Pitch Silent Chain (ANSI B29.2)

STANDARD WIDTHS 2, 2½, 3, 4, 5, 6, 7, 8, 9, 10, 12, 14, 16

No. teeth, small sprocket	Revolutions per minute, small sprocket										
	100	200	300	400	500	700	1000	1200	1500	1800	2000
*17	3.8	6.3	8.8	11	14	18	21	23	—	—	—
*19	3.8	7.5	10	13	15	20	25	26	28	—	—
21	3.8	7.5	11	15	18	23	29	31	33	33	—
23	3.8	8.8	13	16	19	25	31	35	38	38	—
25	5.0	8.8	14	18	21	28	35	39	41	41	41
27	5.0	10	15	19	24	30	39	43	46	46	45
29	5.0	11	16	20	25	33	41	46	50	51	50
31	6.3	11	16	23	28	35	45	50	54	55	54
33	6.3	13	18	24	29	38	49	54	59	59	58
35	6.3	13	19	25	30	40	51	56	61	63	61
37	6.8	14	20	26	33	43	54	60	65	66	—
40	7.5	15	23	29	35	45	59	65	70	—	—
45	8.8	16	25	31	39	51	65	71	76	—	—
50	10	19	28	35	43	56	71	78	—	—	—

Type I Type II Type III

* For best results, smaller sprocket should have at least 21 teeth.
Source: Whitney Chain Operations, Dresser Industries, Inc., Hartford, Conn.

TABLE 2-28 Horsepower Capacity per Inch of Width, 1¼-in Pitch Silent Chain (ANSI B29.2)

STANDARD WIDTHS 2½, 3, 4, 5, 6, 7, 8, 9, 10, 12, 14, 16, 18, 20

No. teeth, small sprocket	Revolutions per minute, small sprocket										
	100	200	300	400	500	600	700	800	1000	1200	1500
*19	5.6	10	15	20	24	26	29	31	34	35	—
21	6.3	11	18	23	26	30	33	36	40	41	—
23	6.9	13	19	24	29	34	36	40	45	46	46
25	7.5	14	20	26	31	36	40	44	50	53	53
27	8.0	15	23	29	35	40	44	49	54	58	58
29	8.6	16	24	31	38	43	48	53	59	63	64
31	9.3	18	26	34	40	46	51	56	64	68	69
33	9.9	19	28	35	43	49	55	60	69	73	74
35	11	20	29	38	45	53	59	64	73	78	78
37	11	21	30	40	48	55	63	68	76	81	—
40	12	24	34	44	53	60	68	74	83	88	—
45	13	26	38	49	59	68	75	81	91	—	—
50	15	29	43	54	65	74	83	90	100	—	—

Type I — Type II — Type III

* For best results, smaller sprocket should have at least 21 teeth.

Source: Whitney Chain Operations, Dresser Industries, Inc., Hartford, Conn.

TABLE 2-29 Horsepower Capacity per Inch of Width, 1½-in Pitch Silent Chain (ANSI B29.2)

STANDARD WIDTHS 3, 4, 5, 6, 7, 8, 9, 10, 12, 14, 16, 18, 20, 22, 24

No. teeth, small sprocket	Revolutions per minute, small sprocket										
	100	200	300	400	500	600	700	800	900	1000	1200
*19	8.0	15	21	28	31	35	39	40	41	43	—
21	8.8	16	24	30	36	40	44	46	49	49	—
23	10	19	26	34	40	45	49	53	55	56	55
25	10	20	29	38	44	50	55	59	61	65	64
27	11	23	31	40	48	54	60	64	68	70	70
29	13	24	34	44	51	59	65	70	74	75	76
31	14	25	36	46	55	64	70	75	79	81	83
33	14	28	39	50	59	68	75	80	85	88	89
35	15	29	41	53	63	71	79	85	90	93	94
37	16	30	44	59	66	76	84	90	96	99	—
40	18	33	48	66	73	83	90	98	105	—	—
45	19	38	54	68	81	93	101	108	113	—	—
50	21	41	59	75	89	101	111	118	—	—	—

Type I Type II Type III

* For best results, smaller sprocket should have at least 21 teeth.

Source: Whitney Chain Operations, Dresser Industries, Inc., Hartford, Conn.

TABLE 2-30 Horsepower Capacity per Inch of Width, 2-in Pitch Silent Chain (ANSI B29.2)
STANDARD WIDTHS 4, 5, 6, 7, 8, 10, 12, 14, 16, 18, 20, 22, 24, 30

No. teeth, small sprocket	Revolutions per minute, small sprocket								
	100	200	300	400	500	600	700	800	900
*19	14	26	36	44	50	54	56	—	—
21	16	29	40	50	53	63	65	—	—
23	17	33	45	55	64	70	74	75	—
25	18	35	49	61	70	78	83	85	85
27	20	38	54	66	78	85	91	94	94
29	21	41	58	73	84	93	99	103	103
31	23	44	63	78	90	100	106	110	110
33	25	46	66	83	96	106	114	118	118
35	26	50	71	88	103	114	121	125	125
37	28	53	75	93	110	124	128	131	—
40	30	58	81	101	118	129	138	141	—
45	34	64	90	113	131	144	151	—	—
50	38	71	100	125	144	156	—	—	—

Type I Type II Type III

* For best results, smaller sprocket should have at least 21 teeth.
Source: Whitney Chain Operations, Dresser Industries, Inc., Hartford, Conn.

TABLE 2-31 Steel Detachable Link Chains, General Dimensions (ANSI B29.6)

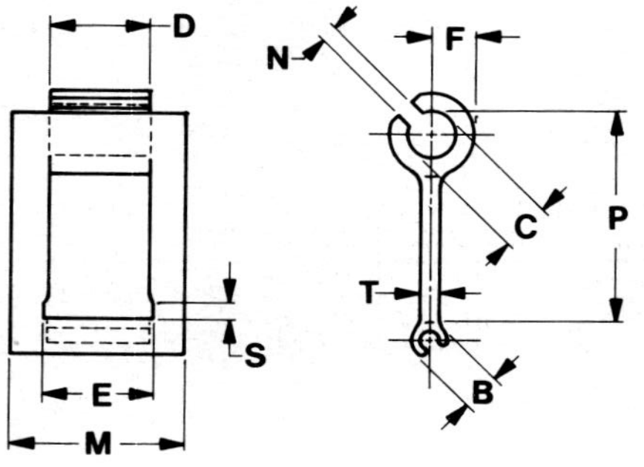

(In Inches)

Chain no.	*B*, max	*C*, min	*D* +0.012 −0.008	*E*, min	*F* +0.000 −0.030	*M* +3/32 −1/32	*N*, max	*P**	*S*, min	*T*	Minimum ultimate tensile strength, lb
25	0.190	0.200	0.422	0.438	0.180	45/64	0.063	0.904	0.078	0.073	760
32	0.240	0.250	0.594	0.610	0.230	15/16	0.080	1.157	0.100	0.090	1320
32W	0.240	0.250	0.594	0.610	0.232	1 1/16	0.085	1.157	0.100	0.095	1320
42	0.295	0.305	0.781	0.800	0.265	1 7/32	0.095	1.375	0.110	0.105	1720
45	0.335	0.345	0.781	0.800	0.303	1 7/32	0.095	1.630	0.110	0.105	1680
51	0.240	0.250	0.703	0.720	0.232	1 3/32	0.090	1.133	0.105	0.100	1680
52	0.320	0.330	0.844	0.860	0.303	1 13/32	0.110	1.508	0.125	0.120	2160
55	0.335	0.345	0.796	0.813	0.320	1 9/32	0.115	1.630	0.130	0.125	2240
62	0.355	0.360	0.984	1.002	0.335	1 9/16	0.138	1.654	0.155	0.148	3520
62A	0.355	0.360	0.984	1.002	0.358	1 15/16	0.160	1.664	0.180	0.170	4000
62H	0.355	0.360	0.984	1.002	0.343	1 7/8	0.145	1.654	0.160	0.155	3600
67	0.500	0.510	1.093	1.110	0.428	1 7/8	0.145	2.313	0.160	0.155	3600
67H	0.500	0.510	1.093	1.110	0.448	1 7/8	0.175	2.313	0.203	0.185	4400
67XH	0.400	0.500	1.093	1.110	0.448	1 7/8	0.190	2.313	0.205	0.200	5500
72	0.445	0.455	1.093	1.110	0.409	1 15/16	0.160	2.025	0.175	0.170	4000

* Assembled chain pitch.

Source: Extracted, with permission, from ANSI B29.6-1972, published by The American Society of Mechanical Engineers.

TABLE 2-32 Malleable Iron Detachable Link Chains, General Dimensions (ANSI B29.7)

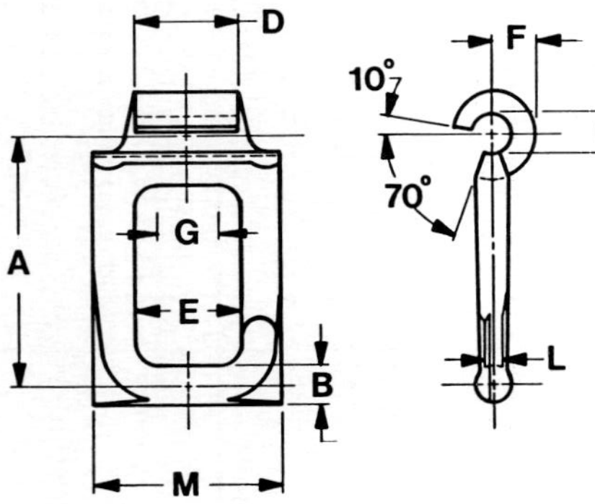

Style 1,
plain hook

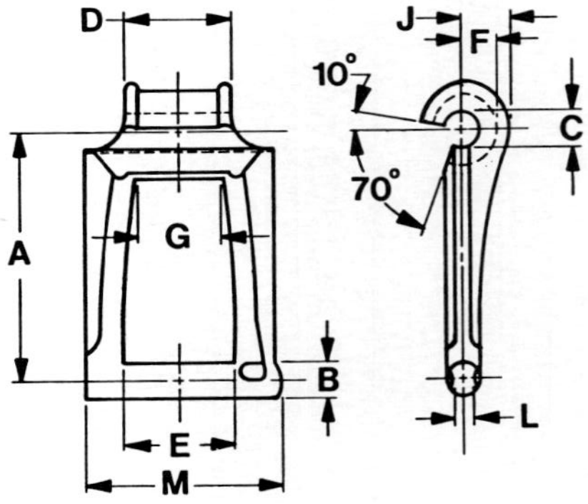

Style 2,
ribbed hook

(In Inches)

Chain no.	*A**	*B*	*C*	*D*	*E*	*F*	*G*†	*J*	*L*	*M*	Style	Finished assembled pitch‡	Chain no.
25	0.894	0.147	0.157	0.400	0.421	0.203	3/8		0.098	0.726	1	0.902	25
32	1.146	0.178	0.188	0.580	0.610	0.250	1/2		0.119	0.938	1	1.154	32
34	1.390	0.194	0.204	0.679	0.713	0.266	1/2		0.127	1.068	1	1.398	34
42	1.366	0.216	0.228	0.763	0.800	0.281	5/8		0.141	1.226	1	1.375	42
45	1.620	0.216	0.228	0.790	0.828	0.297	11/16		0.141	1.242	1	1.630	45
51	1.145	0.268	0.281	0.670	0.704	0.359	9/16		0.179	1.148	1	1.155	51
52	1.496	0.268	0.281	0.826	0.865	0.344	5/8		0.179	1.460	1	1.506	52
55	1.621	0.268	0.281	0.776	0.814	0.359	11/16		0.179	1.290	1	1.631	55
57	2.298	0.268	0.281	1.087	1.138	0.406	11/16		0.179	1.692	1	2.308	57
62	1.643	0.317	0.330	0.963	1.030	0.406	13/16		0.194	1.545	1	1.654	62
67	2.297	0.319	0.334	1.361	1.422	0.406	11/16	17/32	0.212	1.952	2	2.308	67
75	2.598	0.388	0.402	1.116	1.168	0.483	15/16		0.260	1.964	1	2.609	75
77	2.285	0.380	0.397	1.423	1.486	0.359	11/16	39/64	0.260	2.188	2	2.297	77
78	2.596	0.414	0.432	1.618	1.687	0.460	15/16	21/32	0.271	2.570	2	2.609	78
88	2.595	0.485	0.505	1.784	1.858	0.460	15/16	25/32	0.321	2.720	2	2.609	88
103	3.061	0.570	0.591	2.030	2.112	0.609	1 1/8	1	0.379	3.256	2	3.075	103
114	3.232	0.651	0.679	1.937	2.016	0.813	1 1/8	1 11/64	0.439	3.446	2	3.250	114
124	4.043	0.764	0.797	2.290	2.379	0.859	1 1/4	1 5/16	0.505	4.023	2	4.063	124

* Link pitch.
† Maximum sprocket tooth face.
‡ Chain pitch—chain pitch equals link pitch plus one half joint clearance.
Source: Extracted, with permission, from ANSI B29.7-1971, published by The American Society of Mechanical Engineers.

2-10 Detachable-Link Chains (ANSI B29.6,.7)

Steel Detachable-Link Chain (ANSI B29.6) Steel detachable chain is an assembly of identical one-piece formed steel links that hook and interfit together to form a continuous chain belt. When not under tension, this chain can be separated, or detached, by flexing two consecutive links so that they are perpendicular to each other, then sliding or driving the joint apart.

Fifteen standard models are available, ranging in pitch size from 0.904 in to 2.313 in. A wide range of attachments is offered for five of the most popular chain models. See Table 2-31.

Malleable Iron Detachable-Link Chain (ANSI B29.7) This group of chains employs cast malleable iron one-piece links that hook together in a manner similar to the steel detachable chain. Because this malleable detachable chain has fallen into disuse and is rapidly becoming unavailable, it is mentioned here primarily for identification purposes. ANSI B29.7 was formally withdrawn June 24, 1981. There were 18 standard chain models covering a pitch range running from 0.894 in to 4.043 in, as well as a group of attachment links. See Table 2-32 and Fig. 2-16.

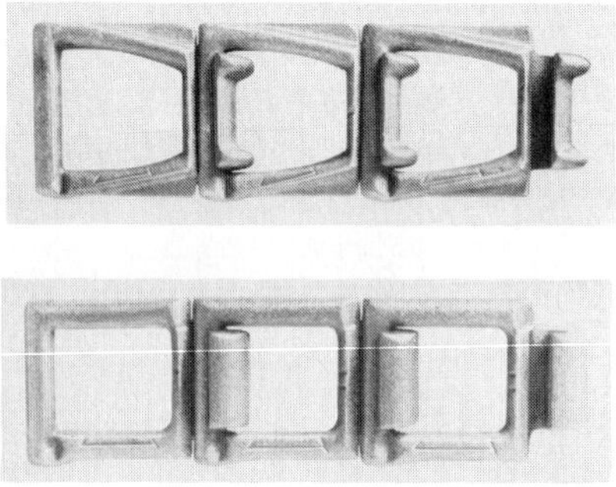

FIG. 2-16 Detachable link chain, malleable iron type. (*Source: Jeffrey Chain Division, Dresser Industries, Inc., Morristown, Tenn.*)

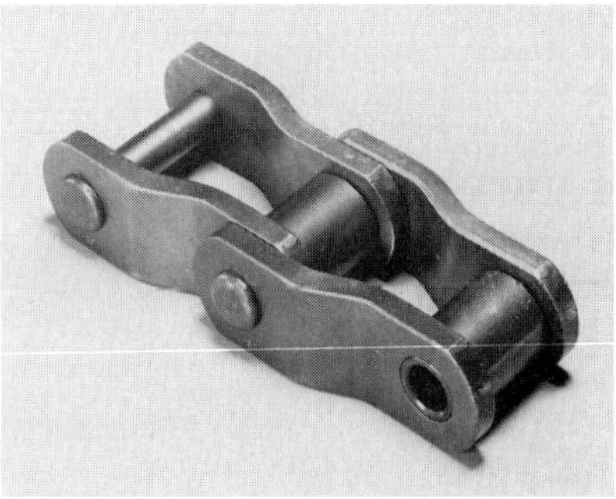

FIG. 2-17 Heavy-duty offset-sidebar roller chain. (*Source: Jeffrey Chain Division, Dresser Industries, Inc., Morristown, Tenn.*)

2-11 Heavy-Duty Offset-Sidebar Power-Transmission Roller Chain (ANSI B29.10)

Offset-sidebar roller chain is an assembly of identical interfitting links with link plates, or sidebars, which are formed to provide one wide end and one narrow end for each link. A pin at the wide end of each link passes through the bore of the bushing in the narrow end of each succeeding link to provide articulation at each joint. All chain models have free-turning rollers surrounding each bushing. See Fig. 2-17.

There are eight pitch sizes available in the standard group of chains; these range from 2½-in to 7-in. Multiple-strand assemblies are not available. See Tables 2-33 through 2-41 and Fig. 2-18.

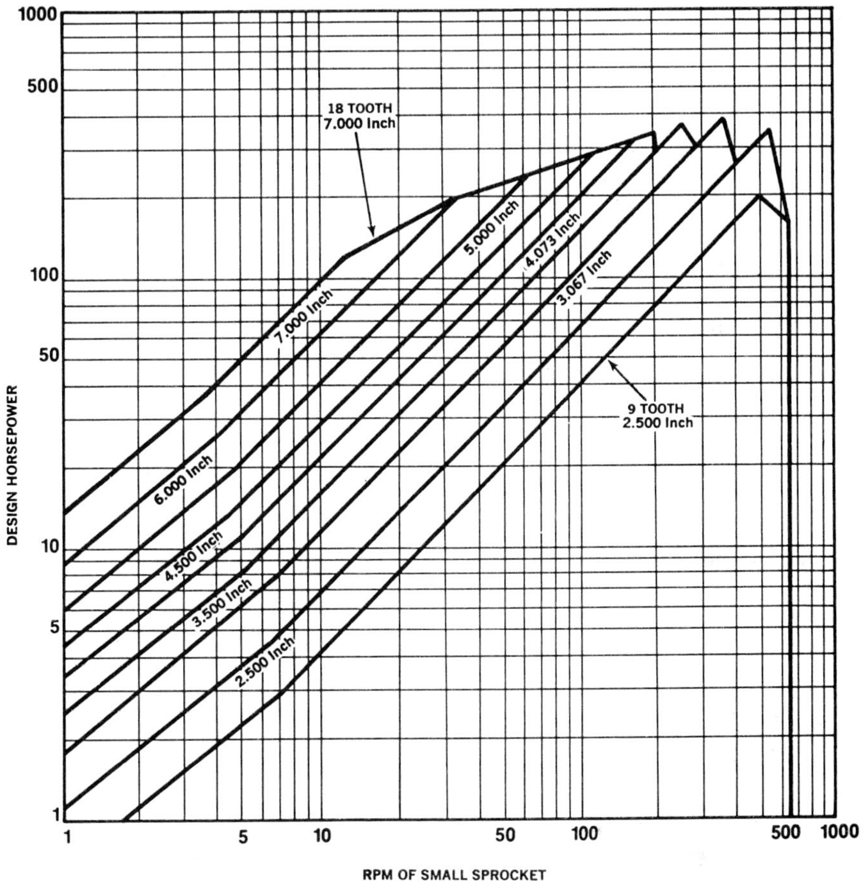

Notes:

Lower line is a 9-tooth **2.500 Inch Pitch**

Top line is for an 18-tooth **7.000 Inch Pitch**

Intermediate lines are approximate mid-points for sprocket-tooth range shown on charts.

Where the HORSEPOWER-RPM intersection takes place near a line both chains on each side should be checked on the charts.

$$S = \frac{PTN}{12}$$

$$HP = \frac{LS}{33000} \text{ or } \frac{NT}{63000}$$

$$T = \frac{63000 \text{ H.P.}}{N}$$

Where
S = Linear Chain Speed in Feet per Minute
T = No teeth in Sprocket
P = Pitch of Chain in inches
N = RPM of Sprocket
L = Chain Load in pounds
H.P. = Horsepower

FIG. 2-18 Heavy-duty offset-sidebar roller chain, quick selector chart. (*Source: Jeffrey Chain Division, Dresser Industries, Inc., Morristown, Tenn.*)

TABLE 2-33 Heavy-Duty Offset-Sidebar Power-Transmission Roller Chain—ANSI B29.10—General Dimensions

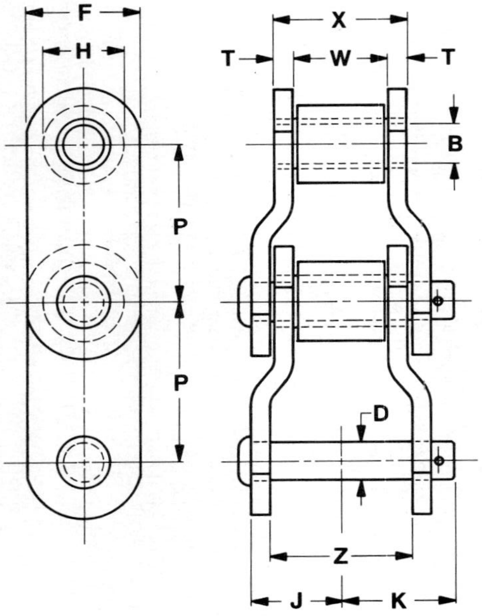

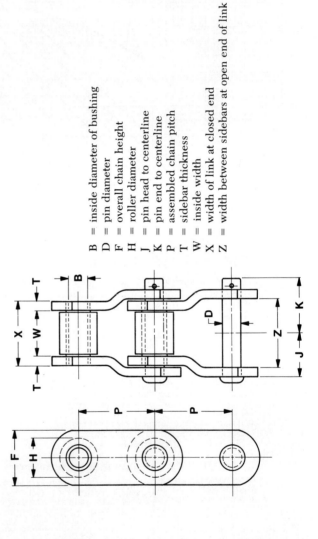

B = inside diameter of bushing
D = pin diameter
F = overall chain height
H = roller diameter
J = pin head to centerline
K = pin end to centerline
P = assembled chain pitch
T = sidebar thickness
W = inside width
X = width of link at closed end
Z = width between sidebars at open end of link

(In Inches)

	Chain number							
	2010	2512	2814	3315	3618	4020	4824	5628
P	2.500	3.067	3.500	4.073	4.500	5.000	6.000	7.000
D	0.625	0.750	0.875	0.038	1.100	1.250	1.500	1.750
T	0.31	0.38	0.50	0.56	0.56	0.62	0.75	0.88
F	1.75	2.25	2.25	2.38	3.00	3.50	4.00	5.00
H	1.25	1.62	1.75	1.78	2.25	2.50	3.00	3.50
W	1.50	1.56	1.50	1.94	2.06	2.75	3.00	3.25
Minimum ultimate strength, lb	57,000	77,000	106,000	124,000	171,000	222,500	287,500	385,000

Source: Extracted, with permission, from ANSI B29.10M-1981, published by The American Society of Mechanical Engineers.

TABLE 2-34 Horsepower Capacity, No. 2010 (ANSI B29.10)

2.500-in Pitch

No. of Teeth	Horsepower Capacity												
	RPM												
	2	3	7	10	20	30	40	100	200	250	350	450	600
9	1.1	1.4	2.7	3.9	7.7	11.6	15.4	38.6	77.2	96.5	135.1	100.1	65.0
10	1.1	1.5	3.0	4.3	8.6	12.9	17.2	42.9	85.8	107.3	150.2	117.2	76.1
11	1.2	1.7	3.3	4.7	9.4	14.2	18.9	47.2	94.4	118.0	165.2	135.2	87.8
12	1.3	1.8	3.6	5.1	10.3	15.4	20.6	51.5	103.0	128.7	180.2	154.1	100.1
13	1.4	1.9	3.9	5.6	11.2	16.7	22.3	55.8	111.5	139.4	195.2	173.7	112.8
14	1.5	2.0	4.2	6.0	12.0	18.0	24.0	60.1	120.1	150.2	210.2	194.2	126.1
15	1.5	2.1	4.5	6.4	12.9	19.3	25.7	64.4	128.7	160.9	225.2	215.3	139.9
16	1.6	2.2	4.8	6.9	13.7	20.6	27.5	68.6	137.3	171.6	240.3	237.2	154.1
17	1.7	2.3	5.1	7.3	14.6	21.9	29.2	72.9	145.9	182.3	255.3	259.8	168.8
18	1.8	2.4	5.4	7.7	15.4	23.2	30.9	77.2	154.5	193.1	270.3	283.1	183.9
19	1.9	2.5	5.7	8.2	16.3	24.5	32.6	81.5	163.0	203.8	285.3	307.0	....
20	1.9	2.6	6.0	8.6	17.2	25.7	34.3	85.8	171.6	214.5	300.3	331.5	....
21	2.0	2.7	6.3	9.0	18.0	27.0	36.0	90.1	180.2	225.2	315.3	356.7	....
22	2.1	2.8	6.6	9.4	18.9	28.3	37.8	94.4	188.8	236.0	330.4	382.5	....
23	2.1	3.0	6.9	9.9	19.7	29.6	39.5	98.7	197.4	246.7	345.4	405.3	....
24	2.2	3.1	7.2	10.3	20.6	30.9	41.2	103.0	205.9	257.4	360.4	414.4	....
	MANUAL LUBRICATION						OIL BATH		OIL STREAM LUBRICATION				

Continuous operation in the shaded area may produce some galling of the live bearing surfaces of the chain joints, even though lubrication is as recommended.

The ratings shown on these charts are for chain which operates over machine-cut tooth sprockets.

Source: Jeffrey Chain Division, Dresser Industries, Inc., Morristown, Tenn.

TABLE 2-35 Horsepower Capacity, No. 2512 (ANSI B29.10)

3.067-in Pitch

No. of Teeth	Horsepower Capacity												
	RPM												
	1	3	6	10	20	40	100	150	200	250	300	350	400
9	1.0	2.4	4.0	6.4	12.7	25.5	63.7	95.6	127.4	159.3	191.1	171.8	140.6
10	1.1	2.6	4.3	7.1	14.2	28.3	70.8	106.2	141.6	177.0	212.4	198.9	164.7
11	1.2	2.7	4.7	7.8	15.6	31.1	77.9	116.8	155.7	194.7	231.3	215.5	190.0
12	1.3	2.9	5.1	8.5	17.0	34.0	85.0	127.4	169.9	212.4	248.6	231.5	216.5
13	1.4	3.1	5.5	9.2	18.4	36.8	92.0	138.0	184.1	230.1	265.3	247.0	232.3
14	1.4	3.3	5.9	9.9	19.8	39.6	99.1	148.7	198.2	247.8	281.4	262.1	246.4
15	1.5	3.5	6.4	10.6	21.2	42.5	106.2	159.3	212.4	265.5	296.9	276.6	260.0
16	1.6	3.7	6.8	11.3	22.7	45.3	113.3	169.9	226.5	283.2	312.0	290.6	273.2
17	1.7	3.8	7.2	12.0	24.1	48.1	120.3	180.5	240.7	300.9	326.5	304.1	285.9
18	1.7	4.0	7.6	12.7	25.5	51.0	127.4	191.1	245.9	318.6	340.5	317.1	...
19	1.8	4.2	8.1	13.5	26.9	53.8	134.5	201.8	269.0	336.3	354.0	329.7	...
20	1.9	4.3	8.5	14.2	28.3	56.6	141.6	212.4	283.2	354.0	367.1	341.9	...
21	1.9	4.5	8.9	14.9	29.7	59.5	148.7	223.0	297.3	371.7	379.2	353.0	...
22	2.0	4.7	9.3	15.6	31.1	62.3	155.7	233.6	311.5	389.4	391.7	364.8	...
23	2.1	4.9	9.8	16.3	32.6	65.1	162.8	244.2	325.6	407.1	403.4	375.7	...
24	2.2	5.1	10.2	17.0	34.0	68.0	169.9	254.9	339.8	424.8	414.6	386.1	...
	MANUAL LUBRICATION					OIL BATH		OIL STREAM LUBRICATION					

Continuous operation in the shaded area may produce some galling of the live bearing surfaces of the chain joints, even though lubrication is as recommended.

The ratings shown on these charts are for chain which operates over machine-cut tooth sprockets.

Source: Jeffrey Chain Division, Dresser Industries, Inc., Morristown, Tenn.

TABLE 2-36 Horsepower Capacity, No. 2814 (ANSI B29.10)

3.500-in Pitch

No. of Teeth	Horsepower Capacity												
	RPM												
	1	3	6	10	20	35	60	100	125	150	200	250	300
9	1.4	3.3	5.5	8.8	17.6	30.8	52.8	88.1	110.1	132.1	176.1	178.7	170.8
10	1.5	3.5	6.0	9.8	19.6	34.2	58.7	97.8	122.3	146.8	195.7	196.1	187.4
11	1.6	3.8	6.5	10.8	21.5	37.7	64.6	107.6	134.5	161.4	215.2	213.0	203.6
12	1.8	4.1	7.0	11.7	23.5	41.1	70.4	117.4	146.8	176.1	234.8	229.5	219.4
13	1.9	4.3	7.6	12.7	25.4	44.5	76.3	127.2	159.0	190.8	254.4	245.6	234.7
14	2.0	4.6	8.2	13.7	27.4	47.9	82.2	137.0	171.2	205.5	273.9	261.2	249.6
15	2.1	4.8	8.8	14.7	29.4	51.4	88.1	146.8	183.4	220.1	292.1	276.3	264.1
16	2.2	5.1	9.4	15.7	31.3	54.8	93.9	156.5	195.7	234.8	307.7	291.1	278.2
17	2.3	5.3	10.0	16.6	33.3	58.2	99.8	166.3	207.9	249.5	322.8	305.5	. . .
18	2.4	5.5	10.6	17.6	35.2	61.6	105.7	176.1	220.1	264.2	337.6	319.4	. . .
19	2.5	5.8	11.2	18.6	37.2	65.1	115.5	185.9	232.4	278.8	351.9	333.0	. . .
20	2.6	6.0	11.7	19.6	39.1	68.5	117.4	195.7	244.6	293.5	365.8	346.1	. . .
21	2.7	6.2	12.3	20.5	41.1	71.9	123.3	205.5	256.8	308.2	379.3	358.9	. . .
	MANUAL LUBRICATION						OIL BATH		OIL STREAM LUBRICATION				

Continuous operation in the shaded area may produce some galling of the live bearing surfaces of the chain joints, even though lubrication is as recommended.

The ratings shown on these charts are for chain which operates over machine-cut tooth sprockets.

Source: Jeffrey Chain Division, Dresser Industries, Inc., Morristown, Tenn.

TABLE 2-37 Horsepower Capacity, No. 3315 (ANSI B29.10)

4.073-in Pitch

No. of Teeth	Horsepower Capacity												
	RPM												
	1	3	6	10	20	30	40	65	80	100	125	150	200
9	2.0	4.7	8.0	12.8	25.5	38.3	51.1	83.0	102.1	127.7	159.6	168.2	166.3
10	2.2	5.1	8.7	14.2	28.4	42.6	56.7	92.2	113.5	141.8	177.3	185.0	182.9
11	2.4	5.5	9.4	15.6	31.2	46.8	62.4	101.4	124.8	156.0	195.0	201.5	199.2
12	2.5	5.9	10.2	17.0	34.0	51.1	68.1	110.6	136.2	170.2	212.8	217.6	215.1
13	2.7	6.3	11.1	18.4	36.9	55.3	73.8	119.9	147.5	184.4	230.5	233.4	230.7
14	2.9	6.6	11.9	19.9	39.7	59.6	79.4	129.1	158.9	198.6	248.2	248.8	246.0
15	3.0	7.0	12.8	21.3	42.6	63.8	85.1	138.3	170.2	212.8	265.9	263.9	261.0
16	3.2	7.3	13.6	22.7	45.4	68.1	90.8	147.5	181.6	227.0	280.7	278.7	275.6
17	3.3	7.7	14.5	24.1	48.2	72.3	96.5	156.7	192.9	241.1	295.3	293.2	289.9
18	3.5	8.0	15.3	25.5	51.1	76.6	102.1	166.0	204.3	255.3	309.6	307.3	303.9
19	3.6	8.4	16.2	27.0	53.9	80.9	107.8	175.2	215.6	269.5	323.5	321.2	317.6
20	3.8	8.7	17.0	28.4	56.7	85.4	113.5	184.4	227.0	283.7	337.1	334.7	...
21	3.9	9.0	17.9	29.8	59.6	89.4	119.2	193.6	238.3	297.9	350.5	347.9	...
	MANUAL LUBRICATION						OIL BATH			OIL STREAM LUBRICATION			

Continuous operation in the shaded area may produce some galling of the live bearing surfaces of the chain joints, even though lubrication is as recommended.

The ratings shown on these charts are for chain which operates over machine-cut tooth sprockets.

Source: Jeffrey Chain Division, Dresser Industries, Inc., Morristown, Tenn.

TABLE 2-38 Horsepower Capacity, No. 3618 (ANSI B29.10)

4.500-in Pitch

No. of Teeth	Horsepower Capacity												
	RPM												
	1	3	6	10	20	30	35	50	65	80	100	125	150
9	2.6	6.0	10.2	16.3	32.6	48.9	57.0	81.5	105.9	130.4	153.8.	156.6	158.8
10	2.8	6.5	11.1	18.1	36.2	54.3	63.4	90.5	117.7	144.9	169.5	172.5	175.0
11	3.0	7.0	12.0	19.9	39.8	59.8	69.7	99.6	129.5	159.4	184.8	188.1	190.8
12	3.3	7.5	13.0	21.7	43.5	65.2	76.1	108.7	141.3	173.9	199.8	203.4	206.3
13	3.5	8.0	14.1	23.5	47.1	70.6	82.4	117.7	153.0	188.3	214.6	218.4	221.6
14	3.7	8.5	15.2	25.4	50.7	76.1	88.7	126.8	164.8	202.8	229.1	233.2	236.6
15	3.9	8.9	16.3	27.2	54.3	81.5	95.1	135.8	176.6	217.3	243.4	247.7	251.3
16	4.1	9.4	17.4	29.0	58.0	86.9	101.4	144.9	188.3	231.8	257.4	261.9	265.7
17	4.2	9.8	18.5	30.8	61.6	92.4	107.8	153.9	200.1	246.3	271.1	275.9	279.9
18	4.4	10.2	19.6	32.6	65.2	97.8	114.1	163.0	211.9	260.8	284.6	289.6	293.8
19	4.6	10.7	20.6	34.4	68.8	103.2	120.4	172.0	223.7	275.3	297.8	303.1	307.5
20	4.8	11.1	21.7	36.2	72.4	108.7	126.8	181.1	235.4	289.8	310.7	316.3	320.9
21	5.0	11.5	22.8	38.0	76.1	114.1	133.1	190.1	247.2	304.2	323.5	329.2	334.0
	MANUAL LUBRICATION						OIL BATH			OIL STREAM LUBRICATION			

Continuous operation in the shaded area may produce some galling of the live bearing surfaces of the chain joints, even though lubrication is as recommended.

The ratings shown on these charts are for chain which operates over machine-cut tooth sprockets.

Source: Jeffrey Chain Division, Dresser Industries, Inc., Morristown, Tenn.

TABLE 2-39 Horsepower Capacity, No. 4020 (ANSI B29.10)

5.000-in Pitch

No. of Teeth	Horsepower Capacity												
	RPM												
	.5	1.0	3	6	10	20	30	35	50	65	80	100	125
9	2.0	3.4	7.8	13.3	21.1	42.2	63.3	73.8	105.5	133.9	139.3	145.3	151.6
10	2.2	3.7	8.5	14.4	23.4	46.9	70.3	82.0	117.2	147.6	153.6	160.2	....
11	2.3	3.9	9.1	15.5	25.8	51.6	77.4	90.3	128.9	161.2	167.7	174.9	....
12	2.5	4.2	9.7	16.9	28.1	56.3	84.4	98.5	140.7	174.5	181.6	189.4	....
13	2.6	4.5	10.3	18.3	30.5	61.0	91.4	106.7	152.4	187.7	195.2	203.7	....
14	2.8	4.7	10.9	19.7	32.8	65.6	98.5	114.9	164.1	200.6	208.7	217.7	....
15	2.9	5.0	11.5	21.1	35.2	70.3	105.5	123.1	175.8	213.4	222.0	231.6	....
16	3.1	5.2	12.1	22.5	37.5	75.0	112.5	131.3	187.5	225.9	235.0	245.2	....
17	3.2	5.5	12.7	23.9	39.9	79.7	119.6	139.5	199.3	238.2	247.8	258.6	....
18	3.4	5.7	13.3	25.3	42.2	84.4	126.6	147.7	211.0	250.4	260.5	271.7	....
	MANUAL LUBRICATION							OIL BATH		OIL STREAM LUBRICATION			

Continuous operation in the shaded area may produce some galling of the live bearing surfaces of the chain joints, even though lubrication is as recommended.

The ratings shown on these charts are for chain which operates over machine-cut tooth sprockets.

Source: Jeffrey Chain Division, Dresser Industries, Inc., Morristown, Tenn.

TABLE 2-40 Horsepower Capacity, No. 4824 (ANSI B29.10)

6.000-in Pitch

No. of Teeth	Horsepower Capacity												
	RPM												
	.5	1.0	3	6	10	20	30	35	40	45	50	60	70
9	3.1	5.3	12.2	20.7	30.3	66.0	96.1	101.5	106.3	110.8	115.0	122.6	129.0
10	3.4	5.7	13.2	22.4	36.6	73.3	106.2	112.1	117.5	122.5	127.1	135.5	....
11	3.6	6.2	14.2	24.2	40.3	80.6	116.1	122.6	128.5	133.9	139.0	148.2	....
12	3.9	6.6	15.2	26.4	44.0	87.9	126.0	133.0	139.4	145.3	150.8	160.8	....
13	4.1	7.0	16.2	28.6	47.6	95.3	135.7	143.2	150.1	156.5	162.4	173.2	....
14	4.4	7.4	17.1	30.8	51.3	102.6	145.3	153.4	160.8	167.6	173.9	185.4	....
15	4.6	7.8	18.0	33.0	55.0	109.9	154.8	163.4	171.3	178.5	185.3	197.5	....
16	4.8	8.2	18.9	35.2	58.6	117.3	164.2	173.3	181.6	189.3	196.5	209.5	....
17	5.1	8.6	19.8	37.4	62.3	124.6	173.4	183.1	191.9	200.0	207.6	221.3	....
18	5.3	9.0	20.7	39.6	66.0	131.9	182.6	192.7	202.0	210.6	218.5	233.0	....
	MANUAL LUBRICATION									OIL BATH		OIL STREAM	

Continuous operation in the shaded area may produce some galling of the live bearing surfaces of the chain joints, even though lubrication is as recommended.

The ratings shown on these charts are for chain which operates over machine-cut tooth sprockets.

Source: Jeffrey Chain Division, Dresser Industries, Inc., Morristown, Tenn.

TABLE 2-41 Horsepower Capacity, No. 5628 (ANSI B29.10)

7.000-in Pitch

No. of Teeth	Horsepower Capacity													
	RPM													
	.1	.5	1	2	4	6	10	15	20	25	30	35	40	45
9	1.3	4.6	7.7	13.1	22.2	30.2	48.1	67.1	76.7	85.0	92.5	99.4	105.7	.0
10	1.4	4.9	8.4	14.2	24.0	32.7	53.5	74.2	84.8	94.0	102.3	109.9	.0	.0
11	1.6	5.3	9.0	15.2	25.9	35.3	58.8	81.2	92.8	103.0	112.0	120.3	.0	.0
12	1.7	5.7	9.6	16.3	27.6	38.5	64.2	88.2	100.8	111.8	121.7	130.7	.0	.0
13	1.8	6.0	10.2	17.3	29.4	41.7	69.5	95.1	108.7	120.6	131.2	140.9	.0	.0
14	1.9	6.4	10.8	18.3	31.1	44.9	74.8	102.0	116.5	129.2	140.6	151.1	.0	.0
15	2.0	6.7	11.4	19.3	32.7	48.1	80.2	108.8	124.3	137.8	150.0	161.1	.0	.0
16	2.1	7.1	12.0	20.3	34.4	51.3	85.5	115.5	132.0	146.4	159.3	171.1	.0	.0
17	2.2	7.4	12.5	21.2	36.4	54.5	90.9	122.2	139.6	154.8	168.5	180.9	.0	.0
18	2.3	7.7	13.1	22.2	38.5	57.7	96.2	128.8	147.1	163.2	177.5	190.7	.0	.0
	MANUAL LUBRICATION													OIL BATH

Continuous operation in the shaded area may produce some galling of the live bearing surfaces of the chain joints, even though lubrication is as recommended.

The ratings shown on these charts are for chain which operates over machine-cut tooth sprockets.

Source: Jeffrey Chain Division, Dresser Industries, Inc., Morristown, Tenn.

2-12 Combination Chain (ANSI B29.11)

This group of chains are referred to as combination chains because they are an alternating assembly of cast one-piece inner links, or block links, and outer links, or pin links, that consist of steel link plates and pins. The cast block links are typically of malleable iron or pearlitic malleable iron, the latter providing approximately 25 percent greater strength.

There are seven standard pitch sizes available, ranging from 1.631 in to 6.050 in. A variety of attachment plates and block links are offered for the more popular chain models. See Table 2-42.

2-13 Steel-Bushed Rollerless Chain (ANSI B29.12)

This group of chains are based on the same dimensional framework as the combination chains of B29.11. In these steel-bushed rollerless chains the inner links, or block links, are assemblies composed of link plates and bushings, while the outer links, or pin links, are assemblies of link plates and pins. The chain is an alternating assembly of inner block links and outer pin links. See Fig. 2-19.

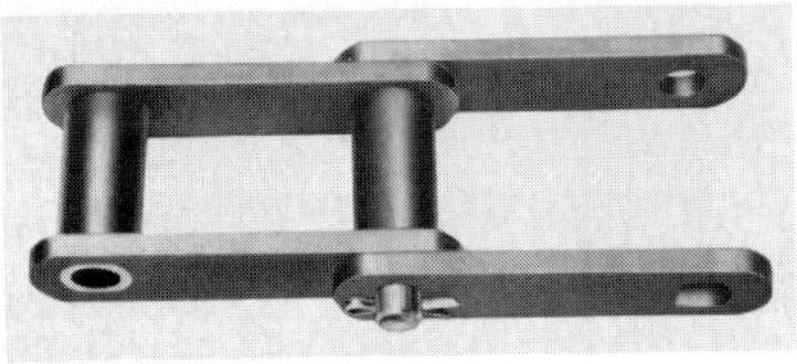

FIG. 2-19 Steel-bushed rollerless chain. *(Source: Jeffrey Chain Division, Dresser Industries, Inc., Morristown, Tenn.)*

Six standard pitch sizes are offered, ranging from 2.609 in to 6.050 in. A variety of attachment links are available for the more popular chain modes. See Table 2-43.

2-14 Pintle Chain (ANSI B29.21)

This group of chains, known specifically as 700 class pintle chains, are assembled as a series of identical cast offset links connected with removable pins. There are two standard types: a straight-sidebar type for use with plain sprockets, and a curved-sidebar type which has curved support surfaces to engage with sprocket flanges.

Five standard chain models are available, all of which have a pitch dimension of 6 in. Additionally, each of the models is produced as class M, malleable iron, and class P, pearlitic malleable iron. See Table 2-44.

A variety of attachment links is available for this group of chains.

2-15 Mill and Drag Chains (ANSI B29.16,.18)

Mill Chains, Welded Type (ANSI B29.16) This mill chain is a consecutive assembly of identical welded-steel offset links joined together with steel pins. The welded links consist of two sidebars joined, at the narrow end, by a bushing. The outside diameter of the bushing interacts with the sprocket teeth, while the pins articulate within the bushing bores. These mill chains can be supplied in two material combinations: with only pins heat-treated, and

TABLE 2-42 Combination Chains, General Dimensions (ANSI B29.11)

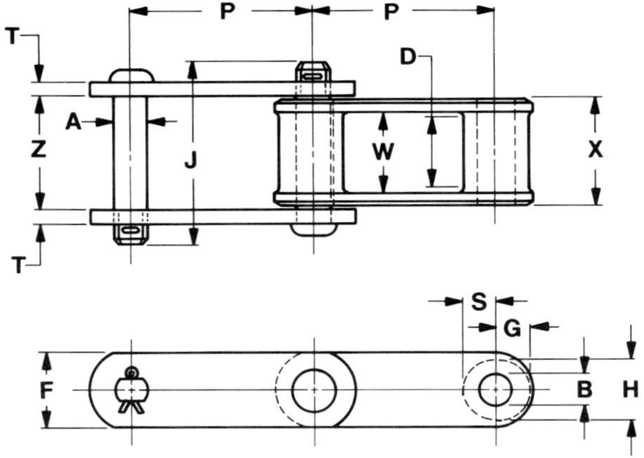

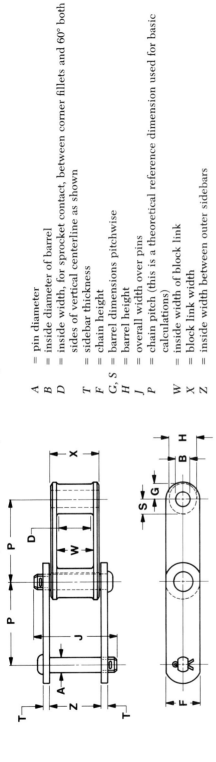

A = pin diameter
B = inside diameter of barrel
D = inside width, for sprocket contact, between corner fillets and 60° both sides of vertical centerline as shown
T = sidebar thickness
F = chain height
G, S = barrel dimensions pitchwise
H = barrel height
J = overall width over pins
P = chain pitch (this is a theoretical reference dimension used for basic calculations)
W = inside width of block link
X = block link width
Z = inside width between outer sidebars

(In Inches)

	Chain no						
	C55	C102B	C110	C111	C131	C132	C188
P(ref.)	1.631	4.000	6.000	4.760	3.075	6.050	2.609
A	0.375	0.625	0.625	0.750	0.625	1.000	0.500
F	0.75	1.50	1.50	1.75	1.50	2.00	1.12
H	0.72	0.98	1.26	1.44	1.22	1.74	0.88
T	0.19	0.38	0.38	0.38	0.38	0.50	0.25
Proof test loads, lb							
Class M	3,600	9,600	9,600	14,400	9,600	20,000	5,600
Class P	4,500	12,000	12,000	18,000	12,000	25,000	7,000

Source: Extracted, with permission, from ANSI B29.11-1974, published by The American Society of Mechanical Engineers.

TABLE 2-43 Steel Bushed Rollerless Chains, General Dimensions (ANSI B29.12)

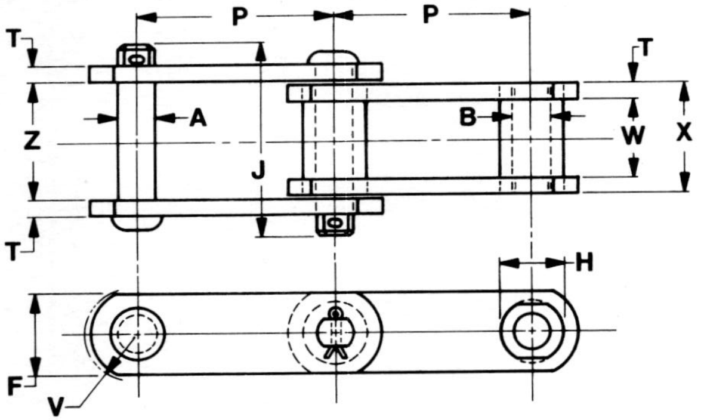

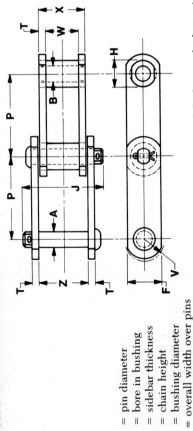

A = pin diameter
B = bore in bushing
T = sidebar thickness
F = chain height
H = bushing diameter
J = overall width over pins
P = chain pitch (this is the theoretical reference dimension used for basic calculations)

V = outer sidebar end clearance radius
W = inside width
X = block link width
Z = inside width between outer sidebars
R = attachment clearance radius

(In Inches)

	Chain no.					
	S102B	S110	S111	S131	S150	S188
P	4.000	6.000	4.760	3.075	6.050	2.609
A	0.625	0.625	0.750	0.625	1.000	0.500
F	1.50	1.50	2.00	1.50	2.50	1.12
H	1.00	1.26	1.44	1.26	1.76	0.88
T	0.38	0.38	0.38	0.38	0.50	0.25
W	2.125	2.125	2.625	1.313	3.313	1.063
Minimum ultimate strength, lb	36,000	36,000	43,000	36,000	85,000	23,000

Source: Extracted, with permission, from ANSI B29.12-1974, published by The American Society of Mechanical Engineers.

with both pins and welded links heat-treated. The all-heat-treated chain provides increased strength and wear resistance.

Eight standard chain models are available, ranging in pitch size from 2.609 in to 6.050 in. A variety of standard attachment links are offered for the more popular chain models. See Table 2-45.

Drag Chains, Welded Type (ANSI B29.18) These welded-steel drag chains consist of a consecutive assembly of identical offset links joined together with steel pins. Drag chains are relatively wide compared with mill chains, and are provided with barrels of shaped cross section to create an optimal pushing and/or scraping action in conveying applications. See Fig. 2-20.

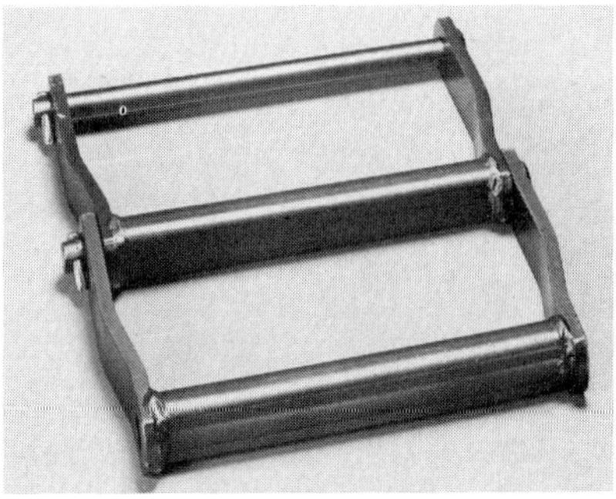

FIG. 2-20 Drag chain, welded type. (*Source: Jeffrey Chain Division, Dresser Industries, Inc., Morristown, Tenn.*)

The barrels contact the sprocket teeth directly, and the chain pins articulate within the barrel bores. As is the case with welded mill chains, all models are available in two material combinations: with only pins heat-treated, and with both pins and welded links heat-treated.

Nine standard chain models are available, ranging in pitch size from 5 in to 8 in. See Table 2-46.

2-16 Drop-Forged Rivetless Chain (ANSI B29.22)

This group of chains is made up of an alternating assembly of a single inner loop link and two outer links, joined by a removable T-headed pin. This allows for simple disassembly of the component parts. These chains are widely used in overhead trolley conveyor applications, where they can be driven either by sprockets or by interengaging drives of precision roller chain fitted with driving dogs. See Fig. 2-21.

There are 15 standard models available, ranging in pitch from 2 in to 9 in. A variety of attachments are available for the more popular models. See Table 2-47.

2-17 Leaf Chain (ANSI B29.8)

Leaf chain is essentially a byproduct derivative of the basic B29.1 precision roller chain. It consists of link plates and pins assembled in various interleaved lacing patterns, with freedom to articulate at each joint. No bushings or rollers are used. Leaf chain is the only type of chain in the B29 group of ANSI standards which does not engage with a sprocket; it articulates over plain round sheaves, and is attached at each end with clevises.

The two styles of leaf chain in most common use in recent years are known

TABLE 2-44 Pintle Chains, General Dimensions (ANSI B29.21)

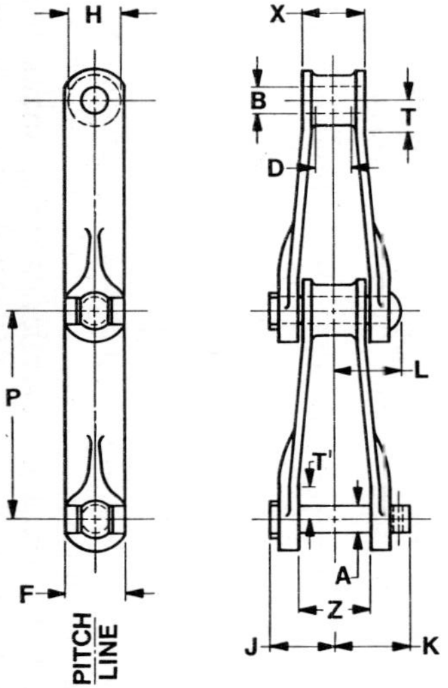

Straight-Sidebar Type
Class 720, 720S, 730

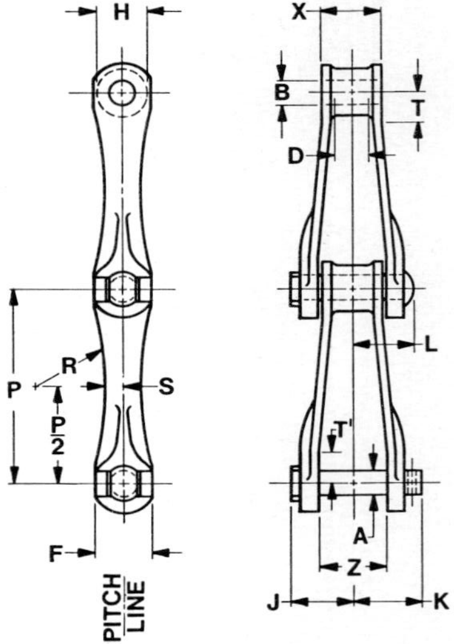

Curved-Sidebar Type
Class CS720S, CS730

A = pin diameter
B = inside diameter of barrel
D = inside width, for sprocket contact, between corner fillets extending 60° both sides of vertical centerline as shown
F = chain height
H = barrel height
J = pin head to centerline
K = pin end to centerline

L = riveted head to centerline
P = chain pitch (this is a theoretical reference dimension used for basic calculations)
R = radius of sidebar sprocket flanges
S = sidebar height at waist, from pitch line
T = straight before bend
T' = straight before bend
X = width of link at barrel end, extending to a point on the pitch line T inches from the centerline as shown
Z = width between sidebars at pin end of link, extending to a point on the pitch line T' inches from the centerline as shown

	Chain no.				
	720	720S	730	CS720S	CS730
P, in	6.000	6.000	6.000	6.000	6.000
A, in	0.688	0.750	0.750	0.750	0.750
F, in	1.50	1.56	1.75	1.56	1.75
H, in	1.38	1.44	1.50	1.44	1.50
Minimum ultimate strength, lb					
Class M	20,000	27,000	27,000	27,000	27,000
Class P	25,000	34,000	34,000	34,000	34,000

Source: Extracted, with permission, from ANSI B29.21M-1981, published by The American Society of Mechanical Engineers.

FIG. 2-21 Drop-forged rivetless chain, X-type. (*Source: Jeffrey Chain Division, Dresser Industries, Inc., Morristown, Tenn.*)

FIG. 2-22 Leaf chain, four-by-four lacing. (*Source: Whitney Chain Operations, Dresser Industries, Inc., Hartford, Conn.*)

as "AL," which utilizes a roller-chain regular-series pin link plate configuration and larger pin diameter. The 1977 revision of ANSI B29.8 took note of current trends in leaf-chain usage in new designs, and dropped the AL series leaf chain. Now all lacing patterns, 2 by 2, 2 by 3, 3 by 4, 4 by 6, and 6 by 6, are available in the BL series configuration. The 8 by 8 assembly is also offered by most manufacturers. See Fig. 2-22. There are 10 pitch sizes listed, ranging from ½ in to 2½ in. See Table 2-48.

2-18 Hinge-Type Flat-Top Conveyor Chain (ANSI B29.17)

As the term "hinge-type" indicates, this style of conveyor chain employs an interfitting curled joint quite similar in appearance to a door hinge. The chain is a consecutive assembly of identically formed steel flat-top links joined together with steel pins. The curls project below the flat-top conveying surface in order to contact the sprocket teeth.

All the standard chain models covered by ANSI B29.17 are of the one basic 1½-in pitch design. Five top plate widths are available as standard, ranging from 3.25 in to 7.50 in. (See Table 2-49.)

Additionally, all standard chain models are offered in carbon steel (24C26) and in austenitic stainless steel (24A26).

2-19 Safety Considerations

In using chains and sprockets to transmit power, to convey material, or to lift loads, the designer should be aware of the possibility of injury to personnel as a result of inadvertent contact with moving chains, sprockets, or shafts.

When chain drives are provided with a lubrication system, the drives usually are enclosed, and the enclosure also functions as a guard. For drives which are lubricated manually, thought should be given to providing guards to protect the unwary operator from physical contact with moving parts. For guidance in this area, two publications of the American National Standards Institute are helpful—ANSI B15.1, Safety Standard for Mechanical Power Transmission Apparatus, and B20.1, Safety Standard for Conveyors and Related Equipment.

For applications in which chain is used for overhead hoisting, only chain designs which are specifically intended for this purpose should be selected. Standards publications which will be helpful to the designer are ANSI B30.16, Safety Standard for Overhead Hoists, and ANSI B29.24, Roller Chains for Overhead Hoists.

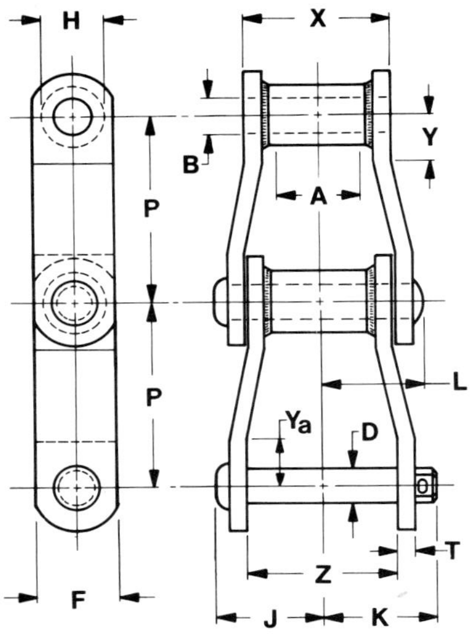

D = pin diameter
B = inside diameter of bushing
A = inside width for sprocket contact
T = sidebar thickness
F = chain height
H = barrel outside diameter
J = pin head to centerline
K = pin end to centerline
L = riveted head to centerline
P = chain pitch (this is a theoretical reference dimension used for basic calculations)
Y = straight before bend—bushing end
Y_a = straight before bend—pin end
X = width of link at bushing end extending to a point on the pitch line Y inches from the centerline as shown
Z = width between sidebars at pin end extending to a point on the pitch line Y_a inches from the centerline as shown

	Chain no.							
	W78	W82	W106	W110	W111	W124	W124H	W132
P, in	2.609	3.075	6.000	6.000	4.760	4.000	4.063	6.050
D, in	0.500	0.562	0.750	0.750	0.750	0.750	0.875	1.000
F, in	1.12	1.25	1.500	1.50	1.50	1.50	2.00	2.00
H, in	0.88	1.22	1.44	1.25	1.44	1.44	1.62	1.75
T, in	0.25	0.25	.38	0.38	0.38	0.38	0.50	0.50
Minimum ultimate strength, lb								
Pin heat-treated	21,000	22,500	38,000	38,000	38,000	38,000	62,000	62,000
All heat-treated	24,000	29,500	50,500	50,500	50,500	50,500	80,000	85,000

Source: Extracted, with permission, from ANSI B29.16M-1981, published by The American Society of Mechanical Engineers.

TABLE 2-46 Drag Chains (Welded Type), General Dimensions (ANSI B29.18)

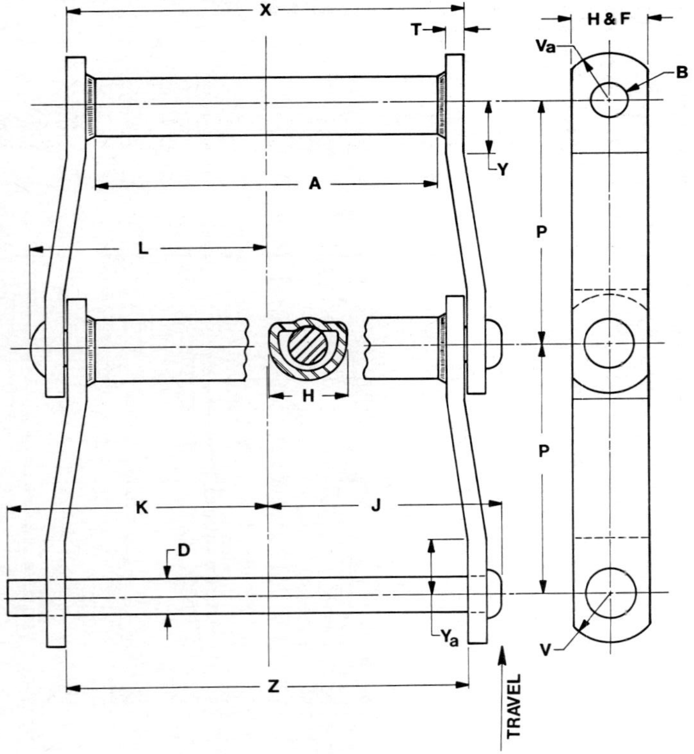

D = pin diameter
B = S.B. hole or barrel I.D.
A = inside width for sprocket contact
T = sidebar thickness
F = chain height
H = barrel height
J = pin head to centerline
K = pin end to centerline
L = rivet head to centerline
P = chain pitch (this is a theoretical reference dimension used for basic calculations)
Y = straight before bend—barrel end
Y_a = straight before bend—pin end
V = sidebar end clearance radius—pin end
V_a = sidebar end clearance radius—barrel end
X = width of link at barrel end, extending to point on the pitch line Y inches from the centerline as shown
Z = width between sidebars at pin end, extending to a point on the pitch line Y_a inches from the centerline as shown

	Chain no.								
	WD102	WD104	WD110	WD112	WD113	WD116	WD118	WD122	WD480
P, in	5.000	6.000	6.000	8.000	6.000	8.000	8.000	8.000	8.000
D, in	0.750	0.750	0.750	0.750	0.875	0.750	0.875	0.875	0.875
F, in	1.50	1.50	1.50	1.50	1.50	1.75	2.00	2.00	2.00
H, in	1.50	1.50	1.50	1.50	1.50	1.75	2.00	2.00	2.00
T, in	0.38	0.38	0.38	0.38	0.50	0.38	0.50	0.50	0.50
Minimum ultimate strength, lb									
Pin heat-treated	38,250	38,250	38,250	38,250	48,000	55,000	70,000	70,000	70,000
All heat-treated	55,000	55,000	55,000	55,000	57,000	59,000	79,000	79,000	79,000

Source: Extracted, with permission, from ANSI B29.18M-1981, published by The American Society of Mechanical Engineers.

TABLE 2-47 Drop-Forged Rivetless Chains, General Dimensions (ANSI B29.22)

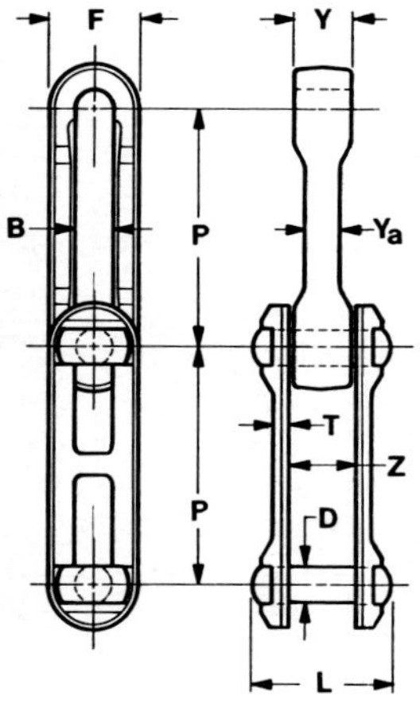

Regular
rivetless chain

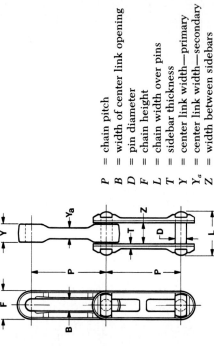

P = chain pitch
B = width of center link opening
D = pin diameter
F = chain height
L = chain width over pins
T = sidebar thickness
Y = center link width—primary
Y_a = center link width—secondary
Z = width between sidebars

	Chain no.						
	458 (100-16)	468 (100-19)	658 (150-16)	678 (150-22)	698 (150-28)	998 (230-28)	9118 (230-35)
Reference pitch, in	4	4	6	6	6	9	9
P	4.031	4.031	6.031	6.031	6.031	9.031	9.031
B (min.)	0.66	0.84	0.66	0.95	1.18	1.18	1.47
D	0.63	0.75	0.63	0.87	1.12	1.12	1.36
F (max.)	1.44	1.93	1.44	2.03	2.69	2.69	3.03
L (max.)	2.31	3.31	2.31	3.03	3.75	3.75	4.88
T	0.31	0.43	0.47	0.47	0.58	0.63	0.75
Y	1.00	1.61	1.00	1.28	1.54	1.54	1.95
Y_a	0.64	1.14	0.64	0.83	1.00	1.00	1.31
Z	1.08	1.70	1.08	1.41	1.64	1.64	2.05
Minimum ultimate strength, lb	42,000	68,000	42,000	72,000	136,000	136,000	187,000

TABLE 2-47 Drop-Forged Rivetless Chains, General Dimensions (ANSI B29.22) (*Continued*)

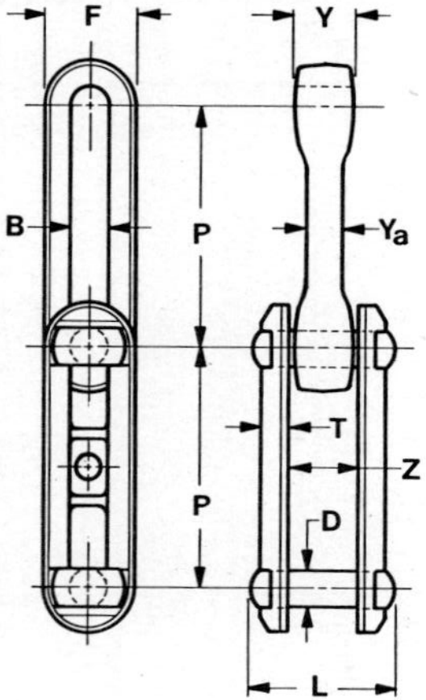

X-type
rivetless chain

P = chain pitch
B = width of center link opening
C = center link mounting dimension
D = pin diameter
F = chain height
L = chain width over pins
T = sidebar thickness
Y = center link width—primary
Y_a = center link width—secondary
Z = width between sidebars

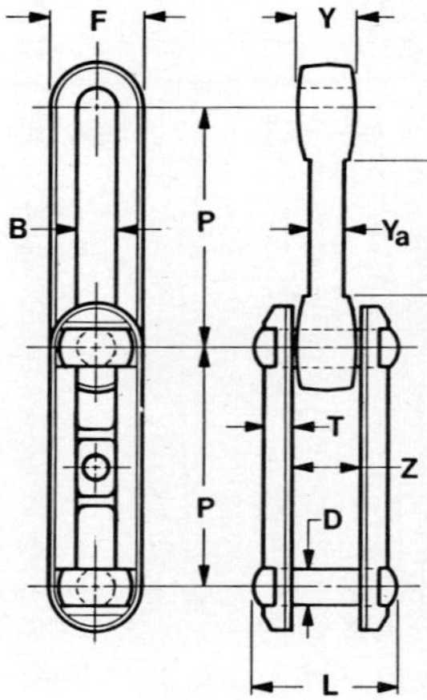

Modified X-type
rivetless chain

	Chain no.			
	X228 (X50-6)	X348 (X75-13)	X458 (X100-16)	X678 (X150-22)
Reference pitch, in	2	3	4	6
P	2.010	3.015	4.031	6.031
B (min.)	0.31	0.53	0.66	0.95
D	0.25	0.49	0.63	0.87
F (max.)	0.71	1.10	1.44	2.03
L (max.)	1.09	1.73	2.25	3.03
T	0.25	0.40	0.48	0.70
Y	0.47	0.74	1.00	1.28
Y_a	0.37	0.51	0.64	0.83
Z	0.51	0.79	1.07	1.35
Minimum ultimate strength, lb	6,000	22,000	42,000	72,000

	Chain no.		
	X348 (X75-13)	X458 (X100-16)	X678 (X150-22)
Reference pitch, in	3	4	6
P	3.015	4.031	6.031
B (min.)	0.53	0.66	0.95
C (min.)	1.59	2.31	3.34
D	0.49	0.63	0.87
F (max.)	1.10	1.44	2.03
L (max.)	1.73	2.25	3.03
T	0.40	0.48	0.70
Y	0.74	1.00	1.28
Y_a	0.51	0.64	0.83
Z	0.79	1.07	1.35
Minimum ultimate strength, lb	22,000	42,000	72,000

Source: Extracted, with permission, from ANSI B29.22M-1980, published by The American Society of Mechanical Engineers.

TABLE 2-48 Leaf Chains, General Dimensions (ANSI B29.8)

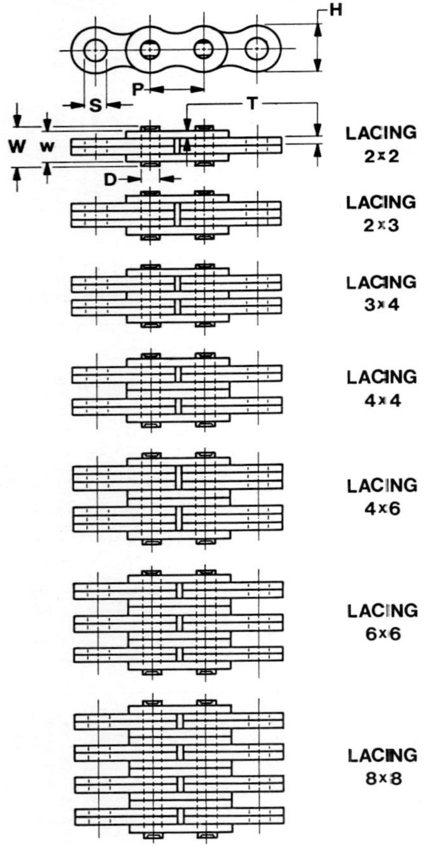

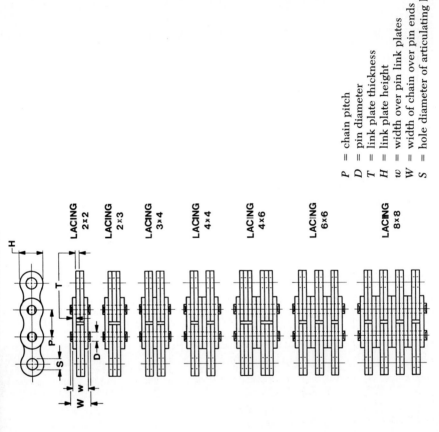

P = chain pitch
D = pin diameter
T = link plate thickness
H = link plate height
w = width over pin link plates
W = width of chain over pin ends
S = hole diameter of articulating link plates

Chain no.	Pitch, in	Lacing, in	w max, in	W max, in	D max, in	S min, in	H max, in	T max, in	Minimum ultimate tensile strength, lb
BL422		2 × 2	0.334	0.434					5,000
BL423		2 × 3	0.418	0.518					5,000
BL434		3 × 4	0.585	0.685					7,500
BL444	0.500	4 × 4	0.668	0.768	0.2005	0.2015	0.475	0.082	10,000
BL446		4 × 6	0.835	0.935					10,000
BL466		6 × 6	1.002	1.102					15,000
BL488		8 × 8	1.336	1.436					20,000
BL522		2 × 2	0.390	0.507					7,500
BL523		2 × 3	0.488	0.605					7,500
BL534		3 × 4	0.683	0.800					11,000
BL544	0.625	4 × 4	0.780	0.897	0.2345	0.2355	0.594	0.096	15,000
BL546		4 × 6	0.975	1.092					15,000
BL566		6 × 6	1.170	1.287					22,500
BL588		8 × 8	1.560	1.677					30,000
BL622		2 × 2	0.528	0.684					11,000
BL623		2 × 3	0.660	0.816					11,000
BL634		3 × 4	0.924	1.080					17,000
BL644	0.750	4 × 4	1.056	1.212	0.3125	0.3135	0.713	0.130	22,000
BL646		4 × 6	1.320	1.476					22,000
BL666		6 × 6	1.584	1.740					33,000
BL688		8 × 8	2.112	2.268					44,000
BL822		2 × 2	0.652	0.840					19,000
BL823		2 × 3	0.815	1.003					19,000
BL834		3 × 4	1.141	1.329					29,000
BL844	1.000	4 × 4	1.304	1.492	0.3755	0.3765	0.950	0.161	38,000
BL846		4 × 6	1.630	1.818					38,000
BL866		6 × 6	1.956	2.144					57,000
BL888		8 × 8	2.608	2.796					76,000

TABLE 2-48 Leaf Chains, General Dimensions (ANSI B29.8) (*Continued*)

Chain no.	Pitch, in	Lacing, in	w max, in	W max, in	D max, in	S min, in	H max, in	T max, in	Minimum ultimate tensile strength, lb
BL1022		2 × 2	0.780	0.999					26,000
BL1023		2 × 3	0.975	1.194					26,000
BL1034		3 × 4	1.365	1.584					41,000
BL1044	1.250	4 × 4	1.560	1.779	0.4375	0.4385	1.188	0.193	52,000
BL1046		4 × 6	1.950	2.169					52,000
BL1066		6 × 6	2.340	2.559					78,000
BL1088		8 × 8	3.120	3.339					104,000
BL1222		2 × 2	0.916	1.166					34,000
BL1223		2 × 3	1.145	1.395					34,000
BL1234		3 × 4	1.603	1.853					55,000
BL1244	1.500	4 × 4	1.832	2.082	0.5005	0.5015	1.425	0.227	68,000
BL1246		4 × 6	2.290	2.540					68,000
BL1266		6 × 6	2.748	2.998					102,000
BL1288		8 × 8	3.664	3.914					136,000

BL1422		2 × 2	1.040	1.321					43,000
BL1423		2 × 3	1.300	1.581					43,000
BL1434		3 × 4	1.820	2.101					71,000
BL1444	1.750	4 × 4	2.080	2.361	0.5625	0.5635	1.663	0.258	86,000
BL1446		4 × 6	2.600	2.881					86,000
BL1466		6 × 6	3.120	3.401					130,000
BL1488		8 × 8	4.160	4.441					172,000
BL1622		2 × 2	1.192	1.536					65,000
BL1623		2 × 3	1.490	1.834					65,000
BL1634		3 × 4	2.086	2.430					99,000
BL1644	2.000	4 × 4	2.384	2.728	0.6875	0.6885	1.900	0.296	130,000
BL1646		4 × 6	2.980	3.324					130,000
BL1666		6 × 6	3.576	3.920					195,000
BL1688		8 × 8	4.768	5.112					260,000
BL2022		2 × 2	1.568	2.037					97,500
Bl2023		2 × 3	1.960	2.429					97,500
BL2034		3 × 4	2.744	3.213					146,000
BL2044	2.500	4 × 4	3.136	3.605	0.9375	0.9385	2.375	0.390	195,000
BL2046		4 × 6	3.920	4.389					195,000
BL2066		6 × 6	4.704	5.173					292,500
BL2088		8 × 8	6.272	6.741					390,000

Source: Extracted, with permission, from ANSI B29.8-1977, published by The American Society of Mechanical Engineers.

TABLE 2-49 Hinge-Type Flat-Top Conveyor Chain, General Dimensions (ANSI B29.17)

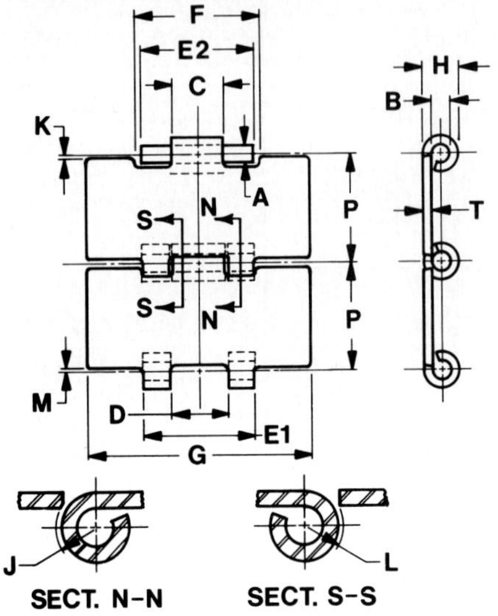

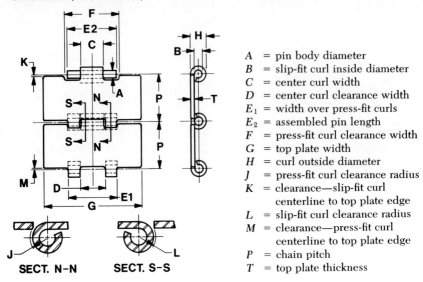

A = pin body diameter
B = slip-fit curl inside diameter
C = center curl width
D = center curl clearance width
E_1 = width over press-fit curls
E_2 = assembled pin length
F = press-fit curl clearance width
G = top plate width
H = curl outside diameter
J = press-fit curl clearance radius
K = clearance—slip-fit curl centerline to top plate edge
L = slip-fit curl clearance radius
M = clearance—press-fit curl centerline to top plate edge
P = chain pitch
T = top plate thickness

	Chain no.	
	24A26 stainless steel	24C26 carbon steel
P, in	1.50	1.50
A, in	0.25	0.25
H, in	0.50	0.50
E_1, in	1.625	1.625
T, in	0.12	0.12
G, in	3.25	3.25
	4.00	4.00
	4.50	4.50
	6.00	6.00
	7.50	7.50
Minimum ultimate strength, lb	2000	2400

Source: Extracted, with permission, from ANSI B29.17-1974, published by The American Society of Mechanical Engineers.

Bulletins covering safe practices for cutting, altering, installing, and repairing chain drives by maintenance and millwright mechanics are available from the American Chain Association, 1133 Fifteenth Street N.W., Washington, DC 20005, or from any of the Association's member companies. These bulletins are entitled *Connect and Disconnect Instructions* and can be obtained for roller chain, engineering steel and cast chains, and silent chains.

2-20 Glossary of Selected Chain Terms

Articulation

The action occurring at each joint when a chain engages with or disengages from a sprocket. The angle of articulation is inversely proportional to the number of teeth in the sprocket.

Attachment

A modification or extension of standard chain links or pins to adapt the chain for purposes such as conveying, elevating, or timing.

Block Chain

A chain type in which the inner link is either a solid block or a block built up of laminations.

Bottom Diameter

The diameter of a circle tangent to the seating curve at the bottom of each tooth space of a sprocket. Also called root diameter.

Break-in

The elongation occurring during the first hours of operation of a new chain.

Caliper Dimension

The measurement across the bottoms of opposite tooth spaces of a sprocket with an even number of teeth, or across the most nearly opposite spaces of a sprocket with an odd number of teeth.

Casing

An enclosure, or housing, for a chain drive which contains the lubricating system and protects the chain and lubricant from contamination. A casing also serves the function of a guard.

Catenary Effect

The catenary is the curve a length of chain assumes between its suspension points. Catenary tension is the tensile load imposed on the chain due to the weight of the chain and the amount of sag between suspension points.

Center Distance

The dimension between the centers of the shafts of a chain drive.

Clevis

A block dimensioned to receive and anchor the end of a chain length used in a tension linkage or hoisting application.

Connecting Link

A pin link, or outer link, in which one plate is readily removable and replaceable to facilitate connecting or disconnecting the ends of a length of chain. For offset-sidebar chains and pintle chains, a link provided with a readily removable and replaceable pin to accomplish connecting and disconnecting.

Design Horsepower

A calculated value arrived at by multiplying the specified power by a service factor which reflects various operating conditions.

Double-Cut Sprocket

A sprocket, for use with extended pitch chains, having two sets of effective teeth, with one set of tooth spaces located midway between the other set.

Double-Pitch Chain

A chain that has a pitch dimension twice that of a standard, or base-pitch, chain, but with otherwise identical pins, bushings, and rollers.

Effective Teeth

In double-cut sprockets, the number of sprocket teeth engaging with the chain in one revolution.

Galling

A form of joint wear in which wear particles are torn from one of the contacting surfaces and adhere to the other surface.

Idler Sprocket

In a drive layout, a sprocket which does not transmit power, but is positioned in the layout to control the chain's travel path, either to avoid contact with the machine's structure or to take up slack in the return strand of chain.

Knuckle

A manufacturer's term for the bushings used in steel-bushed rollerless chain.

Link

The portion of a chain's structure which connects the adjacent joint.

Link Plate

Any of the individual side plates which make up the chain's outer, inner, or offset links. Also known as sidebars.

Measuring Load

The magnitude of the standardized load imposed on a length of chain when it is to be stretched taut for measuring purposes.

Multiple-Strand Chain

In various types of roller chain, an assembly made up of two or more rows of roller links joined into a single structure by pins extending transversely through all rows.

Multiple-Strand Factor

A factor for obtaining the horsepower rating of a multiple-strand chain by multiplying by the rating value of single-strand chain.

Offset Link

A specific type of chain link that utilizes bent, or offset, link plates and is assembled with a pin at one end and a bushing and roller at the other so as to act as a combination link. Used in alternately assembled chain to allow an odd number of links to be connected. Usually the pin is readily removable and replaceable so that the offset link can act as a connecting link.

Offset Section

A factory-assembled section of two or more links which includes a press-fit-assembled offset link. Such a section can be coupled into a drive by the use of two pin-link-type connecting links to avoid a removable-pin-type offset link.

Pitch

The fundamental dimension in chain and sprocket measurement. For chain, it is the center-to-center dimension between chain joints. For a sprocket, it is the chordal dimension between the centers of rollers bedded against the bottoms of adjacent tooth spaces.

Pitch Diameter

The diameter of a circle passing through the chain joint centers when the chain is in wrapped contact on the sprocket.

Prestressing

The practice of subjecting new chain to a tensile loading immediately following assembly for the purpose of seating mating components to minimize break-in elongation, and to control chain length relative to standard length tolerances. In multiple-strand chain assemblies, prestressing promotes uniform load distribution across the width of the assembly.

Proof Loading

The practice of subjecting new chain to a tensile loading to some predetermined percentage of the chain's rated strength. The purpose is to prove that each chain is free of substandard parts which might degrade its capability to withstand the specific load.

Root Diameter

Same as bottom diameter for roller chains. For silent chains it is the diameter of a circle which would pass through the root line surfaces developed by the cutters.

Service Factor

A factor, or multiplier, used to calculate design horsepower for selection purposes. Service factors take into account the relative smoothness of various types of power sources and of driven load.

Sidebar

Alternative term for link plate, more commonly used in connection with engineering-class chain.

Single-Cut Sprocket

A sprocket used with double-pitch chain that has one set of effective teeth.

Take-Up

Provision, in a drive or conveyor layout, for compensating for chain elongation, either by increasing center distance (as with a shaft take-up) or by increasing the chain's travel path (as with an idler sprocket, or shoe, on an adjustable mount).

Tensioner

A take-up device which uses springs, weights, or air or hydraulic pressure to provide constant tension in the slack strand despite changing chain elongation caused by wear or thermal expansion and contraction.

Thimble

A manufacturer's term for bushings used in certain engineering-class roller chains.

Tooth Form

The shape of the sprocket tooth from the bottom of the seating curve up through the working faces to the tip of the tooth.

Tooth Profile

The shape of the cross section of the sprocket tooth taken through the radial center line of the tooth.

Torque

The expression of load on a shaft or sprocket in terms of torsional moment. Usually expressed in inch-pounds, calculated as the product of chain tension (in pounds) and sprocket pitch radius (in inches).

Transverse Pitch

The lateral, or transverse, dimension between the center lines of adjacent strands of a multiple-strand chain assembly. For sprockets, the dimension between the center lines of adjacent tooth flanges, or profiles.

3
BELTS AND PULLEYS

JAMES D. SHEPHERD & DAVID E. ROOS
Gates Rubber Company/Denver, Colorado

Note: Portions of this section are reprinted with permission of the Gates Rubber Company.

INTRODUCTION

The origin of present-day flexible belt drives lies in the rope drives and flat belt drives used in the 1800s and early 1900s. While flat belts were the first type of belt commonly used in early industry as a means of power transmission, other belt types are today receiving greater emphasis. These include the V belt, synchronous belts, and V-ribbed belts. Although each of the many belt types has found its own niche in today's field of power transmission, the fundamentals of drive operation remain the same.

3-1 Where Belts Are Used

Many methods of power transmission are used to drive industrial, domestic, agricultural, and automotive machinery. When the desired speed of the driveN shaft is the same as that of the driveR, direct connection, sometimes through flexible couplings, is the most common method. When the speed of the driveN shaft must be in exact relationship with that of the driveR, gears or chains are most frequently used. However, synchronous belts are increasingly being used in this relatively limited field.

By far the most common type of drive is the type in which synchronization is not required, driveR and driveN speeds are different, and the prime mover and the driveN machinery are separated by some distance. This type of drive lends itself ideally to the use of the various kinds of belts that exist today. With the trend toward higher-speed prime movers, there has been a distinct increase in the use of belt drives, since they offer speed reduction possibilities that allow the design engineer to maintain proper driveN speeds.

3-2 Advantages of Belt Drives

Belts have many advantages over other power transmission methods. One of the most important is overall economics. The high efficiency and reliability of belts as a power transmission medium is well recognized. Belts are clean and require no lubrication. Belts can transmit power between shafts spaced far apart with a wide selection of speed ratios. They can be used for special design purposes, such as clutching and speed variation. They are capable of handling large load fluctuations and have excellent shock-absorbing abilities.

TYPES OF BELT DRIVES

3-3 Heavy-Duty Industrial V Belts

Most industrial drives have load and reliability requirements that cannot be met by single-belt installations. Therefore, 2 to 12 (or more) industrial V belts are normally used together. Multiple-belt drives deliver up to several hundred horsepower continuously and can absorb reasonable shock loads. Temperature limits range from −30 to +140°F.

There are two lines of V belts in this market—the narrow 3V, 5V, 8V line and the classical A, B, C, D, E line. Two standards, published jointly by the Rubber Manufacturers Association (RMA) and the Mechanical Power Transmission Association (MPTA), give the engineering specifications for belts and sheaves for heavy-duty industrial drives. Their titles are:

- "Engineering Standard, Specifications for Drives Using Narrow Multiple V-Belts (3V, 5V, and 8V Cross Sections)." This is RMA number IP-22.
- "Engineering Standard, Specifications for Drives Using Classical Multiple V-Belts (A, B, C, D, and E Cross Sections)." This is RMA number IP-20.

The standards contain specifications for belt and sheave dimensions, along with horsepower ratings and a drive design procedure. As a result, V-belt drives can be developed directly from the standard without assistance from a manufacturer. Industrial drives developed from the standard or from manufacturers' design manuals can be expected to provide 3 to 5 years of service.

Heavy-duty industrial V belts are used in such special applications as the following:

V flat drives

Variable-pitch drives

Crossed drives

Quarter- (and eighth-) turn drives

Vertical-shaft drives

Information on these special drives is given later in this chapter.

3-4 Joined Industrial V Belts

Classical and narrow V belts described in Sec. 3-3 are available in the joined configuration, where several belt strands are connected by a tie band across the top of the belts. The tie band improves lateral stability and solves the problems of belts turning over or coming off the sheaves. The tie band rides above the sheave, so it does not interfere with the wedging action of the individual belt strands. These belts are also described in RMA Standards IP-20 and IP-22.

3-5 Heavy-Duty Industrial Double-V Belts

Double-V belts (sometimes called hexagonal belts) are commonly used in multiple, and are used in certain applications where the regular V belt would have to transmit load to a flat pulley from the back (top) of the belt. These belts operate in the same sheaves as regular classical V belts.

The double V standard is also published jointly by the RMA and the MPTA: "Engineering Standard, Specification for Drives Using Double-V Belts (AA, BB, CC, DD Cross Sections)." The RMA number is IP-21.

3-6 Agricultural V Belts

While agricultural applications require heavy-duty drives, the requirements are significantly different from those of most heavy-duty industrial drives. For instance, industrial drives run at fairly constant loads, whereas agricultural drives are subjected to a wide range of loads, including high, intermittent shock loads. Thus, agricultural V belts tend to have a more durable undercord to prevent the belts from being pulled down into the sheave under these shock loads. In addition, most agricultural drives require the belt to bend in reverse over small-diameter idlers; therefore, agricultural V belts are designed accordingly.

Since load and speed requirements can vary widely from one moment to the next, depending on the field conditions, a weighted average of the effects of these variable conditions must be taken into account in the drive design. Also, many agricultural drives have more than one driveN shaft, and many have idlers to apply the required belt tension. These added sheaves complicate the drive design procedure. Some belt manufacturers publish special design manuals for these applications.

Agricultural V belts are available in the joined configuration for multiple-belt drives and in the hexagonal or double-V style. Double-V belts are used on drives with more than one driveN shaft, where shaft location and direction of rotation make it necessary to drive from both sides of the belt. Again, these belts operate in the same sheaves as regular V belts.

The American Society of Agricultural Engineers (ASAE) has issued a standard covering cross sections for fixed- and adjustable-speed agricultural V belts. This standard, ASAE S211.3, describes the HA, HB, HC, HD, and HE cross sections in the classical V-belt style and the HAA, HBB, HCC, and HDD sections in the double-V style. This standard does not include application data or drive design procedures because of the complexity of agricultural drives. It will be necessary to refer to manufacturers' design manuals.

3-7 Non-Endless Heavy-Duty V Belts

Where it is extremely difficult or impossible to replace worn endless V belts, a non-endless type of belting is used. Such belts do not ordinarily give as long service as the endless belts they replace because of their lower horsepower load capacity. There are basically two types, spliced belting and link belting. They are available in A, B, C, and D sections.

The RMA adopted, many years ago, a "Standard Open End V Belting" which specifies the nominal dimensions of the spliced type of V belting. However, this standard was unpublished, and is not now available. No standard exists for the link type. A variable-speed link-type V belting is also made.

3-8 Light-Duty Industrial V Belts

These belts are similar to classical V belts except that they have slightly thinner cross sections. Thus, they are more adaptable to the small sheave diameters found on light-duty, fractional-horsepower drives.

Most drives designed with light-duty V belts are single- rather than multiple-belt drives. Normally, these drives operate only intermittently; therefore, light-duty belt drives are often designed for lower overall operating lives. Service requirements for these belts vary widely, from as little as 2 or 3 hours per week for power lawn mowers to 40 or more hours per week for office machines.

An RMA/MPTA standard, IP-23, "Engineering Standard, Specifications for Drives Using Single V-Belts (2L, 3L, 4L, and 5L Cross Sections)," lists belt and sheave dimensions for the four standard sizes of light-duty belts. The standard also includes horsepower ratings and a simplified drive design procedure.

Nominal top width for these belts ranges from 1/4 to 5/8 in. The standard numbering system for these belts is based on this width expressed in 1/8-in increments followed by the letter L. For example, the 2L cross section designates a 1/4-in top-width belt.

3-9 Automotive V Belts

Automotive accessory drives require special narrow V belts that can fit within the limited space available in engine compartments. Not only must these belts transmit high horsepower, they also must do so over relatively small-diameter sheaves while being subjected to ambient temperatures of up to 300°F.

The Society of Automotive Engineers (SAE) has standardized several cross sections for driving automotive accessories. SAE Standard J636c describes these belts and the sheaves in which they operate. The most widely used cross sections are the high-capacity or narrow sections designated as SAE sizes 0.380 and 0.500.

These belts are also standardized by the ISO (International Standards Organization) in Standard 2790. This standard covers the AV10 and AV13 cross sections, which are comparable to the 0.380 and 0.500 (nominally 3/8 in and 1/2 in top width, respectively) SAE sizes. Like agricultural drives, the drive design procedure for automotive accessory drives must take into account speed and load variations as well as the effect of several accessories being driven off the same belt. Drive design manuals are available from various belt manufacturers.

3-10 Variable-Speed V Belts

Within certain limits, V belts are well suited to drives that must run at varying input and output speeds. The speed ratio on these drives is controlled by moving one sheave sidewall relative to the other so that the belt rides at different diameters. For industrial applications that require significant speed variation, special variable-speed V belts are available. These belts have wide, thin cross sections to provide a wide range of speed variation. They transmit up to 100 hp per belt and provide speed variations up to 10:1.

An RMA/MPTA standard entitled "Engineering Standard, Specifications for Drives Using Variable-Speed V Belts" gives sheave groove and belt dimensions for twelve of the most widely used cross sections. This is RMA number IP-25. These belts have top widths ranging from 0.875 to 3.0 in. This type of drive is discussed in more detail in Sec. 3-40.

3-11 V-Ribbed Belts

Ribs on the underside of V-ribbed belts mate with corresponding grooves in the pulley. This mating guides the belt and makes it more stable than a flat belt. The ribs fill the pulley grooves completely; therefore, V-ribbed belts do not have the wedging action of V belts and, consequently, must operate at higher belt tensions.

V-ribbed belts have higher lateral stability than do flat belts. In fact, they have the stability advantages of joined V belts on drives where individual belts whip and turn over or come off the sheave. However, the tensile section of V-ribbed belts rides completely out of the pulley grooves and the ribs can be deflected easily, so V-ribbed belts are more susceptible to jumping grooves on misaligned drives than are joined V belts.

While V-ribbed belts are not as flexible as flat belts, they still perform well on small-diameter pulleys, provided that the shafts and bearings can withstand the resulting higher belt tension. In a clothes dryer, for example, a speed ratio of 30:1 is required between the motor and drum; therefore, the motor pulley must be extremely small. V-ribbed belts operate on a small enough motor pulley to produce this speed ratio. In addition, the ribs provide enough traction so that the belt can ride directly on the flat drum without a V-grooved pulley.

"Engineering Standard, Specifications for Drives Using V-Ribbed Belts (H, J, K, L, and M Cross Sections)" is RMA number IP-26. The standard includes belt and pulley dimensions for these five belt sections. The smallest belt operates on pulley diameters down to 0.50 in, and the largest operates as small as 7.00 in.

Three of the belt sections are used primarily on industrial drives. These are the J (0.092-in rib width), L (0.185-in rib width), and M (0.370-in rib width) sections. The smallest section (H, with a rib width of 0.063 in) is used primarily in light, fractional-horsepower applications such as clothes dryers. The K section (0.140-in rib width) was developed mainly for automotive accessory drives.

3-12 Synchronous Belts

V belts and flat belts "creep" somewhat in use (usually about 0.5 percent for V belts), so they are not suitable for drives requiring synchronized input and output (see Sec. 3-21). Synchronous belts were developed to overcome this characteristic. These belts eliminate slip by transmitting power through the positive engagement of belt teeth with pulley teeth. Thus, precise speed ratios and synchronization are possible in such applications as machine-tool indexing heads, automotive camshafts, office machines, and computer peripheral equipment.

Synchronous belt drives have an advantage over gears and chains in that they transmit reasonably high loads with low noise and without lubrication. The shock-absorbing characteristics of the rubber teeth against the metal pulley also can be helpful. However, synchronous belts can wear rapidly and fail prematurely if pulleys are not aligned to close tolerances, especially in long-center-distance drives, where the belts tend to drift sideways against the pulley flanges.

Synchronous belts are sometimes used in applications where synchronization is not required. However, in these applications, V belts often prove to be less expensive. There are two cases in which synchronous belts have an advantage over V belts:

- Synchronous belts incorporate an extremely high-modulus, low-stretch tensile cord to maintain uniform pitch or spacing between teeth. The resultant low growth under load, coupled with the fact that these belts are well sup-

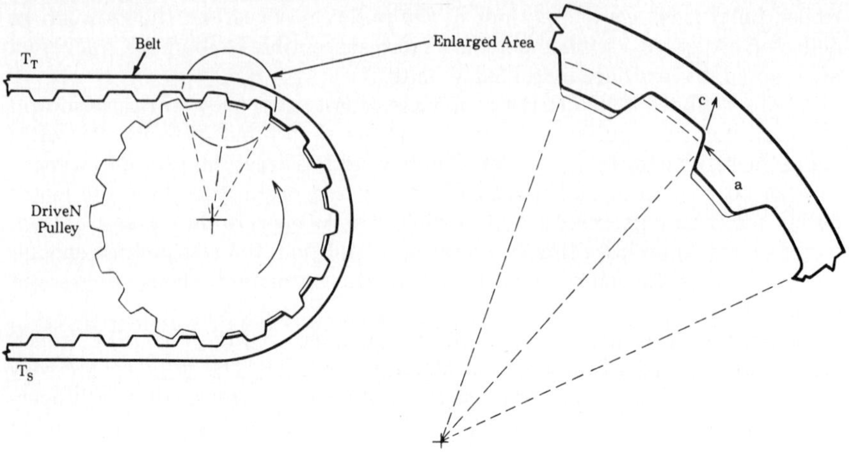

FIG. 3-1 Synchronous belt and pulley forces. Belt must exert circumferential force *a* against pulley tooth to cause pulley rotation. Since tooth face is inclined, belt tries to slide up pulley tooth in direction *c*. This elongates belt, increasing tension.

ported by the pulleys, results in minimal need for center-distance adjustment. Therefore, synchronous belts are often used on drives with limited space for center-distance movement or where the drive is inaccessible.

- Static tension (when the drive is stopped) can be less for synchronous belts than for V belts, imposing lower static loads on shaft bearings. This lower preload reduces starting load, an advantage in drives where the prime mover has a low starting torque.

It is important to remember, however, that the total tension in a synchronous belt increases as torque is applied. The interaction between belt and pulley teeth develops a radial force that increases running tensions. This is illustrated in Fig. 3-1. As a result, dynamic bearing loads for synchronous belts and for V belts are in the same general range.

There are three basic types of synchronous belts: the regular belt with trapezoidal teeth, the high-torque drive (HTD) belt, and the super-torque positive drive (STPD) belt. The last two incorporate curvilinear (round) teeth.

Regular Synchronous Belts These have been used for about 30 years. They are common on many synchronized industrial drives and on automotive camshaft drives.

There are six standard belt sections: MXL, XL, L, H, XH, and XXH. The smallest (0.080-in pitch) can be used on pulley diameters as small as 0.255-in. The largest (1.250-in pitch) can transmit up to 168 hp when operating on pulley diameters of 15.915 in and using an XXH500 belt (5.0-in width).

Belt and pulley dimensions are published in an RMA/MPTA standard entitled "Engineering Standard, Specifications for Drives Using Synchronous Belts." The RMA number is IP-24.

Since the industry standard includes power ratings as well as other application data, it can be used in drive design. However, manuals published by synchronous-belt manufacturers are also available.

HTD Belts These were introduced in 1970. They have deeper, rounded teeth which allow higher torque capacity for a given belt width—hence the HTD nomenclature, which stands for high-torque drive.

The HTD synchronous belt does not cover as wide an application range as regular synchronous belts. Four sizes are available:

- 3-mm pitch
- 5-mm pitch
- 8-mm pitch, which can be used on pulleys as small as 2.206-in pitch diameter
- 14-mm pitch, which can transmit up to 293 hp when the widest stock belt of 6.69 in is used on a 14.036-in pitch diameter pulley.

No standards have been published for this belt line.

STPD Belts These were introduced in 1977. Like HTD belts, they have a curvilinear tooth form. They are available in 2.5-, 4.5-, 6-, 8-, and 14-mm pitch. No standards have been published for this belt line.

3-13 Flat Belts

Flat belts are the simplest and generally least expensive type of belt. They are made of leather, solid rubber, or rubber reinforced with fabric tensile cords. Reinforced flat belts are capable of transmitting high power—up to 500 hp if wide, flat belts are used on large pulleys. Such drives are cumbersome, however, and flat-belt drives have largely been replaced by more compact V-belt drives.

Modern flat-belt drives are generally used for applications where high speed is more important than high power transmission. At speeds near 10,000 r/min, all belts develop centrifugal tension which reduces effective pull, limiting the power transmission capabilities of the drive. In fact, some V belts can develop such high centrifugal tensions that no drive tension remains to transmit power. Flat belts, with their lower mass, do not generally develop significant centrifugal tensions at these high speeds, so they are still able to transmit a reasonable amount of power.

With their thin sections, flat belts also develop less bending tension as they flex around the pulley, especially on small-diameter pulleys. Because bending tension detracts from drive tension, flat belts offer advantages on drives with pulleys that may be too small for V belts. But flat belts depend on high tensions to maintain traction; therefore, the drives must be designed with rugged shafts and bearings to prevent high wear rates or early failure. Alignment also must be controlled closely because flat belts have a tendency to track off their pulleys. Many belt manufacturers can provide data on crowned pulleys, which enhance proper belt tracking.

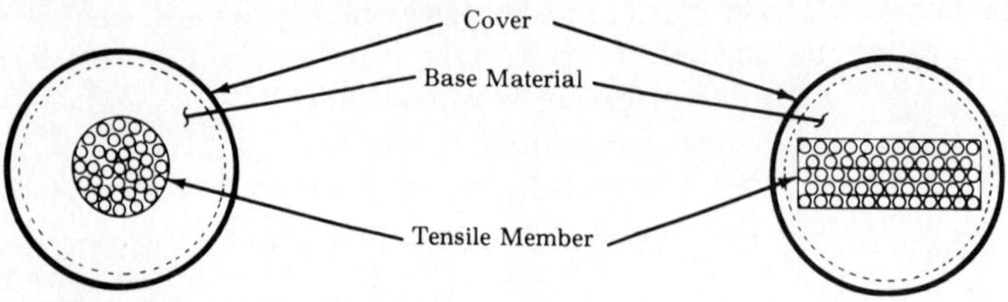

FIG. 3-2 Round belt constructions.

3-14 Round Belts

Sewing machines were the first modern machines on which round belts were used. Round leather belting is still used to some extent for applications of this kind in diameters of 1/4, 9/32, 5/16, 11/32, and 3/8 in. Round belts can be made to length (endless), or the proper length can be cut from rolls of belting and spliced with a metal hook to form an endless product.

The major present-day use of round belts is for agricultural machinery drives and some light-duty or appliance drives, such as vacuum cleaners. These belts are usually similar to V belts, and they run in V sheaves. The component parts of such round endless belts are shown in Fig. 3-2. The rectangular tensile member is the most frequently used type. It is better suited to modern belt production methods than the round tensile member.

Materials used in the various components are similar to those used in V belts. These round belts are available in most fractional diameters from 3/16 to 1 1/16 in.

No industry standards exist for round belts, but there are certain popular lengths which are stocked occasionally by some belt manufacturers. Most of the time, round belts are built to special lengths for specific OEM (original equipment manufacturer) drives.

Sheaves used for round-belt drives are normally those of the 2L, 3L, 4L, 5L, or A, B, C, D, E lines, but there are some special sheaves for the small leather belts previously described. Round grooves do not afford wedging and are seldom used.

3-15 RMA Standards

The list of Rubber Manufacturers Association (RMA) publications for various industrial belts is shown in Table 3-1. Industrial belt sections are illustrated in Fig. 3-3.

3-16 Belt Construction

Figure 3-4 shows the construction of a typical industrial V belt. This type of belt has five basic components:

- Overcord
- Adhesion material

- Tensile cord
- Undercord
- Band or cover (optional)

Since a belt is fundamentally a tension device, the *tensile cord* provides nearly all the tensile strength of the belt. The ideal tensile member must have combined qualities of fatigue strength, tensile strength, shock resistance, adhesion ability, stretch resistance, and resilience.

Dependent on the qualities desired, the most frequently used fibers have included rayon, nylon, polyester, glass, steel, and, today, the new aramid fibers. Polyester has, in recent years, gone far toward replacing rayon and nylon because of its greater strength per unit volume and higher modulus. Thus, it provides greater potential load capacity with low stretch. Glass fibers, while exhibiting excellent strength and low stretch characteristics, are subject to compression failure on some twisted or shock-loaded applications. Steel has exceptionally high strength for some unique applications, but is subject to fretting corrosion, high bending resistance, and cutting problems during manufacturing. The new aramid fibers have an excellent blend of strength and modulus properties and are finding increasing use in belts today.

The *undercord* has to support the tensile member. In the case of V-type belts, it must transfer the load from the tensile member to the sheave groove sidewalls. It must be stiff enough to bridge the groove and keep the center

TABLE 3-1 RMA Standards for Industrial Belts

Publication no.	Subject
IP-20	Specifications: Joint MPTA/RMA/RAC Classical Multiple V-Belts (1977) (A, B, C, D, and E Cross Sections)
IP-21	Specifications: Joint RMA/MPTA Double-V Belts (1984) (AA, BB, CC, DD Cross Sections)
IP-22	Specifications: Joint MPTA/RMA/RAC Narrow Multiple V-Belts (1983) (3V, 5V, and 8V Cross Sections)
IP-23	Specifications: Joint RMA/MPTA Single V-Belts (1968) (2L, 3L, 4L, and 5L Cross Sections)
IP-24	Specifications: Joint MPTA/RMA/RAC Synchronous Belts (1983) (MXL, XL, L, H, XH, and XXH Belt Sections)
IP-25	Specifications: Joint MPTA/RMA/RAC Variable-Speed Belts (1982) (12 Cross Sections)
IP-26	Specifications: Joint MPTA/RMA/RAC V-Ribbed Belts (1977) (H, J, K, L, and M Cross Sections)

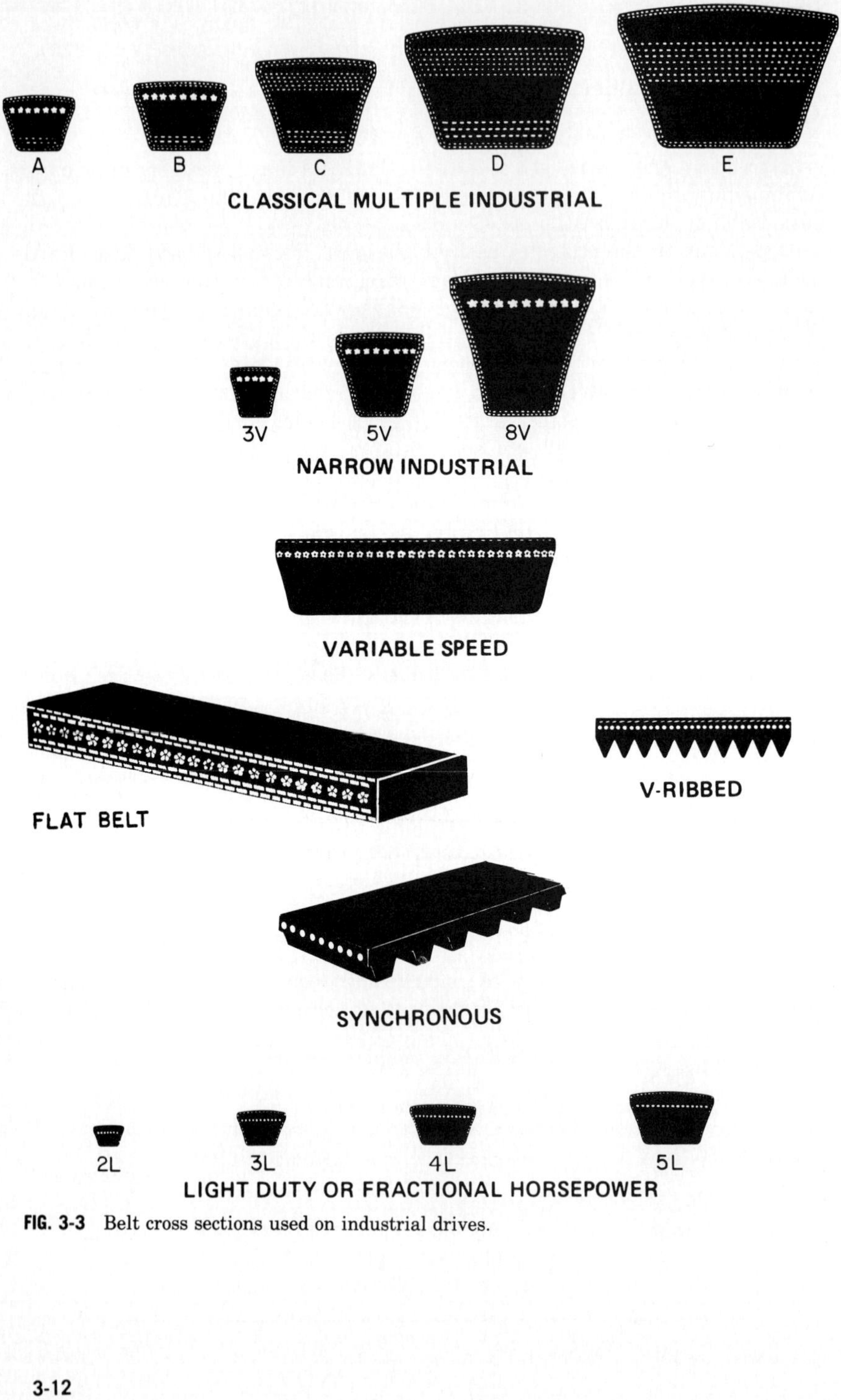

FIG. 3-3 Belt cross sections used on industrial drives.

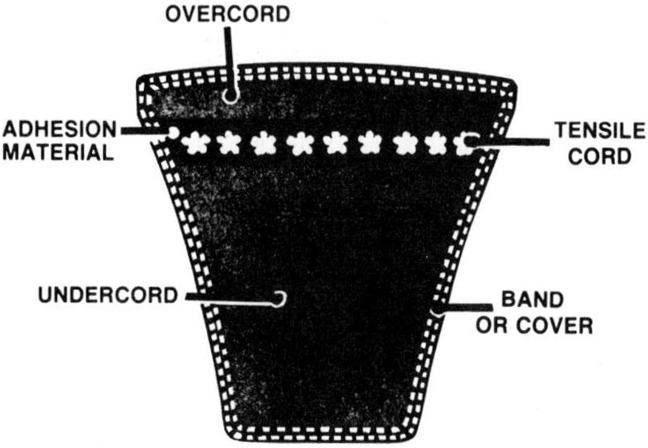

FIG. 3-4 Basic V-belt construction.

tensile cords from sagging under load, yet flexible enough to bend over the sheave through millions of cycles without cracking. Notching or cogging relieves bending stress in the undercord on smaller-diameter sheaves, but it can reduce the support that the tensile member vitally needs.

The *adhesion material* is used to assure complete bonding of the tensile member to the other parts of the belt. It is important to the belt, since a breakdown of the bond in this area will result in premature belt failure.

The *overcord* material locates the tensile member in the belt and adds to the transverse support of the tensile member.

The *band or cover* material is used to protect the rest of the belt from oil, dust, and other destructive elements. The band is optional, and many modern V belts do not include a cover material.

Optimization of the V belt's, or any belt's, construction must take into consideration the specific characteristics of the application or market toward which the belt is directed. The selection of a particular tensile member and of undercord and overcord materials, and the choice between banded or non-banded construction can mean success or failure in any given application. The belt manufacturers' engineering personnel can provide specific details of construction features for any application.

FUNDAMENTALS OF BELT POWER TRANSMISSION

3-17 Work and Power

In order to define the work being performed by a belt drive, a basic understanding of general power transmission is required.

Work can be defined as a force acting through a distance. A formula for work is

$$W = Fl \tag{3-1}$$

where W = work performed, ft·lb
F = force exerted, lb
l = distance moved, ft

Power, then, is defined as the rate at which work is performed, or work per unit time. The most frequently used English unit of power is the horsepower, which is defined as 33,000 ft·lb of work per minute. An equation for horsepower is

$$P = \frac{Fl}{33{,}000t} \tag{3-2}$$

where P = power, hp
F = force exerted, lb
l = distance moved, ft
t = time, min

3-18 Belt Tensions

While considerable time could be spent discussing all the fundamentals of belt power transmission, several basic equations stand out as being most important.

The force that produces work with a belt drive acts on the rim of the pulley, causing it to rotate. When a drive is transmitting power, the belt pulls, or belt tensions, are not equal, as shown in Fig. 3-5.

There is a tight-side tension T_T and a slack-side tension T_S. The difference between these two tensions $T_T - T_S$ is called effective pull or net pull. Substituting in Eq. (3-2) yields the equation

$$T_T - T_S = \frac{33{,}000P}{V} \tag{3-3}$$

where T_T = tight-side tension, lb
T_S = slack-side tension, lb
P = power, hp
V = belt speed, ft/min

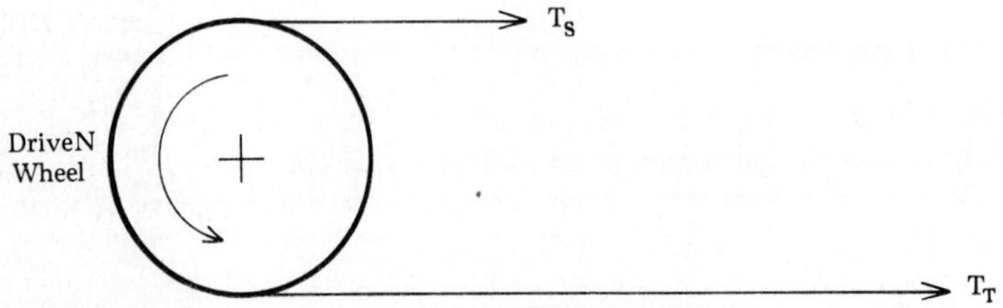

FIG. 3-5 Belt drive tensions while transmitting power.

In this relationship, belt velocity V is defined as

$$V = \frac{D \times N}{3.82} \tag{3-4}$$

where V = belt speed, ft/min
D = pitch diameter, in
N = rotational speed, r/min

If the horsepower to be transmitted and the belt speed are known, Eq. (3-3) can be used to determine the effective pull. A second relationship between T_T and T_S is required to find the value of each. The most commonly used relationship is the tension ratio.

3-19 Tension Ratio

The ratio of the tight-side to the slack-side tension is commonly referred to as the "tension ratio." The higher the ratio between tight- and slack-side tension, the closer a given belt is to slip—the belt is loose. A low tension ratio means there is more slack-side tension in comparison to the tight-side tension, and the belt is less likely to slip. In the latter case, since the slack-side tension is greater, the tight-side tension must also be proportionately greater in order to yield the same effective pull for transmitting the required power. The principle of tension ratio is illustrated in Fig. 3-6. The lower tension ratio creates a greater shaft pull although the effective pull is 40 lb in both cases.

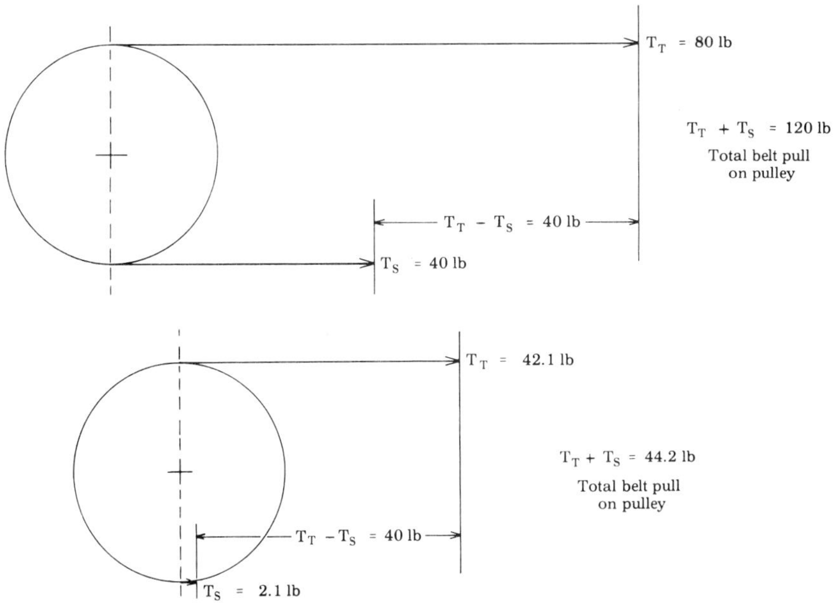

FIG. 3-6 Effect of tension ratio on belt tensions. Tensions shown to scale.

The fundamental tension ratio formula for V belts at the point of *impending* slip is

$$R = \frac{T_T}{T_S} = e^{W\mu\theta} \tag{3-5}$$

where R = tension ratio
T_T = tight-side tension, lb
T_S = slack-side tension, lb
e = a constant, the base of natural logarithms, 2.718
μ = coefficient of friction between the belt and pulley surface
θ = arc of contact of the belt on the pulley, rad
W = wedging factor for V belts; not present in the same formula for flat belts

Note that coefficient of friction, arc of contact, and wedging factor are the factors affecting tension ratio at the point of *impending slip*.

The wedging factor takes into account the multiplication of force between the belt and the groove surface that occurs because of the wedging action of the belt in the groove. The wedging factor increases the exponent of *e*, allowing V belts to operate at higher average tension ratios than flat belts. The higher tension ratio translates directly into lower operating tensions and bearing loads for V belts than for flat belts.

The same is true of the arc of contact (sometimes called belt wrap). At an increased arc of contact, a drive can operate at a higher tension ratio and lower operating tension.

There is a general relationship between tension ratio and belt slip. This can best be illustrated by Fig. 3-7.

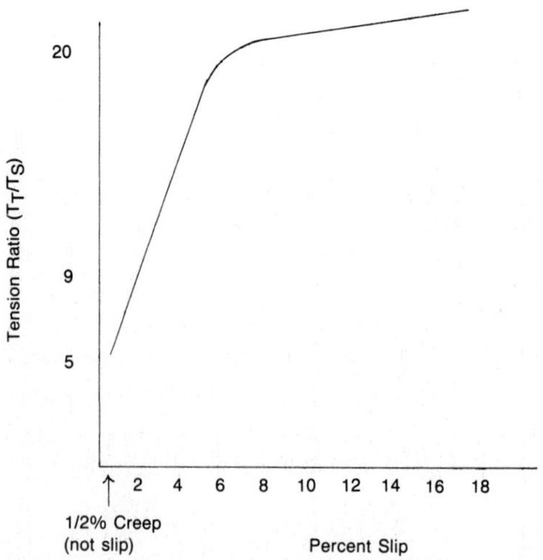

FIG. 3-7 Tension ratio vs. percent slip.

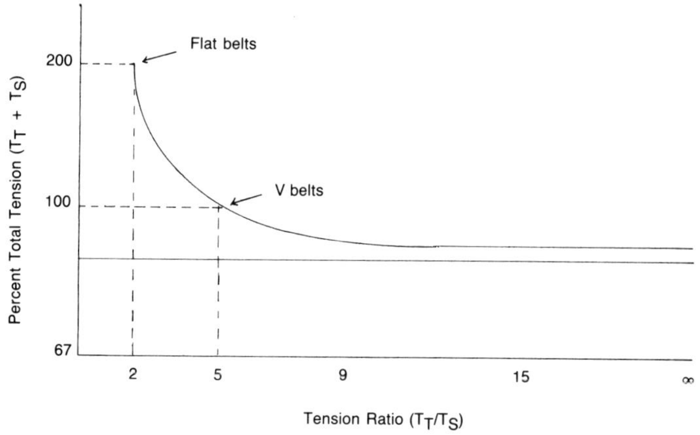

FIG. 3-8 Effect of tension ratio on total tension.

There have been occasions when a design engineer has wanted to design at the point of impending slip. Figure 3-7 shows that this is not practical, because at a tension ratio of about 20, belt slip increases dramatically. It should also be noted that the tension ratio vs. slip characteristics for a drive will depend on the physical condition of the belt and sheave groove surfaces, water or oil contamination, and other factors. The common V-belt *design* tension ratio of 5 (at 180° arc of contact) allows a reserve for nonideal conditions.

Locked-center V-belt drives are designed at tension ratios of 5:1, while design tension ratios of 9:1 are sometimes used with drives which are equipped with automatic tensioning devices. Flat belts are designed at a 2:1 tension ratio. The effect this has on the drive can be seen in Fig. 3-8.

The greatest significance in terms of total tension reduction is seen when going from flat belts to V belts. Designing at tension ratios much beyond 9:1 results in very little reduction in total tension.

In actual design practice, the design tension ratio of 5 is used. Equation (3-5) is modified to reflect this value so that arc-of-contact corrections can be made. Some design procedures use a correction factor (commonly called factor *G*) which facilitates calculation.

Equation (3-5) is generally not used in actual belt applications. For application work the previously discussed "design" tension ratio is—5:1 for a V-belt drive and 2:1 for a flat belt drive, with an arc of contact of 180°. The design tension ratio must then be corrected for the actual arc of contact on the driveR sheave. This arc-of-contact correction factor (factor *G*) is commonly derived from graphs or tables supplied in manufacturers' design manuals. The specific values of factor *G* for a two-sheave drive may be seen in Table 3-2.

Using the arc-of-contact correction factor from Table 3-2, simplified formulas for tight-side and slack-side tension may be derived.

For flat belts:

$$T_T = \frac{66{,}000\,P}{GV} \tag{3-6}$$

TABLE 3-2 Factor *G* for Arc of Contact Correction

$\frac{D-d}{C}$	Arc of contact, degrees	V-belt factor *G*— for belts in V grooves	Flat-belt factor *G*— for belts on flat pulleys
0.0	180	1.00	1.00
0.1	174	0.99	0.98
0.2	169	0.97	0.96
0.3	163	0.96	0.93
0.4	157	0.94	0.91
0.5	151	0.93	0.88
0.6	145	0.91	0.86
0.7	139	0.89	0.83
0.8	133	0.87	0.80
0.9	127	0.85	0.77
1.0	120	0.82	0.74
1.1	113	0.80	0.71
1.2	106	0.77	0.67
1.3	99	0.73	0.63
1.4	91	0.70	0.59
1.5	83	0.65	0.55

$$T_S = \frac{33{,}000(2.0 - G)\,P}{GV} \tag{3-7}$$

For V belts:

$$T_T = \frac{41{,}250\,P}{GV} \tag{3-8}$$

$$T_S = \frac{33{,}000(1.25 - G)\,P}{GV} \tag{3-9}$$

where G = arc-of-contact correction factor
V = belt speed, ft/min
P = power, hp

FUNDAMENTALS OF BELT-DRIVE GEOMETRY

3-20 Defining Sheave or Pulley Diameters

To properly design a drive, the sheave or pulley diameter to be used in calculating tensions and belt lengths must be determined. This is important because a belt has thickness, and thus extends through a diameter range when it is on a pulley or in a sheave. In designing belt drives, two diameters are commonly used for a given belt-sheave combination: the pitch diameter and the effective outside diameter. This is illustrated in Fig. 3-9.

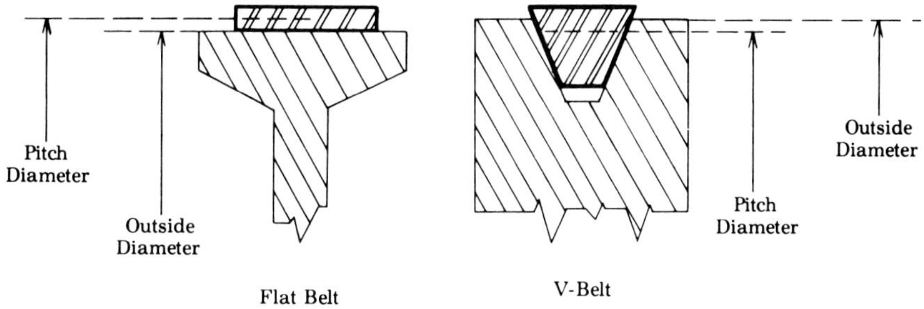

FIG. 3-9 Pitch diameter and outside diameter.

Generally, the effective outside diameter is used to determine belt length, while the pitch diameter is used to determine belt tensions and driveN speed (an exception to this is the classical industrial belts, which are specified by pitch length). The relationship between the outside diameter and the pitch diameter is defined by the standards organizations, such as RMA, SAE, and ASAE.

3-21 Speed Ratio

Most prime movers rotate at a different speed from the one desired for the driveN machine. The ratio between the speed in revolutions per minute of the faster shaft and that of the slower shaft is called the speed ratio.

$$\text{Speed ratio} = \frac{N \text{ of faster shaft, r/min}}{N \text{ of slower shaft, r/min}} \tag{3-10}$$

For design purposes, the speed ratio of a belt drive is always considered to be equal to or greater than 1 whether the drive is a speed-up or speed-down.

The individual speeds in revolutions per minute of the two sheaves or pulleys are inversely proportional to their pitch diameters. In equation form,

$$\text{Speed ratio} = \frac{N \text{ faster shaft}}{N \text{ slower shaft}} = \frac{\text{diameter larger wheel}}{\text{diameter smaller wheel}} \tag{3-11}$$

There are some belt drives for which the speed of the driveN shaft is more critical. In the case of exact synchronization, designing with a synchronous belt should be considered. If the driveN speed must be held close, but not exact, then V belts can be considered.

Both V belts and flat belts will exhibit what is known as "creep." Creep results from the difference in tension from the tight to the slack side of a sheave or pulley. Since a belt elongates and narrows slightly under tension, a greater length of belt leaves the driveN sheave than goes back on it. This makes the driveN shaft rotate at a slightly slower speed than if creep were not a factor. While the degree of creep can vary somewhat based on load and speed, its magnitude is assumed to be ½ percent for most industrial V-belt drives.

Other variables which should be checked in an exact speed situation include:

- Condition of sheave groove or pulley surfaces
- Accuracy of speed measuring equipment
- Belt tension
- Belt construction (cordline location in the belt)

3-22 Belt Length

When the desired driveN speeds of a belt drive have been established by selecting the sheave or pulley diameters, the belt length can then be determined.

For two-sheave drives, the calculation is simple:

$$L = 2C + 1.57(D + d) + \frac{(D - d)^2}{4C} \tag{3-12}$$

where L = belt length, in
D = diameter of large sheave or pulley, in
d = diameter of small sheave or pulley, in
C = center distance, in

If pitch diameters are used in the formula, the belt length is pitch length. If outside diameters are used, the length is outside length. For drives with more than two sheaves it is often neccssary to make a scaled drive layout to determine the correct belt length. Procedures for this method are described in manufacturers' manuals.

Once the correct belt length has been determined, the drive must be checked to make sure that provision has been allowed for installation and take-up of the belt. These values are included in manufacturers' literature or applicable standards.

STRESS FATIGUE

3-23 Working Tensions

In this section the factors that contribute to belt life will be discussed. More specifically, the section will study the process by which a belt fails in normal service, which is often called stress fatigue. V belts will be used as an example throughout this section. However, most of the principles discussed also apply to all power transmission belts.

Section 3-18 discussed working tensions (tight and slack side) as they relate to horsepower requirements, speed, diameters, and other factors, using certain design tension ratios. These working tensions are important from several standpoints, one of these being that they affect the life of the belt.

3-24 Bending and Centrifugal Tensions

In addition to the working tensions in the drive, there are two other tensions that develop when a belt is operating on a drive:

- *Bending tension* occurs when a belt bends around a sheave or pulley. Since the top part of the belt being bent is in tension and the bottom part is in compression, compressive stresses also occur. But the effect of bending is most commonly evaluated in terms of the tension introduced in the tensile member of the belt; hence the term bending tension. The amount of tension incurred in a belt depends on the radius of the bend (diameter of sheave or pulley) and is described in the following equation:

$$T_B = \frac{C_B}{D} \tag{3-13}$$

where T_B = bending tension, lb
C_B = a constant depending on the belt size and construction
D = sheave or pulley diameter, in

- *Centrifugal tension* occurs in a belt because of centrifugal force. The belt is rotating around the drive. It has weight and tries to pull out, away from the sheave or pulley. Tension in the belt results. This tension depends on the belt speed, as shown in the following equation:

$$T_C = MV^2 \tag{3-14}$$

where T_C = centrifugal tension, lb
M = a constant depending on the weight of the belt
V = belt speed, ft/min

Neither the bending nor the centrifugal tension is imposed on the sheaves, pulleys, shafts, or bearings—only on the belts.

3-25 Tension Diagram for Stress Fatigue

A schematic tension diagram of the belt tensions imposed on the upper part of the belt during one revolution of operation is shown in Fig. 3-10.

Starting at point A on the drive, the slack-side tension T_S and the centrifugal tension T_C are imposed on the belt. The centrifugal tension is imposed equally on all parts of the drive. When the belt enters the driveN sheave at point B, the bending tension T_B is imposed. As the belt goes from the slack side of the sheave to the tight side, the working tension increases from T_S to T_T. At point C, when the bending tension is removed, T_T and T_C remain. The opposite occurs over the driveR wheel—the bending tension is imposed at E and removed at F, returning the belt to the original slack-side plus centrifugal tensions.

Visualize this entire cycle occurring extremely fast—10 times per second is

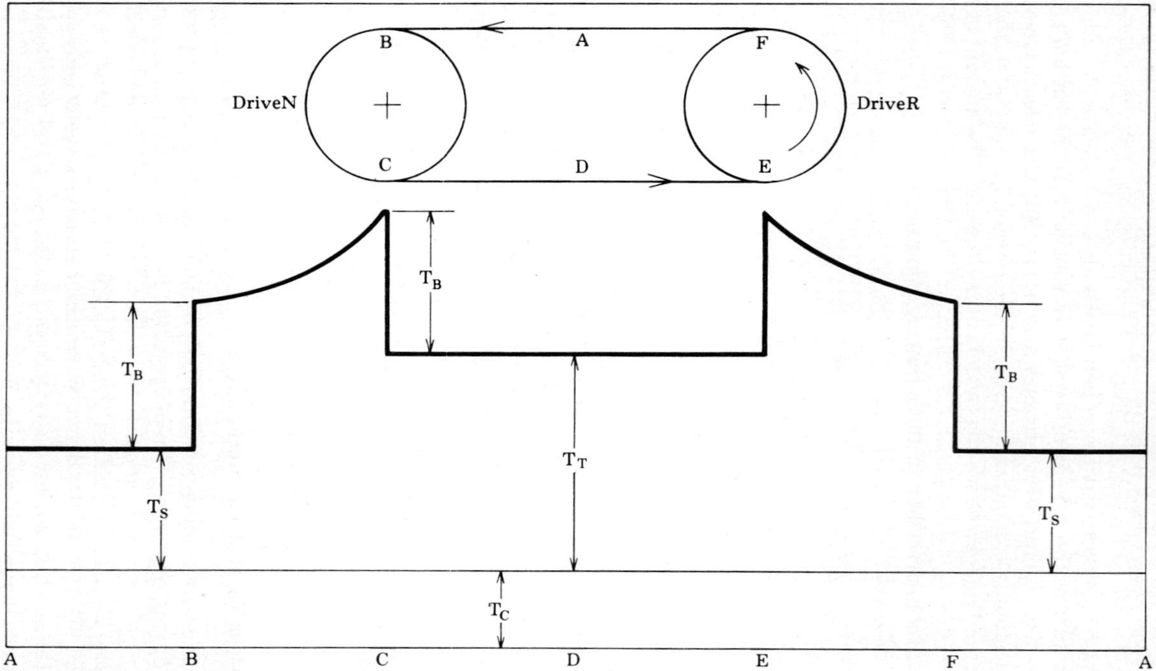

FIG. 3-10 Working, bending, and centrifugal tensions in a belt drive.

not unusual—and you get some idea of the stresses that a belt "feels" when transmitting power.

A belt "wears out" in normal service because of these repetitive stresses incurred as it rotates around the drive. Proper analysis of these stresses should, therefore, make it possible to predict belt life under given conditions. For purposes of calculation, these stresses are expressed in terms of peak tension.

$$T_{peak} = T_T + T_B + T_C \tag{3-15}$$

where T_{peak} = peak tension at a sheave, lb
T_T = tight-side belt tension, lb
T_B = bending tension, lb
T_C = centrifugal tension, lb

3-26 Stress-Cycle Relationships

Peak tension is important because it is directly related to belt life. This relationship has been established through field testing of actual applications and through laboratory testing. The correlation is illustrated in Fig. 3-11.

The curve on the right-hand portion of Fig. 3-11 is the empirical link between peak tensions in a belt and the service life it will give on a drive which imposes those peak tensions. The curve is obtained by running thousands of belts to failure under controlled laboratory conditions, then checking the results with actual controlled field tests.

3-27 Illustration of Effect of Drive Variables

By using the previous information, the effect of changing drive conditions on belt life can be quickly illustrated. Assume first that the sheave diameters on a drive are reduced without changing belt speed, horsepower load, or belt length. Figure 3-12 illustrates these changes.

Since the load, speed, and belt length do not change, T_T and T_C are the same, and T_B is the only factor that changes. T_B increases, increasing the peak tension and thus decreasing the number of cycles to failure. This decreases belt life.

Next, assume that the horsepower load increases without changing diameters, belt speed, or belt length. Figure 3-13 illustrates the changes. The working tensions T_T and T_S both increase because the load goes up. Thus the peak tension increases. The number of cycles to failure therefore decreases, decreasing belt life.

Next, assume that everything is the same except that belt length is reduced by reducing the center distance. Figure 3-14 illustrates the change. All tensions remain the same in this instance, and so peak tension is the same and the cycles to failure are the same. When length decreases, life decreases because of the higher number of cycles per unit of time.

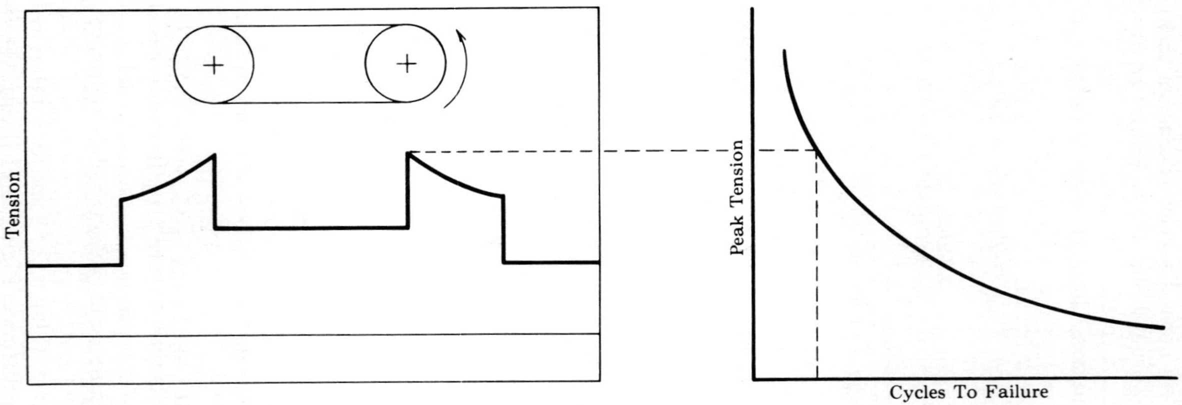

FIG. 3-11 Peak tensions vs. cycles to failure.

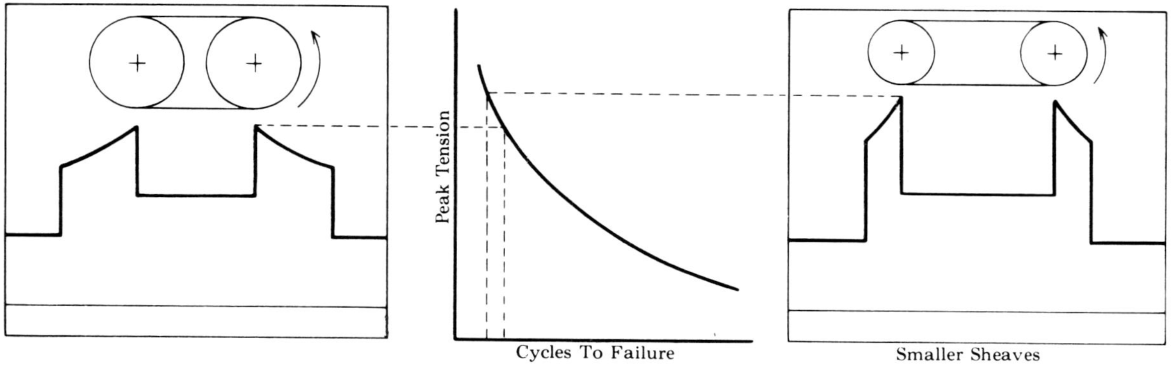

FIG. 3-12 Effect of reducing sheave diameters.

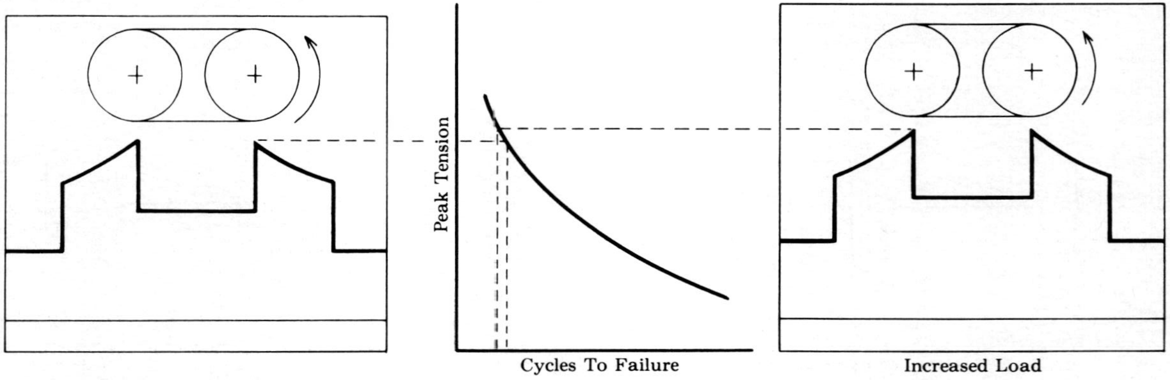

FIG. 3-13 Effect of increasing horsepower load.

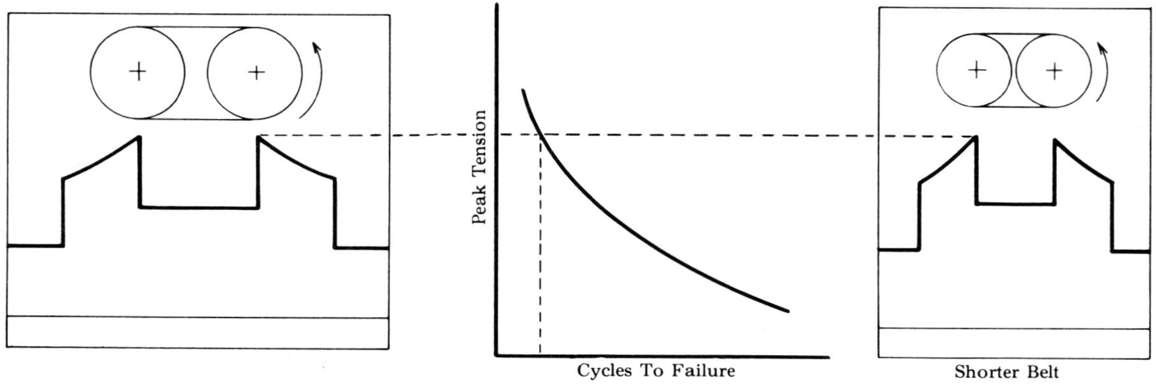

FIG. 3-14 Effect of reducing belt length.

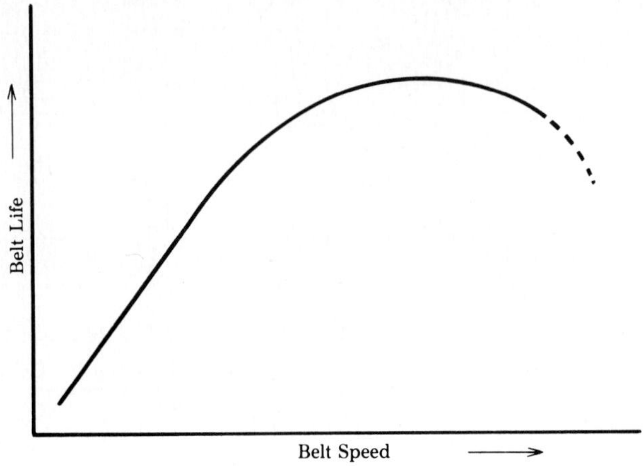

FIG. 3-15 Belt life versus belt speed.

Finally, let's assume that belt speed increases. There are three effects:

- T_T decreases
- T_C increases
- Rate of imposed stresses increases

The combined effect of these changes depends on the original belt speed and on the amount of belt speed change. In general, increasing speed increases belt life. But if the belt speed goes too high, the decrease in life caused by increasing T_C, plus the increased rate of imposed stresses, offsets the increase in life caused by decreasing T_T. The overall effect is a decrease in belt life. This is illustrated in Fig. 3-15.

The above illustrations show the effect of drive parameter changes on belt life. These changes are taken into consideration in the rating of belts.

3-28 Rules of Thumb

The life curve illustrated in the previous figures is actually a complex function. Therefore, it is difficult to generalize about the amount of change in belt life that will occur as a result of a change in one or more drive variables. It is necessary to substitute actual values into the design procedures. These drive design procedures take each of the drive variables into account automatically.

But there are a few rules of thumb for V belts that give general indications of the effect of changes. These rules may help the user to estimate the effect of changing a drive variable.

1. When sheave diameters are reduced 10 percent, the belt life is reduced by approximately half.
2. When horsepower load increases 10 percent, the belt life is reduced by approximately half.

3. When belt length is cut in half, belt life is reduced by approximately half—life is directly proportional to belt length.

3-29 Design Horsepower

Design horsepower is found by multiplying the horsepower requirement by a service factor greater than 1. The reason for increasing the load for design purposes is that the actual horsepower absorbed by a driveN machine varies. Variation can be caused by high starting loads or by high intermittent loads during running. The average of the horsepower absorbed by the driveN machine over a long period of time is generally thought of as the horsepower requirement of the drive. Since drives designed for a peak load value would be overdesigned, the design horsepower selected is something less. Nor is it the actual average (timewise) of the horsepower absorbed by the driveN machine, because the life-load curve is not a straight line. It is an equivalent horsepower that would give the same life, if it were imposed continuously, as the actual application. Service factor tables found in belt manufacturers' catalogs should be used to find the proper design horsepower.

Sometimes an additional service factor is imposed on a drive when load variation cannot be evaluated or for a reason other than load variation. An example of the first situation might be an application with an unusually large amount of acceleration or deceleration involved, for which loads cannot be calculated. An example of the second situation might be an application subjected to abrasive conditions or elevated ambient temperatures.

HOW BELTS ARE APPLIED

3-30 Horsepower Rating vs. Hours Life Calculations

It is not absolutely necessary to know all the principles of belt-drive design to properly apply a belt to a drive. Most manufacturers make available drive-design manuals which allow the belt drive to be designed on a step-by-step basis. It is, however, important to understand the two basic methods of drive design.

When designing a belt drive, the desired result is to select the size and number of belts (or in some cases the width of belt) and the pulley or sheave sizes which will give (1) the desired center distance, (2) correct shaft speeds, and (3) acceptable belt life with minimum maintenance. Two basic methods are used extensively in the industry to achieve this end. They are:

- The horsepower rating method, which assumes a fixed belt life in order to arrive at a horsepower rating per belt
- The life-in-hours method, which assumes a fixed load per belt in order to arrive at a calculated belt life

The same basic philosophy is behind both approaches—the stress fatigue theory. A belt wears out in normal service because of the repetitive stresses incurred as it rotates around the drive. Proper analysis of the tensions makes possible the prediction of load/life relationships under given conditions.

The example curve shown earlier in Fig. 3-11 represents the link between the peak tensions in a belt and the service life it will give on a drive which imposes those peak tensions. Drive design formulas are based on this principle.

The horsepower rating method assumes an acceptable value of belt life for an industrial application and is the preferred method for the simpler two-point drives. The areas in which this method is heavily used are:

- Industrial heavy-duty belt drives
- Industrial light-duty (fractional horsepower) belt drives
- Industrial variable-speed belt drives

In most instances the industrial user is either putting a new piece of equipment into service or replacing the drive on an older piece of equipment. In both cases, the application often requires 3 to 5 years typical industrial service.

In contrast to industrial applications, many original equipment designers design new equipment for a belt life less than standard industrial life, depending on the intended use of the product.

Design engineers, having gained experience from product testing, might determine that belt lives in the 200 to 2000 hour range can be quite adequate. For example, several hundred hours of belt life for a lawnmower or sewing machine can translate into several seasons or several years of use. On a passenger car accessory drive this same life can translate into tens of thousands of miles of driving. Consequently, use of the life-in-hours design method is applicable in the following markets:

- Automotive
- Agricultural
- Appliance

This method can be used with any number of pulleys in a drive and with any geometry. In the initial design stage, it enables the designer to select the best drive from among several alternative drives being considered for a specific power transmission job. Calculated average belt life can be appropriately matched to the specific equipment based on that equipment's normal life cycle.

3-31 Belt Application Range

Belts have been successfully applied over a wide range of speeds and loads. Because of the numerous belt cross sections available, virtually any power transmission application can be considered a potential candidate. While the normal speed range for belts is the typical motor speeds, from 575 to 3450 r/min, applications for speeds from less than 10 to over 20,000 r/min are not uncommon (refer to Sec. 3-32). Similarly, drives have been designed for loads from less than 1/4 hp to over 2000 hp. Often, even the large drives are space competitive with other forms of power transmission.

Belts are often used for non-power transmission applications. Typically these involve conveying of materials, but other applications are possible. An example application is bowling equipment, which consists of many belt drives that convey the ball and pins.

Any application involving extreme speeds or loads, and special non-power transmission applications, should be referred to the belt manufacturer for proper analysis.

3-32 Allowable Belt Speeds

Although for each belt there is a speed beyond which life or horsepower rating begins to decrease, generally speaking, the upper limit on belt speed is determined more by sheaves and pulleys than by belts. This is especially true of today's smaller, more powerful belts, which develop less centrifugal stress because of their lower weight.

With respect to sheaves and pulleys, both safety and imbalance are considerations in determining upper limits on belt speed.

Stock cast iron sheaves and pulleys should not be run above 6000 or 6500 f/min without checking with the manufacturer for possible dynamic balancing requirements. The higher the speed, the more perfect the dynamic balance must be, especially if smooth running is a consideration.

Also, for cast iron sheaves and pulleys, the strength of the material begins to be a consideration at speeds above 6000 to 6500 f/min. Steel sheaves or pulleys, of course, have an upper safety limit in excess of 15,000 f/min, but the dynamic balancing requirements apply just as for cast iron.

V-belt drives are commonly designed with formed steel sheaves for belt speeds up to 9000 to 10,000 f/min, as in automotive drives. Special belt designs can transmit limited loads at speeds up to 20,000 f/min.

At the other end of the range, low belt speed limitations are imposed only by economics. Low-belt-speed applications usually have high torque requirements. Belt operating tensions tend to be rather high. Therefore, low-speed applications usually require large belts or a large number of belts, and often larger-diameter sheaves or pulleys.

3-33 Acceleration Loads

In addition to the normal load requirements, a belt drive is subjected to acceleration loads resulting from rapid speeding up or slowing down of the equipment. For the majority of drives, the acceleration loads are either small enough or occur seldom enough that they do not require special attention during design. In these cases, the extra loading is taken care of by the service factors selected.

When rapid speed changes occur or when acceleration and deceleration is a normal function of the application, it is best to determine if this extra loading affects the design. An example of this is an automobile engine. When the engine is "hot-rodded," the alternator is forced to accelerate quickly. If it is a fairly large alternator, with high inertia, the extra acceleration loads may equal or exceed the normal loading assigned for drive design purposes.

Another example of acceleration loading occurs on the grinders used to finish the surface of large steel rolls used in steel mills and metal shops. These roll grinders use a belt drive to rotate the roll slowly as the grinding wheel operates against its surface. Once the roll is turning, loads on the drive are very light. The torque required to accelerate the roll to grinding speed is the primary load that causes belt failure, and so this is the important factor in design.

The extra loading due to high or frequent acceleration can be accounted for in the design by one of these methods:

- An extra service factor, usually based on experience with the given application. This is less desirable than the following quantitative methods.
- Calculating the acceleration load and adding it to the driveN machine's normal horsepower requirement. This is applicable when the extra load is truly imposed on top of the regular load for the major portion of the life of the drive.
- Calculating the acceleration load and using it as a separate condition, as in multicondition automotive designs. This method is applicable when the times of acceleration do not coincide with other peak driveN loads, so that conditions must be treated separately.

Acceleration loads may be calculated from the following formula:

$$P_a = (4.30 \times 10^{-9})(WR^2)\,(a)\,(\bar{N}) \tag{3-16}$$

where P_a = acceleration horsepower
WR^2 = inertia of rotating component, lb·in^2
a = acceleration of rotating component, r/(min·s)
$\bar{N}$ = average speed of rotating component through the acceleration period, r/min

3-34 Installation and Takeup Allowances

Installation and takeup allowances are usually provided by movement of one or more shafts. Characteristics of each specific belt type or belt line will dictate the amount of allowance required on a belt drive. Specific values can be obtained from belt manufacturers' catalogs.

While belt stretch might be the most easily recognized reason for takeup, there are other factors which influence the need for installation and takeup allowances. Precise statements of what is involved for both allowances follow.

Installation allowances include:

1. The amount of belt length required to get the belt over the edge of the sheave or the flange of a flat pulley without belt damage

2. The amount of elastic change in belt length from measuring (inspection) tension to no tension
3. The negative belt manufacturing tolerance and the plus manufacturing tolerance on the sheaves or pulleys

Takeup allowances include:

1. The amount of inelastic change in belt length that results from permanent elongation of the belt (stretch) during its normal life
2. The amount of elastic change in belt length from measuring tension to drive operating tension
3. The amount of change in center distance that must be made to account for normal wear of the belts and sheaves or pulleys
4. The plus belt manufacturing tolerance and the negative tolerance on the sheaves or pulleys

The frequency of belt takeup for the purpose of applying tension is a maintenance factor; it will vary depending on the characteristics of the application. It is best determined by actual experience on individual drives.

3-35 Idlers

Idlers are sheaves or pulleys on a belt drive that are not loaded (do not transmit power). They are used for several reasons:

- To provide takeup for fixed center drives
- To clear obstructions
- To turn corners (as in mule pulley idlers, Fig. 3-35)
- To break up long spans where belt whip may be a problem
- To provide automatic tensioning (discussed in Sec. 3-37)
- To increase wrap on critical wheels (careful analysis is needed to make sure that the drive is actually benefited)

Idlers may be run inside or outside the drive (Fig. 3-16). An inside idler decreases arc of contact on adjacent wheels, while an outside idler increases arc. An outside idler, however, bends belts backward, and on some drives this nullifies the gain obtained with increased arc as far as belt life is concerned. Also, if an outside idler is used for takeup, the amount of takeup is limited by the belts on the opposite side of the drive.

If at all possible, idlers should be placed in the slack span (or spans) of the drive. Spring-loaded, or weighted, idlers should always be located on the slack span because the spring force or weight can be much less. Such idlers should never be used on a drive where the belt's direction of rotation can be reversed—where the slack side can become the tight side.

A flat idler pulley, either inside or outside, should be located as far as possible from the next sheave or pulley along the path of the belt. This is

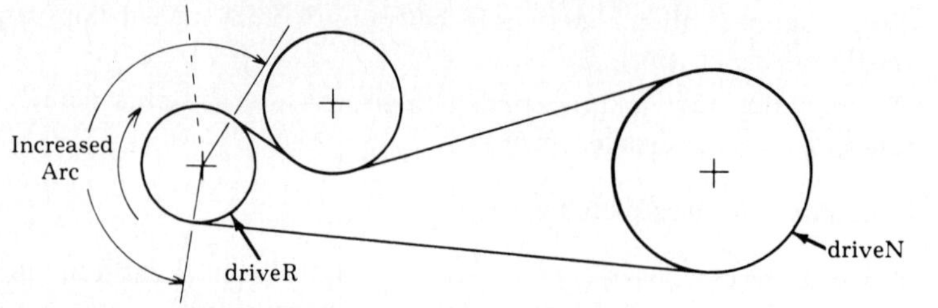

Idler Pulley on the Slack Side of a Drive.

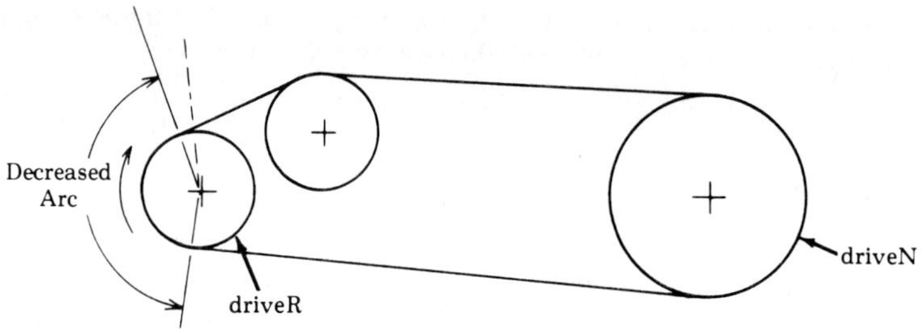

FIG. 3-16 Effect of idler placement.

because belts move back and forth slightly across a flat pulley, and there should be as much span length as possible so that they can be aligned properly for the next wheel. The use of flat idler pulleys on long span drives can sometimes cause severe belt whip and should be avoided, if possible.

The diameter of inside idlers should be at least as large as the diameter of the smallest loaded wheel. The diameter of outside idlers should be at least one-third larger than the diameter of the smallest loaded wheel. One of the most frequent misapplications of belts is their use on drives with too-small idlers. Drives with unusually large sheaves, however, do not always require idlers as large as these rules would indicate. An analysis of the drive will reveal when idlers can be somewhat reduced in size without affecting belt life.

Flat idlers for V-belt drives should not be crowned, but flanging of such idlers is good practice. In the case of flanged idlers, the inside bottom corners of the flange should not be rounded—this may cause a belt to climb over the flange (see Fig. 3-17). The width of flat idlers must be greater than that of the belt (or the total width of the belts).

Brackets for idlers should be sturdily constructed. Frequently, the cause of drive problems described as belt stretch, belt instability, short belt life, belt roughness, belt vibration, and so on can be traced to weak idler bracketry. Idler brackets, bearings, shafts, and other parts must be designed to withstand the forces imposed by belt tensions.

RMA Bulletin IP-3-6 discusses idlers in belt systems.

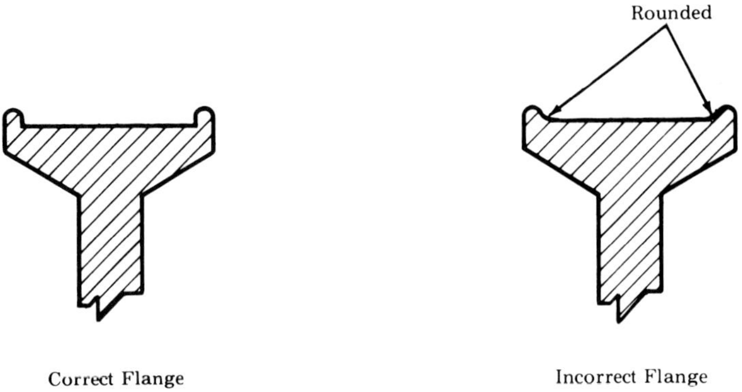

FIG. 3-17 Flanging of flat idlers.

3-36 Clutching Drives

Belt clutches offer a simple and inexpensive way to disconnect a driveN shaft from a driveR shaft, if they are properly designed. Both flat belts and V belts are used on clutching drives. Clutching flat-belt drives are used extensively when many different drives are powered from a line shaft, and must operate independently. V-belt clutching drives are used frequently in power lawn and garden equipment and agricultural machinery.

Belt clutching mechanisms are of four types:

- Clutching sheave, in which the side or sides of a V sheave move outward to declutch, allowing a V belt to run on a free-wheeling center pulley, as in Fig. 3-18
- Clutching idler, in which either an inside or outside idler is moved out of the drive, as in Fig. 3-19

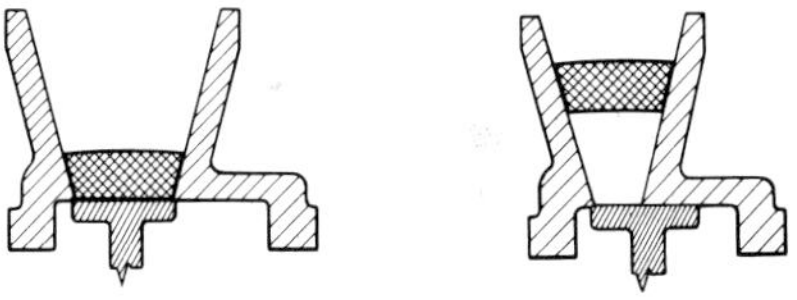

FIG. 3-18 Clutching sheave operation.

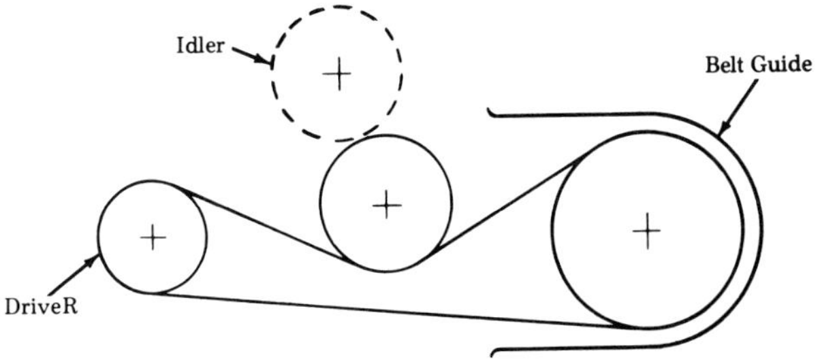

FIG. 3-19 Clutching idler operation.

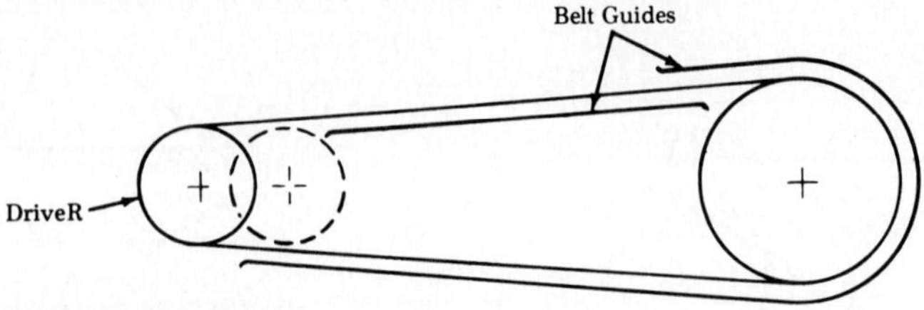

FIG. 3-20 Clutching centers operation.

- Clutching centers, in which the center distance is shortened, as in Fig. 3-20
- Idle pulley, in which the belt is shifted (as with a shifting fork) to a free-wheeling idle pulley, as in Fig. 3-21

On clutching idler and clutching center drives, since the driveR wheel continues to rotate and belt velocity is zero, the belt must be kept from grabbing the driveR wheel and causing the drive to operate sporadically. Under some conditions, belt damage can occur. This is often done by belt guides, as shown in Figs. 3-19 and 3-20, or by guide pins, as shown in Fig. 3-22.

Flanged pulleys are usually required when flat belts are clutched by the clutching idler or clutching center methods, because the belt slip during clutching can cause these belts to come off the pulley.

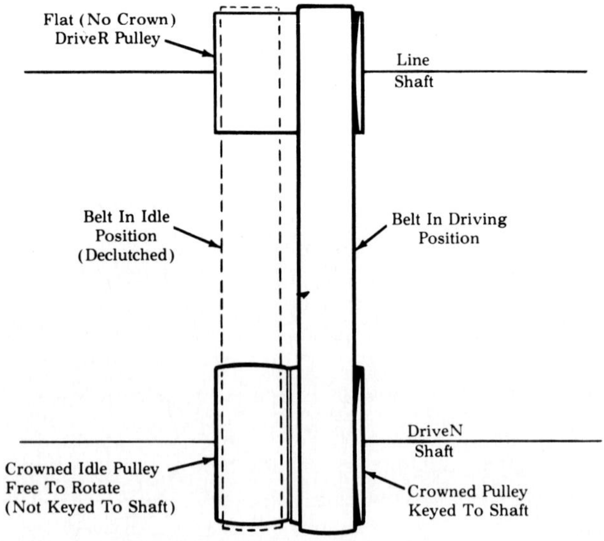

FIG. 3-21 Clutching idle pulley.

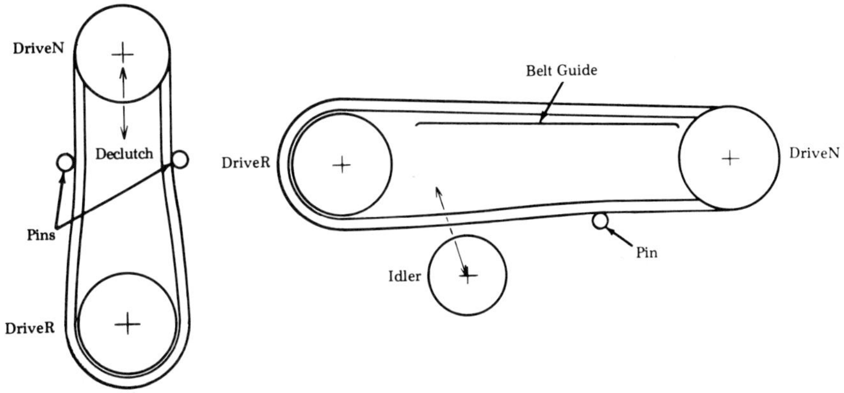

FIG. 3-22 Clutching belt guides.

3-37 Automatic Tensioning

As the name implies, tension is maintained at a proper value by some automatic device, thus minimizing the amount of attention the drive requires. This is sometimes called the constant tension method of tensioning, even though the tension is not always constant. Automatic tensioning can be provided by the various means listed here and discussed in more detail later in this section:

- Spring-loaded (or weighted) idlers
- Spring-loaded prime mover bases
- Gravity or reactive-torque motor bases
- Torque-sensitive sheaves
- Spring-loaded sheaves

Properly designed automatic tensioning systems can:

- Eliminate the need for manual takeup. This usually reduces maintenance costs, and is particularly important for drives which are relatively inaccessible for maintenance purposes.
- Allow the drive to operate at minimum tension. Locked-center drives are slightly overtensioned initially so that adequate tension is provided for a longer time as wear and belt elongation occur. Automatic tensioning can eliminate this initial overtensioning and provide the maximum in belt and bearing life.
- Provide the correct tension at all times. This eliminates the possibility of drive component damage from over- or undertensioning.

Some discretion must be used in applying most of the automatic tensioning devices when shock or pulsating loads occur. Spring-loaded idlers and the

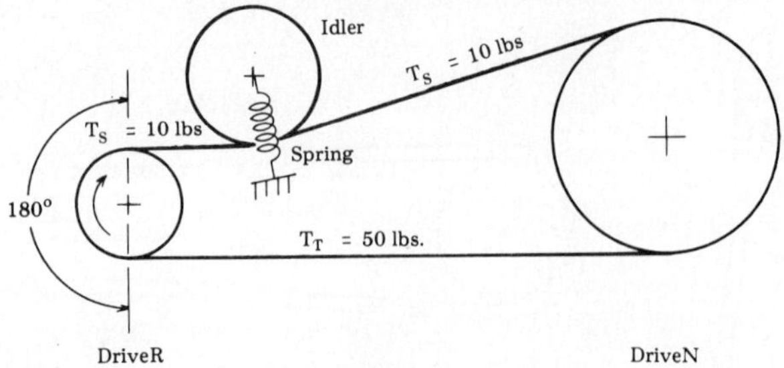

FIG. 3-23 Effect of spring-loaded idler placement.

various types of motor bases can encounter severe bounce and instability. This can result in belts being thrown off the drive or in belt breakage. Although this bounce can sometimes be dampened out with shock absorbers, the best solution may be a locked-center drive.

The following paragraphs deal with each of the various methods of automatic tensioning.

Spring-Loaded Idlers These are used quite extensively on agricultural machinery. They are especially desirable on drives which encounter wide load variations along with high peak loads. A locked-center drive must be tensioned to prevent slip at the peak load, and as the load drops off, the total tension remains nearly constant at the higher value. On the other hand, a spring-loaded idler allows the total tension to decrease as the load drops off, thus increasing belt and bearing life.

This effect is best illustrated by an example. The V-belt drive in Fig. 3-23 uses a spring-loaded idler on the slack side. Arc of contact on the driveR sheave is 180°, and so the design tension ratio T_T/T_S is 5. Assume that the horsepower load is such that an effective pull $T_T - T_S$ of 40 lb is required. Thus the tight-side tension T_T is 50 lb and the slack-side tension T_S is 10 lb, so that the total tension $T_T + T_S$ is 60 lb.

The idler spring must be sized, in this case, so that the idler force against the belt results in a 10-lb slack-side tension. This slack-side tension then remains at 10 lb regardless of whether the load increases or decreases, or even whether or not the drive is rotating.

If the load were taken completely off the driveN shaft, both T_T and T_S would be 10 lb and $T_T + T_S$ would be 20 lb. As load increases, T_T goes up while T_S stays at 10 lb. Assume that one-half the design horsepower is being transmitted, so that $T_T + T_S$ is 20 lb instead of the above 40 lb. Then, since T_S is a constant 10 lb, T_T would be 30 lb, and $T_T + T_S$ would be 40 lb, compared with 60 lb at full horsepower load.

It was mentioned in Sec. 3-35, "Idlers," that spring-loaded idlers should be located on the slack side of the drive. This is because the idler force can be much less in this position, resulting in less costly springs. It was also mentioned that they should not be used on reversing drives. This is because the drive will either slip (if the idler is located on the original slack side) or be severely overtensioned (if the idler is located on the original tight side). For example, if the drive of Fig. 3-23 were to have the driveR and driveN reversed, the spring-loaded idler would be in the tight span. But it imposes only 10 lb on that strand, and the tension on the new slack span would be even less. As a result, the drive would transmit only a fraction of the required horsepower without slipping. On the other hand, with the idler in this position, if the idler force were increased to 50 lb of span tension (so that $T_T - T_S$ would be 40 lb, and the load could be transmitted) and the drive reverted to the original driveR and driveN positions, T_T would be 90 lb ($T_T - T_S = 40$ lb and $T_S = 50$ lb) and $T_T + T_S$ would be 140 lb—a great deal overtensioned.

Weighted Idlers These idlers, sometimes called gravity idlers, are a special case of the spring-loaded idler where the weight of the assembly is the idler force; there is no spring.

Spring-Loaded Prime Mover Bases These are usually called spring-loaded motor bases because they are used more with electric motors than with other types of prime movers. The principle on which they operate is sketched in Fig. 3-24.

Here, the spring force must be such that the motor is pulled back against the drive with a force equal and opposite to the resultant belt pull. The spring must have some extra force, however, to overcome the friction in the base, and, if the mounting is not horizontal, the effect of motor weight must be taken into account.

Spring-loaded motor bases, unlike spring-loaded idlers, provide a nearly constant total drive tension $T_T + T_S$. Most of these bases are used in the industrial market; therefore, each individual drive is not always designed for the correct tension. Instead, standard bases which have an adjustment for the spring tension are available. A good way to set this spring is to install the drive and operate it, adjusting the spring tension until the drive does not slip under the highest normal load condition. Some spring force is lost as the belt seats in and elongates, and so occasional spring adjustments may be necessary.

Gravity Motor Bases These use the weight of the motor to tension the drive, as shown in Fig. 3-25. Standard gravity-type bases are also available. The base plate is usually slotted so that the drive tension can be adjusted by moving the motor closer to or further from the pivot point.

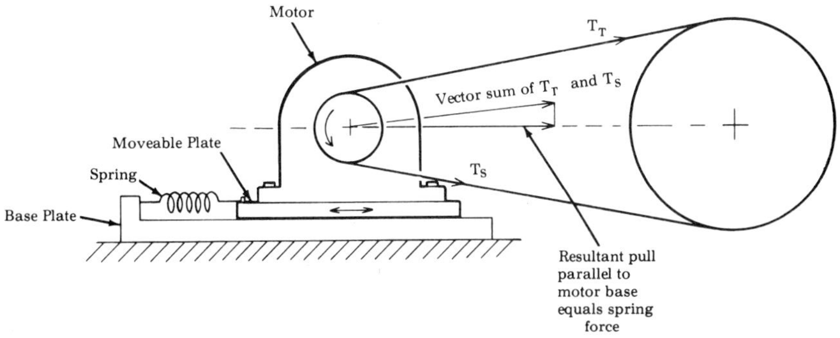

FIG. 3-24 Spring-loaded motor base.

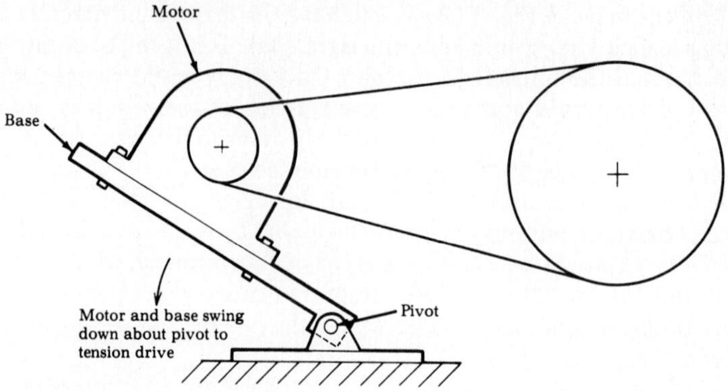

FIG. 3-25 Gravity motor base.

Reactive-Torque Motor Bases These use the reactive torque of the motor to help tension the drive. When a motor is delivering power to a sheave or pulley, its frame tries to rotate in the direction opposite to that of the sheave or pulley. Reactive torque is this tendency to rotate. See the illustration in Fig. 3-26.

By placing the motor in a cradle with a pivot point near the motor shaft, the motor can be allowed to move away from the driveN sheave, tensioning the drive. As torque increases, tension increases. See Fig. 3-27.

This reactive-torque base is the same as the gravity base illustrated in Fig. 3-25 except that the pivot point on the gravity base is quite far away from the motor. In that case, the reactive torque has little effect on drive tension.

The reactive-torque motor base must be installed so that the direction of rotation is correct. If the motor shaft in Fig. 3-27 were rotating in the opposite direction, the reactive torque would actually loosen the drive, rather than tightening it. Operating in the direction shown, both the reactive torque and the motor weight act to tension the drive.

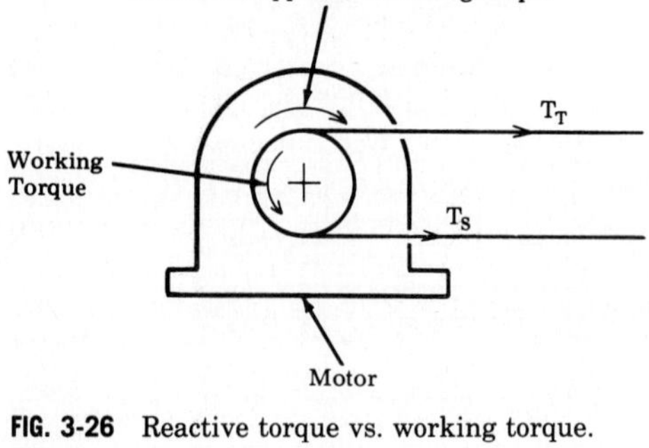

FIG. 3-26 Reactive torque vs. working torque.

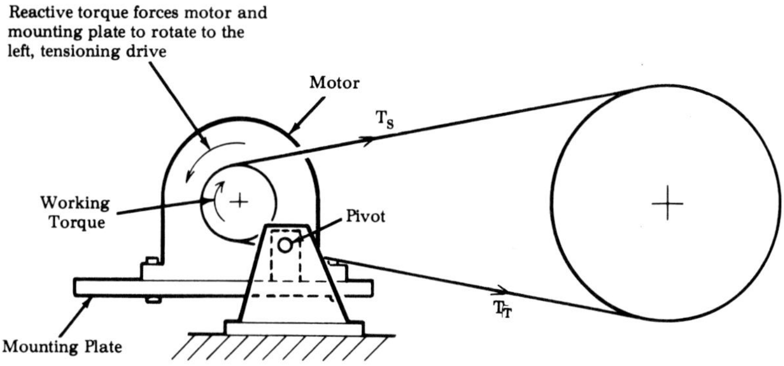

FIG. 3-27 Reactive-torque motor base.

Standard bases are usually available with slots so that the motor can be moved as needed to get the proper tension when the drive is installed. Although specification data usually show a given setting (or a fixed base) for a certain NEMA motor frame size, it is good practice to determine experimentally if the tension in the drive is correct after the drive is installed. This is done by imposing the highest load on the motor after the belts are run in, to see if the drive slips. The pivot point must be adjusted if the tension is not correct.

Torque-Sensitive Sheaves

These are made in separate halves, similar to variable-speed sheaves. In the past, threads or angled slots were used on one of the sheave halves so that it moved toward or away from the other half as torque demand increased or decreased. In more recent years, torque-sensitive variable-speed sheaves have included a combination of springs and cam surfaces which reacted similarly to the torque demands placed on them by the drive. See Fig. 3-28.

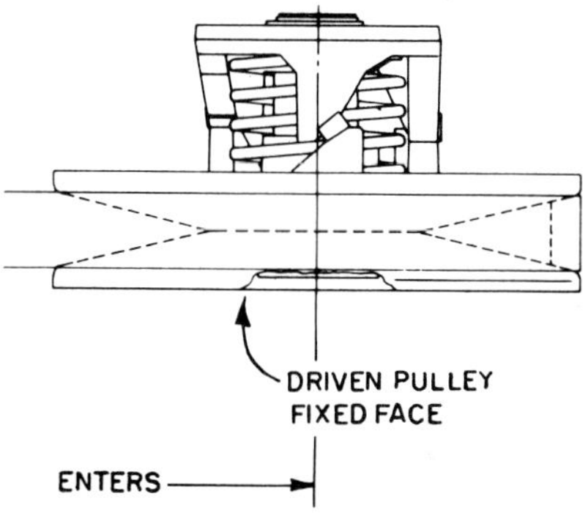

FIG. 3-28 Torque-sensitive sheave operation.

Like torque-reactive motor bases, torque-sensitive sheaves must be installed so that the direction of rotation is correct. If it is not, the sheaves will open up and the drive will not operate. If they are installed properly, such sheaves can provide a nearly constant tension ratio for the drive with various loads. This tension ratio depends on the angle of the inclined plane or threads.

The amount of total takeup that such sheaves can provide is limited by the total possible diameter change. This diameter change can be calculated from Eq. (3-17).

$$\Delta D = \frac{TW}{\tan \alpha/2} - 2(TH) \tag{3-17}$$

where ΔD = total possible change in effective outside diameter, in
TW = top width of groove when closed, in
α = groove angle, degrees
TH = belt thickness, in

If longer belts (which require greater takeup) or belts of small cross section (which have small ΔD) are used, provision for shaft movement for additional takeup must be made.

When torque-sensitive sheaves are used as the driveR on drives with high inertia loads, problems of sporadic speed variation can occur as the drive comes up to speed. The sheaves keep opening and closing slightly as a result of overrun of the driveN unit.

Torque-sensitive sheaves in the light-duty or fractional horsepower range are available as standard items from various manufacturers. They are occasionally built as a special item for other types of V-belt application.

Spring-Loaded Sheaves These are commonly used in variable-speed drives. They provide automatic tensioning even though their main function in those drives is to provide variable speed. These sheaves will be discussed later in this chapter in Sec. 3-40, "Speed Variation by V Belts."

3-38 Sheaves in More Than One Plane

The usual drive design has the belts and sheaves essentially in one plane, as shown in Fig. 3-29. But there are many belt drives in which the belts and sheaves lie in more than one plane—a multiplane drive. One example is shown in Fig. 3.30.

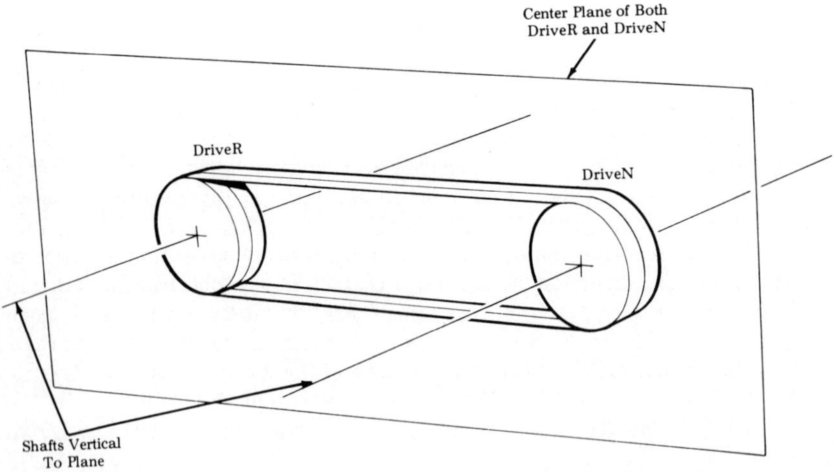

FIG. 3-29 One-plane drive.

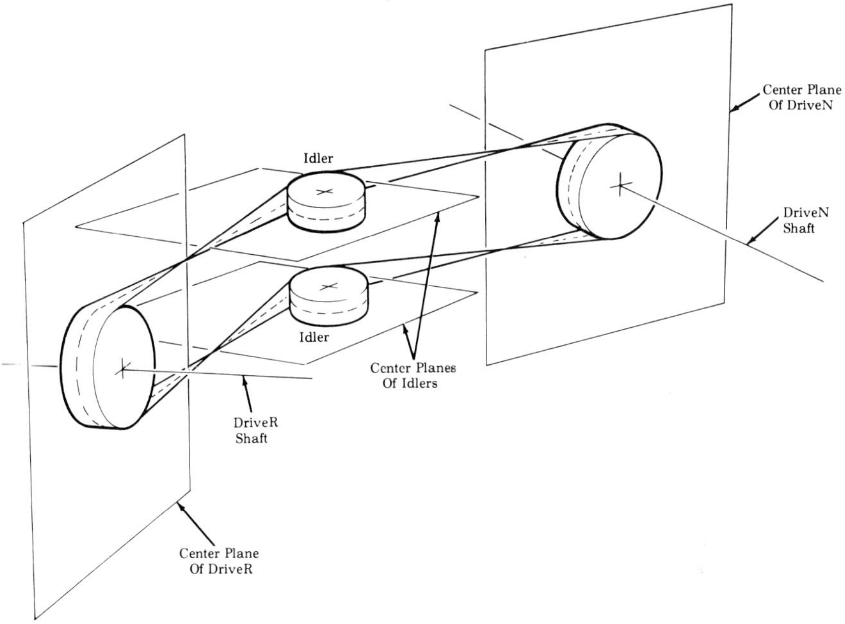

FIG. 3-30 Multiplane drive.

This drive's functions include changing the direction of power transmission, as well as the normal functions of a belt drive. Such drives are used quite frequently on agricultural equipment.

There are two factors common to multiplane drives which must be considered in their design.

- Twist of the belt (all multiplane drives)
- Kink of the belt (most multiplane drives)

Twist of the belt always reduces belt life because the edge of the belt must elongate slightly, as shown in Fig. 3-31. This elongation causes a higher tension in that part of the belt, reducing life.

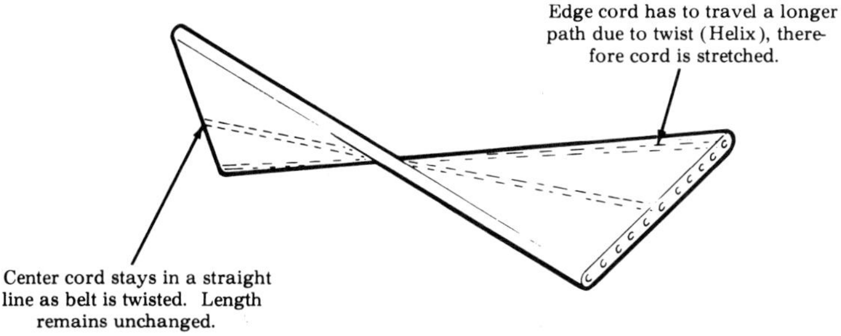

FIG. 3-31 Belt twist.

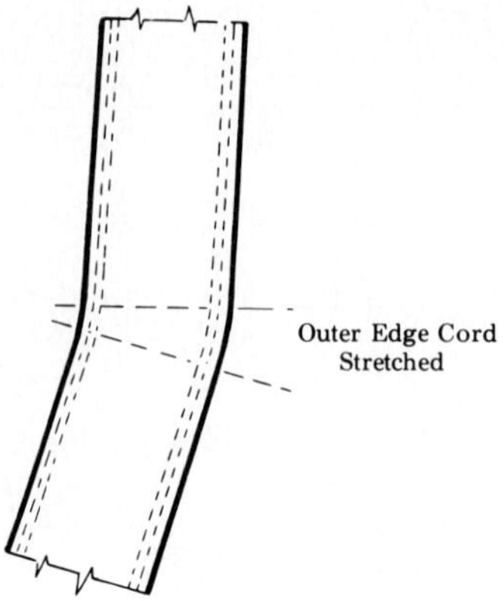

FIG. 3-32 Belt kink.

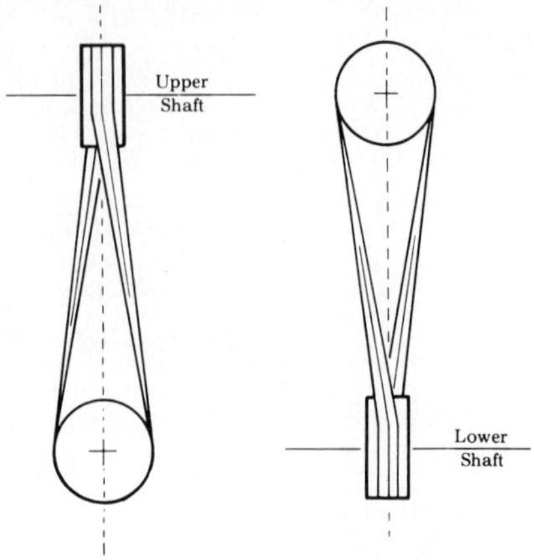

FIG. 3-33 Quarter-turn drive.

Also, twist can cause V-belt turnover if it is too severe. Most manufacturers' design manuals give recommendations for minimum tangent lengths for a given amount of twist. (The tangent length is the length of the belt span between the points of contact on the adjacent sheaves.) These minimum tangent lengths are selected to keep the amount of twist down to a minimum so that it does not affect the stability of the V belt in the groove.

When a belt enters a sheave misaligned, kinking results, and the edge of the belt must elongate. See Fig. 3-32. This elongation creates additional tension in that part of the belt, reducing life.

Kink of the belt reduces belt life even more than twist. Belts do not bend sideways very easily. A belt with a low tensile modulus of elasticity will undergo less increase in tension, and experience a smaller reduction in life, than one with a high modulus. High-modulus steel cable belts, for example, can withstand very little twist or kink without significant life reduction.

Misalignment also causes increased wear of belt sidewalls from rubbing against the side of the sheaves during entry and exit. This can cause excessive tension loss. In severe cases of misalignment, belt stability in the groove can become a problem, causing belt turnover or belts coming off the drive. For this reason, deep-groove sheaves are commonly used for multiplane drives. Specially placed idlers to reduce inherent misalignment are also a possibility.

Three of the most popular multiplane drives are quarter-turn (Fig. 3-33), crossed belt (Fig. 3-34), and mule drives (Fig. 3-35).

All multiplane drives should be referred to the belt supplier in order to ensure optimum design.

3-39 Vertical-Shaft Drives

Vertical-shaft drives are used to a significant extent—a common example is on cooling towers. These drives are designed the same as ordinary drives, but with some special considerations.

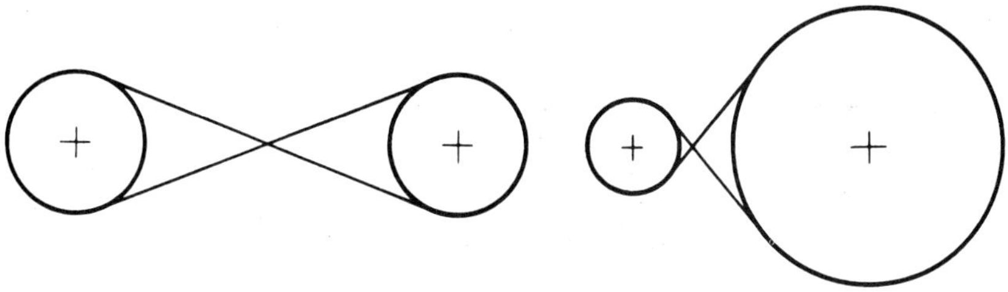

Preferred
(a)

More Wear
(b)

FIG. 3-34 Crossed drives.

Flat belts are seldom used because gravity tends to cause the belt to ride down off the pulleys. This is especially true on drives with long center distances or where tension maintenance is casual. Flanged pulleys help, but the belt continually riding against the flanges can lead to reduced service life. Synchronous belt drives can be used on vertical shafts, but flanged pulleys must be used.

V belts are acceptable, but problems can be reduced by avoiding long center distances and using deep-groove sheaves. Preventing belts from coming off the drive requires either more tension maintenance or automatic tensioning devices.

3-40 Speed Variation by V Belts

Many driveN machines require a change in speed. Some speed variation requirements are minimal, such as a ventilating fan that is adjusted only once after installation to allow for minimal changes of the airflow requirements. Other machines, such as drill presses, may require changes in speed for different kinds of work. Still others require speed changes while operating.

DriveN speed changes can be accomplished by V-belt drives. Regardless of the type of V belt used, the principle is always: The diameter at which the V belt operates on one or more sheaves will vary. This is done by varying the width of the sheave groove. See Fig. 3-36.

The diameter range over which a V belt operates in one or more sheaves gives a speed variation to the driveN shaft. Speed variation can be best expressed as a range, such as variation from 1750 to 875 r/min of the driveN shaft. Most often, however, it is expressed as a ratio, with the term "speed ratio" applied to it. For example, the variation range given can be expressed as 2.0 speed ratio. This can be confusing because "speed ratio" is normally used to indicate a relationship of driveR and driveN shaft speeds, but in this case the term refers to a relationship between the highest and lowest speeds of the driveN shaft, or in other words, a range of driveN speeds. Care must be taken to express the variation or range being considered properly—speed variation or speed range is better than speed ratio when the variation of the driveN shaft is meant.

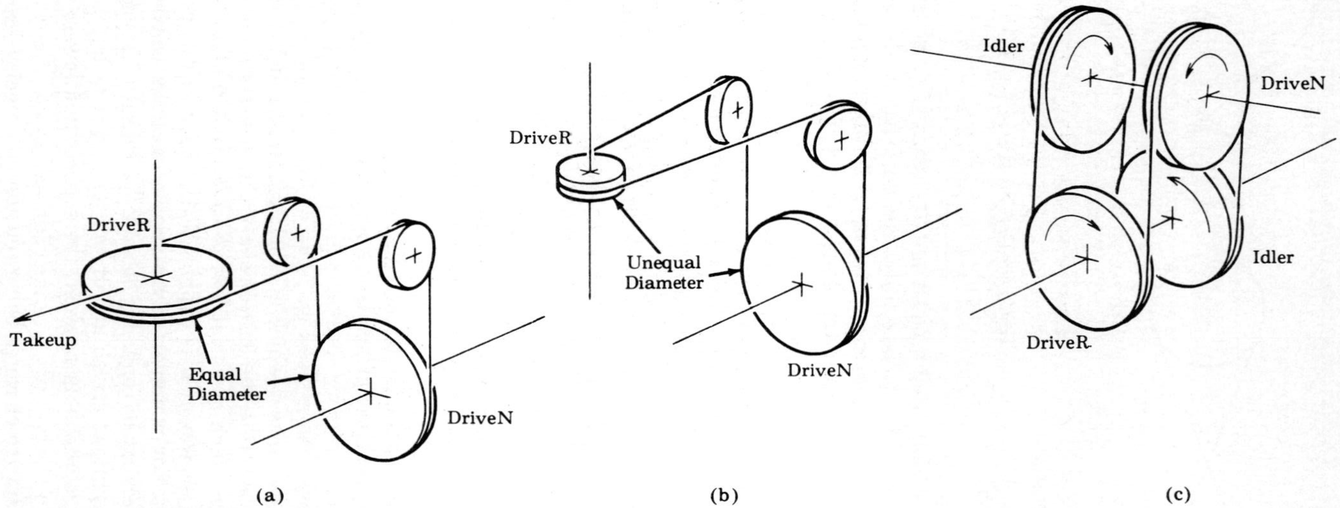

FIG. 3-35 Various types of mule drives.

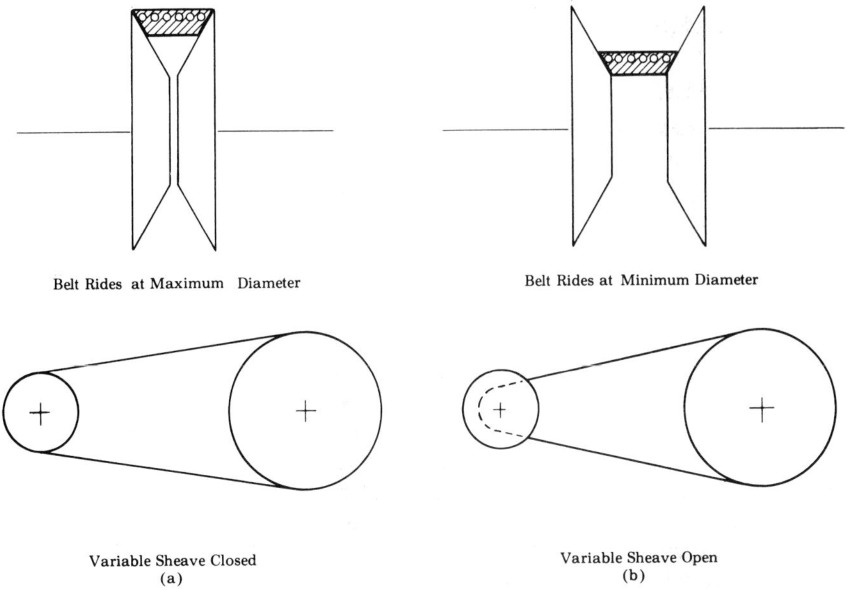

FIG. 3-36 Depiction of closed and open variable sheave.

Various industries use many types of V-belt and sheave combinations for driveN speed variation. Those most commonly referred to include (1) variable-pitch drives, (2) variable-speed drives, and (3) adjustable-speed drives.

Variable-Pitch Drives Standard lines of variable-pitch sheaves are used with both heavy-duty industrial belts (A, B, C, D and 5V, 8V) and light-duty belts (3L, 4L, 5L). The light-duty sheaves are usually single groove; the heavy-duty are usually multiple groove. These variable-pitch drives are used where required speed variation is not great.

In designing variable-pitch drives, either the driveR or the driveN can be the variable sheave—or both, if more speed variation is required. One variable-pitch sheave gives about 20 to 40 percent variation, depending on cross section and diameter. Two variable-pitch sheaves give about 40 to 100 percent.

When only one variable-pitch sheave is being used, the drive requires (with multiple-belt drives) a companion sheave which yields appropriate spacing between the grooves. This is required in order to minimize belt misalignment. There are special companion sheaves available, and under certain application conditions standard sheaves can be used. A belt manufacturer should be contacted.

Variable-Speed Drives For drives that require more speed variation than can be obtained with conventional (A, B, C, D or 5V, 8V, or 3L, 4L, 5L) belts, standard-line variable-speed drives are available. These drives use special wide, thin belts. See RMA Standard IP-25.

Package units of standard-line variable-speed belts and sheaves, combined with the motor and an output gear box (if specified), are available in horsepower ranges from about ½ up through 100 hp. These will be either single- or dual-belt drives, depending on horsepower and supplier.

The speed range of variable-speed drives can be much greater than that of variable-pitch drives. Speed ranges up to 10 to 1 can be obtained on smaller units. Spring-loaded or hand-adjustable motor sheaves are available for the standard variable-speed belts, in packages containing the motor, the sheaves, the belt, and sometimes an adjustable motor base.

Adjustable-Speed Drives Agricultural machinery uses variable-speed drives of both the variable-pitch type (HA, HB, HC, etc.) and the wider, thinner belt type. The latter are called agricultural adjustable-speed drives. They use special belts with top widths from 1 to 2½ in. Propulsion drives for combines (traction drives) and cylinder drives are examples of applications in which adjustable-speed belts are used.

3-41 V-Flat Drives

A V-flat drive is one which uses a V sheave on one shaft and a flat pulley on the other. Usually the sheave is the small wheel and the pulley is the large wheel. When a large number of older flat-belt drives were being converted to V belts, the V-flat drive became popular. It was economical because the large pulley did not have to be discarded.

A flat-belt drive requires more tension (to transmit the same horsepower at the same speed) than a V-belt drive. Therefore, the design of a drive with one sheave and one flat pulley, and a V belt, involves the use of a different factor *G*. V-flat drives are discussed in most manufacturers' manuals.

Special consideration must also be given to crown on the flat wheel (no crown is preferred), pitch diameters, pulley face width, and idlers. These are also discussed in manufacturers' manuals.

3-42 Mechanical Efficiency of V-Belt Drives

When proper V-belt drive design procedures are followed, the mechanical efficiencies are sufficiently high that they are not normally a matter of importance. However, for those drives where system efficiency is of maximum importance, the following sources of belt power losses should be considered:

- Losses resulting from the longitudinal (circumferential) movement of the belt in the sheave groove. This is known as creep, and it results from the difference in belt tension from one side of a sheave to the other, causing a small loss of driveN speed.
- Losses resulting from the vertical (radial) movement of the belt in the sheave groove. This is the seating of the belt in the groove, which is magnified by misalignment.
- Losses resulting from bending of the belt over the sheaves, because of hysteresis losses in the belt materials.
- Losses resulting from windage and miscellaneous effects.

The number of factors operating within each of these major categories is so large that it is difficult to analyze the mechanical efficiency of V-belt drives quantitatively. Many of the factors have more than one effect—sometimes in opposite directions. Some of them change throughout the life of any one drive.

However, some generalizations can be made with respect to the values of a few of the major drive design factors which tend to reduce efficiency. This should not be construed to mean that V-belt drive efficiencies are low. On the contrary, mechanical efficiencies for properly designed drives have been measured consistently over the years at 95 to 98 percent.

Power Transmitted Greatly overloaded or underloaded drives can exhibit poor efficiency, but tests at loads ranging from 50 to 225 percent of rated belt capacities show very little decrease from the 95 to 98 percent efficiency range.

Belt Speed Higher belt speeds reduce required belt tension for given power transmission requirements, thus tending to reduce belt movement in the grooves. However, higher belt speeds also increase hysteresis losses, and tests show that efficiency decreases slightly with increased belt speed.

Sheave Diameter Diameters within the recommended published ranges do not change efficiency significantly. When belts are run over pulleys or sheaves that are well below the established minimum for that particular belt, efficiency can drop sharply.

Tension Ratio When belts are tightened significantly beyond what is needed to transmit the required horsepower, efficiency can decrease. For V-belt drives, decreased efficiency can result from tightening a belt so that its tension ratio is in the range of 2½ to 1 or below (this is within the range of tensions required for flat-belt power transmission). But normal efficiencies are exhibited from this point upward until belt slippage starts.

Arc of Contact (Speed Ratio, Center Distance) Drive geometry affects efficiency only to a limited extent. When speed ratio and center distance combine to form a very low arc of contact on the smaller sheave (70°, for example), somewhat decreased efficiency can result from the higher tensions required to transmit the power (see Table 3-2).

For additional information on the mechanical efficiency of power transmissions for belt drives, refer to RMA Bulletin IP-3-13.

4

SHAFTS AND COUPLINGS

ROBERT O. PARMLEY, P.E.
President/Morgan & Parmley, Ltd./Ladysmith, Wisconsin

SHAFTS

A rotating bar, usually cylindrical in shape, which transmits power is called a shaft. Power is delivered to the shaft through the action of an outside tangential force, resulting in a torsional action set up in the shaft. The resultant torque allows the power to be distributed to other machines or to various components connected to the shaft.

4-1 Usage and Classification

Shafts and shafting may be classified according to their general usage. The following categories are presented here for discussion only and are basic in nature.

Engine Shafts An engine shaft may be described as a shaft directly connected to the power delivery of a motor.

Generator Shafts Generator shafts, along with engine shafts and turbine shafts, are called prime movers. There is a wide range of shaft diameters, depending on power transmission required.

Turbine Shafts Also prime movers, turbine shafts have a tremendous range of diameter size.

Machine Shafts General category of shafts. Variation in sizes of stock diameters ranges from $\frac{1}{2}$ to $2\frac{1}{2}$ in (increments of $\frac{1}{16}$ in), $2\frac{1}{2}$ to 4 in (increments of $\frac{1}{8}$ in), 4 to 6 in (increments of $\frac{1}{4}$ in).

Line Shafts Line shafting is a term employed to describe long and continuous "lines of shafting," generally seen in factories, paper or steel mills, and shops where power distribution over an extended distance is required. Stock lengths of line shafting generally are 12 ft, 20 ft, and 24 ft.

Jackshafts Jackshafts are used where a shaft is connected directly to a source of power from which other shafts are driven.

Countershafts Countershafts are placed between a line shaft and a machine. The countershaft receives power from a line shaft and transmits it to the drive shaft.

4-2 Torsional Stress

A shaft is said to be under torsional stress when one end is securely held and a twisting force acts at the opposite end. Figure 4-1 illustrates this action. Note that the only deformation in the shaft is the rotation of the cross sections with respect to each other, as shown by angle ϕ.

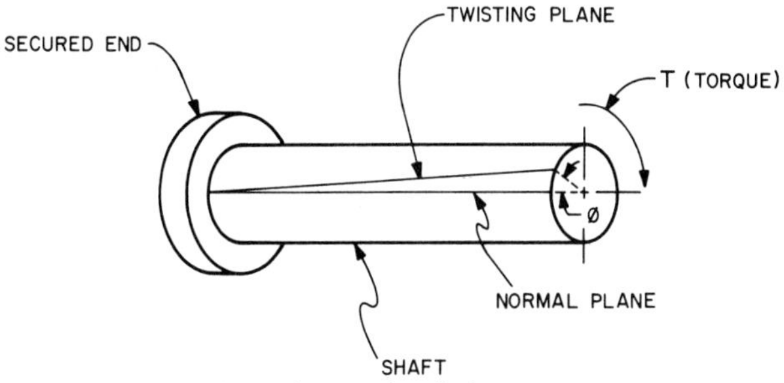

FIG. 4-1 Shaft subjected to torsional stress.

Shafts which are subjected to torsional force only, or those with a minimal bending moment that can be disregarded, may use the following formula to obtain torque in inch-pounds, where horsepower P and rotational speed N in revolutions per minute are known.

$$T = \frac{12 \times 33{,}000\,P}{2\pi N} \tag{4-1}$$

4-3 Twisting Moment

Twisting moment T is equal to the product of the resultant P_r of the twisting forces multiplied by its distance from the axis R. See Fig. 4-2.

$$T = P_r \times R \tag{4-2}$$

4-4 Resisting Moment

Resisting moment T_r equals the sum of the moments of the unit shearing stresses acting along the cross section of the shaft. This moment is the force which "resists" the twisting force exerted to rotate the shaft.

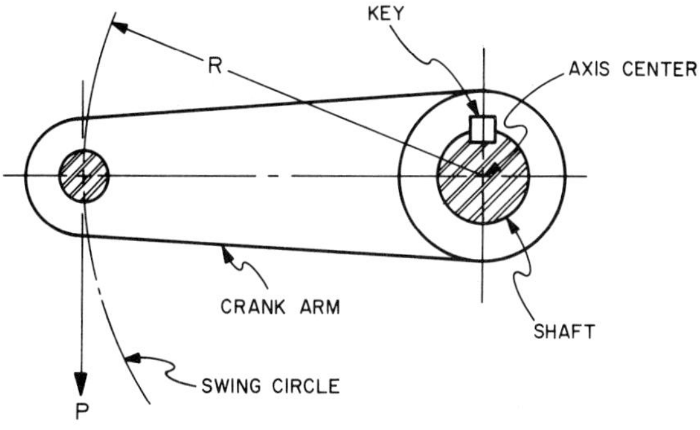

FIG. 4-2 Typical crank arm forces.

4-5 Torsion Formula for Round Shafts

Torsion formulas apply to solid or hollow circular shafts, and only when the applied force is perpendicular to the shaft's axis, if the shearing proportional limit (of the material) is not exceeded.

Conditions of equilibrium, therefore, require the "twisting" moment to be opposed by an equal "resisting" moment. The following formulas may be used to solve the allowable unit shearing stress τ if twisting moment T, diameter of solid shaft D, outside diameter of hollow shaft d, and inside diameter of hollow shaft d_1 are known.

Solid round shafts:

$$\tau = \frac{16T}{\pi D^3} \tag{4-3}$$

Hollow round shafts:

$$\tau = \frac{16Td}{\pi(d^4 - d_1^4)} \tag{4-4}$$

4-6 Shear Stress

In terms of horsepower, for shafts used in the transmission of power, shearing stress may be calculated as follows, where P = horsepower to be transmitted, N = rotational speed in revolutions per minute, and the shaft diameters are those described previously. Maximum unit shearing stress τ is in pounds per square inch.

Solid round shafts:

$$\tau = \frac{321{,}000P}{ND^3} \tag{4-5}$$

Hollow round shafts:

$$\tau = \frac{321{,}000Pd}{N(d^4 - d_1^4)} \tag{4-6}$$

The foregoing formulas do not consider any loads other than torsion. Weight of shaft and pulleys or belt tensions are not included.

4-7 Critical Speeds of Shafts

Shafts in rotation become very unstable at certain speeds, and damaging vibrations are likely to occur. The revolution at which this mechanical phenomenon takes place is called the "critical speed."

Vibration problems may occur at a "fundamental" critical speed. The following formula is used for finding this speed for a shaft on two supports, where W_1, W_2, etc. = weights of rotating components; y_1, y_2, etc. = respective static deflection of the weights; g = gravitational constant, 386 in/s^2.

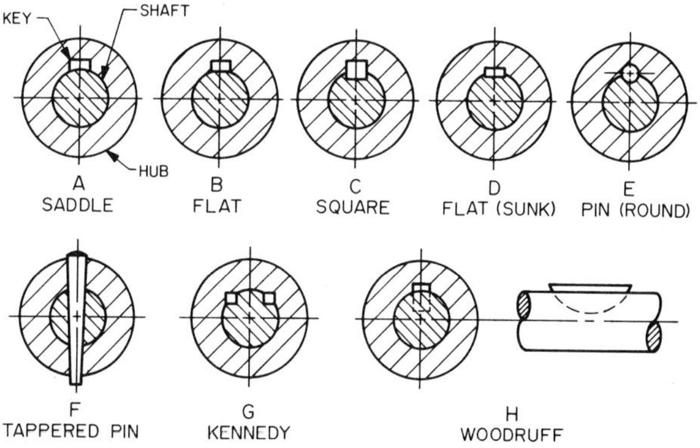

FIG. 4-3 Types of keys.

$$f = \frac{1}{2\pi}\sqrt{\frac{g(W_1y_1 + W_2y_2 + \cdots}{W_1y_1^2 + W_2y_2^2 + \cdots}} \quad \text{cycles/s} \tag{4-7}$$

A thorough discussion of this phenomenon is beyond the scope of this book. Readers should consult the many volumes devoted to vibration theory for an in-depth technical presentation.

4-8 Fasteners for Torque Transmission

Keys Basically keys are wedge-like steel fasteners that are positioned in a gear, sprocket, pulley, or coupling and then secured to a shaft for the transmission of power. The key is the most effective and therefore the most common fastener used for this purpose.

Figure 4-3 illustrates several standard key designs, including round and tapered pins. The saddle key (*a*) is hollowed to fit the shaft, without a keyway cut into the shaft. The flat key (*b*) is positioned on a planed surface of the shaft to give more frictional resistance. Both of these keys can transmit light loads. Square (*c*) and flat-sunk (*d*) keys fit in mating keyways, half in the shaft and half into the hub. This positive holding power provides maximum torque transfer. Round (*e*) and tapered (*f*) pins are also an excellent method of keying hubs to shafts. Kennedy (*g*) and Woodruff (*h*) keys are widely used. Figure 4-4 pictures feather keys, which are used to prevent hubs from rotating on a shaft, but will permit the component part to move along the shaft's axis. Figure 4-4*a* shows a key which is relatively long for axial movement and is secured in position on the shaft with two flat fillister-head matching screws. Figure 4-4*b* is held to the hub and moves freely with the hub along the shaft's keyseat.

A more in-depth presentation of keys will be found in Sec. 12, "Locking Components."

Set Screws Set screws may be used for light applications. A headless screw with a hexagon socket head and a conical tip should be used. Figure 4-5 illustrates both a "good" design and a "bad" design. The set screw must be threaded into the hub and tightened on the shaft to provide a positive anchor.

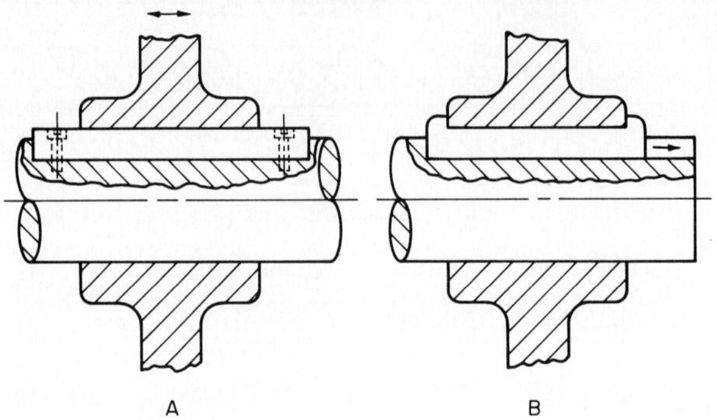

FIG. 4-4 Feather keys.

Pins Round and taper pins were briefly discussed previously, but mention should be made of the groove, spring, spiral, and shear pins. The groove pin has one or more longitudinal grooves, known as flutes, over a portion of its length. The farther you insert this pin, the tighter it becomes. The spring or slotted tubular pin is a hollow tube with a full-length slot and tapered ends. This slot allows the pin's diameter to be reduced somewhat when the pin is inserted, thus providing easy adaptation to irregular holes. Spirally coiled pins are very similar in application to spring pins. They are fabricated from a sheet of metal wrapped twice around itself, forming a spiral effect. Shear pins, of course, are used as a weak link. They are designed to fail when a predetermined force is encountered.

Pins are discussed in Sec. 12, "Locking Components," along with keys. Section 14, "Innovative Design," has some sample applications for pins.

4-9 Splines

Spline shafts are often used instead of keys to transmit power from hub to shaft or from shaft to hub. Splines may be either square or involute.

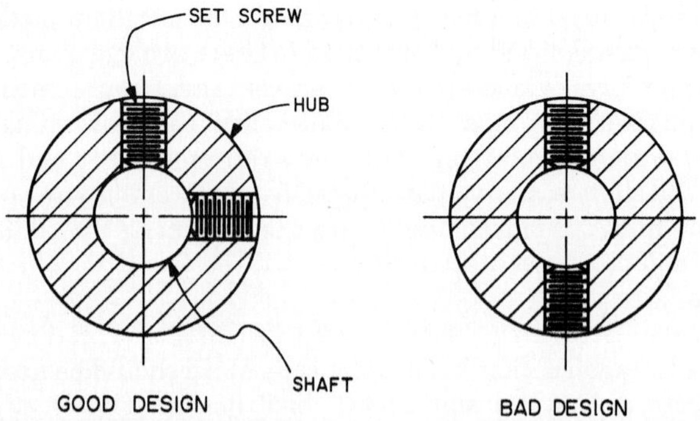

FIG. 4-5 Use of set screws.

One may think of splines as a series of teeth, cut longitudinally into the external circumference of a shaft, that match or mate with a similar series of keyways cut into the hub of a mounted component. Splines are extremely effective when a "sliding" connection is necessary, such as for a PTO (power take-off) on agricultural equipment.

Square or parallel-side splines are employed as multispline shaft fittings in series of 4, 6, 10, or 16. Refer to Sec. 12-17 for a typical square design pattern.

Splines are especially successful when heavy torque loads and/or reversing loads are transmitted. Torque capacity (in inch-pounds) of spline fittings may be calculated by the following formula:

$$T = 1000NrhL \qquad \text{in}\cdot\text{lb} \tag{4-8}$$

where N = number of splines
r = mean radial distance from center of shaft/hub to center of spline
h = depth of spline
L = length of spline bearing surface

This gives torque based on spline side pressure of 1000 lb/in^2. Involute splines are similar in design to gear teeth, but modified from the standard profile. This involute contour provides greater strength and is easier to fabricate. Figure 4-6 shows five typical involute spline shapes.

Section 14, "Innovative Design," illustrates several spline connection designs which will be of interest to the reader.

SHAFT COUPLINGS

In machine design, it often becomes necessary to fasten or join the ends of two shafts axially so that they will act as a single unit to transmit power. When this parameter is required, shaft couplings are called into use. Shaft couplings

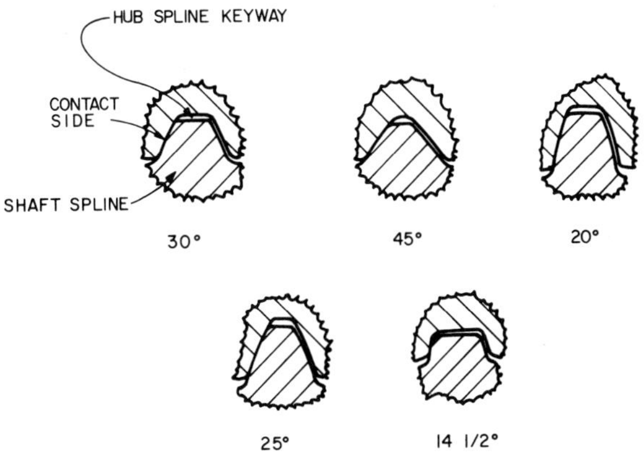

FIG. 4-6 Involute spline shapes.

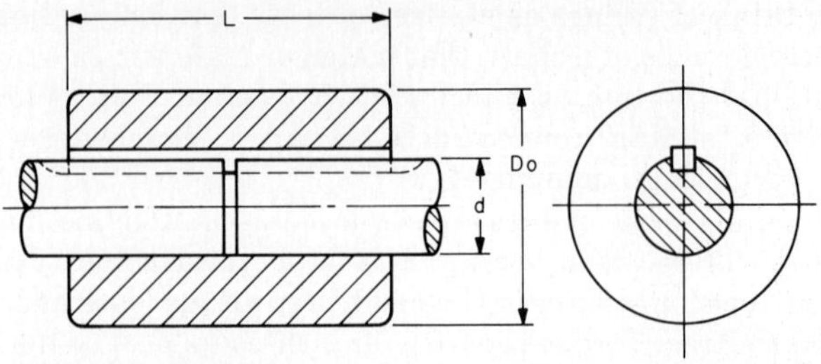

FIG. 4-7 Sleeve coupling.

are grouped into two general classifications: rigid (or solid) and flexible. A rigid coupling will not provide for shaft misalignment or reduce vibration or shock from one shaft to the other. However, flexible shaft couplings provide connection of misaligned shafts and can reduce shock and/or vibration to a degree.

Section 14, "Innovative Design," pictorially presents several unusual designs of shaft connections.

4-10 Sleeve Coupling

Sleeve coupling, as illustrated in Fig. 4-7, consists of a simple hollow cylinder which is slipped over the ends of two shafts fastened into place with a key positioned into mating keyways. This is the simplest rigid coupling in use today. Note that there are no projecting parts, so that it is very safe. Additionally, this coupling is inexpensive to fabricate.

Figure 4-8 pictures two styles of sleeve couplings using standard set screws to anchor the coupling to each shaft end. One design is used for shafts of equal diameters. The other design connects two shafts of unequal diameters.

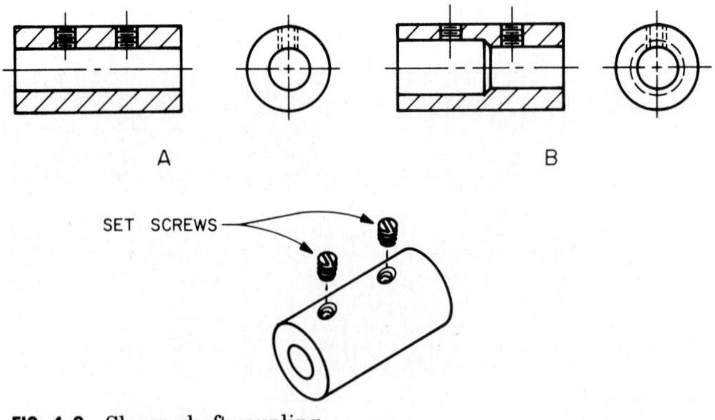

FIG. 4-8 Sleeve shaft coupling.

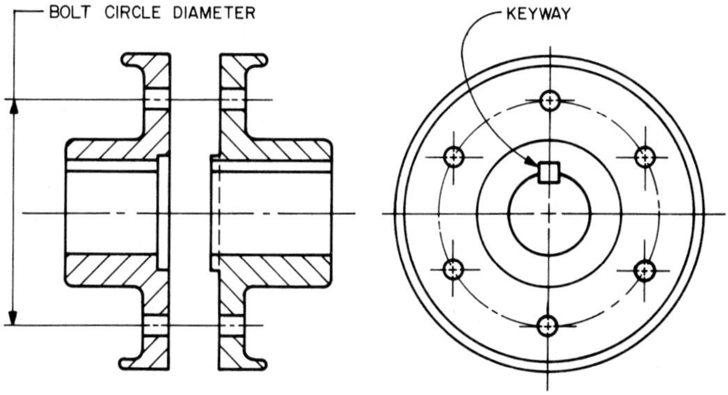

FIG. 4-9 Solid coupling.

4-11 Solid Coupling

The solid coupling shown in Fig. 4-9 is a tough, inexpensive, and positive shaft connector. When heavy torque transmission is required, a rigid coupling of this design is an excellent selection.

4-12 Clamp or Compression Coupling

The rigid coupling shown in Fig. 4-10 has evolved from the basic sleeve coupling. This clamp or compression coupling simply splits into halves, which have recesses for through bolts that secure or clamp the mating parts together, producing a compression effect on the two connecting shafts. This coupling may be used for transmission of large torques because of its positive grip from frictional contact.

4-13 Flange Coupling

Flange couplings are rigid shaft connectors, also known as solid couplings. Figure 4-11 illustrates a typical design. This rigid coupling consists of two components, which are connected to the two shafts with keys. The hub halves

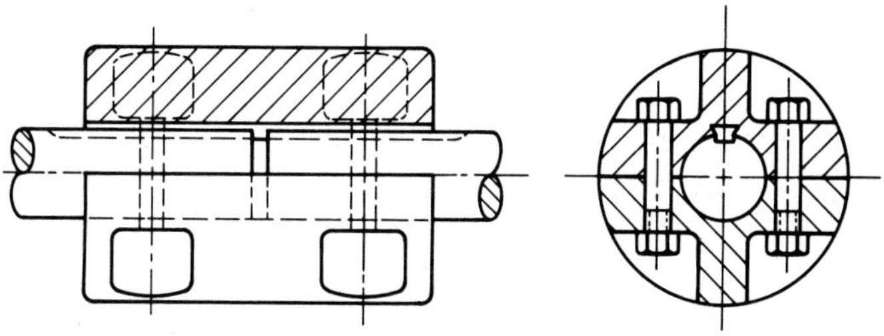

FIG. 4-10 Clamp or compression coupling.

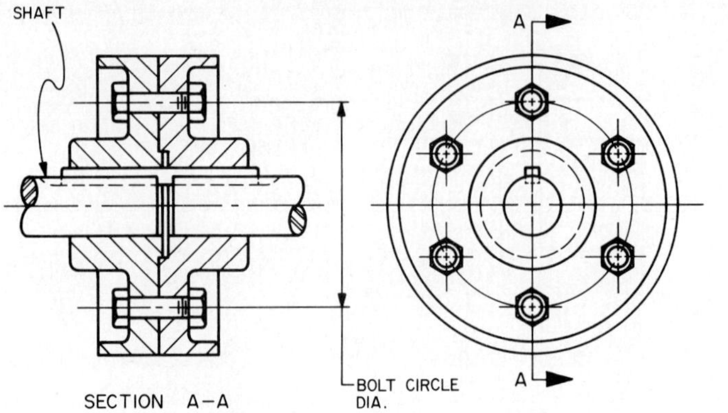

FIG. 4-11 Flange coupling.

are fastened together with a series of bolts arranged in an even pattern concentrically about the center of the shaft. A flange on the outside circumference of the hub provides a safety guard for the bolt heads and nuts, while adding strength to the total assembly.

4-14 Flexible Coupling

Flexible couplings connect two shafts which have some nonalignment between them. The couplings also absorb some shock and vibration which may be transmitted from one shaft to the other.

There are a wide variety of flexible-coupling designs. Figure 4-12 pictures a two-part cast-iron coupling which is fastened onto the shafts by keys and set

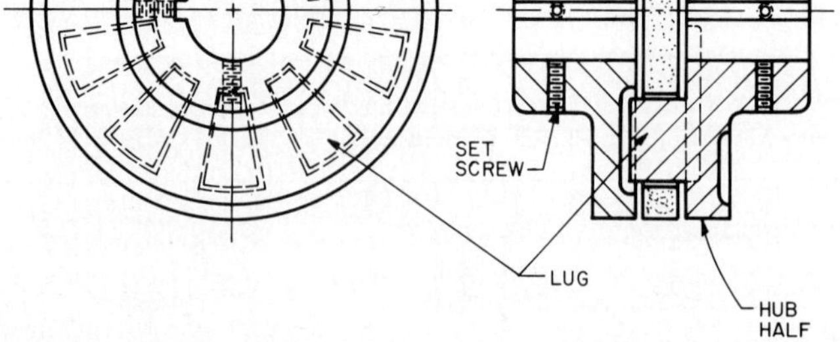

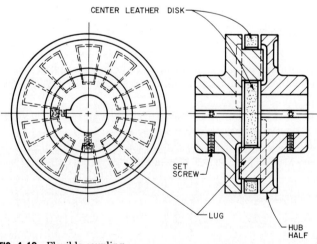

FIG. 4-12 Flexible coupling.

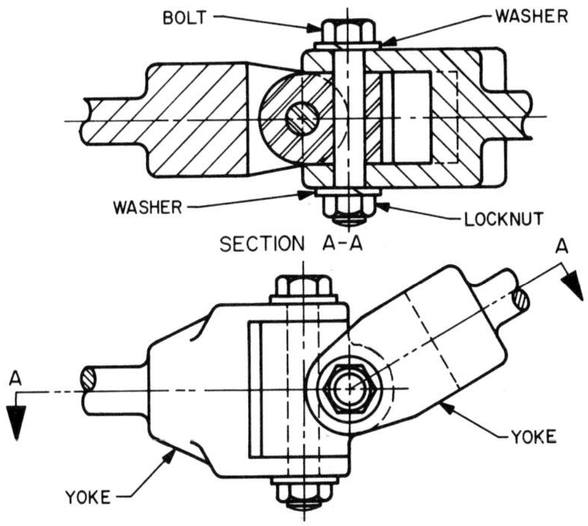

FIG. 4-13 Universal coupling.

screws. The halves have lugs, which are cast an an integral part of each hub half. The lugs fit into entry pockets in a disk made of leather plies which are stitched and cemented together. The center leather laminated disk provides flexibility in all directions. Rotation speed, either slow or fast, will not affect the efficiency of the coupling.

4-15 Universal Coupling

If two shafts are not lined up but have intersecting centerlines or axes, a positive connection can be made with a universal coupling. Figure 4-13 details a typical universal coupling.

Note that the bolts are at right angles to each other. This makes possible the peculiar action of the universal coupling. Either yoke can be rotated about the axis of each bolt so that adjustment to the angle between connected shafts can be made. A good rule of thumb is not to exceed 15° of adjustment per coupling.

4-16 Multijawed Coupling

This rigid-type shaft coupling is a special design. The coupling consists of two halves, each of which has a series of mating teeth which lock together, forming a positive jawlike connection. Set screws secure the hubs onto the respective shafts. This style of coupling is strong and yet easily dismantled. See Fig. 4-14.

4-17 Spider-Type Coupling

The spider-type or Oldham coupling is a form of flexible coupling that was designed for connection of two shafts which are parallel but not in line. The two end hubs, which are connected to the two respective shafts, have grooved

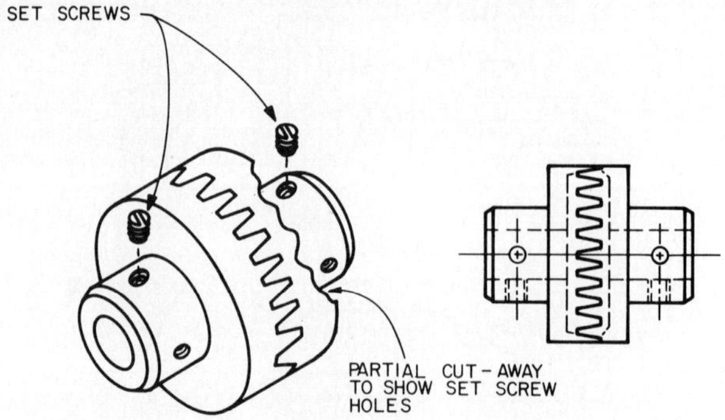

FIG. 4-14 Multijawed coupling.

faces which mate with the two tongues of the center disk. This configuration and slot adjustment allow for misalignment of shafts. Figure 4-15 shows an assembled spider-type coupling.

4-18 Bellows Coupling

Two styles of bellows couplings are illustrated in Fig. 4-16. These couplings are used in applications involving large amounts of shaft misalignment, usually combined with low radial loads. Maximum permissible angular misalignment varies between 5° and 10°, depending on manufacturer's recommendation. Follow manufacturer's guidelines for maximum allowable torque. Generally, these couplings are used in small, light-duty equipment.

4-19 Helical Coupling

These couplings, also, are employed to minimize the forces acting on shafts and bearings as a result of angular and/or parallel misalignment.

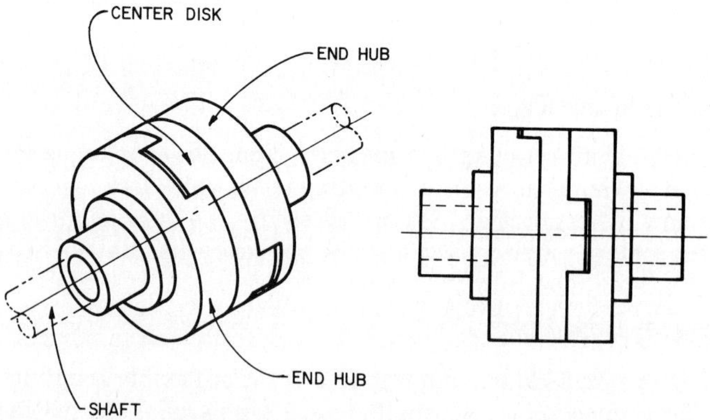

FIG. 4-15 Spider-type coupling.

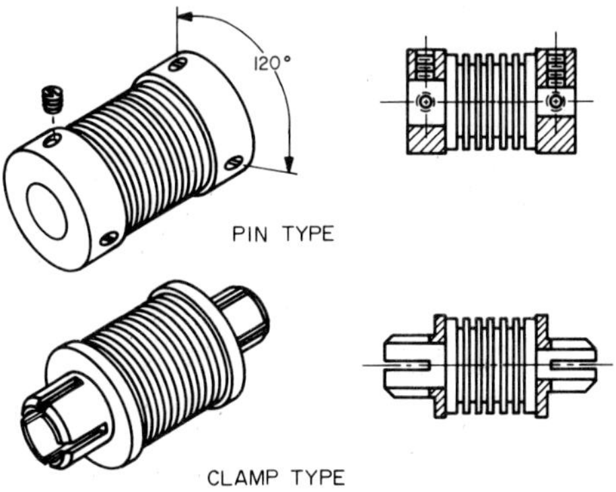

FIG. 4-16 Bellows couplings.

These couplings are used when motion must be transmitted from shaft to shaft with constant velocity and zero backlash.

The helical coupling achieves these parameters by virtue of its patented design, which consists of a one-piece construction with a machined helical groove circling its exterior diameter. Removal of this coil or helical strip results in a flexible unit with considerable torsional strength. See Fig. 4-17, which pictures both the pin- and clamp-type designs.

4-20 Offset Extension Coupling

Figure 4-18 depicts an offset extension shaft coupling. This coupling is used to connect or join parallel drive shafts that are offset ±30° in any direction, with separations generally greater than 3 in. Shafts are secured to the coupling with set screws.

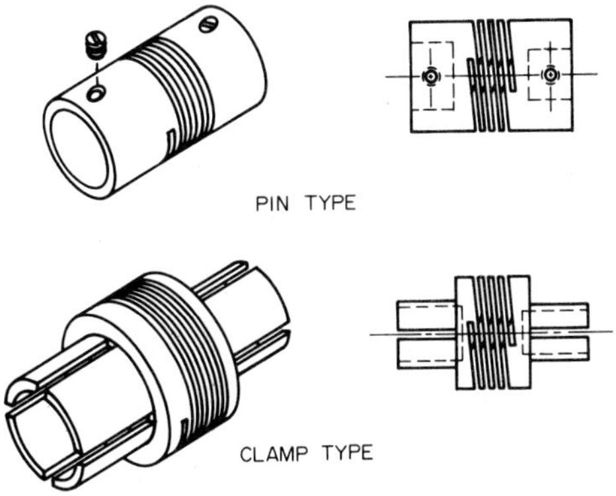

FIG. 4-17 Helical couplings.

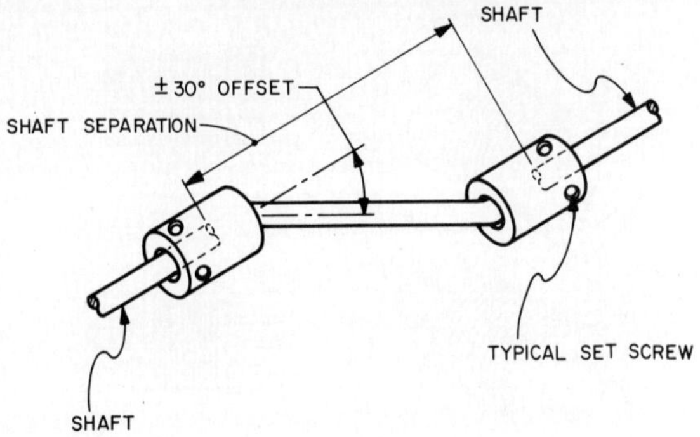

FIG. 4-18 Offset extension shaft coupling.

REFERENCES

Master Catalog 82, Sterling Instrument Division of Designatronics, Inc., New Hyde Park, N.Y.

Levinson, Irving J.: *Machine Design,* Reston Publishing Co., Reston, Va., 1978.

Parmley, R. O.: *Standard Handbook of Fastening and Joining,* McGraw-Hill, New York, 1977.

Spotts, M. F.: *Design of Machine Elements,* 5th ed., Prentice-Hall, Englewood Cliffs, N.J., 1978.

Winston, Stanton E.: *Machine Design,* American Technical Society, Chicago, 1956.

Carmichael, Colin, ed.: *Kent's Mechanical Engineer's Handbook,* 12th ed., Wiley, New York, 1958.

5
Bearings

Members of the Staff

PT Components, Inc./Link-Belt Bearing Division
Indianapolis, Indiana

DESIGN AND MANUFACTURING

Today there are many publications that contain information about bearings, ranging from highly technical treatises for the bearing specialist to standard textbook presentations for the engineering student. The need has persisted, however, for a single presentation that would combine a depth of theoretical and research expertise with a broad scope of experience in practical application. To this end, the technical staff of PT Components have developed this presentation, which is extracted from their 1982 *Bearing Technical Journal.*

Proper selection and application of bearings, including the related areas of product safety, are major concerns of all responsible equipment and bearing manufacturers. Potential operating and performance problems associated with manufactured products will vary appreciably depending on the intended use of, or application of, the equipment.

Bearings are normally a critical part of a machine or mechanical system, and will provide safe and reliable service when properly applied and maintained. It is intended that information in this section, when properly used, will provide clear and reliable guidelines to the proper general selection and application of bearings.

5-1 Rolling Contact

A rolling-element bearing consists of four basic parts: (1) inner ring, (2) outer ring, (3) rollers or balls, and (4) retainer or separator. Three of these parts, the inner ring, the outer ring, and the rollers or balls, support the bearing load. The fourth part, the bearing retainer, serves to position the rolling elements. This simple four-element mechanism is a vital component of practically all forms of dynamic design and assumes a more complex character as we examine it in greater depth.

Some of the factors considered in the design of rolling-element bearings relate specifically to the rolling contact zone, contact areas, contact stress, elastic deformation, and rolling and sliding friction. While rolling friction may approach zero, some degree of sliding friction will exist in all rolling-element bearings. Bearing components, although made of hardened steel, are elastic bodies under load, and react accordingly. This may be visualized by thinking of the bearing elements as made of rubber. In this case, a stationary roller under load would appear as shown in Fig. 5-1. With the introduction of motion, the same roller would appear as shown in Fig. 5-2. Here, there is a continuous

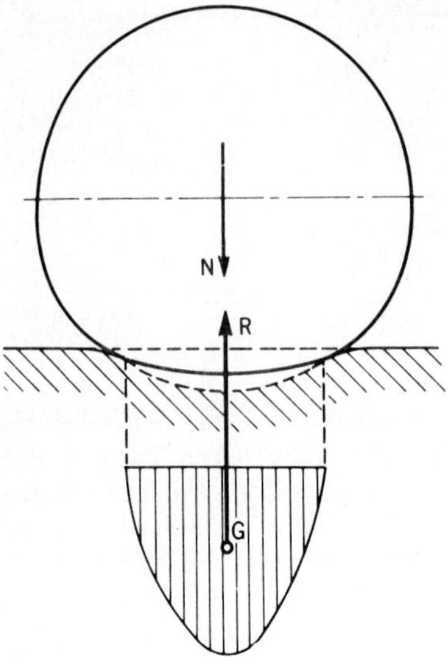

FIG. 5-1 Deformation of a stationary roller and diagram of the elastic reactions.

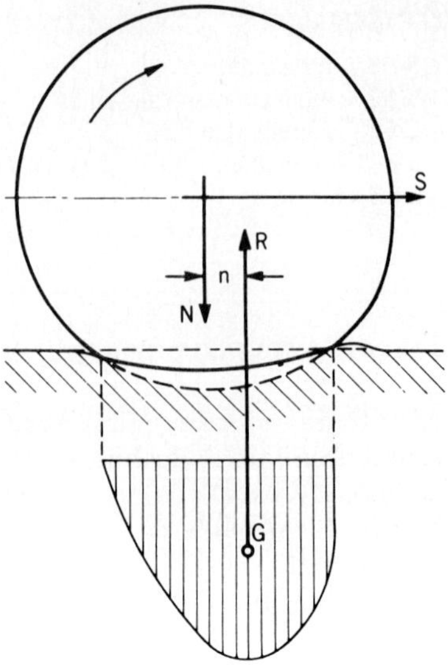

FIG. 5-2 Deformation of a moving roller and diagram of the elastic reactions.

flexing and wave motion of the material in front of and behind the moving contact zone. In this deformed contact zone, some slippage and therefore sliding friction will prevail. This slippage occurs when different surface points on the moving body traverse slightly different arc distances in the same increment of time.

5-2 Bearing Design

Elastic deformation, rolling contact stress, rolling and sliding friction, and contact area are some of the design parameters considered in developing bearings to operate at satisfactory thermal and stress levels. Governing these factors involves proper determination and control of raceway curvature. A line contact of roller and raceway, for example, spreads the load over virtually the entire roller length. In practice, however, line contact can result in the edge-loading phenomenon and severe stress concentrations, as illustrated in Fig. 5-3. Link-Belt roller bearings are designed with an osculation that provides substantially uniform stress across the length of the most heavily loaded roller at design load levels.

Spherical-roller-bearing rollers generally have a true circular crown, while Link-Belt cylindrical roller bearings have polycrowns developed by a polynomial equation, which assure freedom from edge loading under various conditions of load and misalignment. For certain applications where the exact crowning requirements cannot be accurately predetermined, an operation known as contact pattern analysis may be used. In making a contact pattern analysis, production rings with specially treated surfaces are used. Static load-

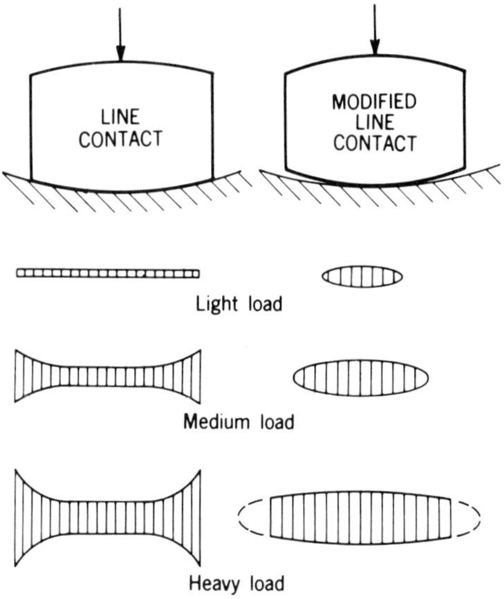

FIG. 5-3 Exaggerated contact patterns illustrate differences between line contact and modified line contact.

ing of the bearing to the critical load level develops the actual contact of the roller on the treated ring, under the influence of the mutual deflection of all system components. Analysis of this contact pattern allows engineers to predict the required crown.

Contact stresses are also influenced by the amount of clearance in a bearing. In a radial ball bearing under a given radial load, the ball which is at the point of maximum load concentration carries much more load than the others. The stress on this ball will vary significantly with varying clearance. For example, a bearing mounted with zero clearance would, under load, have the radial displacement and corresponding distribution of the load indicated by Fig. 5-4.

This same bearing, except with a great amount of initial clearance, would at times have almost the entire load sustained by one ball. These load distributions may be calculated on the basis of the angular distance between rolling elements, the initial clearance before the load is applied, and proportionality between stress and strain.

A very important factor in the design of most rolling-element bearings is the retainer or separator. A retainer provides positive separation of adjacent rolling elements. Without this separation, constantly variable rolling-element velocities in and out of the load zone will cause direct impact, rubbing, and pounding of the rolling elements at high relative sliding velocities. A retainer will limit and absorb these dynamic forces while guiding and spacing the rolling elements. Retainer design becomes increasingly important to bearing

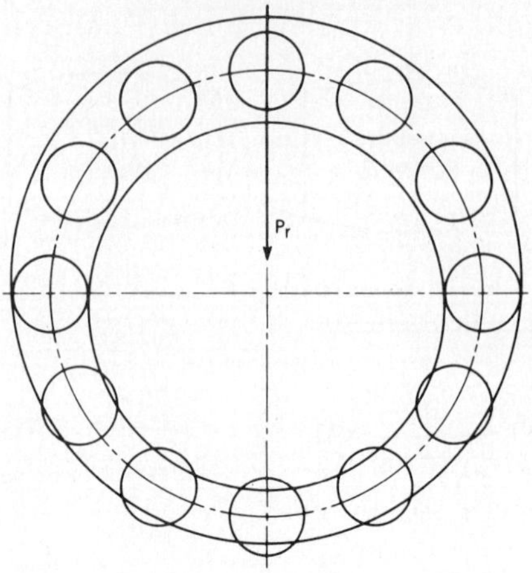

FIG. 5-4 Radial displacement in a ball bearing without clearance under the effect of an externally applied load.

performance at higher speeds and loads, particularly under cyclic speed and load variations.

5-3 Materials

Selection of materials and control of material quality are critical in the manufacture of rolling-element bearings. The theoretical hertzian stress attained in rolling contact may exceed 400,000 lb/in^2, and the stress repetitions endured by many high-speed bearings become astronomical. Bearing steels must possess high strength, toughness, wear resistance, dimensional stability, excellent fatigue resistance, and freedom from internal defects; producing them requires rigid controls. Bearing-quality vacuum-degassed steels are used in the production of Link-Belt ball and roller bearings. The vacuum degassing of bearing steels reduces the inclusion content as well as the hydrogen and nitrogen content. Both carburizing and through-hardening alloys are used, depending on the bearing type, size, configuration of component parts, and application requirements.

Vacuum-degassed steels should not be confused with vacuum-melted steels, which are produced by the vacuum induction melting or consumable-electrode vacuum-remelt process. These two processes are extremely expensive. At present they are used only for manufacture of very specialized bearings.

The selection and control of materials are also vital to facets of bearing engineering other than the bearing raceways and rolling elements. Many different materials may be used in bearing retainers, including bronze alloys, steel, iron, aluminum, polymers, and composites. Other bearing parts, such as seals and housings, also may be made out of a broad range of materials. In these areas as well as in bearing steels, a rigid system of material controls is an essential part of the manufacturing responsibility.

5-4 Manufacturing

The manufacture of ball bearings and roller bearings involves four basic process stages: first-form, heat-treat, final-form, and assembly.

The first-form process stage is the starting point and foundation for final part quality. Rings are formed from solid bars, seamless tubing, or forgings by multiple-spindle automatic bar machines, automatic chucking machines, and numerical-controlled contouring machines. Balls and rollers are cold-headed or turned from wire or bar stock. Adherence to critical bearing corner standards for matching shaft or housing shoulder and fillet standards is directly controlled in first-form processes.

Heat treatment of rings and rolling elements through controlled-atmosphere furnaces, quenching media, and tempering treatments is the second process stage. The hard, tough metallurgical structures which are attained provide each bearing component with the required basic strength. The successful heat treatment of bearing components involves precise control of temperature, atmosphere, and rates of heating and cooling.

With through-hardening steels, parts are held above the austenizing temperature, where the basic structure is carbides in a matrix of austenite. After a certain time the part is cooled in a quenching medium, where the carbon atoms are trapped in a shrinking iron crystal. The resulting crystalline structure is an extremely hard, brittle steel phase called martensite. This material is highly stressed internally and must be tempered to obtain the hard but tough structure required.

With carburizing steels the carbon content is lower than in through-hardening steels, and it is necessary to add carbon to attain full case hardness. To accomplish this, an atmosphere rich in carbon is precisely controlled through the required time and temperature cycles. On quenching and tempering, the metallurgical transformations are similar to those described for through-hardening steels, but the core, remaining lower in carbon content, is at a lower hardness. The result is a part with a fully hardened outer case supported by a tough, ductile core structure.

The final-form process stage involves the generation of contours and sizes to accuracies within millionths of an inch. Some of the techniques utilized are shoe-type centerless grinding, cross-axis grinding, controlled-force grinding, rotary diamond dressing, geometric race honing, vibratory finishing, and rotary barrel finishing. Surface finishes of ball- and roller-bearing raceways are commonly held to AA values as low as 2 to 4 μin (0.05 to 0.10 μm). Raceway curvatures are generated to chordal accuracies as close as 20 μin (0.50 μm). With shoe-type centerless grinding techniques, skilled operators consistently produce large inner rings with race-to-bore runouts of 0.0002 in (5.08 μm) or less. Final honing processes are designed for stock removal of as little as 0.0001 in (2.54 μm) to attain surface and curvature accuracies measurable only by profile tracing instruments.

Bearing components are subject to rigid inspection and statistical sampling plans at every operation before they reach the assembly stage. For matched assembly bearing types, additional measurement of raceway sizes is necessary to complete assemblies at the specified internal bearing clearance. Final sampling inspection of the assembled bearing for sound and vibration characteristics, followed by cleaning, preservation, lubrication, and packaging, completes the manufacturing cycle.

5-5 Measurement

The control of quality in any manufacturing process depends on accurate measurement. High-volume production of precision rolling-element bearings would be literally impossible without instruments which can measure in millionths of an inch. Attainment of superior product quality is engineered and programmed from the design concept stage through completion of the finished product. Quality starts with and is dependent on the attitudes of every individual. Combined with these attitudes must be the process capability of tools and machines, and the skilled use of precise measuring and gauging equipment.

The gauging system is based largely upon a system of masters. Each component at various process stages is directly compared on a suitable gage with a master that has exactly known characteristics. Under this system, thousands of master parts and segments are maintained under the jurisdiction of the chief inspector. These are issued to the appropriate work stations as required, so that the machine operator has the necessary gages and masters pertaining to the particular part or operation being performed.

The master system of gauging is supplemented by many other instruments and techniques. For example, a tracer-type vibration pickup is used in the evaluation of raceway waviness. This instrument will evaluate the magnitude and frequency of deviations from a perfect surface of revolution. With this instrument, the part being checked is rotated, with the tracer pickup in light contact with the surface being checked. Instrument output is produced by an oscilloscope, audio speaker, and electronic meter. Similar instrumentation is used in the evaluation of running accuracy and sound characteristics. In the grinding process stages, eddy current and magnetic flux techniques are utilized for flaw detection.

Regardless of the quality of inspection techniques, the inspection process in itself cannot correct defective parts. Thus, a major objective of inspection is defect prevention as well as detection. Inspection data can pinpoint potential problems. This permits the correction of difficulties before defective parts are produced. The same techniques are applied to materials provided by outside suppliers. A supplier rating system promotes dependable quality of all purchased material.

This entire program of measurement, covering all facets of bearing production, is reflected in the superior performance of quality bearings.

5-6 Standardization

Progress in technology necessitates standardization in all areas of design and manufacture. Rolling-element bearings are no exception. The bearing industry early recognized the need for international bearing standards and has worked continuously for their development.

In the United States, the Anti-Friction Bearing Manufacturers Association (AFBMA) has published industry standards on terminology, bearing boundary dimension plans, tolerances, gauging practices, identification codes, packaging, load ratings, mounting practices, mounting accessories, balls, and instrument ball bearings. The three engineering committees (ABEC, RBEC, and BMEC*) developed these standards; they also work with the American National Standards Institute, Inc. (ANSI) and the International Standards Organization (ISO) in developing national and international bearing standards. In the improvement of communications and understanding, the benefits of such standards to both manufacturers and users are immeasurable.

A dynamic and reasonable standard should never impose a limitation on improvement of bearing quality, precision, or performance. All makes of bearings that meet a given standard cannot be assumed to be exactly equivalent. Where interchangeability is required on severe or critical applications, consultation is recommended. While the appropriate standards will assure interchangeability in dimensions, basic ratings, and other such factors, bearings may differ in operating temperature, lubrication requirements, and other operating characteristics. These differences may be of little concern in most applications, but serious difficulties could develop on more critical applications. For this reason, bearing engineers are usually interested in analyzing application details when substitution of interchangeable bearings is considered.

5-7 Accessories

Bearing accessories are those components necessary for adapting a bearing to a machine. These include housings, seals, tapered adapter sleeves, locknuts, lockwashers, flingers, clamp plates, and so on. These items are as much a part of successful bearing design and application as the four basic bearing parts. A bearing housing that deflects excessively may cause serious malfunctions even though actual bearing performance is acceptable. A housing must provide firm, uniform support, capable of sustaining the design loads without undue deflection. Accuracy equivalent to the accuracies of ground bearing components is commonly obtained in the production of bearing housings. Special housing designs occasionally require grinding and honing operations combined with stress relief treatments to meet critical requirements.

The effective performance of seals is a major factor in extending bearing life. Seal types and configurations are discussed in detail later.

Other bearing accessories are discussed in later sections. They involve a number of design and manufacturing characteristics directly related to bearing performance. Bearing locknuts, for example, must have rigid control of thread to locknut face squareness. Tapered adapter sleeves require accurate taper, taper-to-bore concentricity, and thread-to-taper squareness. Squareness, parallelism, and concentricity, as well as size and roundness, are the

* ABEC, Annular Bearing Engineers Committee; RBEC, Roller Bearing Engineers Committee; BMEC, Ball Manufacturers Engineers Committee.

characteristics most commonly required in design and manufacture of bearing accessories. These factors must be properly specified to achieve the manufacturing accuracy and precision required of rolling-element bearings.

OPERATIONAL CHARACTERISTICS

5-8 Bearing Dynamics

Bearing dynamics encompasses the forces and movements prevailing in operating bearings. Some of these forces and movements were discussed in Sec. 5-1, "Rolling Contact." There it was shown that in all rolling-element bearings, some degree of slip will occur in the rolling contact zone. The frictional forces developed in the slip areas are seldom in balance with the static equilibrium of the rolling elements. A small force couple will generally prevail, creating a ball spin effect in ball bearings and a roller skew effect in roller bearings. Roller skew seeks to divert the roller from its normal raceway path. A means of guidance is therefore provided in the form of flanges, rings, or retainers.

Other forces present within the bearing are gyroscopic and centrifugal effects. Gyroscopic force is usually not significant except in extremely high-speed bearings. Centrifugal force can affect ball or roller motion and guidance forces over a fairly broad range of speeds, but the effect on bearing loading is usually a significant factor only at extremely high speeds.

The orbital speeds of the rolling elements and rotational speed of the retainer can be calculated from the raceway and rolling-element diameters, as illustrated in Fig. 5-5. The motion relations are similar to those of a planetary gear system. Generally the retainer or rolling-element orbital speed will be in the range of 40 to 60 percent of the rotating raceway speed. The ratio of

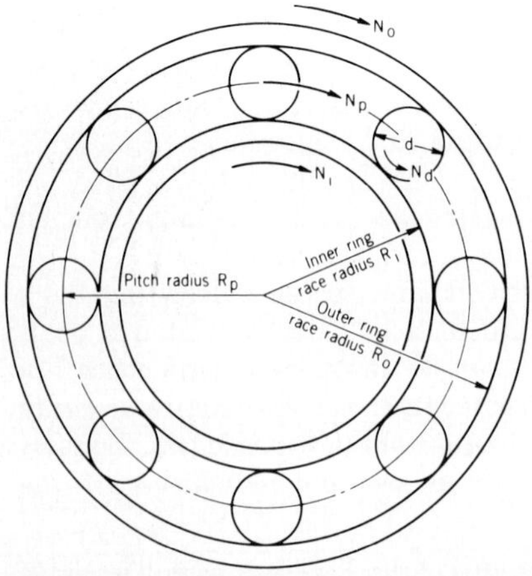

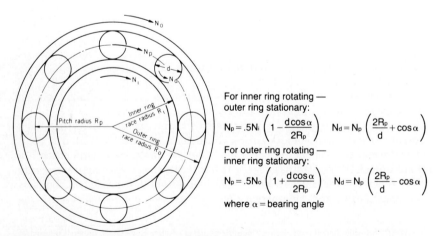

For inner ring rotating — outer ring stationary:

$$N_p = .5N_i\left(1 - \frac{d\cos\alpha}{2R_p}\right) \qquad N_d = N_p\left(\frac{2R_p}{d} + \cos\alpha\right)$$

For outer ring rotating — inner ring stationary:

$$N_p = .5N_o\left(1 + \frac{d\cos\alpha}{2R_p}\right) \qquad N_d = N_p\left(\frac{2R_p}{d} - \cos\alpha\right)$$

where α = bearing angle

FIG. 5-5 Bearing component speeds with pure rolling.

retainer speed to bearing speed is easily observed by noting the angular rotation of the retainer during one full revolution of the rotating raceway. The ratio of angular motion to 360° will then provide the percentage of rotation velocity for a particular case. The rolling-element rotational speed about its own axis may be observed in a similar way or calculated when the raceway diameter relationships are known.

5-9 Bearing Torque

Bearing torque is the frictional resistance of a bearing to rotation; it may be used to estimate the power requirements or heat generation of the bearing. This frictional resistance is influenced by many factors, the most important being lubrication, bearing design, and bearing load. In practice, a constant coefficient of friction can be used for most purposes. Coefficient of friction values shown in Table 5-1 are generally conservative for normal operating ranges of properly lubricated bearings with no seals. The values can vary, however, with grease-packed bearings and with lightly loaded small-size bearings. Also, low-temperature operation or the use of high-viscosity lubricants will significantly increase rolling-bearing friction. Friction is inherently very low in rolling-contact bearings, so far as the load-dependent friction is concerned. Therefore the torque estimates for lightly loaded bearings may not adequately predict the basic zero-load friction characteristics of a particular bearing or application.

The other friction component not included in the value shown is seal friction. Seal torque may vary from a fraction of the bearing torque to several times this level. Seal torques will vary appreciably with type, size, and design.

The total power requirements, or heat generation, are therefore seen to consist of several variables which can seriously affect estimates where critical decisions must be made on horsepower requirements or heat-generation effect. Consultation is recommended to establish more accurate estimates of the system's torque requirements.

TABLE 5-1 Average Bearing Coefficients of Friction Measured at Shaft Surface

Type of load	Coefficient of friction					
	Ball bearings		Spherical roller bearings		Cylindrical roller bearings	
	Starting	Running	Starting	Running	Starting	Running
Radial	0.0025	0.0015	0.0030	0.0018	0.0020	0.0011
Thrust	0.0060	0.0040	0.0120	0.0080	—	—

Bearing torque = coefficient of friction × load × shaft radius.

5-10 Bearing Speed

Most bearing manufacturers specify speed limits for their products. These serve as a useful guide for the majority of normal bearing applications. In addition to catalog speed limits, a bearing may have a thermal speed limit, lubricant speed limit, or design speed limit.

The published speed limits are of practical value only when they are considered along with other factors of bearing operation. Not every application will function satisfactorily at such speed. Load, lubrication, and temperature factors will influence the performance. Bearing operation at the catalog speed limit demands good lubrication, moderate load, and reasonable temperature environment.

The speed beyond which bearing temperatures exceed a critical value is the thermal speed limit. This can be a limit imposed by the bearing user when operating temperature must be limited. In such a case, the thermal speed limit may be less than the catalog limits. Since temperature increases with speed, a limit will be reached at which the lubricant deteriorates excessively or the bearing steel may soften. This may also be classed as a thermal speed limit, but it cannot be evaluated except by experience.

The lubricant speed limit is related to lubricant, temperature, and other application conditions. Different lubricants will have different dynamic shear stability, thermal deterioration limits, lubricity, and so on. The maximum speed for a given bearing application will therefore vary with different lubricants or lubrication methods. The lubricant speed limit may be higher or lower than the catalog limits.

The bearing design speed limit is the ultimate speed limit. It is mainly of interest to the bearing engineer, who seeks to extend it through improved design. Beyond this limit the bearing will not function properly and will destroy itself.

The "speed limit" is a flexible value; however, cataloged limits should be exceeded only after consultation and review of the application.

5-11 Bearing Vibration

By definition, noise is unwanted sound. Sound originates with generated vibrations. The nature and level of the sound depend on the medium through which the vibrations are transmitted to the ear. When noise reduction is required, vibrations in machinery must be isolated, lowered, or eliminated.

A source vibration may transmit a practically imperceptible sound directly to the atmosphere. However, if the same source vibration transmits vibrations through other machine members, sound amplification can occur. Vibrations can be transmitted through shafts, housings, or framework as well as through bearings. Where vibrations reach sheet metal, such as ducts, the sound can be greatly amplified.

Rolling-element bearings generate vibrations. Even ultra-quiet bearings, such as those selected for submarine service, generate vibrational frequencies, but at lower amplitudes. All bearings assembled by major manufacturers meet strict proprietary standards of operational quietness. Nevertheless, an acceptable bearing may vibrate with frequency characteristics that can generate excessive noise on a specific machine. Generation of such excessive noise is always possible in a mechanical system that may include gears, chain drives, other bearings, and other components, each with its own vibration characteristics.

In some instances, noise abatement may require major machinery design changes. In other cases, simple modifications, such as introducing damping materials at the vibration surface or varying the operation speeds, are notably successful. Where it is difficult to dampen the resonance and/or amplification, a selected ultra-quiet bearing may reduce the sound to a tolerable level.

Common causes of noise are often overlooked. A list of contributory factors that have been found repeatedly in the area of sound problems includes:

1. Large areas of unbraced sheet metal
2. Housings insecurely fastened to supports
3. Bearings improperly fitted to shaft or housing
4. Pulleys or gears loose on shaft
5. Belts improperly tensioned
6. Keys loose in keyways
7. Out-of-round shafts
8. Unbalance
9. Pulley groove wobble

A thorough approach to noise control must examine all possible vibration paths from the source to the ear. Thorough examination will ensure selection of the most economical and desirable solution to a noise problem.

5-12 Shaft Control and Preload

Shaft control is a basic requirement of a bearing. In any mechanical arrangement, the static and dynamic position of a shaft is established and held by the support bearings. As might be expected, there are deviations from the optimum condition, wherein a shaft will not rotate precisely about an intended axis. Deviations from perfect trueness of rotation are termed "runout." This term may apply to a radial, axial, or angular deviation from the intended path of movement. In the simple case of a shaft rotating under pure radial load and supported by two bearings, as many as 15 specific factors related to shaft, bearing components, and housing can affect the radial runout of the shaft. These factors and their influence become more complex with different types of loading or rotational conditions. The conditions that may influence shaft positioning are:

1. Shaft—straightness, roundness, size or fit with bearing, bearing seat concentricities, bearing seat squareness, deflection characteristics, and balance

2. Bearing inner ring—raceway-to-bore runout, raceway-to-bore squareness, and roundness
3. Rolling elements—roundness and size variation
4. Bearing outer ring—raceway-to-OD runout, raceway-to-OD squareness, and roundness
5. Housing—roundness, bearing housing fit, housing and frame deflections
6. Bearing assembly—operating clearance and deflection

Precise control of the shaft must be maintained for coating rolls, precision gearing, machine tool spindles, and so on. The other extreme is the type of mechanical equipment in which deflections and runouts visible to the naked eye may be tolerated. Generally, the tolerances held in the manufacture of standard ABEC-1 or RBEC-1 grade bearings (see Tables 5-2 and 5-3) will provide rotational accuracy suitable for the vast majority of bearing applications. When more rigid requirements prevail, closer-tolerance bearings may be necessary. For the most precise requirements, preloaded bearings are frequently used.

Preloading eliminates all free play within the bearing. Deviation of shaft position chargeable to bearings is due only to deflection of the bearing or inaccuracies in bearing components.

A satisfactory technique for preloading has been developed involving the use of tapered-bore spherical roller bearings. These bearings can be designed to be forced up a precision-ground tapered shaft seat a specific distance. This provides a preload at the final installed position on the shaft, in combination with the correct tightness of fit between the shaft and bearing. Such applications must be engineered by the bearing manufacturer in cooperation with the machinery builder to obtain maximum control of variables. A preloaded bearing, while minimizing shaft excursions, becomes very sensitive to thermal gradients. Lockup conditions or loss of preload can occur if speeds are too high or operating temperatures are not properly controlled. Generally, commercial "off-the-shelf" bearings are not intended for preload operation.

Dynamic runout differs from the runout observed during rotation by hand. Usually the important factors in the dynamic state are deflections and unbalanced forces. For precise shaft control, a dynamic balancing of the system is mandatory. Often, bearings have the high point of the inner ring identified or marked. Bearings are then positioned on the shaft seat to counteract runout, minimizing unbalance of the system.

5-13 Thermal Characteristics

Bearing operating temperature is governed by many factors. The heat-generating characteristic of a bearing is purely a function of torque and speed and can usually be estimated to a reasonable degree. At some particular temperature, the heat transfer away from a bearing equals the heat generated by the bearing. This is the steady-state operating temperature, and it is dependent on the heat generated by the bearing and the heat-transfer characteristic of the machine system in its environment.

TABLE 5-2 Tolerance Limits for Inner Rings of Metric Radial Bearings of Tolerance Class ABEC-1, RBEC-1

Basic bore diameter d				Bore†					Width Allowable deviation from B	
				Allowable deviation from d						
mm		Inch		Single mean diameter of bore D_{mp}		Single diameter of bore d_s*		Radial runout K_i	Single width of ring B_s	
Over	Incl.	Over	Incl.	Low	High	Low	High	Max.	High	Low
Tolerance limits in 0.0001 in										
10	18	0.3937	0.7087	− 3	+0	− 4	+1	4	+0	− 50
18	30	0.7087	1.1811	− 4	+0	− 5	+1	5	+0	− 50
30	50	1.1811	1.9685	− 5	+0	− 6	+1	6	+0	− 50
50	80	1.9685	3.1496	− 6	+0	− 8	+2	8	+0	− 60
80	120	3.1496	4.7244	− 8	+0	−10	+2	10	+0	− 80
120	180	4.7244	7.0866	−10	+0	−12	+3	12	+0	−100
180	250	7.0866	9.8425	−12	+0	−15	+3	16	+0	−120
250	315	9.8425	12.4015	−14	+0	−18	+4	20	+0	−140
Tolerance limits in 0.000001 m										
10	18	0.3937	0.7087	− 8	+0	−11	+3	10	+0	−120
18	30	0.7087	1.1811	−10	+0	−13	+3	13	+0	−120
30	50	1.1811	1.9685	−12	+0	−15	+3	15	+0	−120
50	80	1.9685	3.1496	−15	+0	−19	+4	20	+0	−150
80	120	3.1496	4.7244	−20	+0	−25	+5	25	+0	−200
120	180	4.7244	7.0866	−25	+0	−31	+6	30	+0	−250
180	250	7.0866	9.8425	−30	+0	−38	+8	40	+0	−300
250	315	9.4825	12.4015	−35	+0	−44	+9	50	+0	−350

* d_s applies only to bearings of metric diameter series 0, 2, 3, and 4.
In diameter series 0 up to and including d = 40 mm.
In diameter series 2 up to and including d = 180 mm.
For larger sizes in series 0 and 2 and all sizes of series 1, 8, and 9, d_s is not restricted.
† Bore tolerance limits do not apply to tapered bore inner rings.

TABLE 5-3 Tolerance Limits for Outer Rings of Metric Radial Bearings of Tolerance Class ABEC-1, RBEC-1.

Basic outside diameter D				Outside diameter: Allowable deviations from D							
				All bearings		Open bearings		Bearings with shields or seals			
mm		Inch		Single mean diameter of OD D_{mp}		Single diameter of OD D_s*		D_s*		Radial runout K_e	Width limits
Over	Incl.	Over	Incl.	High	Low	High	Low	High	Low	Max.	
Tolerance limits in 0.0001 in											
30	50	1.1811	1.9685	+0	− 5	+ 2	− 7	+ 3	− 8	8	
50	80	1.9685	3.1496	+0	− 5	+ 2	− 7	+ 4	− 9	10	
80	120	3.1496	4.7244	+0	− 6	+ 2	− 8	+ 5	−11	14	Identical
120	150	4.7244	5.9055	+0	− 8	+ 2	−10	+ 5	−13	16	to those
150	180	5.9055	7.0866	+0	−10	+ 3	−13	+ 6	−16	18	of inner
180	250	7.0866	9.8425	+0	−12	+ 3	−15	+ 8	−20	20	ring of

250	315	9.8425	12.4015	+0	−14	+ 4	−18	+ 8	−22	24	the same
315	400	12.4015	15.7480	+0	−16	+ 4	−20	+ 9	−25	28	bearing
400	500	15.7480	19.6850	+0	−18	+ 5	−23	+10	−28	32	
500	630	19.6850	24.8031	+0	−20	+ 6	−26	+12	−32	40	

Tolerance limits in 0.000001 m

30	50	1.1811	1.9685	+0	−11	+ 3	−14	+ 8	−20	20	
50	80	1.9685	3.1496	+0	−13	+ 4	−17	+10	−23	25	
80	120	3.1496	4.7244	+0	−15	+ 5	−20	+13	−28	35	Identical
120	150	4.7244	5.9055	+0	−18	+ 6	−24	+13	−33	40	to those
150	180	5.9055	7.0866	+0	−25	+ 7	−32	+15	−41	45	of inner
180	250	7.0866	9.8425	+0	−30	+ 8	−38	+20	−51	50	ring of
250	315	9.8425	12.4015	+0	−35	+ 9	−44	+20	−56	60	the same
315	400	12.4015	15.7480	+0	−40	+10	−50	+23	−64	70	bearing
400	500	15.7480	19.6850	+0	−45	+12	−57	+25	−71	80	
500	630	19.6850	24.8031	+0	−50	+14	−64	+30	−81	100	

* D_s applies only to bearings of metric diameter series, 0, 2, 3, and 4 (prior to insertion of internal snap rings).
In diameter series 0 up to and including D = 80 mm.
In diameter series 2 up to and including D = 315 mm.
For larger sizes in series 0 and 2 and all sizes of series 1, 8, and 9, D_s is not restricted.

Determination of heat transfer is an extremely complex problem, often requiring digital computer solution. The exact methods are beyond the scope of this text. However, some appreciation of the nature of the problem is of value in judging whether rigorous solutions are necessary. Heat conduction, convection, and radiation are the three modes of transfer involved in the operation of bearings. In the simplest case, a bearing as the only heat source, the major paths of flow will be from the inner ring to the shaft and from the outer ring to the housing. The shaft will conduct heat away from the bearing and transfer it to the atmosphere or to other elements attached to the shaft. The housing will transfer heat to the atmosphere or to the supporting structure. The lubricant will affect both heat generation and heat transfer. The effect of total bearing environment on ultimate heat balance is very critical. This environment includes the total machine, frame, shaft, and so on, and the surrounding atmosphere. Air movement will also have a major effect. External heat sources, such as a hot shaft, may reverse a normally expected heat flow. Each individual bearing application becomes, in effect, a separate system, unique in its heat-transfer characteristics.

The heat-transfer problem is complex; however, background experience on existing equipment often provides a satisfactory basis for simple judgments on new equipment. In marginal situations, consultation with bearing engineers is advisable. A decision as to whether cooling, special lubrication, or special design is required may have serious economic, safety, or reliability impact. Often prototype testing is necessary and may provide more reliable data than a theoretical analysis.

5-14 Summary

The operational characteristics of rolling-contact bearings have been discussed very briefly. Many questions and areas of interest have necessarily been omitted.

This review will, however, instill some awareness of the complexities of bearing operation. It shows that the selection and use of bearings may involve factors of judgment, experience, and in-depth knowledge. Engineering and research facilities are available from the manufacturers, as well as additional information and consultation for special requirements.

RATING AND LIFE

5-15 Rating Standards

A standard method of establishing a bearing's rating permits a valid comparison of various types and makes of bearings. The Anti-Friction Bearing Manufacturers Association (AFBMA) has therefore adopted standards developed by the member companies. The AFBMA standards on ratings were approved as USA Standard B 3.11-1959 by the organization now known as the American National Standards Institute, Inc. (ANSI). PT Components, Inc. is a participating member of AFBMA and ANSI engineering committees and follows the ANSI standards in the establishment of ball- and roller-bearing ratings.

There are two standardized ratings associated with a rolling-element bearing, the basic load rating and the static load rating. For bearings that rotate only occasionally or not at all, selection is usually based on the static load rating C_0. For bearings that rotate, the basic load rating C is used. These ratings are calculated from the ANSI standard rating formulas as follows:

For radial ball bearings:

$$C_0 = f_0 iZD^2 \cos \alpha \tag{5-1}$$

$$C = f_c(i \cos \alpha)^{0.7} Z^{2/3} D^{1.8}$$

For balls larger than 1 in, use $D^{1.4}$ (5-2)

For roller bearings:

$$C_0 = f_0 iZl_{\text{eff}} D \cos \alpha$$

$$f_0 = 6430 \text{ (for inch and pound units) or } 44 \text{ (for millimeter and newton units)} \tag{5-3}$$

$$C = f_c(il_{\text{eff}} \cos a)^{7/9} Z^{3/4} D^{29/27} \tag{5-4}$$

where C = basic load rating, lb or N

C_0 = static load rating, lb or N

f_c = a factor which depends on the geometry of the bearing components, the accuracy to which the various bearing parts are made, and the material, its cleanliness, and its processing

f_0 = a factor which depends on the bearing type, used to determine static load rating

i = number of rows of rolling elements

Z = number of rolling elements per row

α = bearing contact angle, degrees

D = largest diameter of rolling elements, in or mm

l_{eff} = effective length of contact between roller and ring where contact is the shortest, in or mm

d_m = pitch diameter of rolling elements, in or mm

Values of f_c and f_0 are shown in Tables 5-4 and 5-5.

The use of these standard C and C_0 ratings permits an accurate comparison of various bearings. The C rating provides the basic value from which statistical fatigue life is derived for different applications.

5-16 Static Load Rating

As covered in Sec. 5-22, safe static loads must be considered for their effect on bearing structural stability, reliability, and the noise and vibration requirements of the system in which the bearings are installed.

A static load equal to the AFBMA Static Load Rating C_0 will result in contact stresses of 580,000 lb/in^2 (4000 MPa) in bearings under conditions of normal load distribution. Higher loads can be considered in certain applica-

TABLE 5-4 Maximum Values of f_c and f_0 (for Use with Inch-Pound Units Only)

$\frac{D \cos \alpha}{d_m}$	Ball bearings f_c	Ball bearings f_0	Roller bearings f_c	Roller bearings f_0
0.08	4020	2140	7290	6430
0.10	4220	2080	7570	6430
0.12	4370	2020	7760	6430
0.14	4470	1950	7880	6430
0.16	4530	1890	7950	6430
0.18	4550	1820	7980	6430
0.20	4550	1760	7970	6430
0.22	4530	1690	7920	6430
0.24	4480	1630	7850	6430
0.26	4420	1570	7760	6430
0.28	4340	1510	7650	6430
0.30	4250	1460	7520	6430

tions, since measurable permanent deformation is generally not found until contact stresses approach 600,000 lb/in^2.

There are other factors which must be considered in static loading evaluations, such as nature of load, frequency of occurrence, and the nature of bearing supporting structure. An impact or shock load corresponds to a high strain rate and can produce a greater amount of permanent deformation, and a high repetitive cycle can have a cumulative effect. The bearing support structure or shaft design will influence system deflections under static loads such that bearing surface damage and stress concentrations can occur.

TABLE 5-5 Maximum Values of f_c and f_0 (for Use with Millimeter-Newton Units Only)

$\frac{D \cos \alpha}{d_m}$	Ball bearings f_c	Ball bearings f_0	Roller bearings f_c	Roller bearings f_0
0.08	52.8	14.7	81.2	44
0.10	55.5	14.3	84.2	44
0.12	57.5	13.9	86.4	44
0.14	58.8	13.4	87.7	44
0.16	59.6	13.0	88.5	44
0.18	59.9	12.5	88.8	44
0.20	59.9	12.1	88.7	44
0.22	59.6	11.6	88.2	44
0.24	59.0	11.2	87.5	44
0.26	58.2	10.8	86.4	44
0.28	57.1	10.4	85.2	44
0.30	56.0	10.1	83.8	44

5-17 Basic Load Rating and Life Rating

Using the basic load rating C, standard formulas have been developed to predict the statistical life of a bearing under any given set of conditions. These formulas are based on an exponential relationship of load to life which has been established as

$$L = \left(\frac{C}{P}\right)^b \tag{5-5}$$

where L = rating life, millions of revolutions
C = basic load rating, lb or N
P = equivalent radial load, lb or N
b = 3 for ball bearings, $10/3$ for roller bearings

To determine hours of life L_{10}, this formula becomes:

$$L_{10} = \frac{1{,}000{,}000}{N \times 60}\left(\frac{C}{P}\right)^b \tag{5-6}$$

where N = rotational speed, r/min

These formulas reveal the following significant relationships:

1. Doubling the speed of a ball or roller bearing halves the life, while halving the speed doubles the life.
2. Doubling the load on a ball bearing, where $b = 3$, reduces the life to one-eighth, while halving the load increases the life eight times.
3. Doubling the load on a roller bearing, where $b = 10/3$, reduces the life to one-tenth, while halving the load increases the life ten times.

These relationships are frequently helpful in mentally assessing the significance of a change in load or speed.

5-18 Reliability

The term "rating life" is defined as the (fatigue) life that may be expected from 90 percent of a given group of approximately identical bearings operating under equal conditions of load and speed. This is a statistical term that is commonly referred to as L_{10} life, B-10 life, or minimum life. The median life that 50 percent of this same group of bearings would complete or exceed may be from three to five times the L_{10} life. Rating life is really an expression of reliability, because there is a 90 percent probability that a bearing will equal or exceed the given number of revolutions. In modern technology 90 percent reliability is sometimes not acceptable, and other levels of reliability may be considered.

The frequency and dispersion of metal fatigue in rolling bearings correspond to a Weibull distribution. The Weibull function can be represented by a straight line on special logarithmic graphs in which the percentage of surviv-

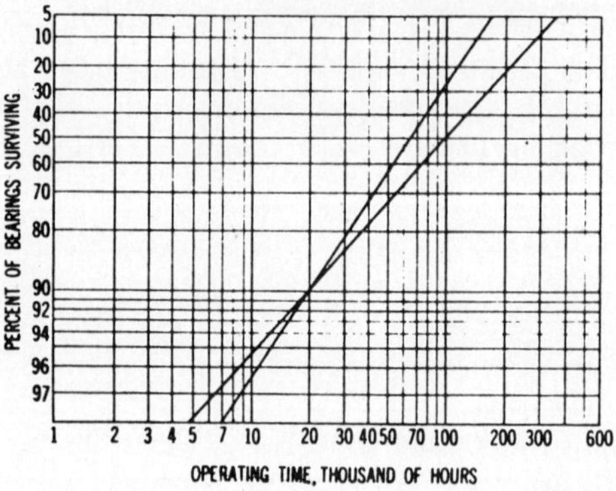

FIG. 5-6 Life reliability chart: relationship of bearings.

ing bearings in a given test group is plotted against the life attained. It is possible to establish a statistical life for a group of bearings at reliability levels either higher or lower than the normal 90 percent base. Figure 5-6 shows the approximate relationship between life and the percentage of surviving bearings for two different slopes of dispersion. In the example of the graph, it is shown that a bearing with an L_{10} life of 20,000 may have from 40 to 60 percent reliability at 80,000 h and 97 to 98 percent reliability at 7000 h.

Continued research on improved designs, materials, and methods is constantly increasing the reliability of rolling-type bearings. For this reason, applications that require high degrees of reliability should be reviewed by the bearing manufacturer. It should always be remembered that the reliability and rating life concepts apply strictly to statistical fatigue life, which can differ appreciably from operational life.

5-19 Life Adjustment Factors

The AFBMA currently recognizes three life adjustment factors which pertain to reliability, material, and application conditions. These factors are:

a_1 = life adjustment factor for reliability
a_2 = life adjustment factor for material
a_3 = life adjustment factor for application conditions

As previously explained, L_{10} fatigue life refers to 90 percent reliability or the life which 90 percent of a group of apparently identical bearings will complete or exceed at a particular loading condition. For some applications, however, life must be defined at a reliability level greater than 90 percent. Table 5-6 lists the life adjustment factors to be used for other reliability levels. The factor a_1 is used such that $L_n = a_1 L_{10}$.

Life adjustment factor a_2, for material, is intended to provide an adjustment to the rating life attributable to the utilization of premium materials and advanced processes in the manufacture of bearing rings and rolling elements.

TABLE 5-6 Life Adjustment Factors

Reliability %	L_n	Life adjustment factor a_1
90	L_{10}	1.00
95	L_5	0.62
96	L_4	0.53
97	L_3	0.33
98	L_2	0.33
99	L_1	0.21

Although research studies and field experience have demonstrated that cleanliness and selection of materials and processes have significant effects on fatigue life of bearings, no specific values for the a_2 factor have been established by AFBMA. This is due to the wide variety of steels and processes available and the anticipated future changes in utilization of various materials and processes. Specific a_2 life adjustment factors are also not recommended, and a_2 factors greater than 1.0 are used only after complete study of the bearing application. Since improvement in reliability and reduction in early failures is a primary design consideration in rolling-bearing applications, it is generally recommended that designers use catalog ratings and predicted life without reference to a_2 adjustments.

Life adjustment factor a_3 relates to application conditions which affect rating life, including lubrication and load distribution, effects of clearance, misalignment, housing and shaft stiffness, types of loading, mounting arrangements, and thermal gradients. Rating life (L_{10}) calculations assume that all these conditions are adequate; if conditions are less than adequate, the a_3 factor must logically be less than 1.0. An a_3 factor greater than 1.0 is considered possible only in connection with thick-film elastohydrodynamic lubrication along with freedom from contaminants or other adverse application conditions. Since lubricant and bearing cavity cleanliness are normally difficult to ensure, a_3 factors greater than 1.0 are seldom used and are recommended only after complete study of the bearing application.

In accordance with the AFBMA load rating standards, the three life adjustment factors are intended to be combined with the rating life to obtain an adjusted life, as follows:

$$L_n = a_1 a_2 a_3 L_{10} \tag{5-7}$$

It is emphasized that indiscriminate use of the life adjustment factors in this equation may lead to serious problems. For example, use of improved bearing steel must not be considered a compensating factor for inadequate lubrication or contamination. The use of bearing ratings and calculations of bearing rating life, along with life adjustment factor modifications, is a complex consideration. Consequently, various bearing manufacturers may respond by selecting or recommending different bearings for a particular application. In some situations the competitive bearings may be essentially equivalent in basic construction, rolling elements, quality, and so on, yet different life estimates are offered. Generally such differences are resolvable among reputable manufacturers, based on differing assumptions or factor selections.

The equipment designer should examine such proposals carefully to identify (and, if necessary, challenge) the various differences in capacities, factors, life calculations, and assumptions.

An L_{10} life calculated from the listed capacities is a theoretical statistical fatigue life, which is then subject to adjustment upward or downward using the life adjustment factors previously described. The resulting adjusted life estimate is still a theoretical statistical fatigue life, not an operating or service life prediction.

5-20 Operational Life

The operational life of a ball or roller bearing is governed by many external factors and influences. These include the machine design, shaft and support design, bearing installation, maintenance, and lubrication. Since it is generally conceded that only a small percentage of bearings removed from service have incurred rolling-contact fatigue, operational life is generally less than fatigue life. Consideration of all other factors which prevent a bearing from achieving its theoretical rating, or fatigue life, is required in order to evaluate the difference between operational life and rating life.

As an example, a bearing with 40,000-h rating life operating in an extremely dirty environment is not likely to achieve 40,000-h life unless special precautions are taken in sealing, housekeeping, and regreasing. A review of the fundamental failure modes of rolling-element bearings will be helpful in understanding the consequences of various operating conditions which influence the service life of a bearing. The basic failure modes are contact fatigue, wear, plastic flow, and fracture. These basic modes are further subdivided as described and illustrated in the following examples.

Contact Fatigue Subsurface fatigue or spalling is the basis of bearing rating life calculations; it is initiated by subsurface cracking at a material inclusion location. See Figs. 5-7 and 5-31. If these types of failures occur prematurely, it is because the loading of the bearing exceeds the design load, either in fact or because of stress or load concentrations resulting from mounting distortions.

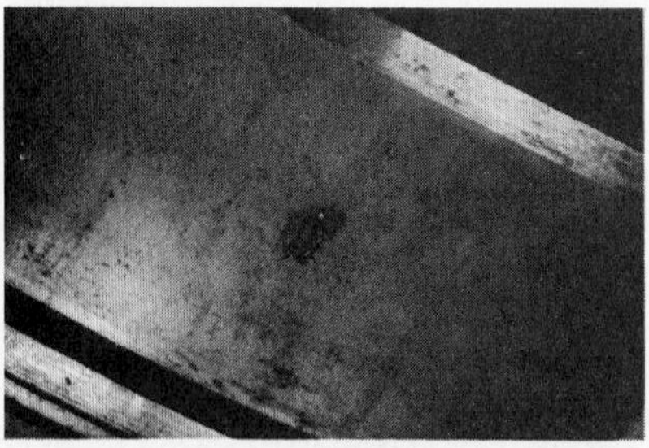

FIG. 5-7 Subsurface fatigue.

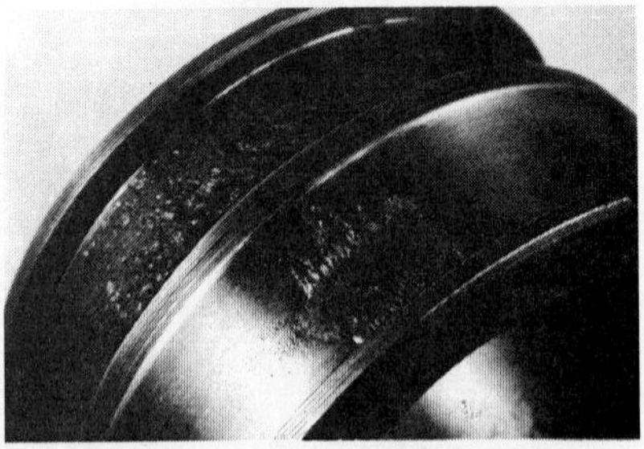

FIG. 5-8 Surface fatigue or peeling.

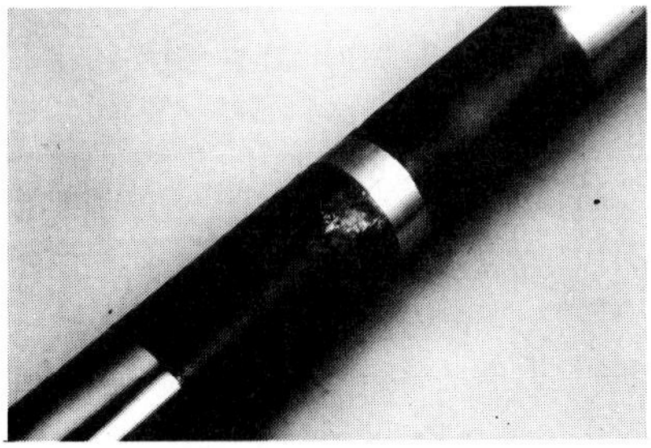

FIG. 5-9 Fretting.

FIG. 5-10 Abrasive wear.

Surface fatigue or peeling is a cracking and peeling of the surface metal. It is essentially the result of poor lubrication or surface damage which interrupts the lubricant film. See Figs. 5-8 and 5-28.

Wear Fret wear or fretting corrosion (sometimes called friction oxidation) is usually associated with improper fitting practices, with resultant damage to bearing bores, ODs, housings, and shafts, as shown in Fig. 5-9. These same phenomena can occur on bearing raceways under conditions of oscillating loads or where machinery vibrates while the bearing is stationary.

Abrasive wear, shown in Fig. 5-10, is the result of contamination. Elimination of this problem may require the use of special seals, auxiliary seals, or shields and/or improved relubrication schedules to aid in the exclusion of abrasive contaminants.

Scores and scratches are generally the result of hard particles being trapped in a bearing. This may also be the result of inadequate sealing, contamination of the lubricant, or improper precautions during installation of the bearing. See Fig. 5-11, showing scoring damage on rollers.

Corrosion is a form of wear by chemical action, caused by water or moisture inside the bearing. See Figs. 5-12 and 5-13. A solution to the problem is again better protection of the bearing by seals and/or improved lubrication practices.

False brinelling is a special form of fretting between a ball or roller and the bearing raceway. This often has an appearance almost identical to true brinelling. It is caused by a bearing being subject to vibration while stationary. See Fig. 5-14.

Smearing is surface damage resulting from unlubricated sliding contact

FIG. 5-11 Scoring.

FIG. 5-12 Corrosion.

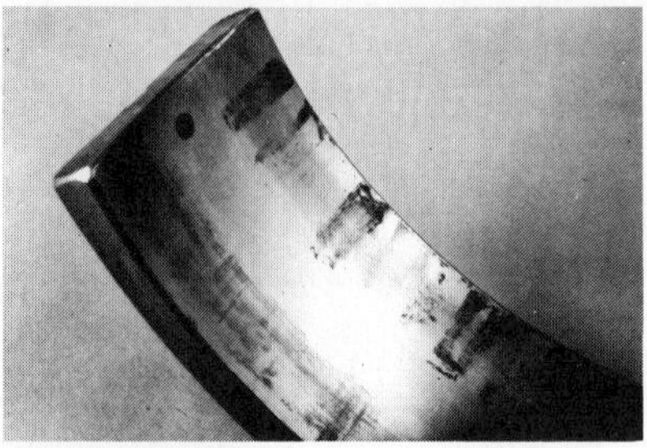

FIG. 5-13 Puddle corrosion.

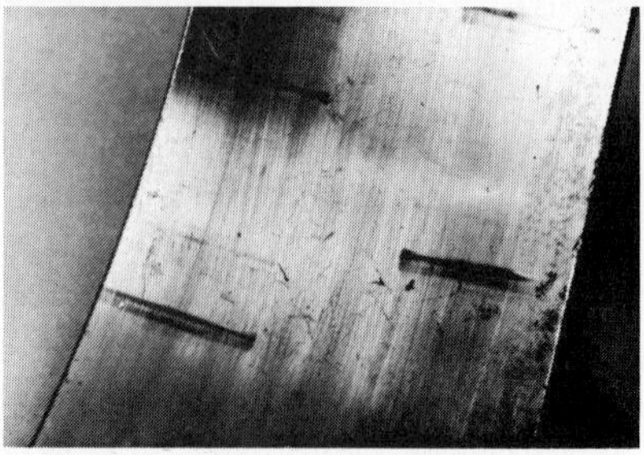

FIG. 5-14 False brinelling.

within a bearing. The plowed furrows and metal transfer are often the result of improper care and use of force or impact during installation of a bearing. See Fig. 5-15.

Electrical pitting wear is a result of electric current passing through a bearing; intermittent breaking of contacts results in electrical discharges and actual pits on the surfaces of bearing races. See Figs. 5-16 and 5-17. A special form of this, shown in Fig. 5-18, is termed fluting; it is the result of electric current passing through the bearing combined with vibration during operation.

Plastic Flow Brinelling is the actual indentation of a rolling element

FIG. 5-15 Smearing.

FIG. 5-16 Electrical pitting.

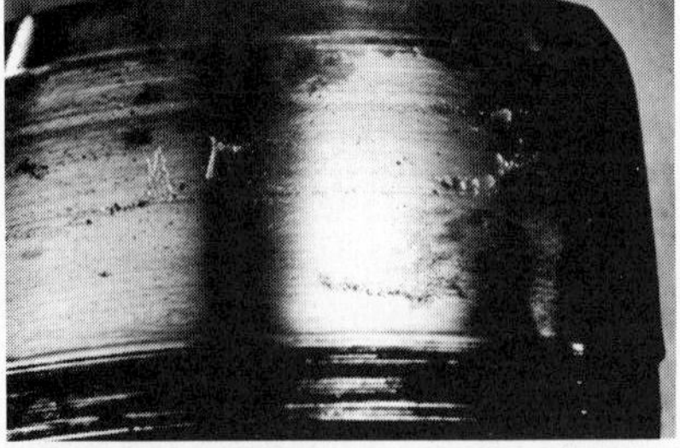

FIG. 5-17 Electrical pitting.

FIG. 5-18 Fluting.

FIG. 5-19 Brinelling.

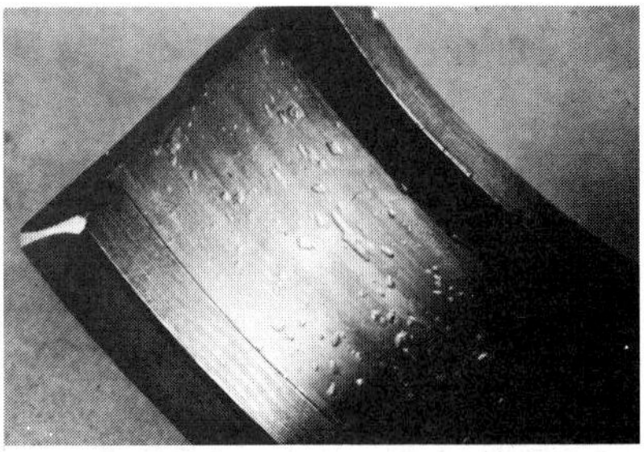

FIG. 5-20 Debris denting.

under excessive load or impact that causes stresses beyond the yield point of the materials. See Fig. 5-19.

Debris denting is essentially the same phenomenon as brinelling except that the indentation is the result of a hard particle being trapped between the raceway and the rolling element, causing a load concentration and stress exceeding the yield point of the material. See Fig. 5-20.

Cold working and bending again is the result of excessive stress beyond the yield point of the material. Bending would normally be associated with a retainer or seal component being damaged during handling or installation of a bearing. See Figs. 5-21 and 5-22.

In hot working, the bearing raceway overheats sufficiently to actually soften the hardened steel to the point where plastic flow or hot rolling occurs. This is a catastrophic failure and usually occurs from problems such as no lubrication, extreme overload, or failure of some other component. See Fig. 5-23.

Fracture Fracture is actual breakage or cracking of a component. It would usually be associated with overload, improper fitting practice, bending fatigue, or component defect. Often the cause can be determined from careful examination of the fractured component. See Figs. 5-24, 5-25, and 5-26.

The information gained from careful examination of failures can be vital in the diagnosis of the problem when bearing failure rates are excessive and occur short of design life. The evaluation of failed bearing components from both R & D testing and field service has been of significant value in the design and development of Link-Belt bearing products. The scanning electron microscope in particular has provided, for some failure analyses, a preciseness never

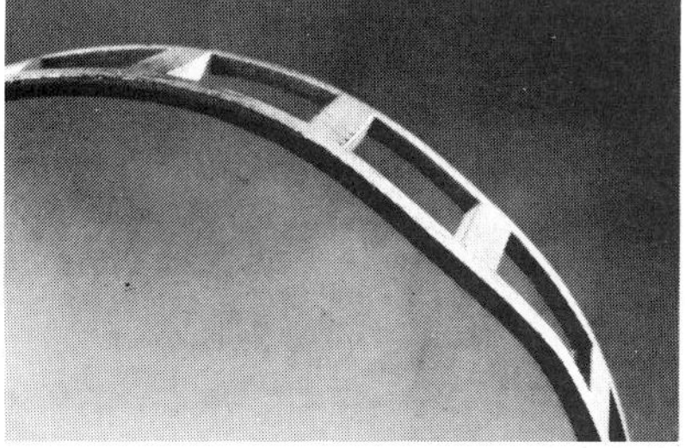

FIG. 5-21 Bending.

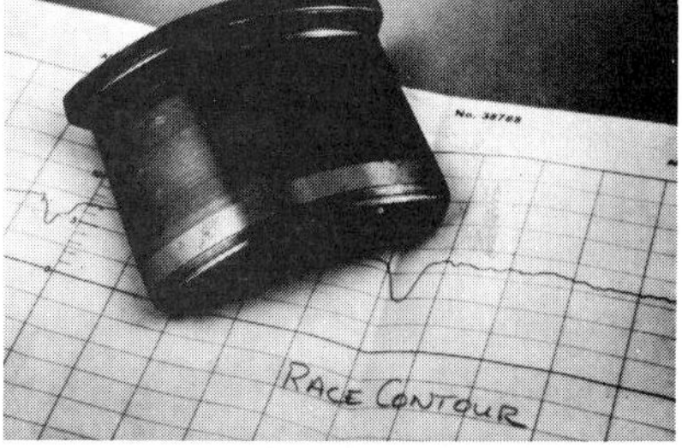

FIG. 5-22 Cold working.

FIG. 5-23 Hot working.

FIG. 5-24 Fracture.

before attainable. The SEM permits examination of surfaces containing various flaws, irregularities, dents, spalls, fractures, and so on, without metallurgical sectioning, polishing, or etching. The magnification of the undisturbed field of examination with 500 times the depth of focus of an optical microscope permits a view of the specimen in excellent detail at magnifications up to 10,000× or greater. Most examinations of bearing components will be in the range of 50× to 1000×, as shown in the examples.

Figure 5-27 shows a bearing raceway surface from which it was determined that the coolant used in grinding had been contaminated. The surface is shown at both 50× and 250×, and the general character of the ground surface as well as the defective portion is clearly seen.

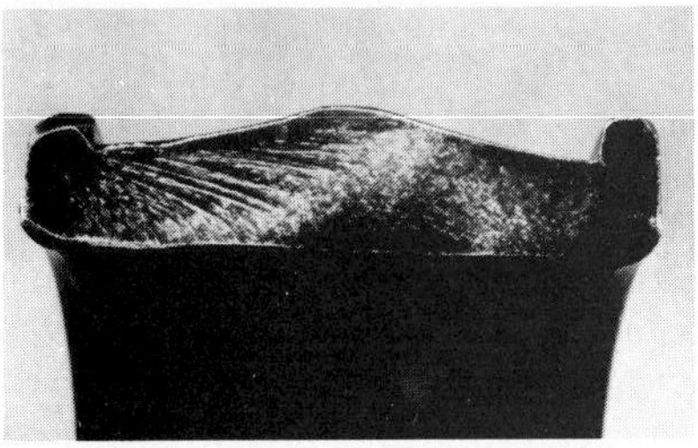

FIG. 5-25 Fracture.

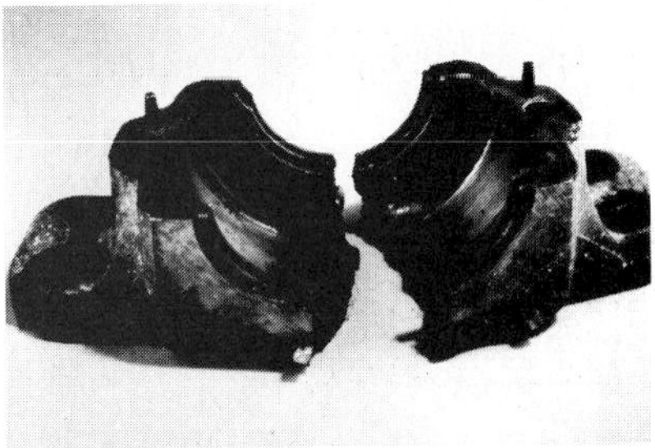

FIG. 5-26 Housing fracture.

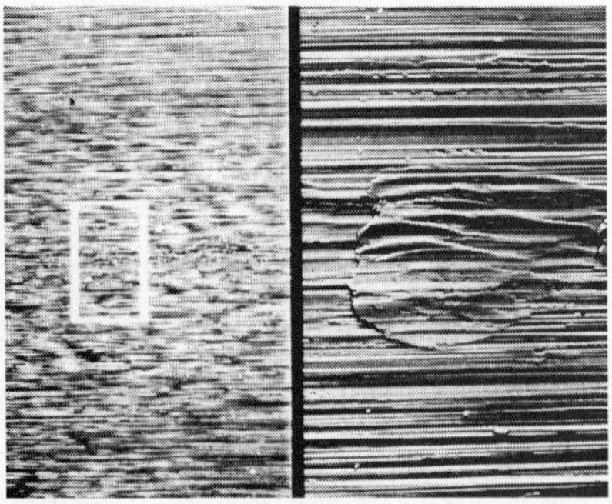

FIG. 5-27 Bearing raceway surface.

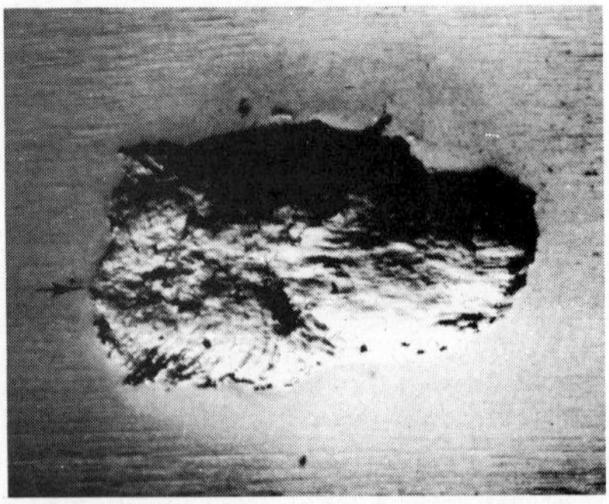

FIG. 5-28 Surface nucleated spall (50×).

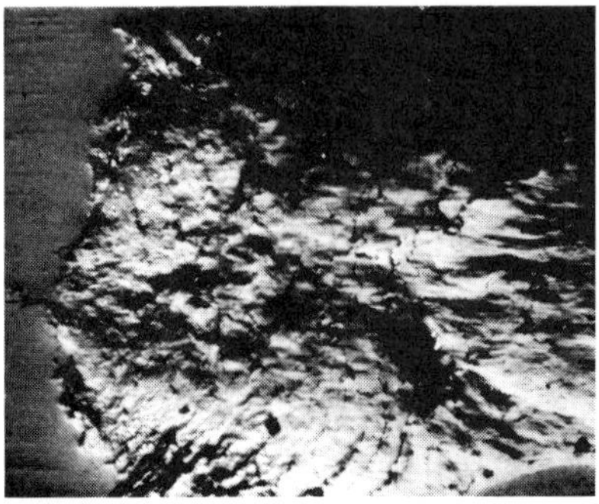

FIG. 5-29 Surface nucleated spall (100×).

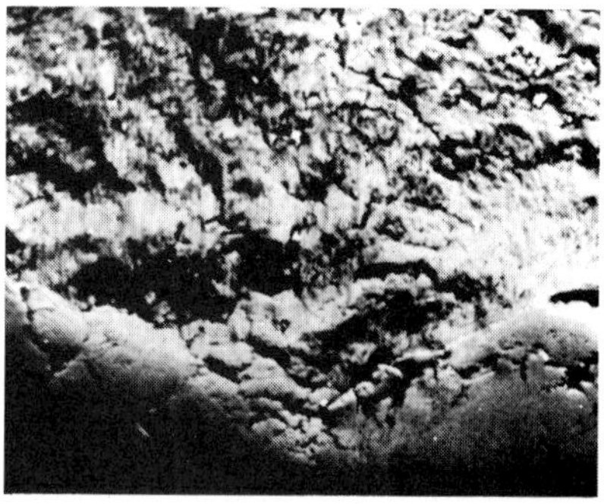

FIG. 5-30 Surface nucleated spall (500×).

A surface nucleated spall is shown in Fig. 5-28 at 50×, and the same spall is seen in Fig. 5-29 at 100× and in Fig. 5-30 at 500×. The arrow in each photo indicates the point of origin.

A subsurface fatigue spall is illustrated in Figs. 5-31 and 5-32.

A galled roller raceway surface is shown in Fig. 5-33 at 100×, and the same part at 1000× is shown in Fig. 5-34.

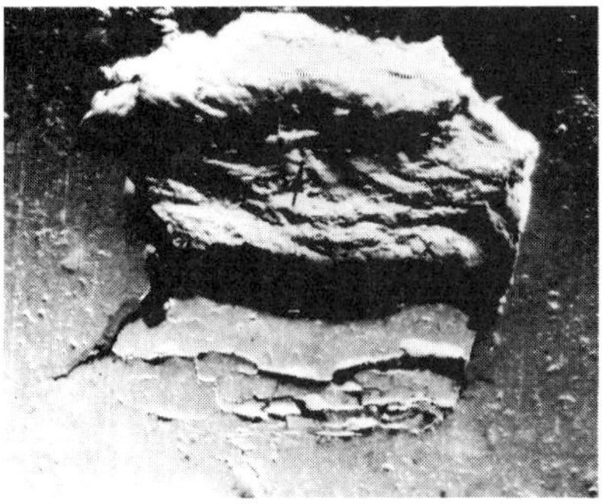

FIG. 5-31 Surface fatigue spall.

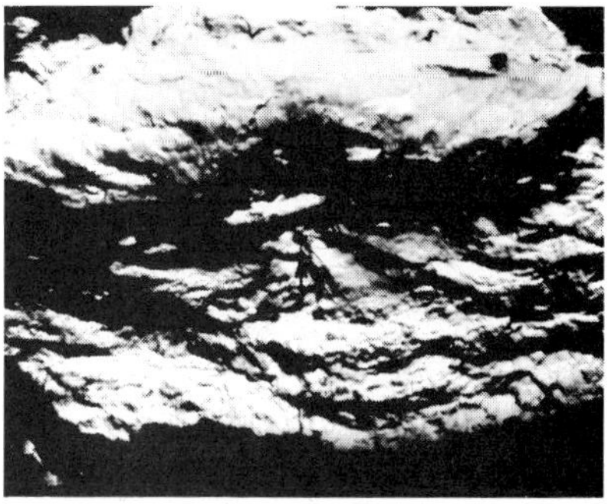

FIG. 5-32 Surface fatigue spall.

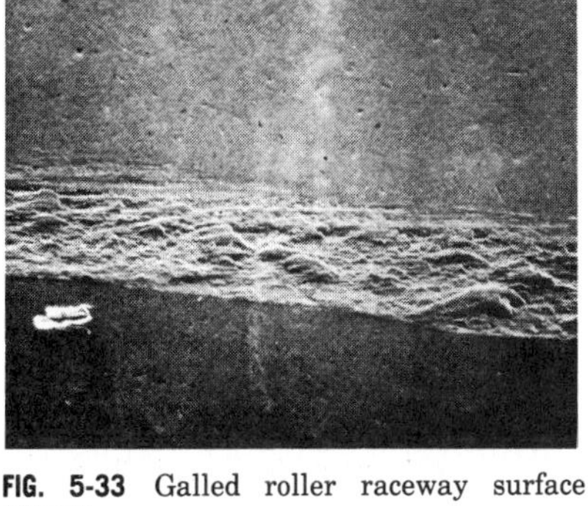

FIG. 5-33 Galled roller raceway surface (100×).

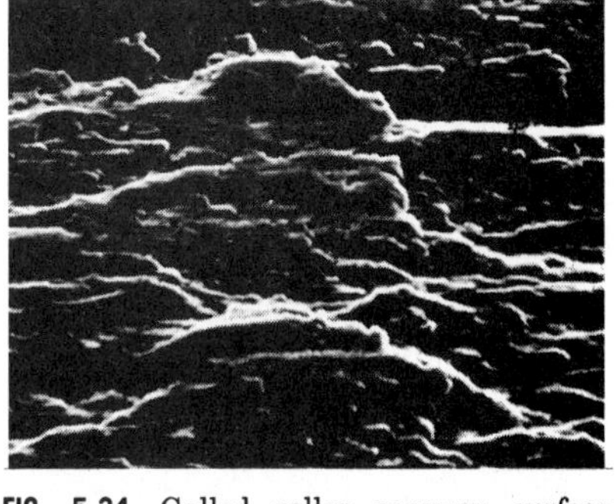

FIG. 5-34 Galled roller raceway surface (1000×).

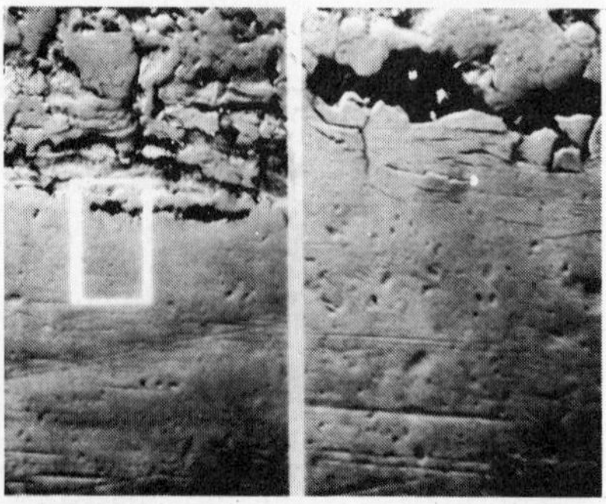

FIG. 5-35 Microspalling of a surface (100× and 500×).

Microspalling of a surface, along with surface debris indentations, is shown in Fig. 5-35 at both 100× and 500×.

These few examples illustrate the capabilities utilized in failure analysis, whether it be for R&D purposes or for the diagnosis of field service problems. If an application has experienced excessive failure rates, adequate evaluation can often make feasible the achievement of service life comparable to the theoretical L_{10} rating lives of bearings by identifying and eliminating the causes of premature failures.

LOADING ANALYSIS

5-21 Equivalent Load

In a loading analysis all bearing loads are converted to an equivalent load P. This conversion permits a direct comparison of the converted loading with the published bearing rating. The equivalent load is the constant stationary radial load that would result in the same fatigue life as that which the bearing will attain under the actual conditions of loading.

The equivalent load P is determined by the following general formula, which is applicable to both ball and roller bearings:

$$P = XF_r + YF_a \tag{5-8}$$

where P = equivalent load
F_r = radial load
F_a = thrust (axial) load
X, Y = radial and thrust factors
C_0 = basic static load rating
α = bearing contact angle, degrees
e = a variable reference value

For ball bearings, the X and Y factors are variable according to the ratio of the radial to the thrust load, as shown in Table 5-7.

For spherical roller bearings, the X and Y factors are derived from the bearing contact angle α, as shown in Table 5-8.

For cylindrical roller bearings, the X factor is 1.0 and the Y factor is zero, and thrust load theoretically has no effect on fatigue life. Cylindrical roller bearings with integral guiding ribs on the inner and outer rings will support limited thrust loads. Maximum allowable thrust under ideal conditions is

TABLE 5-7 Ball-Bearing Factors

		$F_a/F_r \leqq e$		$F_a/F_r > e$	
F_a/C_0	e	X_1	Y_1	X_2	Y_2
0.014*	0.19	1.00	0	0.56	2.30
0.021	0.21	1.00	0	0.56	2.15
0.028	0.22	1.00	0	0.56	1.99
0.042	0.24	1.00	0	0.56	1.85
0.056	0.26	1.00	0	0.56	1.71
0.070	0.27	1.00	0	0.56	1.63
0.084	0.28	1.00	0	0.56	1.55
0.110	0.30	1.00	0	0.56	1.45
0.17	0.34	1.00	0	0.56	1.31
0.28	0.38	1.00	0	0.56	1.15
0.42	0.42	1.00	0	0.56	1.04
0.56	0.44	1.00	0	0.56	1.00

* Use 0.014 if F_a/C_0 is less than 0.014.

given by the formula

$$T_M = 0.052C_A \left(\frac{500}{N}\right)^{0.3} \tag{5-9}$$

where T_M = maximum allowable thrust load
C_A = load rating C of the narrowest series of the same annulus
N = operating speed, r/min

Thrust loading on cylindrical roller bearings that is greater than a nominal locating thrust requires excellent lubrication, a stabilizing radial load, and careful control (limitation) of misalignment. Consultation is recommended when a questionable or unusual condition exists.

Specific formulas for different types of bearings and load conditions are shown in Table 5-9. These are derived from the general formula and values in Tables 5-7 and 5-8.

The equivalent load formulas for pure radial or pure thrust load take the simple form shown in Table 5-9. For simultaneous radial and thrust bearing

TABLE 5-8 Spherical-Roller-Bearing Factors

		$F_a/F_r \leqq e$		$F_a/F_r > e$	
Type of spherical roller bearing	e	X_1	Y_1	X_2	Y_2
Single row	1.5 tan α	1.00	0	0.40	0.40 cot α
Double row	1.5 tan α	1.00	0.45 cot α	0.67	0.67 cot α

Values of e, Y_1, and Y_2 are given for each bearing in the specification section.

TABLE 5-9 Equivalent Load Formulas

Type of bearing	Kind of load: Radial	Thrust	Combined radial-thrust*
Ball	$P = F_r$	$P = Y_2F_a$	$P = 0.56F_r + Y_2F_a$ or $P = F_r$
Single-row spherical roller	$P = F_r$	$P = Y_2F_a$	$P = 0.40F_r + Y_2F_a$ or $P = F_r$
Double-row spherical roller	$P = F_r$	$P = Y_2F_a$	$P = 0.67F_r + Y_2F_a$ or $P = F_r + Y_1F_a$
Cylindrical roller	$P = F_r$	—	$P = F_r$†

* Use whichever resultant is larger. The upper formula, with Y_2F_a, will yield the larger resultant if the value of F_aF_r is greater than e.

† See Sec. 5-21 for allowable thrust on cylindrical roller bearings.

loads, two equations are solved, with the larger resultant being used as the equivalent load P. The alternate solution for P is to select the applicable equation on the basis of the ratio of the thrust load to the radial load and the e factor of the bearing.

5-22 Static Load

Some applications subject nonrotating bearings to high radial, thrust, or combined radial/thrust loads. In these circumstances, permissible loads are governed by the static load rating C_0 or the total permanent deformation of the rolling element and raceway. A static equivalent load is defined as the static radial load which, if applied to the bearing, would produce the same maximum raceway contact stress as that which occurs under the actual condition of loading. The static equivalent load P_0 for radial-type bearings under combined radial thrust load is

$$P_0 = X_0F_r + Y_0F_a \tag{5-10}$$

where P_0 = static equivalent load
X_0 = radial factor (see Table 5-10)
Y_0 = thrust factor (see Table 5-10)
F_r = radial load
F_a = thrust (axial) load

The static equivalent load P_0 is compared directly to the static load rating C_0 of the bearing to determine the suitability of a selection. The static load rating C_0

TABLE 5-10 X_0 and Y_0 Factors

Type of bearing	X_0	Y_0
Ball	0.6	0.5
Single-row spherical roller	0.5	$0.55Y_2$
Double-row spherical roller	1.0	$0.66Y_2$

If P_0 is less than F_r, use $P_0 = F_r$.

is a practical load limit for most applications. A static equivalent load equal to the AFBMA static load rating C_0 will result in contact stresses of 580,000 lb/in^2 (4000 MPa) in Link-Belt bearings under conditions of normal load distribution. No measurable permanent deformation will be found at this stress level, and vibration characteristics will be unaffected in subsequent operation. Careful judgments must be made that the shaft and bearing support systems will not deflect adversely to produce abnormal load distribution under the static load. Also, impact loads, shock loads, and highly repetitive static loads will involve higher strain rates or cumulative damage and produce a greater amount of permanent deformation than a gradually applied or steady load condition. System deflections under high static loads may also produce slight lateral motion between the rolling elements and raceways, resulting in potential bearing surface damage.

Static loads beyond the static load rating C_0 are permissible in many applications where slight permanent deformations are not detrimental to subsequent bearing operation. Measurable permanent deformation is generally not found until contact stresses approach 600,000 lb/in^2 (4137 MPa). In applications such as tank-gun trunnion bearings, heavy hangar doors, rotating display signs, and so on, loads of three to four times the static load rating have been applied successfully. Consultation is recommended when static loads exceed ratings.

5-23 Bearing Selection

The load-speed-life relationship for rolling bearings is expressed by the basic formula discussed in Sec. 5-17, where

$$L = \left(\frac{C}{P}\right)^b \quad \text{or} \quad L^{1/b} = \left(\frac{C}{P}\right) \tag{5-11}$$

in which $b = 3$ for ball bearings or $10/3$ for roller bearings.

From this it is seen that, for a given ratio of bearing rating to bearing load C/P, there is only one resultant life L in millions of revolutions. To select a bearing for a given fatigue life, it is only necessary to provide a bearing C rating that yields the desired C/P ratio.

A required bearing C rating may therefore be computed by the formula

$$C_{req} = P\left(\frac{C}{P}\right) = P(L)^{1/b} \tag{5-12}$$

For example, under an equivalent radial load P of 5000 lb (22,241 N), what C rating will be required to provide a life of 25 million revolutions?

$C_{req} = 5000(25)^{1/3} = 5000 \times 2.92 = 14{,}600$ lb (64,944 N) for ball bearings
$C_{req} = 5000(25)^{3/10} = 5000 \times 2.63 = 13{,}150$ lb (58,494 N) for roller bearings

The tables and nomographs provided in manufacturers' literature show C/P values (same as $L^{1/b}$) for various combinations of life and speed. It must be emphasized that there are separate tables for ball and roller bearings because of the different values of the exponent b, as previously described.

In theory, the bearing selection process is straightforward:

1. Decide what life is appropriate, thereby establishing a required value for C/P.
2. Calculate the required bearing C rating.

$$C_{req} = \text{applied } P \times \text{required } \left(\frac{C}{P}\right)$$

3. Any bearing with a C rating equal to or greater than C_{req} will satisfy the desired life requirement.
4. The exact fatigue life of the bearing selected is calculated from the ratio of its C rating to the applied equivalent radial load P.

$$L = \left(\frac{C}{P}\right)^b \quad \text{in millions of revolutions}$$

or

$$L_{10} = \frac{1{,}000{,}000(C/P)^b}{n \times 60} \quad \text{in hours} \tag{5-13}$$

In practice, the selection process for pure radial loads is exactly as shown. Where thrust loads or combined radial-thrust loads exist, the method is the same, but the problem becomes more complicated. In Table 5-9 it can be seen that the equivalent load P is dependent on F_a/F_r and the e factors of the bearings selected. This means that a trial method or multiple selection process is required.

For ball bearings, the method involves the calculation of a maximum required C rating. Tentative selection of a bearing with a C rating about equal to the maximum C_{req} value is then made. An exact calculation of P and C_{req} is then performed for the tentative bearing selection. The C_{req} calculated for the tentative selection will usually be lower, and may indicate that use of a smaller bearing is possible. The process is then repeated to arrive at the optimum selection.

For spherical roller bearings, the method is identical to that for ball bearings if the bearing selections being considered have widely varying Y_2 factors. If, as in many cases, the bearings being considered have Y_2 factors that vary by only 5 or 10 percent, a single calculation for a required rating is usually sufficient.

An alternative method of bearing selection is provided by the expanded rating tables. These tables show bearing ratings computed for a broad range of hours of life and rotational speeds. A radial equivalent load P can be compared directly with the values in the tables for a quick bearing selection.

Where several bearings are being evaluated as alternative selections, a tabular worksheet is often helpful. Table 5-11 illustrates such a worksheet with some examples filled in. By this method, several alternative bearing selections are entered in the table, along with ratings, factors, etc. An exact calculation of equivalent radial loads P and the L_{10} life for each bearing considered can then be made.

In summary, the selection of a bearing for a desired fatigue life requires the selection of one with a basic load rating C that provides a satisfactory ratio to the equivalent radial load P. Where bearing loads are radial only, a direct comparison and selection may be made from the expanded rating tables.

5-24 Bearing Reaction

Loads exerted on shafts by machine components are translated into bearing reactions by the use of static beam solutions. The shaft is normally considered to be a simple beam with two supports (bearings). Self-aligning-type bearings are considered nonrigid supports, thereby creating no moment reaction. Cylindrical roller bearings may be treated as self-aligning, provided that deflections do not exceed 0.005 in per inch of shaft length. This type of static beam problem is widely covered in textbooks and handbooks. However, a typical example is shown in Fig. 5-36.

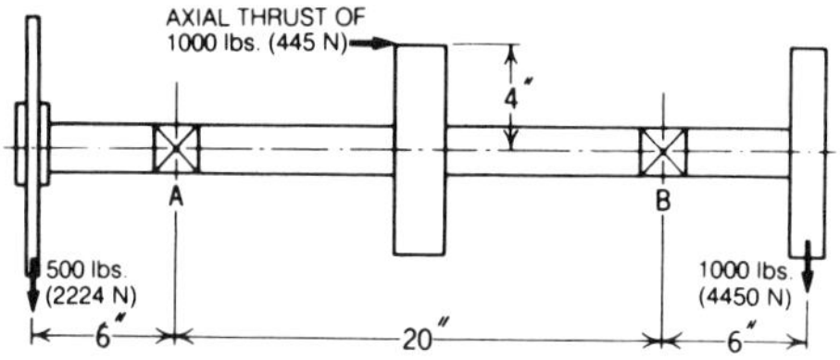

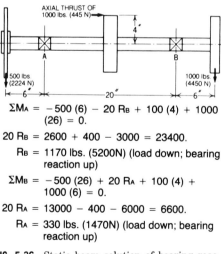

$\Sigma M_A = -500\,(6) - 20\,R_B + 100\,(4) + 1000\,(26) = 0.$

$20\,R_B = 2600 + 400 - 3000 = 23400.$

$R_B = 1170$ lbs. (5200N) (load down; bearing reaction up)

$\Sigma M_B = -500\,(26) + 20\,R_A + 100\,(4) + 1000\,(6) = 0.$

$20\,R_A = 13000 - 400 - 6000 = 6600.$

$R_A = 330$ lbs. (1470N) (load down; bearing reaction up)

FIG. 5-36 Static beam solution of bearing reactions.

TABLE 5-11 Evaluation of Alternative Bearing Selections

Given application conditions:

F_r = 1000 lb (4450 N) L_{10} desired life = 20,000 h Required $L^{1/3}$ = 12.9

F_a = 350 lb (1560 N) n = 1800 r/min $L = 2160 \times 10^6$ revolutions Required $L^{3/10}$ = 10.0

Bearing number	U331D	A22196S	22210LB	22310LB
Bearing type	Single-row ball	Single-row spherical roller	Double-row spherical roller	Double-row spherical roller
Basic load rating C	10,700 lb (47,600 N)	11,100 lb (49,400 N)	15,700 lb (69,800 N)	35,500 lb (157,900 N)
Static load rating C_0	7360 lb (32,740 N)	—	—	—
X_1 = 1.0 for all bearings	1.0	1.0	1.0	1.0
Y_1 = 0 for all ball bearings; 0 for single-row spherical roller bearings; As shown for double-row spherical roller bearings	0	0	2.57	1.73
X_2 = 0.56 for all ball bearings; 0.40 for single-row spherical roller bearings; 0.67 for double-row spherical roller bearings	0.56	0.40	0.67	0.67

$\frac{F_a}{C_0}$	For ball bearings, establishes values of e and Y_2	0.048	—	—	—
e	(See tables in specification section)	0.25	0.40	0.26	0.39
Y_2	(See tables in specification section)	1.78	1.49	3.83	2.57
$\frac{F_a}{F_r}$	If greater than e, use X_2 and Y_2 values to calculate P. Otherwise, use X_1 and Y_1	0.35	0.35	0.35	0.35
$P =$	$X_1F_r + Y_1F_a$ or $X_2F_r + Y_2F_a$	Use X_2 & Y_2: $0.56 \times 1000 = 560$ $1.78 \times 350 = \underline{623}$ 1183 lb (5260 N)	Use X_1 & Y_1: $1.0 \times 1000 = 1000$ $0 \times 350 = \underline{0}$ 1000 lb (4450 N)	Use X_2 & Y_2: $0.67 \times 1000 = 670$ $3.83 \times 350 = \underline{1340}$ 2010 lb (8940 N)	Use X_1 & Y_1: $1.0 \times 1000 = 1000$ $1.73 \times 350 = \underline{605}$ 1605 lb (7140 N)
$\frac{C}{P}$	Must equal or exceed required $L^{1/b}$	9.04	11.1	7.81	22.1
L_{10}	From nomograph, table, or calculation: $L_{10} = \left(\frac{C}{P}\right)^b \frac{1{,}000{,}000}{n \times 60}$	6840 h	28,200 h	8755 h	281,000 h

In the example shown, the solution uses the two moment equations with moments taken about the bearing positions. Often it is convenient to use the summation of forces and one moment equation. For computer programs, moments are often taken about points outside the shaft system. Approaches may vary for different problems or analytical techniques. The primary point is to be careful to use consistent sign terminology for clockwise and counterclockwise moments and force directions. Also, force directions and reaction directions must be accurately determined. Where loads and forces occur in different planes, they are usually converted to perpendicular planes or to horizontal and vertical planes.

5-25 Three-Bearing Shafts

A shaft is considered indeterminate when it is supported by three bearings. The third bearing affects the load on the other two, inducing leverage forces caused by deflections of the shaft. The equations for static equilibrium do not yield a solution.

Methods are available for approximating bearing reactions on shafts supported at three or more points. The procedure for determining reactions involves (in addition to the equations for static equilibrium) an assumed deflection at some point of support. An expression for this deflection is obtained by methods of double integration, area-moment, or superposition. The reactions calculated are subject to the accuracy of the bearing installation. An error in the bearing position assumed in calculation can lead to serious errors in calculated bearing loads. With three bearings on short centers, an alignment error of a few thousandths of an inch can impose destructive loads on the bearings. Such applications should be examined very carefully for validity of the solutions.

5-26 Variable Loads

Many bearing applications involve variable loading conditions. Usually loads and speeds will change in accordance with a known pattern. A typical example of such variable loading is a multiple-speed transmission. Here it is necessary to establish an equivalent radial load P as defined in Sec. 5-21.

If a bearing operates under two or more applied loads $F_{r1}, F_{r2}, F_{r3}, \ldots, F_{rn}$ at speeds of $N_1, N_2, N_3, \ldots, N_n$ and periods of time $T_1, T_2, T_3, \ldots, T_n$, respectively, the following equation is used to calculate the equivalent radial load P:

$$P = \left[\frac{T_1N_1(F_{r1})^b + T_2N_2(F_{r2})^b + \cdots + T_nN_n(F_{rn})^b}{N(T_1 + T_2 + T_3 + \cdots + T_n)}\right]^{1/b} \qquad (5\text{-}14)$$

$b = 3$ for ball bearings; $10/3$ for roller bearings.

The N term in the denominator may be any selected speed, but $33\frac{1}{3}$ r/min is normally chosen. This permits direct comparison with bearing ratings.

In using the variable load equation, the load terms F_{r1}, F_{r2}, etc., require modification of thrust loading, as each F_r term is considered a radial load. For example, in a machine where a portion of the cycle involves thrust or combined radial-thrust loading, the appropriate $(F_r)^b$ term will become $(XF_r + XF_a)^b$ in accordance with Sec. 5-21 (treatment of combined radial-thrust loads).

On occasion, variable loading analysis must be based on assumptions because of insufficient knowledge of the actual dynamic forces involved. Logical assumptions and use of the variable load formula will provide an analysis that is superior to that obtained by using an arbitrary overall factor.

5-27 Shock Loads

The effect of shock loads on fatigue life is difficult to assess. With precise knowledge of the rate of load change and the peak duration, the variable-load equation (Sec. 5-26) can be used with discretion. Because of the time factor in this equation, medium shock loads of short duration will in theory have little effect on equivalent load and bearing life. Heavier shock loads may result in serious plastic deformations, impairment of smooth operation, and reduction of bearing life. When the bearing C_o rating is exceeded, a careful evaluation of the application is necessary. For example, an explosive shock that occurs once in the life of a machine may cause damage. Such damage can often be tolerated for a limited period of time until repairs can be made. If such a shock occurs on a more frequent basis, a heavier bearing would be necessary to avoid damage.

5-28 Oscillating Loads

Applications involving oscillating motion sometimes make bearing selection difficult because they present variations in stress cycling. These applications should be referred to the bearing manufacturer for a thorough analysis.

Figure 5-37 illustrates schematically four significant positions in the angular movement of an inner ring in a rolling bearing. Position 1 is the starting position. Position 2 shows the rotation of the inner ring through an angle A, during which the rolling elements have rotated 180°. In position 3, further rotation of the inner ring through angle B shows point a on the inner ring at the first point of contact with the previously adjacent roller 2. Position 4, with the inner ring rotated through angle C, shows point b on the outer ring at the first point of contact with the previously adjacent roller 3.

These three angular movements represent three distinct and important half-angles of oscillation. Angle A is the minimum half-angle at which each roller will be stressed through the same plane four times in each full oscillation cycle. Angle B is the single roller overlap angle at the inner ring raceway. Angle C is the half-angle at which single roller overlap occurs at the outer ring raceway. A similar set of half-angles exists for outer ring oscillation (inner ring fixed).

These critical angles vary for each size and type of bearing. In some instances angle A may be greater than angle B or C, depending on the bearing dimensions and the arc between the rolling elements. These rotational angles determine the pattern of repetitive stress on the rings and rolling elements and, in turn, determine the proper oscillating load factor to be used in computing an equivalent radial load.

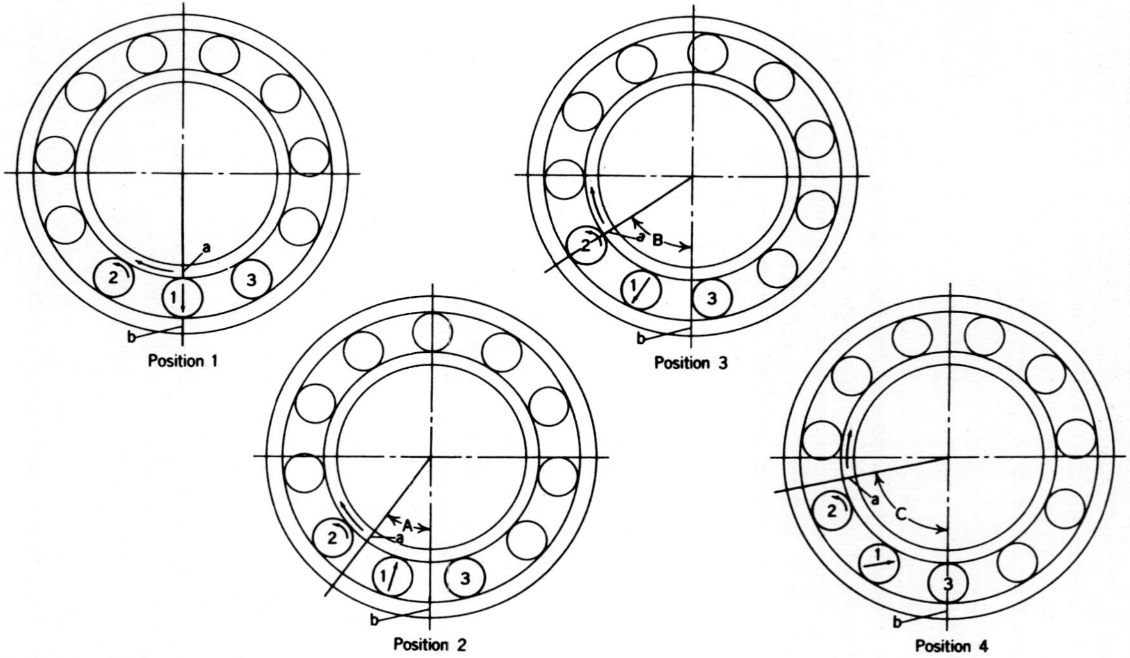

FIG. 5-37 Positions of rolling elements (outer ring fixed).

Where angles of oscillation are less than the single overlap angles of the inner or outer ring, lubrication is more critical than the selection of load-modifying factors. With small-angle oscillation, the rolling elements, in effect, wipe the raceways clean of lubricant, with resultant damage by false brinelling. In such cases, the elements can partially bury themselves in the raceways of the inner or outer ring. Keeping the bearing cavity full of a soft EP-type grease or, even better, an EP oil has been found effective in minimizing this action. Where extremely small angles of oscillation prevail, a rolling-element bearing is not suitable. Alternative methods of support, such as torsion bars or rubber joints, are suggested.

5-29 Unbalanced Loads

Unbalanced forces may prevail as a condition of machine design or may exist as an undesirable but predictable condition of operation. In either case, these forces must be considered in an analysis of bearing loads in a machine system. Usually these forces move in phase with a rotating shaft. The effect on the bearing is similar to, but more severe than, that produced by a stationary load and shaft and a rotating outer ring.

Vibratory action is coming into increasing use in process equipment. Much of this equipment functions on the principle of eccentric weights. Rotating at high speed about a center of rotation, these weights create a known path and frequency of vibration. The force and motion involved are calculated from the well-known centrifugal force formulas:

$$F = \frac{MV^2}{r} \quad \text{or} \quad F = Mr\omega^2 \tag{5-15}$$

In other cases, the unbalanced forces are not designed into the machine, but must be calculated by using assumptions. A case in point is a fan operating in a dirty environment. Dirt builds up on the fan blades and is then thrown off in large masses, thereby creating variable degrees of unbalance. At some point the fan blades must be cleaned so that the fan is again well balanced. This phenomenon causes a variable unbalanced load cycle. It is assumed that the degree of unbalance gradually increases to a peak value at a fairly uniform rate, then returns to a low level when the blades are cleaned.

Rotating loads induce a vibratory effect on all parts of a machine. All parts of the bearings, especially the retainers, may have their normal paths of motion altered. This vibratory type of loading is considered the most severe loading normally encountered by a rolling-element bearing; it is used extensively in development and testing of bearings.

5-30 Inertia Loads

Changes in velocity or direction of a moving body are caused by external forces. These may be the force of gravity, collision with another body, or the force exerted by a connecting body. Bearings are frequently subject to these inertial forces. Generally, any mechanism using a reciprocating motion or a rotary mechanism subject to fluctuating loads should be examined for inertial bearing load effects.

The simple crank is one of the widely used mechanisms producing inertial forces. The crank produces simple harmonic motion, and the forces are easily calculated:

$$F = Mr\omega^2 \cos \omega t \tag{5-16}$$

where F = force
$M = W/g$ = mass
r = crank radius
ω = angular velocity
t = time, where ωt defines the angular position of the crank

There are more complex linear motions, such as those generated by cams, linkages, and so on. Here a specific mathematical description of the motion is necessary to accurately analyze accelerations and force.

One aspect of loading analysis which is not always recognized is the influence of cyclic or random torsional variations. In systems utilizing diesel engines, hydrostatic drives, or other such devices, or those involving the rotation of randomly oriented masses, significant load and speed variations may be encountered. In extreme cases the weighted average load on bearing may be affected. More often the influence adds to stresses and forces on bearing retainers and mounting systems. Retainer pocket wear and even retainer fracture have been observed, and normal shaft fits or mounting arrangements may be inadequate. Heavier shaft fits, alternative bearing retainer types, or equipment modification may be necessary where such speed and load variations are found or expected.

Purely rotary mechanisms produce inertial forces, termed a "flywheel effect." By virtue of the flywheel effect, a machine may operate with a 10-hp (74.6-kW) drive but produce work through a portion of its operating cycle at a 500-hp (373-kW) rate. Flywheel effect calculations can be made on the basis of flywheel "slip," whereby an assumed or known reduction in speed occurs over a certain period of time. With an assumed or known slip and a known rotational inertia, an energy calculation is made as follows:

$$E = \frac{k^2 W}{2g} (\omega_1^2 - \omega_2^2) \tag{5-17}$$

where k = radius of gyration
W = weight of rotating body
g = acceleration of gravity
ω = angular velocity
$\omega = 2\pi N/60$, where N = rotational speed, r/min
E = energy transfer

For example, a rotor weighing 10,000 lb (44,500 N) runs at 1800 r/min and slows to 1600 r/min in 1 min. Its radius of gyration is 2 ft (0.61 m). What is the generated power?

$$E = \frac{(2)^2 \times 10{,}000}{2 \times 32.2}\left[\left(\frac{2\pi \times 1800}{60}\right)^2 - \left(\frac{2\pi \times 1600}{60}\right)^2\right] \tag{5-18}$$

$$E = 4{,}630{,}000 \text{ ft-lb } (6{,}277{,}000 \text{ J})$$

The generated power is

$$\frac{4{,}630{,}000}{33{,}000} = 140 \text{ hp } (104 \text{ kW})$$

This power may be transmitted through gears, pulleys, or other elements to result in direct bearing loads. These inertial forces are inherently intermittent and are usually treated as variable loads, as discussed in Sec. 5-26.

5-31 Power Transmission Loads

Mechanical devices transmit and/or consume power. The transmission of power to or from a rotating shaft is by means of the torque created by belts, chains, gears, wheels, cranks, and so on. An analysis of forces resulting from power transmission by these components is always based on the power-torque relationship. A tangential force applied to the periphery of a wheel, sprocket, pulley, and so on, is multiplied by the radius of the component, creating torque. When rotary motion is added, we have the concept of power.

$$F = \frac{cH}{D_m N} \tag{5-19}$$

where F = tangential force
H = power transmitted
D_m = diameter of machine component at which tangential force acts
N = rotational speed in r/min
c = constant depending on units used
= 126,000 (using lb, in, and hp)
= 19,100 (using N, m, and kW)

In the analysis of various kinds of machine components, the tangential force is created by different types of pressures and contacts. In gears, for example, the total contact pressure between teeth does not act in a purely tangential direction. In belt drives, the tangential force is developed by virtue of friction between the belt and pulley. To develop the required frictional and tangential force, the actual tension in the belt must be higher than the tangential force. The following specific examples of power transmission force analysis provide the basic information from which most systems may be analyzed.

Chains Chain drives produce a radial force on a driving or driven shaft through the power transmitted. The formula for calculating the radial force F produced by the chain is

$$F = \frac{cH}{D_m \times N} \tag{5-20}$$

where H = power transmitted
D_m = sprocket pitch diameter
N = rotational speed, r/min
c = constant

The force F acts along the line between the driving and driven sprockets and tends to pull the shafts toward each other. This calculation is valid for all but very long drive centers with tight chains.

Belts Belts produce a radial force on the shaft through the torque transmitted. The formula is essentially the same as that used for chains, except that a tension factor must be used to compensate for initial tensioning of the belt, which is necessary to prevent belt slippage. Thus, the formula for calculating the radial force F is

$$F = \frac{cHG}{D_m N} \tag{5-21}$$

where G = tension factor
c = 126,000 when using lb, in, and hp
c = 19,100 when using N, m, and kW

Tension factors for belts vary from 1.5 to 5.0, depending on the transmitted power, the number and size of the belts, the sheave center distance, and the drive ratio. For greatest accuracy, designers using belt drives should calculate the tension factors from the belt manufacturer's catalog.

Information necessary for the complete calculation of belt drives includes:

1. Power transmitted
2. Center distance between shafts
3. Sheave diameters
4. Number, size, and type of belts
5. Shaft speeds
6. Spacing of components along shaft
7. Direction of driving force

If this information is unavailable, the tension factors in Table 5-12 are used for approximate calculations.

Gears Forces exerted on bearings by gear drives are tangential, separating, and in some cases thrust, depending on the type of gears. The direction and magnitude of the forces from helical, worm, bevel, and spiral bevel gears are dependent on the nature of the tooth contact. These forces are grouped into radial and thrust components for use in moment equations and in determining the bearing reactions. Application details necessary for the complete engineering solution of a gear drive are as follows:

1. Gear pitch diameter
2. Shaft speeds, r/min

TABLE 5-12 Tension Factors for Belt Drives

Type of drive	Tension factor G*
V-belts	2.0
Leather belts, 1, 2, or 3-ply	2.0
Leather belts, short centers	3.0
High ratios, low speeds	3.0
Flat rubber belts, 3 to 12 ply	2.0

* As a point of caution, it is important to know that drive belts are often improperly tensioned. This can drastically affect the life of the bearings that support the drives as well as the life of the belts.

3. Tooth pressure angle, degrees
4. Spiral or helix angle, degrees
5. One-half pitch cone angle, degrees
6. All shaft center distances
7. Directions of shaft rotation
8. Horizontal and vertical locations of shafts
9. Power transmitted
10. Worm-gear lead (single, double, or triple)

When calculating gear drive forces, load factors from Sec. 5-33 should be used to compensate for inaccuracies in gear-tooth engagement or tooth impact at high speeds.

The following symbols are used in the analysis of gear forces:

D_m = pitch diameter of gear
F = tangential force
F_w = tangential force of worm
F_r = radial load on bearing
F_a = thrust load on bearing
H = power transmitted
N = shaft speed, r/min
S_f = gear separating force
T = thrust force (from bevel, helical, or worm gear)
W = total radial gear-tooth force
θ = tooth pressure angle, degrees
α = spiral or helix angle, degrees
β = one-half pitch cone angle, degrees

The total radial force produced by a spur gear drive is the resultant of the tangential force F and the separating force S_f, where:

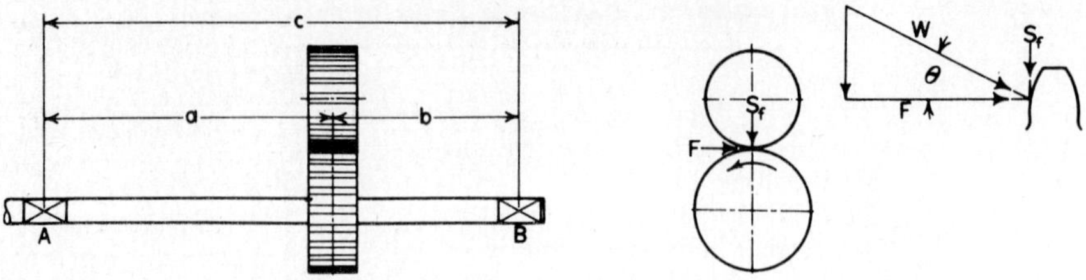

FIG. 5-38 Forces produced by a spur gear.

$$F = \frac{cH}{D_m \times N} \quad \text{and} \quad S_f = F \tan \theta \tag{5-22}$$

Since these forces act at right angles, the resultant or total radial force W is

$$W = \sqrt{F^2 + S_f^2}$$
$$c = 126{,}000 \text{ when using lb, in, and hp} \tag{5-23}$$
$$c = 19{,}100 \text{ when using N, m, and kW}$$

(See Fig. 5-38.)

The total radial force produced by a helical gear drive consists of a tangential force F, the separating force S_f, and the thrust T, where

$$F = \frac{cH}{D_m \times N} \qquad S_f = \frac{F \tan \theta}{\cos \alpha}; \tag{5-24}$$

$$F_a = T = F \tan \alpha \tag{5-25}$$

For this type of application, it is usually convenient to work with forces in only two planes at right angles to each other, designated as vertical and horizontal planes. All forces are resolved into vertical and horizontal components in these two planes. Notice that thrust force T produces a $TD_m/2$ moment, contributing to the radial loading of the bearings. (See Fig. 5-39.)

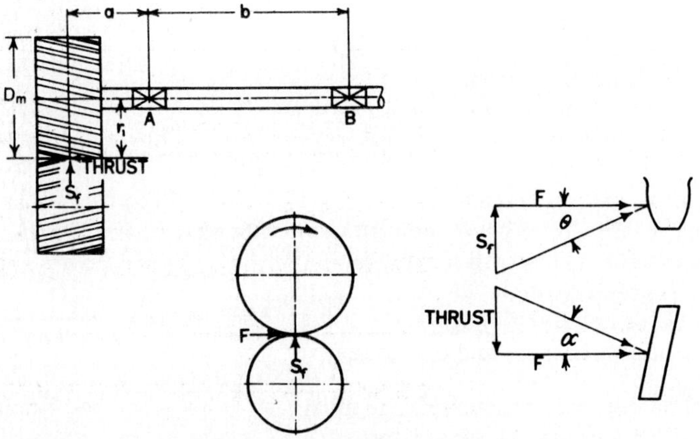

FIG. 5-39 Forces produced by a helical gear drive.

Bearing load determination for worm-gear drives may be simplified by using the three perpendicular components of the normal tooth load. The components are:

F_w = tangential driving force of worm
S_f = separating force caused by tooth pressure angle
T_w = worm thrust at pitch radius due to helix angle α causing worm gear to turn

The formulas for determining these forces are:

$$F_w = \frac{cH}{D_m \times N} \qquad S_f = \frac{F_w \times \tan \theta}{\tan \alpha} \tag{5-26}$$

$$T_w = \frac{F_w}{\tan \alpha^1} \tag{5-27}$$

$(\alpha^1 = \alpha + 3°$ for friction loss)
c = 126,000 when using lb, in, and hp
c = 19,100 when using N, m, and kW

(See Fig. 5-40.)

It is most convenient to work with forces in only two planes at right angles to each other, designated as the vertical and horizontal planes. All forces are resolved into vertical and horizontal components in these two planes. From these components, the radial loads acting in the two planes on each bearing can be determined and then combined to obtain the total radial load F_r on the bearing.

Bearing load determination for straight bevel-gear drives may be simplified by using the three perpendicular components of the normal tooth load. The components are:

F_w = tangential force perpendicular to plane of shafts
S_f = separating force caused by tooth pressure angle; this breaks down into axial forces: pinion thrust T_p and gear thrust T_g

The formulas for determining these forces are:

$$F = \frac{cH}{D_m \times N} \qquad S_f = F \times \tan \alpha$$

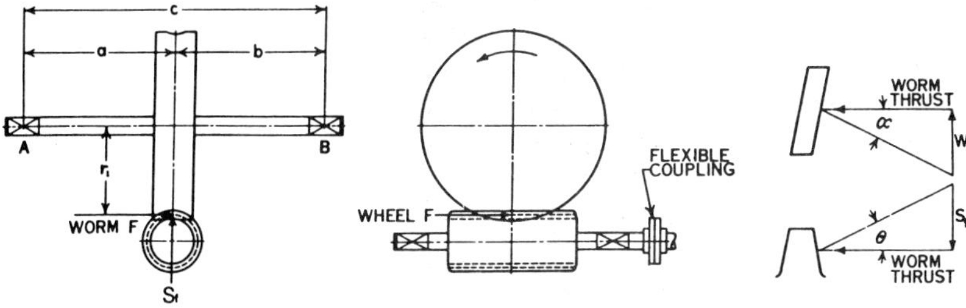

FIG. 5-40 Forces produced by a worm-gear drive.

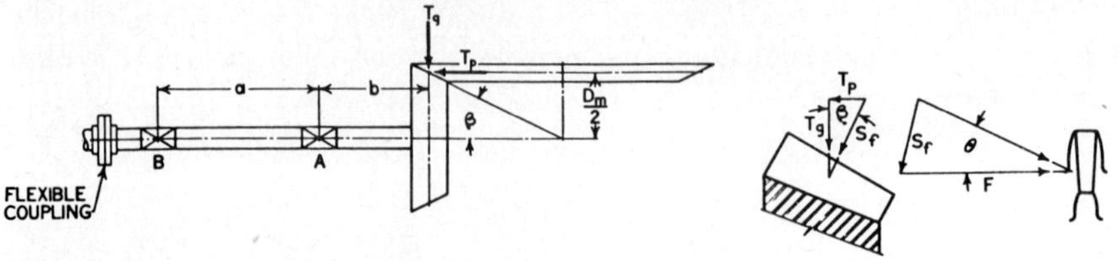

FIG. 5-41 Forces produced by a straight bevel gear drive.

$$T_p = S_f \times \sin \alpha;\ T_g = S_f \times \cos \alpha$$
$$c = 126{,}000 \text{ when using lb, in, and hp}$$
$$c = 19{,}100 \text{ when using N, m, and kW}$$

(See Fig. 5-41.)

It is most convenient to work with forces in only two planes at right angles to each other, designated as the vertical and horizontal planes. All forces are resolved into vertical and horizontal components in these two planes. From these components, the radial loads acting in the two planes on each bearing can be determined and then combined to obtain the total radial load F_f on the bearing.

Bearing load determination for spiral bevel gear drives requires complete information regarding the spiral or helix angle, pitch cone angle, tooth pressure angle, and direction of rotation. (See Fig. 5-42.)

The determination may be simplified by using the three perpendicular components of the normal tooth load. The components are:

F = tangential force perpendicular to plane of shafts
T_p = pinion thrust
T_g = gear thrust

The formulas for determining these forces are:

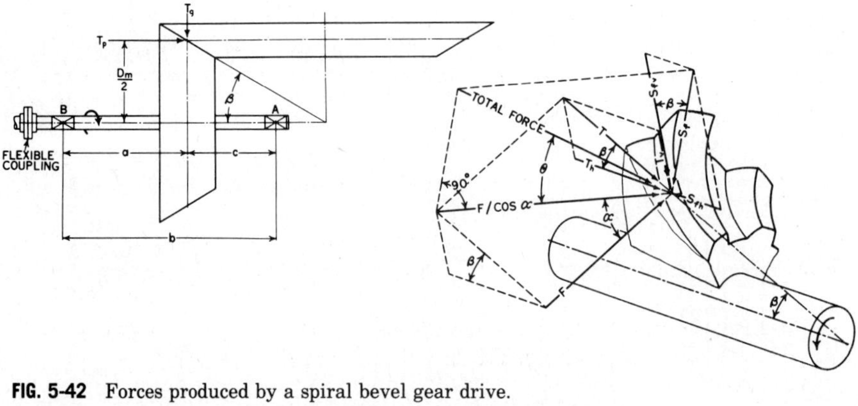

FIG. 5-42 Forces produced by a spiral bevel gear drive.

TABLE 5-13 Spiral Bevel Gear and Pinion Thrust Formulas

Pinion rotation	Spiral	Formulas
Clockwise	Right hand	T_p Pinion thrust $= F\left(\dfrac{\tan\theta \sin\beta}{\cos\alpha} - \tan\alpha\cos\beta\right)$
Counterclockwise	Left hand	T_g Gear thrust $= F\left(\dfrac{\tan\theta \cos\beta}{\cos\alpha} + \tan\alpha\sin\beta\right)$
Clockwise	Left hand	T_p Pinion thrust $= F\left(\dfrac{\tan\theta \sin\beta}{\cos\alpha} + \tan\alpha\cos\beta\right)$
Counterclockwise	Right hand	T_g Gear thrust $= F\left(\dfrac{\tan\theta \cos\beta}{\cos\alpha} - \tan\alpha\sin\beta\right)$

Note: Since the parts of the above formulas expressed by $\tan\theta \sin\beta/\cos\alpha$ and $\tan\theta\cos\beta/\cos\alpha$ are derived from the separating force that acts away from the apex, a positive answer from the formulas indicates the force is away from the apex; a negative answer indicates the force is toward the apex.

c = 126,000 when using pounds, inches, and horsepower
c = 19,100 when using newtons, meters, and kilowatts

$$F = \frac{cH}{D_m \times N}$$

T_p and T_g—see Table 5-13

The direction of rotation and the "hand" of the spiral angle are defined as observed from the large end of the gear. In the following example, the pinion has a right-hand spiral and rotates in a clockwise direction, while the mating gear has a left-hand spiral and rotates in a counterclockwise direction.

5-32 Bearing Thrust Reaction

With single-row angular contact bearings, any radial load on the bearings develops a thrust reaction. This reaction must be opposed by an equal force to prevent separation of the inner and outer rings of the bearing. The opposing force is provided by the opposite bearing in a two-bearing shaft system. This thrust reaction is treated simply as an external thrust load on the opposite bearing, and vice versa, where the opposite bearing is also a single-row angular contact bearing. The magnitude of the thrust reaction T_F is calculated from the bearing thrust factor Y_2 and the radial load F_r.

$$T_F = \frac{0.5F_r}{Y_2} \tag{5-28}$$

The thrust reaction T_F should be computed for each single-row angular contact bearing on the shaft, then combined with all the other external thrust forces.

In the example shown in Fig. 5-43, a shaft is supported by two A22275S single-row spherical roller bearings. The Y_2 factor for these bearings is 1.59. Bearing loadings in this example may be calculated as follows:

For bearing A:

$$F_r = \frac{10{,}000 \times 16}{20} = 8000 \text{ lb } (35{,}585 \text{ N}) \tag{5-29}$$

and

$$T_F = \frac{0.5F_r}{Y_2} = \frac{0.5 \times 8000}{1.59} = 2515 \text{ lb } (11{,}187 \text{ N}) \tag{5-30}$$

For bearing B:

$$F_r = \frac{10{,}000 \times 4}{20} = 2000 \text{ lb } (8896 \text{ N}) \tag{5-31}$$

and

$$T_F = \frac{0.5F_r}{Y_2} = \frac{0.5 \times 2000}{1.59} = 629 \text{ lb } (2798 \text{ N}) \tag{5-32}$$

The maximum thrust reaction governs, and the resultant thrust of the shaft is 2515 lb (11,187 N) in a direction toward bearing B. The total bearing loads are therefore:

On bearing A: 8000 lb (3558 N) (radial)
On bearing B: 2000 lb (8896 N) (radial)
2515 lb (11,187 N) (thrust)

The supporting structure between the two bearings must sustain a thrust of 2515 lb (11,187 N) with sufficient rigidity to maintain proper bearing clearance adjustment.

As shown in Fig. 5-43, the direction of the thrust reaction T_F always separates the bearing or causes the shaft to move toward the opposing bearing. When the opposing bearing is a double-row bearing, it will not produce a thrust reaction beyond that required to oppose the single-row bearing. The maximum T_F value, after being combined with all other thrust forces, is used in the selection as a thrust load F_a on the bearing.

5-33 Load Factors

Previous sections of this chapter have illustrated methods for calculating the machinery-induced forces that act on a bearing. In practice, these forces may

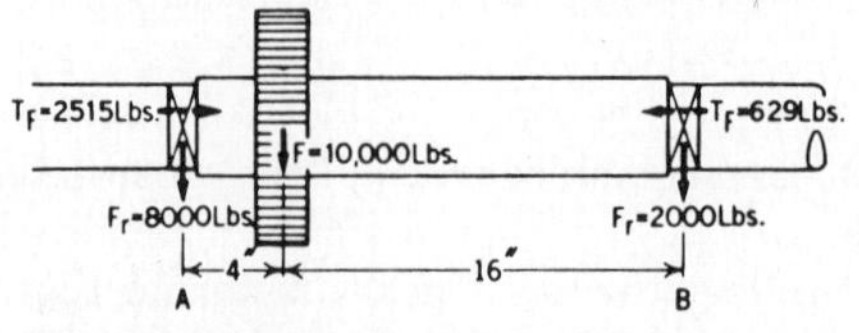

FIG. 5-43 Thrust reaction in a single-row angular contact roller bearing.

be of greater magnitude and duration than calculated. For example, because of variations in manufacturing tolerances, installation practices, operating conditions, and so on, machines may develop forces that are appreciably different from those theoretically expected. Load factors are used to compensate for the deviations of the actual force systems from the theoretically calculated values. These factors may be applied to individual forces or to a composite bearing load that is the resultant of many individual forces. A more precise method is to apply separate load factors for each questionable force acting upon the bearing. As an example, a bearing load may consist partially of a dead weight, which may be accurately determined. However, the force exerted by a belt drive is subject to adjustment by the operator. In this case, a load factor is applied to modify the belt force only.

When an accurate analysis of bearing loads is not possible, a careful selection of load factors is essential. It must be emphasized that load factors for bearings are not the same as safety factors for bolts, beams, shafts, and other such components. The bearing load factor provides an exponential margin of safety in its effect on bearing life. This effect is realized by examination of the load factor curves in Fig. 5-44.

These curves show the modifications necessary to provide for overloads of specific magnitude and duration. For example, a bearing might be called upon to operate at 90 percent overload for 20 percent of its operating life. Under these circumstances a load factor of 1.3 would be used to modify the loading calculation for bearing selection. If the overload occurs as anticipated, the

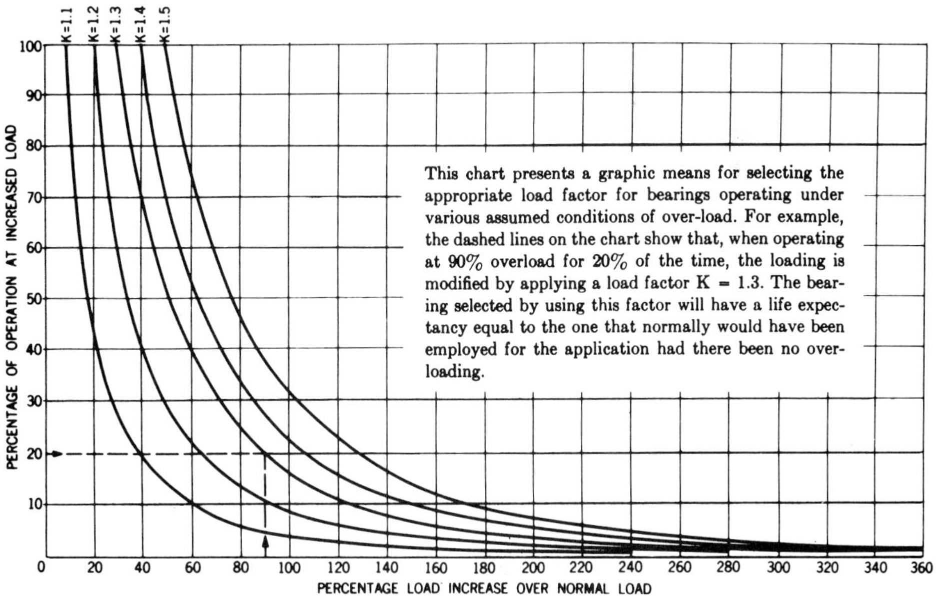

FIG. 5-44 Curves for load factor selection.

bearing is adequate. If the overload does not occur, increased life and reliability will be the result.

The use of a load-modifying factor is not expected to compensate for serious stress concentration due to mounting distortions, excessive deflections, and so on. Such conditions cannot be compensated for except by correction of the problem. For example, a sliver of metal a few thousandths thick trapped between mounting surfaces may induce distortions in a bearing raceway sufficient to cause rapid failure. An excessive shaft deflection or machine frame deflection can produce serious edge-loading conditions or distortions of bearing geometry. Such conditions can be serious machine design problems for which a higher-capacity bearing may not be a satisfactory solution.

In a competitive and ever-changing market, the intelligent selection of load factors can be of appreciable significance in maintaining competitive costs. For example, unnecessarily large bearings are often selected because of unrealized compounding of load factors. Selection of these large and uneconomical bearings occurs when a machinery manufacturer submits data to the bearing producer that already include some unmentioned load factors. The bearing application engineer, not knowing this, may apply additional load factors before recommending a specific bearing for the application. Such errors can increase machine costs drastically by causing the specification of bearings, shafts, and housings larger than necessary for a given application.

LUBRICATION

5-34 Basic Principles

In a correctly operating rolling-element bearing, a thin film of lubricant separates the rolling elements from the raceways. This film should be of sufficient thickness to prevent asperity contact of the metal surfaces. Contact of these asperities, or high points of the raceway surfaces, will result in wear, scoring, and possible seizure. Providing this lubricating film is the primary function of the lubricant in a rolling-element bearing. Lubricants also provide (1) protection from corrosion, (2) dissipation of heat, (3) exclusion of contaminants, and (4) flushing away of wear products. A good lubricant is expected to have high film strength; chemical, thermal, and mechanical stability; and corrosion-prevention properties. These and other specialized properties are provided in modern lubricants by various formulations, refining techniques, and additives.

The requirements for a lubricant for rolling-element bearings are often more severe than realized. In a rolling-element bearing there are conditions of both rolling and sliding with extremely high contact pressures. The lubricant must withstand high rates of shear and mechanical working not generally prevalent in other mechanical components. For these reasons, proper attention to equipment lubrication is vital from design to operation. The following sections discuss types of lubricants, their selection and application.

5-35 Greases

Grease is a combination of petroleum oil or synthetic fluid and a suitable thickener. The thickener in grease may range from 3 to 30 percent or more. Grease consistency or stiffness is determined primarily by the percentage of thickener and the base oil viscosity. A grease of given consistency may be compounded in many ways by varying the thickener percentage and oil viscosity. Because of this, greases of equal stiffness cannot be assumed to be equal in performance.

In the past, the most widely used thickeners were either sodium or calcium soaps; however, today's greases contain many types of thickeners. Each grease has certain characteristics, but categorizing them for use on the basis of type of thickener can be very misleading. For example, a sodium-base grease is not necessarily best for a given type of service. For general industrial use, sodium, calcium, lithium, aluminum, and mixed soap-base greases in the metallic soap types, along with various synthetic and non-soap-base greases, are widely used. Considerable effort has been directed to the development of a multipurpose lubricant; however, no grease yet satisfies all requirements.

The performance requirement of a grease lubricant has led to the development of many standardized laboratory tests for measuring the properties or characteristics of a grease. Grease will deteriorate as a result of time, temperature, shear (mechanical working in the bearing), and contamination. These characteristics are therefore measured by standard tests of oxidation resistance, water resistance, mechanical stability, oil separation, dropping or melting point, and evaporation. The stiffness property of the grease is measured by a hardness or penetration test. A change in this property during operation can be caused by excessive working or churning, oil separation or vaporization, change in oil viscosity due to oxidation, and so on. The multitude of standard laboratory tests, while important, have not always been found to correlate with field performance. In some cases greases have been found excellent in laboratory evaluation and completely unsuitable in field performance. In other instances the reverse has been true. For this reason, field testing and field development of lubrication requirements for a particular equipment installation are often necessary.

Bearings and bearing units are designed for service ranging from nonregreasable (lubricated for life) to units designed for almost continuous relubrication by means of automatic grease lubrication systems. The properties and performance characteristics required of greases are not the same for all bearings. In high-speed applications, oil separation or bleed rates are critical and useful grease life is limited because of a gradual reduction in the oil content of the mixture. With slow speed, long grease life, or nonregreasable applications, a high bleed rate would be undesirable.

In summary, grease is critical to bearing performance. It must be properly selected to suit the application requirements, and must be utilized in a properly developed maintenance program to make possible the expected performance of any particular mechanical system.

5-36 Oils

Oil is a fluid lubricant. It can be pumped, circulated, filtered, cleaned, heated, cooled, atomized, and so on. It is more versatile than grease and is suitable for many severe applications involving extreme speeds and high temperatures. On the other hand, because oil is a fluid, it is more difficult to seal or retain in bearings and housings. Oil level or oil flow in high-speed bearings is extremely critical and must be adequately controlled.

Fluid lubricants may be classified into two broad categories: petroleum and synthetic. Petroleum oils are lower in cost and have excellent lubricating properties. The synthetic fluids have been developed largely for improved heat stability and wider temperature range. Animal and vegetable oils are seldom considered for rolling-bearing lubrication because they are more subject to deterioration and formation of corrosive acids.

Viscosity, the measure of an oil's thickness or consistency, is related to film strength for a given type of oil. The selection of proper oil viscosity is essential; it is based primarily on expected operating temperatures of bearings. In rolling-element bearings, excessive oil viscosity may cause skidding of rolling elements and undue lubricant friction. The result can be severe overheating and raceway damage. Insufficient oil viscosity may result in metal contact and possible bearing seizure. Other properties of oils, such as viscosity index, flash point, pour point, neutralization number, carbon residue, corrosion protection, and so on, are significant and should be understood by those involved in the selection of oils. Information on lubricant selection is provided in Sec. 5-42.

5-37 Synthetic Lubricants

The development of a wide range of synthetic lubricants has been prompted largely by the extreme environmental demands of military and aerospace activities. Commercial applications of synthetic lubricants are usually associated with extreme temperature conditions or special environmental requirements such as fire hazard. Prominent classes of current synthetic lubricants are as follows: (1) dibasic—acid esters (diesters), (2) phosphate esters, (3) silicone polymers, (4) silicate esters, (5) polyglycol ethers (including polyalkylene glycols), (6) halogenated compounds (including fluorinated hydrocarbons), (7) polyphenyl ethers, (8) silanes. These classes are commercially available; however, costs range from double to over one hundred times the cost of petroleum lubricants.

Synthetic lubricants usually permit a broader operating temperature range and provide good lubrication at high or low temperatures beyond the limits of petroleum lubricants. Operating temperature limits of synthetics are often misunderstood. For example, in various aircraft and space applications, successful operation at extremely high temperatures is essential, but life requirements may be extremely short. Industrial and commercial service requirements are usually for much longer periods of time, which means that temperature limits for a given synthetic in industry may be much lower than for the more exotic applications.

Synthetic lubricants may also have other limiting characteristics, such as load-carrying ability and high-speed operation. As with more conventional lubricants, it is not possible to assign specific load and speed limitations, because these factors must be considered in combination with bearing type, operating temperatures, and so on.

5-38 Dry Lubricants

Dry lubricants in either powder or colloidal suspension form have been used for many years. Notably successful applications include the lubrication of kiln car wheels, conveyor wheels, furnace roll bearings, and so on. These high-temperature applications involve operations at extremely slow speed where ample torque is available to rotate the bearing at a relatively high coefficient of friction. Usually, the solid is placed in a liquid carrier, which burns off completely, leaving the solid lubricant. In more recent years, dry lubrication has been effected by bonding dry film coatings to bearing raceways. Some of the more promising materials used are graphites, molybdenum disulphide, cadmium iodide, and fluorinated polyethylenes. Common bonding agents are phenolic resins, silicone ceramics, and sodium silicate. Another method incorporates the lubricant into one or more of the bearing components. Typically, a bearing retainer may be constructed partially from lubricating materials. In such cases, the dry lubricant is deposited on the rolling elements of the bearing and, in turn, is transferred to the raceway surfaces. In both this and the bonded coating design, bearing life is governed by the wearout life or depletion of the lubricant. Such special bearings are usually quite expensive. Often, a more economical and practical solution for industrial use is to design equipment so that conventionally lubricated bearings can be used. Application engineers are available for consultation on these problems.

5-39 Lubricant Compatibility

Compatibility is always a concern when lubricants are mixed. The compatibility of lubricants is the solubility or the affinity of lubricants for each other. Because analysis of compatibility failures is difficult and positive identification has been infrequent, commercial products are generally not classified for this characteristic. Manufacturers of lubricants usually recommend exclusive use of their product, with no mixing with other products. This recommendation is based on the premise that the complex petroleum additives used by different producers may react adversely when mixed. Nevertheless, mixing of lubricants is often unavoidable. Problems of compatibility are most likely to arise when mixing greases with different bases and when mixing different types of oils, particularly those with complex additives. With greases, incompatibility between two products may be observed as either a thickening or a thinning of the resultant grease mixture. With oils, incompatibility may be indicated by coagulation and separation of the additive contents. Such coagulation increases the chances of clogging oil lines. Experience has indicated that a period of time may pass before incompatibility becomes apparent.

Hence, simple mixing tests are not considered reliable in evaluating lubricant compatibility. The reliability of lubricant mixes can be established only after observation under actual or simulated operating conditions.

With the large number of lubricants available and the increasing complexity of lubricant structure, there is little likelihood that reliable procedures to consistently predict compatibility among lubricants will be developed. If different lubricants are mixed, the application should be closely observed to permit correction of any lubrication problem before damage to the bearing. In some cases of lubricant incompatibility, it is necessary to flush and clean the bearing units. However, in most instances, simple purging by regreasing or reoiling should be effective.

When a switch of oil or grease within a system is required, a temporary increase in the frequency of relubrication or oil change is advisable. It is recommended that the technical representative of the lubricant suppliers be consulted prior to the selection and use of a different lubricant.

5-40 Corrosion Prevention

Corrosion damage may occur at any time in the life of a bearing—even before it is placed in service. Corrosion marks or pitting on bearing raceways cause stress concentrations in service and premature raceway fatigue. Such damage most frequently results from oxidation of parts by moisture. The basic remedy is to protect the bearing parts from moisture.

A common source of corrosion is condensation of moisture in bearings or lubricants. Changing temperatures can cause free air space within bearings or lubricant containers to breathe moisture-laden air. Bearings and lubricant containers should be stored in controlled-temperature and low-humidity areas.

For low-speed applications under adverse environmental conditions, bearings should be kept full of lubricant to minimize air space. For high-speed applications, this is not recommended because of possible overheating from excess lubricant. When equipment is shut down or stored for long periods, bearings may be particularly vulnerable to corrosion damage. Breathing and condensation of moisture may occur, while oil or grease film protection of bearing surfaces may be interrupted.

Provision should be made for periodic rotation to reestablish oil film. Where scheduled rotation is not practical, other procedures should be implemented, such as 100 percent fill of cavities with lubricant or preservative, or spraying or flushing of cavities with appropriate preservatives. With preservative treatment, some system rotation or vibration should be used at application to assure preservative penetration of any oil miniscus at points of rolling-element contact. It must be emphasized that lubricating oils cannot be expected to provide corrosion protection for idle equipment.

For both low- and high-speed applications, an increased frequency of relubrication is often very effective in flushing away water-contaminated lubricants. Automatic lubrication systems can provide very effective corrosion control.

Most ball- and roller-bearing lubricants contain rust and oxidation inhibitors. However, there are great differences in the corrosion protection provided by various lubricants. In severe moisture conditions, the lubricant selection can be critical.

There are limits to the protection afforded by bearing seals, lubricants, and maintenance procedures. Water flow or heavy splash against bearings may require custom designs. Seal systems are available or can be designed to function reliably under extreme conditions. Where corrosion problems cannot be solved by standard components and procedures, the more expensive custom designs can be well justified.

5-41 Lubricant Temperature Limits

Temperature is the major factor affecting the functional life of a correctly selected bearing lubricant. Lubricant temperature is influenced primarily by bearing speed, bearing load, and ambient temperature. The lower the temperature, the longer the lubricant will last, excluding low-temperature limits.

Whether or not a bearing will operate successfully at a particular combination of load, speed, and temperature depends on the lubricant. Two different greases used on identical applications may produce entirely different results. Base oil viscosity, thickeners, and chemical structure can all contribute to different operating temperatures. Some greases will churn in high-speed bearings and cause overheating, whereas a channeling type grease may function satisfactorily at a much reduced temperature.

Practical temperature limits for a specific grease will vary widely, depending on bearing type, speed and load, ambient temperature, regreasing procedures, and other such factors. Speed and temperature limits for a specific grease are not fixed values. The most valid limits are the boundaries within which the system performs to the satisfaction of the user. Figure 5-45 shows a general comparison of petroleum greases and a number of synthetic greases. The validity of such data, however, is largely dependent on the factors mentioned.

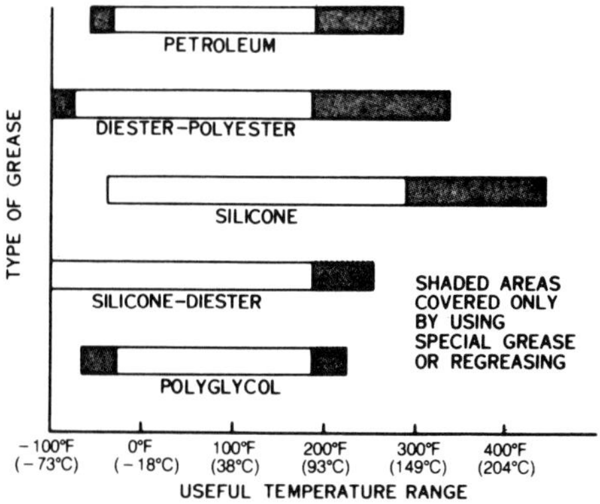

FIG. 5-45 Useful temperature ranges of grease.

The multipurpose grease has gained popularity because it is designed to function over a broad range of operating conditions. Conversely, this type of grease may not be the best choice for a specific set of conditions. Ideal operation of a machine under extreme conditions of speed, temperature, loads, vibration, and so on will usually depend on careful selection of a lubricant designed for these conditions. With oil lubrication, limitations are more specific. Fluid lubricants can be circulated, cooled, filtered, and atomized. In a dynamic oil system, these processes can provide for cooling of the bearings. Oil flow rates can be regulated, and constant lubricant temperature and viscosity can be maintained. Most oil lubricants have a high-temperature limit beyond which deterioration, varnish, and sludge formation will occur. When this limit is exceeded, frequent oil changes and maintenance will be required.

Continuous operation at over 200°F (93°C) is a problem with petroleum lubricants. A limit of 150°F (66°C) must be observed with some lubricants containing active EP additives. Figure 5-45 shows rather broad shaded areas at the high-temperature ends and considerable overlap of various types of lubricants. This chart does not reflect the time effect of temperature. Figure 5-46 shows typical effects of time and temperature. The time effect is often overlooked in discussion of lubricant high-temperature limits. In aerospace applications, lubricant life requirements may range in duration from minutes to a few hundred hours. These life requirements are radically different from those in typical industrial applications, where several thousand hours of service may be required. High temperatures destroy lubricants by causing accelerated oxidation, catalytic decompositions, evaporation, melting, and so on, which result in sludge, burnt carbon deposits, lubricant caking, and varnished raceways. With some active EP additives the compounds are broken down above 150°F (66°C) and may become corrosive, particularly if moisture or other

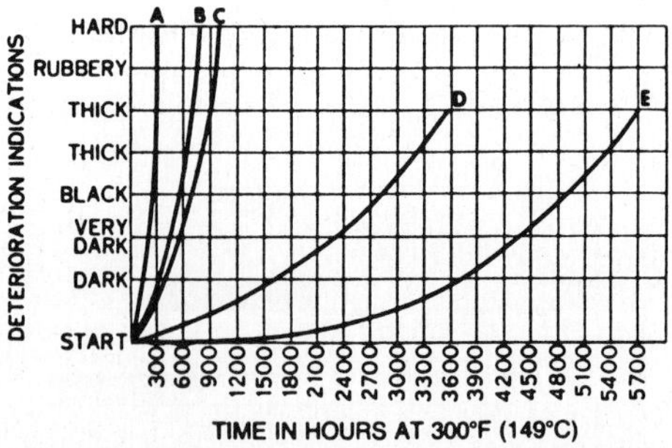

FIG. 5-46 Deterioration of greases at high temperatures. This chart shows some typical results obtained from a study of various lubricant samples held in an oven at 300°F (149°C) with periodic observation of visible changes. A, B, and C show the approximate deterioration of three different types of petroleum base lubricants, while D and E show comparable results with two types of synthetic lubricants. In general the deterioration will be greatly accelerated when lubricants are in actual service.

contaminants are present. Temperature also reduces the effective viscosity of the lubricant, thus reducing the film strength. The difficulties of lubrication over 200°F (93°C) can significantly increase equipment operating cost. A higher initial cost may often be justified if equipment can be designed for reduced operating temperature. Significant temperature reductions can be accomplished by water cooling, insulation, heat flingers, circulating oils, and so on. Conversely, the costly synthetic lubricants may provide satisfactory high-temperature operation if properly selected for compatibility with equipment design.

Lubrication at low temperatures must permit a reasonable starting torque. Lubricants must not freeze or become too stiff. Selecting a lubricant to meet the starting torque requirements is relatively simple. Also to be considered is the total temperature range in which the equipment must operate. The lubricant must permit equipment turnover at the lowest temperature, and must also have adequate viscosity at the higher operating temperatures to provide sufficient oil film strength. For example, a petroleum type lubricant with very low-viscosity oil, considered suitable for startup at −40°F (−40°C) and operation at 100°F (38°C), may be unsuitable for operation at 180°F (82°C). In such cases, a synthetic oil or grease may be required to function satisfactorily at both the high and low operational limits. Lubricant manufacturers usually provide oil pour-point values and grease pumpability temperatures; however, the low-temperature bearing torque specification is used for determining suitability of a lubricant for low-temperature application. When this is unavailable, a simple stir test may suffice. In a sample of lubricant lowered to the lowest expected temperature under which it will operate, the resistance to stirring will assist in judging its suitability. Where there is still doubt, the measurement of lubricant torque in a test bearing at low temperature may be required.

Most low-temperature bearing applications operate satisfactorily with proper lubricant selection, although in some cases equipment modification may be necessary.

Problems associated with high-temperature or high-speed bearing applications usually result from lubricant failure. Increased frequency of relubrication and flushing out of deteriorated lubricant will usually improve bearing performance. In other cases, special lubricants will be required. In some unusual applications, an experimental approach is often the only way to develop successful lubrication procedures. Under such conditions, close observation of the lubricant's operational performance is required, and modifications can be made before trouble develops. An examination of the color and consistency of the lubricant compared with a fresh sample can lead to the establishment of correct lubrication intervals for highest performance. Along with the selection of a correct lubricant, relubrication procedure can be the most vital point in the prevention of lubrication failures.

5-42 Lubricant Selection

The selection of lubricants is relatively straightforward in most cases. A high percentage of applications are in the normal range of load, speed, and tempera-

TABLE 5-14 Operating Ranges, Normal and Critical

	Load	Speed	Temperature
Low	0 to 8% of bearing C rating	Up to 10% of maximum cataloged speed*	Less than 20°F (−7°C)*
Medium	8 to 18% of bearing C rating	10 to 75% of maximum cataloged speed	20 to 200°F (−7° to 93°C)
High	Over 18% of bearing C rating or shock and vibration*	Over 75% of maximum cataloged speed*	Over 200°F (93°C)*

* In critical range.

ture, where high-quality ball- and roller-bearing lubricants can be expected to provide good performance. The balance of applications may be in the critical range of lubrication. Table 5-14 illustrates the normal and critical ranges of bearing operation. The critical range conditions of extreme load, speed, or temperature, where special lubrication provisions may be required, are indicated. Operation in the other areas but at upper limits in two or all three of the factors may also be categorized as critical.

Whether an application is normal or critical, optimum performance is attained when lubricant viscosity is selected for full film separation of the metal surfaces. Elastohydrodynamic theories provide equations for solution of the film thickness problem. In these theories, it is seen that rotational speed is by far the most significant factor. For Link-Belt bearings, an approximation based on relative speed has been found to be more suitable for practical use than the extremely complex equations of the basic theory. In this approximation, the recommended viscosity for all bearings at their maximum cataloged speed and lubricated with a naphthenic-type oil is a constant 70 SUS (Saybolt universal seconds) at operating temperature. The deviation from the exact solution is found to be less than the normal viscosity tolerance in a purchased lubricant. The recommended viscosity for speeds other than maximum cataloged speed is calculated by the following method:

$$\text{SUS viscosity at operating temp.} = 70\left(\frac{\text{maximum catalog speed}}{\text{actual speed}}\right)^{0.7} \tag{5-33}$$

This relationship is held valid for the entire cataloged speed range of a given bearing except for some adjustment in the heavy load or low-speed range (see Table 5-15). Modification of the viscosity calculation is necessary when synthetic or nonconventional lubricants are used. The latest analytical methods and computer capability for the study of theoretical EHD films, are available,

TABLE 5-15 Critical Range Lubrication

	Oils	Greases	Synthetics (fluid or grease)
Load—high Over 18% of bearing *C* rating or shock and vibration	Increase calculated viscosity by factor of 1.5 to 2.0. EP types desirable	Extreme pressure types recommended. Viscosity of oil in grease is chosen on same basis as for oil lubrication	Use with caution, as some products may not provide adequate film strength. Consultation recommended
Speed—high Over 75% of maximum cataloged speed	Oils may be subject to foaming and high temperatures, and should be selected carefully. Circulation or frequent replacement may be required. Oil levels must be controlled.	Grease not normally recommended. Some products suitable depending on other operating conditions. Consultation advisable	Use with caution. Consultation recommended
Speed—low Up to 10% of maximum cataloged speed	Use EP type of highest viscosity possible. Computed viscosity may be increased by factor of 2.0 or more	Use EP types having highest-viscosity oil content. Prefer softer grade than used on medium-speed applications	High viscosity and EP characteristics desirable. Consultation recommended
Temperature—high Over 200°F (93°C)	Deterioration rate increases rapidly above 200°F (93°C). Circulation or frequent replacement indicated. Select high-quality oils with low carbonizing characteristics	Deterioration rate increases rapidly above 200°F (93°C). Regreasing frequency must be increased accordingly and must often be determined experimentally	Superior in heat resistance and operating temperature range. Prefer medium speeds and medium or light loads
Temperature—low Under 20°F (−7°C)	Requires oil viscosity to permit cold start and correct viscosity at subsequent operating temperature. Many conventional oils suitable to −30°F (−34°C)	Requires grease to permit cold start and not thin excessively at subsequent operating temperature. Many conventional greases suitable to −20°F (−29°C)	Usually required at temperatures lower than −20°F (−29°C). Excellent temperature range characteristics

as is broad practical experience on a wide variety of unique bearing applications.

Table 5-15 is a general guide to lubricant selection for bearings operating in the critical range. The information provided is of necessity very general. Bearing maintenance procedures can play a very important part in the suitability of a lubricant. Where reliable maintenance procedures are followed, a standard petroleum lubricant may perform satisfactorily in some critical-range applications. On the other hand, an apparently identical application where maintenance procedures are poor may require a special lubricant.

Various lubricant additives have been developed to enhance specific properties of an oil or grease. Viscosity, viscosity index, apparent viscosity, neutralization number, flash point, pour point, carbon residue, corrosion protection, oxidation resistance, resistance to foaming, film strength, oiliness, temperature stability, shear stability, and dropping point are some of the characteristics measured by standardized tests. Additives can improve these properties, but, in some cases, at a sacrifice in some other characteristics of the lubricant.

A mild EP (extreme pressure) additive is common in ball- or roller-bearing lubricants. This type of additive has been very beneficial in reducing wear in higher-load applications. Rust and oxidation inhibitors form another very important additive class; in some applications, these are mandatory.

Any selection of lubricant, even with the use of Tables 5-14 and 5-15, must be tempered with judgment. For example, in these guides many other environment factors and maintenance procedures are presumed to be normal. Moisture, dirt, radiation, chemicals, vacuum, oxidizing atmospheres, and other such factors also influence lubricant selection.

Lubricant selection must also consider the materials used in a given bearing design, particularly cages and seals. Ball- and roller-bearing cages and seals made of nonmetallics may be subject to deterioration in service with certain types of lubricants. One of the recent developments in bearing technology has been the polymeric cage, usually made of nylon 66 or similar material. This material is suitable for usage with all conventional lubricants, including most synthetics; however, dibasic ester types of synthetics may result in deterioration over a period of time. The elastomeric seals used in bearings and bearing units are most commonly found with a Buna-N type material, which again is generally suitable for use with conventional lubricants, but should not be used with the phosphate ester type fire-resistant fluids.

While the results of a compatibility problem can be serious, such problems are encountered primarily on the rarer applications that require various types of synthetic lubricants. Consequently, more careful attention to selection of lubricants is necessary. The lubricant supplier should have data or information available on the restrictions, if any, for a particular lubricant.

Consultation with a reputable lubricant manufacturer is recommended when unusual environment or maintenance conditions exist.

5-43 Relubrication

The proper relubrication of bearings is often of equal or greater importance than the initial selection of lubricant. The establishment of proper relubrication procedures is also one of the most difficult aspects of lubrication. The bearing manufacturer can provide some general guidelines; however, exact recommendations can be established only from experience and/or a thorough study of a particular application. Bearing application requirements will vary from a lubricated-for-life approach to installations requiring continuous relubrication, such as those with circulating oil systems, oil mist systems, or automatic greasing systems. The decision as to relubrication requirements for a particular bearing application will depend on the nature of the application, the environmental conditions, the type of bearing, operating temperatures, and accessibility. With regard to accessibility, for instance, a remote or inaccessible bearing location will typically be either lubricated for life or equipped with an automatic lubrication system.

The basic concept of relubrication is to provide the right amount of lubricant to a bearing at appropriate intervals so that the bearing will always have a sufficient amount of undeteriorated lubricant available for satisfactory operation under the load, speed, temperature, and environmental conditions involved. Accomplishing this is sometimes easier said than done.

In Sec. 5-41, lubrication temperature limits were discussed, and it was seen that temperature is the factor which most governs the rate of deterioration of a lubricant. In relubrication, the basic concept is to replenish the lubricant at a rate which will eliminate the effects of deterioration due to time exposure at a particular temperature. Concurrent with this requirement, we are concerned with the loss of lubricant through leakage or evaporation, and lubricant replenishment rates must be established to offset such losses. The third basic factor to consider is contamination, such as by entrance of dirt, abrasives, water, chemicals, or even exposure to radiation. Lubricant replenishment rates must be established to offset this type of contamination at a rate which will prevent damage to the bearing. Usually one of these three basic factors will govern or limit the useful life of a lubricant in the bearing. The factors of load and speed are also important and tend to govern the operating temperature of the bearing. Speed and vibration can also have an effect on grease from a standpoint of mechanical working, as under certain conditions grease products may break down, generally to a liquid form. In this situation, the operating conditions must be considered too severe for the grease product used. With all the variables involved in various types of operating machinery, it is not possible to define precise relubrication procedures, and optimum procedures must generally be developed through experience and careful observation of operating equipment.

With oil lubrication, the methods are fairly direct. Oil losses through leakage or evaporation must be observed and replaced periodically. This is generally accomplished by observance of oil levels or utilization of constant-level oilers. The condition of oil can be monitored through periodic sampling and analysis, and when the oil approaches a marginal condition as a result of contamination or deterioration, a drain and replace procedure is implemented.

With grease lubrication, typical bearing constructions do not permit observation of the amount of grease remaining in a bearing or its condition. Therefore, the addition of new grease to an operating bearing is the primary method of observation. The condition of used grease and the amount of grease added to

TABLE 5-16 Grease Lubrication Schedule

Link-Belt® Spherical Roller Bearings—Series B22400 and B22500 for Use on Horizontal Shaft Equipment

Shaft sizes		Amount of grease		Operating Speed (r/min)									
				500	1000	1500	2000	2200	2700	3000	3500	4000	4500
Inches	mm	in^3	cm^3	Lubrication cycle (months)									
3/4–1	25	0.39	6.4	6	6	6	4	4	4	2	2	1	1
1 1/8–1/4	30	0.47	7.7	6	6	4	4	2	2	1	1	1	1
1 7/16–1 1/2	35	0.56	9.2	6	4	4	2	2	1	1	1	1	1/2
1 5/8–1 3/4	40	0.80	13.1	6	4	2	2	1	1	1	1	1/2	
1 15/16–2	45–50	0.89	14.6	6	4	2	1	1	1	1	1/2		
2 3/16–2 1/4	55	1.09	17.9	6	4	2	1	1	1	1/2			

2 7/16 – 2 1/2	60	1.30	21.3	4	2	1	1	1	1/2
2 11/16–3	65–75	2.42	39.7	4	2	1	1	1/2	
3 3/16–3 1/2	80–85	3.92	64.2	4	2	1	1/2		
3 11/16–4	90–100	5.71	93.6	4	1	1/2			
4 3/16–4 1/2	110–115	6.50	106.5	4	1	1/2			
4 15/16–5	125	10.00	163.9	2	1	1/2			

These guidelines are for use on applications approved by PT Components, Inc.

Lubricate with a multipurpose roller bearing NLGI Grade 1 or 2 grease having rust inhibitors, anti-oxidant additives, and a minimum oil viscosity of 400 SSU at 100°F. For operation requiring a monthly or less cycle, the grease should also be suitable for temperatures up to 250°F continuous, dynamically stable, and must not churn or whip.

If bearings are subjected to temperatures below 32° or above 200°F, consult equipment manufacturer for proper lubrication.

Conditions of vibration exceeding 1 to 2 mils, moisture, or dirt will require a more frequent lubrication cycle or special lubricant selection. Rotate bearings during relubrication where good safety practice permits.

Lubricate bearings prior to extended shutdown or storage and rotate shaft monthly to aid corrosion protection.

start purging of the old grease can be observed. It is then possible to make logical judgments as to whether the amount of grease added is sufficient or excessive. Table 5-16 shows data for use on applications. This provides a guideline as to the amount and frequency of grease addition under various operating conditions. While similar information can be developed for any known application, it must still be emphasized that these are guidelines, and, in most cases, more optimum procedures could be developed through an appropriate maintenance program based on careful observation and experience.

5-44 Lubricating Systems

Manual lubrication places on an individual the responsibility for using the correct lubricant and for maintaining the correct frequency of lubrication. This requires training and supervision of lubrication personnel. Even with adequate training and supervision, manual lubrication of bearings is still subject to human error. A large variety of lubricating devices are available, including drip feed from cups, constant-level oil cups, wick feeds, and pressure-feed grease cups. These devices essentially permit longer intervals between manual servicing, but still require trained personnel and rigid observance of servicing schedules. A lubrication system can alleviate such problems and improve reliability. There are three basic types of centralized lubrication systems: centralized greasing, centralized circulating oil, and oil mist or fog systems. Variations of these systems are often custom-designed for individual applications. The design and installation of a lubrication system should be considered in the initial planning of any machine or machinery complex. An automatic or semiautomatic lubrication system can often be justified by reduced maintenance costs and increased effectiveness of lubrication procedures. Grease lubrication systems generally can be adjusted for amount and frequency of lubrication. In each separate centralized grease lubrication system, the same type of grease is distributed to all bearings. With such systems, visual gauges are usually provided to facilitate checking for a continuous lubricant supply to all bearings in the system. In cases where separate bearings within one system operate under different temperature, speed, and load conditions, it may be necessary to use more than one system to meet the correct lubrication needs of the individual bearings. For automatic greasing systems, it is mandatory that all grease lines and bearings be filled properly before the equipment is operated.

Circulating-oil lubrication systems furnish a controlled flow of cool, clean oil to the bearings. These systems are most beneficial when bearings must be cooled continuously and when abrasive materials must be flushed away to assure safe operation of the bearing. Controlling temperature at cooler operating levels extends oil life and maintains excellent control of operating oil viscosity. Circulating-oil lubrication systems nearly always have filter and heat exchange elements in addition to their oil reservoir and pump. They may also have a centrifuge or a sump for separating and removing foreign material from the oil. Unlimited variations are possible by using such equipment as remote controls, warning devices, automatic cutoff switches, and so on. These are particularly useful in meeting the special requirements of paper mills, lumber mills, steel mills, coal processing plants, and other similar applications.

Oil mist lubrication systems use an air stream to provide oil to the bearings. Generally, the oil is not recirculated. Because the system uses air pressure to provide oil to the bearings, it maintains a positive pressure within the bearing chamber. This pressure can effectively prevent foreign matter from entering the chamber. In addition, the air flow of an oil mist system can be regulated to produce minimum lubricant friction and the concomitant lubricant friction temperature effect. The air flow will not, however, provide significant cooling effect.

Air flowing out of a mist-lubricated bearing may impart a fine oil vapor to the surrounding atmosphere. This vapor may be objectionable in some working areas, especially in the food and textile industries. In such cases, it is necessary to vent to other areas or provide air cleaning systems.

Oil mist systems are sensitive and exact devices. Drainage of bearing reservoirs, provision for proper oil levels during bearing start-up, and timing of the mist flow must meet precise specifications. For this reason the system manufacturer should be relied upon to adjust the system for correct operation.

5-45 Cleaning, Preservation, and Storage of Bearings

Bearings operating over a period of time may develop deposits of lubricant varnish, hard residues, and external contaminants that may detrimentally affect bearing performance. Unused bearings may also become contaminated as a result of improper preservation, packaging, or storage. Such deposits or contaminants should be removed to ensure that their accumulation is not harmful. The cleaning procedure for such bearings will depend on the particular circumstances. In some instances, reasonably effective cleaning may be obtained by flushing the bearing with clean lubricant while it is in operation. In more severe cases, extended soaking in kerosene, mineral spirits, or special commercial solvents will be necessary.

Some solvents could be detrimental to nonmetallic cages, seals, or other components. Alcohols, cresols, phenols, fluoropropanol, and other similar chemicals or mixtures are unacceptable for bearings using nylon cages. Some recommended solvents would be mineral spirits, kerosene, Freons, Chlorothene, and Stoddard solvent. After removal from cleaning solvents, bearings are extremely vulnerable to corrosion or mechanical damage and should be protected or lubricated immediately.

Handling with bare hands may be harmful to bearing metals. Without use of fingerprint neutralizer, the acidic moisture on hands will severely corrode the bearing surfaces. A clean, dry, unlubricated bearing is particularly susceptible to mechanical damage from movement between rolling elements and raceways. Such bearings should never be spun or subjected to shock or vibration before they are coated with lubricant or preservative.

Bearings that are to be stored after cleaning should be coated with a preservative compound and wrapped in a neutral greaseproof paper, foil, or plastic film. Various preservative compounds are available to provide protection from several months to several years. Packaged bearings should be cleaned and stored in a dry, low-humidity area. Packages should be identified to avoid unnecessary opening and exposure of the bearing. Packages should be date-marked and used on a first-in, first-out basis. When long-term storage is not anticipated, bearings may be preserved in their lubricating grease or oil.

Economic justification for the cleaning and storage of used bearings in anticipation of their eventual reuse should be carefully considered. Usually such procedures are feasible only for the larger, more expensive bearings. When operating bearings need to be cleaned too frequently, a review of maintenance procedures is recommended. The solution may center on such things as sealing effectiveness, lubricant selection, or relubrication procedures.

5-46 Summary

The three major factors that determine the suitability of a lubricant for a particular application are the rotational speed of the bearing, the load it must support, and the temperature at which it operates. For most applications in which speed, load, and temperature are not extreme, a standard petroleum lubricant will suffice. However, under conditions in which any of the three conditions is extreme, special attention must be given to lubricant selection. In most instances, there is petroleum with special additives or a synthetic available to meet the needs of difficult applications. However, the lubricant user must always keep in mind that lubricants developed to meet the needs of special applications may be restricted in their range of use. For example, synthetics that are suitable for very high temperature operations may not have the load-carrying capacity of standard petroleums used in lower-temperature applications. In many instances, the needs of difficult applications can be met by increasing the frequency of relubrication.

Consultation with a reputable lubricant manufacturer is recommended when unusual environmental or maintenance conditions exist.

SEALING

5-47 Seal Functions

The functions of a seal are containment of lubricant and exclusion of contaminants. These functions are accomplished by designs ranging from a simple baffle or a strip of felt to complex systems of flingers, lips, labyrinths, springs, shields, and so on. In some cases, a seal may be required to perform only one of the two functions; for example, in perfectly clean surroundings, containment of lubricant is the only requirement.

The design of a bearing seal is governed by what the seal is expected to do. This depends on the nature of the material to be excluded, such as liquid, solid, viscous, or powdery; shaft surface speed; temperature conditions; type of lubricant; permissible leakage; and many other conditions of environment and operation.

5-48 Contact Seals

Contact seals vary from simple felt stripping that encircles and rubs the shaft to axial face seals with precision-honed faces flat to within twenty millionths of an inch. In all cases, an intimate contact between moving and nonmoving surfaces provides a barrier against lubricant loss and/or contaminant penetration. Contact seal designs vary widely because of the range of materials available and the broad latitude of seal configurations.

Sliding friction between the seal and the shaft limits the uses of contact-type seals. In general, the maximum limit of rubbing velocity between the shaft and a contact seal varies from 500 to 1000 ft/min (2.54 to 5.08 m/s) for felt seals and from 1500 to 3000 ft/min (7.62 to 15.24 m/s) for lip seals. Precision axial contact seals can be designed for considerably higher speeds.

Shaft surface has a critical effect on contact seal performance. Shaft running accuracy directly affects seal efficiency. Slight surface imperfections should be avoided, especially spiral characteristics such as those produced by through-feed grinding.

For effective application of lip-type contact seals, the shaft surface finish should be 10 to 20 μin AA (0.25 μm to 0.50 μm). Seals should be positioned concentric with the bearing. Nonconcentric seal positioning leads to reduced contact pressure or even to loss of contact.

Wool felt in various densities has long been used as an excellent and reliable sealing medium. In recent years, synthetic fiber felts have come into wider use because of their resistance to high temperatures and corrosive environments along with excellent fiber strength. The felt seal excludes solids effectively unless contact is lost through excessive shaft misalignment. The felt-type seal is generally not suitable for exclusion of liquids or retention of oil because of its wicking property. For this reason, felt is often used in combination with other seal materials, as shown in Fig. 5-47.

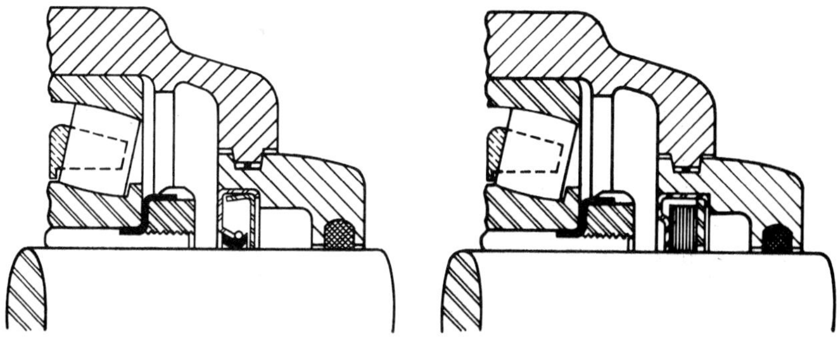

FIG. 5-47 Felt in combination seals.

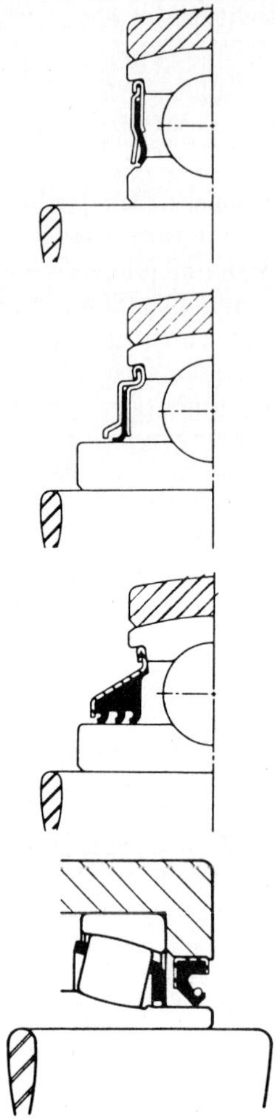

FIG. 5-48 Lip Seals in Link-Belt bearings.

Lip-type contact seals are excellent for sealing solids, liquids, and gases at reasonable pressures. Lip seals are made of leather, synthetic elastomers, plastics, or laminates. Individual lip seals are designed to operate under specific conditions of contact pressure, temperature, and deleterious bearing environment. Because of this, they are available with a broad range of capabilities and operating characteristics.

Commercial cartridge-type lip seals generally consist of a molded synthetic elastomer lip member with a mechanical spring controlling lip pressure against the shaft. These seals usually have high torque characteristics. Sometimes the torque is greater than that required to rotate the bearing, and it will significantly increase the operating temperature of the system. Seal torques should not be overlooked when designing equipment or estimating power requirements.

Many variations of lip seals are available for low temperature, low speed, high speed, and other special conditions. Figure 5-48 illustrates lip seals used in various Link-Belt bearing products. Selection of the optimum seal type should be given full consideration at the equipment design level, as this factor is often over-

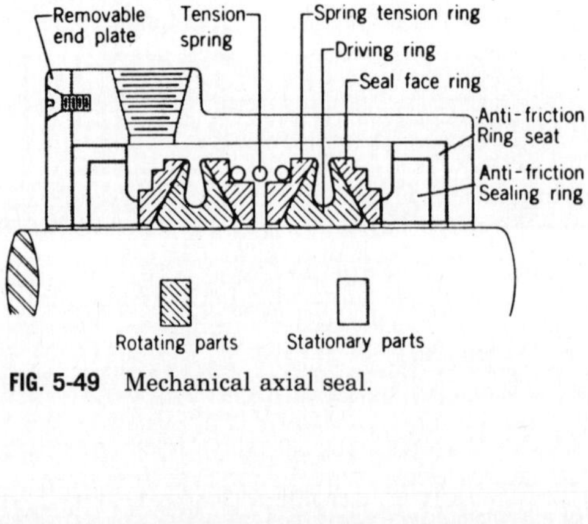

FIG. 5-49 Mechanical axial seal.

looked in the application of lip seals. Most designs are dependent on some lubrication at the lip contact area, as lip wear is the normal life determinant.

Although the mechanical axial seal is a most effective design, it is limited in application. This type of seal is usually recommended when the application requires sealing liquids or gases at medium to high pressures and high speeds. Axial-type seals are expensive; they also require an appreciable amount of space and often demand very precise alignment. Figure 5-49 shows a section of a balanced pressure type of mechanical axial seal.

5-49 Labyrinth or Clearance Seals

In contrast to contact seals, labyrinth or clearance seals provide a running clearance between moving and nonmoving surfaces. Since there is no inherent rubbing contact, frictional drag and heat generation are minimal. High speed, therefore, presents no problem and will increase the effectiveness of some designs. Labyrinth seals are seldom considered for sealing against pressure or for operations that require submersion of the bearing in either liquids or granulated solids. A labyrinth-type seal may range from a simple low-clearance shaft opening to a complex system of labyrinths, flingers, and baffles.

Figure 5-50 illustrates a few types of labyrinth seals, with items *d*, *e*, and *f* showing standard designs used in Link-Belt bearings. Components of labyrinth seals are generally of rigid construction. A broad choice of metals and plastics provides great leeway in designing labyrinth seals to withstand service under corrosive environments, extreme temperature conditions, exposure to abrasion, and other such situations.

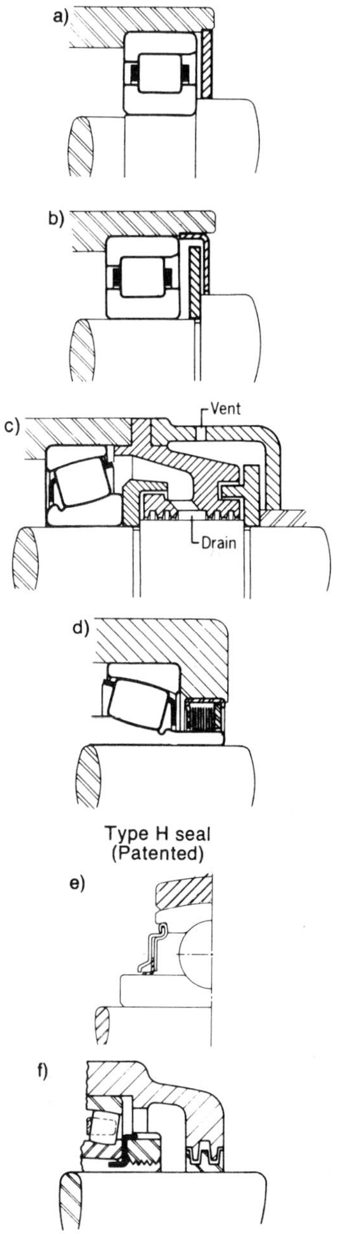

FIG. 5-50 Variations of labyrinth seals.

The design of labyrinth or clearance seals incorporates the following aspects and considerations:

1. Contaminant exclusion. Seal effectiveness depends on maintenance—housekeeping, lubrication, and so on. The probability of adequate maintenance is a design consideration.
2. Lubricant retention. Oil lubrication requires a return drain from an appropriate part of the seal to the oil reservoir. Venting provisions are sometimes necessary to eliminate pressure gradients across a seal. (Note Figure 5-50*c*.)
3. Grease pressure. High-pressure or high-volume greasing equipment can damage seal parts unless relief is provided or the design is adequate to resist such forces.
4. Seal clearance. Seal clearance must be sufficient to avoid fouling in service. Tolerances on concentricity, squareness, and parallelism must be considered. Operating deflections and movements must be fully evaluated to assure running clearance.
5. Installation. Seals are often ruined before operation because of installation damage. Some designs minimize this danger by providing husky outer members which resist pressure or impact.

To summarize, a labyrinth or clearance seal has the inherent advantage over a contact seal of having minimum friction. Excluding the obvious conditions demanding a contact seal, the labyrinth seal is suitable for a broad range of applications—especially for higher speeds.

5-50 Extreme Environments

For extreme conditions, custom seal designs must be created. The following environmental classifications may be considered extreme, depending on the nature of the application and the degree of exposure:

1. Liquid environment. High humidity, water vapor, steam, water splash or spray, rain, snow, corrosive chemicals, chemical vapors, and so on. Volatile or solvent chemicals that may remove or dilute preservatives or lubricants
2. Dirt environment. Dust and dirt of all types, from micro-sizes to large particles, chemically active to chemically inert, and highly abrasive to nonabrasive. Also vegetable or synthetic fibers with wrappage tendencies
3. Atmospheric environment. Gases, heat, cold, pressure and vacuum, and so on
4. Radiation environment. Nuclear and ultraviolet radiation
5. Mechanical environment. Vibration deflection, impact, shock, and abuse

These categories are listed without regard to degree of exposure. For example, exposure to rain may be of no concern in arid or temperate climates, but it could be a severe problem in the tropics. Along with the degree of exposure, the nature of the application will determine whether a given environment must be considered extreme.

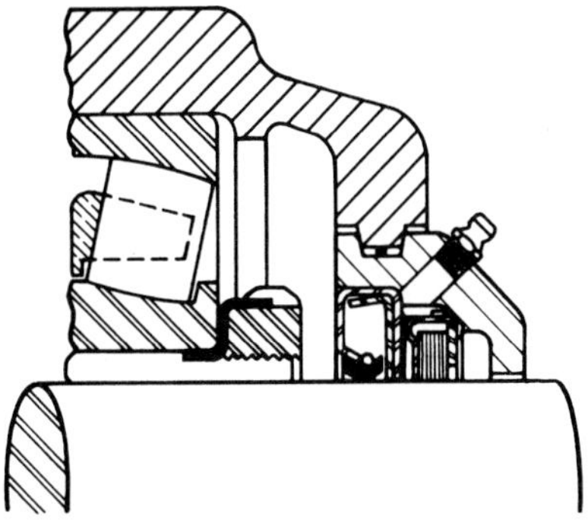

FIG. 5-51 D8 independently flushable seal (patented).

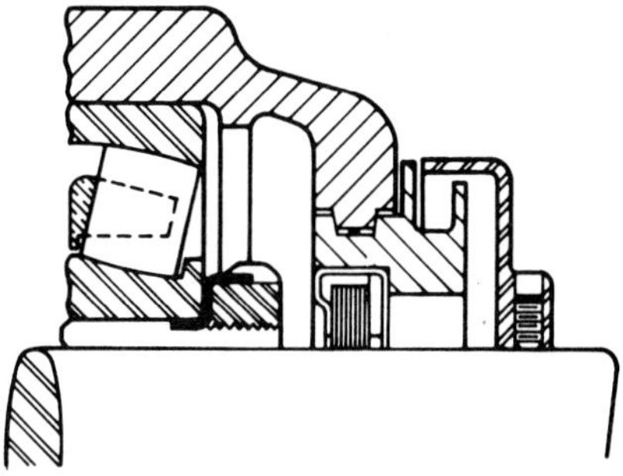

FIG. 5-52 Liquid seal.

The evaluation of extreme conditions and the selection of appropriate seals are unfortunately not clear scientific decisions. There is no substitute for engineering judgment backed by experience. The basic decision relating to a given seal is whether or not it will function as desired in the anticipated environment. When custom sealing is deemed necessary, combination or auxiliary seal designs are often considered. For example, Fig. 5-51 illustrates the patented Link-Belt type D8 seal, which combines a lip seal and a clearance seal along with a proportioned grease-flushing action. Lubricant, introduced to the space between the two seals, purges through and around the labyrinth seal. The configuration illustrated in Fig. 5-52, developed by PT Components, Inc., protects against severe liquid splash or spray. The outer flinger member of this arrangement blocks most of the splash liquid. Any residual fluid that penetrates beyond the flinger is trapped in the channel member and returned by gravity to the exterior.

Seals of the general style illustrated in Figs. 5-51 and 5-52 will solve the majority of dirt or liquid problems, excluding submerged conditions. Submergence in liquid or dirt requires a contact-type seal, usually of custom design.

The third and fourth listed environments (atmospheric and radiation) will usually govern the materials of construction. For example, conventional lip-type seals may be adversely affected by extreme temperatures or radiation. Sealing of high pressure or vacuum will usually require a mechanical face contact seal.

The fifth factor (mechanical environment) is the easiest to design seals for. These conditions will usually affect the seal design more than the seal type. For example, a rock crusher might utilize a basic seal type, either contact or labyrinth, but it may require a husky outer member or guard to withstand the repeated impacts of rocks and earth.

The vast majority of bearing applications are successfully sealed by standard seals. These "off-the-shelf" units provide an opportunity to match the seal type to the functional requirements. Outstanding performance in severe conditions of mud, sand, rain, snow, and other such conditions has provided application engineers with valuable experience in practically every conceivable combination of conditions and equipment. Application engineers will assist in determining when extreme conditions demand a custom seal design.

5-51 Seal Life

An accurate prediction of seal life is seldom possible because of the many variables associated with seal performance. The life of a seal could be defined as the length of service obtained before the seal ceases to provide the expected function of lubricant containment and/or contaminant exclusion. The definition of when a seal ceases to perform its intended function may vary considerably with type of application. For example, if the leakage of an oil seal increases slightly above a normally accepted value, this may be classed as a failure on certain applications. On other applications life may be appreciably longer because a higher leakage rate is acceptable.

The variables which affect seal performance life include shaft finish, runout, misalignment, temperature, lubrication, contamination, proper installation, and, often most important, the maintenance practice. Seals can be damaged by excessive greasing pressure or volume or by lack of relubrication. They can be damaged by excessive exposure to contaminants. Most conventional seals are not designed to function in a submerged condition or with dirt piled against any portion of the seal.

Bearing units are equipped with seals that are designed to function effectively over a reasonable service life, provided the bearings are properly selected and maintained. With contact-type seals some amount of wear is eventually predictable if any abrasive contaminants reach the rubbing contact area. Generally, relubrication at appropriate intervals will help to exclude contaminants and promote a lubricated rubbing contact to minimize or eliminate wear. Clearance-type seals will generally develop wear only when exposed to abrasive contamination. Again, proper relubrication procedures will often prevent wear deterioration.

In many applications seals run directly on the shaft, utilizing the shaft surface as one element of the seal assembly; therefore, the shaft finish and hardness must be adequate for the service life and operating conditions anticipated. To reduce shaft wear, flushable seals are employed in contaminated environments, with the frequency and amount of flushing developed empirically. Experience in such applications has demonstrated long seal and shaft life in highly contaminated atmospheres. Consultation with respected manufacturers is recommended when seals for contaminated environments are considered.

BEARING MOUNTING

5-52 Machine-Bearing Function

A bearing is a part of a mechanical system in which the various parts are usually interdependent to some degree. The function of a bearing in a machine is directly affected by the mating components—primarily the shaft and housing. The effects on a bearing of shaft and housing size, shape, strength, and so on are predictable.

Both the inner and outer rings of a rolling-element bearing are relatively elastic. Imperfections in a shaft or housing often can be detected by a distortion of the raceway geometry. A bearing that is tightly fitted to an out-of-round shaft will operate with the inner ring stretched out of round to approximately the same degree. Incorrect taper, squareness of shoulders and slots, or any other condition of the housing or shaft that deviates from true cylindrical form will affect running accuracies, vibration level, bearing life, and so on.

Manufacturers' data contain fitting practice tables that give recommendations for proper fits between the shaft and housing for various conditions of load, speed, application, and so on. A study of these tables reveals that an assembly will fall between two limits of fit if the tolerances are controlled within recommended limits. Actually, the mean of a fit recommendation is usually the optimum fit for a given assembly. The extremes of a recommended fit represent a compromise with manufacturing tolerances. When a manufacturing facility or process is known to produce heavily at one extreme of manufacturing tolerances, resultant assemblies will predominate at one extreme of the recommended fit. Some compensation may be desirable in such cases. With one-of-a-kind assembly, the statistical distribution and random assembly concepts are worthless. In some cases, manufacturing to fit specifications may be justifiable even at considerable expense.

To facilitate assembly, most bearings are fitted loosely to either the shaft or housing. A loose fit is subject to wear and fretting from differential rolling or "creep." This rolling occurs when the loose-fitted ring is rotating with respect to load direction. Figure 5-53 shows a bearing loosely fitted to the shaft with the inner ring rotating and the load direction fixed. It can be seen that the inner ring is gripped between the shaft and the rolling elements. It is,

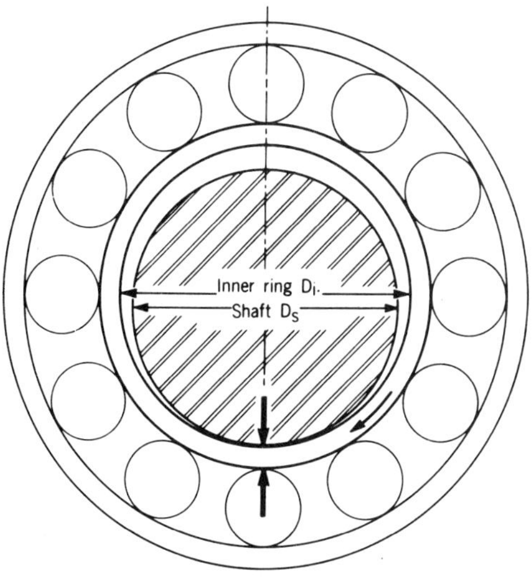

FIG. 5-53 Differential rotation.

in effect, pulled through the load zone much as a plate is moved through rollers in a rolling mill operation. The linear velocity of the shaft surface is equal to the linear velocity of the (bore) surface of the inner ring if the motion is pure rolling. Since the shaft circumference is less than the inner ring bore circumference, the inner ring will revolve slightly less than one revolution for each full revolution of the shaft. The relative linear movement per shaft revolution is π times the amount of looseness in the fit. For example, a bearing fitted 0.003 in (0.08 mm) loose on a shaft rolls a distance of 0.003π in (0.08π mm) each revolution. With a 3600-r/min shaft speed, the inner ring can creep around the shaft a distance of over 3 mi (4.83 km) in 100 h of operation. With sparse lubrication, if any, between the shaft and inner ring, rapid wear can destroy the system. An equivalent phenomenon occurs with a loose-fitted outer ring if there is relative rotation between the load direction and the outer ring. Outer ring creep in a housing is often encountered in applications with unbalanced loads.

In the design of a machine, it is essential to consider the effect of the shaft and housing on bearing function. Whether the housing fit should be loose or tight must be determined. The required rigidity and permissible deflections are also considerations. The effect of shaft and housing inaccuracies on bearing and machine performance should be evaluated. Bearing applications vary appreciably in their mounting requirements. Under some conditions, slightly out-of-round shafting can cause intolerable vibration, noise, or faulty performance. In other cases, running accuracies may be of no particular concern. An objective evaluation of the machine/bearing interdependence will contribute to optimum design and performance.

5-53 Shafting

Shaft components of a machine system serve to transmit power or torque and to sustain axial or radial loads. Shaft size and design may be dictated by strength, deflection, or critical speed requirements of the application. In other cases, bearings or other machine components may govern the size of the shaft, and stress or deflection will not be a major factor. Specific data on shaft design, stress analysis, deflections, critical speed, and so on, may be found in many textbooks devoted to these subjects.

Assuming correct design, the aspects of shaft quality that affect bearings are geometric and dimensional accuracy, surface finish, deflections, material, and hardness. Geometric and dimensional inaccuracies, as noted in Sec. 5-52, may directly affect bearing and machine performance.

The geometric accuracy (concentricity and squareness) of the various working surfaces with each other, including bearing seats, shoulders, and seal surfaces, must be held within reasonable limits. In certain applications, tolerances less than 0.001 in (0.25 μm) are necessary. Out-of-roundness in the shaft can affect the dynamic accuracy of bearing rotation and the vibration characteristics of a machine system. A shaft that is not straight can develop destructive forces, particularly when bearings are not internally self-aligning.

Dimensional accuracy requirements apply to both shaft diameters and axial locations along the shaft. Out-of-tolerance conditions between bearing location shoulders can result in cross-loading of the bearing, while locating shoulder off-squareness can induce both misalignment and cross-loading. Oversize diameters invite overheating or preloading, while undersize shafting may contribute to fretting, loosening, or excessive internal clearance.

The surface finish of the shaft, if too rough, may cause loss of press fits and excessive wear and fretting of the bearing seat. A 63-μin AA (1.6 μm) surface finish on bearing seats is usually the maximum limit of roughness. If a seal contacts the shaft surface, a finish between 10 and 20 μin AA (1.6 μm) surface finish on bearing seats is usually the maximum limit. Turning or through-feed centerless grinding may cause leakage of the seals.

Retention of a bearing on a shaft is affected by the material, hardness, finish, and deflection of the shaft. Torsional or bending deflections involve minute relative movement at the shaft-bearing interface. Such deflections may dictate use of heavier press fits to reduce or eliminate fretting: Bearings mounted with a slip fit on the shaft but held by spring-locking collars, eccentric cam locking collars, or other devices are also vulnerable. Provision of proper shaft tolerance is critical to this type of bearing mounting. Excessive shaft deflections may wear or destroy the clamping contact or the grip of setscrews on the shaft. With hardened steel shafting, mounting accessories that use setscrews are often ineffective—there is little or no penetration of the setscrew into the shaft.

Nonferrous shaft materials require special attention. The difference in the thermal coefficient of expansion of ferrous and nonferrous metals means that the fit of a bearing on the shaft varies with the temperature. The specific effects of varying the fit should be determined, and then the correct shaft tolerances can be established.

A shaft surface is sometimes substituted for the inner ring of a cyclindrical roller bearing. The shaft, used with an outer ring and roller assembly, then becomes an integral part of the bearing. The serviceability and rating of such a bearing are directly dependent on shaft quality and precision. A bearing-quality steel, proper finishes and geometry, dimensional accuracy, and adequate hardness become essential to attainment of full bearing rating and reliable performance. Rockwell C59 minimum hardness, a maximum roughness of 15 μin AA (0.38 μm), and freedom from objectionable lobing and waviness are usually required unless reduced performance is acceptable.

In summary, the correct design and manufacture of a shaft contribute to the reliable performance of bearings and the machine system they serve. Careful attention to application requirements is essential in the establishment of shaft specifications. The shaft chosen must be appropriate to the bearing/shaft mounting method.

5-54 Housings

The housing should provide a rigid support for the bearing. A variable housing stiffness created by webs, ribs, and reliefs may provide distorted bearing support under load, thereby affecting bearing load distribution. Housings may be separate components fastened to a machine frame or foundation, or they may be an integral part of the machine. In the latter case, the machine frame is the bearing housing. In addition to supporting the load, the housing protects the bearing and often provides a lubricant reservoir, a lubricant flow system, cooling, seals, and so on.

In the case of cylindrical roller bearings, a housing may be used as the outer ring for an inner ring and roller assembly. The housing is then an integral part of the bearing and must be manufactured to the same rigid standards as rolling-element-bearing components. Where full bearing rating and performance are required, the housing bore (bearing race) should be hardened to Rockwell C59 or higher and should have a surface finish of 15 μin AA (0.38 μm) or less. In addition, bearing-quality steel and proper dimensional accuracy are necessary, as in the case of shafting used with outer ring and roller assemblies. Manufacturers' data provide housing bore tolerances and other engineering information on the use of cylindrical inner ring and roller assemblies.

Standard housings offer a relatively simple means of adapting bearing and shaft systems to machine design. Manufacturers provide a broad selection of basic mounting styles. These standard units are complete and include bearings, seals, mounting accessories, lubrication fittings, and so on. Occasionally, a custom-designed housing is required. Engineers have designed a great variety of housed units with special features tailored to a broad spectrum of application requirements.

As is the case with bearings, a housing is selected for a specific service. Primary consideration is given to the loading, to the stability and accuracy it must provide, and to environmental conditions and servicing requirements. Table 5-17 lists the factors that affect housing design or selection.

A housing is only as stable as its supporting structure. To provide effective support, the housing itself should rest upon or with a stable foundation. If it is bolted to a weak machine frame which undergoes excessive deflection, it will be subject to twisting and bending loads for which it may not be designed. Even though such distortions may not harm the housing, they may be transmitted to the bearing and cause damage. With a distorted, inaccurate frame or foundation, housing breakage may occur as a result of bolting stresses even before dynamic load is applied.

The ultimate strength of a bearing housing is the maximum load it can sustain without fracturing. Ultimate strength varies, depending on the direction of loading. For this reason, housings must be selected and installed with regard to both the degree and the direction of the forces they will incur. The maximum capacity of a housing is usually attained with the load direction toward and perpendicular to the mounting surface. Where the load is parallel or oblique to the mounting surface, supplementary support is often required to assure a safe and reliable installation. Where the load is directed away from the mounting surface, the installed tension of mounting bolts must exceed the applied load to maintain contact between the housing and the mounting surface.

The safety and reliability of housings are dependent on good design, materials, installation, and so on. Safety factors must be used in accordance with

TABLE 5-17 Factors Affecting Housing Design or Selection

Loading
Magnitude of load: variable or constant
Direction of load: variable or constant
Shock
Vibration
Accuracy
Axial control of shaft
Radial control of shaft
Bearing—housing fit
* Squareness and concentricity
Environment
Corrosion resistance
Radiation resistance
Heat and cold resistance
Magnetic permeability
Servicing and maintenance
Installation problems
Removal: frequent, or only at failure
Relubrication: regreasing or changing oil
Accessories and auxiliaries
Lubrication: grease or oil
Lube method: circulating, reservoir, mist
Seals and sealing
Controls: thermocouples, switches, sensors
Styling, appearance, and cost
Housing: solid or two-piece design
Construction: casting or fabrication
Weight: massive or light

* Refers to accessories (seals and flingers, etc.), as well as to bearings and shafts.

accepted engineering practice. Any bearing mounting subject to heavy load (over 18 percent of the C rating) should be checked for an adequate margin of safety. Even when the loading is below the 18 percent level, housings loaded in an unfavorable direction should be checked for a margin of safety, depending on the risks involved in the event of breakage. Appropriate safety factors in machine design range from 1.5 to 20, depending on materials, type of equipment, risks of injury or equipment damage, and so on. Specific values of ultimate strength cannot be provided for commercial bearing housings because of the variability of strength with the direction of the load. In general, the various cataloged housing styles will sustain loads equal to 18 percent of the C rating in the design direction. The margin of safety would be at least 3.0—and usually over 5.0—under such conditions. The cast steel housings of mill bearing pillow blocks, for example, will sustain more than the bearing C rating when adequately supported.

One of the frequent omissions in the planning of machines is selecting a housing design that facilitates removal of the bearing. Often bearings must be

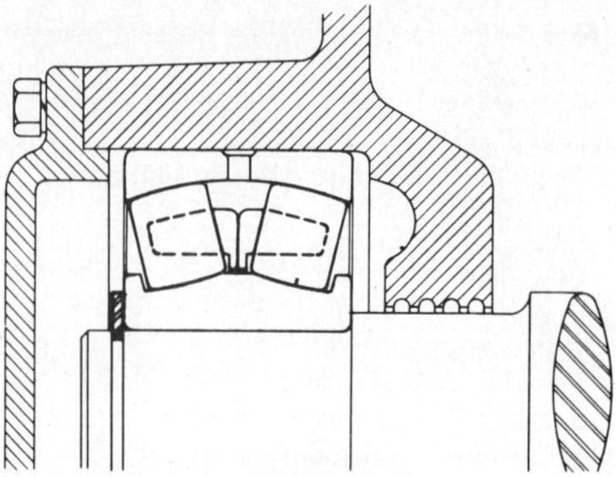

FIG. 5-54 Double-row spherical roller bearing with inner ring press fitted to shaft.

burned off to be replaced. The result may be damage to an expensive shaft or rotor.

Figure 5-54 illustrates a bearing that is press fitted on the shaft and is destroyed on removal. The advantage of this design is simplicity and ease of installation when bearing replacement is known to be rare. A number of designs make installation removal, or servicing difficult. Usually a review of the cross-sectional layouts will reveal such problems, and alternative designs can be developed.

Sealing, lubrication, and control requirements often affect the size of a bearing housing. For example, special seals may require more space along the shaft. Lubrication and cooling may demand oversize reservoirs. An alternative here may be a compact housing with an external circulating oil system. The influence of these factors on bearing centers, and in turn on shaft or machine frame design, is obvious. Failure to provide adequate space for proper housing design can create chronic maintenance problems for the life of the machine.

Housing materials are selected from a wide variety of ferrous alloys. Nonferrous housing materials are seldom used; if they are to be considered, consultation with the bearing manufacturer is recommended. The selection of the housing material is an integral part of design. For example, a well-designed malleable iron housing may be superior in performance to a poorly engineered housing of high-strength alloy steel. In the manufacture of standard housings, the basic materials used are high-test iron, ductile iron, malleable iron and steel in castings, and low-carbon steel strip in some of the fabricated housings. Each housing type is designed with the specific material that provides appropriate strength, rigidity, and cost in relation to the general application of the product.

5-55 Expansion Bearings

A fixed bearing is one that is secured axially both in the housing and on the shaft. An expansion bearing is free to move axially, within fixed limits, in relation to the shaft or the housing or both; it is not expected to sustain external axial loads. Most shafts are supported by two bearings. Both bearings may be fixed, or one may be fixed and the other expansion, or both may be expansion. With two expansion bearings, the axial positioning of the shaft is controlled by some means other than a bearing.

The purpose of an expansion bearing is to eliminate the thrusting of one bearing against the other bearing at the opposite end of the shaft. The cross-loading of two fixed bearings may result from differential thermal expansion between the shafting and machine frame or from installation inaccuracies. The cross-loads developed on the bearings are in addition to the design loads of a machine. They reduce machine life, and in some cases may become totally destructive.

A cylindrical roller bearing can provide an excellent method of accommodating expansion. Various configurations are available in which either the inner ring or the outer ring permits axial shift of the roller set in either or both directions. If a housing bore or shaft is substituted for one of the cylindrical-roller-bearing rings, an almost unlimited allowance for differential expansion can be provided.

Expansion designs for ball bearings and spherical roller bearings are accomplished by a free fit of the bearing in its housing or on the shaft. The location of this loose fit is usually governed by the nature of the loading. For example, with a rotating inner ring and stationary load, the inner ring is tightly fixed to the shaft. The expansion feature is therefore made a function of the loose-fitted outer ring. The reverse holds for a rotating outer ring and stationary load or a rotating load and stationary outer ring. Here the outer ring fit is tight in the housing, and expansion is best accommodated on the shaft.

Whether a bearing must slide on the shaft or in its housing, there is sliding friction involved. A force is necessary to cause the required movement, and this force bears on both the fixed and expansion bearings as a cross-load effect. Such an effect is generally temporary, although in some instances it is considered in the loading analysis. The continuous flexing action in a rotating ball or spherical roller bearing results in a relatively low coefficient of friction (0.05 to 0.15) between the bearing and the housing or shaft. With a rotating cylindrical roller bearing providing expansion, the effect is insignificant. A nonrotating bearing will require a higher expansion force, with coefficients of friction ranging from 0.25 to 0.35, even with a cylindrical roller bearing. The cylindrical roller bearing is susceptible to damage with static expansion and is usually not recommended. If sliding surfaces have deteriorated because of rusting, fretting, galling, and so on, the force needed for axial movement may increase to the point where the bearing is not effective as an expansion unit.

The necessity for expansion bearings is well established, but their use is not always justified. Much mechanical equipment involves short bearing centers and differential expansion potential of a very low order. Installation can often be modified to permit the use of fixed bearings.

5-56 Adjustable Bearings

Single-row adjustable-clearance bearings are almost always used in pairs, and the adjustment of operating clearance is made at the time of installation.

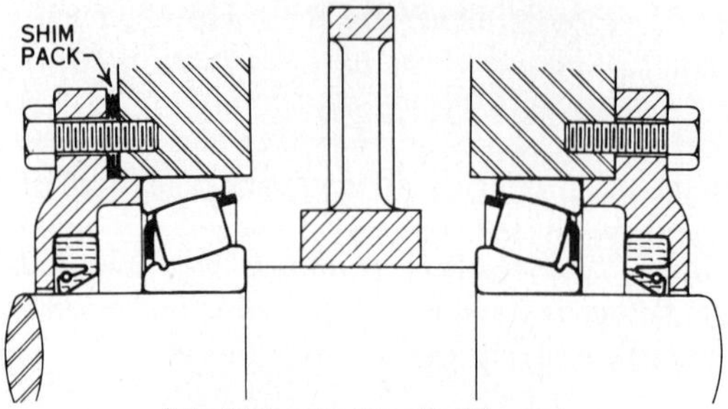

Adjustable outer ring with shim pack

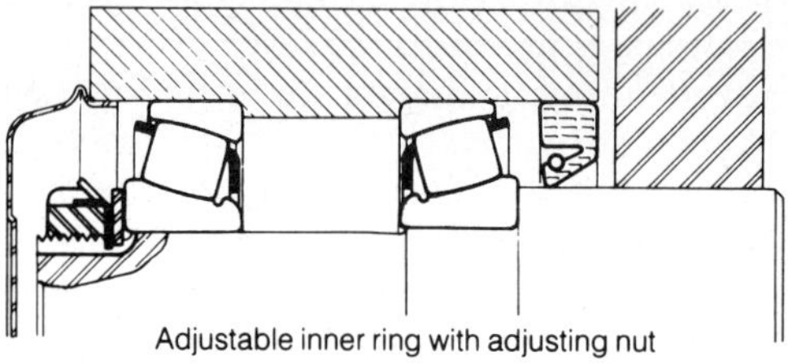

Adjustable inner ring with adjusting nut

FIG. 5-55 Methods of adjusting single-row angular contact bearings.

Provision is made in the mounting arrangement for the axial movement of one bearing toward the other. The adjustment is done by means of shim packs or various threaded adjusting devices. The method may involve movement of the inner ring on the shaft or movement of the outer ring in the housing, depending on the fitting practice used in the application. Figure 5-55 illustrates two of the methods used.

With the adjustable outer ring mounting, the shaft is rotating and the load direction is constant; therefore, the inner ring is press fitted. The outer ring is loose fitted in the housing and thus is easily adjustable by means of a shim pack arrangement. An alternative method uses a threaded cover to engage a threaded housing, eliminating the shims. With the adjustable inner ring mounting, a reverse arrangement prevails: the outer ring and housing are rotating, the shaft is stationary, and the load direction is constant. Here the outer ring is a tight fit in the housing and the inner ring is a slip fit on the shaft. The bearing clearance is set by means of an adjusting nut on a threaded shaft. The final adjustment of the bearings is usually done by first observing the condition of zero clearance, or zero end play in the system. Then the adjustment is backed off to provide the specific operating clearance desired. With the shim pack arrangement, a measurement of the gap at zero clearance is made, then a shim pack is provided that is equal to the measured gap plus the operating clearance. An adjustment is often made by "feel" or by measuring direct end play or torque. A system of adjusting by feel is acceptable if it is done by trained and experienced personnel. Many other procedures may be

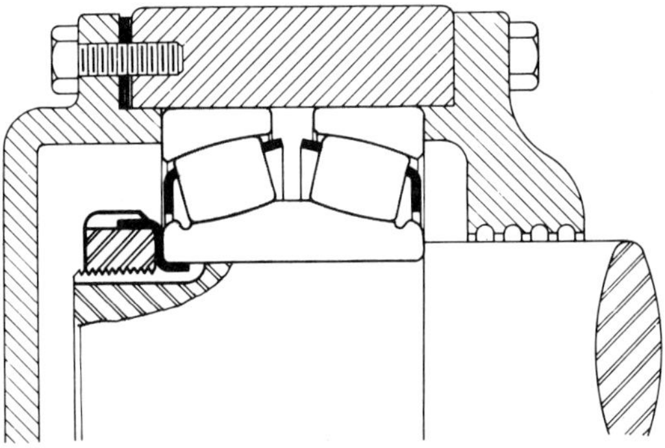

FIG. 5-56 Double-row spherical roller bearing with adjustable outer ring.

used on this type of bearing, depending on the nature of the application. The design of the mounting arrangement should provide a means for adjusting the clearance during the operating life of the bearing. A thorough education of personnel is recommended to achieve uniform results and optimum performance.

A differential thermal expansion between the shaft and the machine frame will have a direct effect on bearing adjustment. With long bearing centers involving temperature differentials, single-row adjustable bearings may not be suitable. Excessive deflection of bearing supports, machine frame, or shaft may have the same detrimental effect as a thermal expansion differential. The single-row bearing will develop a thrust reaction (described in Sec. 5-32) that must be sustained by the frame or shaft without excessive deflection. The magnitude of this thrust reaction is calculated in accordance with the procedures described in Sec. 5-32.

Double-row adjustable-type bearings, as illustrated in Fig. 5-56, must also have provision for the adjustment of operating clearance. With the double-row bearing, however, each bearing in the system must be individually adjusted. When the double-row bearing is used in conjunction with a single-row bearing at opposite ends of a shaft, both bearings must have provision for clearance adjustment. In this case, the double-row bearing must be adjusted before the single-row bearing.

5-57 Misalignment

Misalignment of the shaft with the bearing housing or machine frame can be anticipated in all except the most precisely manufactured equipment. Even in precision equipment, deflections in operation will exist and will result in some degree of misalignment. This misalignment is ideally compensated for by internally aligning rolling-element bearings. In such a design, the function of the bearings is unaffected by dynamic or static misalignment. All other types of bearings must be accurately aligned within close limits or must be provided with a means for external alignment, often with a ball-and-socket arrangement. In Link-Belt ball-bearing units, for example, the bearing outer ring is ground to a spherical contour, and a matching spherical curvature is provided in the bearing housing.

The basic function of the ball-and-socket arrangement is to compensate for static misalignment occurring as a result of simple deflection or mounting inaccuracies. With the ball-and-socket concept, moment loads may prevail during operation if the magnitude of the moment is less than the force required to overcome the friction of the ball-and-socket joint. In most instances, a slight jar or impact will relieve any twist load that is not relieved by normal vibrations that occur when the equipment is operating.

Installation should be as accurate as possible, even when self-aligning bearings are used. There are limits to the self-aligning construction, and misalignment beyond the limits may damage some of the components, such as the seals. Generally, the alignment feature in Link-Belt bearings, other than cylindrical roller bearings, provides ample allowance for a completely visual alignment of the bearings. Standard Link-Belt cylindrical roller bearings are designed to accommodate misalignments in the range of 0.0015 in/in (m/m). Methods of verifying the mutual deflection effects in mechanical systems have been developed. Utilizing these methods, the true misalignment and deflection effect on a bearing is evaluated to establish special bearing design requirements when necessary.

In the installation of cylindrical roller bearings care must be taken to align the rings on shafts and in housings to face runouts generally not exceeding 0.0004 in/in so as to minimize installation effects.

5-58 Mounting Accessories

Mounting accessories aid in the mounting of a bearing on a shaft or in a housing. Some of these devices have been standardized by the bearing industry, while others are exclusive or are designed for the bearings of a given manufacturer.

The three standardized mounting accessories are tapered adapter sleeves, bearing locknuts, and bearing lockwashers. Extended inner ring bearings utilize a locking collar of either the spring or eccentric cam type. Both types serve to clamp a loose-fitted inner ring to the shaft and thus provide axial location and concurrent rotation of the shaft and inner ring. Each of these accessories is effective when properly applied; however, recommended tolerances must be observed.

Many other accessories are used in the mounting of bearings. Spacers or sleeves are often used for axial location or axial clamping of the rings. Many axial clamping arrangements are possible where bearing locations are fixed by gear hubs, pulleys, clamp plates, snap rings, and other devices. With an axial clamping device, differential rotation of the ring is not necessarily prevented. The friction exerted against the face of the ring is often insufficient to prevent creep. An extremely lightly loaded bearing is the exception, but even here deflections and wear may eventually eliminate the clamping pressure exerted against the face of the ring. Extremely tight axial clamping of the rings should be avoided to prevent distortion in the raceway geometry.

Mounting accessories also include devices for bearing removal. Various aids to installation and removal can be incorporated in a machine. A spacing sleeve, for example, may serve as a removal tool. Housing bolts may serve as jackscrews for housing removal.

Ball and roller bearings are especially susceptible to damage during installation and removal. Often damage is caused when the rolling elements transmit the force used to install or remove the bearing. Sharp blows from tools can cause great damage. Mounting accessories that avoid this type of damage therefore contribute to reliability and service life of bearings.

5-59 Bearing Clearance

The establishment of correct bearing clearance is essential for reliable performance of rolling-element bearings. Excessive bearing clearance will result in poor load distribution within the bearing, decreased fatigue life, and possible excessive dynamic excursions of the rotating system. Insufficient bearing clearance may result in excessive operating temperature or possible thermal lockup and catastrophic failure.

Bearings are manufactured with an initial bearing clearance which can be varied by the manufacturer and is generally part of the bearing purchase specification. This initial clearance, or unmounted bearing clearance, is then altered (generally reduced) by the fits utilized on the shaft and/or housing and by the thermal gradients established in operation of the equipment. The clearance remaining in the bearing in the operating mode is termed "operating clearance." Operating clearance requirements will vary with nature of application, type of bearing, load, and speed. Generally, higher-speed applications will be designed for a greater amount of operating clearance, to allow for less predictable thermal differentials. Lower-speed applications often involve heavier loads and are designed for reduced operating clearance to obtain optimum load distribution and maximum fatigue life. In some applications, such as printing presses and precision coating rolls, bearings are engineered for negative operating clearance or preload to maximize control of shaft position.

The evaluation of clearance change from the initial bearing clearance to the final operating clearance is generally straightforward where iron or steel housings are utilized with solid steel shafting and the application does not involve extreme speeds or external thermal effects. When any of these normal conditions change, the design should be evaluated very carefully.

5-60 Mounting Precautions

After selecting bearings suitable to the application, finalizing equipment designs, and establishing specifications for housings, shafting, and so on, the actual mounting or installation of the bearing must be considered. Proper planning and consideration of bearing installation can actually be the most critical factor in a successful bearing application. The training of personnel involved in the installation and maintenance of bearings or bearing units is essential to reliable and safe performance of the equipment. Bearings must be properly handled, protected, and maintained during the critical installation procedure and subsequent operation. Service instructions are available for all Link-Belt bearings, and they should be carefully followed. In specific problem situations, field consultation may be requested.

Common problems that can be encountered during installation are abusive handling, exposure to contaminants, bearings cocked on shaft or in housings, improper tightening of mounting sleeves, improper tightening of fasteners, bearing damage from other operations on equipment (such as welding), improper shimming, improper positioning, and so on. Operational and maintenance problems often encountered are improper lubrication or omission of lubrication, stray electric current through bearings, exposure to steam cleaning or wash-down, exposure to shock or abusive handling of equipment, exposure to external vibration, exposure to excessive heat, and so on. The adverse conditions or events which can lead to problems ranging from reduced life to sudden catastrophic failure are almost endless and beyond the scope of this text. However, bearings are essentially safe products and can be expected to provide reliable, safe service under normal conditions.

Since bearings are used to support large inertia systems and masses, their failure can release forces that can cause personal injury or property damage. Therefore, all equipment should be evaluated with regard to potential risk or hazard in the event of a bearing malfunction. Problems resulting in bearing failure are generally obvious, and they can largely be avoided by proper planning, training, and instruction of personnel, including equipment users.

SLEEVE BEARINGS

5-61 Introduction

A sleeve bearing is a simple device for providing support and radial positioning while permitting rotation of a shaft. It is the oldest bearing device known. In the broad category of sleeve bearings can be included a great variety of materials, shapes, and sizes. Materials used include an untold number of metallic alloys, sintered metals, plastics, wood, rubber, ceramics, solid lubricants, and composites. Types range from a simple hole in a cast-iron machine frame to some exceedingly complex gas-lubricated high-speed rotor bearings.

The sleeve bearing is inherently quiet in operation because it has no moving parts. With proper selection, installation, and lubrication, it does not fail suddenly. Wear is gradual, and replacement of worn bearings can be scheduled when equipment is normally idle. These bearings are uniquely suited to conditions involving oscillating movement of the shaft. Excessive moisture, submersion in liquids, and extreme temperatures can be accommodated with the proper selection of materials.

Sleeve bearings, like rolling-element bearings, must be properly selected and used, with control of shaft tolerances, housings, mounting or installation procedures, and lubrication.

5-62 Application and Selection

The application and selection of sleeve bearings, like rolling-element bearings, is ideally based on a thorough analysis of all the conditions of operation. These conditions include load, speed, temperature, environment, and service requirements. Under the category of load, one should know the variations in loading, peak loads, shock loads, vibration, load direction, and so on. Speed involves starting, stopping, frequency, acceleration, deceleration, and variations while running. The consideration of temperature must include maximum and minimum expected temperatures as well as operating temperatures and expected variations. The external environment may be gas or liquid, with solid contamination, corrosive or abrasive, explosive, ultra-clean, and so on. The quality of maintenance and service expectations in terms of life and reliability can influence bearing selection. At the initial design stage, there are often unknowns among the various conditions mentioned. Assumptions and decisions can be made based on best judgments; however, the risks in such decisions must be evaluated.

The load and speed capabilities of the various cataloged bearings were outlined previously. The values provided are straightforward, but, as with all physical ratings of machine elements, engineering judgments are necessary. Sometimes ratings can be exceeded, and, conversely, situations exist in which use of a product within the catalog ratings could be unsafe. For example, a load may be well within the capability of a given bearing, but vibration or shock consideration may rule the bearing inadequate. A given bearing might be satisfactory for continuous operation at a particular speed but prove unsuitable in the start-up condition. Because of the complexity of operating conditions in many applications, it is not feasible to reduce the bearing selection technique to a precise engineering formula, and engineering judgment is a prime requisite.

Assuming that a particular application is best served by a sleeve bearing rather than a rolling-element bearing, decisions are required as to the type. The section on bearing materials provides a brief review of the various materials and their characteristics. The analysis of operating conditions will generally point to a particular material as the most suitable for an application. If the selection of material is a clear-cut decision because of temperature or environmental requirements, the selection procedure then establishes the bearing size that will satisfy the load and speed conditions. Along with size determination, the style of mounting and installation arrangement should also be considered. For example, an uplift load condition may require the inverted mounting of a pillow block. This in turn could complicate any gravity feed arrangements for lubrication.

The service life of a sleeve bearing is generally not predictable. With all conditions of selection, installation, lubrication, and maintenance in accordance with theory, bearing failure will develop only gradually, as wear conditions exceed allowable amounts. In practice the wear rate cannot be precisely predicted, except from the history of performance on a particular application.

5-63 Bearing Materials

Babbitt Babbitt metal bearings are universally accepted as providing reasonable capacity and dependable service, often under adverse conditions. It is a relatively soft bearing material, and for this reason minimizes the danger of scoring or damage of large, expensive shafts or rotors. Babbitt can often be repaired quickly in the field by rescraping, pouring of new metal, and so on. The actual bearing operating temperature must not exceed 200°F (93°C), and maximum ambient temperatures should not exceed 130°F (70°C). Babbitt is used for applications involving light to moderate loadings, with bronze required for heavier loadings or higher temperatures.

Bronze Bronze bearings are suitable for heavier loads than babbitt, with capacities ranging from 75 to 200 percent higher, depending on the specific ranges of loads and speeds. Bronze withstands higher shock loads than babbitt, permits slightly higher-speed operation, and can be used up to 300°F (149°C) ambient temperature. Bronze is a harder material than babbitt and has a greater tendency to score or damage journals in the event of malfunction. Field repair of a bronze bearing generally requires replacement of the bushing.

Sintered Metal An increasingly popular bearing material is the sintered porous metal bushing, which is usually an alloy of bronze. This is the material used in the type T1G Flex-block® bearing (consult manufacturer). In the fabrication of this type of bearing, the powdered metal alloy is pressed in dies to a controlled density, then sintered at high temperature in a reducing atmosphere. Any subsequent machining of the part must be controlled with proper tooling and machining techniques so that the open pore structure is maintained.

The porous bearing material functions somewhat as a sponge, with the lubricant retained in the voids. These voids normally make up about 20 percent of the volume of the bushing. Oils used for impregnation should be a nongumming type resistant to oxidation. In operation the development of heat or pressure between the shaft and the bushing initiates a capillary action, bringing oil to the surface for lubrication. This type of bearing will function until the supply of lubricant contained within the bushing is exhausted. With the Link-Belt type T1G bearing, an additional supply of lubricant is provided in a wicking-type material surrounding the bushing, with provision for additional replenishment through the oil supply hole.

Carbon-Graphite A carbon-graphite material is used for applications at temperatures up to 700°F (371°C), where the use of conventional bearings and lubricants is next to impossible. This type of bearing is typified in the type T3 Flex-block® bearing (consult manufacturer), which utilizes a carbon-graphite bushing. This design is entirely self-lubricating; in essence, the bearing itself is the lubricant. In service this solid lubricant is gradually consumed, with the shaft/bushing clearance gradually increasing as the solid lubricant is used. These bearings have been particularly useful in ovens, driers, furnaces, and other such applications, where loads are light and speeds are relatively low. They are often used in alien environments because of the basic inertness of the carbon-graphite material.

Occasionally this type of bearing will develop a high-frequency vibration in resonance with the operating system, producing a high-pitched squeal or noise condition. Correction usually requires damping or modification of the resonant frequency of the shaft.

Cast Iron Cast iron is very suitable for many slow-moving shafts and oscillating arms supporting relatively light loads. The wear-limiting characteristics of cast iron are attributed to the free graphite flakes present in the alloy. When cast-iron bearings are used, higher shaft clearance is usually provided. Thus any large wear particles or extraneous debris will not jam the clearance spaces and seize the bearing. These bearings have been applied at temperatures as high as 1000°F (538°C) under light loads and slow-speed intermittent operation.

Special Bearing Materials In some situations special bearing materials are required for an application. When feasible, custom designs can be provided; however, full application information is necessary for evaluation. Custom designs generally involve high engineering and preparation costs, extended deliveries, and occasionally development and prototype testing. A minor modification of the equipment design to allow standard bearings to be used is often a more desirable approach.

5-64 Lubrication

The three basic conditions of bearing/shaft interface shown in Fig. 5-57 illustrate full film, boundary (or thin film), and extreme boundary lubrication conditions. Of these, full film lubrication is obviously the ideal operating condition for a sleeve bearing; however, many bearings are designed to operate in the other two modes. The Link-Belt sleeve-bearing product

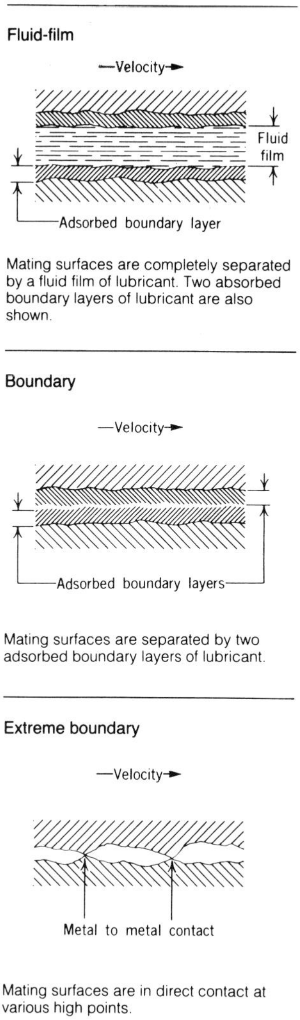

FIG. 5-57 Basic types of lubrication.

lines, with some exceptions, are designed and rated for operation essentially in the boundary lubrication regime, although some operating conditions will develop full film lubrication. The type T3 bearing unit is intended for operation without lubrication, and might therefore be placed in the extreme boundary lubrication category.

In the classic hydrodynamic theory of lubrication, the motion of the shaft forces the lubricant to flow in the direction of rotation. In the area of greatest load, the shaft is displaced from the geometric center of the bearing by the applied loads. As the distance between shaft and bearing decreases at the load zone, the well-known wedge-shaped clearance is created, ranging from the high clearance space opposite the load zone to the minimal clearance space at the load zone. The lubricant entrained by the shaft and carried into the diminishing clearance wedge is squeezed severely. Since the lubricant is essentially incompressible, a pressure is developed with the oil film which is sufficient to support the applied loads. The development of the fluid film wedge is largely dependent on relative velocity between shaft and bearing. For this reason, starting and stopping conditions are often critical for bearings designed for full fluid film operation. In some cases a hydrostatic principle of lubrication is introduced, where the oil pressure necessary to support the applied load is provided from an external source.

Boundary lubrication, thin film lubrication, and partial film lubrication are synonymous terms applied to a prevalent condition of sleeve-bearing lubrication. These terms describe a lubrication regime in which the separating film of lubricant exists over only part of the load-bearing surface, so that part of the load may be carried by direct contact of the shaft and bearing material. The characteristics of the lubricant and the sliding surfaces become deciding factors in determining overall friction and wear of the bearing. Link-Belt rigid sleeve bearings are designed for grease lubrication, which is retained in the bearing longer than oil, primarily because of a reduced side leakage effect. In many cases grease will provide hydrodynamic full film lubrication, but because of its high apparent viscosity and its other inherent characteristics, it will function more effectively than oil in a boundary lubrication regime. This protects bearings in start-up and shut-down operations or in slow-speed operation where velocities are not sufficient to develop a hydrodynamic film. In many instances greases fortified with solid lubricant additives such as molydisulfide and graphite have proven beneficial, particularly in conditions approaching extreme boundary lubrication. Oil may be used in the rigid sleeve bearings; however, the expected side leakage would normally be greater than with grease. An adequate oil supply must be provided to avoid risk of oil starvation. Oil viscosity should be chosen between 100 and 200 SUS (20 and 43 cSt) at the estimated operating temperature. Grease consistency is normally chosen ranging from National Lubricating and Grease Institute (NLGI) no. 0 at operating temperatures below freezing up to NLGI no. 3 consistency at operating temperatures above 100°F (38°C).

Extreme boundary lubrication is defined as metal-to-metal contact or shaft-to-bearing-material contact, at least at various high points or asperities. This condition is encountered in bearings under conditions of lubricant starvation, overloading, or improper lubricant, and will generally result in rapid wear or failure if the bearing is not designed or intended for this type of service. Bearings designed for dry operation, such as the Link-Belt type T3, will still wear; however, it is a programmed wear which is anticipated, and with proper bearing selection the functional life requirement is satisfied. Most bearings designed for this type of service are dry lubricant types or self-lubricating. The type T3 bearing, for example, is a solid lubricant carbon-graphite bushing. The bearing itself is the lubricant, which is gradually consumed as shear and wear occur. This material is utilized effectively in extremely high-temperature applications, normally up to 700°F (371°C).

In summary, lubricant determinations are usually governed by the type of bearing selected, since most sleeve bearings are designed for a specific type of lubricant. The method of applying lubricants may vary from simple oil cups, grease cups, or fittings to completely automatic systems. The preferred method relates to the nature of the application and the economics of providing the lubricant to the bearings at the required rate.

5-65 Mounting

The proper mounting of a sleeve bearing on the shaft and machine frame is a critical factor in the successful application of such a unit. The bearing can be very sensitive to distortions, deflections, misalignments, and surface imperfections of the mating machine elements. Such imperfections due to design, manufacture, or installation will have their effect on load distribution, lubrication, or other design characteristics of the bearing. The Link-Belt sleeve-bearing product line is basically a mounted bearing series, with a properly designed outer support provided for the various types. When these products are used, the primary concerns are the shaft element of the machine, installation and alignment, and the accuracy of the bearing-support surface of the machine.

Housings The bearing housing provides the basic support and location of the bearing. The housing must provide adequate back-up or support so that there will be no linear or radial distortion induced in the bore. There are two general classes of housing with respect to alignment of bearings: the rigid mounting and the alignable type. With the rigid type, the required careful alignment of the bearing must be obtained by proper shimming and adjustment of the housing position on the machine. With the alignable style, designs generally involve an intermediate cartridge housing with a spherical OD supporting the basic bearing sleeve. This cartridge then fits into the main housing, which is provided with a spherical seat, forming a ball-and-socket joint.

Housing styles are further subdivided into one-piece and two-piece housings. The two-piece housing provides a simplified mounting or installation arrangement but is more restrictive than the one-piece housing with respect to direction of applied loads. Several of the basic types of housings are illustrated in Fig. 5-58. Each housing design fulfills certain requirements of various applications and types of bearings.

Shafting The primary requirement for bearing shafting is to provide the proper size and finish. The shaft surface finish requirements differ for various

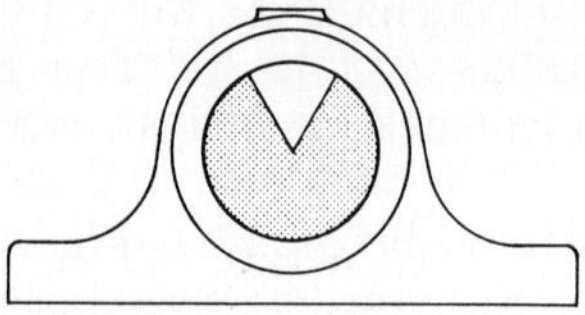

One-piece housings

Pillow blocks and flanged units for service where removable caps are not required. Available with babbitt, bronze, or plain iron bearings. Recommended loading range, 300°.

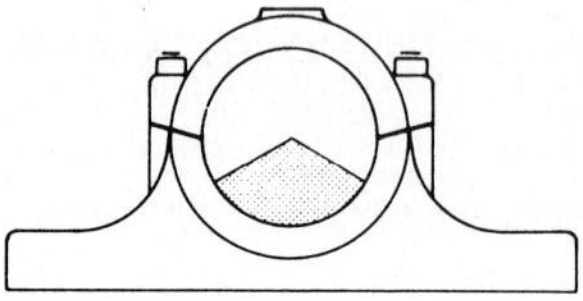

Two-piece housings

Pillow blocks for loads involving no side pressure. Available with babbitt or bronze bearings. Recommended loading range, 120°.

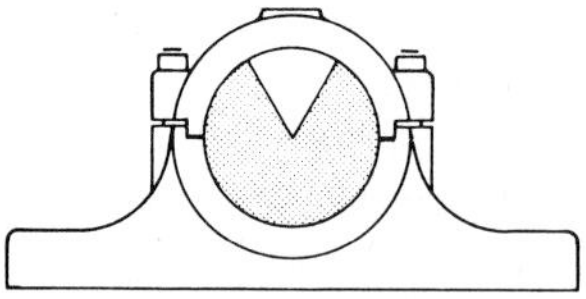

Two-piece housings with gibbed joints

Ideal for installations involving heavy side pressure. Available with babbitt or bronze bearings; grease lubricated. Recommended loading range, 300°.

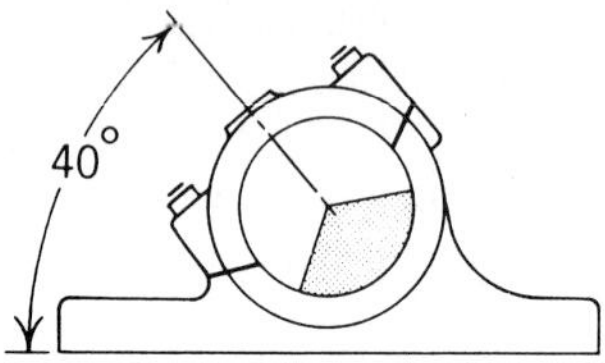

Angular two-piece housings

Pillow blocks adapted to angular loads. Recommended loading range, 120°.

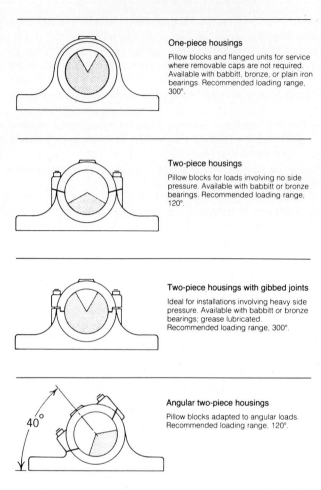

FIG. 5-58 Recommended loading range for basic types of sleeve bearing.

bearing materials. Type T1G and T3 bearings require shaft finish of 12 μin AA (0.30 μm) max., while the rigid babbitt and bronze sleeve bearings may have shaft finish up to 32 μin AA (0.81 μm). Shafting should be straight and free of seams, nicks, scratches, or burrs. Imperfections in shaft surfaces can disrupt the lubricant film, thereby promoting a condition of localized scoring. Permanent shaft bending creates a dynamic misalignment action within a bearing, which can be destructive to the system. Shafting deflection may also distort the pressure distribution within the bearing. An alignable-type bearing will compensate for shaft deflection, provided the deflection is not a dynamic condition. A rigid bearing mounting may sometimes be shimmed at installation to provide alignment of the bearing with a deflected shaft if the deflection does not vary significantly in operation.

Installation Assuming proper bearing selection, the installation and lubrication of the bearing are usually the most critical factors in its serviceability. Each type of bearing has unique characteristics and installation requirements. Service instructions are provided for various types of Link-Belt bearings; however, installation procedures for each application may differ to some degree. To a great extent, common sense and consideration of the way a bearing functions in a given application will dictate installation and maintenance procedures. With the rigid-type bearing, for example, very careful alignment is an absolute necessity. With the alignable bearing styles, exact alignment is of little concern. With most bearings the factors affecting lubrication should be considered. For example, lubricant entry points or grooves generally should be opposite the load direction, except in bearings operating at less than 50 ft/min (0.25 m/s). With low-speed bearings it is often desirable to have lubricant entry points in the load zone. Any factor which may disrupt or remove the lubricating film should be eliminated. For example, sharp edges on the shaft or the bearing surface may act as scrapers to destroy or remove lubricant films. Grease grooves in bearings must have blended edges. A shaft keyway extending into the bearing would be very poor practice because any oil film effect would be destroyed with every revolution.

Dirt or other contaminants can rapidly destroy a bearing through abrasive wear. In such cases sealing arrangements may be required. Felt seal washers may be used to seal directly against the ends of bearings.

6
SEALS AND PACKINGS

DR. LESLIE A. HORVE, P. E.
Vice President of Technology/Chicago Rawhide Manufacturing Company/Elgin, Illinois

INTRODUCTION

6-1 Seal Requirements

Modern machinery is highly dependent upon seals to retain liquids, solids, or gases. A secondary but equally important function of sealing devices is to prevent foreign particles from entering the sealed cavity.

There are many seal designs and sealing principles used in industry. Selecting a sealing principle and ultimately a seal design for a given application is not an easy task. The application must be studied to correctly identify the nature of the sealing problem. The application conditions must be considered to further refine the selection (Fig. 6-1). Some important conditions include system pressure, temperature, medium to be sealed, medium to be excluded, static or dynamic conditions, and type of motion if dynamic. Economics is also a factor. In general, the complexity of seal designs and the cost of sealing increases as the tolerance for leakage decreases.

The most frequently used seals are packings, which are usually the simplest and least expensive solution to sealing problems. Packings are shown as the bottom of the sealing pyramid in Fig. 6-2. The complexity and cost of sealing devices increase as one moves up the pyramid. The horizontal axis represents quantities of seal types used.

6-2 Allowable Leakage Rates

Many sources will argue that a condition of zero leakage does not exist. Virtually leakproof static seals can be obtained by plastically deforming metal. The

Seal Variables	Effects Causing Static or Dynamic Leakage	Application Variables
Material	Thermal expansion & contraction	Static or dynamic
Design	Seal material degradation	Sealed medium properties
	Corrosion	Type of shaft speed & motion
	Fatigue	Shaft eccentricity
	Vibration	Shaft diameter
	Wear	Shaft finish
	Shaft & bore defects	Shaft material
	Lubricant breakdown	Housing
	Case leakage	Finish
	Misalignment	Eccentricity
		Interference tolerance
		Operation
		Cycles (pressure, temperature)
		Run time
		Ambient
		Temperature & range
		Ozone
		Dust/mud
		Pressure & range

FIG. 6-1 Application conditions.

leakage obtained with this gasket type is less than 10^{-8} cm^3/s (atm.) gaseous nitrogen at 300 lb/in^2 (gage) at ambient temperatures. Leakage of fluids is typically measured in terms of drops (approximately 0.05 cm^3) per day. Different applications will tolerate different leakage rates. Allowable leakage rates range from zero to many cubic centimeters per hour.

It has always been necessary to minimize the leakage of expensive, toxic, corrosive, or noxious gases and liquids. Pressures to reduce energy losses, conserve fuels and lubricants, and protect the environment have increased. It has been estimated that 100 million gallons of fluid are lost to the environment each year from hydraulic machines and systems alone. Additional losses occur from engines, transmissions, and gearboxes.

It is possible to reduce and even stop external leakage. Aircraft hydraulic systems are expensive, but they are virtually leak-free. Leakage in industrial machinery and vehicles has been substantially reduced.

These reductions are due in part to the availability of better seal designs,

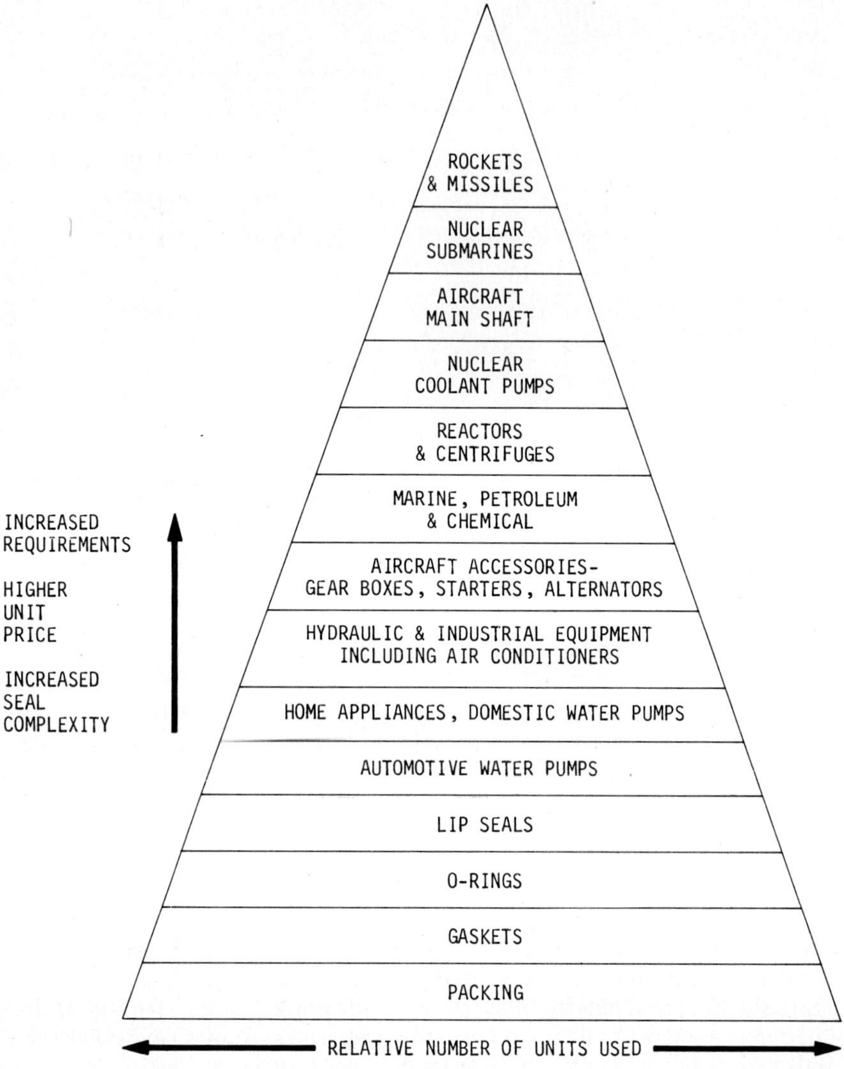

FIG. 6-2 Sealing pyramid.

materials, and technology. They are also due to the great care seal users take in selecting the proper sealing concept for their application.

6-3 Seal Classification

Seals are classified into two major categories, static and dynamic. Static seals are used in applications in which there is no movement between mating surfaces. Deformable gaskets and sealants are used extensively in static applica-

SEAL CLASSIFICATION

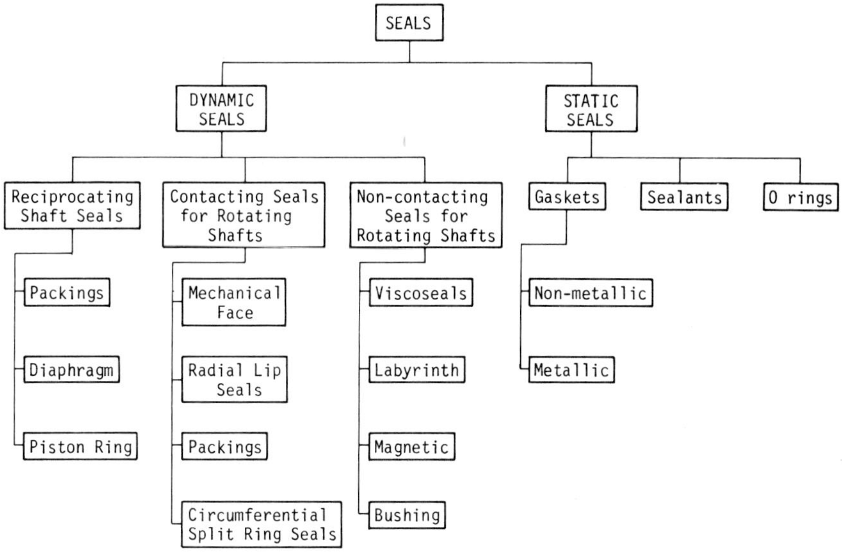

FIG. 6-3 General classification chart.

tions. Dynamic seals are used in applications in which shaft rotation, reciprocation, and oscillation are present. A wide variety of seal designs can be used in dynamic applications. A general classification chart for seals appears in Fig. 6-3.

6-4 Sealing History

The earliest seal consisted of leather straps arranged to form a labyrinth at the end of a wheel axle to hold grease or animal fat in place. The industrial revolution of the eighteenth century created demands for more sophisticated lubrication and sealing systems. Compression-packed stuffing boxes were used to seal both reciprocating and rotating shafts. Early packings were made from available fibers, such as flax, hemp, cotton, and wool. Animal fats were used as lubricants and fillers. Then, in the early 1900s, asbestos increased the temperature range and chemical resistance of packings. Synthetic fibers later became available; these are now replacing asbestos because of its recently recognized carcinogenic characteristics. Lubricants and fillers were improved with petroleum-base lubricants, fish oils, waxes, and synthetic lubricants such as mica, graphite, polytetrafluoroethylene (PTFE), and molydisulfide-base materials. Various fabrication techniques, new arrangements of braided fibers, and various reinforcing techniques using wire or metal foils all serve to extend the capability of compression packing.

As the speeds of rotating shafts increased, it became necessary to consider an alternative to packings. In the mid-1920s, radial lip seals made of leather were used in automotive applications to follow the eccentricities of high-speed rotating shafts. These assembled seals had advantages that could be readily appreciated: They were compact, self-contained units, and they were easily press fitted into a housing bore.

The mechanical face seal was also developed in the 1920s to provide low leakage rates for the mechanical refrigeration industry. The early refrigeration compressors used a variety of gases, such as ammonia, sulfur dioxide, and methyl chloride. These compressors used stuffing boxes, were difficult to adjust, and permitted significant leakage that drained off the refrigerant gas supply. In addition, the refrigerant gases used at that time were obnoxious or toxic to the personnel in the area. Mechanical seals replaced the packings and lip seals that had been used, solving these problems.

As World War II approached, rotating machinery turned rapidly to antifriction bearings. Shaft speeds increased, and operating temperatures rose. The existing seal designs and materials were no longer adequate. The newly developed oil-resistant synthetic elastomers replaced leather in radial lip seal designs for rotating shaft applications. Upper temperature limitations were extended from 200°F to 300°F. High-pressure hydraulic systems were used in more and more applications. A host of packings made from synthetic elastomers were introduced. These packings (V rings, U rings, O rings, chevrons, and so on) helped to extend the operating pressure range of hydraulic systems while reducing leakage. The jet and space age that started in the 1950s spurred the development of new materials and sealing concepts. Sophisticated labyrinths and bushings were developed for jet engines. All-metallic mechanical seals and new elastomeric materials extended the available temperature ranges for new applications.

Bonding cements were improved, and the assembled rotary lip seal was replaced by simpler designs with the elastomer bonded directly to the metal case. Internal leakage through the assembled components was eliminated. The zero leakage requirements of the space and nuclear applications of the 1970s and 1980s have led to the development of magnetic sealing systems. New sealing systems that will economically ensure that there is little or no leakage must be developed to protect the the environment and sophisticated equipment from the inadvertent leakage of harsh chemicals and lubricants.

STATIC SEALS—GASKETS

6-5 Gasket Applications

Gaskets are used to develop and maintain a barrier between mating surfaces of mechanical assemblies when the surfaces do not move relative to each other. The barrier is designed to retain internal pressure, prevent liquids and gases from escaping from the assembly, and prevent contaminants from entering the assembly. It is essential that joint design and gasket design be considered together. The gland-type joint captures the gasket (Fig. 6-4) and locks it in

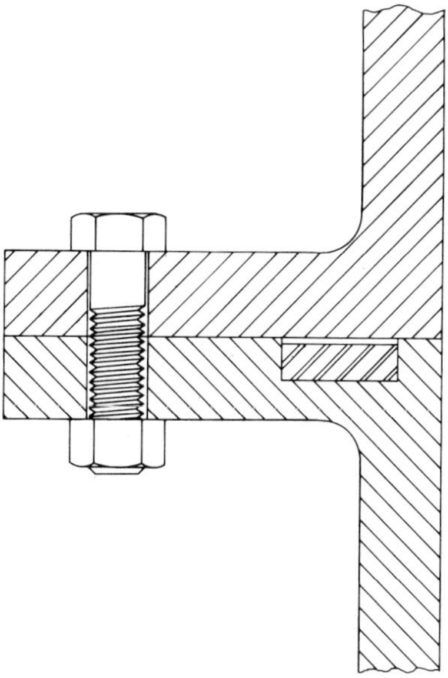

FIG. 6-4 Gland-type joint.

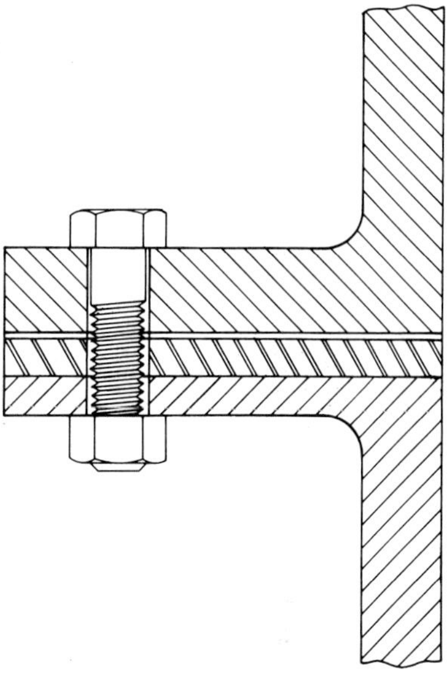

FIG. 6-5 Glandless joint.

place. It is more expensive to machine than the glandless type (Fig. 6-5). Application parameters (liquid or gas to be sealed, internal pressure, temperature, exterior contaminants, types of surfaces to be joined, surface roughness, and others) must be considered carefully before the gasket type and material are selected. The success of the gasket is dependent on the flange pressure developed at the joint. Flange pressure is defined as the effective compressive load per inch or centimeter of gasket area, expressed in pounds per square inch or pascals. Flange pressure compresses the gasket material and causes the material to conform to surface irregularities in the flange. Flange pressure is developed by tightening bolts that hold the assembly together. About half of the initial tightening torque on a bolt is used to overcome collar friction, and 40 percent is required to overcome mating screw-thread friction. The remaining 10 percent creates the gasket load in flanged joints. The flange pressure can be estimated using Eq. (6-1).

$$F_p = \frac{N_B T}{A_r K D} = \frac{N_B T}{A_r} \qquad (6\text{-}1)$$

where F_p = flange pressure, lb/in^2 (bars)
N_B = number of bolts in the assembly
T = initial bolt torque, lb·in (N·cm)
K = torque friction coefficient
A_r = gasket area, in^2 (cm^2)
D = nominal bolt diameter, in (cm)
P = bolt clamp load developed by tightening, lb (N)

TABLE 6-1 Clamp Loading Chart

	Tensile		SAE Grade & Bolts			Tightening torque		
Size	Bolt diam. D, in	Stress area, A_s, in	Tensile strength, min lb/in²	Proof load, lb/in²	Clamp† load P, lb	Dry $K = 0.20$	Lub. $K = 0.15$	Tensile strength, min lb/in²
						lb·in	lb·in	
4-40	0.1120	0.00804	74,000	55,000	240	5	4	120,000
4-48	0.1120	0.00661			280	6	5	
6-32	0.1380	0.00909			380	10	8	
6-40	0.1380	0.01015			420	12	9	
8-32	0.1640	0.01400			580	19	14	
8-36	0.1640	0.01474			600	20	15	
10-24	0.1900	0.01750			720	27	21	
10-32	0.1900	0.02000			820	31	23	
¼-20	0.2500	0.0318			1,320	66	49	
¼- 8	0.2500	0.0364			1,500	76	56	
						lb·ft	lb·ft	
5⁄16-18	0.3125	0.0524			2,160	11	8	
5⁄16-24	0.3125	0.0508			2,400	12	9	
⅜-16	0.3750	0.0775			3,200	20	15	
⅜-24	0.3750	0.0878			3,620	23	17	
7⁄16-14	0.4375	0.1063			4,380	30	24	
7⁄16-20	0.4375	0.1187			4,900	35	25	
½-13	0.5000	0.1419			5,840	50	35	
½-20	0.5000	0.1599			6,600	55	40	
9⁄16-12	0.5625	0.1820			7,500	70	55	
9⁄16-18	0.5625	0.2030			8,400	80	60	
⅝-11	0.6250	0.2260			9,300	100	75	
⅝-18	0.6250	0.2560			10,600	110	85	
¾-10	0.7500	0.3340			13,800	175	130	
¾-16	0.7500	0.3730			15,400	195	145	
⅞- 9	0.8750	0.4620	60,000	33,000	11,400	165	125	
⅞-14	0.8750	0.5090			12,600	185	140	
1- 8	1.0000	0.6060			15,000	250	190	
1-12	1.0000	0.6630			16,400	270	200	
1⅛- 7	1.1250	0.7630			18,900	350	270	105,000
1⅛-12	1.1250	0.8560			21,200	400	300	
1¼- 7	1.2500	0.9690			24,000	500	380	
1¼-12	1.2500	1.0730			26,600	550	420	
1⅜- 6	1.3750	1.1550			28,600	660	490	
1⅜-12	1.3750	1.3150			32,500	740	560	
1½- 6	1.5000	1.4050			34,800	870	650	
1½-12	1.5000	1.5800			39,100	980	730	

Notes:

* Torque-tightening values are calculated from the formula $T = KDP$, where T = torque tightening, lb·in; K = torque-friction coefficient; D = nominal bolt diameter, in; and P = bolt clamping load developed by tightening, lb.

† Clamp load is also known as preload or initial load in tension on bolt. Clamp load (lb) is calculated by arbitrarily assuming usable bolt strength is 75% of bolt-proof load (lb/in²) times tensile stress area (in²) of threaded

	SAE Grade 5 Bolts			SAE Grade 7‡			SAE Grade 8¶		
		Tightening torque			Tightening torque			Tightening torque	
Proof load, lb/in^3	Clamps† load P, lb	Dry $K = 0.20$	Lub. $K = 0.15$	Clamps† load P, lb	Dry $K = 0.20$	Lub. $K = 0.15$	Clamp† load P, lb	Dry $K = 0.20$	Lub. $K = 0.15$
		lb·in	lb·in		lb·in	lb·in		lb·in	lb·in
85,000	380	8	6	480	11	8	540	12	9
	420	9	7	520	12	9	600	13	10
	580	16	12	720	20	15	820	23	17
	640	18	13	800	22	17	920	25	19
	900	30	22	1,100	36	27	1,260	41	31
	940	31	23	1,160	38	29	1,320	43	32
	1,120	43	32	1,380	52	39	1,580	60	45
	1,285	49	36	1,580	60	45	1,800	68	51
	2,020	96	75	2,500	120	96	2,860	144	108
	2,320	120	86	2,860	144	108	3,280	168	120
		lb·ft	lb·ft		lb·ft	lb·ft		lb·ft	lb·ft
	3,340	17	13	4,120	21	16	4,720	25	18
	3,700	19	14	4,560	24	18	5,220	25	20
	4,940	30	23	6,100	40	30	7,000	45	35
	5,600	35	25	6,900	45	30	7,900	50	35
	6,800	50	35	8,400	60	45	9,550	70	55
	7,550	55	40	9,350	70	50	10,700	80	60
	9,050	75	55	11,200	95	70	12,750	110	80
	10,700	90	65	12,600	100	80	14,400	120	90
	11,600	110	80	14,350	135	100	16,400	150	110
	12,950	120	90	16,000	150	110	18,250	170	130
	14,400	150	110	17,800	190	140	20,350	220	170
	16,300	170	130	20,150	210	160	23,000	240	180
	21,300	260	200	26,300	320	240	30,100	380	280
	23,800	300	220	29,400	360	280	33,600	420	320
	29,400	430	320	36,400	520	400	41,600	600	400
	32,400	470	350	40,100	580	440	45,800	660	500
	38,600	640	480	47,700	800	600	54,500	900	680
	42,200	700	530	52,200	860	660	59,700	1000	740
74,000	42,300	800	600	60,100	1120	840	68,700	1280	960
	47,500	880	660	67,400	1260	940	77,000	1440	1080
	53,800	1120	840	76,300	1580	1100	87,200	1820	1360
	59,600	1240	920	84,500	1760	1320	96,800	2000	1500
	64,100	1460	1100	91,000	2080	1560	104,000	2380	1780
	73,000	1680	1260	104,000	2380	1780	118,400	2720	2040
	78,000	1940	1460	111,000	2780	2080	126,500	3160	2360
	87,700	2200	1640	124,005	3100	2320	142,200	3560	2660

section of each bolt size. Higher or lower values of clamp load can be used depending on the application requirements and the judgment of the designer.

‡ Tensile strength (min lb/in^2) of all Grade 7 bolts is 133,000. Proof load is 105,000 lb/in^2.

¶ Tensile strength (min lb/in^2) of all Grade 8 bolts is 150,000 lb/in^2. Proof load is 120,000 lb/in^2.

(Reprinted by permission of the copyright owner *Machine Design*.)

Courtesy: Armstrong Gasket Corporation.

The effect of load loss due to bolt stretching and other factors must be considered. For many applications a rigid tightening sequence is necessary to ensure that the gasket loading is uniformly applied. The amount of initial bolt torque T required to give a desired bolt clamp load P for dry and lubricated conditions is given in Table 6-1 for various bolt sizes and grades.

6-6 Nonmetallic Gaskets

The success of nonmetallic gaskets in a given application is highly dependent on the material selected. The permeability, deformation and flow characteristics under load, resistance to fluids, and cost of the material must be considered when the selection is made. The extent to which gaskets must be compressed to obtain a seal depends on the finish of the faces and the internal pressure. Materials with high compression are generally used for low-pressure applications where a good face finish is not available. Low-compression materials seal high-pressure applications. Materials used for gaskets include Teflon, rubber, asbestos, paper, cork, and various combinations of materials. The amount of compression expected for these materials as a result of various flange pressures is given in Fig. 6-6.

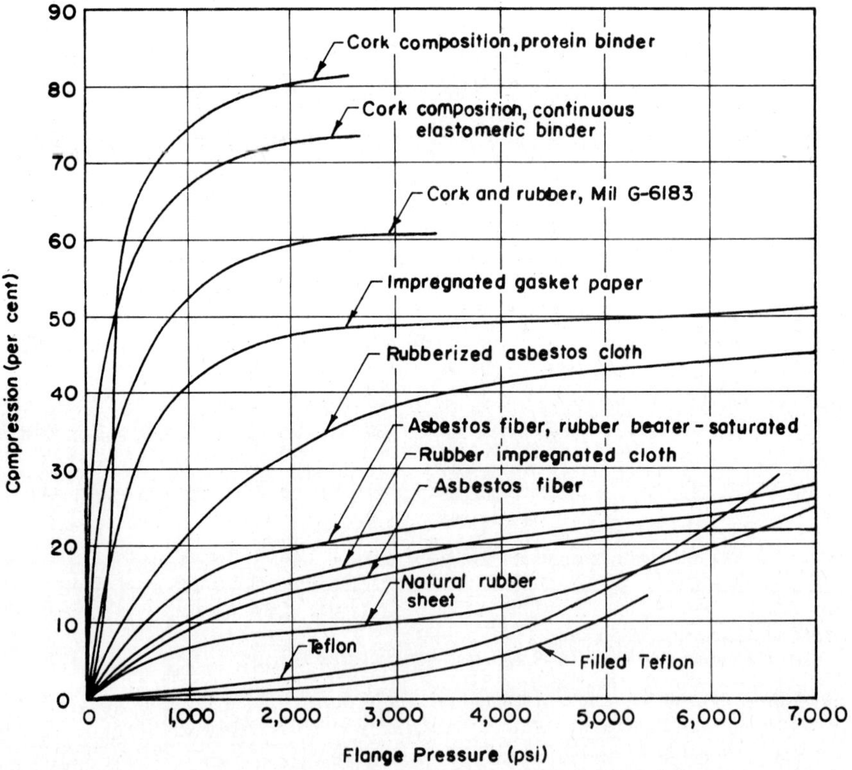

FIG. 6-6 Percent compression vs. flange pressure. (*Courtesy: Machine Design.*)

Paper gaskets can be made from organic or mineral (asbestos) fibers. Untreated paper gaskets are highly permeable and are not typically used to contain fluids. They are used to prevent contaminants such as dust or water splash from entering a cavity. The permeability is reduced by saturating the paper with latex or glue. Saturated papers are used to hold oils and gasoline at moderate pressures (0 to 20 lb/in^2) at temperatures less than 160°F. About 500 to 1000 lb/in^2 is required to seat the flange to the paper gasket.

Elastomeric materials can be compounded to produce a great variety of rubber gaskets. Since rubber is incompressible, the designer must provide room for flow or deformation when the rubber sheet is loaded. Rubber gaskets can be molded to virtually any shape to meet special application needs. The material can be stretched over projections during assembly. The hardness and compression set of the rubber materials can be adjusted by compounding. Rubber materials for gaskets are usually compounded to give low compression. There is a wide variety of elastomeric materials to choose from. The desired temperature range and the fluid in the application must be considered when selecting a rubber material (Table 6-2).

Asbestos fibers are low in strength and high in porosity. Rubber and plastic material are usually mixed with asbestos to make a composite material. Rubber-asbestos sheets are used where bolting pressures are high. The material is compacted under high pressure and becomes impermeable. Compressed asbestos requires relatively high loads to achieve intimate contact with the flange faces. It is used in relatively heavy construction where rigid flanges with adequate bolting capacity are used.

Compressed asbestos has little elasticity. After the bolts have pulled up to the point where the gasket is well seated and has conformed to the irregularities of the joint, it will yield little more than the metal components themselves.

Cork compositions combine granulated cork with a binder to form a widely used gasket material. It is used to mate irregular surfaces where temperatures are moderate (less than 160°F) and pressures are low (less than 30 lb/in^2). Cork compositions have high compression under load and are used as shock absorbers for joining glass to metal. They are unaffected by oils and aromatic solvents, but are not recommended for use with acid or alkaline solutions. Cork has a tendency to dry, which can cause shrinking and hardening.

When rubber is mixed with cork and vulcanized, the compression and flow characteristics of the resulting gasket can be controlled. Gaskets which are nearly as compressible as cork or almost as incompressible as rubber can be made. The temperature limitations of the material depend on the elastomer. The material die-cuts well and is used in a wide variety of applications. It is not used for gasket steam lines or high-temperature combustion chambers.

PTFE materials are used in applications where broad temperature ranges, resistance to harsh fluids, and chemical inertness are required. The gaskets can be cut from sintered billets, die-cut from sheets, or molded from powder.

Laminated gaskets are often made by combining two or more materials. The usual reason for laminating is to combine the properties of a strong, incompressible material with those of a weak, highly compressible one. Cork or asbestos is sometimes attached to both sides of a steel sheet to form a highly compressible, strong gasket.

TABLE 6-2 Properties of Elastomers Commonly Used for Gasket Materials

Properties	Styrene-butadiene	Ethylene propylene	Polychloroprene	Silicone	Nitrile-butadiene	Chlorosulfonated polyethylene	Fluorocarbon	Polyacrylate
Useful temperature range	−70 to +250°F	−65 to +350°F	−50 to +250°F	−120 to +600°F	−65 to +300°F	−60 to +250°F	−40 to +600°F	−30 to +400°F
Water	E	T	T	T	E	E	E	F
Acid	G	E	G	G	G	E	E	P
Alkali	E	G	E	F	G	E	G	P
Gasoline	P	P	F	P	E	F	E	G
Petroleum oil	P	P	G	P	E	G	G	E
Animal and vegetable oil	P-G	G	G	G	E	F	G	E
Hydrocarbon solvents	P	P	G (except aromatics)	P	G-E	F	E	G
Oxygenated solvents	F	G	P	P	P	F	F	P
Ozone	P	E	G	E	P	E	E	G

Key: E = Excellent, G = Good, F = Fair, P = Poor.

The properties of nonmetallic gasket materials are given in Table 6-3. Treatments and coatings that are sometimes applied to gaskets are given in Table 6-4.

6-7 Metallic Gaskets

Gaskets made from soft materials, such as cork, rubber, and asbestos, are used to seal low-pressure applications. High-pressure gases at high temperatures are usually contained with metallic gaskets. If the product of the internal pressure in pounds per square inch and the operating temperature in degrees Fahrenheit exceeds 250,000, only metallic gaskets can be used. Sealing the head of internal combustion engines is an example of a typical application for metallic gaskets.

The material selected for metallic gaskets depends on operation conditions. Soft metals, such as lead, brass, and copper, are selected if the mating surfaces are rough. Stainless steels are used to prevent corrosion due to oxidation. Aluminum is selected for its light weight and good corrosion resistance. Gal-

TABLE 6-3 Gasket Materials

Material	Properties	General usage
Untreated paper	Highly permeable, low cost, noncorrosive	To exclude dust and water splash
Treated paper	General purpose, up to 160°F at 0 to 20 lb/in²	Seal oil, gasoline
Rubber	Incompressible, impermeable, can vary in hardness, can be stretched, can be molded to special shapes	Used when stretching is required for installation. Seal alkalis, hot water, and some acids
Rubber-asbestos	Tough, durable, dimensionally stable, resistant to steam and hot water	Water and steam fittings to 500°F
Cork-compositions	Lightweight, inert, high compression, will not deteriorate, low cost, excellent oil and solvent resistance, high friction, poor resistance to acids and alkalis	Mates irregular surfaces up to 160°F. Used to join glass and ceramic to metal
Cork-rubber	Can compound material to give various levels of compression, high coefficient of friction	General-purpose gaskets. Temperature limits depend upon elastomer
PTFE	Chemically inert, wide temperature range, resistant to most solvents. Expensive	Used in special applications where heat and fluid resistance is required

TABLE 6-4 Treatments and Coatings

Description of treatment	Method of application	Purpose
Synthetic rubber–neoprene	Dipped or sprayed on exterior of cork-composition gaskets	Resists oil penetration
Fungicides (betanaphthol, pentachlorophenol, salicylailide, copper/mercury compounds)	Compounded into materials, dusted onto materials	Resist mold growth
Reflective coating	Dip or spray aluminum paint or lacquer on exterior or gasket	Reflects heat
Graphite	Applied to exterior of gasket. Dusted as a dry flake	Prevents adhesion of gasket to metal surfaces
Adhesives	Applied to exterior of gasket	Bond gasket to flange

vanic corrosion may occur when two dissimilar metals contact an electrolyte. Temperature of the application is also important (Table 6-5) and will dictate what material must be used.

There are many types of metallic gasket designs. Corrugated gaskets (Fig. 6-7) are made from thin metallic sheets (0.010 to 0.031 in) and are used for smooth-faced, low-pressure (500 lb/in^2) applications such as valve bonnets and fuel and combustion lines. In some cases, the corrugated material is coated

TABLE 6-5 Temperature Limitations for Metallic Gasket Materials

Material	Maximum temperature, °F
Lead	212
Common brasses	500
Copper	600
Aluminum	800
Stainless steel	1600
Soft iron, low-carbon steel	1000
Titanium	1000
Nickel	1400
Monel	1500
Inconel	2000
Hastelloy	2000

FIG. 6-7 Corrugated gasket.

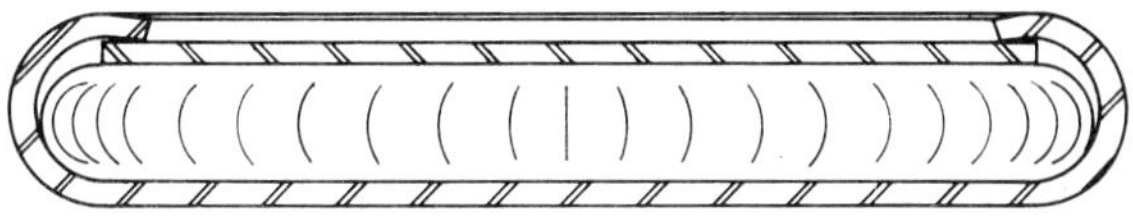

FIG. 6-8 Metal-jacketed soft-filler gasket.

with a sealing compound to extend the pressure limit to 1000 lb/in^2. Asbestos cord is cemented in the corrugations if the surfaces are uneven. A soft compression filler (usually asbestos) can be partially or totally encased in a metal jacket (Fig. 6-8) to provide more compression than corrugated types. These are used to compensate for flange irregularities in applications up to 1200°F. Spiral-wound gaskets (Fig. 6-9) have the best resilience of all metal asbestos-type gaskets. They are made by winding a preformed V-shaped strip of metal into a spiral. The metallic layers are separated by an asbestos filler. Flat metallic gaskets of various shapes and sizes are relatively inexpensive to produce and perform satisfactorily in many types of applications. Flat gaskets seal because the compression loads of the joining bolts exceed the tensile strength of the gasket metal and flow the material into the flange surface irregularities. Some flat gaskets have serrated grooved surfaces to promote deformation (Fig. 6-10). Flat plain metal gaskets should be used with flanges that are grooved or prepared with a surface finish of 80 μin rms.

Solid metal rings with various cross sections are used to provide gas-tight seals at relatively low flange pressure. These gaskets usually seal by line contact or wedging action that causes surface flow. Some designs are pressure-activated. Higher pressures generate greater closing force. Design and uses of solid metal ring gaskets are shown in Fig. 6-11.

Metallic rings of various cross sections (Fig. 6-12) can be designed to expand when internal pressures increase. The rings are usually coated with a soft metallic or plastic material that provides the actual sealing. A typical application is shown in Fig. 6-13. Hollow metallic O rings made from tubing are

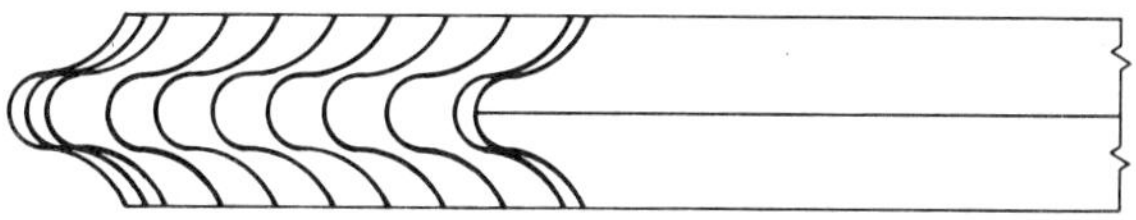

FIG. 6-9 Spiral-wound gasket.

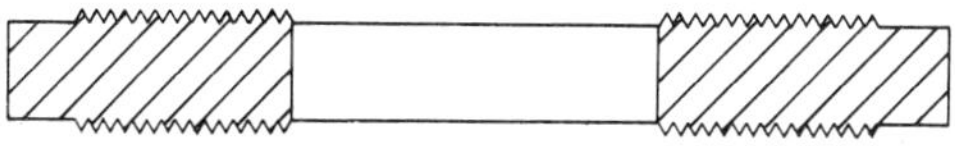

FIG. 6-10 Serrated flat metallic gasket.

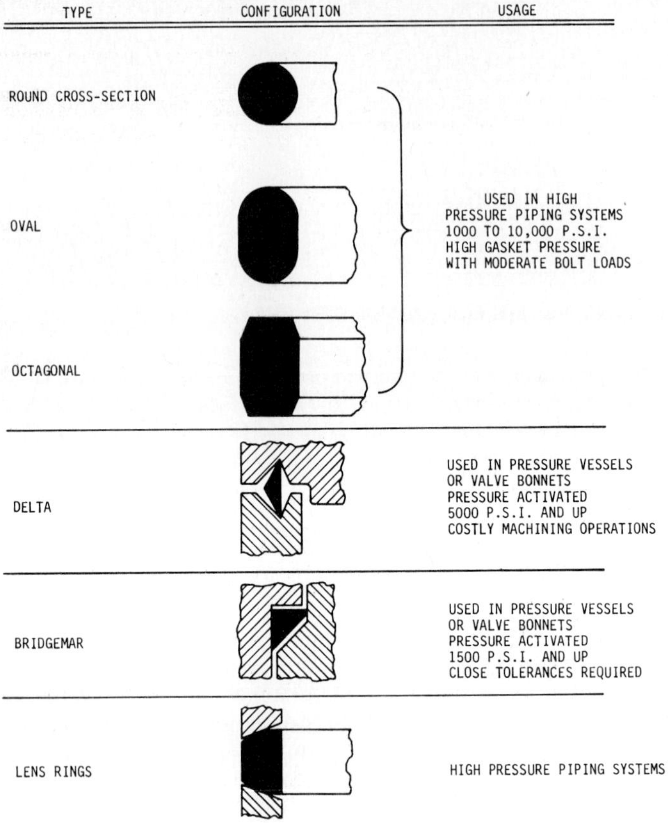

FIG. 6-11 Design and uses of solid-metal ring gaskets.

compressed between parallel faces to form a seal. Vent holes are often provided at the ID for pressure activation (Fig. 6-14).

6-8 Formable Gaskets—Sealants

Sealants are used to form gaskets for applications that have relatively low pressure and temperature requirements. The high-viscosity liquids flow over the flange surfaces, filling voids and leveling the contact areas. These properties enable sealants to tolerate surface irregularities that cannot be overcome with preformed gaskets. Formable gaskets can be applied directly to intricate and complex seal-area geometries with relatively simple automatic metering equipment. Relatively low clamping forces are required to seal the joints. Metal-to-metal contact takes place at the high spots, and the sealant fills the void area with a continuous film.

Sealants can be classified into hardening and nonhardening types. The hardening sealants require cure times that range from 15 min to 2 weeks,

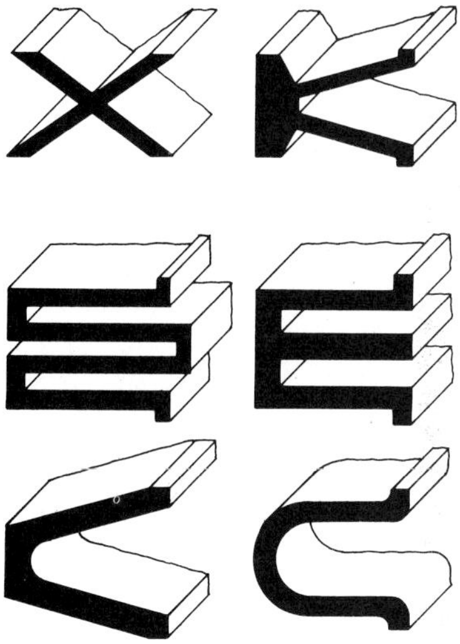

FIG. 6-12 Pressure-actuated metal rings.

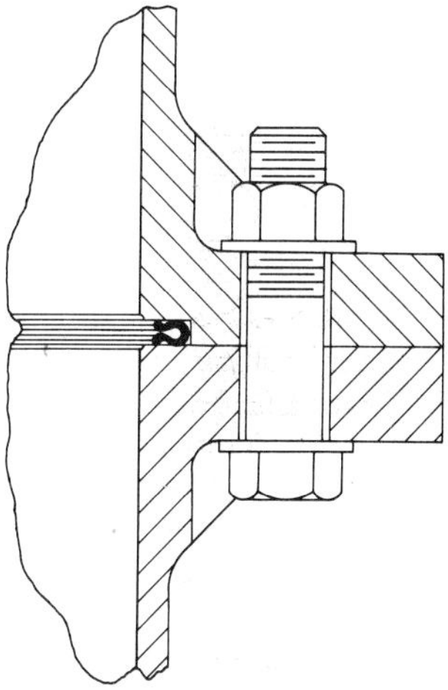

FIG. 6-13 Application of pressure-actuated metallic gasket.

depending on sealant type, humidity, and temperature. Some sealants will cure to a rigid state, and others will remain flexible. Nonhardening sealants do not cure. There is no chemical change after application. Solvents may be added to the nonhardening sealants to improve handling. The solvent will evaporate after application to change the consistency of the gasket.

The mechanical properties of sealants are sensitive to composition; minor changes in composition can result in major property alterations. Other factors that must be considered include chemical compatibility, permeability, weather resistance, abrasion resistance, adhesion, electrical characteristics, color, resistance to heat, toxicity, flammability, and method of application.

Sealants are used for static pressure applications. Pipe threads, flange joints, tanks, and pan gaskets are examples of applications for sealants.

Characteristics of typical sealants are given in Table 6-6.

STATIC SEALS—ELASTOMERIC

6-9 Elastomeric Static Seals

A wide variety of shapes can be molded to make an elastomeric static seal. Some special shapes are designed to expand under pressure to increase the rubber-to-metal contact (Figs. 6-15 and 6-16). Cross sections of other seal types include square lathe cut, rectangular lathe cut, oval

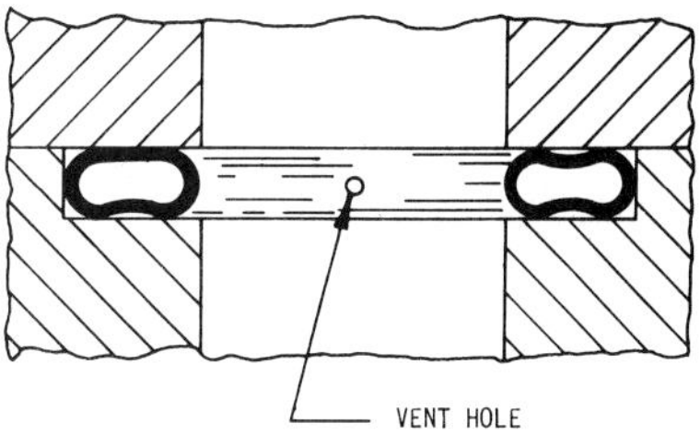

FIG. 6-14 Hollow metallic O ring.

TABLE 6-6 Sealant Properties

Sealant base	Sealant form	Curing time and temperature range*	Pot life,† h	Relative cost/gal	Uses
Epoxy	One-part liquid	1–16 h @ 180–350°F	½–8		Potting electrical connectors, encapsulating miniature components, coating circuit boards, and cable splicing. Caulking and pipe sealing.
	Two-part liquid	1–8 h @ 75°F; some formulations require 4 to 6 days	½–3	7–12	
	Powder	½–3 h @ 150°F 24 h @ 250°F 2 h @ 320°F 5 min @ 500°F	Unlimited		Used as a sealer and abrasion-resistant coating for concrete
Modified	Two-part liquid modified with polysulfide polymer	½–8 h @ 75°F	⅙–2		
Polyester	Two-part liquid	8–24 h @ 75°F. Less time at higher temperature	⅙–4	6–12	Potting, molding and encapsulating. Gasket and pipe thread sealants
	One-part liquid	¼–24 h @ 75°F	Unlimited if kept in contact with oxygen		
Polysulfide	One-part liquid	14–21 days @ 75°F	1 year	17–25	General construction-type sealing, caulking, and glazing. Sealing between dissimilar metals. Deck caulking and sealing of refracting mirrors
Polyurethane	Two-part	16–24 h @ 75°F	3–6	12–15	Potting and molding of electrical connectors; encapsulating of hydrophones, transducers, and circuit boards. Caulking where compatibility with LOX is required

Polyurethane/ silicone	One-part	14–21 days @ 75°F	3–9 months	12–18	General construction-type sealing, caulking and glazing
	Two-part	5–10 h @ 180°F	¼–10	45–100	Potting and molding of electrical connectors. Potting firewall connectors and coating of umbilical cables. Sealing of heat shields
Silicone	One-part	1–14 days @ 75°F	6 months	22–35	General construction-type sealing, caulking, and glazing. Potting and molding
Acrylic	Made in one-part release and two-part	Varies, depending on formulation and curing system	—	8–10	Same general uses as nonhardening formulations. Hardening-type materials should be used where pressure limits exceed the limitations of the nonhardening formulations
Oleo-resin	One-part putty or paste	14 days or more @ 75°F to reach specific hardness	Very long	2–6	
Asphalt and bituminous	One-part putty or paste	Several weeks @ 75°F	Very long	1–4	
	Two-part flow-type paste	1–8 h @ 75°F	¼–¼		
Hardening sealants—nonrigid types					
Polysulfide	Two-part liquid	16–72 h @ 75°F	1–6	10–22	Sealing integral fuel tanks and pressure tanks. Sealing faying surfaces and channel. Potting, molding, and sealing of Plexiglas. General construction sealing and caulking of metal, wood, masonry joints, and glass
Modified epoxy	Two-part	2 h @ 75°F	½–1	7–12	Potting, molding, encapsulating; sealing transformers; high-voltage splicing; capacitor sealing. General construction caulking

TABLE 6-6 Sealant Properties (*Continued*)

Sealant base	Sealant form	Curing time and temperature range*	Pot life,† h	Relative cost/gal	Uses
Acrylic-Viton	One-part	21 days @ 75°F	3–9 months	8–10	General construction-type sealing, caulking and glazing
	Two-part	24 h @ 75°F	3–4	80–115	For high-temperature service where fuel and oil resistance is required. Sealing fuel tanks, channels, and faying surfaces
Solvent-Release Systems					
Neoprene	—			8–12	Sealing between dissimilar metals. Caulking and general sealing.
Hypalon	—	7–14 days @ 75°F	6–12 months	8–12	Similar in use to Neoprene.
Butyls	—			5–10	Glazing and caulking of metal, glass, and masonry-type joints

		Nonhardening sealants			
Oleo-resin	One part mastic, putty, or paste	—	—	2–6	Sealing of concrete joints, masonry copings, and tile. Glazing. Sealing of cable pressure splices, electrical conduit, and glass-to-metal meter cases
Asphalt and bituminous		—	—	1–4	Sealing faying surface metal joints, silos, and air conditioners. Caulking for expansion and contraction joints
Butyl		—	—	5–10	Caulking expansion and contraction joints. Metal-to-glass seals and metal-to-metal sealing to separate dissimilar materials. Sealing electrical conduit
Acrylic		—	—	8–10	Pipe joints, glazing, masonry, and metal caulking. Special compounds used as liquid gaskets and pipe dope
Polybutene		—	—	3–7	General construction-type caulking and glazing. Seal between dissimilar metals

* Ranges given are representative. Formulation and addition of solvents or thinners can alter these values.

† Values given are for standard conditions—75°F and 50 percent relative humidity. Times represent limits obtainable by formulation or addition of solvents. Pot life for specific formulations varies with temperature and humidity.

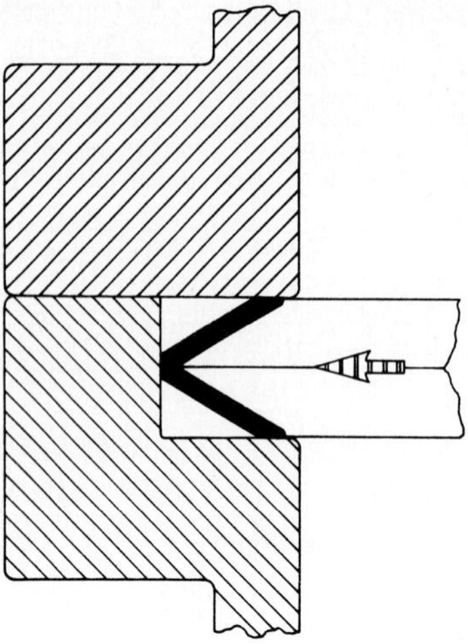

FIG. 6-15 Internal-pressure expanding V ring.

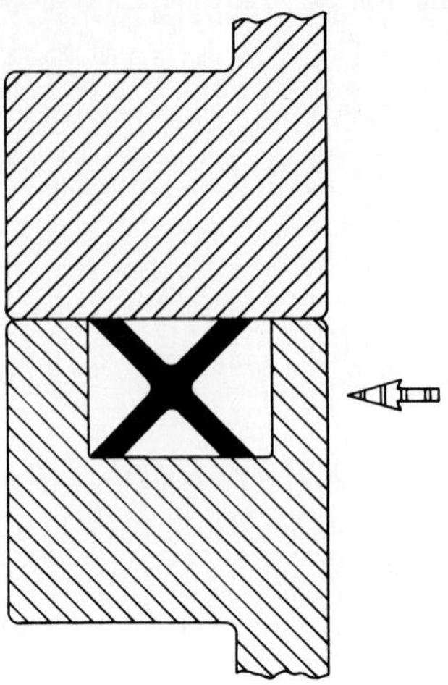

FIG. 6-16 Quad lip seal expands as pressure increases.

octagon, pyramid, round, and others (Fig. 6-17). These types of seals are typically installed in machined grooves in one of the surfaces to be sealed. When the surfaces are brought together to form a gland, the rubber ring is squeezed and deformed. This deformation creates a pressure that effectively blocks the leak path in the gland and prevents fluid leakage. The most common shapes are square-cut, rectangular-cut, and round. The round cross section is better known as the O ring.

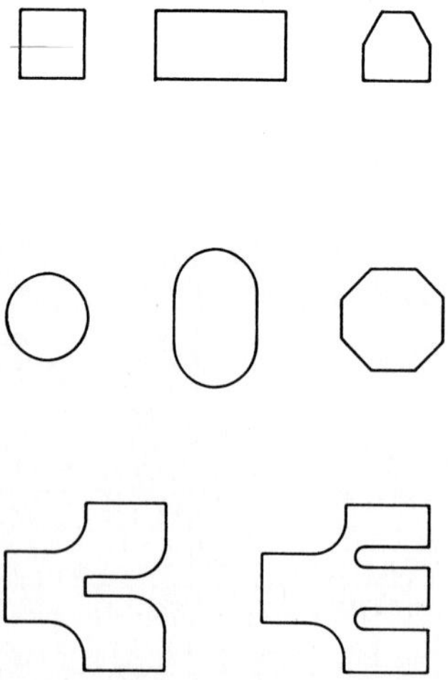

FIG. 6-17 Various cross sections for elastomeric static seals.

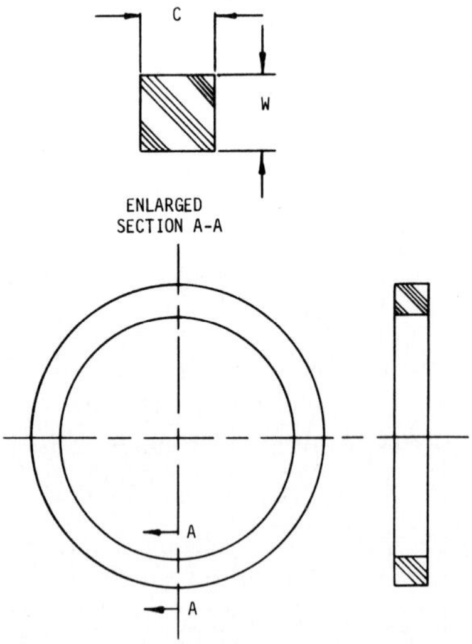

FIG. 6-18 Rectangular lathe-cut rings.

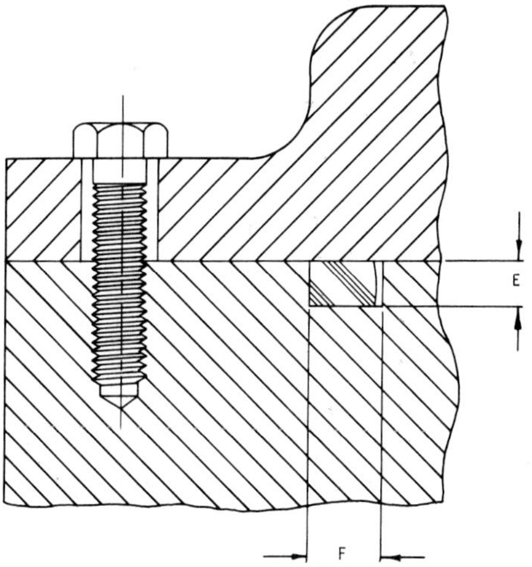

FIG. 6-19 Axial-squeeze application.

The usual method of manufacturing rubber rings with square or rectangular cross sections (Fig. 6-18) is to cut them on a lathe to the desired size from vulcanized rubber billets. These rings are recommended only for static applications up to 1500 lb/in^2 fluid pressure. Typical applications for these rings include axial squeeze to form gaskets (Fig. 6-19) and radial squeeze (Fig. 6-20) to seal cylinder walls. The detail of the groove for the lathe-cut rings is given in Fig. 6-21.

The O-ring design (Fig. 6-22) is perhaps the most widely used, since its

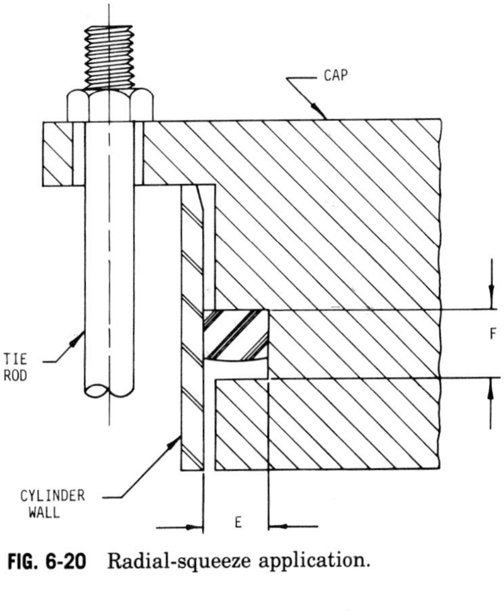

FIG. 6-20 Radial-squeeze application.

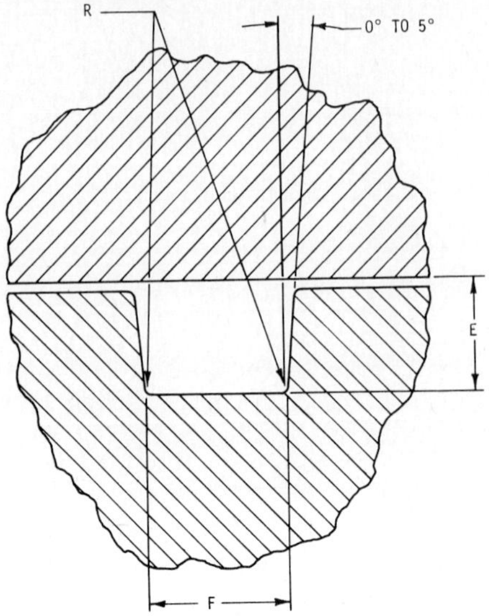

FIG. 6-21 Groove detail for lathe-cut rectangular and square rings.

round cross section readily fits many applications. The depth of the gland must be less than the cross-section diameter of the O ring. The microfinish of the gland should be 32 to 63 μin.

Three O-ring gland shapes suggested for many applications are the rectangle, the vee, and the undercut, Fig. 6-23. The rectangular groove may be used for every type of application, including high-pressure static situations. Its sides slope toward 7° with edges broken to about 0.005-in radius.

The vee groove is suitable for low-pressure applications and low-temperature use, where it resists cold by increasing squeeze on the seal. This gland is the simplest to machine in either male or female configuration, and, therefore,

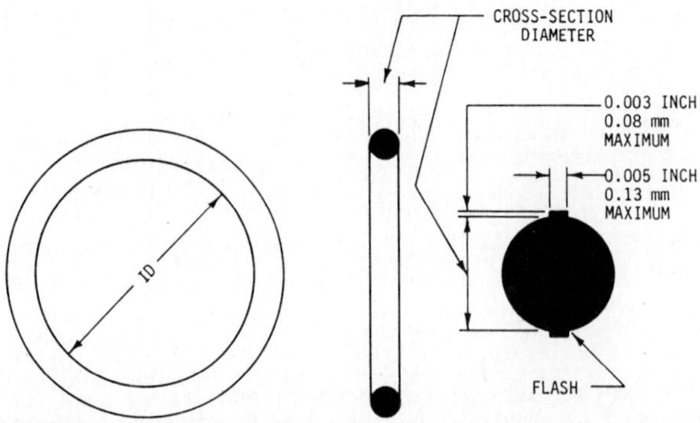

FIG. 6-22 O-ring details.

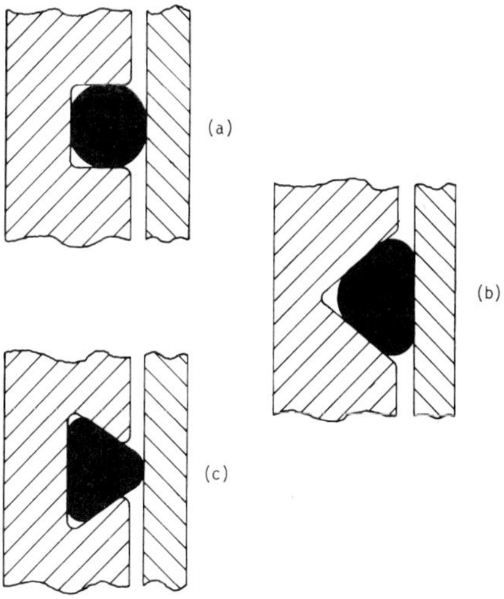

FIG. 6-23 Common O-ring gland shapes.

is least expensive. However, a seal in a vee gland generates the highest friction of all three shapes.

The undercut or dovetailed groove prevents extrusion, makes a tighter dynamic seal, and reduces starting and operating friction. Unfortunately, it is the most difficult to machine and, therefore, most expensive.

The minimum squeeze must be 0.006 in, regardless of O-ring cross-section diameter. This will account for differences in the shrinkage of rubber and metal as temperatures vary. The maximum recommended squeeze is 35 percent of the O-ring cross-section diameter.

Selection of materials for elastomeric static seals depends on the operating temperature, the internal pressure, and the fluid to be sealed. Table 6-7 provides a guide for material selection.

Static seal leakage can be caused by a variety of reasons ranging from installation damage to extrusion. Extrusion and "nibbling" results when pulsating pressure forces the rubber into the clearance between the metal faces (Fig. 6-24). Increasing the hardness of the elastomer will increase resistance to extrusion and larger clearances between the metal surfaces can be tolerated (Fig. 6-25).

Table 6-8 can be used as a guide when analyzing static seal leaks in the applications.

DYNAMIC SEALS—PACKINGS

6-10 Packing Classification and Usage

Packings are used in reciprocating and rotating applications to seal fluids and exclude contaminants. Packings are usually divided into three categories: compression, molded (or automatic), and floating.

TABLE 6-7 Elastomeric Capabilities Guide

	Nitrile (Buna-N)	Ethylene-propylene	(Chloroprene) neoprene	Fluorocarbon (Viton, Fluorel)	Silicone	Fluoro-silicone
			General			
Hardness range, A scale	40–90	50–90	40–80	70–90	40–80	60–80
Relative static ring cost	Low	Low	Low/Moderate	Moderate/High	Moderate	High
Continuous high-temperature limit	257°F 125°C	302°F 150°C	284°F 140°C	437°F 225°C	482°F 250°C	347°F 175°C
Low-temperature capability	−67°F −55°C	−67°F −55°C	−67°F −55°C	−40°F −40°C	−103°F −75°C	−85°F −65°C
Compression set resistance	Very good	Very good	Good	Very good	Excellent	Very good
			Fluid compatibility summary			
Acid, inorganic	Fair	Good	Fair/good	Excellent	Good	Good
Acid, organic	Good	Very good	Good	Good	Excellent	Good
Aging (oxygen, ozone, weather)	Fair/poor	Very good	Good	Very good	Excellent	Excellent
Air	Fair	Very good	Good	Very good	Excellent	Very good
Alcohols	Very good	Excellent	Very good	Fair	Very good	Very good
Aldehydes	Fair/poor	Very good	Fair/poor	Poor	Good	Poor
Alkalis	Fair/good	Excellent	Good	Good	Very good	Good
Amines	Poor	Very good	Very good	Poor	Good	Poor
Animal oils	Excellent	Good	Good	Very good	Good	Excellent
Esters, alkyl phosphate (Skydrol)	Poor	Excellent	Poor	Poor	Good	Fair/poor
Esters, aryl phosphate	Fair/poor	Excellent	Fair/poor	Excellent	Good	Very good
Esters, silicate	Good	Poor	Fair	Excellent	Poor	Very good
Ethers	Poor	Fair	Poor	Poor	Poor	Fair
Hydrocarbon fuels, aliphatic	Excellent	Poor	Fair	Excellent	Fair	Excellent
Hydrocarbon fuels, aromatic	Good	Poor	Fair/poor	Excellent	Poor	Very good
Hydrocarbons, halogenated	Fair/poor	Poor	Poor	Excellent	Poor	Very good
Hydrocarbon oils, high aniline	Excellent	Poor	Good	Excellent	Very good	Excellent
Hydrocarbon oils, low aniline	Very good	Poor	Fair/poor	Excellent	Fair	Very good
Impermeability to gases	Good	Good	Good	Very good	Poor	Poor
Ketones	Poor	Excellent	Poor	Poor	Poor	Fair/poor
Silicone oils	Excellent	Excellent	Excellent	Excellent	Good	Excellent
Vegetable oils	Excellent	Good	Good	Excellent	Excellent	Excellent
Water/steam	Good	Excellent	Fair	Fair	Fair	Fair

Styrene-butadiene (SBR)	Poly-acrylate	Poly-urethane	Butyl	Polysulfide (Thiokol)	Chloro-sulfonated poly-ethylene (Hypalon)	Epichloro-hydrin (Hydrin)	Phospho-nitrilic fluoro-elastomer (PNF)
General							
40–80	70–90	60–90	50–70	50–80	50–90	50–90	50–90
Low	Moderate	Moderate	Moderate	Moderate	Moderate	Moderate	High
212°F	347°F	212°F	212°F	212°F	257°F	257°F	347°F
100°C	175°C	100°C	100°C	100°C	125°C	125°C	175°C
−67°F	−4°F	−67°F	−67°F	−67°C	−67°C	−67°C	−85°F
−55°C	−20°C	−55°C	−55°C	−55°C	−55°C	−55°C	−65°C
Good	Fair	Fair	Fair/good	Fair	Fair/poor	Fair/good	Good
Fluid compatibility summary							
Fair/good	Poor	Poor	Good	Poor	Excellent	Fair	Poor
Good	Poor	Poor	Very good	Good	Good	Fair	Fair
Poor	Excellent	Excellent	Very good	Excellent	Very good	Very good	Excellent
Fair	Very good	Good	Good	Good	Excellent	Good	Excellent
Very good	Poor	Poor	Very good	Fair/good	Very good	Good	Fair
Fair/poor	Poor	Poor	Good	Fair/good	Fair/good	Poor	Poor
Fair/good	Poor	Fair/good	Excellent	Poor	Excellent	Fair	Good
Fair	Poor	Poor	Good	Poor	Poor	Poor	Good
Poor	Excellent	Good	Good	Poor	Good	Good	Fair
Poor	Poor	Poor	Very good	Poor	Poor	Poor	Poor
Poor	Poor	Poor	Excellent	Good	Fair	Poor	Excellent
Poor	Fair/poor	Poor	Poor	Fair/poor	Fair	Good	Excellent
Poor	Fair/poor	Fair	Fair/poor	Good	Poor	Good	Poor
Poor	Very good	Good	Poor	Excellent	Fair	Very good	Excellent
Poor	Poor	Fair/poor	Poor	Good	Fair/poor	Very good	Excellent
Poor	Fair/good	Fair	Poor	Good	Fair	Excellent	Fair
Poor	Excellent	Excellent	Poor	Very good	Excellent	Excellent	Excellent
Poor	Excellent	Very good	Poor	Good	Very good	Excellent	Excellent
Fair/good	Very good	Fair	Excellent	Very good	Very good	Excellent	Fair
Poor	Poor	Poor	Excellent	Good	Fair	Fair	Poor
Excellent	Excellent	Excellent	Excellent	Excellent	Excellent	Excellent	Excellent
Poor	Good	Fair	Good	Poor	Good	Excellent	Fair
Fair	Poor	Poor	Excellent	Fair	Fair	Good	Fair

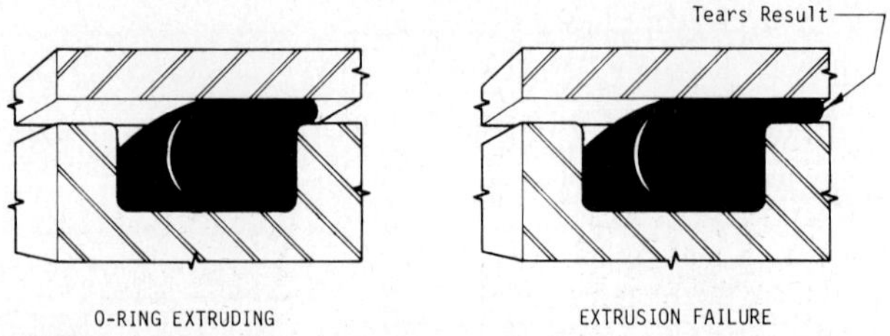

FIG. 6-24 Tears can result from nibbling as a result of pressure pulses.

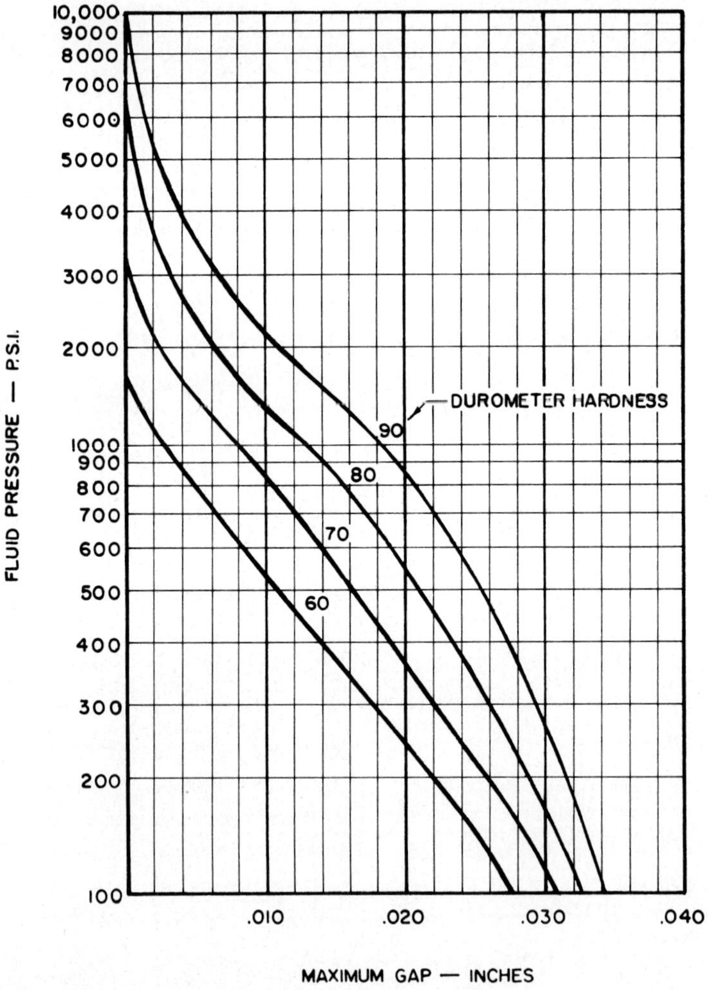

FIG. 6-25 Plot of maximum clearance vs. system fluid pressure for various durometer hardness. (*Courtesy: Hydraulics & Pneumatics.*)

TABLE 6-8 Static-Seal Leakage Analysis

Possible source of trouble	Suggested remedy
Seal has extruded or been nibbled to death	Replace seal and check the following: • Sealing surfaces must be flat within 0.005 in; replace part if out of limits • Initial bolt torque may have been too low; check manual for proper torque • Pressure pulses may be too high; check for proper relief valve setting • If normal operating pressures exceed 1500 lb/in^2, backup rings are required
Seal is badly worn	Replace seal and check the following: • Sealing surface too rough; polish to 16 μin if possible, or replace part • Undertorqued bolts permit movement; check manual for correct setting • Seal material or durometer may be wrong; check manual if in doubt
Seal is hard or has taken excessive permanent set	Replace seal and check the following: • Determine what normal operating temperature is, check system temperature • Check manual to determine that correct seal material is in use
Sealing surfaces are scratched, gouged, or have spiral tool	Replace faulty parts if marks cannot be polished out
Seal has been pinched or cut on assembly	Use petrolatum to hold seal in place with blind assembly; use protective shim if seal must pass over sharp threads
Seal leaks for no apparent reason	Check seal size and parts size; get correct replacement parts

Compression packing is used primarily in rotary shaft applications. It is squeezed between the throat of a box and a gland to effect a seal. The packing flows outward to seal against the bore of the box and inward to seal against the moving shaft or rod. Compression packing requires periodic tightening to compensate for wear and loss of volume.

The automatic or molded packing relies on operating pressures to create a seal; therefore, very little gland adjustment is required. These designs use flexible lips to seal against one or both surfaces in a stuffing box. One lip seals against the stationary bore and the other lip against the moving part. In some cases, O rings, quad rings, and other molded static seals are used in dynamic applications. These seals are called "squeeze" seals and depend on internal pressure to effect a seal. More molded packings are used in reciprocating applications. Floating packings include piston rings and segmental rod packings that fit into grooves. These rings are used in rotary and reciprocating motion.

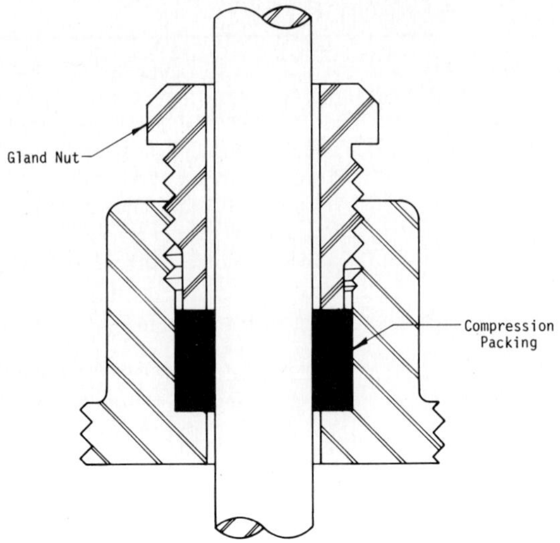

FIG. 6-26 Compression packing installed in stuffing box.

6-11 Compression Packings

Compression packings create a seal by being squeezed between the throat of a stuffing box and its adjustable gland (Fig. 6-26). The squeeze forces push the material against the throat of the box and the reciprocating or rotating shaft. When leakage occurs, the gland is tightened further.

Materials are extremely important in selecting the proper packing for an application. The packing must be able to withstand the conditions outlined in Table 6-9. Packing material is made of fabric or metal. The packing cross section can be round, square, or rectangular.

Fabric packings are made from plant fibers, mineral fibers, animal products, or artificial fibers. Plant fibers include flax, jute, ramie, and cotton. Cotton is usually used in cloth form, and the other materials are braided to form high-strength, water-resistant packings. All plant fibers are restricted to low-temperature applications. Asbestos is the most versatile mineral fiber. It is resistant to strong mineral acids. Asbestos loses strength at about 800°F.

Leather strips are used to make braided packings. Leather is restricted to applications of 200°F or less. Artificial fibers are used in applications to seal

TABLE 6-9 Packing Resistant Conditions

Must withstand the temperature of the applications.
Must be plastic to conform to bore and shaft under moderate pressure.
Must resist application fluid.
Must be elastic to absorb shaft vibrations.
Must be noncorrosive.
Must lose volume slowly to minimize adjustment.

corrosive liquids at temperatures up to 500°F. PTFE and graphite are used to make yarns for packings. The simplest yarn-type packing consists of strands of material twisted together to form a rope (Fig. 6-27). Braided packings can be obtained in round, square, or interlocked configurations (Fig. 6-28). The packings are impregnated with mineral oil, grease, or graphite. This lubricates the moving parts of the application to reduce wear, retain packing flexibility, and help effect a seal. Cloth packings are made from sheets of asbestos or cotton duck that are rolled, folded, or laminated with rubber to form the desired cross section (Fig. 6-29).

FIG. 6-27 Twisted-yarn fabric packing.

Metallic packings are used in high-temperature applications. Lead is used in applications up to 450°F. Soft pure copper is flexible; it is used in hot oil, tar, or asphalt pump applications. Aluminum foil is more flexible than copper and is resistant to sour crude oil. Shafts for copper and aluminum packings must be hardened to 500 Bhn; copper and aluminum can handle 1000°F application temperatures. Pure nickel is used to handle steam or caustic alkali materials at temperatures of up to 1500°F. Metallic packing designs are shown in Fig. 6-30. Metal is sometimes combined with other materials to form semimetallic designs (Fig. 6-31).

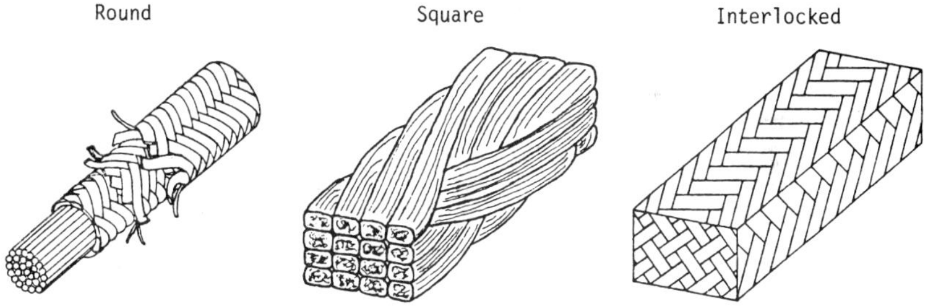

FIG. 6-28 Braided-yarn fabric packing types.

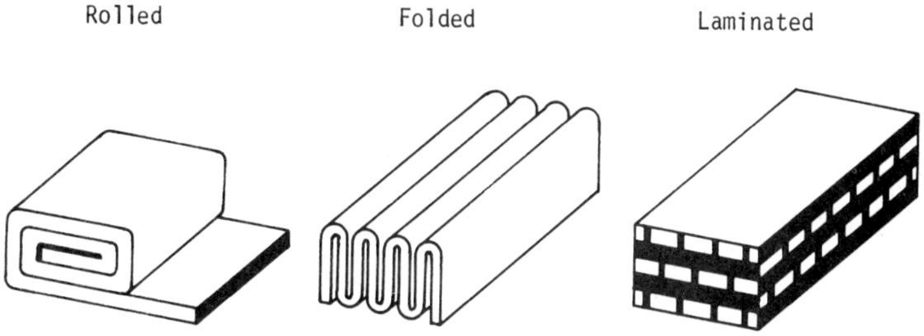

FIG. 6-29 Cloth fabric packing types.

TABLE 6-10 Compression Packings for Various Service Conditions

Fluid medium	Service condition			
	Reciprocating shafts	Rotating shafts	Pistons or cylinders	Valve stems
Acids and caustics	Asbestos (blue) Metallic Semimetallic TFE fluorocarbon resins and yarns	Asbestos (blue) Semimetallic TFE fluorocarbon resins and yarns	TFE fluorocarbon resins	Asbestos (blue) Semimetallic TFE fluorocarbon resins and yarns Graphite yarn
Air	Asbestos Metallic Semimetallic	Asbestos Semimetallic	Leather Metallic	Asbestos Semimetallic
Ammonia	Duck and rubber Metallic Semimetallic	Asbestos Semimetallic	Duck and rubber	Asbestos Duck and rubber Semimetallic
Gas	Asbestos Metallic Semimetallic	Asbestos Semimetallic	Leather Metallic	Asbestos Semimetallic
Cold gasoline and oils	Asbestos Semimetallic	Asbestos Semimetallic	Leather	Asbestos Semimetallic

Low-pressure steam	Asbestos Duck and rubber Metallic Semimetallic	Asbestos Metallic Semimetallic	Duck and rubber Metallic	Asbestos Duck and rubber Semimetallic
High-pressure steam	Asbestos Metallic Semimetallic	Asbestos Metallic Semimetallic	Metallic	Asbestos Metallic Semimetallic
Cold water	Duck and rubber Flax, jute, or ramie Leather Semimetallic	Asbestos Flax, jute, or ramie Semimetallic	Duck and rubber Semimetallic	Asbestos Flax or cotton
Hot water	Duck and rubber Leather Semimetallic	Asbestos Semimetallic	Duck and rubber	Asbestos Duck and rubber Semimetallic

Courtesy: *Machine Design*.

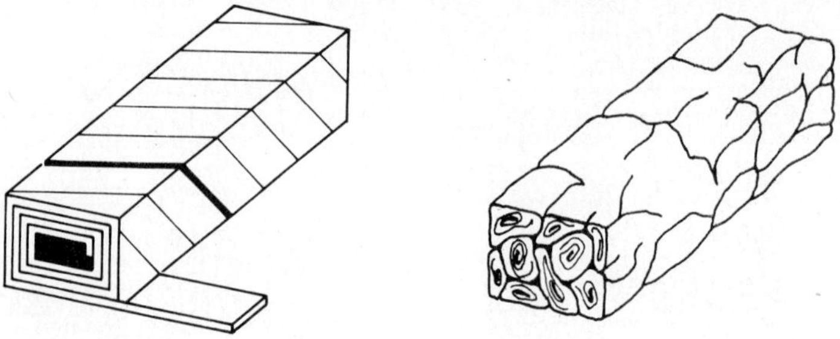

Spiral Wrapped Metal Foil   Folded and Twisted Metal Foil

FIG. 6-30 Metallic packings.

The types of packings that should be used for various service and fluid conditions are summarized in Table 6-10.

6-12 Molded (Automatic) Packing

Molded packings are sometimes referred to as automatic, hydraulic, or mechanical packings. They rely on the fluid pressure of the application to press the packing material against the wear surfaces. Lip-type packings are used primarily for reciprocating shafts. V- and U-ring packings can be used for both high- (50,000 lb/in^2) and low-pressure applications. Multiple V rings are often installed in sets; they are primarily packed on the outside of a reciprocating rod (Fig. 6-32). U rings are usually used to seal a piston (Fig. 6-33). Cup packings have a single lip and are used to seal pistons (Fig. 6-34). Flanged packings seal on the ID (Fig. 6-35) and are used for low-pressure, outside packed installation where there is not enough room for V or U rings.

Exclusion seals are used in conjunction with packings to keep out solid and liquid contaminants. Rod wipers (Fig. 6-36) are usually molded from tough materials that resist abrasion. Carboxylated nitriles and polyurethanes are usually used. They are hard (85 to 95 Shore A) with high (6000 lb/in^2) tensile strength. V- and U-cup packings are sometimes combined with a rod wiper to

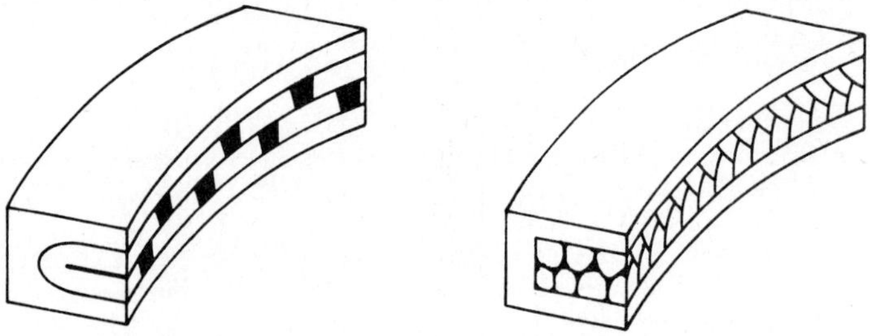

Metal Core   Braided Core

FIG. 6-31 Semimetallic packings.

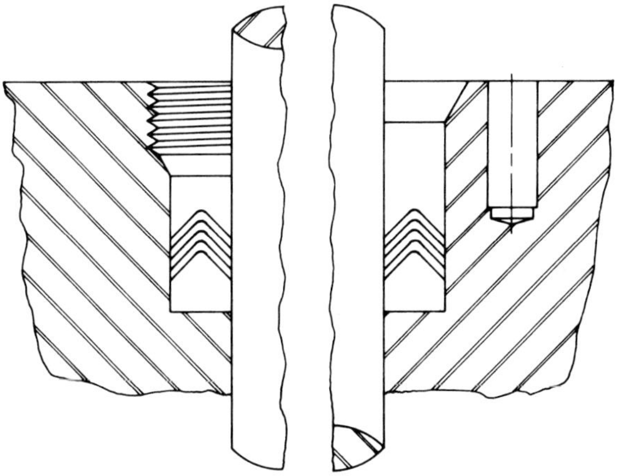

FIG. 6-32 V-ring packing in a gland.

form a utilized seal (Figs. 6-37 and 6-38). Wiper scraper seals (Fig. 6-39) have a metal ring made from copper, aluminum, bronze, or brass that scrapes debris from the shaft. The metal used must have low corrosion and low friction and be soft enough to resist scoring the shaft. Boots (Fig. 6-40) are used to prevent contamination from getting to the packings. Many shapes of boots are available. Specialized boots for hydraulic and pneumatic cylinders accommodate various stroke lengths.

Packings are made from rubber, fabric-reinforced rubber, and leather. Leather is usually impregnated with a synthetic rubber such as polyurethane

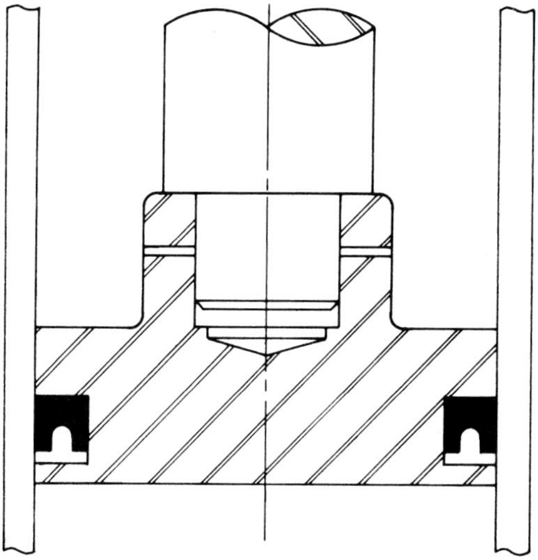

FIG. 6-33 U-ring packing used to seal a piston.

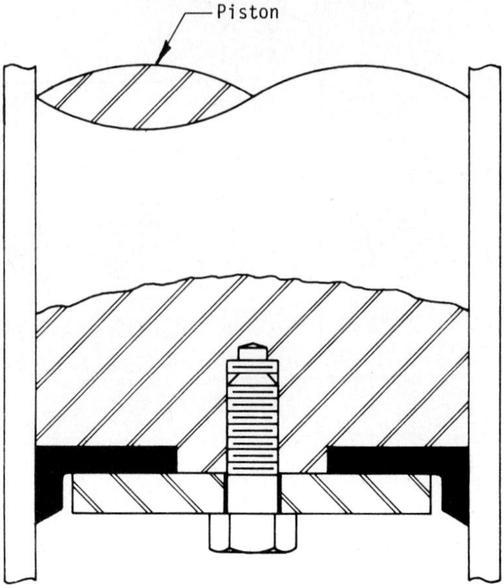

FIG. 6-34 Cup packing.

to fill voids between the fibers. Leather is quite flexible at temperatures down to −65°F. It deteriorates rapidly at temperatures above 200°F. Leather should not be used for pressurized steam or for strong acids and alkalis. In general, leather can be used if the pH of the liquid lies between 3 and 8.5. Leather will conform to rough surfaces up to 60 μin rms.

Packings are also made from synthetic rubber. The upper pressure limit of homogeneous rubber packings is 5000 lb/in^2. This limit can be extended to 8000 lb/in^2 if the material is reinforced with fabric. Cotton duck is commonly

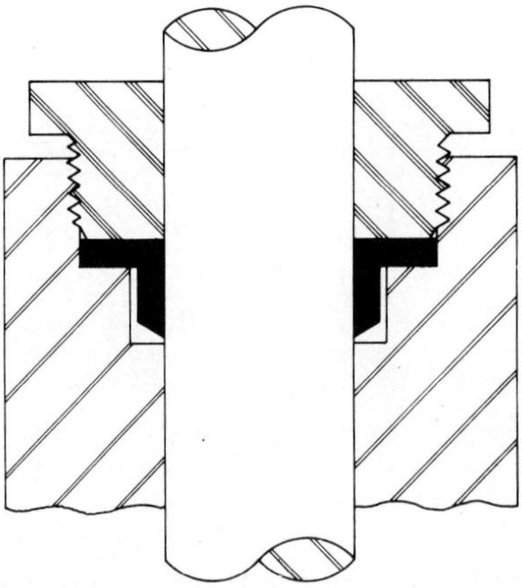

FIG. 6-35 Flange packing.

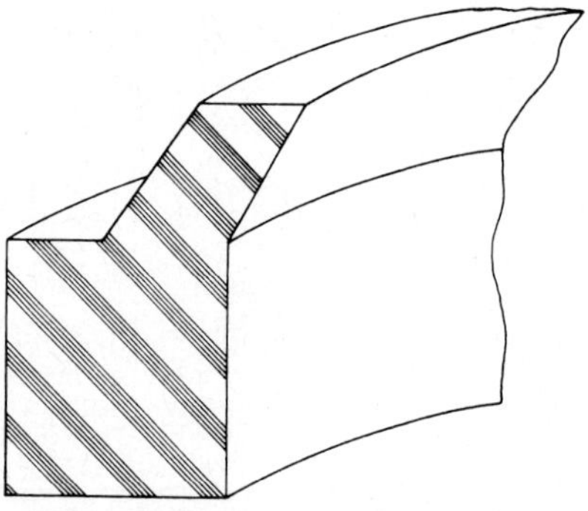

FIG. 6-36 Rod wiper.

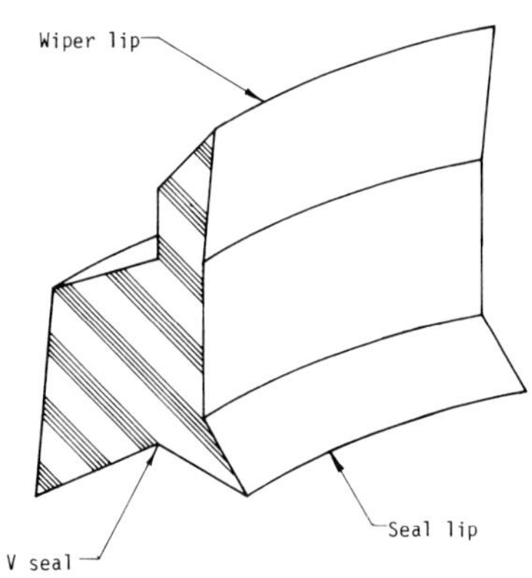

FIG. 6-37 Rod wiper combined with V packing.

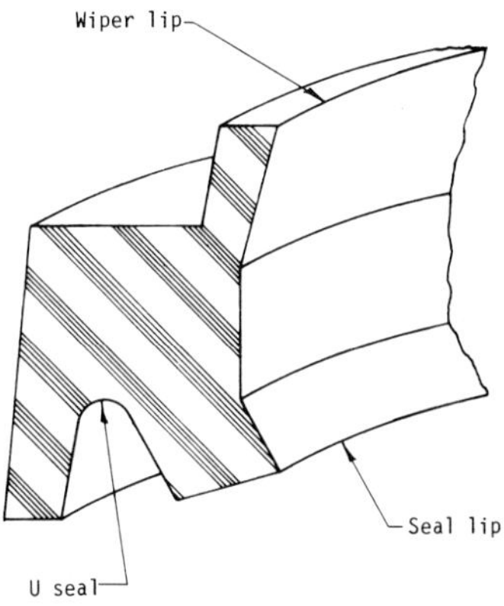

FIG. 6-38 Rod wiper combined with U packing.

used to reinforce rubber when the temperature is less than 250°F. Asbestos is used for temperatures above 250°F. Nylon is used when strength and flexibility are required.

The synthetic rubber used depends on the type of fluid and the temperature. The most common base polymers are polychloroprene, Buna-N, Buna-S, butyl, and fluoroelastomer. Polychloroprene and Buna-N are used for oil service, Buna-S for water, and butyl for phosphate esters. Fluoroelastomer is used for high-temperature applications.

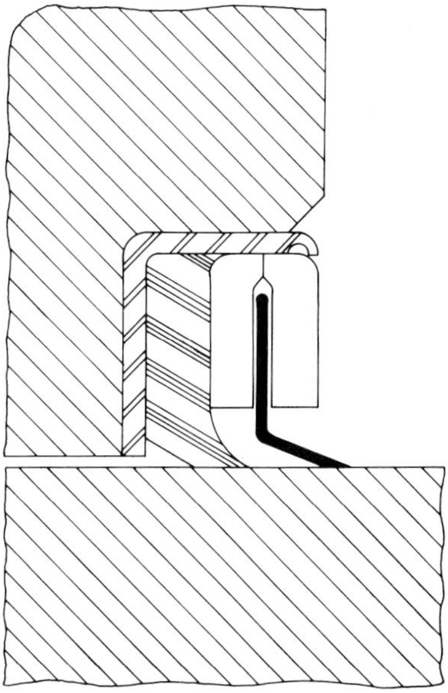

FIG. 6-39 Wiper scraper.

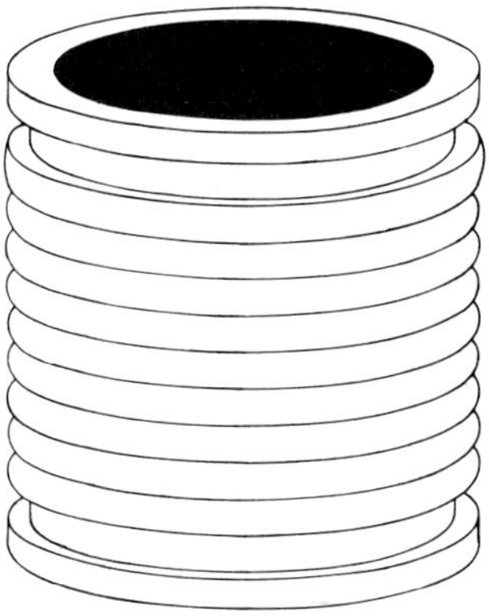

FIG. 6-40 Boots.

Metal surface finish for fabric-reinforced rubber packings should be a maximum of 32 μin rms with a preferred value of 16. On a rough surface, the packing fabric abrades quickly, causing early failure. Homogeneous rubber seals require a surface finish between 8 and 16 μin rms. PTFE packings have very little flexibility, but they are resistant to practically all chemicals and solvents. They are used to seal very corrosive fluids at pressures up to 3000 lb/in^2 and temperatures of 300°F or less.

Table 6-11 can be used as a guide for selecting molded packing materials.

6-13 Floating Packings (Split-Ring Seals)

Expanding split-piston-ring seals are used to seal gases in internal combustion engines and fluids in pumps. Contracting split-ring or rod seals are used in linear actuators whenever space, high temperature, or excessive pressure prohibit the use of packings. In all applications, the application pressure forces the rings against the surfaces that require sealing. Expanding split-ring seals must mate on the ID of a cylindrical bore and the side of a piston (Fig. 6-41).

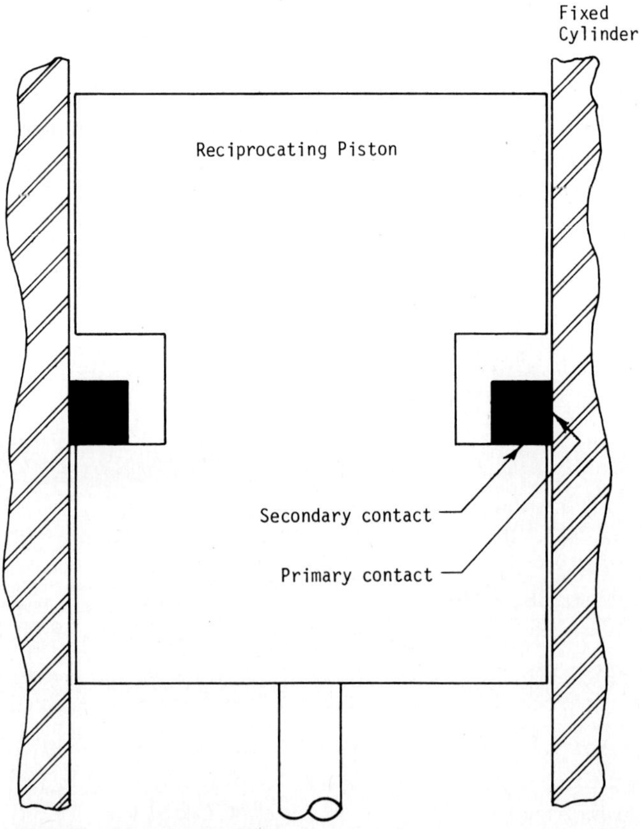

FIG. 6-41 Expanding split ring.

TABLE 6-11 Guide for Selection of Molded Packing Material

Condition	Leather	Homogeneous rubber	Rubber fabric-reinforced
Oil	Good	Good	Good
Air	Good	Good	Good
Water	Good	Good	Good
Steam	Not recommended	Good	Good
Solvents	Not recommended	Good	Good
Acids	Not recommended	Good	Good
Alkalis	Not recommended	Good	Good
Fire-resistant fluids:			
Phosphate ester	Wax or polysulfide impregnation	Butyl base polymer	Butyl base polymer
Water-glycol	Not recommended	Buna-N base polymer	Buna-N base polymer
Water-oil emulsion	Wax, polyurethane, or polysulfide impregnation	Buna-N base polymer	Buna-N base polymer
Temperature range	−65 to +180°F	−65 to +400°F	−40 to +400°F*
Types of metal	Ferrous and nonferrous	Chrome-plated steel and nonferrous alloys with hard, smooth surfaces	Chrome-plated steel and nonferrous alloys with hard, smooth surfaces
Metal finish, rms (max)	60	16	32
Clearances	Medium	Very close	Close
Extrusions or cold flow	Good	Poor	Fair
Friction coefficient	Low	Medium and high	Medium
Resistance to abrasion	Good	Fair	Fair
Maximum pressure, lb/in²	125,000	5000	8000
Concentricity	Medium	Very close	Close
Side loads	Fair	Poor	Fair
High shock loads	Good	Poor to fair	Fair

* Depending on specific formulation or combination of materials.
Courtesy: *Machine Design*.

Contracting-type seals must contact the side surface of a fixed housing and the OD of a reciprocating rod (Fig. 6-42). The simplest and least expensive split ring has a straight-cut joint (Fig. 6-43). It is used as a low-pressure piston seal where joint leakage is not critical. Step joints (Fig. 6-44) of various types are used to reduce joint leakage. Rings are balanced when hydraulic fluid pressure is high. This is accomplished by machining small-circumference grooves in the wear surface of the ring (Fig. 6-45). System fluid flows into the groove and forms a very thin dam, which acts as a pressure reducer. Balancing usually increases leakage. Multiring systems are used to seal high internal pressure. Table 6-12 gives the number of rings required for various pressure levels.

Metal rings are usually used in lubricated applications. If the application is not lubricated, PTFE rings impregnated with carbon-graphite or molybdenum disulfide can be used. Table 6-13 gives the temperature limitations for materials used in ring applications. The limitations of various plating treatments appear in Table 6-14.

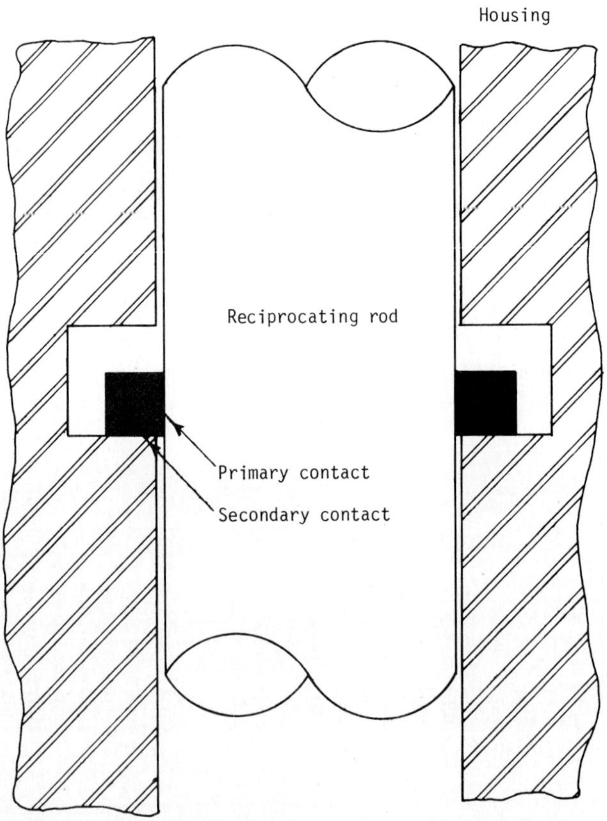

FIG. 6-42 Contracting split ring.

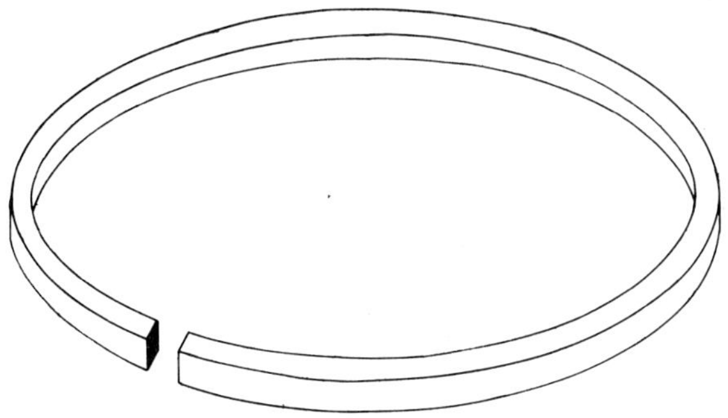

FIG. 6-43 Simple straight-cut split ring.

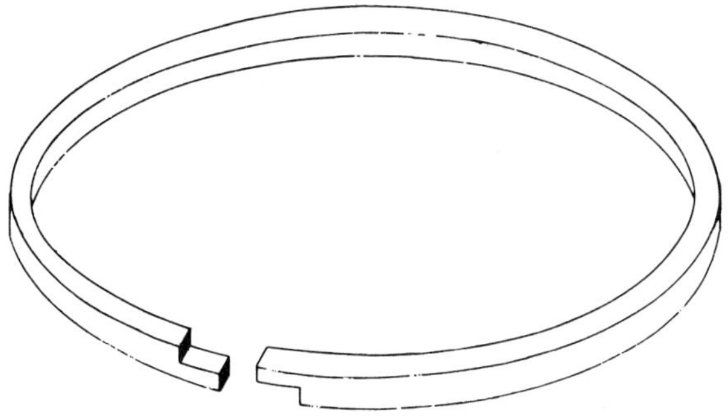

FIG. 6-44 Step-cut split ring.

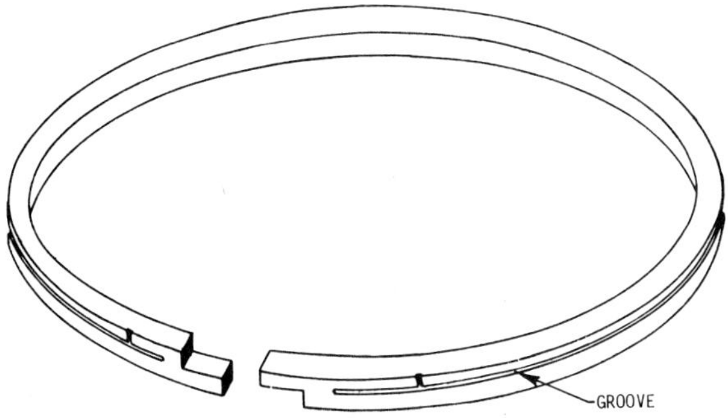

FIG. 6-45 Balanced ring.

TABLE 6-12 Ring Pressure Limitations

Pressure, lb/in²	No. of rings required
Up to 300	2
300 to 900	3
900 to 1500	4 plain face
	5 balanced
1500 to 3000	5 plain face
	6 balanced
3000 and up	6 balanced

TABLE 6-13 Material Temperature Limitations

Material	Temperature, °F
Low-alloy gray irons	650
Malleable iron	700
Ductile iron	700
Ni-Resist	800
Ductile Ni-Resist	1000
410 stainless steel	900
17-4 pH stainless steel	900
Bronze	500
Tool steel, Tc 62-65	900
Carbon (high temperature)	950
K-30 (filled Teflons)	450 to 500
S-Monel	950
Polyimide	750

Courtesy: *Machine Design.*

TABLE 6-14 Plating Temperature Limitations

Surface treatment	Temperature, °F
Chromium plate	500
Tin plate	700
Silver plate	600
Cadmium-nickel plate	1000
Flame plate	1000–1600
Flame plate LC-1A	1600
Flame plate LA-2	1600

Courtesy: *Machine Design.*

6-14 Diaphragm Seals

Diaphragms are membranes that are used to prevent movement of fluid or contamination from one chamber to another (Fig. 6-46). There are static diaphragms that merely separate fluids. These diaphragms are subject to very little displacement. Dynamic diaphragms are attached to the stationary and moving members and usually transmit force or pressure. Dynamic diaphragms can be flat or rolling. A flat diaphragm has no convolutions or convolutions that are less than 180°. Most heavy-duty flat diaphragms have molded convolutions to allow flexibility (Fig. 6-47). They are attached to the stationary housing at the edges. Plates are used to attach the diaphragm to the pushrod. Springs are used to return the diaphragm to the neutral position; they ride in the central disk. Positive mechanical stops are used to prevent overstroking the diaphragm. Diaphragm motion should be less than 90 percent of the maximum possible stroke. The diaphragm must be thin enough to prevent wrinkling.

Rolling diaphragms are used for medium- and high-pressure long-stroke applications (Fig. 6-48). As the piston moves down, the diaphragm rolls off the piston sidewall onto the cylinder sidewall. The diaphragms are usually from 0.010 to 0.035 in thick. The pressure is supported by the cylinder head, and it holds the diaphragm against the piston and cylinder walls. Rolling dia-

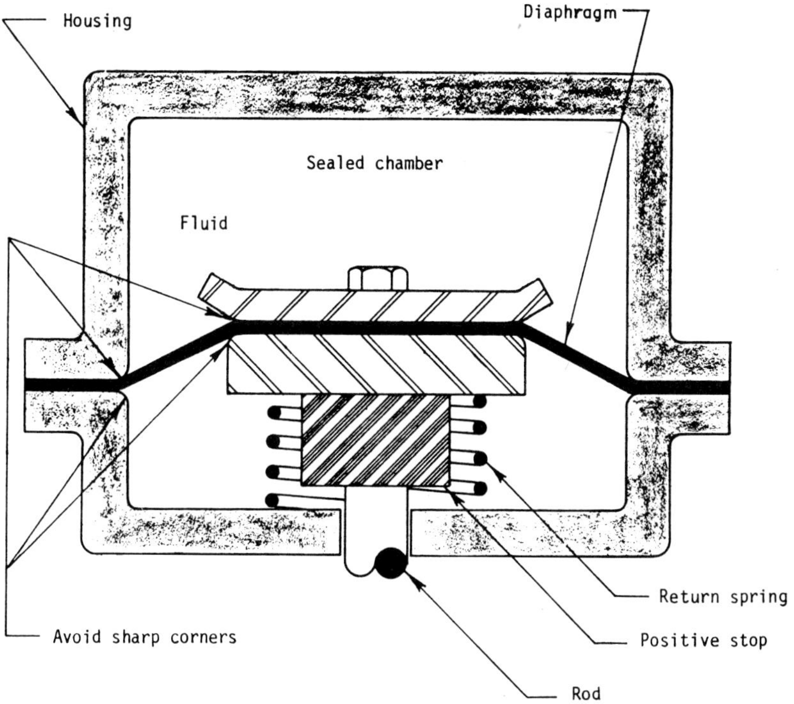

FIG. 6-46 Typical diaphragm application.

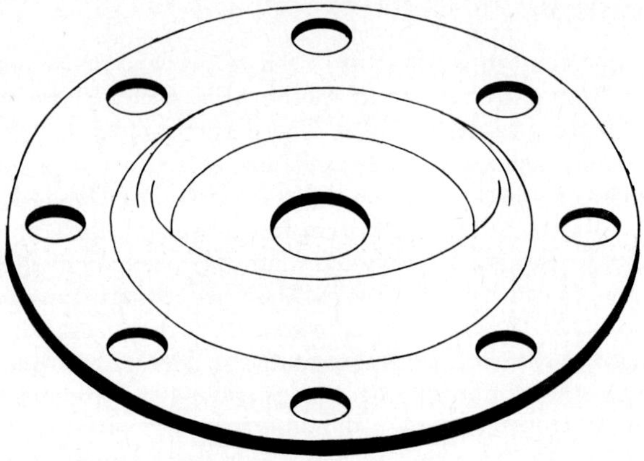

FIG. 6-47 Flat convoluted diaphragm.

phragms are usually molded in the shape of a top hat, and the convolute is generated by inversion during assembly. In some installations, the resiliency of the material may cause it to revert to its original position (Fig. 6-49). Sidewall scrubbing and high wear will result. This effect is prevented by using retained plates with curved lips. Rolling diaphragms cannot be used in applications where the low- and high-pressure chambers can be reversed. Pressure reversal will cause the diaphragm to wrinkle and scuff against the sidewall. High wear and premature failure can result.

Materials used to make diaphragms must have a high burst strength. This is the pressure that will rupture the material. Material modulus and tensile

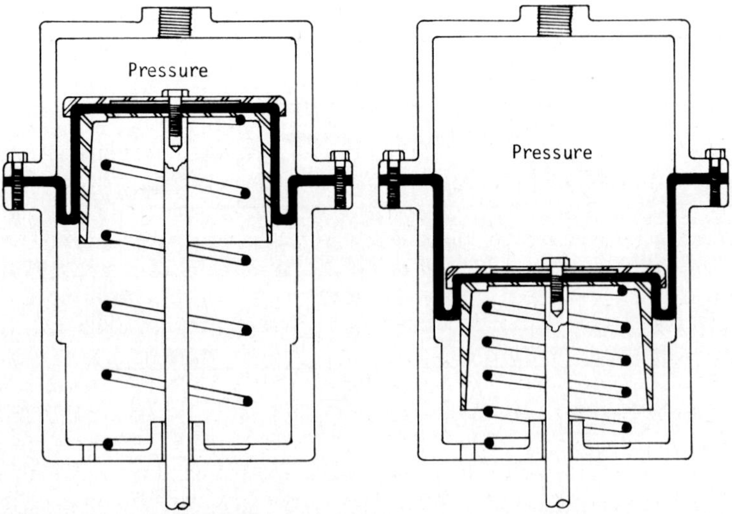

FIG. 6-48 Rolling diaphragm.

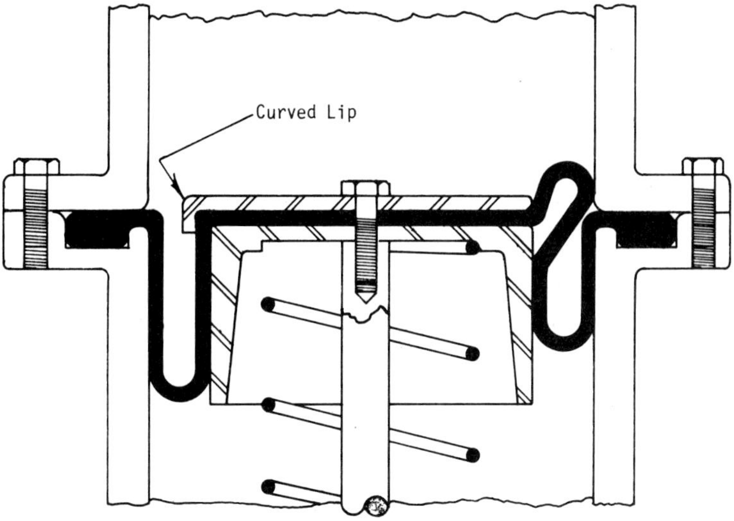

FIG. 6-49 Rolling diaphragm inversion is prevented by curved lip.

strength must be high. Blends of fabrics and elastomers are used in diaphragms. Cottons and nylons are used in flat diaphragms at temperatures less than 250°F. Nylon resists abrasion and fatigue, but will creep under pressure and take a set. Davron can be used at temperatures up to 350°F, and Nomex is used for prolonged exposures at 500°F. All fabrics must be impregnated with elastomers to render the diaphragm impermeable. Elastomeric properties appear in Table 6-15.

DYNAMIC SEALS—CONTACTING TYPE

6-15 Mechanical Face Seals

Mechanical face seals are used in applications where low leakage rates and long life are essential. Standard face-seal designs are ideal for high pressure [up to 300 lb/in^2 (absolute)], high shaft speeds (up to 50,000 r/min), and broad temperature ranges (−425 to 1200°F). Special design arrangements are required for conditions that exceed these limits. In some cases, tandem seals are required to handle large pressure differentials.

Mechanical seals employ two wearing faces. The seal head is usually spring-loaded and is pressed against a mating ring or seal. Rotating spring-loaded heads (Fig. 6-50) made from graphite or plastic are used with shafts that are machined to close tolerances from high-quality materials. The stationary ring, which is mounted in the housing, is made of ceramic or metal. If comparatively high speeds (5000 r/min or more) are encountered, stationary heads and rotating seals (Fig. 6-51) give the best results. The stationary seal head requires less critical dynamic balancing. Close bore tolerances with a

TABLE 6-15 Elastomeric Properties

Elastomeric type	Air permeability rating	Operating temperature limits, °C	Properties
Silicone	170 to 260	−80 to 260	General-purpose, low-temperature resistant; high permeability
Fluorosilicone	50	−60 to 230	Oil-resistant, high temperature
Nitrile	0.25 to 1.00	−40 to 120	Oil-resistant, low cost, attacked by ozone
Neoprene	1.40	−35 to 120	Weather-resistant; fair oil-resistance
Ethylene propylene	9.60	−40 to 150	Steam-, ozone-, acid-, and alkali-resistant
Fluorocarbon	0.32	−20 to 280	Oil-, fuel-, and chemical-resistant
Polyacrylate	1.50	−30 to 175	Hot-oil- and ozone-resistant
Epichlorohydrin	0.15 to 0.70	−40 to 150	Low permeability, oil-resistant

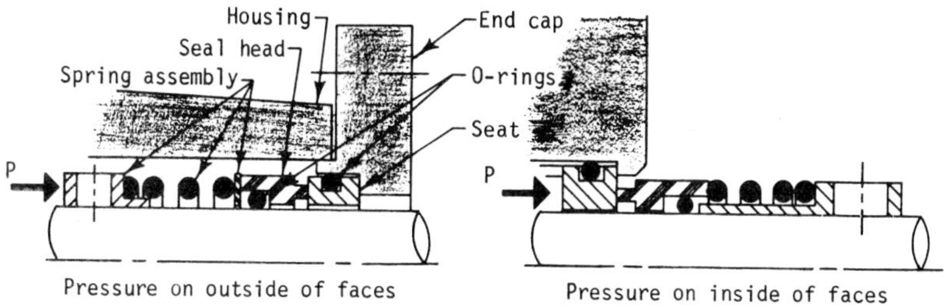

FIG. 6-50 Rotating seal heads.

high-grade finish must be maintained if a stationary head is used. The method of sealing depends on the direction of the pressure. The configurations shown in Figs. 6-50 and 6-51 must be adjusted to ensure that internal pressure will keep the seal faces closed.

It is necessary to provide a static seal between the wear members and the shaft or housing the member is attached to. This prevents leakage between the members. Elastomeric elements are typically used in pusher-type seals to prevent leakage between the rotating shaft and the rotating head. As the sealing face wears, the static seals are pushed along the shaft by the spring. These elements include the O ring, V ring, U cup, and wedge (Fig. 6-52).

Some seals, known as nonpusher types, use a bellows-shaped element to form a static seal between itself and the shaft. The bellows arrangement is used to compensate for axial movement. Elastomeric bellows (Fig. 6-53) can be used. The upper temperature limitation of the seal is determined by the material used for the static seal or the bellows. In general, most synthetics are limited to −40 to 225°F. TFE fluorocarbon can be used from −400 to 500°F. Metal bellows are used to extend the upper temperature limit to 1200°F. Metal

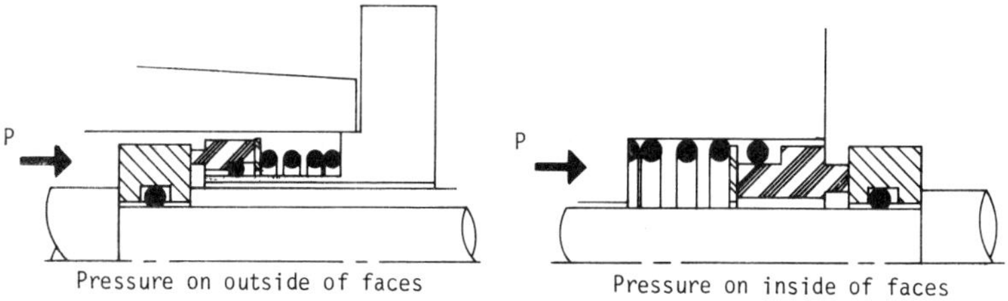

FIG. 6-51 Stationary seal heads.

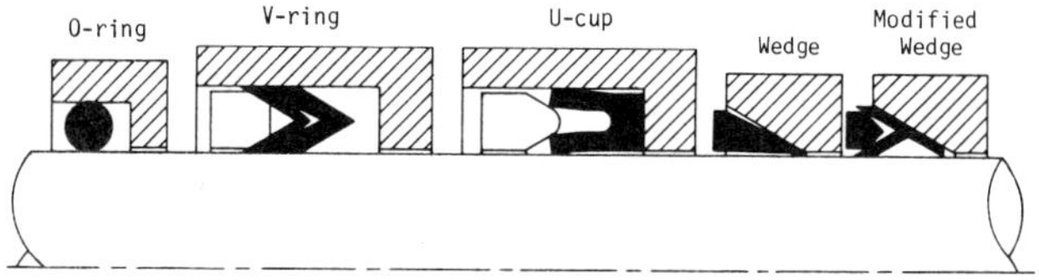

FIG. 6-52 Pusher types.

FIG. 6-53 Elastomeric bellows.

bellows can be produced from tubing by hydraulic forming in dies or by roll forming (Fig. 6-54). A more expensive method of making the bellows is to weld disks together (Fig. 6-55). The welded bellows requires less axial space than the formed metal bellows.

Metal bellows seals are suited for use in harsh environments that would destroy elastomers. They are used with liquid oxygen, fluorine, acids, and a wide variety of caustic chemicals.

Face Loading The sealing faces are kept closed under all operating conditions by providing an axial load. The load must be high enough to overcome friction, but not high enough to accelerate seal wear and shorten life.

Belleville, finger, slotted, and wavy washers provide a loading force in a small amount of axial space. These spring types must be heat-tempered to be

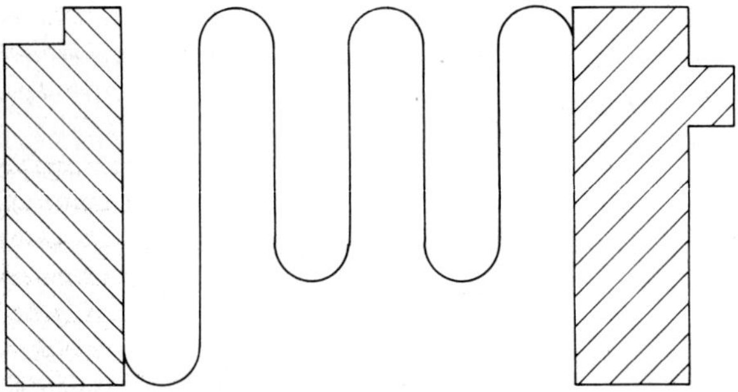

FIG. 6-54 U-shaped convolutions.

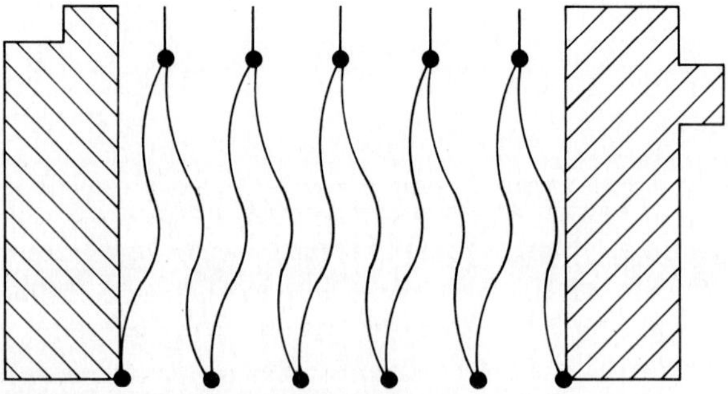

FIG. 6-55 Bellows welded from disks.

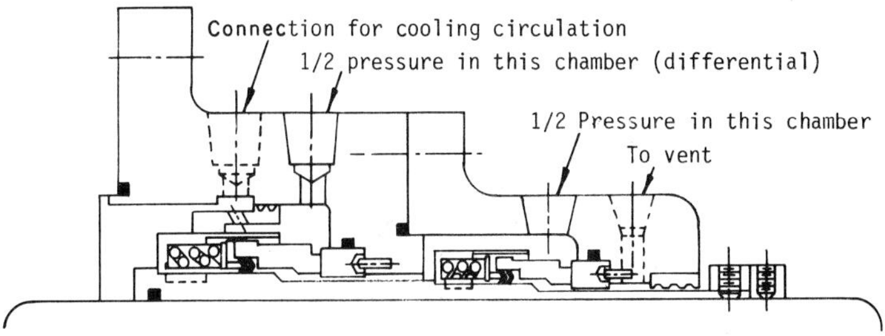

FIG. 6-56 Tandem pusher seal arrangement with V-ring static seals and multiple springs.

effective. This requirement limits the selection to materials that are not as corrosion-resistant as stainless steel.

A single spring with a heavy coil that surrounds the shaft (Figs. 6-50 and 6-51) can withstand a relatively high degree of corrosion. This design generally requires a great deal of axial space. Centrifugal forces will attempt to unwind the space. It is generally difficult to provide uniform face loadings with single springs.

Seals with multiple springs require less axial space than single-coil springs. The same springs can be used in various combinations for seals of many sizes (Fig. 6-56). Multiple springs resist unwinding from centrifugal force. A small amount of axial movement can generate the desired force changes.

The small cross section of the wire used to make the springs will corrode rapidly unless corrosion-resistant materials are used.

Rubber elements are sometimes used to load the faces of seals that are used in relatively low-pressure, low-speed applications. These seal types (Fig. 6-57) are used for track and idler wheels on earth-moving equipment.

Metal bellows that are used to replace static seal elements are also used to supply the force to load the sealing faces. In addition to extending the upper temperature limits of the seal, the metal bellows serves a dual purpose and reduces the number of components in the seal (Table 6-16).

Seal Balancing As the internal pressure increases, the force pushing the wear faces together can increase dramatically if the seal is unbalanced (Fig.

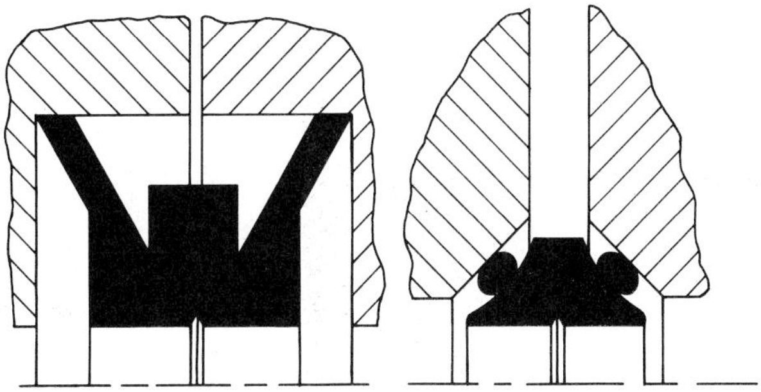

FIG. 6-57 Rubber-loaded faces.

TABLE 6-16 Common Construction Materials for Mechanical Seals

Primary ring	Mating ring	Secondary seal	Spring-housing components
Carbon	Ceramic	Nitrile Ethylene propylene	18-8 stainless
Tungsten carbide	Cast iron	Fluorocarbon resin	316 stainless
Bronze	Cobalt-base alloy		Nickel-base alloys
Silicon carbide	Silicon carbide Tungsten carbide	Fluoroelastomer Perfluoroelastomer	Titanium

6-58). The wear on the face surfaces will increase as the hydraulic pressure increases. The hydraulic pressure can be balanced by designing a step in the seal head. Any portion of the face load can be cancelled by changing the step dimensions (Fig. 6-59). The balance ratio is defined as the ratio of the amount of face area above the balance line to the amount of face area below. A 55/45 percent ratio is usual with poor lubricants, and an 85/15 percent ratio is usual with good ones. Completely balanced seals (Fig. 6-60) depend only on the spring force to load the faces. The load is independent of internal pressure.

PV Relationships An important criterion in determining the limitations of seal-face materials is the *PV* factor. The factor is the product of the unit pressure acting on the seal-face junction (in pounds per square inch) and the rubbing velocity (in feet per minute). Unit pressure results from spring, bellows, or diaphragm tension and the unbalanced portion of the hydraulic load.

For any given combination of seal-face materials, the limiting *PV* value is set by factors such as the tenacity of the film on the face surfaces, rate of heat conduction away from the heat-sensitive elements of the seal, and quality of the surfaces.

At high rubbing speeds, the quality of the seal-face surfaces becomes very important. Surface quality is evaluated by roughness, expressed as an rms

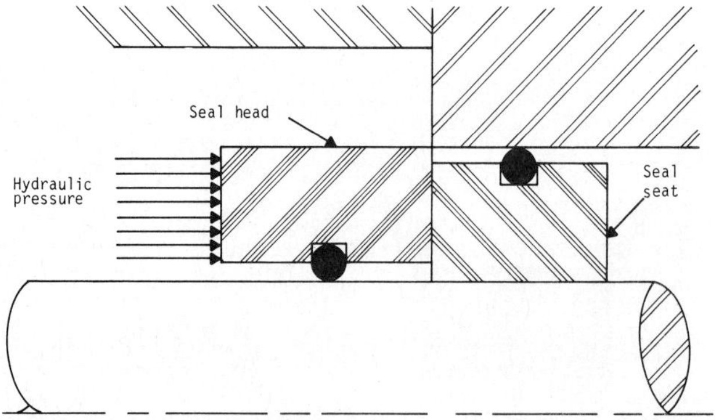

FIG. 6-58 Unbalanced seal. Total pressure pushes head and seat together; heavy wear can result.

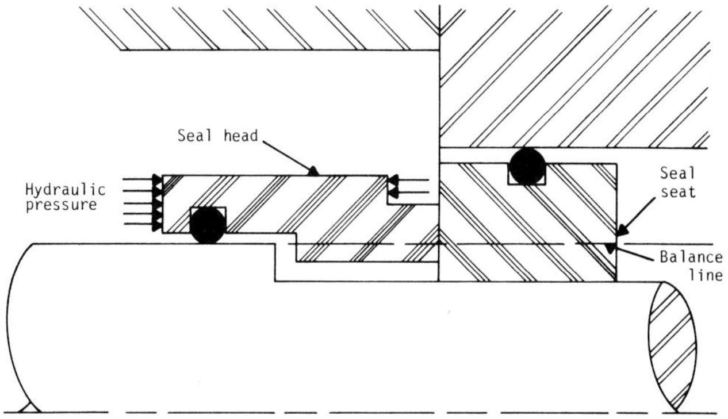

FIG. 6-59 Partially balanced seal. Part of pressure is relieved by step in head.

value, and by flatness, usually given in light bands. As speeds increase, the degree to which planes of the mating surfaces approach the true normal to the shaft axis becomes very important.

The factors that are most significant to seal performance are leakage, power consumption, and life or wear. The wear rate, and thus life, is highly dependent on the maintenance of a lubricant film between the seal faces. Theoretical analyses that examine the factors that influence leakage, power, and life generally define the pressure distribution across the seal face. This pressure distribution is used to predict the volumetric flow rate and the torque. Most face-seal designs have the high-pressure side of the seal on the outer diameter of the interface. Centrifugal force opposes any leakage that will flow radially inward. The radial pressure distribution for laminar flow between the smooth parallel surfaces defined by the geometry of Fig. 6-61 is given by Eq. (6-2). The amount of seal leakage appears in Eq. (6-3), the theoretical condition for zero leakage is given by Eq. (6-4), and the power consumed is given by Eq. (6-5).

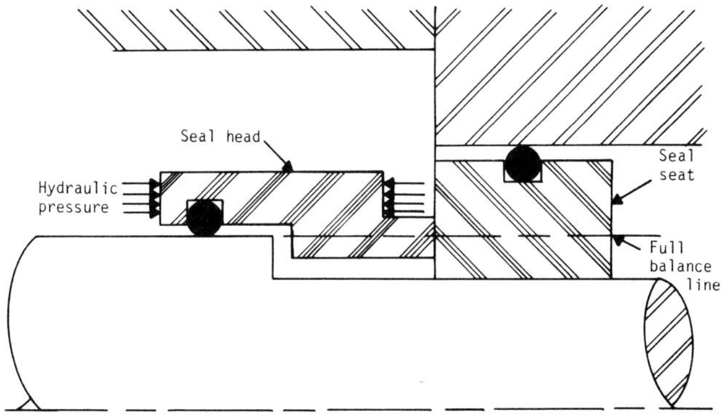

FIG. 6-60 Completely balanced seal. Force between head and seat is independent of pressure.

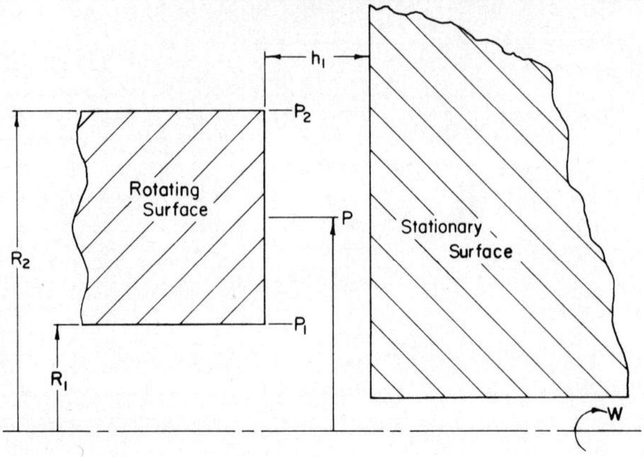

FIG. 6-61 Face seal geometry.

$$p - p_1 = \frac{3\rho\omega^2}{20g}(r^2 - R_1^2) - \frac{6\nu}{\pi h^3}\ln\frac{r}{R} \qquad (6\text{-}2)$$

$$Q = \frac{\pi h^3}{6\nu \ln(R^2/R^1)}\left[\frac{3\rho\omega^2}{20g}(R_2^2 - R_1^2) - p_2 - p_1)\right] \qquad (6\text{-}3)$$

$$p_2 - p_1 = \frac{3}{20}\rho\omega^2(R_2^2 - R_1^2) \qquad (6\text{-}4)$$

$$P = \frac{\pi\nu\omega^2}{13{,}200h}(R_2^4 - R_1^4) \qquad (6\text{-}5)$$

where p = pressure at radial position r, lb/in^2
p_1 = pressure at seal ID, lb/in^2
p_2 = internal hydraulic pressure, lb/in^2
r = radial position, in
R_1 = ID of rotating member, in
R_2 = OD of rotating member, in
h = thickness of fluid between members, in
Q = leakage rate, in^3/s
ν = kinematic viscosity, lb·s/in^2
ρ = fluid density, lb/in^3
ω = rotational speed, rad/s
P = power loss, hp

Troubleshooting Face Seals Leaking face seals often show telltale contact patterns that help identify the cause of failure. Table 6-17 lists things to look for.

6-16 Radial Lip Seals

Radial lip seals are used primarily to retain lubricants in equipment with rotating, reciprocating, or oscillating shafts. The secondary purpose is to ex-

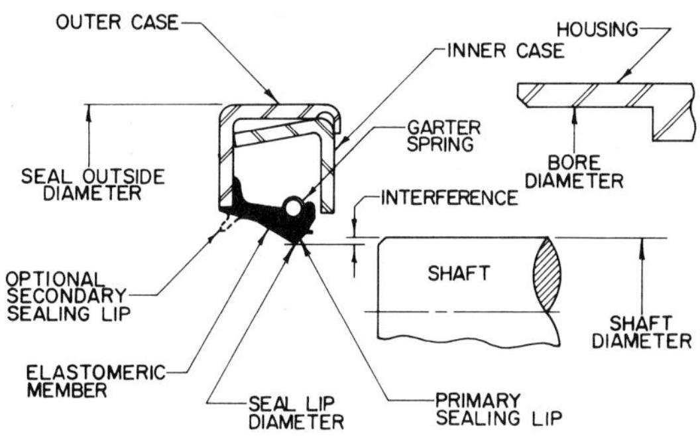

FIG. 6-62 Radial lip seal terminology.

clude foreign matter. Application conditions may vary from high-speed rotation up to 7500 r/min in a clean environment to low-speed shaft reciprocation in very dusty environments. The seal must operate under conditions of extreme cold (−50°F). A typical seal design with standard terminology is shown in Fig. 6-62. Sealing is dependent on maintaining an interface between the shaft and the elastomeric member, and between the bore and the seal outside diameter. The basic advantages of radial lip seals include low cost for effective sealing, small space requirements, easy installation, and an ability to seal multivariable applications.

Seal Types Non-spring-loaded seals (Fig. 6-63*a*) are used to retain highly viscous materials like grease at shaft speeds less than 2000 ft/min (609 m/min). If dust exclusion is desired, a secondary lip can be added (Fig. 6-63*b*). Typical applications include conveyor rollers, vehicle wheels, and so on. For maximum effectiveness as a dirt excluder, the seals are installed with the primary lip facing away from the bearings. Heavy-duty multilip seals (Fig. 6-63*c*) provide good lubricant retention under severe dust conditions, such as in disk harrows. The shaft speed is usually limited to 500 r/min. External multilip seals (Fig. 6-63*d*) are used with a fixed shaft and rotating bore. The bore or a wear ring at the bore is used as a sealing surface.

Spring-loaded seals are used to retain low-viscosity lubricants such as oils at speeds up to 3600 ft/min (1098 m/min). The single-lip seal is the most economical general-purpose oil seal (Fig. 6-64*a*).

Typical applications include engines, drive axles, transmission pumps, electric motors, and speed reducers. If medium-duty dirt exclusion is required, a dual-lip seal is used (Fig. 6-64*b*). These seals are commonly used in automotive, farm, and industrial applications where they are protected from external abuse during operation.

Dual-case seals are also available (Fig. 6-64*c* and *d*) for heavy-duty applications. The secondary case is added to provide additional strength and protection during assembly or operation. These seals are commonly used in construction equipment, farm, and industrial applications. External spring-loaded seals are used in applications with rotating housing and stationary shafts (Fig. 6-65).

TABLE 6-17 Troubleshooting for Face Seals

Symptoms	Appearance	Causes
Seal leaks steadily despite apparent full contact of mating rings	Full 360° contact patterns on mating ring; little or no measurable wear	Secondary seal leakage caused by: • Nicked, scratched, or porous seal surfaces • O-ring compression set • Chemical attack
Steady leakage at low pressure; little or no leakage at high pressure	Heavy contact on mating ring OD; fades to no visible contact at ID. Possible edge chipping on primary ring OD	Deflection of primary ring from overpressurization. Seal faces not flat because of improper lapping
Seal leaks steadily when shaft is rotating; little or no leakage when shaft is stationary	Heavy contact pattern on mating ring ID; fades to no visible contact at OD. Possible edge chipping on primary-ring ID	Thermal distortion of seal faces. Seal faces not flat because of improper lapping
Seal leaks steadily whether shaft is stationary or rotating	Two large contact spots; pattern fades away between spots. Contact through about 270°; pattern fades away at low spot. Contact spots at each bolt location	Mechanical distortion caused by: • Overtorqued bolts • Out-of-square clamping parts • Out-of-flat stuffing box faces • Nicked or burred gland surface • Hard gasket

Seal leaks steadily whether shaft is stationary or rotating. Noise from flashing or face popping	High wear or thermal distress on mating ring. High wear and carbon deposits on primary ring. Possible edge chipping on primary ring. Thermal distress over one-third of mating ring, located 180° from inlet of seal flush. High wear and possible carbon deposits on primary ring. Thermal distress at 2 to 6 locations on mating ring. High wear and possible carbon deposits on primary ring	Sealed liquid vaporizing at seal interface, caused by: • Low suction or stuffing box pressure • Improper running clearance between shaft and primary ring • Insufficient cooling • Improper bushing clearance • Circumferential flush groove in gland plate missing or blocked
Seal leaks steadily whether shaft is stationary or rotating	High wear and grooving on mating ring	Poor lubrication from sealed fluid. Abrasives in fluid
Steady leakage when shaft is rotating; no leakage when shaft is stationary	Contact pattern on mating ring is slightly larger than primary ring width. Possible high spot opposite drive pin hole	Out-of-square mating caused by: • Nicked or burred gland surfaces • Improper drive pin extension • Misaligned shaft • Piping strain on pump casing • Bearing failure • Shaft whirl
Damaged mating ring; seal leaks whether shaft is stationary or rotating	Eccentric contact pattern, although width equals that of primary ring. Possible cracks on mating ring	Misaligned mating ring caused by: • Improper clearance between gland plate and stuffing box • Lack of concentricity between shaft OD and stuffing box ID

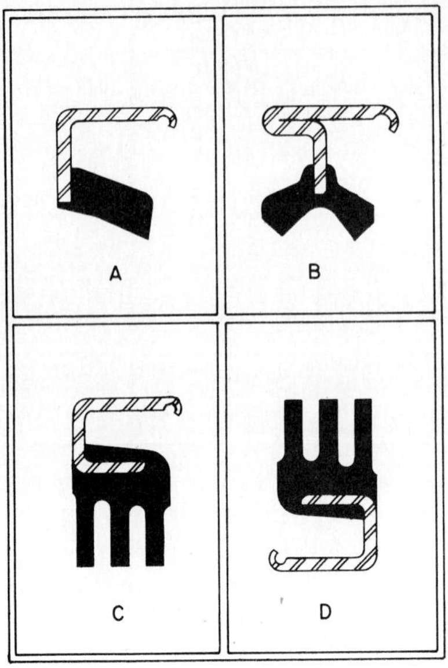

FIG. 6-63 Non-spring-loaded seals.

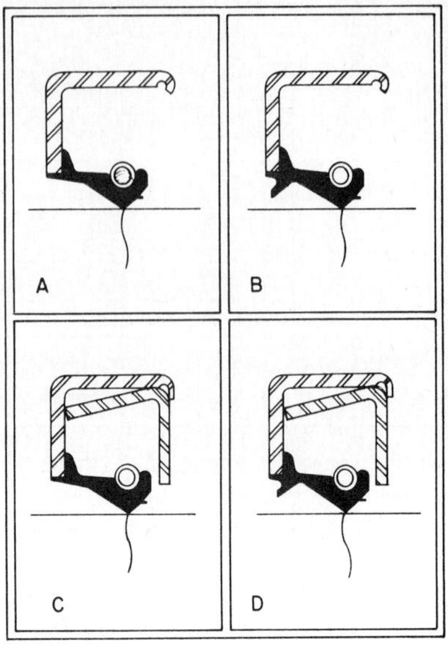

FIG. 6-64 Spring-loaded seals.

In general, pressure reduces seal life. For conventional seal designs, the pressure limitations are found in Table 6-18.

If the pressure requirement is greater, special designs must be considered. These unique designs are usually developed for specific applications. A seal designed for intermittent operation at 1200 to 1500 lb/in^2 (8280 to 10,350 kPa) is shown in Fig. 6-66.

In some cases, a rubber OD seal (Fig. 6-67) is preferred. The rubber OD is used when bore finish recommendations are not followed. If the bore material is different from the seal case material, rubber ODs can be used to eliminate bore leakage resulting from differential thermal expansion.

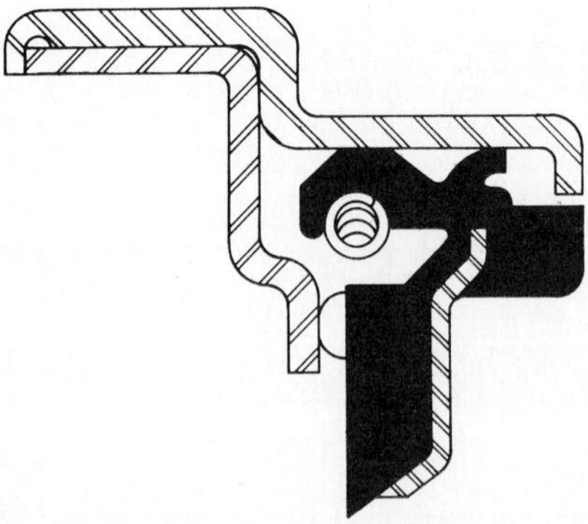

FIG. 6-65 Seal for rotating OD stationary shaft.

TABLE 6-18 Operating Pressure Limits

Shaft speed		Maximum pressure permissible	
ft/min	m/min	lb/in²	kPa
0–1000	0–304.8	7	48.2
1000–2000	305–609	5	34.5
2000–3600	610–1098	3	20.7

For applications where the total eccentricity exceeds the recommended levels, special seal designs are used (Fig. 6-68). The elastomer member is convoluted to provide added flexibility, and the ability to follow the shaft is improved. Eccentricity motion as great as 0.060 in (1.524 mm) can easily be handled with these designs. The complexity of the seal design results in a cost penalty.

Some seals are made with lip materials that cannot be bonded to the case. The lip material is tightly clinched between metal stampings to prevent the packing from rotating (Fig. 6-69). An internal seal must be provided to prevent fluid from leaking through the seal. The cost of assembled seals is typically more than that of bonded ones.

Some seals have ribs molded on the air side on the rubber element. Any leakage that seeps past the lip is pumped back into the sump. These seals typically run hotter than plain lip seals because there is an increased amount of rubber contacting the shaft. They do not function well in dusty environ-

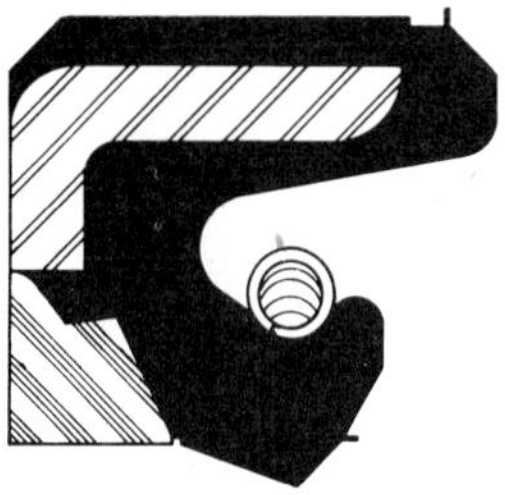

FIG. 6-66 High-pressure seal.

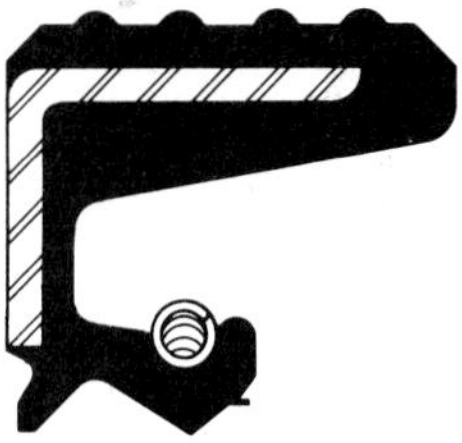

FIG. 6-67 Rubber OD seal.

FIG. 6-68 Seal for high shaft runout.

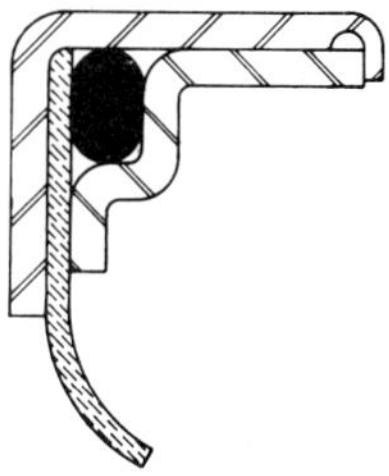

FIG. 6-69 Assembled seal.

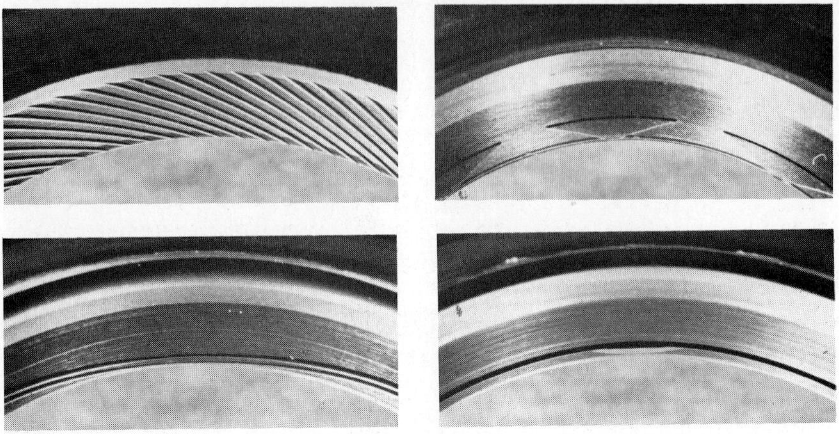

FIG. 6-70 (*a*) Helix seal—unidirectional. (*b*) Triangular seal—birotational. (*c*) Sine wave—birotational. (*d*) Combination pad and rib—birotational.

ments, since the ribs will sweep dust into the sump. A helix seal is shown in Fig. 6-70*a*. It will function only if the shaft rotates in a single direction. Seals for birotational operation have triangular pads (Fig. 6-70*b*), sine waves (Fig. 6-70*c*), and combinations of ribs and pads (Fig. 6-70*d*).

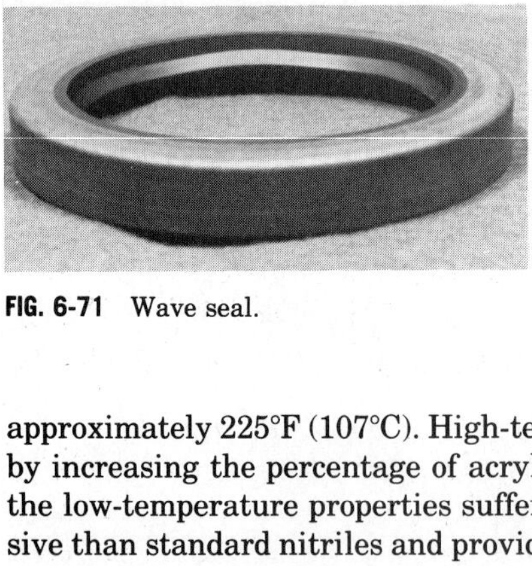

FIG. 6-71 Wave seal.

One seal type has the entire element formed in a sine-wave configuration. This provides a mild hydrodynamic (Fig. 6-71) action without ribs to create heat and pump dust.

Shaft Seal Materials Nitrile is a copolymer of butadiene and acrylonitrile. It is the most extensively used shaft seal material because of its good oil resistance, wear resistance, good low-temperature properties, and low cost. The upper temperature limit is approximately 225°F (107°C). High-temperature performance can be improved by increasing the percentage of acrylonitrile in the material. Unfortunately, the low-temperature properties suffer. Carboxylated nitriles are more expensive than standard nitriles and provide better wear resistance at high temperatures. The oil resistance is identical to that of standard nitriles, and low-temperature flexibility is lower. Nitriles are particularly suited for sealing hydrocarbon fluids at moderate temperatures. If low-molecular-weight fluids like gasoline or highly aromatic fluids have to be sealed, some sacrifice in low-temperature properties must be made. Nitriles can be used in EP-type oils at low or moderate temperatures. When sump temperatures get higher than 215°F (102°C), the additives in the oil (i.e., chlorinated paraffins and sulfonated olefins) will harden and crack the material. Nitriles are not recommended for use with low-molecular-weight aromatics such as benzene or polar aliphatics.

The polyacrylic polymers offer improvements over nitrile in high-temperature applications. Polyacrylic compounds function well at 275°F (135°C) in engine or transmission fluids. They resist EP-type additives and should replace nitriles in those applications where hardening and cracking of the nitrile is a problem. Polyacrylate is an ester and is vulnerable to water, acids, bases, and all types of polar solvents like ketones, esters, and so on. The major limitations of polyacrylates have been wear resistance and low-temperature properties. Many polyacrylate compounds will become brittle around 0 to 5°F (−18 to −15°C). New polyacrylate polymers are capable of handling temperatures down to −40°F (−40°C). When selecting a polyacrylate compound for an application, extreme care must be exercised if cold-temperature properties are important.

Silicone is an expensive material that will operate well from −100° to 300°F (−73 to 149°C) in typical engine and transmission oils. Silicones function in water, inorganic acids and bases, diesters, and non-petroleum-based brake fluids. They are not recommended for use in chlorinated or aromatic solvents, gasoline, and other such applications. They are also adversely affected by some EP additives, such as lead naphthanate and zinc dithiocarbamate. Oxidized lubricating oil will also cause the material to revert. Although silicones show fair wear resistance during normal operation, their dry-running characteristics are quite poor. It is recommended that silicone seals be presoaked in oil before installation to prevent dry running during initial break-in. Silicones also have poor tear characteristics, and special care must be exercised during installation to prevent lip damage.

There are two types of fluoroelastomers, which are quite different chemically. Fluorocarbons are used more extensively than fluorosilicone. Both materials are expensive and are applied in critical applications (high temperature, high speed). Fluorocarbons have excellent chemical and high-temperature resistance. The material is very tough and is a popular choice in heavy-duty applications. At low temperatures, the material becomes very stiff, and the sealing lip cannot follow the shaft. Leakage may occur. Fluorosilicone combines the chemical and high-temperature resistance of fluorocarbon with the excellent low-temperature behavior of silicone. Unfortunately, it also has the poor tear strength and wear characteristics of silicone.

Tetrafluoroethylene (TFE) has been used as a sealing material for years, but only within recent times has it been seriously considered for typical rotational shaft seal applications. TFE is resistant to virtually all fluids and has a temperature range from −120 to +400°F (−84 to +204°C). TFE is not an elastomeric material and will not follow shaft motion as well as other seal materials. At temperatures of 100°F (38°C) and below, the TFE material is stiff. The material cannot be easily bonded; thus most TFE seals are of the costly assembled type. The material has a high cost and is easily damaged. Extreme care must be exercised during installation.

Several other materials are used in specialized shaft seal applications. Urethanes are tough materials that are used when the utmost in wear resistance at lower temperatures is required. Butyl and ethylene propylene diene monomer (EPDM) are used to seal non-petroleum-based brake fluids and polar solvents. Epichlorohydrin is an oil-resistant material that lies between nitrile and polyacrylate in cost and properties. In general, these materials represent a small fraction of the shaft seal population because of processing difficulties that create cost penalties.

The chemical resistance of shaft seal materials to various fluids appears in Table 6-19. Even though chemical resistance is of extreme importance, other factors, such as cost, temperature range, and wear resistance, must be considered before the material is selected. These factors appear in Table 6-20. The final material choice is often a compromise that will satisfy several of the most important application parameters.

6-17 Sealing-System Requirements and Recommendations

The seal is only part of the sealing system. The efficiency and operating life is often affected by the condition of the two mating surfaces between which it is installed—the bore and the shaft. Table 6-21 provides the recommended bore sizes for a given shaft size.

Under normal conditions, the portion of the shaft contacted by the seal should be hardened to Rockwell C30 minimum to minimize shaft scoring. There is no conclusive evidence that additional hardening will increase the wear resistance of the shaft.

Two types of eccentricity affect seal performance. (1) Shaft-to-bore misalignment (STBM). This is the amount (in inches) that the center of the shaft is offset with respect to the center of the bore. STBM usually exists to some degree as a result of normal machining and assembly inaccuracies. To ensure proper seal performance, the STBM should be as small as possible. Too much STBM increases friction and creates abnormal wear on one side of the seal. The STBM is measured by attaching a dial indicator to the shaft and indicated off the seal bore as the shaft is slowly rotated. (2) Dynamic runout (DRO). This is the amount (in inches) that the sealing surface of the shaft does not rotate about the true center. DRO is caused by misalignment, bending of the shaft, lack of shaft balance, shaft lobing, and other manufacturing inaccuracies. It is measured by the total movement of an indicator held against the side of the shaft while the shaft is slowly rotated.

Sealing performance in eccentric conditions depends largely on the flexibility of the sealing element. As a general rule, typical spring-loaded seals will operate satisfactorily if the total eccentricity (combined indicator readings) does not exceed the maximums shown in Table 6-22.

Steel or stainless steel shafts are recommended for best sealing performance. Nickel-plated surfaces are acceptable. Brass, bronze, aluminum alloys, zinc, magnesium, and similar materials should not be used except under unusual circumstances. The level of interference between the sealing lip and the shaft surface must be precisely controlled to ensure optimum seal performance. The shaft diameter should be held within the tolerances shown in Table 6-23 unless the sealing requirements are not critical.

The corner of the shaft can damage the sealing lip during installation if it is not properly machined. A burr-free chamfer or radius is recommended (Fig. 6-72). The shaft finish is critical to the proper functioning of the lip seal. The

TABLE 6-19 Seal Element Selection Chart
(Chemical Resistance to Various Fluid Media)

Medium	Nitrile	Polyacrylate	Silicone	Fluoroelastomer		Fluorocarbon TFE
				Fluorosilicone	Fluorocarbon	
Engine oil	Good	Good	Good	Good	Good	Good
ATF-A	Fair	Good	Good	Good	Good	Good
Grease	Good	Fair	Fair	Fair	Good	Good
EP lube	Fair/poor	Good	Poor	Fair	Good	Poor
SAE90	Good	Good	Good	Good	Good	Good
Fuel oil	Good	Fair	Poor	Good	Good	Good
Kerosene	Good	Fair	Poor	Good	Good	Good
Gasoline	Good	Fair	Poor	Good	Good	Good
Petroleum-based hydraulic oil	Good	Good	Good	Good	Good	Good
Brake fluid	Poor	Poor	Poor	Fair	Fair	Good
Skydrol 500	Poor	Poor	Good	Poor	Poor	Good
MIL-L-7808	Fair	Poor	Good	Good	Good	Good
MIL-L-23699	Fair	Poor	Good	Good	Good	Good
MIL-L-6082-A	Good	Good	Good	Good	Good	Good
MIL-L-5606	Good	Good	Poor	Good	Good	Good
MIL-G-10924	Good	Good	Poor	Good	Good	Good
Butane	Good	Good	Fair	Good	Good	Good
Ketone	Poor	Poor	Poor	Poor	Poor	Good
Ammonium gas, cold	Good	Poor	Fair	Poor	Poor	Good
Fresh water	Good	Poor	Good	Good	Good	Good
Salt water	Good	Poor	Good	Good	Good	Good

TABLE 6-20 Seal Element Selection Chart

Material	Approximate relative cost (cpd.)	Approximate relative cost (seals)	Sump oil temperature °F	Sump oil temperature °C	Advantages	Disadvantages
Nitrile	100	100	−50 to +225	−45 to +107	Low cost, low swell. Good wear and oil resistance at moderate temperatures	Poor resistance to EP additives. Poor high-temperature resistance
Polyacrylic	250	115	−40 to +275	−40 to +135	Good oil resistance. Low swell. Generally resistant to EP additives	Fair wear properties. Poor dry running. Poor water resistance
Silicone	625	130	−30 to +300	−34 to +149	Very broad temperature range.	Poor dry running properties. Poor resistance to oxidized oil and some EP additives. Poor tear characteristics

Fluoroelastomer (fluorocarbon)	2250	200	−40 to +350	−40 to +177	Excellent oil and chemical resistance. Good wear properties. Low swell	Becomes stiff at low temperatures. Poor followability at low temperatures
Fluoroelastomer	2250	200	−80 to +300	−62 to +149	Excellent oil and chemical resistance. Good low-temperature properties	Poor tear strength. Poor dry running. Poor wear resistance
Fluorocarbon (TFE)		300	−140 to +400	−95 to +204	Excellent oil and chemical resistance. Excellent temperature range	Easily damaged. Becomes stiff at low temperature. High cost. Poor followability at low temperature

TABLE 6-21 RMA Oil Seal Standard Sizes (in)
(Single- and Dual-Lip Spring-Loaded Bonded Seals)

Standard bore sizes		Standard shaft sizes																		
		in																		
		0.500	0.625	0.750	0.875	1.000	1.125	1.250	1.375	1.500	1.625	1.750	1.875	2.000	2.125	2.250	2.375	2.500	2.625	2.750
		mm																		
in	mm	12.7	15.9	19.05	22.2	25.4	28.6	31.8	34.9	38.1	41.3	44.5	47.6	50.8	54.0	57.2	60.3	63.5	66.7	69.9
0.999	25.4	×	—	—	—	—	—	—	—	—	—	—	—	—	—	—	—	—	—	—
—																				
1.124	28.6	×	×	—	—	—	—	—	—	—	—	—	—	—	—	—	—	—	—	—
1.250	31.8	×	×	×	—	—	—	—	—	—	—	—	—	—	—	—	—	—	—	—
1.375	35.0	—	×	×	×	—	—	—	—	—	—	—	—	—	—	—	—	—	—	—
1.499	38.1	—	×	×	×	×	—	—	—	—	—	—	—	—	—	—	—	—	—	—
1.624	41.3	—	—	×	×	×	×	—	—	—	—	—	—	—	—	—	—	—	—	—
1.752	44.5	—	—	—	×	×	×	×	—	—	—	—	—	—	—	—	—	—	—	—
1.874	47.6	—	—	—	—	×	×	×	—	—	—	—	—	—	—	—	—	—	—	—
2.000	50.8	—	—	—	—	—	×	×	×	—	—	—	—	—	—	—	—	—	—	—
2.125	54.0	—	—	—	—	—	—	×	×	×	—	—	—	—	—	—	—	—	—	—
2.250	57.2	—	—	—	—	—	—	—	×	×	×	—	—	—	—	—	—	—	—	—
2.374	60.3	—	—	—	—	—	—	—	×	×	×	×	—	—	—	—	—	—	—	—
2.502	63.6	—	—	—	—	—	—	—	—	×	×	×	—	—	—	—	—	—	—	—
2.623	66.6	—	—	—	—	—	—	—	—	—	×	×	×	×	—	—	—	—	—	—
2.750	69.9	—	—	—	—	—	—	—	—	—	—	×	×	×	×	—	—	—	—	—
2.875	73.0	—	—	—	—	—	—	—	—	—	—	—	×	×	×	—	—	—	—	—
3.000	76.2	—	—	—	—	—	—	—	—	—	—	—	×	×	×	×	—	—	—	—
3.125	79.4	—	—	—	—	—	—	—	—	—	—	—	×	×	×	×	×	—	—	—
3.251	82.6	—	—	—	—	—	—	—	—	—	—	—	—	—	×	×	×	×	—	—
3.371	85.6	—	—	—	—	—	—	—	—	—	—	—	—	—	—	×	×	×	×	—
3.500	88.9	—	—	—	—	—	—	—	—	—	—	—	—	—	—	—	×	×	×	×
3.623	92.0	—	—	—	—	—	—	—	—	—	—	—	—	—	—	—	—	×	×	×

Standard bore sizes		in																			
		2.625	2.750	2.875	3.000	3.125	3.250	3.375	3.500	3.625	3.750	3.875	4.000	4.250	4.500	4.750	5.000	5.250	5.500	5.750	6.000
		mm																			
in	mm	66.7	69.9	73.0	76.2	79.4	82.6	85.7	88.9	92.0	95.3	98.4	101.6	107.9	114.3	120.0	127.0	133.4	139.7	146.0	152.4
3.623	92.0	×	×	×	—	—	—	—	—	—	—	—	—	—	—	—	—	—	—	—	—
3.751	95.3	×	×	×	×	—	—	—	—	—	—	—	—	—	—	—	—	—	—	—	—
3.875	98.4	—	×	×	×	—	—	—	—	—	—	—	—	—	—	—	—	—	—	—	—
4.003	101.7	—	—	×	×	—	—	—	—	—	—	—	—	—	—	—	—	—	—	—	—
4.125	104.8	—	—	—	×	×	—	—	—	—	—	—	—	—	—	—	—	—	—	—	—
4.249	107.9	—	—	—	—	×	×	×	—	—	—	—	—	—	—	—	—	—	—	—	—
4.376	111.2	—	—	—	—	×	×	×	×	—	—	—	—	—	—	—	—	—	—	—	—
4.500	114.3	—	—	—	—	×	×	×	×	—	—	—	—	—	—	—	—	—	—	—	—
4.626	117.5	—	—	—	—	—	×	×	×	×	×	—	—	—	—	—	—	—	—	—	—
4.751	120.7	—	—	—	—	—	—	—	×	×	×	—	—	—	—	—	—	—	—	—	—
4.876	123.9	—	—	—	—	—	—	—	—	×	×	×	—	—	—	—	—	—	—	—	—
4.999	127.0	—	—	—	—	—	—	—	—	×	×	×	×	—	—	—	—	—	—	—	—
5.125	130.2	—	—	—	—	—	—	—	—	—	—	×	×	—	—	—	—	—	—	—	—
5.251	133.4	—	—	—	—	—	—	—	—	—	—	×	×	×	—	—	—	—	—	—	—
5.375	136.5	—	—	—	—	—	—	—	—	—	—	—	×	×	—	—	—	—	—	—	—
5.501	139.7	—	—	—	—	—	—	—	—	—	—	—	—	×	×	—	—	—	—	—	—
5.625	142.9	—	—	—	—	—	—	—	—	—	—	—	—	×	×	—	—	—	—	—	—
5.751	146.1		—	—	—	—		—			—	—	—	—	×	×	—	—	—	—	—
6.000	152.4	—	—	—	—	—	—	—	—	—	—	—	—	—	—	×	×	—	—	—	—
6.250	158.8	—	—	—	—	—	—	—	—	—	—	—	—	—	—	—	×	×	—	—	—
6.375	161.9	—	—	—	—	—	—	—	—	—	—	—	—	—	—	—	×	×	—	—	—
6.500	165.1	—	—	—	—	—	—	—	—	—	—	—	—	—	—	—	—	×	×	—	—
6.625	168.3	—	—	—	—	—	—	—	—	—	—	—	—	—	—	—	—	×	×	—	—
6.750	171.5	—	—	—	—	—	—	—	—	—	—	—			—	—	—	—	×	×	
6.875	174.6	—	—	—	—	—	—		—	—	—	—	—	—	—	—		—	×	×	—
7.000	177.8	—	—	—	—	—	—	—	—	—	—	—	—	—	—	—	—	—	—	×	—
7.125	180.9	—	—	—	—	—	—	—	—	—	—	—	—	—	—	—	—	—	—	×	×
7.500	190.5	—	—	—	—	—	—	—	—	—	—	—	—	—	—	—	—	—	—	—	×

Courtesy: Rubber Manufacturers Association (RMA).

TABLE 6-22 Shaft Eccentricity Chart

Maximum total eccentricity (STBM plus DRO)		Shaft speed, r/min
In	mm	
0.025	0.635	100
0.020	0.508	200
0.018	0.457	500
0.015	0.381	1000
0.013	0.330	1500
0.010	0.254	2000
0.009	0.229	2500
0.008	0.203	3000
0.007	0.178	4000

TABLE 6-23 Shaft Diameter Tolerance Chart

Nominal shaft diameter		Shaft tolerance	
in	mm	in	mm
Up to and including 4.000	Up to and including 101.6	±0.003	±0.08
4.001–6.000	101.63–152.4	±0.004	±0.10
6.001 and up	152.43–254	±0.005	±0.13

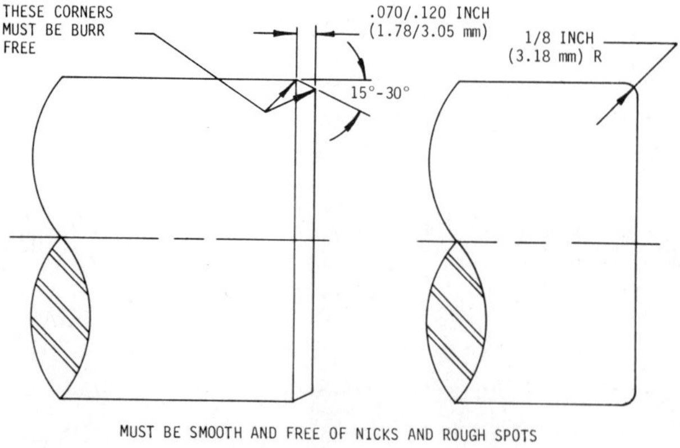

FIG. 6-72 Shaft lead corners.

shaft surface must be smooth enough to provide continuous contact between the sealing lip and the shaft surface. It must also be rough enough to provide pockets for lubricant. This lubricant reduces the coefficient of friction between the seal lip and the shaft and provides coolant for the interface. The shaft surface must not cause excessive seal wear or be uneconomical to produce.

Machine lead must be avoided, since the spiral marks on the shaft may cause lip damage and pump oil out of the sump.

In many applications, wear sleeves or rings of mild steel are pressed over shafts of cast iron or other soft materials. They permit easy replacement of the sealing surface and should be replaced whenever the seal is changed.

The best method of providing all requirements is to plunge-grind the shaft to a finish of 10 to 20 μin (0.25 to 0.50 μm) with no machine lead. Shafts should be ground with mixed-number rotational speed ratios. The grinding wheel should be allowed to spark out to prevent machine lead.

Ferrous materials are most commonly used in the housings that form seal bores. The conventional seal case material is steel; thus thermal expansion is not usually a problem unless other materials (such as aluminum) are used for the housing bore.

Differential thermal expansion rates can cause a sealing problem at the bore, particularly for large-diameter applications. When this problem exists, it can be solved by using special seals with the case material identical to the bore. This usually results in a cost penalty. A steel mating ring pressed into the bore is an alternative solution. In some cases, rubber OD seals are recommended.

To ensure proper sealing at the bore, the proper press fit between the seal OD and the bore must be maintained. The tolerances and press-fit requirements for ferrous bores appear in Table 6-24.

If the bore is stepped, the depth of the bore should exceed the seal width by at least 1⁄64 in (0.397 mm). No specific bore hardness is recommended. It need only be high enough to maintain interference with the seal OD.

Whenever lubricant pressure is present at the outside diameter of the seal, a bore finish of 125 μin (3.15 μm) or less is recommended to prevent bore leakage. If this is not possible, coatings which will effectively fill minor bore imperfections can be applied to the seal OD. If the bore roughness is extreme and tool marks exist, rubber OD seals are sometimes used.

The leading or entering edge of the bore should be chamfered, as shown in Fig. 6-73, to prevent seal damage during installation. The inside corner of the bore should have a maximum radius of 3⁄64 in (1.190 mm).

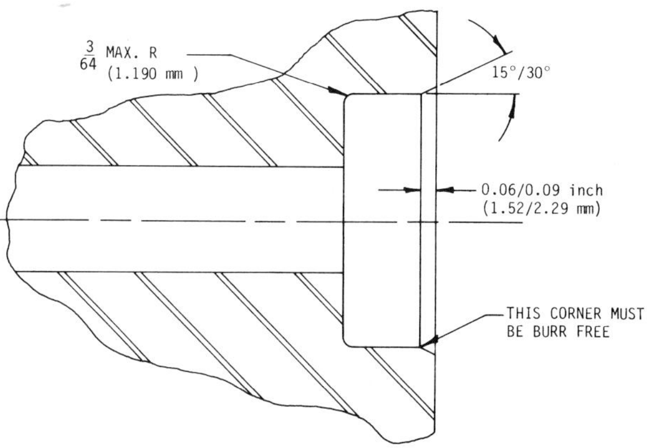

FIG. 6-73 Recommended bore lead corner.

TABLE 6-24 RMA Oil Seal Standard Tolerance
(Metal Outside Diameter Seals)

Bore diameter		Bore tolerance		Nominal press fit		OD* tolerance		Out of round	
in	mm	in	mm	in	mm	in	mm	in	mm
Up to 1.000	Up to 25.4	±0.001	±0.025	0.004	0.102	±0.002	±0.051	0.005	0.127
1.001– 3.000	25.43– 76.2	±0.001	±0.025	0.004	0.102	±0.002	±0.051	0.006	0.152
3.001– 4.000	76.23–101.6	±0.0015	±0.038	0.005	0.127	±0.002	±0.051	0.007	0.178
4.001– 6.000	101.62–152.4	±0.0015	±0.038	0.005	0.127	+0.003/−0.002	+0.076/−0.051	0.009	0.229
6.001– 8.000	152.42–203.2	±0.002	±0.051	0.006	0.152	+0.003/−0.002	+0.076/−0.051	0.012	0.306
8.001– 9.000	203.23–228.6	±0.002	±0.051	0.007	0.178	+0.004/−0.002	+0.102/−0.051	0.015	0.381
9.001–10.000	228.63–254	±0.002	±0.051	0.008	0.203	+0.004/−0.002	+0.102/−0.051	0.015	0.381
10.001–20.000	254.03–508	+0.002/−0.004	+0.051/−0.102	0.008	0.203	+0.006/−0.002	+0.152/−0.051	0.002	0.051
20.001–40.000	508.03–1016	+0.002/−0.006	+0.051/−0.152	0.008	0.203	+0.008/−0.002	+0.203/−0.051	in/in of	mm/mm of
40.001–60.000	1018.03–1524	+0.002/−0.010	+0.051/−0.254	0.008	0.203	+0.010/0.002	+0.254/−0.051	seal OD	seal OD

* The average of a minimum of three measurements to be taken at equally spaced positions.

Preventing Sealing-System Failures Sealing failures can result when one or more deviations from accepted practice are permitted. In many cases, the seal is blamed for leakage when some other component in the sealing system is the culprit. It is difficult to predict or obtain the optimum life of a lip seal, but it is necessary to define and pinpoint reasons for early or premature failure.

Improper installation is one of the main causes of premature leakage. An elastomeric lip seal is a precision mechanical component; it must be assembled properly if the reliability expected for the application is to be obtained. In high-volume production applications, to assemble the seal is desirable. These tools will minimize assembly variables that may affect sealing efficiency.

Installation Procedure The following installation procedure should be followed to ensure proper seal life.

1. Check dimensions. If the shaft and bore dimensions do not match those specified for the selected seal, leakage is more likely to occur. The seal should be replaced with one that has the proper shaft size and outside diameter.
2. Check seal. The seal should be examined for damage that may have occurred prior to installation. A sealing lip that is turned back or nicked will leak. The seal outside diameter should be checked to ensure that there are no dents, scores, or cuts. If there is any damage, the seal should be replaced. Never reuse a seal. Always use a new seal in the application.
3. Check housing bore. The entering edge must be deburred to prevent damage to the seal outside diameter. A chamfer or rounded corner should be provided whenever possible.
4. Check shaft. Remove surface nicks, burrs, and grooves that may damage the sealing lip. Examine the shaft for machine lead. Remove burrs or

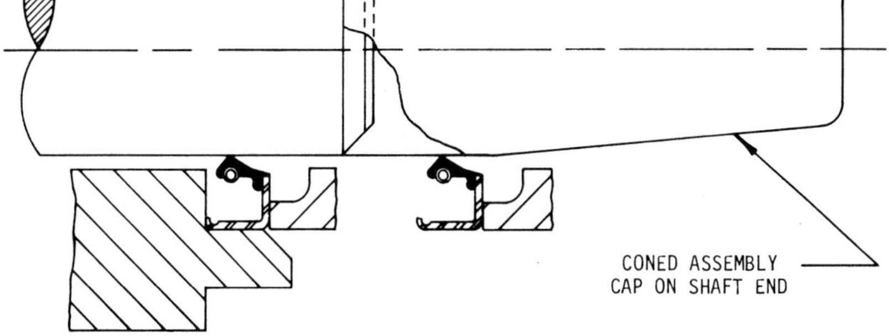

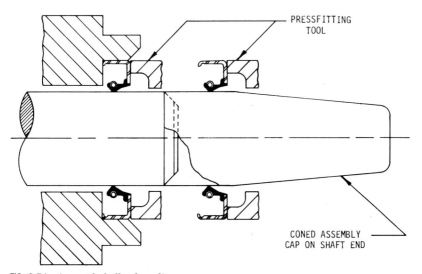

FIG. 6-74 Assembly bullet for splines.

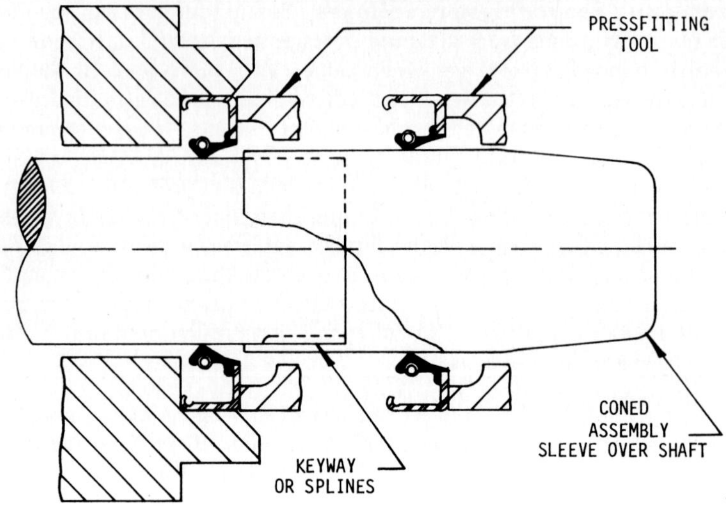

FIG. 6-75 Assembly bullet.

sharp edges from the shaft end. The shaft end should be chamfered. If shaft burrs cannot be removed, a coned assembly tool must be used (Fig. 6-74).

5. Check splines and keyways. When installing a seal over splines and keyways, an assembly sleeve should be used (Fig. 6-75).
6. Check seal lip direction. Make sure that the new seal faces the same direction as the original. Generally, the lip faces the lubricant or fluid to

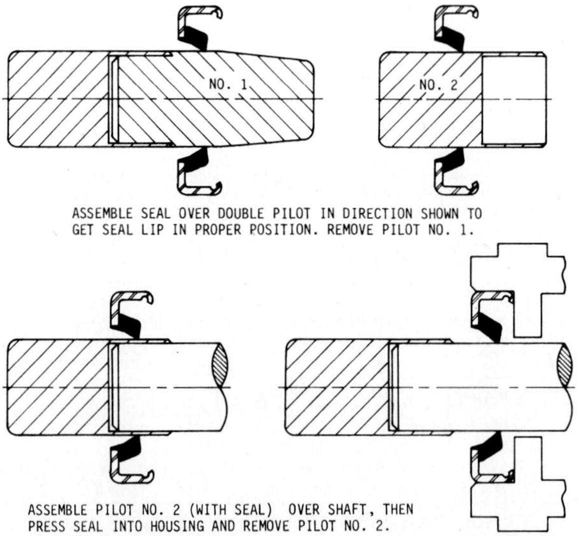

FIG. 6-76 Piloting tools.

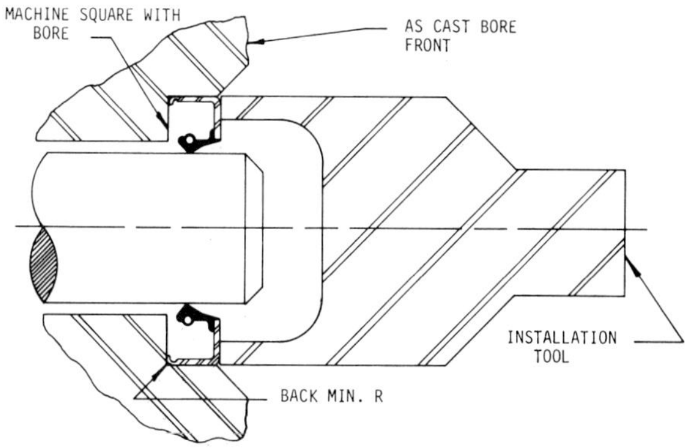

FIG. 6-77 Bottom seal in bore.

be sealed. When the shaft is installed against the lip, use piloting tools shown in Fig. 6-76.

7. Prelubricate the sealing element. Carefully wipe the seal lip with lubricant immediately before installation.
8. Use correct installation tools. Press-fitting tools should have an outside diameter 0.010 in smaller than the bore diameter. Apply pressure only at the seal outside diameter to prevent seal distortion.
9. Use proper driving force. An arbor press is recommended, but a soft-faced mallet can be used with the installation tool. Never hammer directly on the seal face.
10. Bottom out the tool or seal. To avoid cocking the seal in the bore, the installation tool must be designed to bottom the seal in the bore (Fig. 6-77), against the shaft (Fig. 6-78), or against the bore face (Fig. 6-79). If the shaft has been used before, the seal should be positioned to prevent the seal lip from running in the old wear track.
11. Check for parts interference. After the installation is complete, check for other machine parts that may rub against the seal and cause friction and heat.

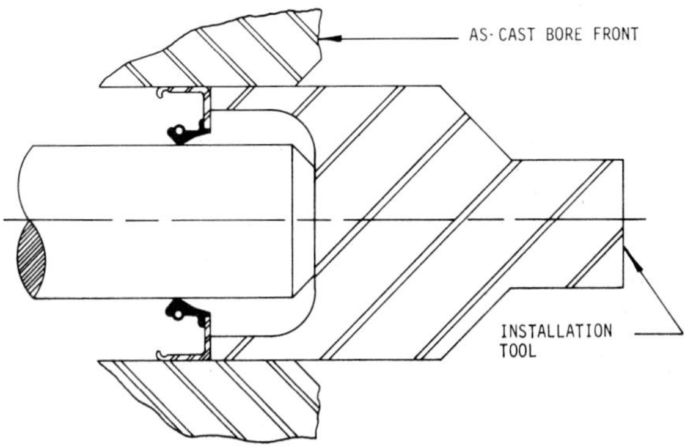

FIG. 6-78 Bottom seal against shaft face.

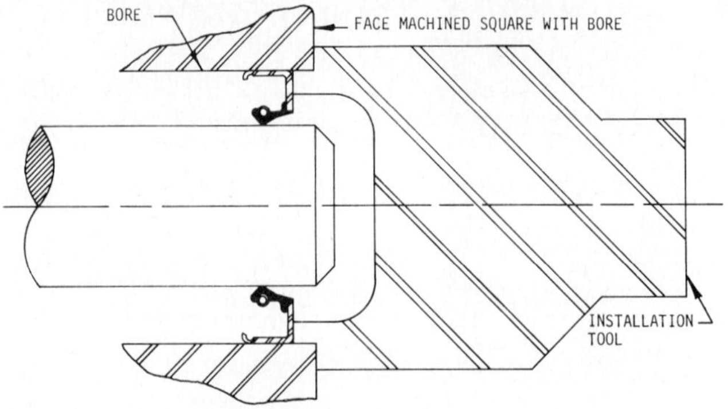

FIG. 6-79 Bottom seal against bore face.

After Installation Perfect installations of seals are sometimes ruined by mishandling of the sealing region during normal maintenance. When machinery is painted, the seal area should be masked to avoid getting paint on the lip or the shaft where the lip rides. If the sump is vented to prevent pressure buildup, mask the vents to prevent clogging. If the paint is to be baked or the mechanism exposed to other outside heat sources, care must be taken to keep the seal temperature below material limitations. When cleaning or testing, do not allow cleaning fluids to contact seals. When testing or breaking in machinery, do not subject seals to conditions that exceed design recommendations. Seal damage may not be evident until much later.

The most important factor in the performance of a lip seal is its ability to maintain a radial force between the sealing element and the shaft. Anything that affects this sealing force will shorten seal life and reduce effectiveness. Excessive heat or lubricant additives may harden or soften the rubber lip and affect seal performance. Rapid wear of the sealing lip can be caused by shaft irregularities, lack of lubrication, and excessive temperatures. Typical radial-seal leakage problems, possible causes, and suggested cures appear in Table 6-25. This table is provided as a guide only and cannot be considered all-inclusive.

DYNAMIC SEALS—NONCONTACTING SEALS FOR ROTATING SHAFTS

6-18 Introduction to Noncontacting Seals

Noncontacting seals have a small clearance gap between the rotating shaft and the stationary housing. Direct contact of the mating parts does not occur; therefore, frictional wear is eliminated. The seals are durable, reliable, simple in design, and easy to maintain. The design principle allows for some leakage to occur. The pressure differential between the internal sump and the external atmosphere is maintained by throttling the leaking fluid with the seal. The leakage can be minimized by controlling the seal design and the magnitude of the clearance gap. A bushing seal has a constant clearance gap. Leakage is minimized by reducing the gap clearance or increasing the bushing length. A labyrinth seal consists of a series of small bushings joined together. Dynamic leakage is virtually eliminated when screw threads on the shaft or housing pump the fluid back into the sump. Some noncontacting seals use magnetic components. A magnetic fluid is placed and held in the clearance gap. The ferrofluidic seal effectively eliminates leakage.

6-19 Bushings

Fixed-Bushing Seals A fixed-bushing seal consists of a close-fitting stationary sleeve or pipe that fits within a housing. The shaft rotates within the bushing (Fig. 6-80).

Fluid seeps through the narrow gap that is formed. The leakage rate of an incompressible fluid flowing in the laminar region through a perfectly concentric and aligned bushing with a small clearance can be estimated with Eq. (6-6). If the center of the bushing does not coincide with the housing center (Fig. 6-81), then Eq. (6-7) must be used to account for eccentricity.

$$Q_L = \frac{\pi R h^3 g}{6\nu\rho}\frac{\Delta P}{L} \tag{6-6}$$

$$Q_{Le} = Q_L(1 + 3/2 N^2) \tag{6-7}$$

where Q_L = laminar volumetric flow rate, cm^3/s
R = mean radius of annulus, cm
h = mean radial clearance, cm
ν = kinematic viscosity, cm^2/s
ρ = fluid density, g/cm^3
p = pressure drop, g/cm^2
N = eccentricity/radial clearance = e/h
g = gravitation constant, 980 cm/s^2

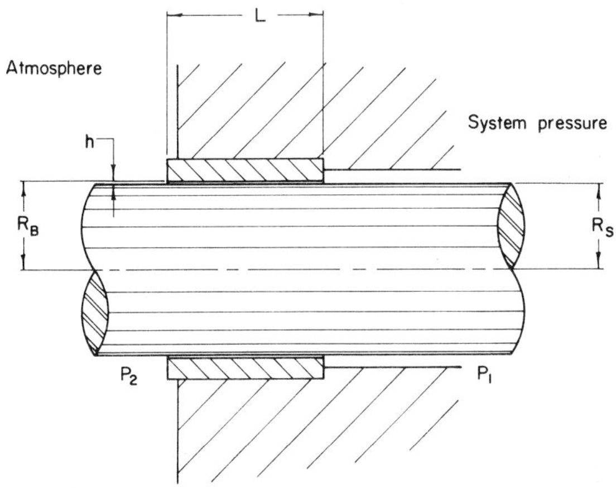

FIG. 6-80 Fixed bushing seal.

TABLE 6-25 Troubleshooting Lip Seals

Symptom	Possible cause	Corrective action
1. Early lip leakage	(*a*) Nicks, tears, or cuts in seal lip	Examine shaft. Eliminate burrs and sharp edges. Use correct mounting tools to protect seal lip from splines, keyways, or sharp shoulders. Handle seals with care. Keep seals packaged in storage and in transit
	(*b*) Rough shaft	Finish shaft to 10–20 μin rms or smoother
	(*c*) Scratches or nicks on surface	Protect shaft after finishing
	(*d*) Lead on shaft	Plunge-grind shaft surface
	(*e*) Excessive shaft whip or runout	Locate seal close to bearings. Ensure good, accurate machining practices
	(*f*) Cocked seal	Use correct mounting tools and procedures
	(*g*) Paint on shaft or seal element	Mask seal and adjacent shaft before painting
	(*h*) Turned-under lip	Check shaft chamfer for roughness. Machine chamfer to 32 μin or smoother, blend into shaft surface. Check shaft chamfer for steepness. Use recommended lead chamfer. Use correct mounting tools and procedures.

	(*i*) Damaged or "popped out" spring	Use correct mounting tools and procedures to apply press-fit force uniformly. Protect seals in storage and transit
	(*j*) Damaged or distorted case	Use correct mounting tools and procedures to apply press-fit force uniformly. Protect seals in storage and transit
	(*k*) OD sealant on shaft or lip element	Use care in applying OD sealant. Purchase precoated seals
2. Lip leakage, intermediate life	(*a*) Excessive lip wear	Check seal cavity for excessive pressure. Provide vents to reduce pressure. Provide proper lubrication for seal. Check shaft finish. Make sure finish is 10–20 μin
	(*b*) Element hardening and cracking	Reduce sump temperature, if possible. Upgrade seal material. Provide proper lubrication for seal. Change oil frequently. Change seal design.
	(*c*) Element corrosion and reversion	Check material-lubricant compatibility. Change material
	(*d*) Excessive shaft wear	Check shaft hardness. Harden to Rockwell C30 minimum. Change oil frequently to remove contaminants. Use dust lip in dirty atmosphere
3. OD leakage	(*a*) Scored seal OD	Check housing machining. Use 125 μin rms. Check edges on housing bore. Use recommended chamfer. Remove burrs.
	(*b*) Damaged seal case	Use correct mounting tools and procedure to apply press fit uniformly. Protect seals in storage and transit

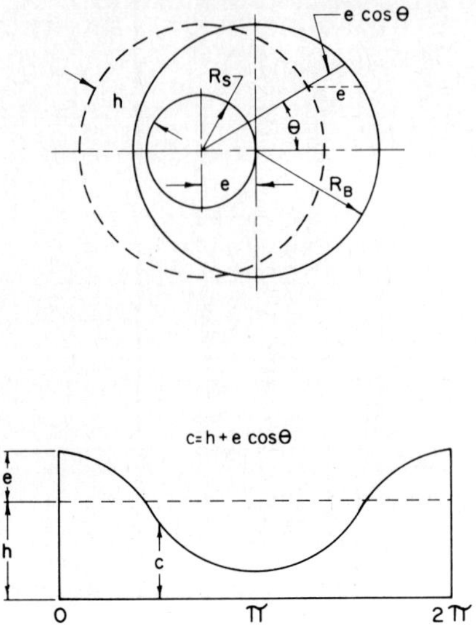

FIG. 6-81 Eccentric fixed bushing.

If the flow rate becomes large, turbulence will result. Equation (6-8) will then apply for the concentric bushing, and Eq. (6-9) must be used for the eccentric case.

$$Q_T = 2\pi R \left(\frac{1}{0.0665} \frac{h^3 gP}{\nu^{1/4}\rho L}\right)^{4/7} \tag{6-8}$$

$$Q_{Te} = 1.315 Q_T \tag{6-9}$$

The leakage rate is reduced as the radial clearance h and the pressure drop from sump to atmosphere are reduced. Leakage rate is also reduced as the bushing length and the fluid kinematic viscosity and density increase.

Leakage rates can be estimated from Table 6-26 and Figs. 6-82 and 6-83.

The geometric and fluid factor F, Eq. (6-10), is calculated and the Reynolds number, Re, Eq. (6-11), is determined from Fig. 6-83. The leakage flow velocity

TABLE 6-26 Leakage Rates

Liquid	°C	Specific weight, g/cm³
Alcohol	20	0.800
Fresh water	15	1.000
Sea water	15	1.024
Lubricating oil	15	0.881–10.946
Glycerin	0	1.260
Fuel oil	15	0.897–1.026

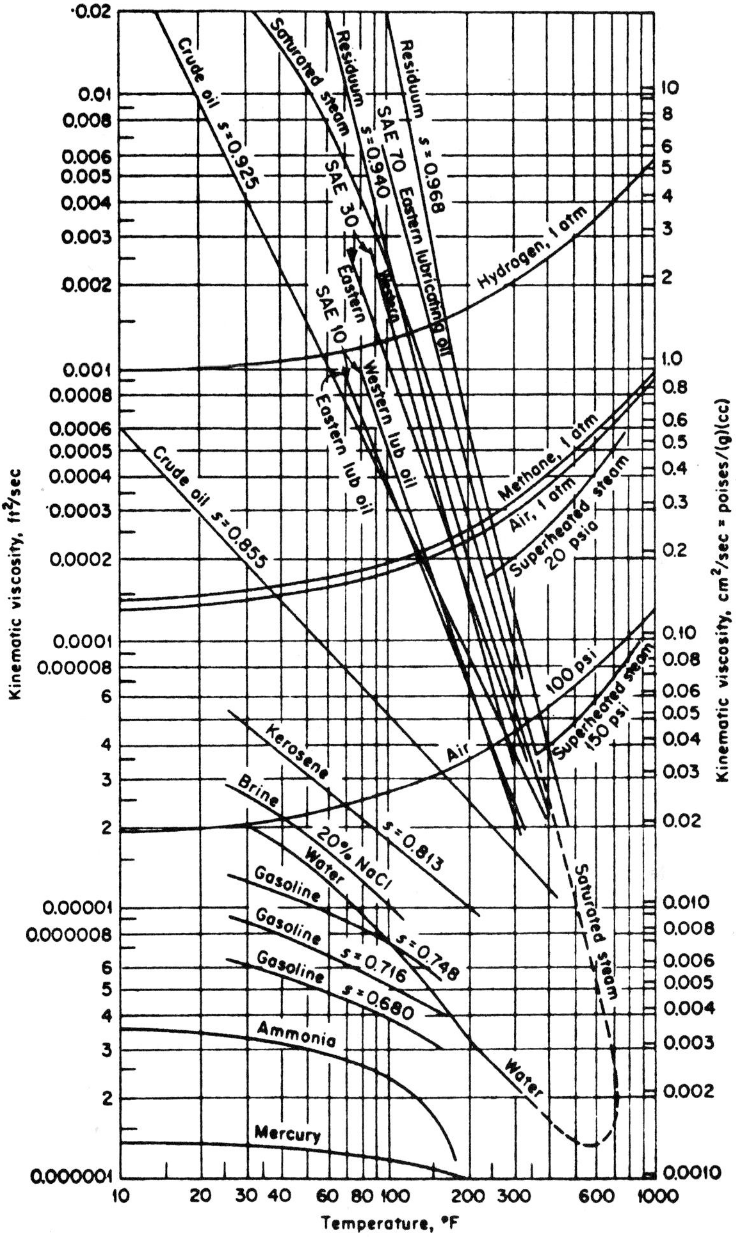

FIG. 6-82 Kinematic viscosity of fluids. (*Courtesy: McGraw-Hill Book Co.*)

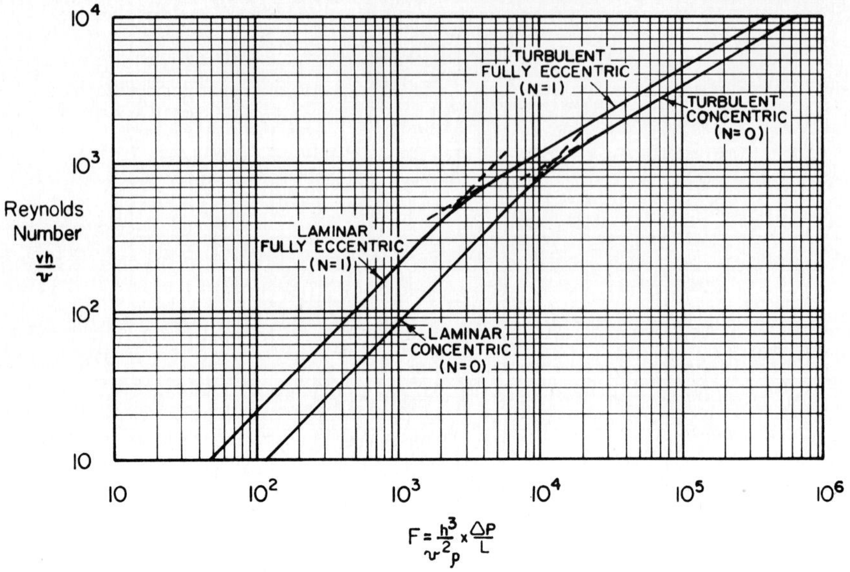

FIG. 6-83

V is calculated with Eq. (6-12), and the volumetric flow rate is obtained from Eq. (6-13).

$$F = \frac{h^3 g \Delta p}{V^2 \rho L} \tag{6-10}$$

$$\mathrm{Re} = \frac{Vh}{\nu} \tag{6-11}$$

$$V = \frac{\nu \mathrm{Re}}{h} \qquad \mathrm{cm/s} \tag{6-12}$$

$$Q = 2\pi RhV \qquad \mathrm{cm^3/s} \tag{6-13}$$

In most cases, the fixed-bushing seal is used for sealing liquids. The viscosity of gases is too low to provide enough friction to keep leakage within tolerable limits.

Since fixed-bushing seals are firmly attached to the housing, the clearance must be large enough to prevent rubbing during operation. Rubbing can occur as a result of vibration, shaft bowing, shaft deflection, thermal growth, shaft-to-bore misalignment, shaft eccentricity, and bore eccentricity. With inadequate clearance, rubbing can induce excessive bearing loads, excessive heat, serious vibration, and seizure. Serious damage to machinery can result.

If large clearances and thus high leakage rates can be tolerated, fixed-bushing seals offer a simple, low-cost, easy to install and maintain solution to sealing problems.

Floating-Bushing and Ring Seals Operating clearances with fixed bushings may often be too large to obtain desired leakage rates. Floating

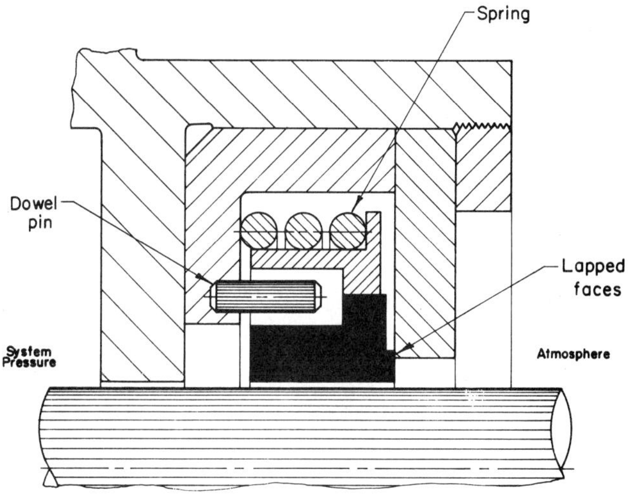

FIG. 6-84 Floating bushing seal.

bushings with smaller clearances can be used to seal gases and reduce leakage (Fig. 6-84). Rotation is prevented with a dowel pin, and the bushing is free to move radially. A helical spring is used to provide a positive axial force to load the bushing face against the housing face to prevent static leakage.

In some cases, multiple bushings or rings are used (Fig. 6-85). The length of each ring in the tandem arrangement is less than the length that would be required if a single bushing was used. This design allows larger shaft misalignments without affecting seal performance.

Instead of using solid rings, it is also customary to design a floating-ring seal as a segmented archbound ring, where the segments are held together by a garter spring (Fig. 6-86). The garter spring provides the contact force re-

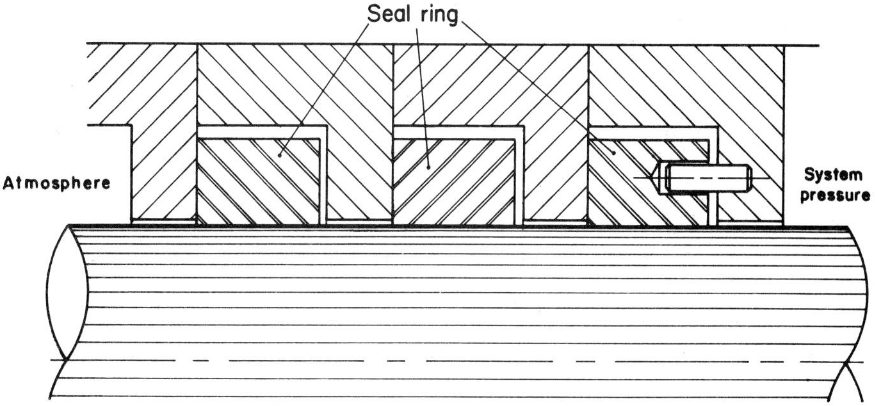

FIG. 6-85 Multistage floating bushing.

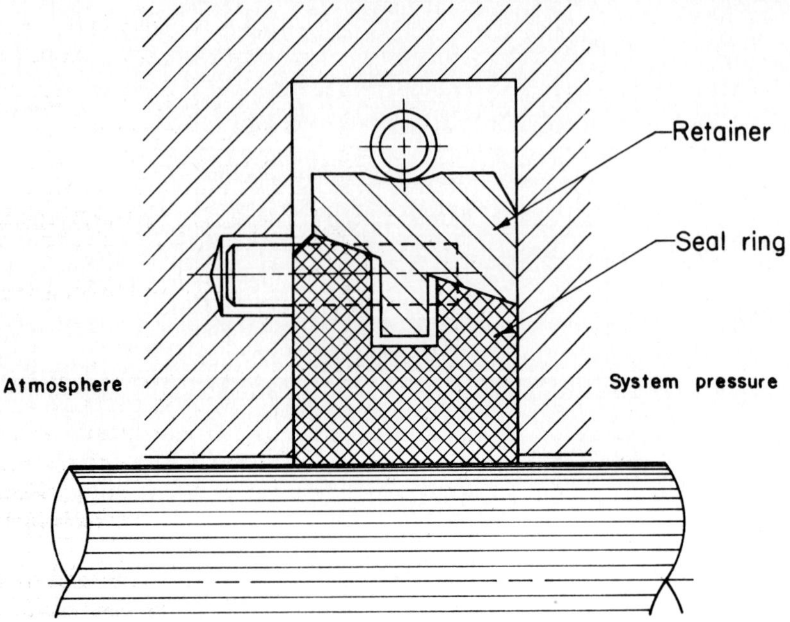

FIG. 6-86 Segmented floating ring seal.

quired to establish a seal with the shaft surface. The metal retainer is not always necessary.

In some cases, the bushing seal can be balanced by providing a step in the sleeve (Fig. 6-87). The step size can be adjusted to provide the desired sealing force at the bushing/housing interface.

The calculation of leakage rates for gases must compensate for density changes in the clearance space. If the flow is steady, laminar, and isothermal and the bushing and shaft are concentric, Eq. (6-14) can be used to estimate leakage rates. If the ring and shaft are eccentric, Eq. (6-15) must be used.

$$G_L = \frac{\pi R h^3 g}{12 \nu L} \left(\frac{P_i^2 - P_o^2}{P_s} \right) \frac{T_s}{T_i} \tag{6-14}$$

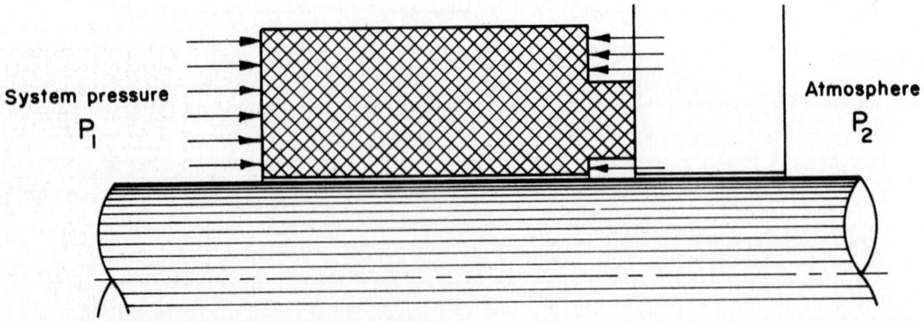

FIG. 6-87 Balanced floating bushing seal.

$$G_{Le} = G_L(1 + 3/2 N^2) \tag{6-15}$$

where G_L = leakage rate of laminar compressible concentric fluid, g/s
G_{Le} = leakage rate of laminar compressible eccentric fluid, g/s
P_i = upstream pressure, g/cm·s^2
P_o = downstream pressure, g/cm·s^2
T_s = standard temperature, K
T_i = upstream temperature, K

Calculations become more complicated if the flow is turbulent. The concentric leakage rate is given by Eq. (6-16). The fully eccentric case is approximately 1.5 times more than the concentric case (see Table 6-27).

$$G_T = 2\pi Rh\alpha\sqrt{gP_1\rho_i} \tag{6-16}$$

where ρ_i = gas density at the upstream condition, g/cm^3
α = flow resistance coefficient

The flow resistance coefficient α is a function of seal geometry, pressure ratio, and shaft speed. The following procedure is used to determine α. The velocity ratio β is calculated from Eqs. (6-17) through (6-19). The multiring factor m is found in Fig. 6-88. The ring resistance factor R_r is determined with Eq. (6-20), and the pressure ratio γ is defined by Eq. (6-21). The values of γ, R_r, and β are used with Fig. 6-89 to determine the flow resistance coefficient. The highest value of β given on the chart is 0.1. If the calculated value of β exceeds 0.1, the data presented for β = 0.1 should provide a satisfactory approximation for α.

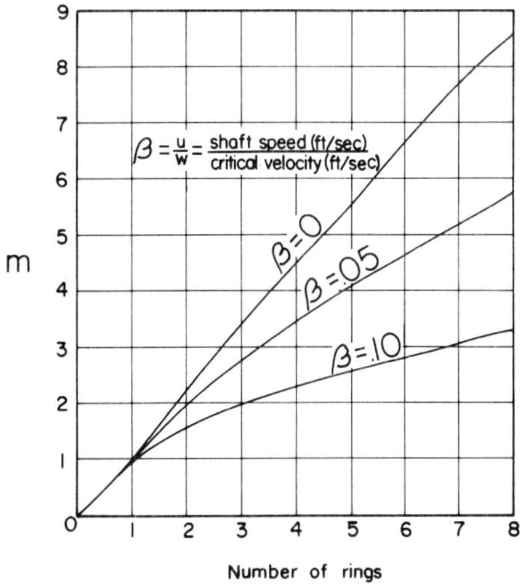

FIG. 6-88 Determination of m for multigland seal rings.

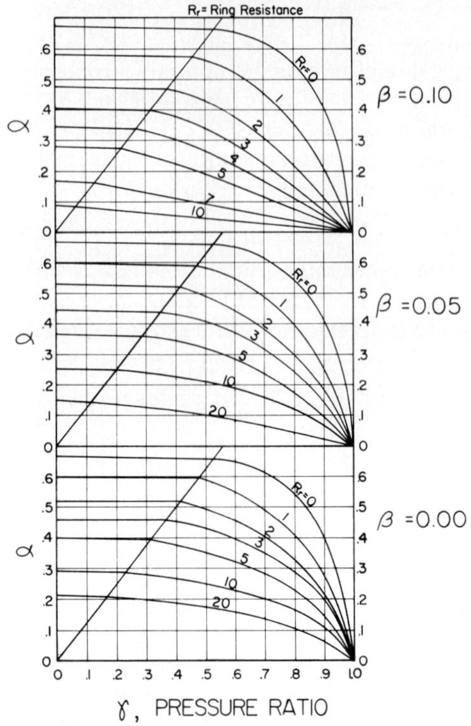

FIG. 6-89

$$V_c = 2g\,\frac{k}{kT_i}\,\frac{P_i}{\rho_i} \qquad (6\text{-}17)$$

$$V_s = 2\pi RN \qquad (6\text{-}18)$$

$$\beta = \frac{V_s}{V_c} \qquad (6\text{-}19)$$

$$R_r = m\left(\frac{w}{h}\right) \qquad (6\text{-}20)$$

$$\gamma = \frac{P_o}{P_i} \qquad (6\text{-}21)$$

where w = ring width, in
k = exponent of adiabatic expansion
V_c = critical fluid velocity, cm/s
V_s = linear shaft velocity, cm/s
N = shaft rotational speed, r/s
γ = fluid resistance coefficient (0.5 for steam and 0.02 for air)

For fixed bushings, the best material is an antifriction type, such as carbon-graphite, molybdenum disulfide, or a combination of both with PTFE resins or

TABLE 6-27 Bushing Equations

	Incompressible	Compressible
Laminar-flow leakage rate (concentric)	$Q_L = \dfrac{\pi R h^3 g \Delta P}{6\nu\rho L}$ in^3/s	$G_L = \dfrac{\pi R h^3 g}{12\nu L}\left(\dfrac{P_i^2 - P_o^2}{P_s}\right)\dfrac{T_s}{T_i}$
Laminar-flow leakage rate (eccentric)	$Q_e = Q_L(1 + 3/2 N^2)$	$G_{Le} = G_L(1 + 3/2 N^2)$
Turbulent-flow leakage rate (concentric)	$Q_T = 2\pi R\left(\dfrac{1 h^3 g \Delta P}{0.0665\nu^{1/4}\rho L}\right)^{4/7}$	$G_T = 2\pi R h \alpha \sqrt{g P_1 \rho_i}$
Turbulent-flow leakage rate (eccentric)	$Q_{Te} \approx 1.315 Q_T$	$G_{Te} \approx 1.5 G_T$

soft metals used as liners. Where temperatures permit, babbitt is a good candidate. For higher temperatures, bronzes and aluminum alloys are well suited. Any material chosen must be compatible with the system fluid.

Floating-bushing and ring seals are subject to the same basic requirements. Differences in thermal expansion are essential and should be carefully evaluated. With carbon-graphite as a major candidate, it is important to consider the low coefficient of expansion compared to the steel of the shaft. By shrinking it into a metallic retainer, the problem can be minimized. Plasma coatings have proven favorable for high-temperature applications.

The system fluid is an important factor and must be carefully evaluated. It must be clean and should not tend to polymerize or crystallize. When water is used, it should be "soft." High surface velocities may produce erosion effects along the bushing surface. A materials selection chart appears in Table 6-28.

6-20 Labyrinth Seals

Labyrinth seals consist of a series of circumferential strips of material that may extend from the shaft or the bore to form a series of annular orifices. Labyrinths are used primarily to seal gases in compressors and steam turbines. They have leakage rates that are higher than those of bushings or face seals, but they are simple, reliable, and maintenance-free. There are a multitude of labyrinth designs. The simplest is the straight-through design (Fig. 6-90). High fluid velocities are generated at the throat of each constriction, and the kinetic energy is dissipated by turbulence in the chamber beyond each throat. There is some velocity carryover with a straight labyrinth. This results in efficiency losses. The labyrinth can be stepped (Fig. 6-91) or staggered (Fig. 6-92) to cause the expanding jet of gas to impinge upon a solid transversal surface. This minimizes the velocity carryover losses. The housing of the staggered labyrinth must be split to allow for assembly. Combinations of labyrinth types are also used (Fig. 6-93). The combination labyrinth shown uses a buffered inlet to prevent flow of the internal gas to the atmosphere. The barrier

TABLE 6-28 Floating-Bushing, Ring-Seal Materials and Environment Combinations

Environment	Seal ring material	Shaft material
Oil	Babbitt Bronzes Aluminum Carbon graphite	Hardened steel Shafting Chrome plate Nitrided steels
Water	Bronzes 416 stainless steel Stellite Carbon graphite Ceramics	440-C Chrome-plated Chrome plate
Gas (Air, CO_2), H_2, He, N_2, O_2)	Carbon graphite	Tool steels, hardened Tungsten carbide plate Ceramic plate Chrome carbide Chrome plate Stainless steel (300) Stainless steels (400) Stainless steels ≈ 50 Rockwell C

fluid is maintained at a pressure higher than that of the process gas. Another variation of the buffered system (Fig. 6-94) uses an aspirating port at the outlet. The outlet pressure is generally equal to or less than the pressure at either end.

The leakage of steam through a labyrinth may be approximated by Eq. (6-22).

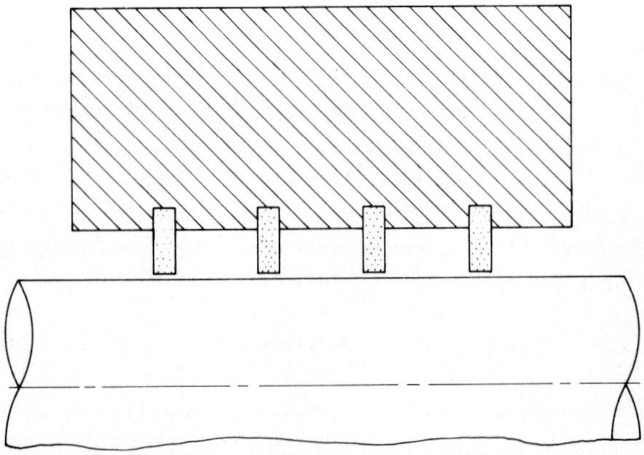

FIG. 6-90 Straight labyrinth.

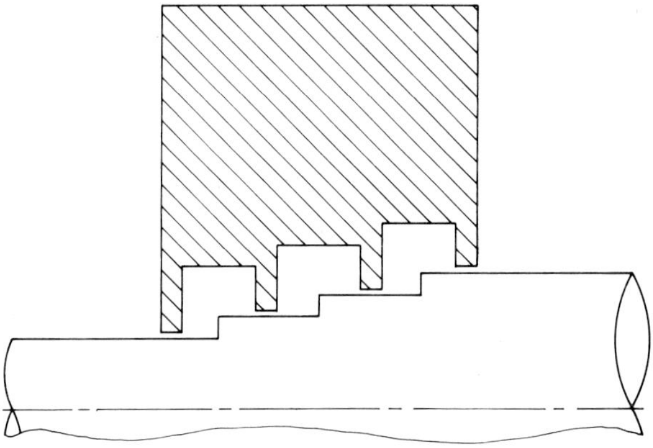

FIG. 6-91 Stepped labyrinth.

$$G = 25KA \sqrt{\frac{P_i}{V_i}\left[1 - \left(\frac{P_o}{P_i}\right)^2\right] \Big/ \left(N - \ln \frac{P_o}{P_i}\right)} \qquad (6\text{-}22)$$

where G = flow rate of steam, lb/h
A = area through packing clearing space, in^2
P_i = upstream pressure, lb/in^2 (absolute)
V_i = initial specific volume of steam, ft^3/lb
P_o = final pressure, lb/in^2 (absolute)
N = number of stages in labyrinth
K = experimentally determined flow coefficient

The coefficient K varies for different types of seal teeth. For simple tooth forms—straight through the inclined teeth—K varies from 60 to 120, with the simplest form yielding the highest coefficient. For staggered tooth forms, K will vary from 30 to 65, with deep high-low forms giving the lowest coefficient.

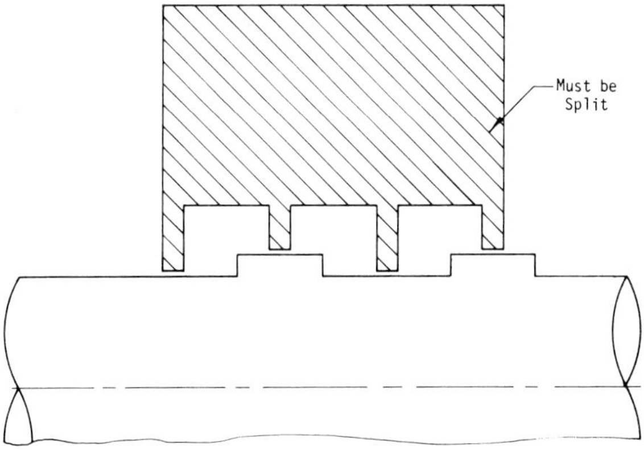

FIG. 6-92 Staggered labyrinth.

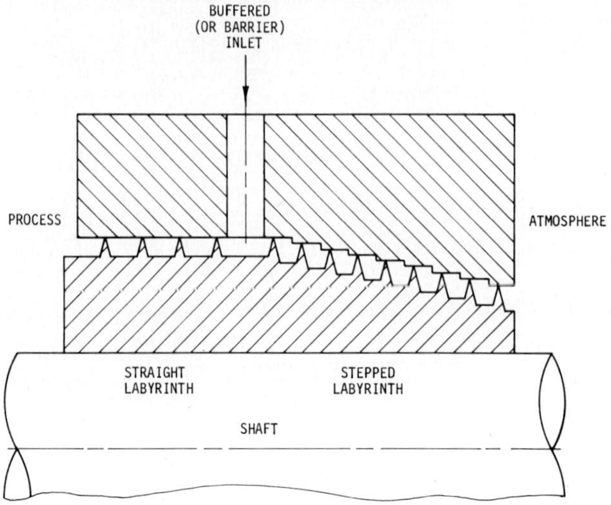

FIG. 6-93 Combination labyrinth.

When designing a labyrinth system, the fluid density, fluid temperature, flow direction, inlet pressure, outlet pressure, amount of leakage that can be tolerated, and available space must all be considered. The simplest forms should be considered first.

The radial clearance at which a labyrinth can operate is probably the most important single factor and the one that most directly affects the effectiveness of the seal; therefore, the clearance selected is usually based on allowing a very slight rub (interference) under the worst conditions. If possible, line-to-line is recommended. In figuring the minimum clearance, machining tolerances on parts together with the fits with control concentricity must be taken into account for static assembly conditions. In addition, however, the effect of differential expansion and of deflection of the parts as a result of the forces during operation must be considered. One must remember that no rotating

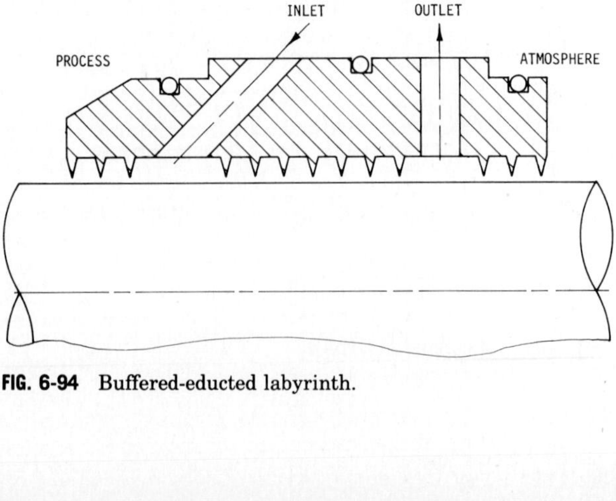

FIG. 6-94 Buffered-educted labyrinth.

shaft is free from wobble or movement in some orbit, and that very often at least one critical speed must be gone through. Prior experience usually determines the amount of clearance which a particular machine will produce in rotating about a center other than the shaft center. One must also consider that the amount of clearance required increases as the span between bearings increases. There is really no merit in providing a very small static clearance when it will be worn to a relatively large value on the first run. In some instances, the labyrinth is designed to "run in." Experience dictates this allowance, but in any case it will be a function of the rubbing speed and the materials used. As an example, aluminum or bronze seal strips or knives running against a steel surface will usually tolerate several thousandths of an inch wear without causing appreciable damage either to the shaft or to the strips or knives.

After the clearance is established, the number of strips or knives must be considered. This selection will be influenced by the allowable leakage and space considerations. Concurrently, the proportions of the strips of knives must be considered. No hard-and-fast rule concerning knife form, knife pitch, or knife depth can be given. As a general guide, one might say that the tip of the knife or strip usually has a width of 0.005 to 0.015 in and that the included angle on the knife or strip is 8° to 12°; the pitch will be in the range of 0.150 to 0.220 in, and the depth of the knife or strip will approximate the pitch. In aircraft service, the number of throttlings for a single pass usually does not exceed four. If more throttlings are desired, this is accomplished in two or more tiers. In steam turbine applications, the end seals may have as many as 30 throttlings at lesser pitch.

The materials used in the rubbing surfaces are a very important consideration. The material combination is chosen on the basis of rubbing qualities and the ability to stand up under the particular atmospheres and temperature existing in the service. With reference to wearing qualities, it is preferable that the end of the tooth remain narrow and clean-cut so as to obtain the minimum flow coefficient. Materials that crumble and break away cleanly are desirable. Materials that mushroom are not desirable, since the flow coefficient, and consequently the leakage, increases because of the rounding at the orifices. Materials that ball up as a result of melting and air hardening are undesirable because they will tear or gouge out a larger clearance than would be expected from normal wear, and thus will cause large leakage. For low-temperature work a soft brass may work well. In steam, which can be considered a protective atmosphere, a leaded nickel-tin bronze is very satisfactory and may be used at temperatures up to 950°F. In the presence of oxygen or air, however, it will oxidize rapidly, and a limiting temperature of 400 to 450°F is indicated. 25 Chrome–20 nickel stainless steel appears to work reasonably well, although it does have a high coefficient of friction. Some of the aluminum bronzes seem to have both wear resistance and oxidation resistance at high temperature. Aluminum is a likely candidate for some applications, but it should not be anodized, since aluminum oxide is formed, and this may score the adjacent part.

Stationary seal elements that have a greater coefficient of expansion than the casing (e.g., rolled in strips), should be split into segments with joint

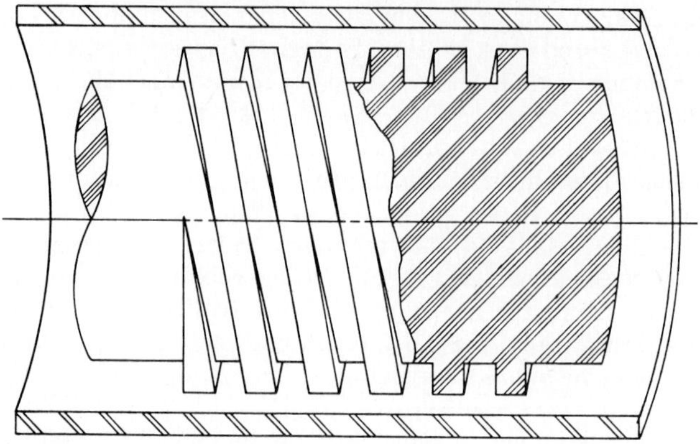

FIG. 6-95 Screw threads machined on shaft.

clearance to allow for minimum radial clearance under all conditions. A method of reducing clearance is to flexibly mount a labyrinth sleeve to permit radial motion relative to the fixed housing. Care must be exercised and pressure balancing considered to assure minimum frictional restraint due to axial forces induced by high pressure. Another method is to use soft or abradable material. In this category, aircraft use silver plate, steel honeycomb, fiber metal, and abradable plasma sprays. Industrially, carbon-graphite labyrinths have found some acceptance. Generally, in steam turbines the clearances are adjusted so that rubbing is not problematical.

6-21 Visco Seals

Visco seals have screw threads machined on the shaft (Fig. 6-95) or into the housing (Fig. 6-96). They are used primarily to screw liquids from the outside toward the interior of a sump. This pumping prevents fluid leakage.

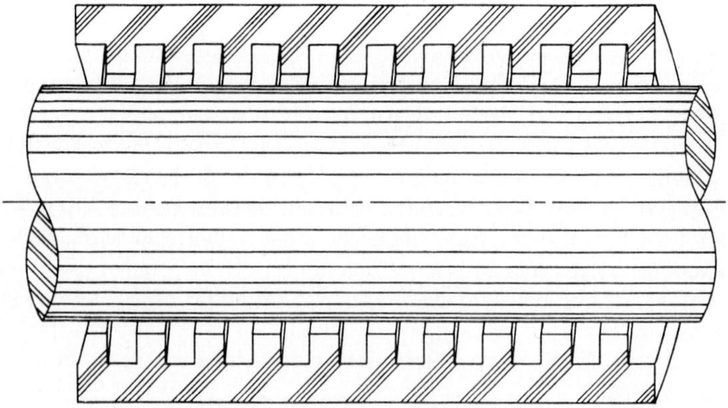

FIG. 6-96 Screw threads machined on housing.

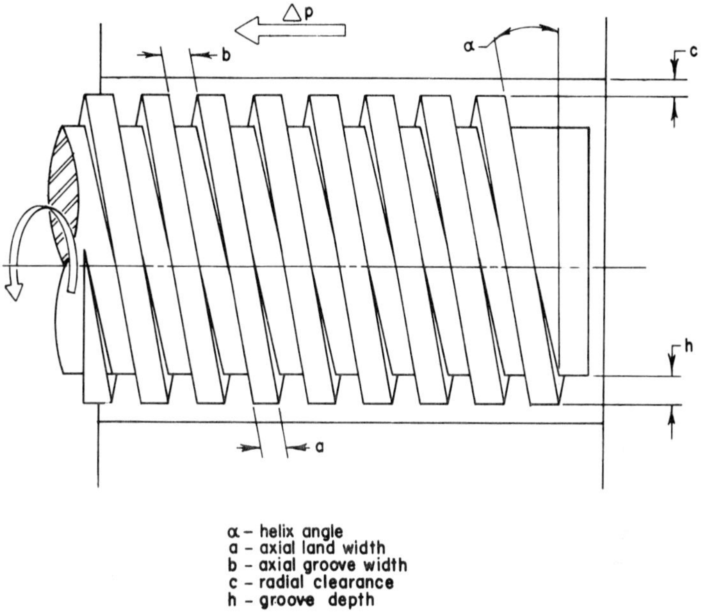

FIG. 6-97 Geometrical factors.

Visco seals are designed to function properly only when the shaft operates at a certain minimum rotational speed. Thus, for very low or zero shaft rotation, a secondary sealing device must be provided. Several designs for industrial visco seals utilize the centrifugal force during rotation to keep built-in lips from establishing contact with the shaft, but once the shaft stops rotating, the lips contact the shaft and establish the necessary seal. This provides a satisfactory seal as long as temperature is not a problem for elastomeric lip materials.

There is no contact between the rotating and the stationary components of the seal; therefore, visco seals have long reliable life.

The key design parameters appear in Fig. 6-97. The optimum sealing performance for laminar flow occurs when the helix angle α is 15 to 20°, γ is 0.5, and β is 3.6 to 4.1. The parameter γ is defined to be the ratio of groove width b to land and groove width at b. The parameter β is the ratio of groove depth h and radial clearance c to the radial clearance, as shown in Eqs. (6-23) and (6-24).

$$\gamma = \frac{b}{a + b} \tag{6-23}$$

$$\beta = \frac{h + c}{c} \tag{6-24}$$

These rules do not apply if the flow becomes turbulent. Helix angles of 10 to 15°, a γ of 0.62, and a β of 4.1 to 6.5 give good results in the turbulent region.

TABLE 6-29 Summary of Physical Properties of Ferrofluids

Carrier fluid	Saturation magnitization, gauss	Viscosity, cP, 27°C	Evaporation rate, g/s·cm², 240°C	Density, g/mL	Thermal conductivity, mW/m·K	Initial susceptibility	Pour point, °C	Permeability	Electrical resistivity, Ω·cm
Ester	450	450	1.39×10^{-6}	1.490	209	2.10	−28	1.0–1.4	1.60×10^{9}
Ester	450	450	1.07×10^{-5}	1.410	168	2.20	−44	1.0–1.4	0.27×10^{9}
Synthetic petroleum	300	120	3.69×10^{-6}	1.195	170	1.49	−54	1.0–1.3	0.94×10^{9}
Petroleum	200	200	3.17×10^{-6}	1.080	—	1.14	−51	1.0–1.2	1.32×10^{9}
Ester	300	75	5.95×10^{-6}	1.258	185	1.37	−54	1.0–1.3	11.30×10^{7}
Ester	450	100	1.27×10^{-5}	1.440	186	2.06	−63	1.0–1.4	9.30×10^{7}
Fluorocarbon	300	3500	2.24×10^{-6}	2.245	117	1.36	−27	1.0–1.2	13.20×10^{9}
Polyphenyl ether	450	4500	9.92×10^{-7}	1.665	—	2.13	−12	1.0–1.4	0.18×10^{9}
Synthetic petroleum	200	1000	6.94×10^{-6}	1.125	158	0.67	−38	1.0–1.1	9.90×10^{9}

Source: Ferrofluidics Corporation, Nashua, N.H.

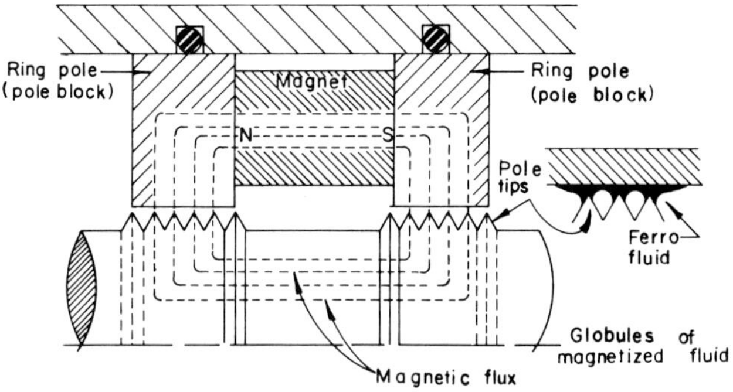

FIG. 6-98 Principle of magnetic seal.

6-22 Ferrofluidic Seals

Ferrofluidic seals use a magnetic fluid to seal the clearance gap in a labyrinth-type seal. This magnetic fluid prevents any leakage from occurring and is thus ideally suited for applications where any leakage is intolerable.

A ferrofluid consists of a carrier liquid that contains ultramicroscopic particles of a magnetic solid, such as magnetite. They are colloidally suspended, then stabilized by physiochemical means. To prevent flocculation even under the influence of a magnetic field, the particles are coated, and random collisions (brownian motion) with the molecules of the carrier liquid keep the particles in colloidal suspension for an indefinite period of time.

At the present time, all commercially available ferrofluids are electrically nonconductive in carrier liquids, such as fluorocarbons, hydrocarbons, polyphenyl ether, and aqueous solutions. Chemically and mechanically, the magnetic fluid offers the same characteristics as those provided by the carrier liquid in which the magnetic particles are colloidally suspended. Physical properties of typical ferrofluids appear in Table 6-29.

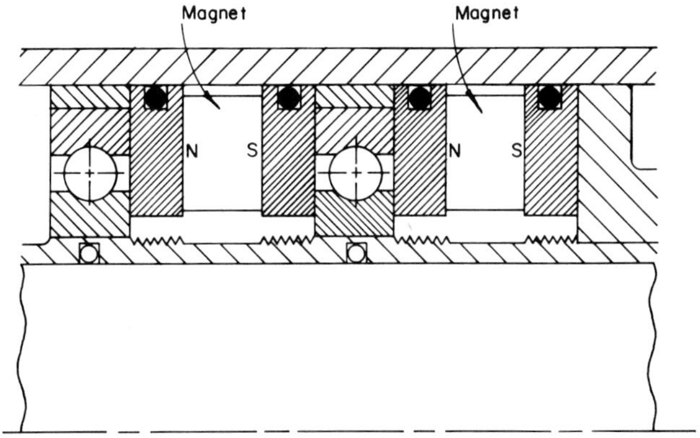

FIG. 6-99 Double seal arrangement.

TABLE 6-30 Noncontacting Seals

Problem	Causes	Corrective action
Excessive leakage (all seals leak at all times)	Excessive radial clearance	Decrease radial clearance
	Excessive pressure	Improve or add vent
	Excessive lubricant fill	Decrease fill or increase cavity
	Excessive temperature	Decrease fill or increase cavity
	Excessive vibration	Increase lube cavity
	Excessive end play	Increase lube cavity
Excessive leakage (some seals leak at all times)	Excessive radial clearance	Improve quality control
	Radial contact	Decrease misalignment
	Bypass leakage	Fill scratches and voids
Excessive leakage (all seals leak after a period of time)	Loss of lubricant viscosity	Improve lubricant
	Excessive relubrication	Decrease relubrication
	Contaminant ingress	Add contaminant shield
Water ingress	Static leakage	Add static seal
	Partial vacuum	Improve or add vent
	No lube leakage	Increase lube fill
	Excessive wet environment	Add water shield
Dirt ingress	Excessive radial clearance	Decrease radial clearance
	Excessively dirty environment	Add dirt seal
	Partial vacuum	Improve or add vent
	No lube leakage	Increase lube fill

The operating principle of magnetic seals is shown in Fig. 6-98. A ring magnet is placed over the shaft contacting a ring pole on either side. With this arrangement, a magnetic field is produced; it is enforced in its effect by placing thread-type serrations either on the shaft surface or on the ID of the ring poles. The magnetic flux path is complete through the clearance gap between the ring poles and the shaft is filled. The fluid bridges the passageway, blocking any trace of leakage flow. The shaft is free to rotate without frictional disturbances by the ferrofluid. Without solid mechanical contact, no wear is possible, and the friction generated in the fluid film is negligible. Even at extreme rotational shaft speeds, maintenance is not required.

Manufacturers claim successful operations at speeds of the order of 10,000 r/min in gaseous atmosphere under pressure or vacuum without leakage. Clearance gaps are typically 0.002 to 0.005 in. Temperatures must be maintained at 225°F or below to prevent sufficient carrier fluid evaporation. A typical double-seal arrangement is shown in Fig. 6-99. At the present time, the magnetic seal is applied to equipment sealing gases only.

6-23 Troubleshooting Noncontacting Seals

Table 6-30 may be helpful when correcting leakage that occurs with noncontacting seals.

7
HOSE FITTINGS

JOSEPH F. BRIGGS
Aeroquip Corporation/Jackson, Michigan

7-1 The Hose Connection

Hose fittings are a key mechanical component in modern design, and choosing the correct type is very important. The designer must realize the full life expectancy of the hose, as well as selecting the proper fitting.

There are many different ways of attaching a fitting to a hose. This section deals with the many methods of hose attachment and proper selection of fittings. Some factors in selecting a hose fitting are the material requirements. Hose fittings are normally available in the following materials:

Cadmium-plated steel
Chrome-dipped
Stainless steel
Brass
Bronze
Monel
Aluminum

Additionally, this subject involves means and methods for attaching fittings to hose. This consists of squeezing, clamping, pressing, and compressing the hose wall between two metal parts, whose recesses, irregular surfaces, serrations, barbs, and so on are forced around and into the inner and outer hose walls. To accomplish this, there are five basic types of hose fittings:

1. Nipple and band-clamp fittings
2. Crimped fittings
3. Swaged fittings
4. Detachable, reusable fittings
5. Built-in fittings

All of these employ a shank, stem, or nipple as part of the fitting. See Fig. 7-1.

The nipple is inserted into the bore of the hose, and the wall of the hose is compressed onto the nipple by various methods. Which method is chosen depends on various factors, which will be discussed along with the description of each method.

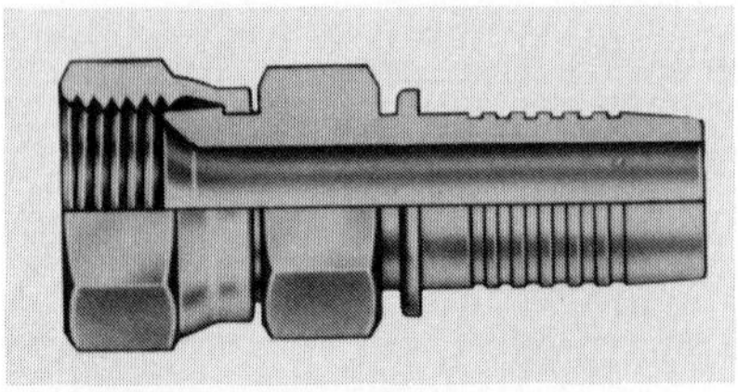

FIG. 7-1 Typical hose fitting.

7-2 Band Clamps

The band clamp works on a compression sealing principle. This is a low-pressure connection. It has been in use for many years and may be the oldest type of piping hose seal. It consists of a band clamp tightened over a rubber hose which has been installed

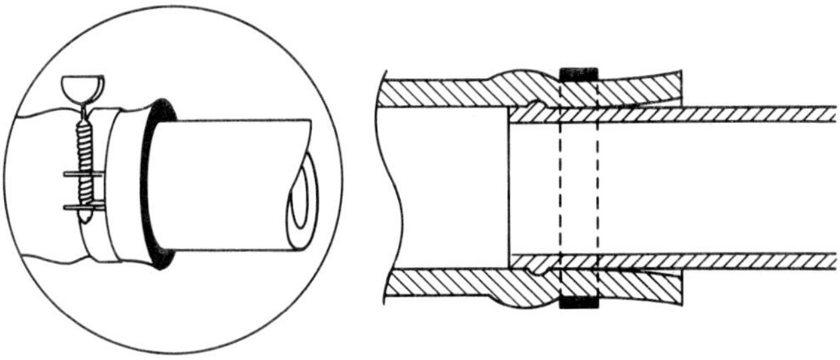

FIG. 7-2 Common band clamp.

over a barbed or beaded nipple, as shown in Fig. 7-2. Applications range from air lines and radiator lines to suction and return lines in some hydraulic systems.

7-3 Worm-Gear Type

Probably the most common band clamp is the worm-gear type (Fig. 7-3). Very high compression-to-torque ratios can be achieved with this design.

7-4 Wire Clamps

A common clamp found under the hood of many automobiles is the wire-spring clamp (Fig. 7-4). It is a semi-self-adjusting clamp, since the wire is spring steel. Other styles of wire clamps which can be applied with special tooling are available.

7-5 Thumb Screw

These clamps are ideal for clamping lightweight hoses, and each clamp will cover a large range of diameters. They all have a minimum diameter capability of from 1⁄4 in (6 mm) up to as high as 28 1⁄8 in (714 mm), depending on the

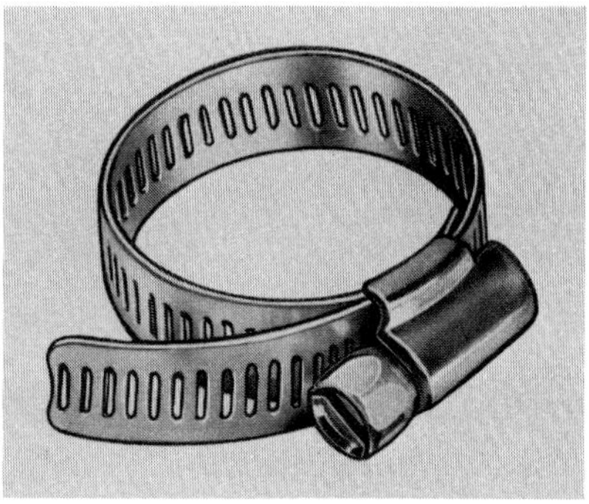

FIG. 7-3 Worm-gear-style band clamp.

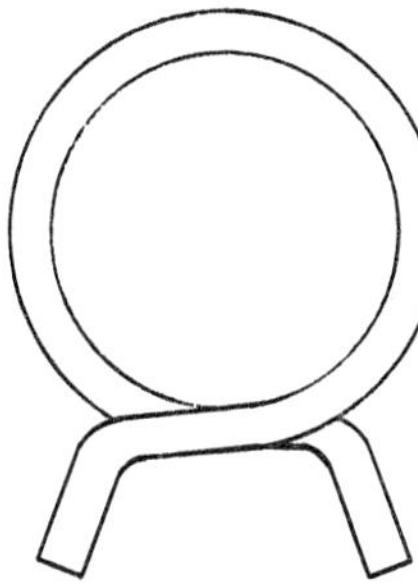

FIG. 7-4 Wire-spring clamp.

part number. See Fig. 7-5. For example, the smallest clamp offered ranges from 0.25-in (6 mm) to 2.125-in (54-mm) diameters. Thumbscrew widths vary from 2.25 in (57.2 mm) to 1.8 in (45.7 mm). This style of clamp is available in three different strengths: light-duty, up to 320 lb (145.2 kg); medium-duty, up to 630 lb (294.8 kg); and heavy-duty, up to 1400 lb (635 kg).

7-6 Heavy-Duty Band Clamps

The heavy-duty band clamp is designed to provide high-performance clamping on heavy-duty applications. Use of this style of clamp can eliminate the need for double clamping to secure heavy rubber hose connections wherever high reliability is required. Refer to Fig. 7-6. Sizes range from 1.38 in (35 mm) to 8.10 in (205.7 mm). The T bolt is curved for some sizes. These clamps are designed for specific diameters.

7-7 Light-Duty Band Clamps—Single Bolt

These light-duty clamps are for use on applications of specific diameters. They can be used for low-pressure applications and thin-walled hose, Fig. 7-7. They are also available in double-bolt configurations for heavy-duty applications.

7-8 Clinched Clamps

These clamps are offered in various materials, sizes, and strengths. The strapping may be carried in bulk lengths along with appropriate buckles, and applied with a special tool that tensions the strap, clinches the buckle, and cuts the strap (see Fig. 7-8). Where desirable, a buckle with a set screw that is reusable may be used. The strap is tensioned with a special tool. The set screw is tightened with a hex wrench. Some buckles are clinched onto the strap by center punching, using a special tool (Fig. 7-9).

7-9 Advantages and Disadvantages of Clamps

One of the problems that may be encountered when using clamps to secure the fitting in the hose is loss of compression force due to cold flow. (Rubber, when compressed, will deform, and over a period of time will assume its deformed

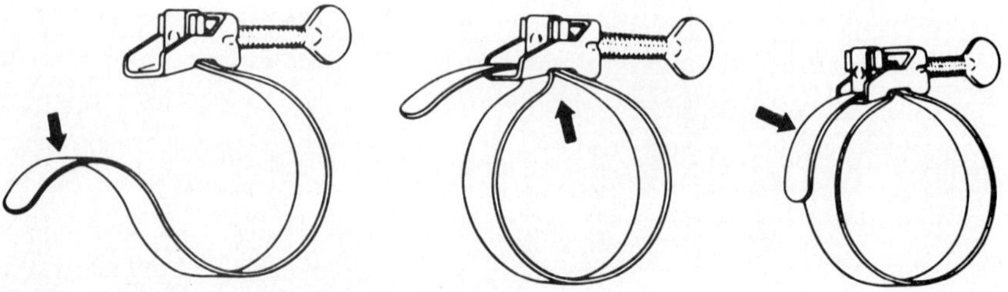

FIG. 7-5 Universal thumb screw band clamp.

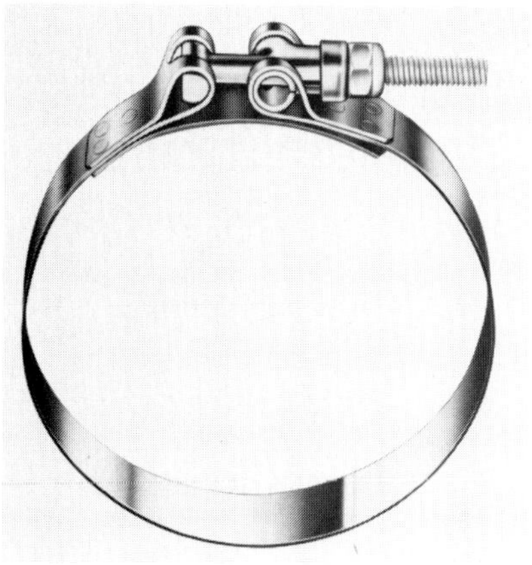

FIG. 7-6 Tee bolt heavy-duty band clamp.

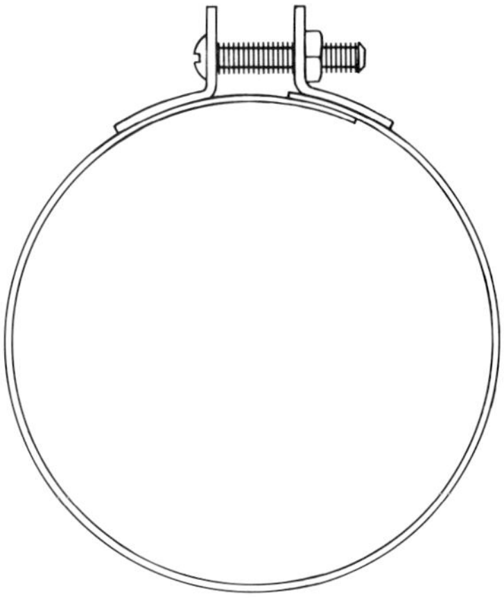

FIG. 7-7 Light-duty single-bolt clamp.

shape when the compression force is removed. This phenomenon is referred to in the industry as cold flow.) In the case of a hose clamp, the compression force directly under the clamp diminishes as the rubber moves to the sides of the clamp and builds up. This possible problem can be controlled by periodically retightening the clamp to make up for this compression loss. Depending on hardness (durometer) of the rubber and other factors, this may or may not present a problem.

The main advantages of clamps are ease of application and economy.

7-10 Crimping

With this method a metal socket is used in conjunction with the hose nipple. See Fig. 7-10. A cavity is formed between the inside diameter of the socket and the outer diameter of the nipple large enough to accommodate the wall of the hose so that the hose can be pushed into the hose fitting by hand. The hose

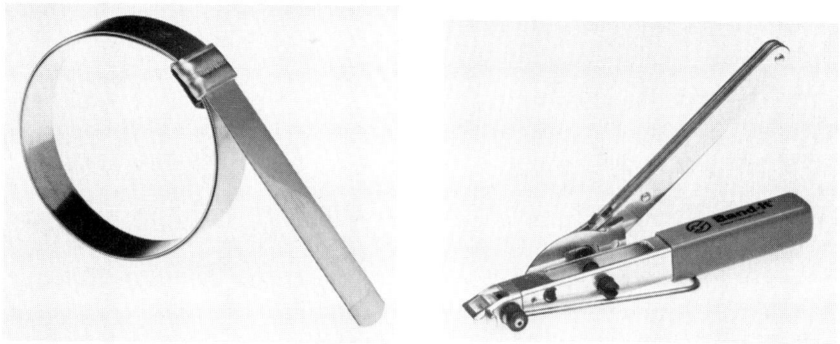

FIG. 7-8 (*a*) Bandit clamp; (*b*) crimping tool. (*Courtesy: Band-It Company.*)

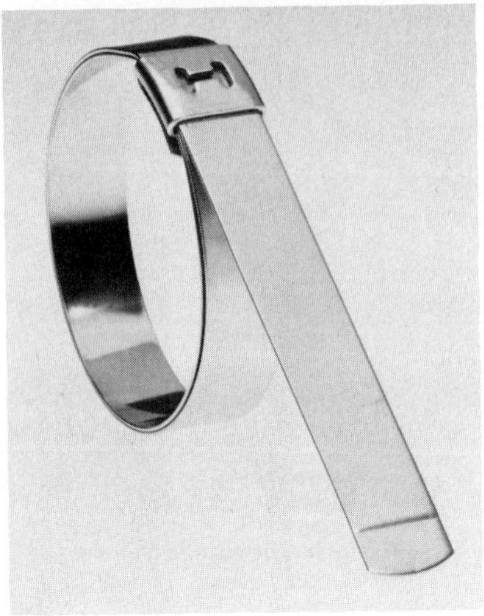

(*a*)

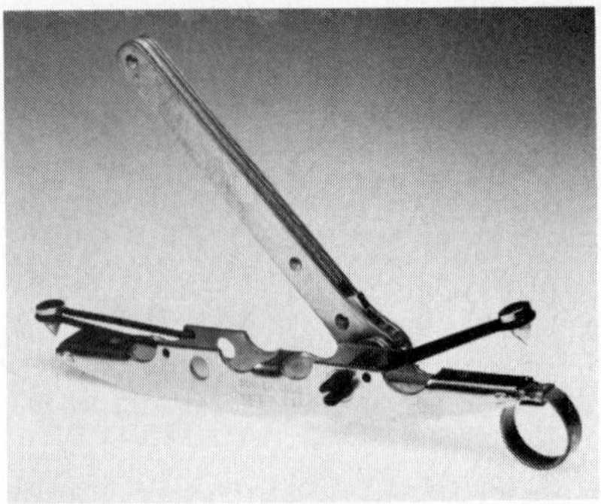

(*b*)

FIG. 7-9 (*a*) Bandit clamp with buckle. (*b*) Center punch tool. (*Courtesy: Band-It Company.*)

fitting is then placed in a machine which has a number of jaws that move radially toward the center of the hose when the machine is activated. This deforms the metal socket, reducing the cavity between it and the nipple, thus introducing a compression force on the walls of the hose. Upon completion of the crimp, the outer surface of the socket may appear dimpled, be corrugated into alternate ridges and furrows, or have annular ridges, depending on the contacting surface finish of the jaws. Refer to Fig. 7-11. Some manufacturers refer to these fittings as "permanent" fittings. This description is used to indi-

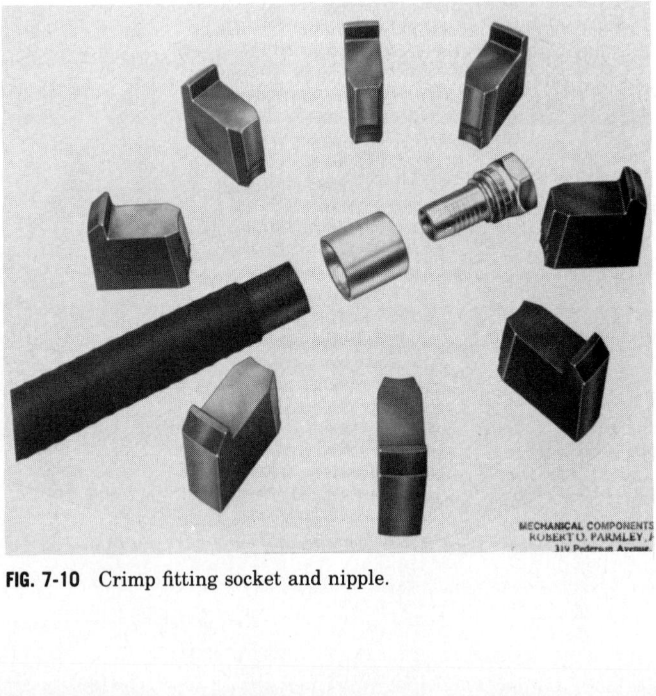

FIG. 7-10 Crimp fitting socket and nipple.

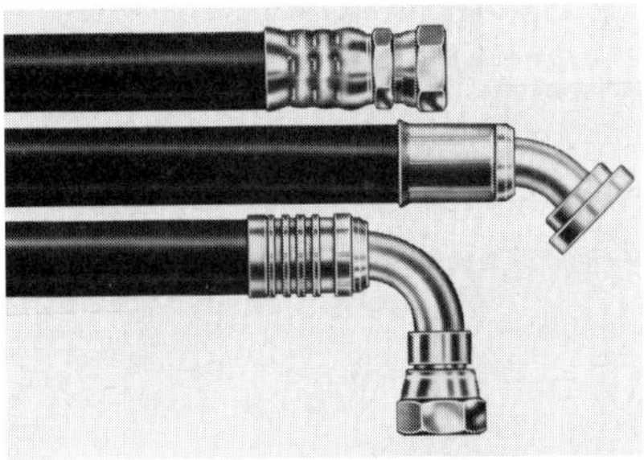

FIG. 7-11 Corrugated flat crimp (top). Dimpled barrel crimp (center). Annular radial crimp (bottom).

cate that the fittings cannot be removed from the hose and reused as detachable, reusable fittings can be. There is a great variety of crimping machines available to the industry, from small field crimpers to large, automated, high-production factory machines. See Fig. 7-12. In many cases, especially with very high pressure, factory-crimped lines, a portion of the hose cover is stripped, or skived, down to the wire braid reinforcement before crimping. A small portion of the inner tube may also be removed so that the wire-braided reinforcement as well as the wall of the hose is locked between the socket and the nipple (Fig. 7-13). This arrangement provides an optimum method of retaining the hose fitting to the hose under very high pressure conditions.

In most field crimping applications, the hose is not skived. In these cases, the hose must be manufactured with a thinner outer cover than normal. Hoses manufactured this way are designated by SAE as T hoses. For example, an SAE 100-R2A hose has two layers of wire-braided reinforcement, a synthetic inner tube, and a heavy synthetic cover. An SAE 100-R2AT hose has exactly the same construction, except that the outer cover is approximately half the thickness. The T-style hoses are for crimping directly over the cover. The standard hose requires skiving (removal) of the cover where the crimping is to take place. Since the outside diameter of the T-designated hose is less than the outside diameter of the standard-style hose, it is important to secure the proper hose for the style of crimping to be used; otherwise, premature failure can occur.

Note also that the heavier cover provides more protection from physical abuse to the reinforcement than does the thinner cover. For example, a 1-in hose under this specification would have a nominal cover thickness of 0.1875 in, while a T hose would have a nominal cover thickness of 0.0685 in.

It is important to follow the crimp machine manufacturer's instructions very closely to assure good-quality results. Either overcrimping or undercrimping can cause early hose failure. It is best to use the same manufacturer's hose and fittings with the machine. Failure to do this could release the manufacturer from all warranty claims. Most manufacturers match their hoses and hose fittings dimensionally to their machinery. The most common cause for failure of this style of fitting is poor hose preparation and/or improper crimping.

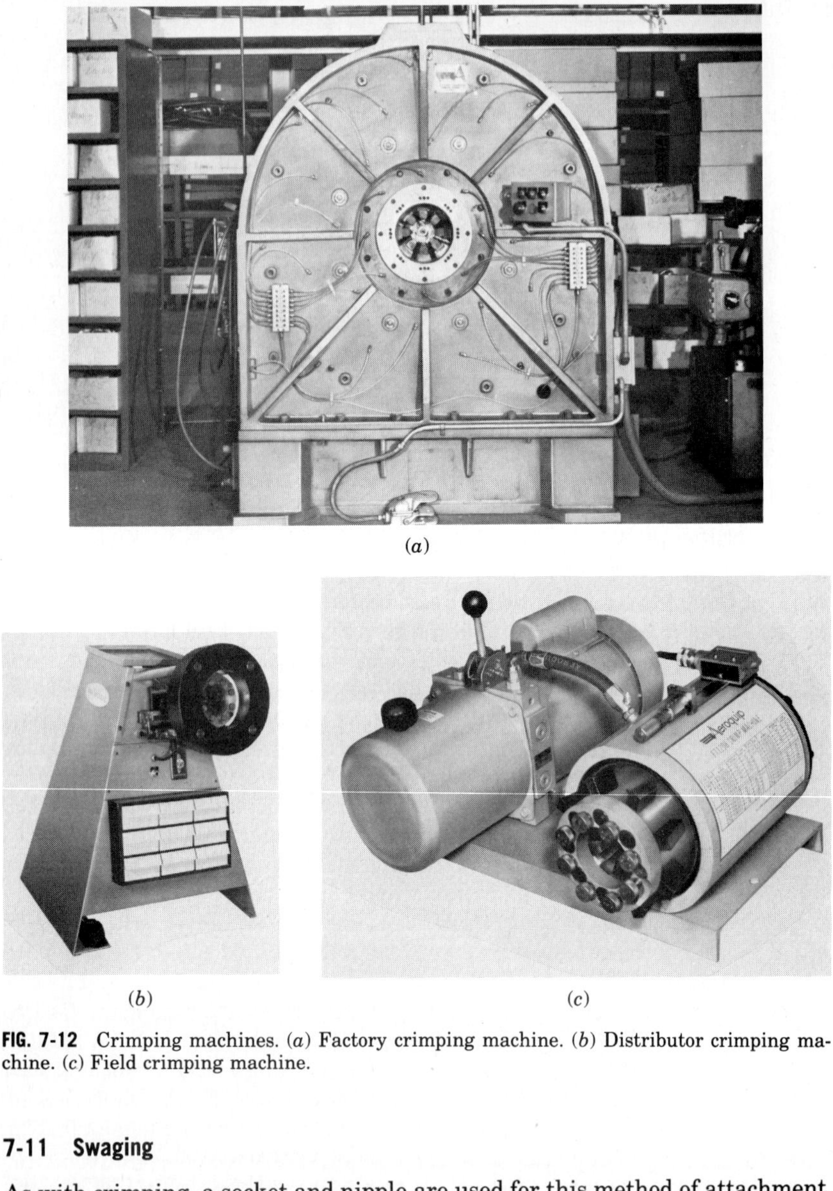

(*a*) (*b*) (*c*)

FIG. 7-12 Crimping machines. (*a*) Factory crimping machine. (*b*) Distributor crimping machine. (*c*) Field crimping machine.

7-11 Swaging

As with crimping, a socket and nipple are used for this method of attachment. The difference between the two lies in the method of reducing, by deformation, the outside diameter of the socket. This is accomplished by using a pair of dies and forcing the fittings through the die forms with pressure.

The two die halves form a tapered hole when they are locked together. The fitting is pushed onto the hose, then, using either mechanical or hydraulic force, the hose fitting is pushed through the dies, reducing the outer diameter of the socket and compressing it onto the hose. Refer to Fig. 7-14.

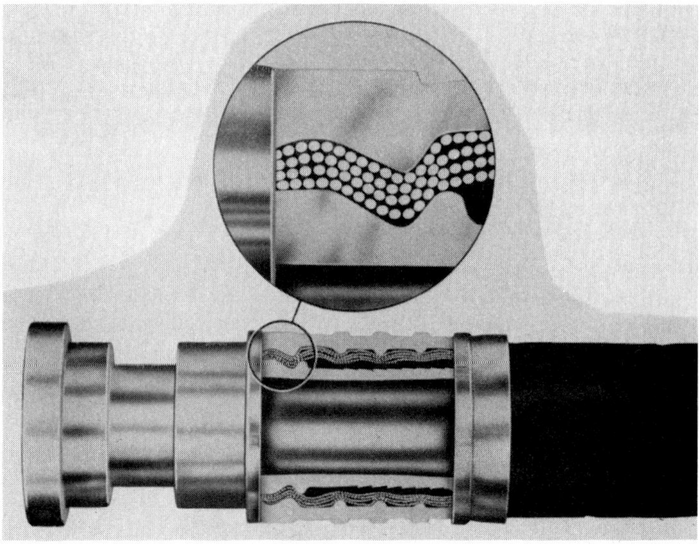

FIG. 7-13 Internal and external skive-type fitting.

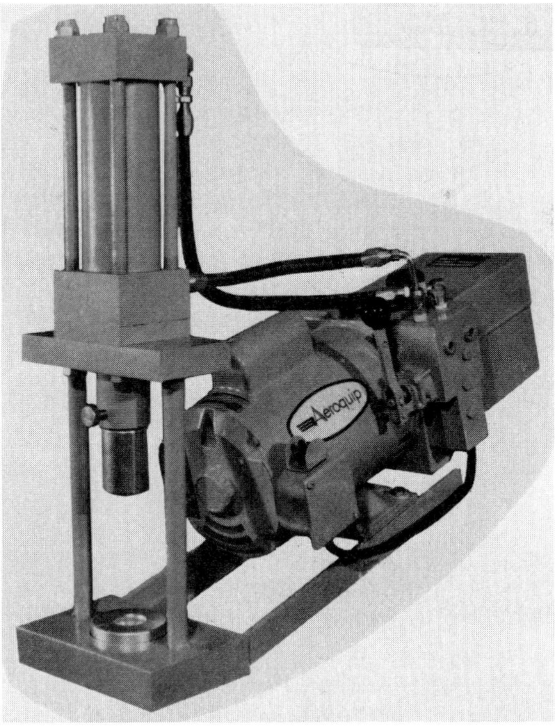

FIG. 7-14 Power swage machine.

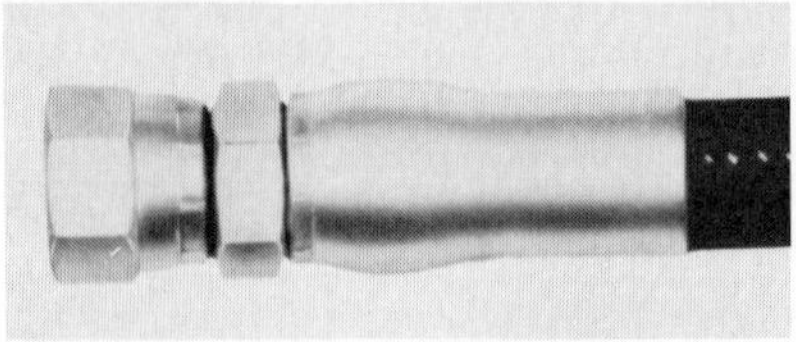

FIG. 7-15 Swaged hose fitting.

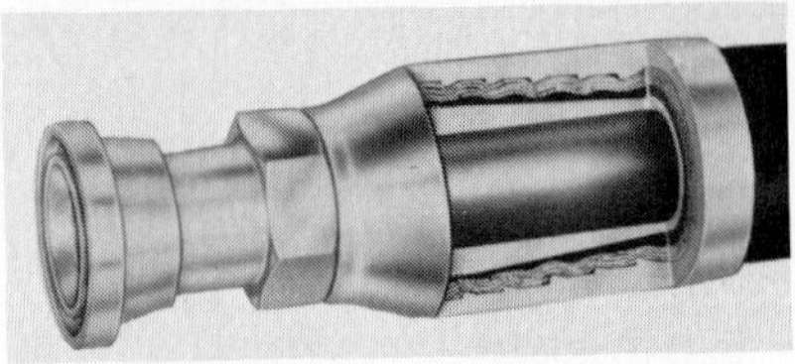

FIG. 7-16 Screw-together reusable fitting.

After swaging, the outside of the socket has a burnished appearance. See Fig. 7-15. As with crimping, it is recommended that the same manufacturer's machine, hose, and fittings be used. Intermixing different manufacturers' equipment or parts can void warranties and may result in premature hose failures. Swaging is limited to T-style hoses and thermoplastic hoses.

7-12 Screw-Together Reusable Fittings

One of the first styles of reusable fittings introduced to the marketplace was the screw-together fitting which was developed by a young German engineer shortly before World War II. He formed a company in the United States to manufacture this fitting. It subsequently became the military standard for all aircraft during World War II and was used on over 300,000 aircraft.

The fitting consists of an internally threaded socket which is installed over the hose, and a tapered, threaded nipple which is screwed into the socket and hose, compressing the wall of the hose as it is threaded, because of its wedged shape (Fig. 7-16).

For high-pressure hose, such as SAE 100-R2A, which has a synthetic rubber tube, two layers of braided wire reinforcement, and a thick synthetic rubber cover, skiving (removing part of the rubber cover) is necessary in order to assure good fitting retention and to eliminate cold flow of the outer cover.

When using 100-R2AT (or any other T-style hoses), the socket becomes the skiving tool. It threads through the cover down to the wire braid reinforcement as it is installed. The stripped cover is simply stored in the internal grooves of the socket. These fittings are referred to as "through the cover" (TTC) fittings or "no skive" fittings.

For medium-pressure hoses, such as SAE 100-R5, which has a synthetic rubber tube, a single layer of braided-wire reinforcement, and a fabric cover,

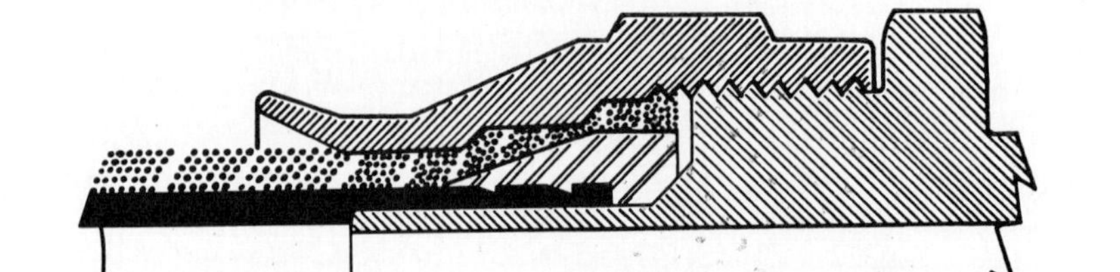

FIG. 7-17 Reusable fitting for Teflon hose.

the socket is installed directly over the fabric cover, eliminating the skiving requirement, and the nipple is screwed into the hose and socket.

A variation of this design developed specifically for Teflon hose is shown in Fig. 7-17. With this arrangement the threaded socket is installed on the outer diameter of the hose. The sleeve is inserted between the stainless steel braided reinforcement and the Teflon inner tube. The nipple is then threaded into the socket, compressing the stainless steel wire braid between the socket and the sleeve. At the same time, the outer diameter of the Teflon tube is pushed into the sealing barbs on the inner diameter of the sleeve. The sleeve seats on the nipple to form a metal-to-metal seal. The barbs on the inner diameter of the sleeve provide a lip seal which seals the agent in the hose. It is advantageous to have this design, which does not compress the Teflon tube, because the tube becomes soft at high temperatures and would cold-flow under compression. Hoses of Teflon can operate successfully at temperatures from −100°F (−73°C) to +450°F (232°C) and with a great variety of fluids.

This design has been used on military and commercial aircraft because of its unique properties.

The reliability of screw-together fittings has been established over many years. Each manufacturer generally will guarantee that the hose and fitting combination will perform as a unit to the established specifications for impulse life and burst pressures of the particular style of hose being used.

The degree of reusability with over-the-cover hose is pretty much unlimited. Reusability is somewhat limited with skived hose because as the socket is threaded over the exposed wire reinforcement, some wear takes place on the inside of the socket. Because of this, many users will notch or prick-punch the socket each time it is reused, then discard it after it has been used from 6 times for ¾ in and larger sizes and up to 12 times for the smaller sizes. In the event of a fitting blow-off, because of excessively high pressure or a misapplication, the socket should be replaced and the old socket scrapped. The nipple can be reused indefinitely as long as it is not physically damaged.

When reusing Teflon hose fittings, it is good practice to discard the old sleeves and install new ones.

Although reusable fittings initially cost more, the higher cost may be justified by the fact that they are reusable and many can be installed with ordinary hand tools. In many cases, from 50 to 80 percent of the cost of a hose assembly is represented by the hardware on the ends of the hose. With the great variety of standards and configurations being used today, it is virtually impossible for the user to stock all the sizes, shapes, and standards required to service the equipment.

If the equipment is originally equipped with reusable fittings, then the manufacturer has provided the customer with a built-in inventory of hose fittings for future hose replacement.

7-13 Bolt-On Reusables

This style of fitting consists of a hose nipple, two outer clamps, and bolts and nuts. The two basic styles are the two-bolt design and the four-bolt design.

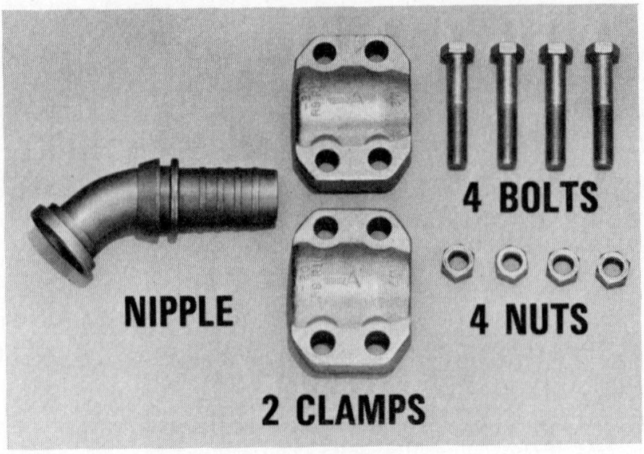

FIG. 7-18 Bolt-on reusable fitting.

Which type is used depends on the pressure requirements. The four-bolt design is shown in Fig. 7-18. With this design, no skiving of the cover is required. The inside and outside of the hose are lubricated with a high-viscosity oil or a light grease. The nipple is pushed into the hose, the socket halves are hooked into

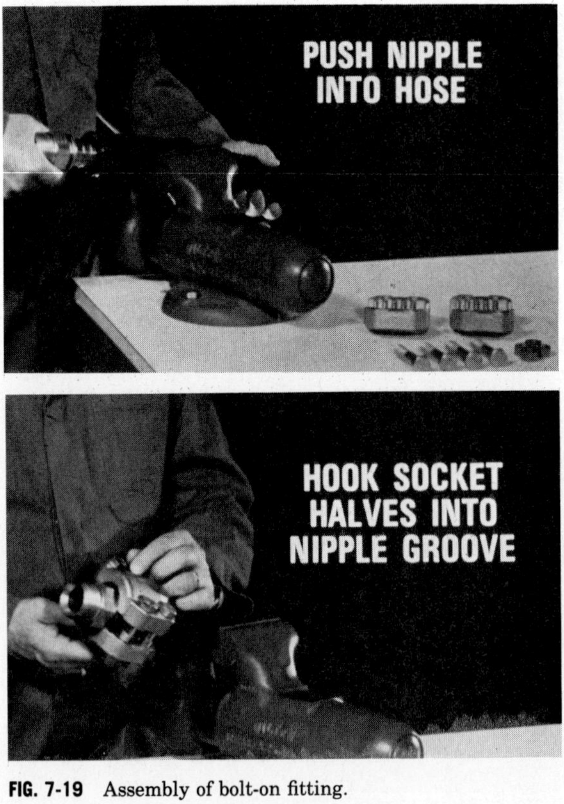

FIG. 7-19 Assembly of bolt-on fitting.

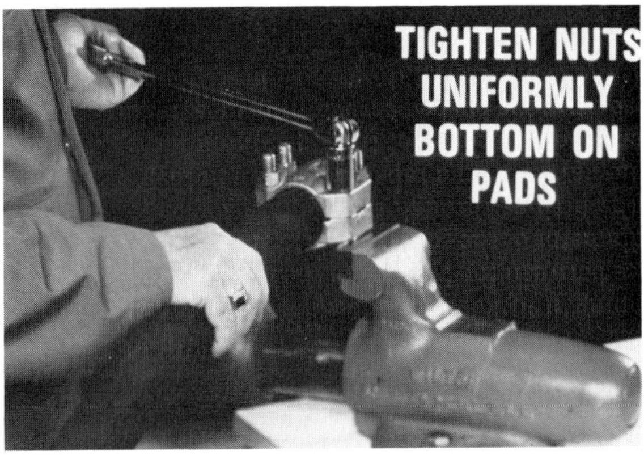

FIG. 7-20 Final assembly of bolt-on fitting.

the nipple groove, and the nuts and bolts are attached (Fig. 7-19). The bolts are then uniformly tightened to the manufacturer's recommended torque. See Fig. 7-20. When installing the clamps it may be necessary to orient them so as to provide clearance where required. Design engineers should consult the SAE Handbook for maximum allowable envelope sizes for this style of fitting. Most manufacturers adhere to these SAE specifications.

7-14 Reusable Segmented Hose Fittings

A segmented fitting consists of a nipple, a nose ring, three (or more) segments, and a retaining ring, as shown in Fig. 7-21. A special tool is required to service this fitting. Figure 7-22 depicts the fitting as installed on a very high-pressure hydraulic hose. With this design, no skiving of the hose is required. The procedure for installing this fitting is as follows: Slip the retaining ring on the hose, then place the nose ring over the nipple and insert the nipple into the hose as far as it can go. Refer to Fig. 7-23. Hook the segments under the nose ring and

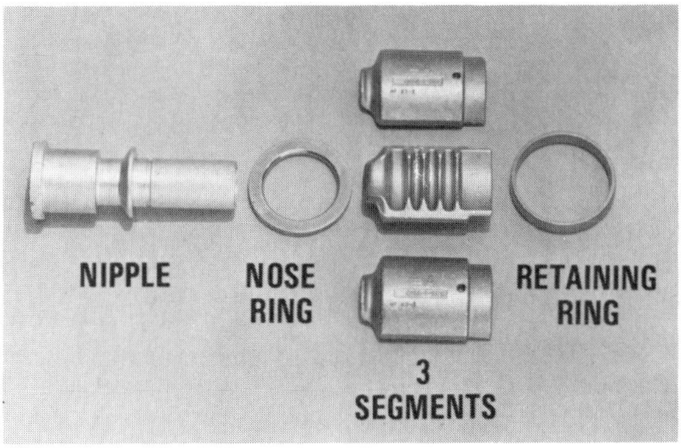

FIG. 7-21 Segmented-fitting component parts.

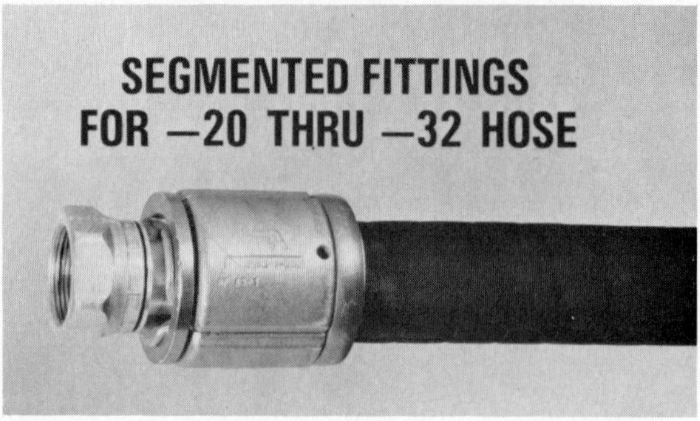

FIG. 7-22 Installed segmented fitting.

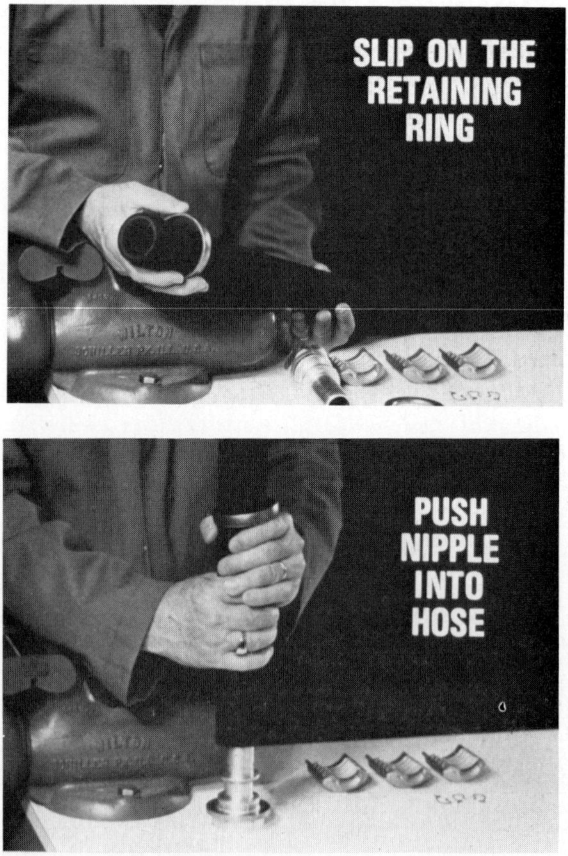

FIG. 7-23 Segmented fitting installation instructions—steps 1 and 2.

attach the tool, making sure the pins on the tool are in the holes on the outside of the segments. See Fig. 7-24. Oil the threads on the assembly tool and wrench-tighten the tool sections uniformly around the fitting. Slide the retaining ring over the ends of the compressed segments, then remove the tool. Refer to Fig. 7-25. This completes the assembly. The fitting is removed from the hose in the reverse order.

7-15 Caterpillar XT-3, XT-5, and XT-6 Reusable Fittings

Another style of reusable fitting is the press-together type, developed by Caterpillar Tractor Co.

It consists of a collet-type nipple (or stem) with multiland fingers which are tapered at their ends and a hardened steel sleeve (or socket). This fitting has been expressly designed and matched to fit Caterpillar-manufactured XT-3, XT-5, and XT-6 hose (Fig. 7-26).

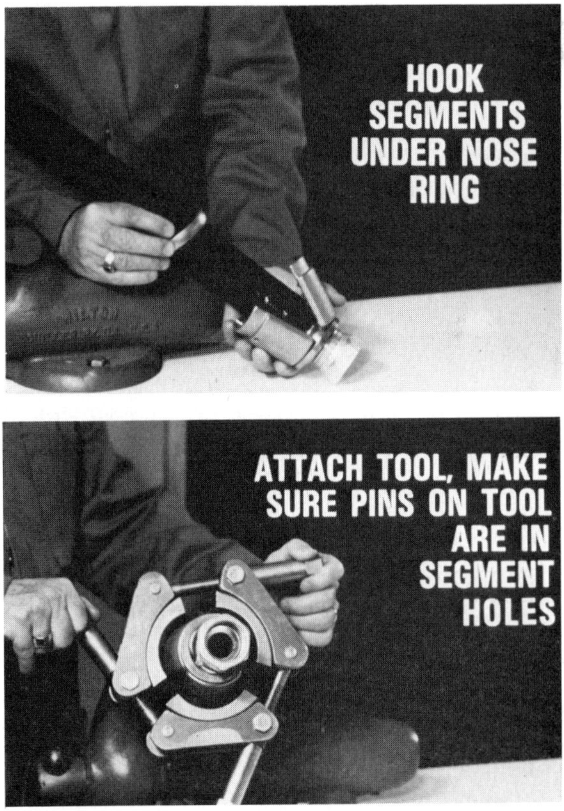

FIG. 7-24 Segmented fitting installation instructions—steps 3 and 4.

FIG. 7-25 Segmented fitting installation instructions—steps 5 and 6.

Care should be taken to match the proper fitting with the proper hose. To facilitate this, the three hoses are color-coded and branded with the symbols Δ, +, or UHP. The fittings are branded in the same manner as the corresponding hose. Intermixing hose and couplings which do not have the same marking can result in a hose failure, along with related damage.

Detailed assembly instructions are available from the manufacturer. Special tooling has been developed to facilitate installation and/or removal of the fittings; this is also available from the manufacturer.

These reusable fittings are designed for working pressures up to 6000 lb/in^2 (413.7 bars), depending on type, and have a minimum safety factor of 4 to 1.

7-16 Push-On Reusable Fittings

The fitting shown in Fig. 7-27, unlike other reusable fittings, consists of the nipple only. Instead of clamping the wall of the hose to achieve fitting retention, the fitting is held in the hose by the design of the hose itself. The hose is manufactured with a special braid angle that causes it to lock onto the nipple

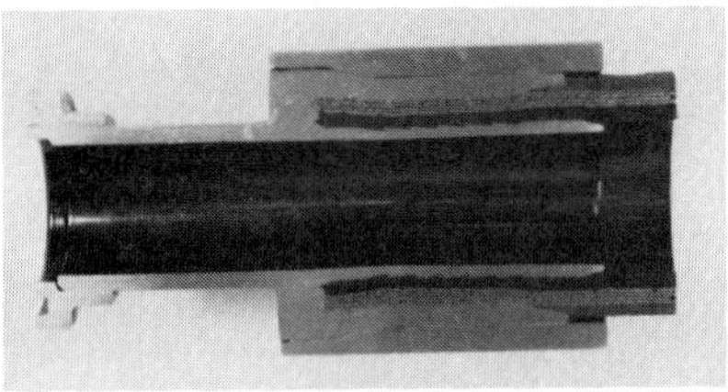

(*a*)

FIG. 7-26 Caterpillar "press on" reusable fittings and tooling. (*a*) Cutaway of installed hose fitting. (*b*) Socket and nipple assemblies. (*c*) Fitting assembly equipment. (*Courtesy: Caterpillar Tractor Company.*)

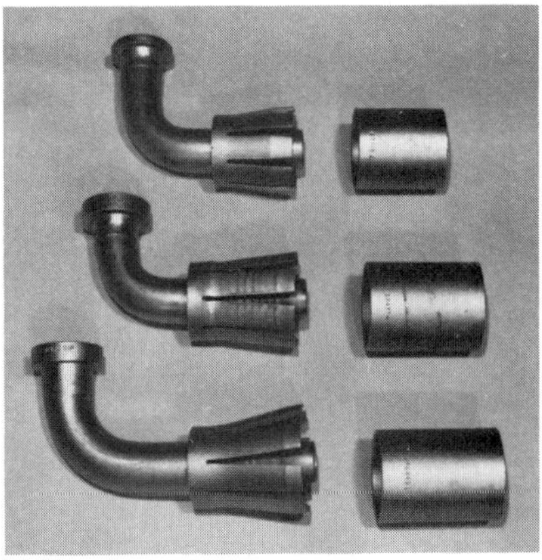

(*b*)

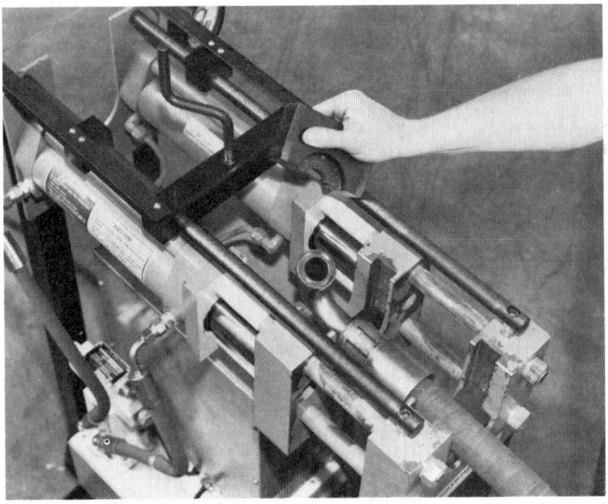

(*c*)

as it is inserted into the hose. The nipple outside diameter is slightly larger than the hose inner diameter, which makes this possible. No clamps, bands, wires, or sockets are used on the outside of the hose. One simply pushes the fitting into the hose and it stays in. The maximum operating pressure for this design is 250 lb/in^2 (17.3 bars) with a 1000-lb/in^2 (69-bars) minimum burst. These fittings are manufactured in sizes from ¼ in (6.35 mm) to ¾ in (19.05 mm). The use of band clamps with this design can cause premature failure

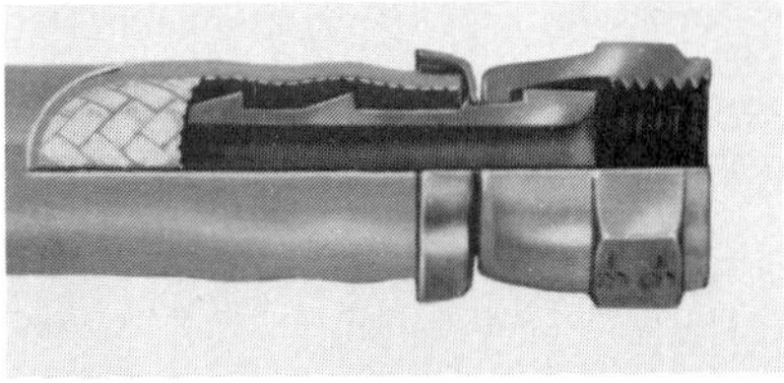

FIG. 7-27 Socketless™ Push-on Fittings. (*Courtesy: TM Aeroquip Corp.*)

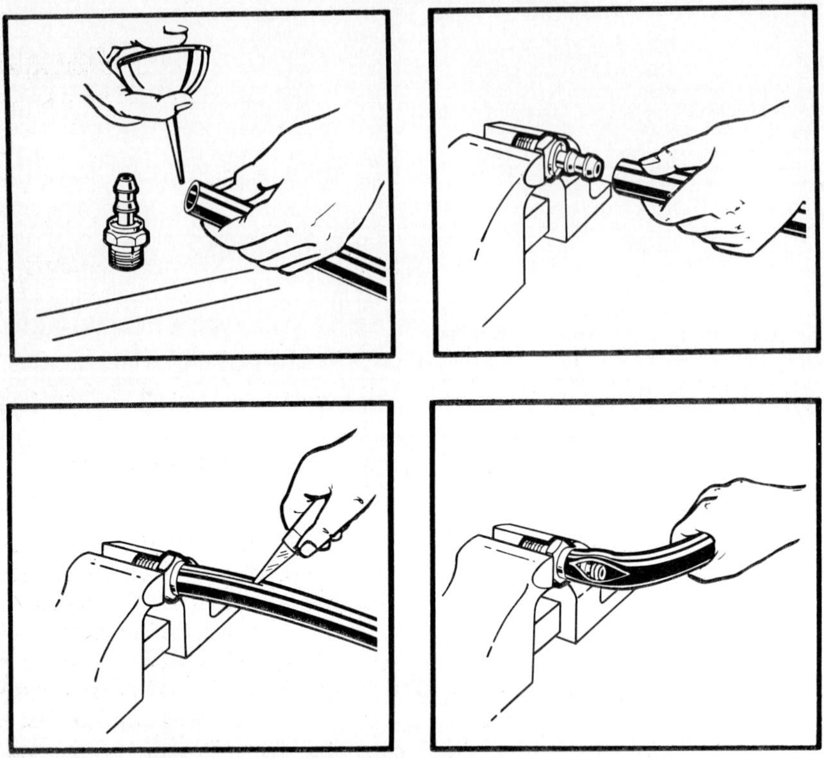

FIG. 7-28 Assembly instruction for push-on fittings.

because it will force the barbs into the inner tube. It is important to match the proper hose with the proper fitting for this design to work. This design is not suitable for hydraulic impulse applications and is not approved for air-brake applications.

To reuse, simply slit the hose lengthwise at the fitting, bend it, and snap it off. See Fig. 7-28.

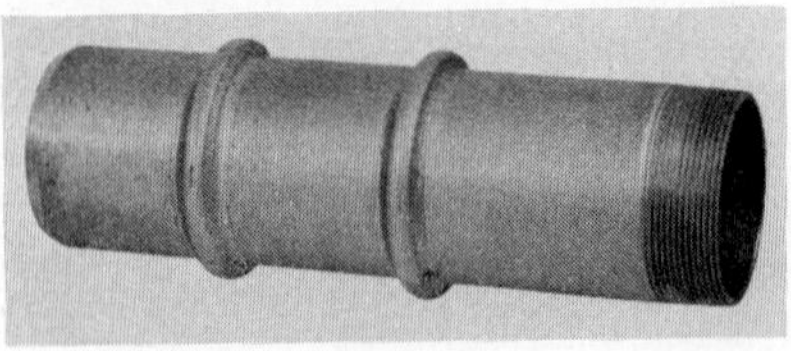

(*a*)

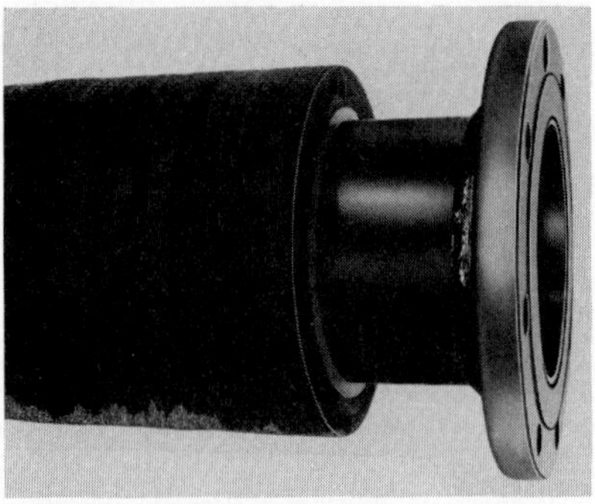

(*b*)

FIG. 7-29 Built-in hose fittings. (*a*) Beaded nipple. (*b*) Completed assembly. (*Courtesy: Goodyear Tire & Rubber Co.*)

7-17 Built-In Fitting

In this design steel bands are welded to the nipple and the hose is built around it by a special process that vulcanizes hose and nipples into one complete unit. This design is used on heavy-duty or high-pressure hand-built hose for oil suction and discharge service, sand suction, and material handling service. The size range is from ½ in (1.27 cm) to 18 in (45.72 cm). See Fig. 7-29.

8

CLUTCHES AND BRAKES

EDITED BY

ROBERT O. PARMLEY,* P.E.

President & Consulting Engineer
Morgan & Parmley, Ltd.
Ladysmith, Wisconsin

* This section was developed by the editor from articles originally published in the 1981–1982 Power Transmission Design Handbook. Permission has been granted from the publisher, Penton/IPC, Inc., for inclusion in this Handbook. Credit to the individual authors has been given at the end of the section under "Acknowledgments." Superscript numbers after major headings refer to individual authors.

Clutches and brakes are often similar in design. Clutches are used to connect shafts to a driving mechanism such as an engine, motor, or line shaft, and may be disconnected at the operator's will. Brakes are used to slow or stop a machine or an individual component.

Clutches may be classified in the following general categories: (1) friction clutches, where the rotating force is transmitted by friction of surfaces in contact; (2) centrifugal clutches, engaged or activated by centrifugal force after a certain driven speed is achieved; (3) air-actuated clutches, pneumatically engaged; (4) electromagnetic clutches, where torque is transmitted across an air gap via the creation of an electromagnetic field; (5) magnetic fluid clutches that use iron particles suspended in machine oil to control the torque; (6) jaw clutches that use jaws, dogs, or teeth within the mechanism to interlock; and (7) freewheeling or overrunning clutches that permit torque transmission in one direction only.

A clutch becomes a brake if it can decelerate and/or stop the driven component. The braking effect converts the kinetic energy of rotation by friction into heat, which must be dissipated to avoid damage to the brake mechanism. There are many types of brakes, from common friction designs to more complex mechanisms such as pneumatic, hydraulic, and electrical.

While space does not allow a full presentation of all clutch and brake designs, the following material has been culled from technical sources to serve as a reasonable coverage.

CENTRIFUGAL CLUTCHES[1]

8-1 Function

Centrifugal clutches connect a prime mover to a load. They have these characteristics:

1. Automatic engagement at a predetermined speed
2. Torque-carrying ability that is a function of the square of prime mover speed
3. Soft starting (high slip during start-up)
4. No slip at rated speed

8-2 Usage

Centrifugal clutches are at their best when:

1. Automatic engagement and disengagement is desirable or tolerable.
2. Motor speed is an adequate clutch control factor.
3. Low to moderate first cost is desired.
4. Gentle starting is required.
5. Essentially no energy waste at load is important.
6. Isolation of shock spikes between prime mover and load is desired.
7. High reliability and maintainability are wanted.

You can select centrifugal clutches which will perform well in the power range from 1 hp to 10,000 hp.

8-3 Advantages

The major advantages of centrifugal clutches are that they:

1. Let the prime mover carry the load only at acceptable (if not optimum) torque and speed. The prime mover "commands" the clutch.
2. Provide automatic engagement and release without peripheral sensors or other equipment.
3. Permit slow or rapid load pickup, depending upon the clutch.
4. Permit slippage during extreme shock or overload to protect equipment.
5. Don't slip during normal running.

Figures 8-1, 8-2, and 8-3 show the behavior of centrifugal clutches with electric motors. The graphs are based on a NEMA Design B motor, but operation is similar for Designs A and C, and shaded-pole motors. Plotting clutch characteristics on a Design D or series-wound motor curve quickly shows the problems that might arise.

Figure 8-1 stresses the relationship between clutch performance and the motor torque-speed characteristics. It shows how the clutch must be selected to pick up the load and carry rated power without slip, but must slip (to destruction if necessary) without forcing an overloaded motor back to pull-out speed, and possibly burn out.

Figure 8-2 demonstrates that clutched motors need to withstand high currents for much shorter time periods than unclutched motors, and therefore can provide the same performance with lighter-duty windings and lower-temperature insulation. The time-delay clutch usually allows the motor to reach synchronous speed before engaging, so all load current is from the portion of the curve after the pull-out condition.

Figure 8-3 depicts the torque-time relationship and shows (in conjunction with Fig. 8-2 and the previous paragraph) that the unclutched motor must be rated considerably higher than the clutched motor. The maximum torque it can provide during most of the acceleration period is the pull-up torque, while the clutched motor can deliver nearly pull-out torque. Therefore the clutched motor spends less time in low-speed high-current operation.

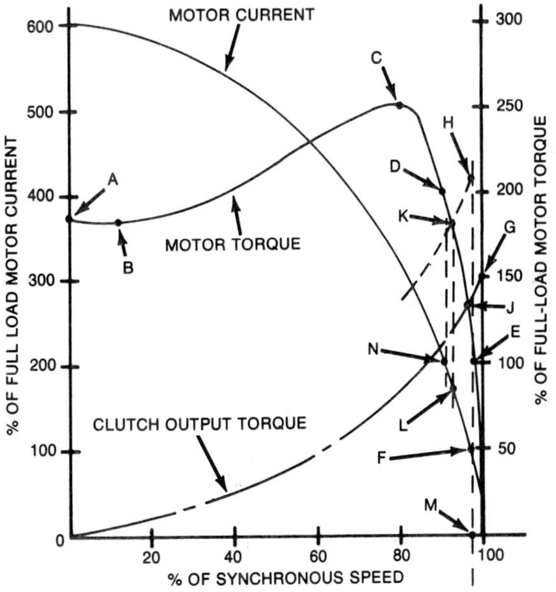

A = LOCKED-ROTOR (STARTING) TORQUE
B = PULL UP (MIN STARTING) TORQUE
C = PULL-OUT (MAX OR BREAK-DOWN) TORQUE
D = RECOMMENDED MAX OVERLOAD TORQUE*
E = FULL-LOAD MOTOR TORQUE
F = FULL-LOAD MOTOR CURRENT
G = CLUTCH DESIGN (MIN STATIC) TORQUE CAPACITY*
H = CLUTCH DESIGN (MAX STATIC) TORQUE CAPACITY*
J = CLUTCH ACCELERATING TORQUE CAPACITY (MIN DYN)
K = CLUTCH SLIP (OVERLOAD) TORQUE CAPACITY*
L = MOTOR CURRENT DRAW AT CLUTCH SLIP*
M = FULL-LOAD MOTOR SPEED
N = MOTOR SPEED CUT-OUT (CURRENT)*

**APPROXIMATE*

FIG. 8-1 Performance characteristics of a NEMA design. B squirrel-cage motor and centrifugal clutch.

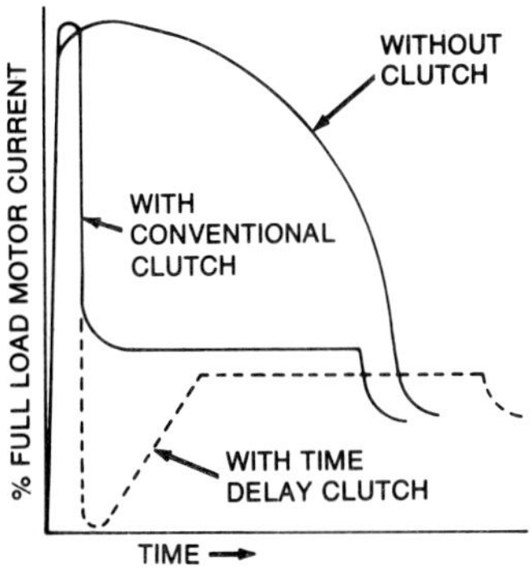

FIG. 8-2 Current-time relationship. Clutched motors need to withstand high current for shorter time periods than unclutched motors.

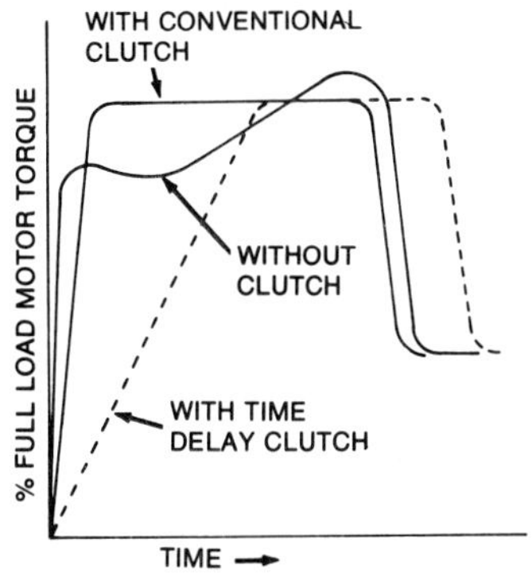

FIG. 8-3 Torque-time relationship.

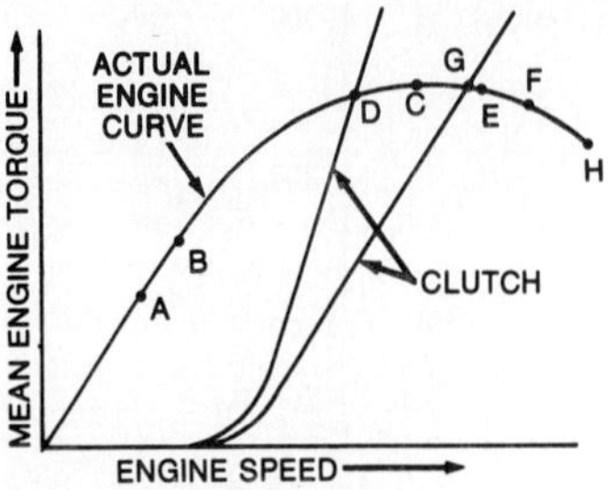

A—MAXIMUM IDLE
B—MINIMUM ENGAGEMENT
C—MINIMUM PULL-DOWN
D—CLUTCH/ENGINE DYNAMIC MATCH
E—RATED ENGINE OPERATION
F—MAXIMUM RATED ENGINE OPERATION
G—MINIMUM STATIC CLUTCH/ENGINE MATCH
H—MAXIMUM ENGINE OVERSPEED CONDITION

FIG. 8-4 Internal combustion engine and centrifugal clutch relationships.

The relationship of centrifugal clutches and internal combustion engines is less complex, but often less understood. Figure 8-4 shows performance curves for a typical gasoline engine and a centrifugal clutch. Good economics dictates that the *actual* use data be considered. Most engines are rated at 70 to 80 percent of maximum torque. Many applications are actually at around 50 percent torque. To specify at 100 percent and operate at 50 percent invariably results in too expensive clutches. Probably the most important clutch function is to allow the engine to start and idle until it reaches minimum operating temperature before full load is applied. Cycle timers and feedback controls from the load can disconnect the engine from the load by throttling the speed below clutch release speed. A manual control can provide the same results. In each condition, changing engine speed (a simple manipulation of the throttle) activates the clutch. This requires little force and can be done either automatically or manually.

Engine idle speed should be as far below the clutch engagement speed as possible to permit practical tolerances in engine idle and clutch engagement speeds. This prevents premature wear caused by clutch slippage during idle. On the other hand, operating and minimum pull-down speeds must be as far above the engagement speed as possible so that the smallest rated clutch will suffice. Minimum pull-down speed is a transient condition, during which the load pulls the engine below minimum rated clutch speed. An example is a chain saw getting stuck in a log. The clutch must survive the minimum pull-down condition. It must not slip, or it must slip at a rate that will not cause unacceptable failure. This condition is often ignored and results in clutch failure. Similarly, the clutch must function during maximum or overspeed conditions.

Nonslip characteristics make these clutches attractive because, once fully engaged, there is no slippage energy loss.

Some systems transmit strong shock pulses between the prime mover and load. A properly sized centrifugal clutch will slip and prevent shock-load passage. This reduces loads on bearings, mountings, shafting, and vibration absorbers. Naturally, slippage causes clutch wear. Clutch life, in spite of this type of slippage, is often tens of thousands of hours. Energy loss due to slip during high-amplitude transients is usually very small because of short duration. Cost of clutch wear and energy loss due to slippage should be compared with the cost of larger clutches and more shock-resistant equipment required if no slippage is allowed.

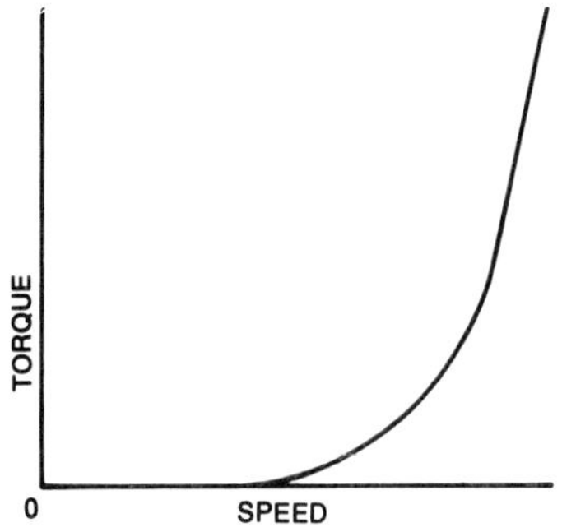

FIG. 8-5 Spring-biased centrifugal clutch.

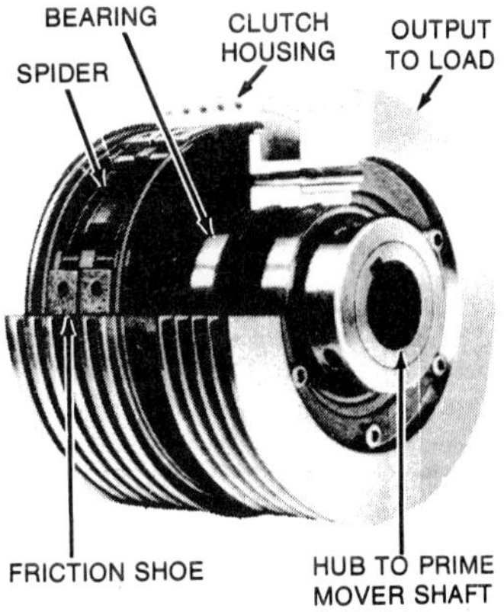

FIG. 8-6 Friction-type centrifugal clutch.

8-4 Design

Special designs of springs and levers give centrifugal clutches a range of special characteristics. The curve of Fig. 8-5 is provided by spring-biased friction shoes which form part of a rotor coupled to the prime mover shaft. Centrifugal force drives the friction shoes into a housing that connects to the load (Fig. 8-6). Friction between shoes and housing generates torque that causes the load to accelerate to shoe spider speed and finally lock-up between the rotor and the housing.

Figure 8-7 shows a proprietary delayed action clutch. It provides delay by metering mercury through an orifice. Mercury is most of the shoe system mass. It permits electric motors to reach synchronous speed before engagement. Remove the mercury and the clutch is a conventional spring-biased centrifugal clutch. Now remove the springs and the clutch becomes an even simpler unbiased clutch. There are many variations of this basic clutch.

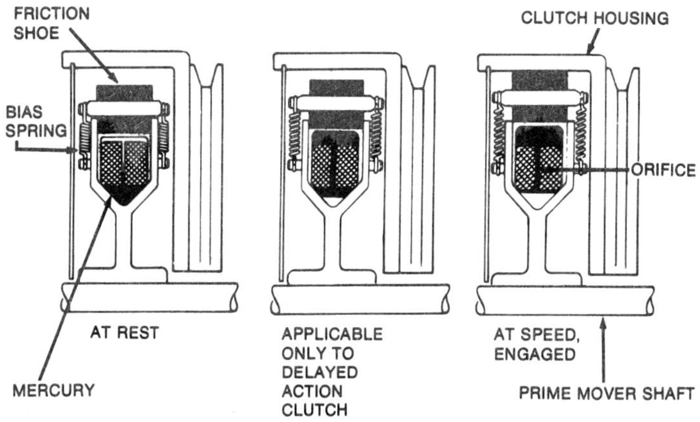

FIG. 8-7 Phases of operation of shoe type friction clutch.

TABLE 8-1 Load Classifications

Class A	Class B	Class C	Class D
Dead load Low inertia Nonpulsating	High inertia (flywheel) requiring prolonged acceleration period Heavy-duty dead load	High breakaway Torque—medium Pulsating loads Unbalanced loads	High breakaway Torque—heavy Pulsating loads

Classifications of various applications

Driven load	Class	Driven load	Class
Agitators:		Generators	B
Vertical	A	Lawn mowers	A
Horizontal	B	Laundry machines	B
Bakery machinery	B	Line shafts	B
Blowers:		Mangles	A
Centrifugal	B	Machine tools:	
Calenders	B	Light	A
Clarifiers	A	Heavy	B
Centrifuges	B	Mills:	
Classifiers	B	Rotary	B
Compressors:		Motor scooters	A
Rotary	A	Presses	B
Multicylinder	C	Propeller shafts	A
Single-cylinder	D	Pulverizers	B
Conveyors:		Pumps:	
Belt	A	Centrifugal	A
Oven	A	Rotary-gear	A
Screw	A	Multicylinder	C
Reciprocating	C	Single-cylinder	
Shaker	C	Sanding machines	B
Crushers:		Screens:	
Coal	B	Rotary	C
Iron	B	Vibrating	C
Stones	B	Stokers:	
Hammermill	B	Screw feed	A
Dryers	B	Ram feed	B
Elevators:		Washing machines:	
Bucket	C	Centrifugal	B
Agitator	Exzcitors B	Textile-paper	A
Fans:		Winding machines:	
Centrifugal	A	Wire	B
Propeller	B	Woodworking machinery	B
Garden tractors	A		

Torque capacity of friction-type clutches can be changed by:

1. Changing the number of shoes, or effective shoe mass
2. Changing the coefficient of friction
3. Changing spring forces
4. Changing the degree of self-energization

How to Analyze Them Most centrifugal clutches have the speed and torque variation of Fig. 8-5. This can be expressed mathematically as:

$$T = T_b(U_o^2 - U_r^2) \tag{8-1}$$

where T = operating torque of clutch at U_o
T_b = basic torque of clutch. This is the torque that the clutch will provide at 1000 r/min without bias springs.
U_o = the operating (r/min)/1000
U_r = the clutch release (r/min)/1000. This is the point at which the shoes no longer contact the housing as the clutch slows down.

The mean torque of the prime mover is

$$T_m = \frac{5252P}{U} \tag{8-2}$$

where U = output shaft speed, r/min
P = horsepower of prime mover, hp

The peak torque is always greater than the mean torque, and an appropriate service factor from 1.5 to 10 is applied depending upon the prime mover, the load, and whether T_b is a static or dynamic value. Tables 8-1 and 8-2 provide typical service factors. The T_m of Eq. (8-2) is multiplied by the service factor.

The equation for a body rotating about an axis is

$$T = I\alpha$$

where T = torque, lb·ft
I = polar moment of inertia, slug·ft²
α = angular acceleration, rad/s²

From this equation we can derive the equation for a combined clutch-load rotating system

$$t = \frac{WR^2U}{307.2(T_m - T_1 - T_2)} \tag{8-3}$$

where t = time to bring load from speed 0 to U, s
U = speed at which clutch locks up, r/min. This speed should equal U_o.
T_m = mean torque, lb·ft. This is often the same as T in Eq. (8-2) when $U_o = U$.
WR^2 = inertia of load and clutch, lb·ft²
T_1 = system torques such as unspecified pumping torques
T_2 = constant torque loads such as friction torques
W = inertial weight
R = radius of gyration

It is difficult to determine T_m and U, as well as T_1 and T_2, although these torques can often be measured. The prime mover usually accelerates the clutch rotor rapidly to U, at which speed the dynamic clutch torque matches

TABLE 8-2 Service Factors

Class load	Power service	Service factors
A	Electric motor	1.5
	Multicylinder engine	1.5
	Single-cylinder engine	2
B	Electric motor	2
	Multicylinder engine	2
	Single-cylinder engine	2.5
C	Electric motor	2.5
	Multicylinder engine	2.5
	Single-cylinder engine	3–5
D	Electric motor—ac	3–5
	Multicylinder engine	3–5
	Single-cylinder engine	5–8
	Electric motor—dc	5–10

the prime mover torque. The clutch continues to slip until the load comes up to speed, and clutch lock-up occurs at some speed between U and U_o. For most applications, a good first shot is to consider U and T_m as the conditions represented at point J on Fig. 8-1 and point D on Fig. 8-4.

Friction-surface temperature usually limits the clutch's ability to accelerate the load. This temperature is difficult to determine. Experience and an acceptable housing bulk temperature rise are often used. This determination belongs more to the clutch designer than user.

8-5 Pitfalls

You must match clutches to their tasks even more carefully than most other drive components. Oversizing a clutch can cause as many system problems as undersizing it. The undersized clutch slips, overheats, scores, and fails prematurely. Less obvious is what happens with the oversized clutch. It provides rapid start-ups that may:

1. Damage machine components because of high dynamic loading
2. Pull motors driving high inertial loads below their pull-out condition and contribute to premature motor failure
3. Pull gasoline engines below their minimum operating speed when cold, or into an auto-ignition condition when hot
4. Cause belt slippage and failure

8-6 Speed Conditions

You must determine the operating conditions. These are important:

Engagement and Release Speeds Specify these speeds such that minimum clutch release speed is greater than maximum idle speed in the case of IC engines. There are some exceptions in special cases. For electric motors, these speeds must permit the clutch to provide the proper torque at points J and K of Fig. 8-1.

Minimum Pull-Down Speed This is comparable to point K on Fig. 8-1 for electric motors. For IC engines, it is the minimum speed anticipated in system operation. Whether this is a slip or no-slip condition depends upon system requirements, such as frequency of occurrence, duration, effect on system components such as belts, and alternatives. This speed should usually be as far above engagement speed as possible. Clutch size is sometimes based on this condition.

Operating Speed This is the full load condition of Figs. 8-1 and 8-4. The clutch must operate without slip at this rated full load condition for electric motors and for IC engines. On small, lightweight, high-output engines, this torque may be as much as seven times the mean torque value on the engine curve. Therefore, operating speed conditions are important.

Maximum Speed This is synchronous speed (Fig. 8-1) for ac motors. It is of little importance in selecting clutches. However, gasoline engines, and series-wound dc motors can present maximum speeds far above rated speed. Overspeed can cause clutch housing failure.

AIR-ACTUATED DISK CLUTCHES AND BRAKES[2]

Here, we introduce various clutch-brake performance data, suggest application classifications to help the designer decide which data are important, and discuss how to evaluate machine requirements for those data pertinent to each classification.

The discussion refers to single-faced, circular disk units which, in our opinion, best demonstrate the following principles. Also, since our experience has been with pneumatically actuated units, we write with them in mind. In most cases, the principles would apply to other friction-type units with mechanical, hydraulic, or electrical actuation.

8-7 Clutch-Brake Performance

Transmitted horsepower and torque, of all variables, are perhaps the ones that you must always consider in one way or another for all applications. Their relation to each other is:

$$T = 63{,}000\ P/N \tag{8-4}$$

where T = torque, lb·in
P = power, hp
N = shaft speed, r/min

Many selections can be made successfully based on transmitted horsepower alone, so a chart like that of Fig. 8-8 is popular in manufacturers' catalogs.

Such a chart is prepared by evaluating Eq. (8-4) for horsepower using the torque value of various clutch models. With typically recommended air pressure of 80 lb/in^2, this chart provides a built-in 1.6 service factor in that it assumes only 50 lb/in^2 to the clutch. For situations that require real values of clutch or brake torque, the manufacturer usually provides a graph like that of Fig. 8-9.

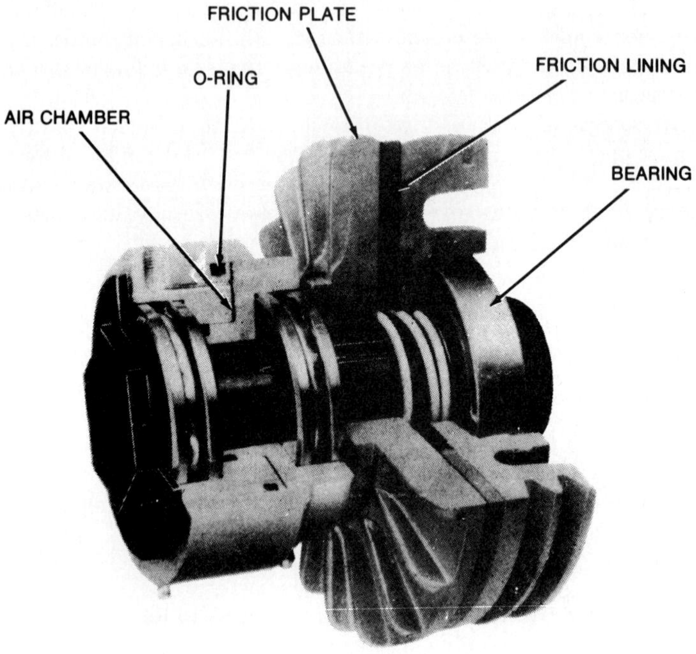

Typical air-actuated, single-faced clutch.

RPM/100

HORSEPOWER	1	2	3	4	5	6	7	8	9	10	11	12	13	14	15	16	17	18	22
1	10	8	6	6	6	6	4½	4½	4½	4½	4½	4½	4½	4½	4½	4½	4½	2¾	2¾
2	10	10	8	8	8	6	6	6	6	6	6	6	6	6	4½	4½	4½	4½	4½
3	10	10	10	8	8	8	8	8	6	6	6	6	6	6	6	6	6	6	6
5	11	10	10	10	10	8	8	8	8	8	8	8	8	8	6	6	6	6	6
7½		11	10	10	10	10	10	10	8	8	8	8	8	8	8	8	8	8	8
10		11	10	10	10	10	10	10	10	10	10	8	8	8	8	8	8	8	8
15			11	11	10	10	10	10	10	10	10	10	10	10	10	10	10	10	8
20				11	11	10	10	10	10	10	10	10	10	10	10	10	10	10	10
25					11	11	11	10	10	10	10	10	10	10	10	10	10	10	10
30						11	11	11	11	10	10	10	10	10	10	10	10	10	10

FIG. 8-8 Typical pneumatic clutch selection by diameter (in inches) for power transmitted at various speeds. Based on 50-16/in^2 air pressure. (*Courtesy: Horton Manufacturing Co., Inc.*)

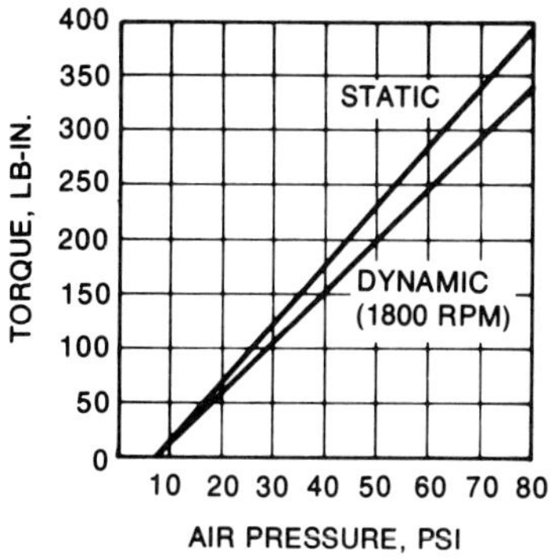

FIG. 8-9 Typical torque as a function of air pressure for 6-in-diameter pneumatic clutch.

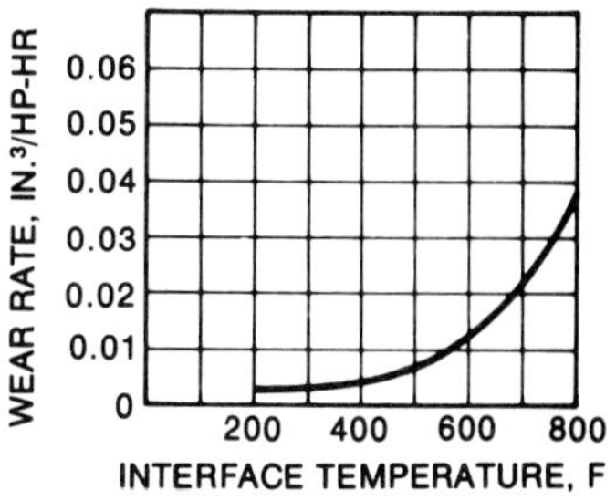

FIG. 8-10 Typical facing wear rate as a function of interface temperature.

8-8 Thermal Ratings

Thermal capacity, next to torque, is probably the most important clutch-brake characteristic, yet it is often the least understood. Some confusion results from the three distinctly different types:

- Peak input rate
- Heat sink
- Continuous dissipation

Peak input rate is the greatest rate at which heat can be generated at the friction interface without raising the interface temperature high enough to damage the friction plate or the lining permanently. Its allowable value depends on rubbing interface surface area and friction-plate conductivity. If heat is generated faster than it can be removed, extreme temperature gradients cause distortion or checking of the friction plate and excessive facing wear.

Figure 8-10 shows that facing wear increases dramatically as interface temperature approaches 400 to 500°F. Temperature differential affects conductive heat flow, so acceptable facing life also becomes a factor in setting the allowable clutch or brake input rate. See Fig. 8-11.

Heat sink values represent the total heat that a clutch or brake can absorb during a given time. It is a measure of the heat required to raise the friction-plate temperature from ambient to some safe level below that where checking or distortion would occur. Again, 400 to 500°F is acceptable, if the peak input rate is not exceeded during any warm-up interval, and if other components like rubber O rings are satisfactorily isolated. The true heat sink value depends on the weight of the friction plate and its specific heat:

$$Q = cW\,\Delta t \tag{8-5}$$

FIG. 8-11 Heat checking of a cast-iron friction plate subjected to excessive peak thermal input rate. (*Courtesy: Horton Manufacturing Co., Inc.*)

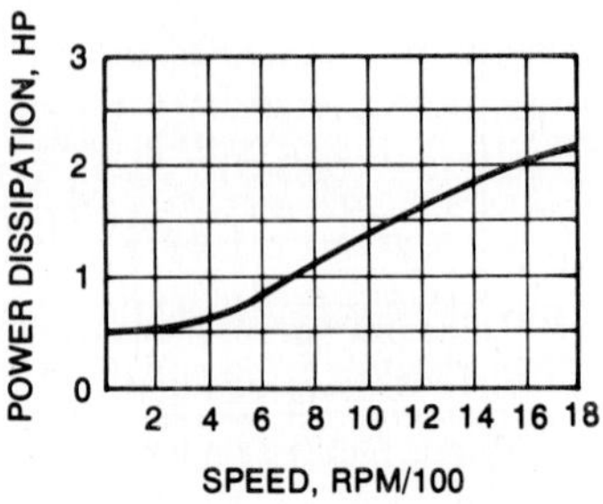

FIG. 8-12 Typical continuous thermal horsepower as a function of speed for 8-inch-diameter pneumatic clutch in 80°F ambient air.

where Q = heat quantity, Btu
W = weight of that part of the friction plate under the rubbing surface, lb
c = specific heat of friction plate material, Btu/lb·°F
Δt = allowable temperature rise, °F

Generally, Q is multiplied by 778 and published in catalogs as work done in foot-pounds.

Continuous dissipation indicates the average rate at which heat can be generated at the friction interface without damaging clutch-brake components or seriously shortening design life. It is generally stated as a horsepower value or an equivalent number of foot-pounds per minute. As with peak input rate, it is based on the critical temperature of components within the unit. For example, a popular bearing seal material is nitrile rubber with a maximum rating of 250°F. A clutch or brake with such a seal cannot have a continuous heat input that would subject the bearing to a greater temperature. Because the unit's ability to expel heat depends on radiation and convection, both ambient temperature and air circulation become important. Figure 8-12 is a way of expressing continuous thermal capacity.

The three types of thermal ratings all have to do with the unit's ability to transfer heat from the friction interface and eventually to surrounding air.

Rotational inertia, commonly called WK^2, is a measure of the unit's resistance to rotational speed change. It is normally expressed in pounds per square foot and is usually important only in cyclic applications. Then it is added to the inertia of the machine itself and becomes part of the start-up or stopping load that confronts the clutch and brake.

Facing life may not be presented directly in manufacturers' catalogs because, unless the application is severe or one of constant slip, it is of no interest. However, it is usually available upon request. A popular way of expressing it is in horsepower-hours.

Response time is that time increment from when power is switched on or off at the control valve until the clutch or brake responds with some given change in torque output. Because the type of control valve contributes to the value, it is conveniently expressed with empirical data.

Air consumption with pneumatic clutches and brakes takes place only when the unit is cycled. While it is engaged, assuming that there is no leak, no air is used. Engagement is maintained by static air pressure. Consumption becomes important only if extremely high cyclic duty is required, or if an installation has no continuous air supply. From multiplying the cyclic rate by the quantity of air used per cycle, you can determine the necessary control valve orifice size. If the orifice were too small, the clutch or brake would lose torque and overheat.

Limiting speed is primarily a function of the unit's diameter. The larger the unit, the lower the limiting speed. Catalogs often relate the unit to standard motor speeds like 1200, 1800, or 3600 r/min. However, many units can operate successfully up to their materials' peripheral velocity limits. Class 40 cast iron, for example, is considered safe at 6200 ft/min. The relationship between speed and peripheral velocity is:

$$V = \pi d(N)/12 \tag{8-6}$$

where V = peripheral velocity, ft/min
d = diameter, in
N = rotary speed, r/min

8-9 Classifying Applications

The designer should classify the application into one of these:

- Occasional start or stop
- Cyclic starts and stops
- Continuous slip
- High inertial start or stop

Occasional start-stop may be the most common application and, fortunately, it involves the fewest complications. Here, the clutch merely disconnects the prime mover from the machine at irregular intervals. The brake is used only to avoid long coast-down periods. The main concern is that the clutch have enough torque to drive the machine at all loading conditions and that the brake bring rotating elements to a stop in an acceptable time. "Required cycle rates of less than four or five times per min" is an approximate rule of thumb when placing a machine in this classification.

Many times, the clutch mounts directly on an electric motor shaft. Then, selection can be made from charts like Fig. 8-8, using the motor's rated horsepower and its operating speed. On the other hand, if the clutch is downstream of the motor, clutch shaft speed must be calculated, and that, along with motor horsepower, is used in the chart.

When a table is unavailable, calculate required torque at the clutch shaft by using Eq. (8-4) and the motor's rated horsepower. Selection is then made from torque graphs like Fig. 8-9 after applying a service factor of 1.2 to 1.5, depending on how conservatively the motor has been sized for the machine.

If possible, for occasional start-stop, the clutch or brake should be on the high-speed shaft, where torque is lowest. In that way, you can use smaller, less expensive units.

A caution for this type of application: Account for torque pulsations from the prime mover and the load. If the prime mover is a gasoline engine or some other piston device, or if the load is reciprocating, apply a service factor of 2 to 3 to the required clutch torque if you calculate it from rated prime mover horsepower. If peak torque can be estimated, that then times the lesser normal service factor would be the required clutch torque.

Any application truly falling into this classification should require no further computation except that of peripheral velocity limits.

Cyclic start-stop is surely the most challenging of all selection tasks because torque, peak thermal input, continuous thermal input, machine and clutch-brake inertia, response time, and facing life may all need evaluation. Even air consumption can become important. Typical examples are process machines where stock is fed into a precise location and then cut, stamped, or formed. The important thing is that the designer be prepared with reasonably accurate shaft and pulley dimensions, so that machine inertia can be estimated.

Any rotating body has kinetic energy as defined by:

$$\mathrm{KE} = 0.00017\,(N^2)WK^2 \tag{8-7}$$

where N = speed, r/min
WK^2 = rotational inertia, lb·ft^2

Each time the rotating parts are brought from rest to speed, heat equivalent to this energy is generated at the clutch interface. Likewise, each time the components are stopped, this amount of energy in the form of heat is generated at the brake interface. For the sake of mechanical completeness, machine frictional drag also contributes to clutch heat on start-up, but reduces the heat load to the brake during a stop.

Frictional drag of any given machine is hard to evaluate and, unless a great amount is known to exist, it is expedient to account for it by applying a service factor to the clutch thermal requirement.

Continuous slip applications appear frequently in the paper industry, where material is pulled from large rolls supported by a center shaft. A clutch or brake connects to the shaft to provide paper tension. The primary concern here is continuous thermal dissipation of clutch and brake rather than torque capabilities, because most units can transmit substantially more power than they can dissipate as heat. Facing wear life is important.

High inertial start-stop applications are characterized by heavy rotating rolls or flywheels. Though only occasional starts and stops are made, they are

Rating Code

VI — very important
MI — may be important
SI — seldom important
NA — not applicable
SC — should be checked

	Application Classification			
	occasional start-stop	cyclic start-stop	constant slip	high inertial start-stop
transmitted horsepower	VI	MI	NA	SI
static torque	VI	MI	NA	SI
dynamic torque	SI	VI	NA	VI
peak thermal rate	SI	SI	NA	VI
heat sink capacity	SI	NA	NA	VI
continuous thermal rate	NA	VI	VI	NA
rotational inertia	SI	VI	NA	MI
response time	NA	MI	NA	VI
air consumption	NA	MI	NA	NA
facing wear rate	SI	MI	VI	MI
limiting speed	SC	SC	SC	SC

FIG. 8-13 Relative importance of clutch and brake performance data based on application type. (Author's opinion based on experience.)

distinguished from that classification by the time required for the load to come to speed or to be stopped. Start-stop periods of more than 0.1 s are suggestive of this type, and any over 0.5 s should surely be considered as such. Here, the designer can take advantage of clutch-brake heat sink values, because the units have time to cool between starts and stops. Since heat transfer to the sink (friction plate) takes place by conduction (compared to radiation and convection for transfer from heat sink to air), input rates far in excess of the continuous value can be tolerated until the sink is full. This conduction rate is the peak input rate.

Figure 8-13 summarizes the classifications along with a suggested importance rating of each clutch-brake performance characteristic. Those rated as very important (VI) are what the engineer should focus attention on first. If he selects a unit with these factors properly sized, it is probable that the other factors will also be adequate or not applicable. For example, if a 1.2 service factor, as recommended, is used in selecting static torque for an occasional start-stop application, Fig. 8-9 shows that even the dynamic torque for the same unit would exceed the true requirement. Likely then, if dynamic torque in the end result turned out to be a limiting factor, the application should have been classified as cyclic or high inertial start-stop instead. By similar reasoning, peak thermal rate is seldom important for cyclic start-stop jobs. However, if the machine's inertia is high, so that start-up times exceed 0.1 s, for example, then the application really suggests both cyclic and high inertial start-stop classifications and requires evaluation of the "VI" ratings for both, except heat sink capacity.

For safety, check the limiting speed (Fig. 8-13) for all applications.

8-10 Machine Requirements

Dynamic torque becomes important in cyclic applications, where start and stop times must be quick or precise. Here, the engineer must first evaluate the WK^2 of all rotating parts that the clutch must bring to speed. Tables are usually available. For commercial sheaves, the WK^2 is in the manufacturer's catalog. Parts driven by the clutch and not rating at clutch speed must be referred to clutch speed by:

$$\text{Referred } WK^2 = (N/n)^2 WK^2 \tag{8-8}$$

where WK^2 = rotational inertia of driven component, lb·ft^2
n = clutch shaft speed, r/min
N = speed of driven part, r/min

With all referred WK^2 summed, including that of the clutch or brake, required dynamic torque is then calculated from:

$$T = 0.039 WK^2 N/t \tag{8-9}$$

where T = dynamic torque, lb·in
N = driven speed of clutch or brake, r/min
t = time required to start or stop, s
WK^2 = rotational inertia, lb·ft^2

If start and stop time requirements are short, the torque from Eq. (8-9) may exceed transmitted torque as selected for occasional starts. Therefore, it is a good idea to check prime mover torque. Most electric motors, however, have "pull-out" torque approximately double their running torque, which generally accounts for quick-start requirements.

Continuous thermal dissipation requirement is calculated as follows: Determine the kinetic energy per start and stop from Eq. (8-7), using the same WK^2 as used to calculate dynamic torque. Then find required thermal capacity for the clutch or brake from:

$$\text{Thermal hp} = (\text{KE})(C)/33{,}000 \tag{8-10}$$

where KE = kinetic energy from Eq. (8-7), ft·lb
C = number of starts or stops per minute

The clutch and brake selected must each have a minimum continuous dissipation equal to or greater than this value.

Heat sink requirement for high inertial loads is merely the kinetic energy of one start or stop as figured from Eq. (8-7). The proper choice is simply a unit with a rated value greater than this result.

Peak thermal input requirement may be a limiting factor in high inertial start-stops, though the heat sink is sufficient. To appreciate this, look at the slip speed of a clutch interface during a start-up interval. By solving Eq. (8-9) for N, it becomes an expression for clutch output speed as a function of time

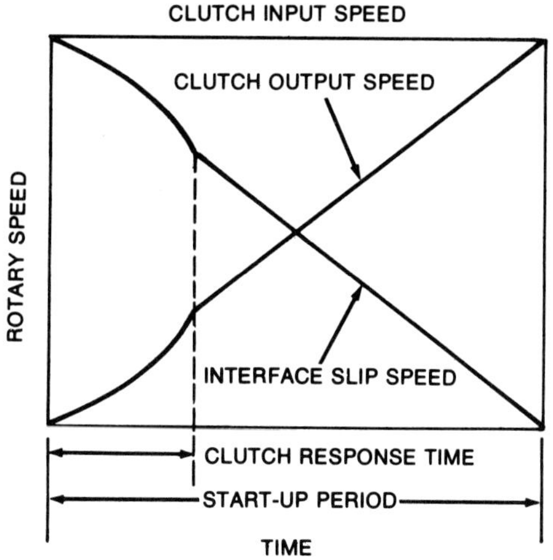

FIG. 8-14 Relationship of clutch input speed, output speed, and slip speed over a start-up period with high inertial load.

and clutch torque. It is valid, however, only *after* clutch torque has stabilized, that is, risen to full value at regulated pressure. Therefore, it can be used only to give the instantaneous value of output speed from the end of the response period until full output speed is reached.

During the response time, while air pressure in the clutch air chamber is still building, clutch torque is changing. If this change is taken to be linear (laboratory results suggest that, except at the very beginning and end of the interval, it is nearly so), and if the linear expression for torque rise is put into the foregoing expression for N, it can be integrated to give an instantaneous value of N over the response period.

Figure 8-14 shows a typical graphic profile of this result as clutch output speed during the response period. Obviously, slip speed at the interface is simply clutch driven speed minus clutch output speed.

Knowing the slip speed and torque at any one time, you can go back to Eq. (8-4) and solve for an instantaneous power input to the clutch by taking speed as instantaneous slip speed and T as instantaneous torque. Figure 8-15 shows a graphical profile of this result.

The dashed line in Fig. 8-15 represents a theoretical condition and shows the peak input that would exist if full torque occurred simultaneously with full slip speed. The other two profiles compare a high inertial start-up, where the total start time is far in excess of the response time, and a small inertial start-up, where full speed is reached before full torque is realized. In the two latter "real-life" cases, you can see that peak input would be reduced by an increase in clutch response time.

Heat generated, which is represented by the area under the input-time curve for a pure inertial load (as in Fig. 8-15), must be the same as the kinetic energy of rotating parts at full speed. Obviously, an extension of the time base

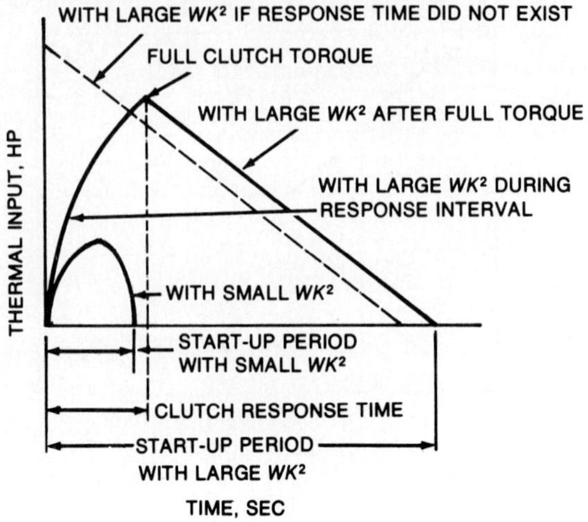

FIG. 8-15 Comparison of thermal input to a pneumatic clutch with large inertial load and with small inertial load.

would require a decrease in the peak input rate to maintain constant area. On the other hand, an increase in response time also increases the total start-up time, but for high inertial applications, this usually does not matter. It provides a way of doing the job without causing surface cracks in the clutch plate. For the sake of simplicity, Figs. 8-14 and 8-15 refer only to clutches. However, the same principles hold for brakes by considering clutch driven speed as brake speed at the beginning of a stop.

Response time requirements, in the light of potentially damaging peak input rates, take on a new dimension. Generally, clutch-brake response time should be as quick as possible. Manufacturers design accordingly and publish results with quick-response control valves. To maximize a machine's cyclic rate, this, of course, is beneficial; but by referring to Fig. 8-15 again, we see that the quick response requirement must be tempered with consideration of the peak input rate.

Increasing the response time of a pneumatic clutch or brake is simply a matter of using a control valve with a smaller orifice or adding additional air cavity between valve and unit. The latter is often done merely with an extra-long air line. Further, Fig. 8-15 suggests that peak input can be controlled by regulating clutch torque to the lowest possible level without causing slippage during the machine's work cycle.

Facing wear rate can finally be evaluated, once the engineer knows how much work the clutch or brake is doing during slip. In cyclic situations, this can be taken as the kinetic energy obtained from Eq. (8-7) or the area under the appropriate curve in Fig. 8-15, depending on which is evaluated.

You can find the number of days that a facing will last by taking the product of work done per cycle and cycles per day and dividing it into the total work that the facing can do before requiring replacement. For continuous slip applications, work done per unit time can be evaluated from Eq. (8-4) by taking N as the slip speed at the clutch or brake, and P as the rate of doing work. If slip speed is not constant, you may have to integrate or use average speeds during a period. Again, peak input rate becomes a factor in the cyclic applications because, as Fig. 8-11 suggests, the total work that one facing can produce depends on interface temperature, and that is directly related to peak input rate. Similarly, for continuous slip, the greater the service factor that can be used for a selection based on continuous thermal capacity, the longer the lining life.

If the application does not require some performance data, you can reduce selection time significantly by merely disregarding it. There is no magic formula for all situations—only experience can give a feeling as to how involved the evaluations must be. The purpose of Fig. 8-13 (and the purpose of this article) is to help guide the selection engineer as to what data must be evaluated and what is most important to the given application.

ELECTRICALLY ACTUATED DISK CLUTCHES AND BRAKES

For most applications where a shaft transmits mechanical power, an electric friction clutch or brake can control that power. Government regulations, particularly Occupational Safety and Health Administration (OSHA) and consumer protection laws, have added applications to which electric clutches and brakes are well suited. Applications include automotive air conditioning, trailer brakes, computer peripheral equipment, office equipment, standby power generators, elevators, escalators, cranes, hoists, conveyor systems, and machine tools.

8-11 General Characteristics

Electric clutches and brakes come in a wide variety of styles and sizes. Compared to other electric clutches, electric friction clutches have a favorable size-to-torque ratio. And their cost can be attractive, as evidenced by the use of electric friction clutches in the highly competitive automotive and agricultural equipment fields.

The electric friction clutch is used in many applications because its torque can be varied by altering applied voltage. Because the torque of these units varies with friction coefficient, they are not used for precise torque control, unless a feedback circuit is included.

Friction surface wear normally does not allow the electric friction clutch to be used for constant slip applications. However, when properly applied, life can be sufficient in some constant slip applications. Electric friction clutches and brakes are ideally suited for start-stop applications.

Electrically operated, friction clutches and brakes have two characteristics in common: They use friction to transmit motion, and an electromagnet to control the normal force applied to friction surfaces. The units can be ac or dc operated. They can be electrically set or electrically released. Some units have single-surface friction faces; others, multiple-disk designs. Also, teeth can replace conventional friction surfaces. Electric clutches and brakes may be operated in dry or wet (usually oil) environments.

The popularity of the electric friction clutch is due primarily to its favorable cost, torque-size relationship, and ease of control. These units are presently manufactured in sizes with a ⅞- to 42-in diameter, and with torque capacities ranging from a few ounce-inches to over 100,000 lb·ft. Normally, they can be supplied for any voltage required by the user.

Operating an electric clutch is usually as easy as opening or closing a switch. Electric power requirements are relatively low, with a typical requirement of less than 1 A.

8-12 AC Motor Brakes

The ac-operated units are available primarily as electric motor brakes (Fig. 8-16). The electric motor brake mounts on a C-flange electric motor with through shaft. The brake solenoid is wired to the motor leads. Because the brake is spring-set, the brake engages when you turn off power to the motor. Conversely, the brake releases when you apply electric power to the motor.

Ac motor brakes are usually multiple-disk, dry designs. They are available in many sizes to fit various motor frame sizes. A wide range of torque ratings suit application needs from a few ounce-inches to 1250 lb·ft. Enclosures come in dust-tight, waterproof, UL explosion-proof, Navy, commercial marine, and Bureau of Mines styles.

In addition, you can get modified versions with special friction disks, wear indicator devices, thermal switches, and a manual release. Typically, the brakes are available for floor mount, integral shaft application, or as a coupling between motor and load.

One characteristic of the ac unit is a high inrush current, which may be 10 times the holding current. The motor brakes also come with a dc solenoid. When operated on dc, a switch changes high energizing current to low current after the solenoid energizes.

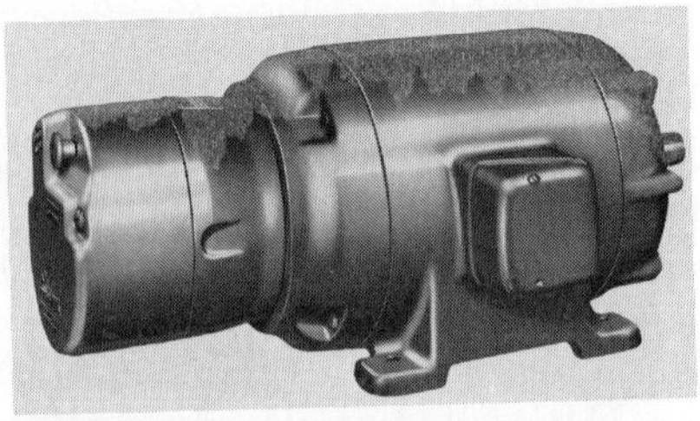

FIG. 8-16 Motor brake, mounted on C face of electric motor. (*Courtesy: PT Components, Inc., Stearns Division.*)

FIG. 8-17 Clutch-coupling, shown on common shaft only for illustrative purposes. (*Courtesy: PT Components, Inc., Stearns Division.*)

8-13 DC Clutch-Brake

The dc-operated units are available in a broader variety of styles than ac units. Although the dc units must operate on dc to the coil, a relatively low-cost rectifier can provide dc when only ac power is available. A dc unit has relatively high inductance and a slower reaction time than a comparable ac unit.

The most common dc clutch is a single-surface, magnetically set unit (Fig. 8-17). This device has high torque capacity for its size and a steel-to-steel friction surface. Such units normally operate dry. They come in a variety of configurations that include shaft mount, floor stand mount, clutch-brake combination, integral sheave, and motor-mounted designs.

Another common dc clutch is the multiple-disk unit (Fig. 8-18). This type is available in magnetically set and magnetically released models. Usually, multiple-disk units are small in diameter for their torque ratings. This characteristic can be advantageous in machine tool and related applications.

Multiple-disk units are available in "wet" configuration. An advantage of multiple-disk styles is easy replacement of worn disks. The single-surface clutch, by contrast, is often replaced with a new unit when the friction surface wears out.

The tooth clutch is a special type of electric friction clutch (Fig. 8-19). The tooth configuration gives the unit high torque capacity for its size, but engagement speeds must be low. However, the tooth clutch can be disengaged at high speeds. The tooth arrangement can be modified so the clutch engages in only one position. This feature is often used for indexing or synchronizing. Tooth clutches can be used in wet or dry environments. Like ac units, dc clutches and brakes can be ordered with a variety of modifications, special friction material, choice of operating voltages, and temperature ratings. If the demand warrants, many manufacturers will design a clutch or brake for a special application.

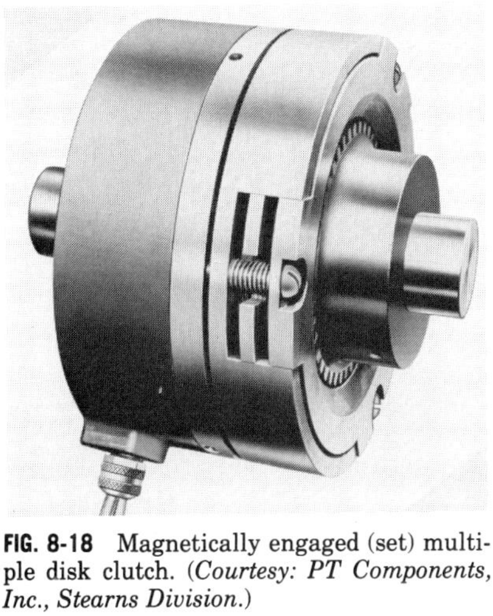

FIG. 8-18 Magnetically engaged (set) multiple disk clutch. (*Courtesy: PT Components, Inc., Stearns Division.*)

FIG. 8-19 Tooth clutch—clutch coupling. (*Courtesy: PT Components, Inc., Stearns Division.*)

8-14 Calculations for Selection

When you must calculate size requirements, it usually makes little difference what clutch or brake type you select for an application. The application analysis is always performed in the same manner. Calculate the torque requirement first.

Sometimes, you can size a clutch or brake to match the motor. An example would be a lathe spindle, where stopping time is not important. A reasonable decision is to stop the spindle at a deceleration rate equal to the acceleration rate. Then, the required brake torque can be calculated:

$$T = (P \times 5250)/N \tag{8-11}$$

where T = brake torque, lb·ft
P = motor power, hp
N = brake shaft speed, r/min

For a 3-hp, 1750-r/min motor,

$$T = (3 \times 5250)/1750 = 9 \text{ lb·ft} \tag{8-12}$$

Thus, we often find a 9- or 10-lb·ft motor brake mounted on a 3-hp, 1750-r/min motor. The calculation to size a clutch is similar.

In some cases a specific minimum or maximum time for starting or stopping may be required. For example, to select a brake, use

$$T = (WK^2 \times N)/(308 \times t) \tag{8-13}$$

where T = brake torque, lb·ft
WK^2 = rotating mass inertia reflected to the brake shaft, lb·ft²
N = brake shaft speed, r/min
t = stopping time, s

Assume you must stop a lathe spindle in a maximum time of 1 s. The spindle inertia is 2 lb·ft^2; spindle speed, 875 r/min. The motor and gear train have an inertia of 0.5 lb·ft^2. Motor speed is 1750 r/min. If you must mount a brake on the motor, you must know the total inertia reflected to the brake shaft. The spindle inertia reflected to the motor shaft is proportional to the square of the speed ratio, so that:

$$\frac{WK_1^2}{WK_2^2} = \left(\frac{N_2}{N_1}\right)^2 \tag{8-14}$$

and

$$\begin{aligned} WK_1^2 &= 2 \text{ lb·ft}^2 \times (875/1750)^2 \\ &= 0.5 \text{ lb·ft}^2 \end{aligned} \tag{8-15}$$

The total inertia reflected to the brake shaft is 0.5 lb·ft^2 + 0.5 lb·ft^2 or 1.0 lb·ft^2. The speed of the brake shaft is 1750 r/min. Thus, the braking torque requirement is

$$\begin{aligned} T &= (WK^2 \times N)/(308 \times t) \\ &= (1.0 \times 1750)/(308 \times 1) \\ &= 5.7 \text{ lb·ft} \end{aligned} \tag{8-16}$$

Another frequently encountered brake application is the overhauling load, as in a hoist application. Assume a hoist has a 3-hp, 1800-r/min motor, the maximum load is 2000 lb and the hook speed is 0.5 ft/s. The load must stop in 0.25 s. The hoist drum has a 6-in diameter and 19.1 r/min speed. Motor and speed reducer inertia is 0.5 lb·ft^2.

In this case, the load inertia is

$$\begin{aligned} WK^2 &= WR^2 = 2000 \times (0.25)^2 \\ &= 125 \text{ lb·ft}^2 \end{aligned} \tag{8-17}$$

The load inertia reflected to the brake is WK^2 = 125 lb·ft^2 × (19.1/1800)^2 = 0.014 lb·ft^2. The total inertia is 0.5 + 0.014 = 0.514 lb·ft^2. Thus, the torque required for deceleration is

$$\begin{aligned} T &= (WK^2 \times N)/(308 \times t) \\ &= (0.514 \times 1800)/(308 \times 0.25) \\ &= 12 \text{ lb·ft} \end{aligned} \tag{8-18}$$

In addition to deceleration torque, you must also consider the holding torque. Holding torque at the drum is 2000-lb load × 0.25-ft drum radius = 500 lb·ft. At the motor, the equivalent torque is 500 lb·ft × (19.1/1800) or 5.3 lb·ft. The required brake torque is the sum of the deceleration torque and the holding torque. Thus,

$$T = 12 + 5.3 = 17.3 \text{ lb·ft} \tag{8-19}$$

Although it may not be obvious at first, the necessity for adding the holding torque and deceleration torque is clear if you consider the effect a brake has with only a 5.3-lb·ft rating. This 5.3-lb·ft brake would only hold the descending speed constant. Any additional brake torque will decelerate the load.

In our example, the stopping time for an ascending load is faster. Stopping time can be calculated as:

$$t = (WK^2 \times N)/308 \times T \tag{8-20}$$

When ascending, the torque used in Eq. (8-20) is 17.3-lb·ft brake torque rating plus the 5.3-lb·ft torque effect of the load, or 22.6 lb·ft total. Thus, when ascending:

$$\begin{aligned} t &= (0.514 \times 1800)/(308 \times 22.6) \\ &= 0.133 \text{ s} \end{aligned}$$

The same calculation procedure applies for a clutch, when the load consists of steady friction or drag load and an inertia that must be accelerated.

In an application where a clutch or brake cycles frequently, you must calculate the required thermal capacity. The required heat dissipation rate of the brake (hp-sec/min) is:

$$\text{HP}\,\frac{\text{sec}}{\text{min}} = \frac{WK^2 \times N^2 \times n}{3{,}220{,}000} \tag{8-21}$$

where WK^2 = the inertia of the rotating mass reflected to the brake shaft, lb·ft²
N = the brake shaft speed, r/min
n = the cycles per minute

This equation is valid only if n is greater than 1.

When infrequent but heavy inertia loads must be stopped, you must also calculate the "crash stop" requirement, a measure of the required brake's heat absorption capacity. This thermal requirement can be calculated with either of the following equations:

$$\text{KE} = \frac{(WK^2 \times N^2)}{5875} \tag{8-22}$$

or by

$$\text{KE} = 0.0524 \times N \times t \times T \tag{8-23}$$

where KE = kinetic energy, ft·lb
WK^2 = rotating mass inertia reflected to the brake, lb·ft²
N = the brake shaft speed, r/min
t = stopping time, s
T = brake torque, lb·ft

8-15 Selection Criteria

Once you know the torque and thermal requirements, you can select a brake or clutch. In the previous examples, the torque requirement was for dynamic (slipping) torque. Although clutches and brakes are rated commonly in static torque, you can approximate dynamic torque. For a dry multiple-disk unit, dynamic torque is 80 percent of static torque. For a single-surface steel-on-steel unit, dynamic torque is 60 percent of static torque.

Actually, dynamic torque varies with slipping speed, as shown in Figs. 8-20 and 8-21. Such curves are typically available from the manufacturer.

When a clutch or brake must start or stop rapidly, the reaction time must be considered, especially in dc magnetically set units (Fig. 8-22). The reaction time of spring-set, ac solenoid-released units is fast enough so that it can be neglected (Fig. 8-23). The reaction time of spring-set, magnetically released dc units varies considerably, depending on the unit and suppression circuit type. Contact the manufacturer for specific details if set or release characteristics are important.

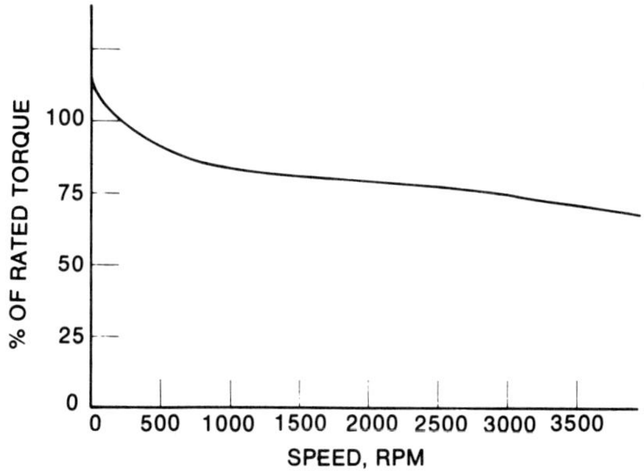

FIG. 8-20 Typical dynamic torque-steel on steel and friction material.

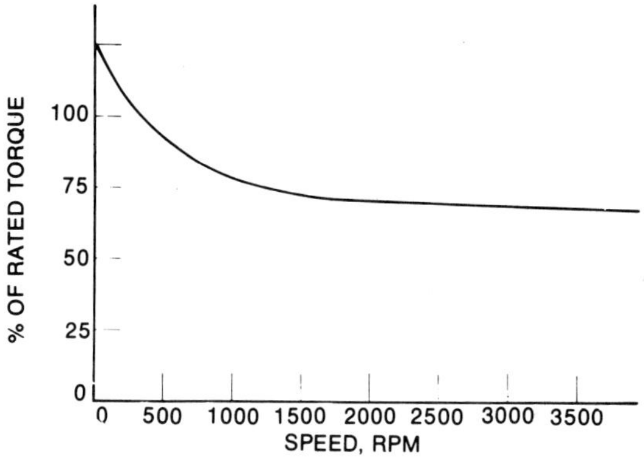

FIG. 8-21 Typical dynamic torque-friction material on cast iron.

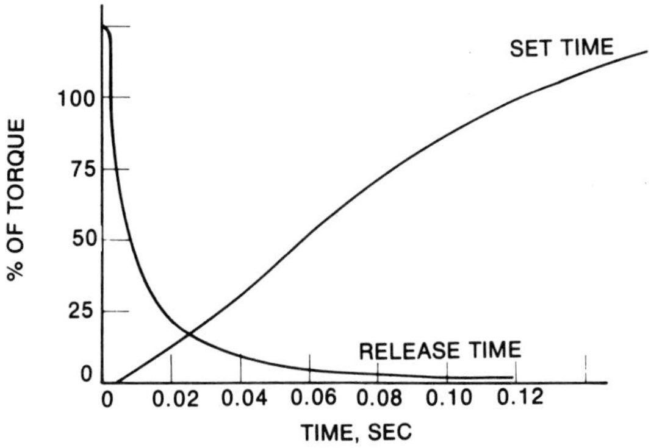

FIG. 8-22 Typical torque vs. time—4-in dc clutch.

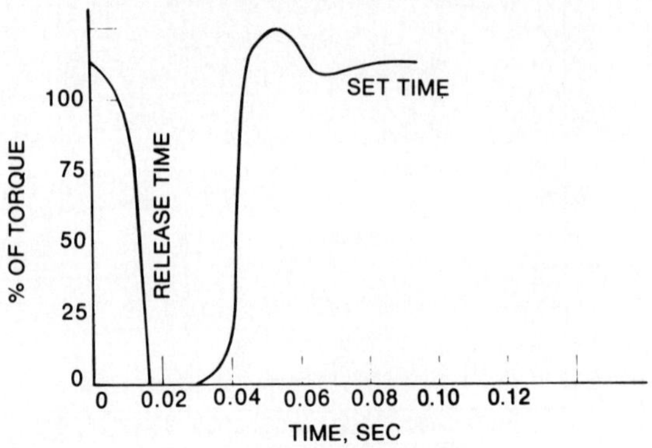

FIG. 8-23 Typical torque vs. time—ac brake.

Commonly, manufacturers give clutch and brake thermal ratings for average room temperature conditions at 1800 r/min. They base their ratings on arbitrarily selected duty cycles. However, the units' effective ratings vary when application conditions vary.

Friction material wear life is a function of temperature (Fig. 8-24). The life of a unit operating near its thermal capacity is considerably less than one-half the life of the same unit operating at one-half its thermal capacity.

Clutch and brake selection should not be made without considering type of service and applicable service factors. Service factors (as in Table 8-3) make an adjustment for differences in operating conditions. For example, a 10-hp single-cylinder engine at 3600 r/min has an average torque of 15 lb·ft. However, even with a flywheel, the engine's peak torque is much higher. Applying a service factor adjusts for the difference between average torque and peak torque.

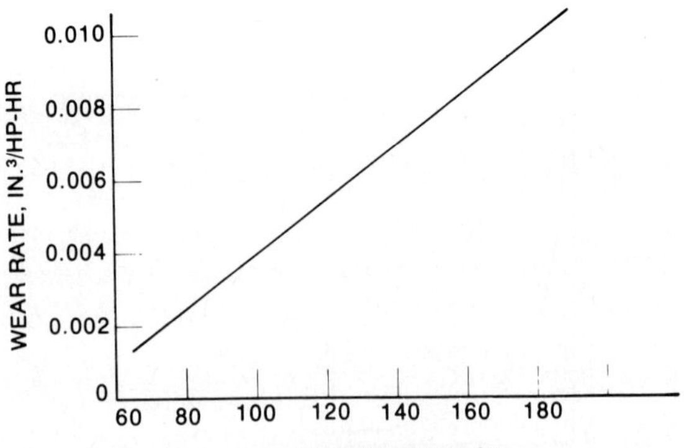

FIG. 8-24 Typical wear rate vs. temperature—friction material on cast iron.

TABLE 8-3 Service Factors for Clutch and Brake Selection

Source of power	Type of driven load*	Number of clutch engagements per hour			
		3 or less	4 to 9	10 to 18	19 to 30
Uniform drive Electric-motor, steam-turbine, steam-engine (belt drive), lineshafts	Steady	1.5	1.8	2.25	2.85
	Light shock	1.75	2	2.5	3.1
	Severe shock	2	2.3	2.75	3.4
Uneven drive, moderate impulses Gas or diesel engine, 4 or more cylinders	Steady	2	2.3	2.75	3.4
	Light shock	2.5	2.8	3.25	3.9
	Severe shock	3	3.3	3.75	4.4
Uneven drive, severe impulses Gas or diesel engine, 1 to 3 cylinders	Steady	2.5	2.8	3.25	3.9
	Light shock	3	3.3	3.75	4.4
	Severe shock	3.5	3.8	4.25	4.9

* **Type of driven load.** *Steady*—starting torque is low, running load is fairly smooth and uniform, without shocks. *Light shock*—starting torque is somewhat greater than running torque, but peak loads are comparatively low and infrequent. *Severe shock*—starting torque and peak loads are 200 to 250 percent of running load. Fluctuating running load, heavy shocks, or reversals occur frequently.

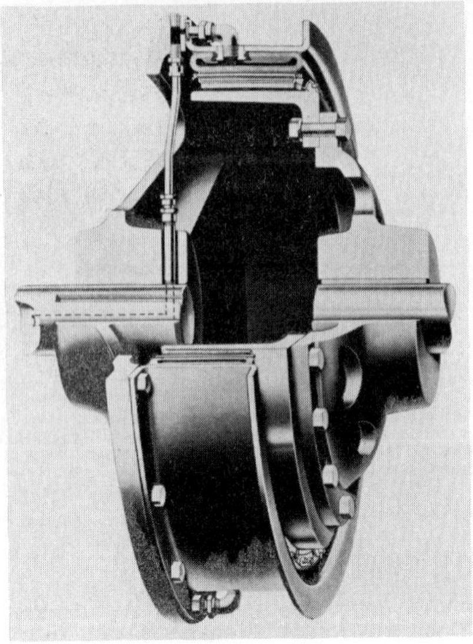

FIG. 8-25 Drum-type clutch. (*Courtesy: Eaton Corp., Industrial Drives Operations—Airflex Division.*)

Also, consider the consequences should a clutch or brake unit lose torque capacity completely or partially. In the lathe example, it might not matter if the spindle took 1.5 s to stop, instead of the planned 1 s. However, in a hoist application, if the brake failed to hold the load as planned, property damage or injury could result. Obviously, a conservative safety factor should be applied in proportion to the consequences of partial or complete failure. In many applications, it is wise to use a redundant or back-up braking system. In passenger elevator applications, this is required by law.

AIR-ACTUATED DRUM-TYPE CLUTCHES[4]

The drum-type clutch, with an air actuating tube, is one of the simplest clutches on the market. It has only one moving part—the air tube. The drum design can accommodate oversized shafts and is readily adapted to any type of drive arrangement (Fig. 8-25).

Drum-type clutches consist of two major components: a drum, which is usually the output side, and the friction shoes with actuating means, which are usually the input side. They can be of the constricting or the expanding type. In the constricting type (Fig. 8-26), the friction shoes grip the outside surface of the drum. This design takes advantage of centrifugal force acting on the friction shoes, preventing unintentional engagement and friction wear from drag.

The shoes grip the inside surface of the drum on the expanding type (Fig. 8-27). Here, depending on operating speed, centrifugal force tends to keep the shoes engaged with the drum.

The simplest way to make the shoes engage the drum is by an air-actuated, rubber-and-cord tube similar to an auto tire. The tube is bonded to a rim's inside surface, if constricting; outside surface, if expanding (Figs. 8-26 and 8-27). Pressurized air expands the tube, making the shoes engage the drum. The tube cords transmit torque. The cords also add strength and firmness to the rubber. The rubber's resiliency keeps the shoes free of the drum, cushions clutch engagement, and compensates for parallel and angular misalignment. The tube design compensates automatically for friction shoe wear.

8-16 Torque Capacity

You can calculate torque capacity of drum-type clutches from the basic equation in Fig. 8-28.

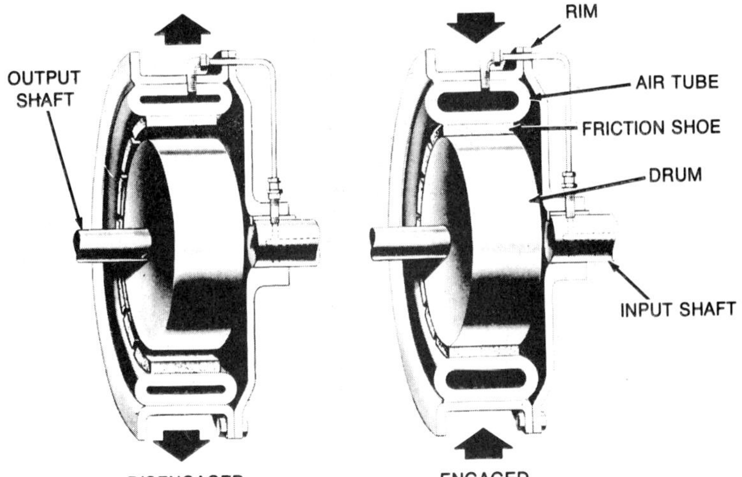

FIG. 8-26 Constricting-type clutch. (*Courtesy: Eaton Corp., Industrial Drives Operations—Airflex Division.*)

Capacity depends on drum radius, tube dimensions, applied pressure, and friction coefficient. For a given clutch size, torque is proportional to applied pressure. By fixing the maximum applied pressure at a given value, the clutch can serve as an overload device, slipping when a specified torque is exceeded. Or, by regulating the rate at which pressure builds up in the tube, you can get a very short or very long acceleration period.

Figure 8-28 shows that in a drum-type clutch, contact velocity across the friction material is constant, resulting in even friction lining wear. Also, each friction shoe carries its share of the load. Because the engaging force is applied radially, there is no axial thrust.

In any application, the clutch must

- Transmit required torque
- Dissipate heat generated during acceleration

You can calculate the torque required to accelerate from:

$$T = 0.0391(WK^2)(N)/t \quad (8\text{-}24)$$

where T = average torque, lb·in
WK^2 = total inertia, lb·ft^2
N = speed, r/min
t = time, s

There are applications where the regular running torque is larger than

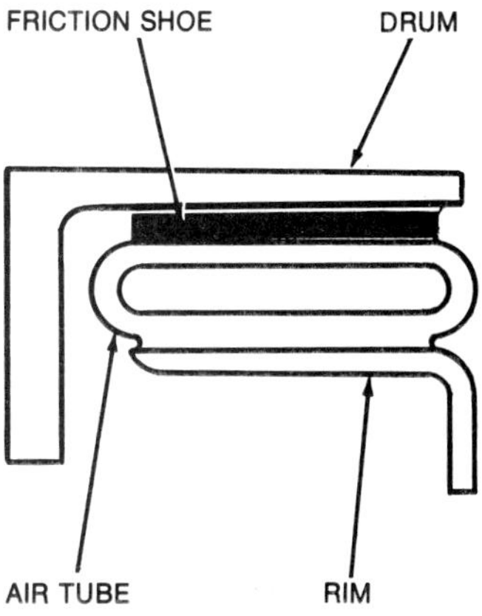

FIG. 8-27 Expanding-type clutch.

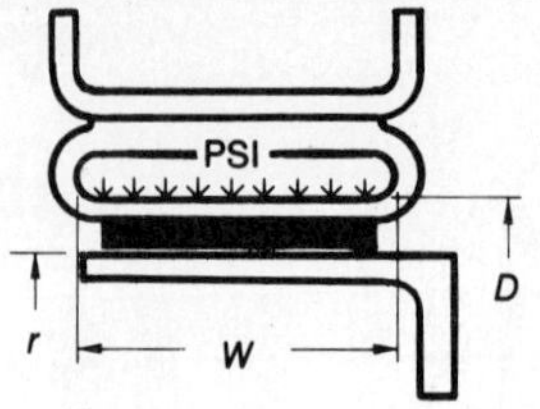

D = TUBE DIAMETER, IN.

f = COEFFICIENT OF FRICTION

PSI = APPLIED PRESSURE, psi

r = DRUM RADIUS, IN.

W = TUBE WIDTH, IN.

F = FRICTIONAL FORCE, LB

N = NORMAL FORCE, LB

T = TORQUE, LB-IN.

$N = \text{AREA} \times \text{PSI}$
$= \pi DW \times \text{PSI}$

$F = fN$
$= f\pi DW \times \text{PSI}$

$\text{TORQUE} = Fr$
$= f\pi DW\ \text{PSI}\ r$

FIG. 8-28 Constant contact velocity across friction material means even wear.

the torque to accelerate. In these situations, determine the torque from

$$T = (P)(63{,}000)/N \tag{8-25}$$

where P = running horsepower

8-17 Heat

The performance of a clutch is affected by temperature more than any other condition. High temperature reduces friction coefficient, which results in loss of torque and increased wear rate.

The heat generated in accelerating a drive to running speed equals the energy acquired by the drive. Energy acquired is

$$E = WK^2(N)/5873 \text{ ft·lb} \tag{8-26}$$

This is also the energy that the clutch must dissipate per start. If starts are cyclic, the clutch must have enough cooling capacity to prevent temperature build-up. If the starts are very infrequent, a temperature should not be reached that would cause fading or loss of friction coefficient while accelerating.

Although both clutches have excellent heat-dissipating characteristics, the outside drum surface of the expanding type exposes greater surface area for radiation. For this reason, the expanding type is used in continuous slip or drag situations, like providing back tension for unwind or rewind applications.

You can approximate drum temperature rise from:

$$\Delta T = E/778MC \tag{8-27}$$

where ΔT = temperature rise, °F
E = energy input, ft·lb
M = mass of drum, lb
C = specific heat of drum, Btu/lb·°F

Keep temperature rise as low as possible. For cyclic applications, it should not exceed 250°F. For infrequent starts, limit the temperature to 400°F. If these temperatures are exceeded, consider external cooling. You can direct an air blower at the clutch, or change the drum configuration to accept water cooling.

8-18 Mounting

Figure 8-29 shows several clutch mounting arrangements. Arrangements *A* and *B* couple two shafts. Here, besides performing its regular function, the clutch also replaces a coupling. The *X* dimension in arrangement *A* permits clutch servicing without moving the driving and driven shafts. The clutch element and the drum will slide through the gap between the shafts. Arrangement *B* conserves space and requires moving one shaft for servicing.

The through-shaft mounting, arrangement *C*, can locate between shaft-support bearings or on an overhanging shaft extension. A sheave, sprocket, or gear can mount on the bearing-mounted hub. Arrangement *D* can mount only outboard of a shaft-support bearing.

8-19 Control

You can control the clutch in a number of ways. Figure 8-30 shows three control systems.

System *A* uses a modulating valve that lets the operator hand-control the rate of pressure build-up in the clutch and, hence, the smoothness of engagement. When the drive is up to speed, the valve can lock open to full line pressure.

The flow control valve in system *B* determines the rate of pressure build-up through a presetting. The operator controls only the *on-off* valve. In system *C*, a solenoid replaces the manual *on-off* valve. The valve can be actuated by the manual push-button station shown, or through a signal received from a timer or control panel.

8-20 Applications

Air-actuated, drum-type clutches are used to good advantage by the petroleum industry on rotary drilling rigs. Rigs must be moved continually from one drilling location to another. The rubber actuating tube compensates for misalignment resulting from the move.

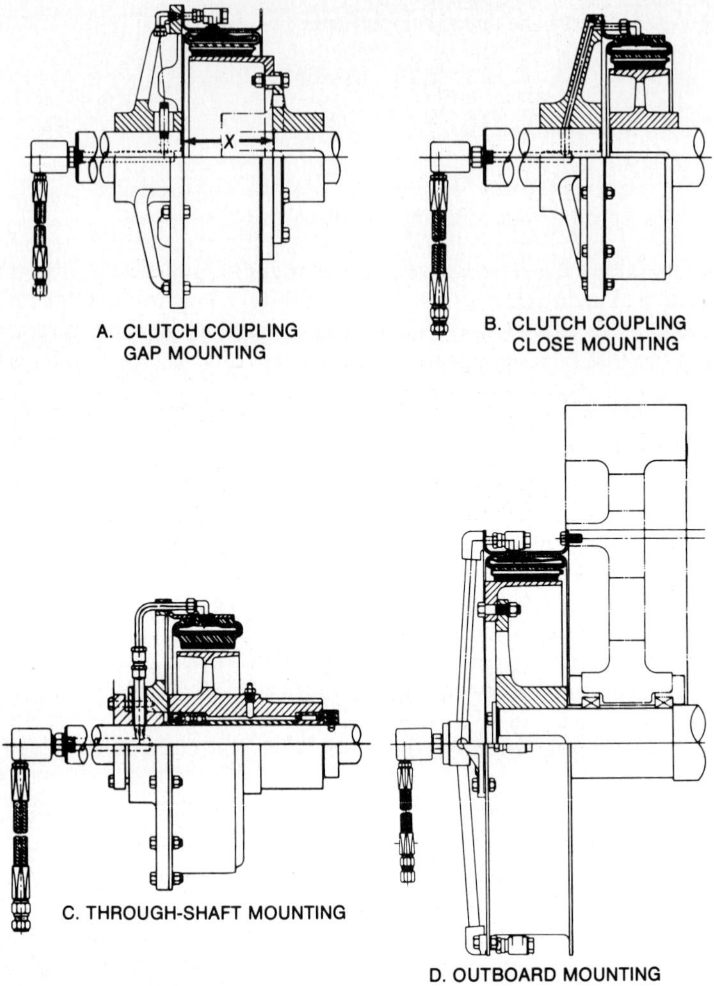

FIG. 8-29 Clutch mountings.

Internal-combustion engines power almost all rigs. The damping characteristics of the rubber tube let it absorb most of the impulse torque of the engine. The tube protects the driving and driven members, resulting in longer life for shafts, bearings, chains, and gears.

The Super 7-11 drill rig by Ideco Div., Dresser Industries, Inc., for example, is equipped with 10 drum-type clutches. They serve on high and low cable drums, engines, rotary tables, sandlines, and pump drives.

The air-actuated, drum-type clutch is widely used in the metal-stamping industry. The shock-absorbing ability of the rubber tube allows higher operating speeds, and prevents broken crankshafts and press frames. At the same

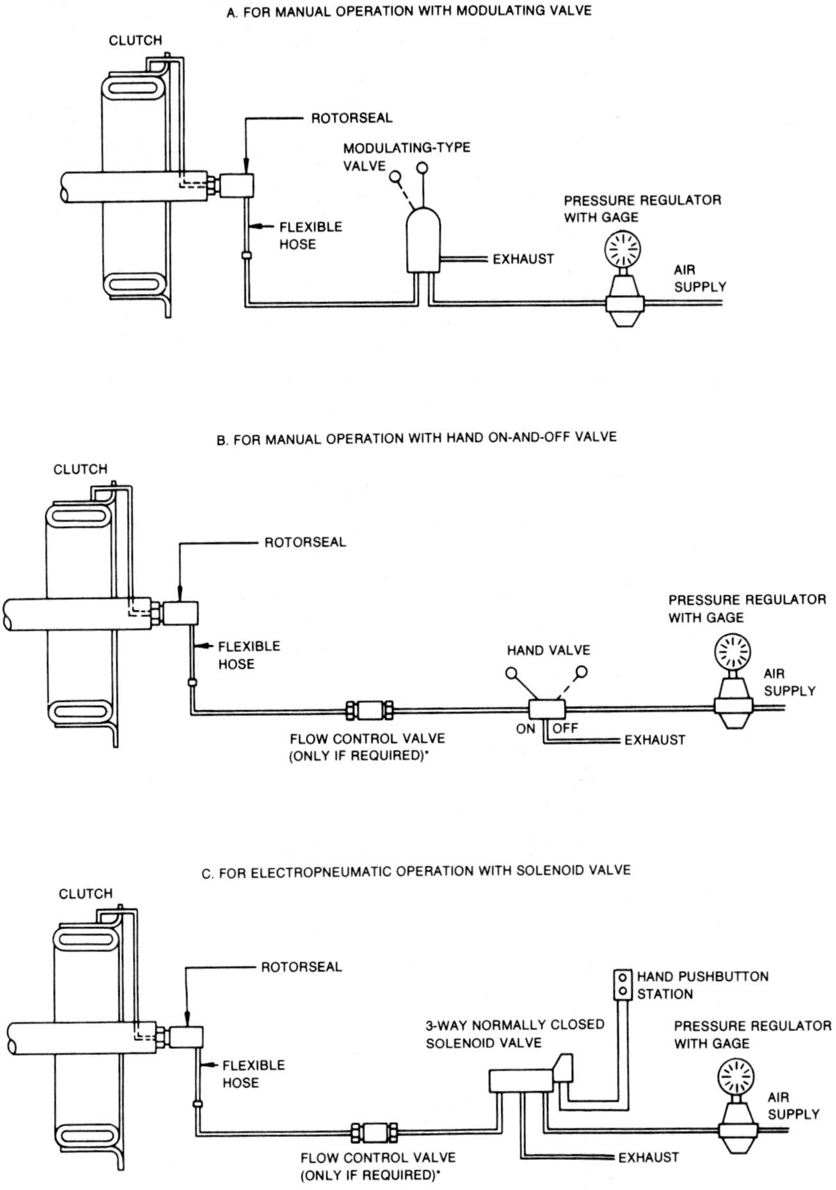

*USE VALVE WITH FREE FLOW IN EXHAUST DIRECTION.

FIG. 8-30 Typical clutch controls.

time, there is no problem from gear backlash. This style of clutch is furnished as original equipment on presses by many manufacturers.

MAGNETIC-PARTICLE CLUTCHES AND BRAKES[5]

Because of rising material and labor costs and the need for higher productivity, power-transmission systems are requiring more control. In the past, "control" meant an operator watching a process and making manual adjustments. Today, magnetic-particle clutches and brakes provide automatic torque control which is linear, repeatable, and independent of speed.

The magnetic-particle clutch is not a "new" device. The principle of torque control between two members by means of iron particles was developed in 1948. In May of that year, Jacob Rabinow disclosed work he had done on the "magnetic fluid clutch."

Essentially, Rabinow's clutch was a disk running in a mixture of iron powder and light machine oil. This clutch had a linear torque-current relationship which was nearly independent of speed. Rabinow and others in this period used oil or lubricant mixed with finely divided iron powder (carbonyl iron). While this mixture produced torque quite well, the oil would disintegrate because of heat generated under slipping conditions. As a result, fluid-mixture clutches were a short-lived phenomenon.

8-21 Dry-Magnetic-Particle Clutch

The dry-magnetic-particle clutch was developed in the 1950s, largely by work sponsored by the U.S. Government. This clutch contains either iron particles or iron particles plus a dry lubricant, but contains no oil.

More recent developments in magnetic-particle devices have been for high torque and heat dissipation units for industrial power-transmission systems. These units dissipate more than 5 hp continuously.

8-22 Principles of Operation

In a magnetic-particle clutch (Fig. 8-31), the rotor, a smooth, solid cylinder, is surrounded by the drive cylinder, a smooth, hollow cylinder. Each element is supported by a pair of ball bearings. The magnetic powder is between the two rotating cylinders, and is contained by the end shield arrangement. The size, shape, and chemical composition of the powder is carefully controlled to ensure proper clutch operation.

The rotating members and their bearings are supported by the stator, which also contains the coil. The narrow space between the stator and drive cylinder is the "air gap," and the space between the drive cylinder and rotor is the "powder gap."

The drive cylinder is connected to the prime mover, and the rotor is coupled to the load. With no coil excitation, the powder moves freely around in the drive cylinder, and there is no coupling effect to the rotor. A flux field is

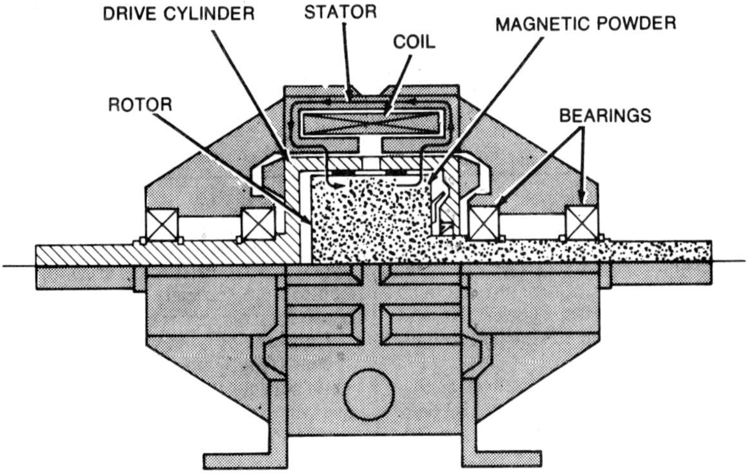

FIG. 8-31 Magnetic-particle clutch.

established when the coil is energized. The flux path is around the coil, through the stator, and across the air gap. Then, the flux path is through the powder gap and powder and into the rotor. Here the flux path is through the powder, drive cylinder, and stator, completing the circuit.

When the flux penetrates the powder, it causes the powder to "link up," binding rotor to drive cylinder. The greater the flux, the greater the bonding action. The bond between rotor and drive cylinder is the torque-coupling effect. Since the rotor and drive cylinder are symmetrical, this torque coupling is independent of the motion of the parts; that is, torque is related only to coil current and is not affected by rotation of either member.

The magnetic particle clutch shows no stick-slip, cogging, or slip-speed dependency. It produces the same torque under forward, reverse, and stationary conditions and is capable of 100 percent lock-up with no slip.

Magnetic-particle brakes operate on the same basic principles as clutches, except there is no drive cylinder, and braking action takes place between rotor and stator.

8-23 Advantages

Since the torque-producing action takes place among many small particles, magnetic-particle clutches and brakes have these characteristics:

Torque Proportional to Current This allows system control by a low-level electric signal (Fig. 8-32).

Torque Independent of Speed This means a system can run at low speed for set-up; then speed can be increased with no torque change. Acceleration is linear and repeatable, with no cogging or jerking at lock-up. Also, application calculations are exact (Fig. 8-33).

Torque Repeatability This means the same results in processing time after time.

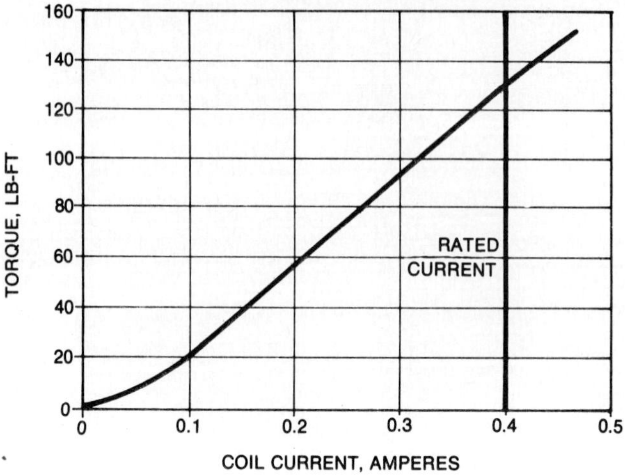

FIG. 8-32 Typical torque vs. current (90 V dc).

Soft Starts and Stops Since torque builds up gradually, smoother acceleration and less inertial loading mean longer mechanical life.

Silent Operation The clutch is silent, since only a barely audible sound is heard when the clutch is slipping, so OSHA noise regulations can be met, particularly on low-speed slipping applications where friction surfaces tend to squeal.

Long Life Since torque action takes place in the powder, the small amount of wear also takes place in the powder. Therefore, there is never a rapid or complete loss of torque. The torque-vs.-current curve merely rotates slightly downward, and by increasing current, you can get the original torque desired.

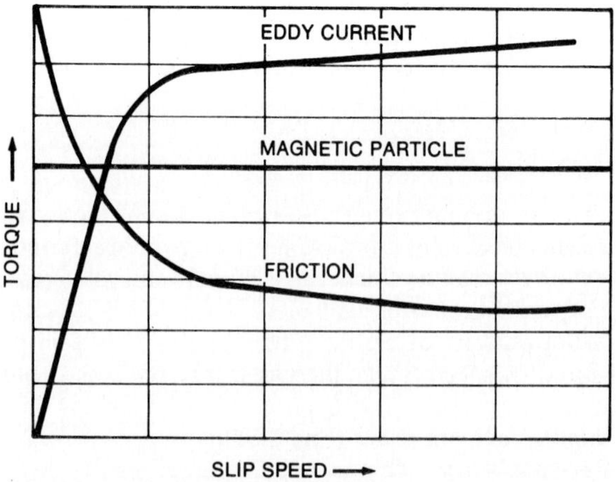

FIG. 8-33 Torque variation with slip speed.

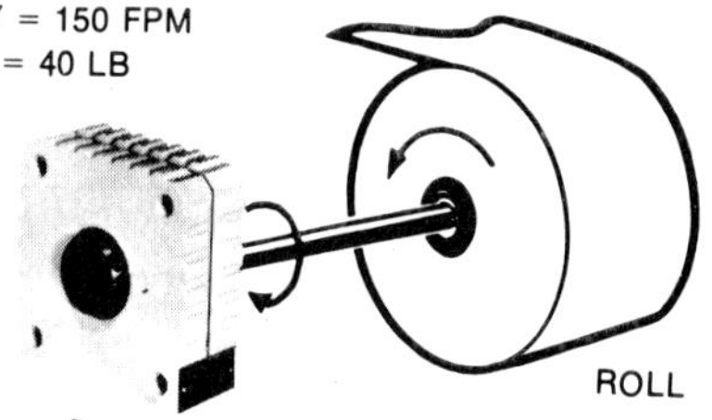

FIG. 8-34 Unwind.

8-24 Typical Applications

Here are some torque-control applications for magnetic-particle clutches and brakes:

Tension Control One of the largest applications is tension control in web-handling equipment. This includes materials like cord, thread, textiles, wire, film, paper, tape, and foil. On an unwind (Fig. 8-34), for example, a brake provides constantly decreasing torque as the size of the roll decreases. The magnetic-particle brake can be easily controlled to adjust torque manually or by a roll follower, dancer roll, or strain-gage sensor feeding the appropriate controller.

Speed is also increasing during unwinding, but these controls do not need to compensate for a changing torque-vs.-speed characteristic. Therefore, these controls are generally much simpler than those for non-magnetic-particle brakes.

Constant or controlled taper tension rewinds (Fig. 8-35) are also simple using magnetic-particle clutches. Typically, an ac gearmotor drives the clutch, which in turn drives the rewind core. Clutch speed is slightly greater than web speed. Thus, tension is controlled by clutch torque, and speed is always perfectly matched and doesn't have to be synchronized to the main machine speed. This allows web tension to be maintained when the main drive stops. The

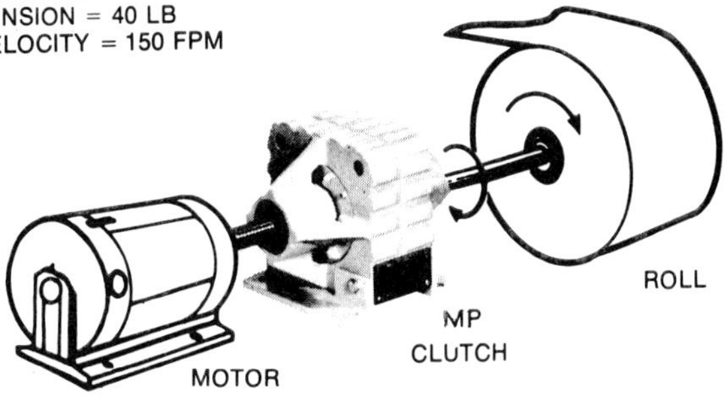

FIG. 8-35 Rewind.

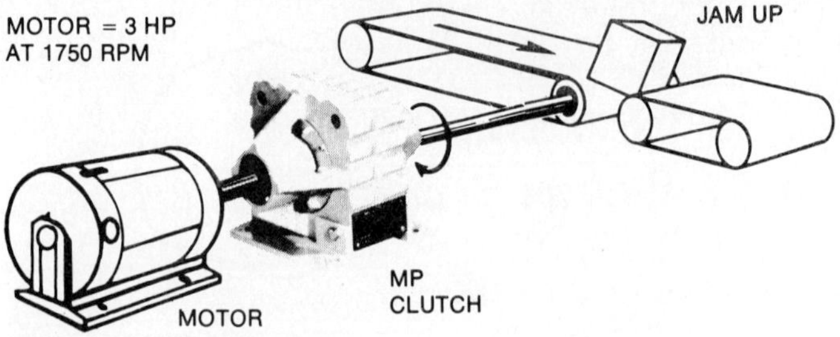

FIG. 8-36 Overload protection.

same type of power supplies may be used for both rewind and unwind tension control.

Torque Limiting The clutch's ability to slip at a torque value set by the coil current, and to do this regardless of the speed at which it is running, makes the magnetic-particle clutch the perfect torque limiter. This torque-limiting ability is also completely adjustable while running.

A typical application is a conveyor drive (Fig. 8-36). Torque can be programmed to a level just high enough to run the conveyor, but low enough so that any overload, no matter how small, will stop the conveyor.

Frequently, zero-speed switches are used in this application, so that when the conveyor jams, the clutch slips and output stops. The zero-speed switch then shuts down the drive system to provide personnel protection and to prevent start-up when the jam is cleared.

Soft Starts and Stops Since torque engagement is a build-up of "links" in the powder, there are no hard, jarring torques to damage machine components. The load is picked up smoothly and continuously, with no jerking or cogging when the load reaches synchronization.

Conveyors are typical soft-start-and-stop applications. Controlled acceleration eliminates swinging of parts on hanging conveyors and spilling of products from uncapped containers. Also, when used as a soft-start device on a motor (Fig. 8-37), the clutch will reach 100 percent of motor speed and then drive without slip, which means no power waste.

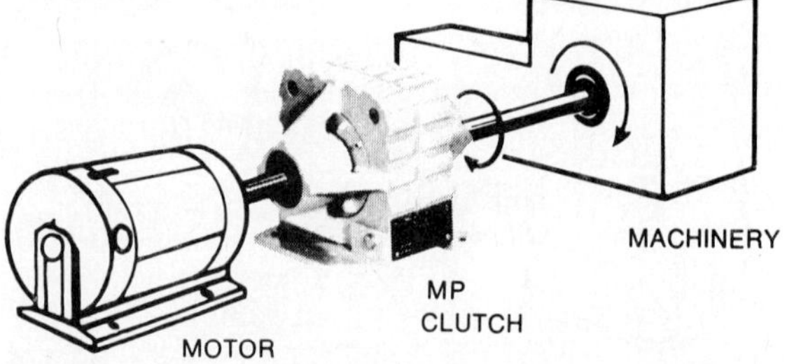

FIG. 8-37 Soft start.

Dynamometers and Test Equipment Repeatability, controllability, and smooth torque delivery make the magnetic-particle clutch or brake an excellent load for a dynamometer. In a typical application on electric motors, the motor drives the brake, which is restrained by a load cell. By manually increasing torque, the motor is loaded from no load (full speed) to full load, then to stall. With an X-Y recorder, the torque-vs.-rotational speed curve can be drawn in seconds.

8-25 Selection Procedure

You should use the same equations selecting a magnetic-particle clutch as you would use for selecting any type of clutch. One exception: Since torque is not related to speed, true rather than average torque values are used. Only torque and slip heat need to be calculated.

Slip heat is generated in the clutch when the clutch slips under load. The heat is power input minus power output, and the calculations are the same for all clutches and brakes, no matter what type. The clutch's ability to dissipate heat is a measure of its usefulness. Most applications for magnetic-particle clutches are ones in which there is appreciable or continuous slip, so consider carefully the size and type of cooling.

To select a clutch or brake, only a few equations are needed:

$$T = FR \tag{8-28}$$

where T = torque
F = force
R = radius

$$N = V/\pi D \tag{8-29}$$

where N = speed, r/min
V = velocity, ft/min
πD = circumference, ft

$$Q = TS/7 \tag{8-30}$$

where Q = slip heat, W
T = torque, lb·ft
S = slip, r/min

The third equation can be rewritten as:

$$P = TN/5252 \tag{8-31}$$

where P = power, hp
T = torque, lb·ft
N = speed, r/min

but 746 W = 1 hp, so multiplying both sides of the equation by 746 W/hp, the constant becomes approximately 7.

Example 1. Select an unwind tension brake for the conditions of Fig. 8-34.

$$T = FR$$
$$T = (40)1 = 40 \text{ lb·ft}$$
$$N = V/\pi D$$
$$N = 150/2\pi = 24 \text{ r/min}$$
$$Q = TS/7$$
$$Q = (40)24/7 = 137 \text{ W}$$

Therefore, the brake must produce up to 40 lb·ft torque and dissipate 137 W continuously.

Example 2. Select a clutch for the constant tension rewind conditions of Fig. 8-35. The gearmotor runs at fixed speed.

$$T_{max} = (40)1 \text{ lb·ft}$$
$$\text{Clutch input (motor) rpm} = 150/(0.75)\pi = 72 \text{ r/min}$$
$$N_{min} = 150/2\pi = 24$$
$$Q = 274 \text{ W}$$

Therefore, a clutch with at least 40 lb·ft torque and 274 W heat dissipation capacity is required.

Example 3. (Fig. 8-38) Select a brake to stop an 8-lb·ft² inertial load in 2 s. The load rotates at 1750 r/min.

$$I = I(N)/308t$$

where T = brake torque, lb·ft
I = load inertia, lb·ft²
t = braking time, s

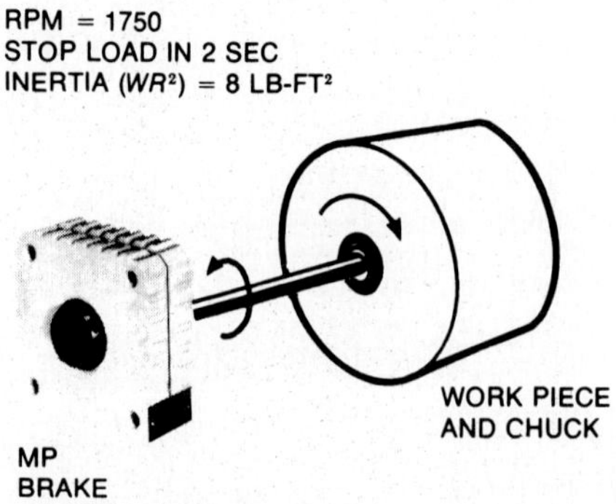

FIG. 8-38 Soft stop.

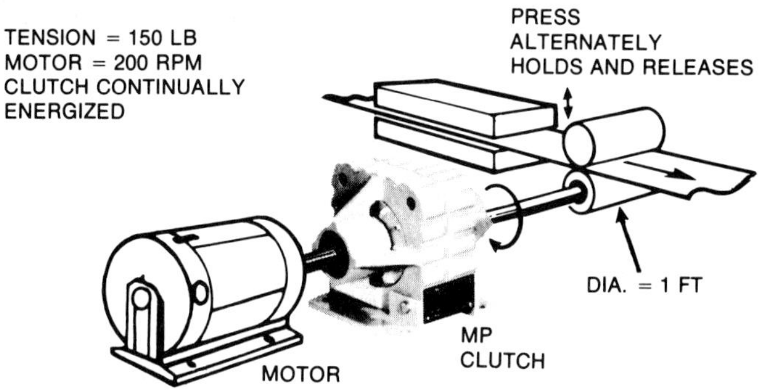

FIG. 8-39 Cycling.

$$T = 8(1750)/308(2) = 22.7 \text{ lb·ft}$$

Therefore, a brake with at least 23 lb·ft torque is required.

Example 4. Select a clutch to satisfy the overload condition caused by Fig. 8-36 conditions.

$$T = P(5250)/1750 = 9 \text{ lb·ft}$$

Therefore, select a clutch that will produce 9 lb·ft torque.

Example 5. Select a clutch for the cycling process of Fig. 8-39.

$$T = FR$$
$$T = (150)(0.5) = 75 \text{ lb·ft}$$
$$Q = TS/7 = 2142 \text{ W}$$

But the clutch only slips half the time, therefore $Q = 1071$. Select a clutch with torque = 75 lb·ft that can dissipate 1071 W of heat.

Example 6. Select a clutch that satisfies Fig. 8-37 conditions. A soft start is desired

$$T = P(5250)/N$$
$$T = 2(5250)/1800 = 5.8 \text{ lb·ft}$$

A clutch producing 5.8 lb·ft torque is required.

8-26 Cycling

For cycling applications (Fig. 8-38), the torque can be determined from either the motor power or the load inertia. Slip heat is calculated during the slip period and averaged over the total cycle time. This is safe, since the clutch torques are exact, not approximate.

8-27 Maintenance

If the torque level becomes too low and increasing coil current doesn't restore it to a satisfactory level, the unit must be rebuilt. Repair kits consisting of bearings, seals, snap rings, and powder are available. Disassembly and kit installation will restore the clutch in about 1 h.

8-28 Controls

Since the coil operates on 0 to 90 V dc, most controls are simple. Usually an SCR power supply is used, and control selection is based on application rather than torque rating.

OIL-SHEAR CLUTCHES[6]

Although the basic principle of oil-shear drives dates back to the nineteenth century, heavy-duty industrial oil-shear drives are less than 20 years old. The earliest drives were clutches that were immersed in oil and applied to limited cyclic duty. Primary application was in the automotive field. Modern automotive transmissions still use oil-shear principles, but with considerable refinement in design.

The first heavy-duty industrial applications were on large stamping presses. High inertia and frequent cycling had made other clutches fail. Success on the presses soon led to applications elsewhere.

8-29 Theory

The flow of any fluid gives rise to tangential friction forces, called *viscous forces*. The action of internal shearing forces causes a degradation of mechanical energy into shear or unavailable thermal energy.

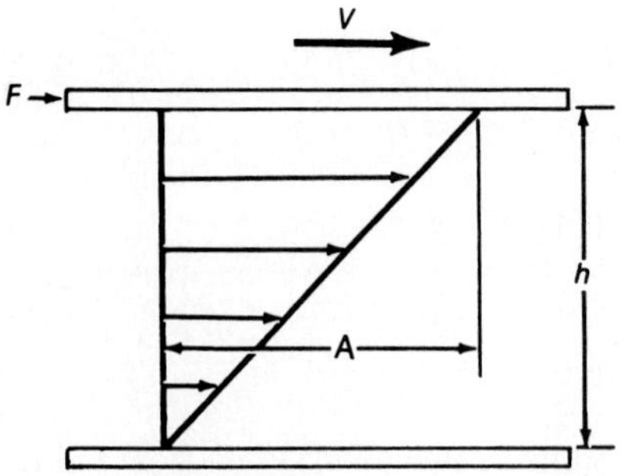

FIG. 8-40 Flow between parallel plates. Lower plate is stationary.

The upper plate in Fig. 8-40 moves with velocity V. The lower plate is stationary. A very thin layer of fluid moves with velocity V. We will assume that the fluid flows in parallel layers or laminas, and that no secondary irregular fluid motion is superimposed on the main flow. This kind of flow is called *laminar*.

F is the force required to maintain the flow and slide the fluid layers relative to each other by overcoming the internal fluid resistance. If A is the area of the plate in contact with the fluid, then the shear stress is F/A. Figure 8-40 shows the linear velocity distribution in the fluid. The rate of shearing of the fluid is V/h. During each unit of time, there is an angular change equal to V/h. Viscosity of the fluid is defined as dynamic viscosity, u.

$$u = \frac{\text{shearing stress}}{\text{rate of shearing strain}} \tag{8-32}$$

$$u = \frac{F/A}{V/h}$$

Rearranging we find

$$F = \frac{uVA}{h}$$

You see that force F (or, in the case of rotary motion, torque) is directly proportional to viscosity, relative velocity of the surfaces, and the area in contact, and is inversely proportional to the thickness of the oil film.

Not all wet clutches are oil-shear clutches. In many drives, the oil is used only to cool the friction surfaces. Unless the clutch is designed to maintain an oil film, it will exhibit the characteristics of a friction-type clutch. Torque will be transferred mechanically from one friction surface to another.

In a true oil-shear drive, torque is transmitted through the shearing of the oil film and there is no metal-to-metal contact between the surfaces until relative velocity approaches zero; and, at this point, there is little wear.

Another difference: In the friction-type clutch, heat is generated in the working surface and then dissipated through air, oil, or water cooling. In the oil-shear drive, the heat is generated in the oil film itself. The heat is easily carried away by displacing the hot oil between the surfaces with cooler oil. The hot oil may then be cooled by contact with the housing, which is cooled by convection and radiation, or water jacketing. For an extremely high rate of dissipation, circulate the oil through an external oil-and-water or oil-and-air heat exchanger.

8-30 Basic Components

Figure 8-41 shows the basic components of an oil-shear drive. A rotary prime mover supplies input power at nominally constant speed. Power is transmitted by the fluid shear to the load.

In its simplest form, the oil-shear drive consists of an axially movable disk on the input shaft and a fixed disk on the output shaft. A thin oil film between the two transmits torque by means of the oil shear. In practice, oil-shear drives

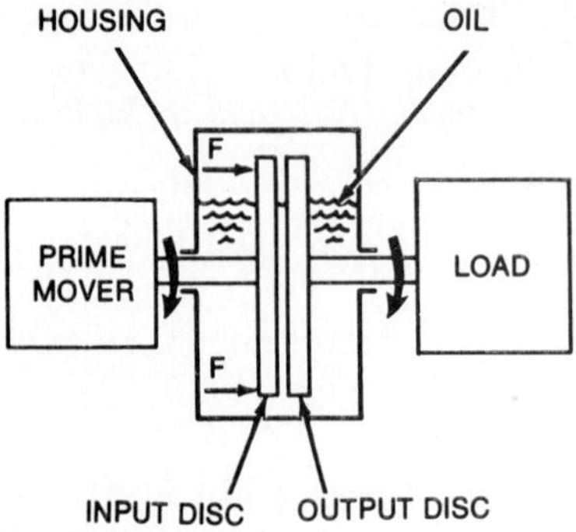

FIG. 8-41 Basic components of oil-shear drive. Oil film between disks transmits power.

use from 10 to 40 working surfaces to transmit maximum torque for a given diameter. Unlike friction drives, which must be partially derated for multiple-plate construction, the oil-shear drive rating increases almost in proportion to the number of working surfaces.

A fringe benefit of the multiple-disk construction is the high torque-to-inertia ratio. Torque diminishes as the cube of the diameter, while inertia shrinks as the fifth power of the diameter.

8-31 Clutch-Brakes

Figures 8-42 and 8-43 show cutaway and cross-section views of a typical oil-shear clutch-brake.

Compressed air or hydraulic fluid introduced in the clutch air-input port makes the nonrotating, centrally located piston clamp the clutch disk pack. The clutch pack consists of steel disks keyed to the input shaft and alternate bronze-faced disks splined to the output shaft. During acceleration, the torque from input to output shafts is transferred through the viscous shear of an oil film between the friction surfaces. As input and output shafts approach syn-

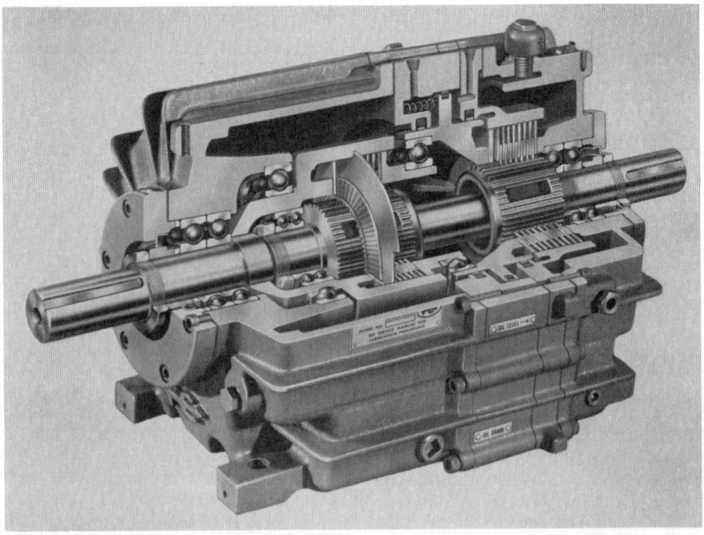

FIG. 8-42 Cutaway of typical oil-shear clutch-brake. (*Courtesy: Force Control Industries, Inc.*)

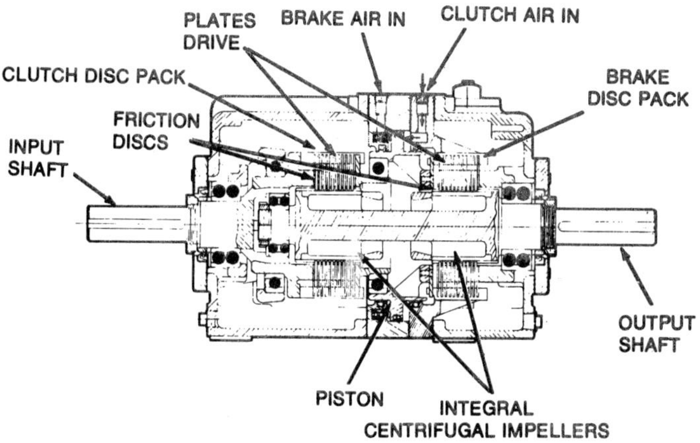

FIG. 8-43 Cross section of typical oil-shear clutch-brake, clutch engaged.

chronous speed, the film breaks down, allowing metal-to-metal contact and 100 percent efficiency in transmission.

For braking action, compressed air, hydraulic fluid, or springs provide a force against the piston toward the output end of the unit as air, or hydraulic fluid, is exhausted from the clutch piston chamber. The piston thrust member clamps the brake disk pack, which is identical to the clutch pack except for the steel plates being keyed to the unit housing.

You can adjust air or hydraulic pressure independently, to provide rapid or cushioned response for clutching and braking.

Integral cast impellers on the output shaft provide positive oil circulation through the clutch and brake packs to remove any residual heat of engagement and to lubricate all bearings.

The springs shown on the left side of the piston are light-duty return springs, for units that use external actuation (compressed air or hydraulic pressure) for brake operation. In the fail-safe brake, heavy-duty springs replace these springs. The springs engage the brakes when actuation pressure is relieved on the clutch side of the piston. The advantage of this model is that failure of the electric power or actuating pressure automatically releases the clutch and sets the brake. The load stops.

This clutch and brake thrust mechanism makes any overlap of the clutching and braking modes a mechanical impossibility.

8-32 Torque Adjustment

We said that torque transmitted by an oil-shear drive varies inversely with film thickness. Direct control of oil-film thickness is impractical with multiple-disk assemblies. Film thickness can be controlled indirectly, however, by controlling clamping pressure. Figure 8-44 shows that torque varies proportionately as actuation pressure. Thus, torque can be adjusted by using a simple, inexpensive pressure regulator.

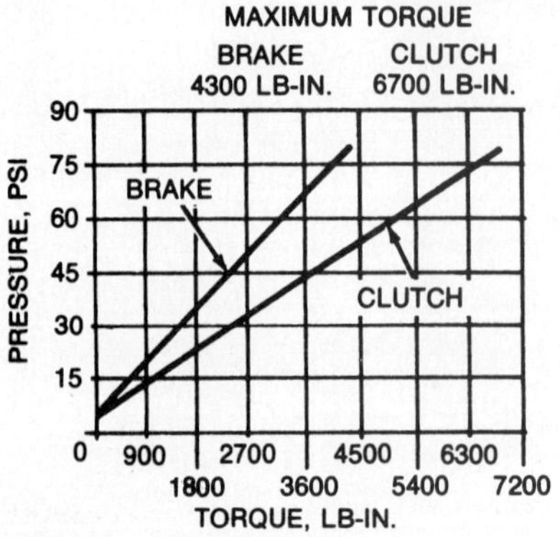

FIG. 8-44 Torque vs. actuation pressure for a typical air-actuated oil-shear clutch-brake. Changing actuation pressure changes torque.

8-33 Advantages

The limitations of oil-shear drives are residual drag and cost. The advantages are low wear, high thermal rating, low inertia, repeatability, tight enclosure (totally enclosed), rapid response, and suitability for vertical mounting.

The very high torque-to-inertia ratio of the multiple-disk stack is advantageous in high cycling applications in which the clutch-brake inertia is many times that of the load. This inertia-to-load relationship usually prevails in a system that contains a speed reducer. The load inertia is reduced by the square of the reduction ratio when referred to the drive. In such a system, 50 percent of the motor output may well go toward accelerating the drive. Thus, 50 percent of the heat generated would come from starting and stopping the rotating portion of the clutch-brake.

It is well known that the coefficient of friction for dry surfaces will vary widely at identical speeds and temperatures, and will vary even more with changing speeds and temperatures. It is not so well known that oil-shear drives are quite consistent under uniform conditions and are relatively consistent at varying speeds and temperatures.

Oil-shear drives must be enclosed to contain the oil. This enclosure also keeps out foreign matter, permitting the drives to be used in dusty, dirty, oily, and hazardous atmospheres.

The factor which has most limited wide use of oil-shear drives has been cost. Although the initial cost is usually somewhat higher, it is easily compensated for by longer life, reduced downtime, and lower maintenance costs. As a result, most oil-shear drives have been introduced to a plant by maintenance personnel rather than by machine manufacturers.

A limitation which is significant only in constant-slip applications is residual drag. With the clutch or brake disengaged, the oil film (which is always present) will continue to transmit some torque. The amount is a function of viscosity, disk size, and relative slip speed. Where residual drag is expected to be a problem, steps can be taken to minimize it.

8-34 Cycling Drives

Applications of oil-shear drives can be grouped into two main categories: cycling drives and constant-slip drives.

You can use oil-shear drives in any application in which conventional clutches and brakes are normally used. They are found most often in cycling applications. Frequent starting and stopping generates considerable heat. Several million cycles a month is common duty. The long wear-life of oil-shear drives suits them for this type of duty. Typical examples are indexing drives, stamping presses, conveyors, packaging machines, and machine tools.

Oil-shear drives can normally dissipate more heat than friction drives. With fan cooling or water cooling, the rate can be increased tenfold. Inertia and frequent cycling of a machine can make a dry clutch burn up. Because an oil-shear drive is completely self-contained with its own bearings, housing, and input and output shafts, conversion is relatively simple. Most oil-shear drives in cycling applications are direct-driven by the motor, then belted to the machine. Simple modifications make this drive suitable for vertical operation, which seems to be required for many machine-tool installations.

One of the most difficult clutch-brake applications is starting and stopping high-inertia loads like flywheels.

8-35 Constant Slip

In start-stop applications, the objective is to bring the load to full engagement with the clutch or brake. In constant-slip applications, the objective is to slip continuously without engagement. Typical installations include unwind brakes, dynamometers, constant-tension drives, and adjustable-speed drives.

Unwind brakes are common in the paper and metal industries. The brake maintains a fairly constant tension in the material, even though the moment arm, and thus the torque required, is changing constantly as roll size diminishes. In noncritical installations, the actuating pressure on the oil-shear drive (and thus, the torque output) is adjusted manually by adjusting a pressure regulator, or by using a follower arm to decrease pressure as roll diameter decreases. These methods will control torque and tension within 5 or 10 percent. Better results with open-loop control are not possible, because torque varies also with the viscosity of the oil and with relative slip speeds. Closed-loop control provides much finer accuracy. A load cell or a similar device measures tension and, through a servo control, constantly adjusts actuating pressure. Tension control is ¼ percent, or better.

The drawworks oil-shear brake used in the petroleum industry for lowering pipe during the drilling process is merely a variation of the unwind brake. The brake converts potential energy of the string into heat. Controls also are similar to those of unwind brakes.

Oil-shear drives work well in most dynamometer applications, provided certain conditions are met. An oil-shear drive that has a multiple-disk stack can develop very high torque with low inertia. Extremely smooth transmission of torque is possible as long as speeds are kept high enough to maintain the oil film. You must also consider drag torque, which limits the minimum torque that can be delivered. Drag torque can be held to a minimum with proper design, but you cannot overlook it. Despite minimum slip and drag torque, oil-shear drives make excellent dynamometers, because they can develop both high and low torque over a wide speed range. Most important, they can de-

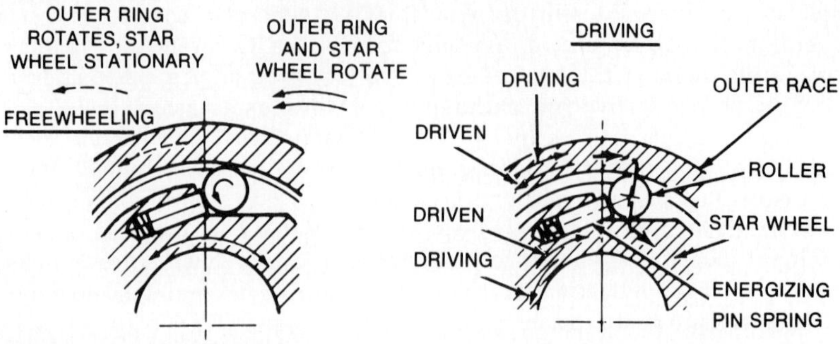

FIG. 8-45 In driving mode, spring holds roller against ramp. When freewheeling, spring compresses, roller turns.

velop very high torque at the low speeds at which other dynamometers cannot function.

The newest and most important use for constant-slip clutches is in adjustable-speed drives. Because torque can be varied by changing the actuation pressure, speed can be controlled by a servo control. This is such an important application that oil-shear drives are covered in a separate article.

ROLLER AND CAM OVERRUNNING CLUTCHES[7]

A freewheeling clutch permits free rotation in one direction and prevents rotation in the other. The roller clutch and the cam clutch are two types of freewheeling clutch. The freewheeling roller clutch has an inner race (or star wheel) and an outer race, enclosing a set of rollers which are held in spring tension against ramps (see Fig. 8-45). In a cam clutch (Fig. 8-46), springs hold segmented rollers in contact with inner and outer races (clutch engaged) until centrifugal force lifts the rollers off (clutch disengaged).

In a roller clutch, power is transmitted from one race, through the rollers which are wedged up the ramps, to the outer race.

In the overrunning mode (Fig. 8-45), the outer race moves faster than the inner race, keeping the ball from wedging. Many roller sizes and shapes are available, as are variously tensioned springs. Also available are bearings to maintain concentricity, a large variety of mounting flanges, seals, and lubrication systems. Roller clutches prevent reverse rotation, separate a driven member from the driven element, or convert reciprocating motion to rotary motion.

8-36 Theory

How does the roller clutch freewheel in one direction and transmit power in the other? The normal force P (Fig. 8-47), acting at the contact points between the roller and the inner and outer races, produces a friction force R, given by

$$R = P\mu \tag{8-33}$$

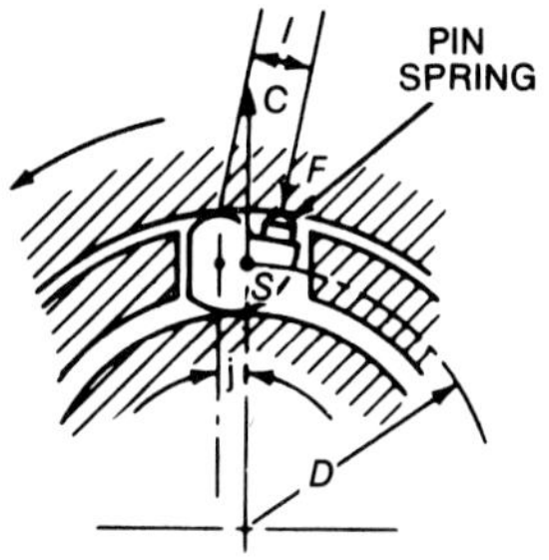

FIG. 8-46 Cam clutch has roller segments that lift by centrifugal force when freewheeling, stop heat generation.

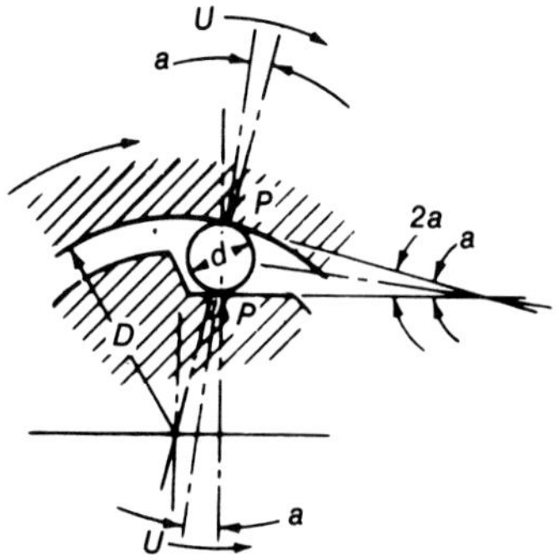

FIG. 8-47 When friction force R is greater than circumferential force U, power flows.

where μ = coefficient of friction.

The tangential vector of P is the circumferential force, given by

$$U = P \tan a \tag{8-34}$$

where $2a$ = angle between locking surfaces of inner and outer races.

For power transmission, R must be greater than U; and, at the instant when slip ceases and power transmission begins,

$$R = U \tag{8-35}$$

Therefore,

$$\mu = \tan a$$

The torque moment M transmitted by the roller is given by

$$M = DU/2 \tag{8-36}$$

where D is the inner diameter of the outer race.

By substitution,

$$M = DP(\tan a)/2 \tag{8-37}$$

The roller-pressure equation gives normal force:

$$P = 2rbk \tag{8-38}$$

where r = roller radius
b = roller length
k = roller pressure

By substitution,

$$M = Drbk(\tan a)$$

which you can use to calculate torque capacity during power transmission.

When the roller clutch is in the freewheeling condition (Fig. 8-48), the spring force F on the roller causes reactive forces F_1 and F_2 at the contact points with the outer and inner races. If the inner race is rotating, then centrifugal force C will supplement F_1. For free running, R must be less than U. Because friction force is

$$R = (F_1 + C)\mu$$

heat and wear caused by friction increase as the speed of rotation of the inner race increases (or as the relative speed between rings increases).

8-37 Strength

Large-diameter rollers carry more torque than small rollers. To maximize strength capacity, some roller clutches are designed with segments of rollers that have exceptionally large diameters. Often, these are called *cam clutches.* Their torque capacity can be increased by further increasing the clutch diameter, the number of roller segments, or both.

To calculate lift-off speed of a cam clutch, consider that the cam has a center of gravity S outside the point of contact (Fig. 8-46). A free-body analysis shows that there is a moment Cj acting counterclockwise, while the moment FI acts clockwise. For lift-off,

$$Cj = FI$$

Applying the equation for centrifugal force:

$$C = (W/g)(2N/60)^2(D/2) \tag{8-39}$$

and solving for N, we find

$$N = 76.9(FI/jWD)^{1/2} \tag{8-40}$$

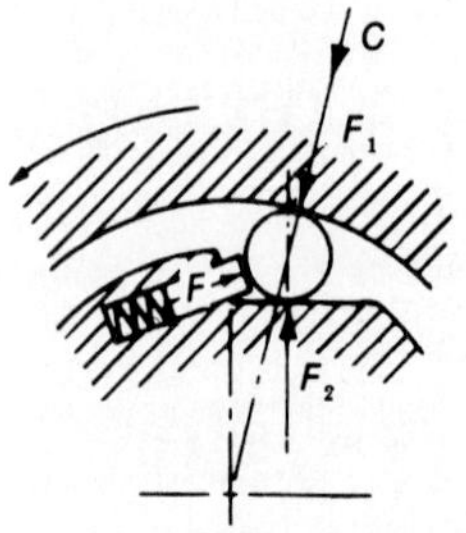

FIG. 8-48 Freewheeling roller clutch has friction forces which generate heat.

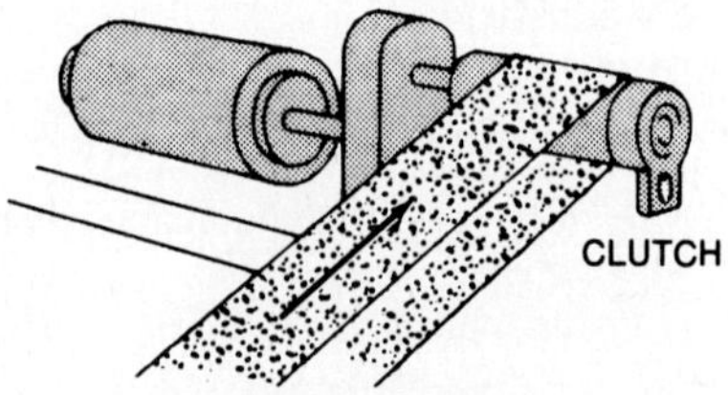

FIG. 8-49 Holdback clutch keeps inclined conveyor from dumping load during power failure.

8-38 Torque and Service Factors

A free-body analysis of the clutch, based on a steady, gradual load, will reveal the torque value. This torque, multiplied by a service factor, gives *design torque,* which you should use to select the roller clutch for optimum value-cost relationship.

For overrunning, indexing, and backstopping, torque is given by

$$T = 5250P/N \tag{8-41}$$

where T = torque, lb·ft
P = power, hp
N = speed, r/min

In each case, you must consider other load factors—friction or drag, torsional or linear vibration.

8-39 Backlash

Indexing accuracy is inversely proportional to backlash. Backlash is caused by penetration of the lubrication film and elastic deformation of materials.

There is no slip in the usual sense in a roller clutch because spring tension holds the rollers in constant contact with the ramp as zero speed in approached. By proper design, roller clutches can achieve an accuracy of 0.004 in per feed at 1000 indexes per minute.

While engaging the roller and ramp, contact surfaces must penetrate the lubrication film until they make dry contact. This causes a delay of several angular minutes in the solid engagement. Proper selection of the lubricating oil and the spring element can minimize the delay. Stronger springs can reduce backlash.

Also in the engagement process, load application causes minor elastic deformation of all elements. This deformation is proportional to the amount of stress, which is, in turn, determined by the materials of the ring, rollers, and ramps.

8-40 Backstopping

Roller clutches can backstop conveyors, lifts, and speed reducers. Usually, backstopping is done by restraining or mounting one ring of a clutch on a stationary member and letting the other ring freewheel the output shaft. When the drive stops or power fails, the load tends to reverse direction, causing the clutch to engage. Backstopping prevents reverse rotation.

On a typical inclined-conveyor application (Fig. 8-49), the clutch freewheels as long as the conveyor moves forward. Should power fail, the clutch will engage, thereby preventing the conveyor from reversing and dumping its load. This is called holdback.

In many applications, a backstop clutch on the opposite end of a speed reducer (Fig. 8-50) prevents reverse rotation. In most situations, the roller

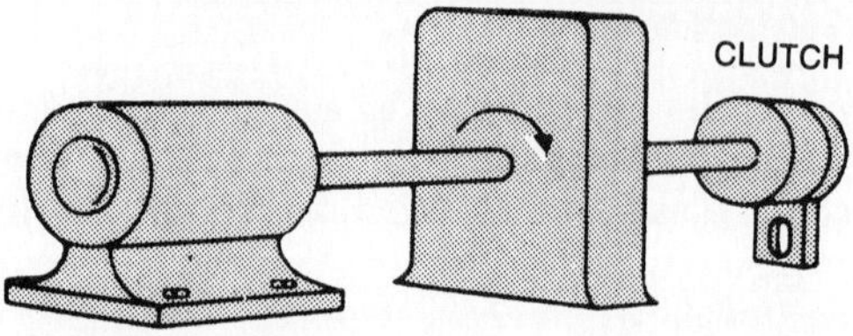

FIG. 8-50 Backstopping clutch on outboard side of speed reducer prevents reverse rotation by engaging when shaft attempts to reverse.

clutch mounts on the reducer high-speed shaft. When the shaft attempts to reverse, the clutch automatically engages, thus preventing damage in the system.

8-41 Overrunning

In overrunning uses, the roller clutch discriminates between the rotary speeds of the races and disengages one race from the other when the speed of the first exceeds that of the second. The clutch engages when the speed of the first race becomes less than that of the second. Many roller-clutch applications are in generators, fans, and starter drives, where the driven member must separate from the driver.

In Fig. 8-51, the roller clutch provides the connection between two motors and their load. The low-speed motor drives the load and armature of the high-speed motor until the high-speed motor starts. Then the clutch disengages so that the low-speed motor does not rotate at high speed.

Energy dissipation is another application for roller clutches. In Fig. 8-52, a motor drives a large fan through an engaged clutch. When the motor is turned off, the fan tends to continue rotating because of its momentum. The clutch freewheels, letting the fan coast to a stop after the motor stops.

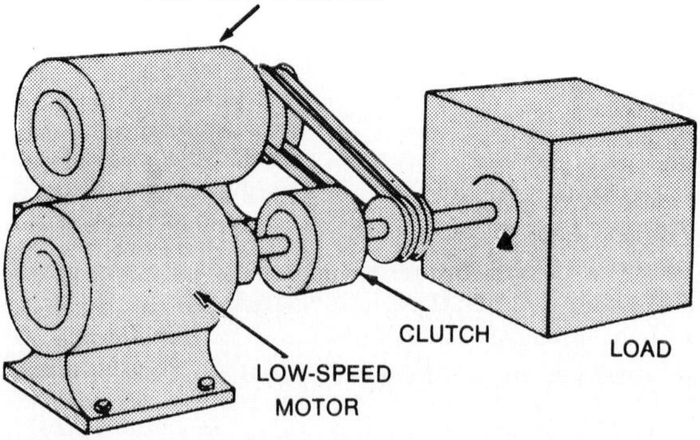

FIG. 8-51 Over-running roller clutch connects low-speed motor to load and to high-speed armature until high-speed motor comes on. At high speed, low-speed motor disengages.

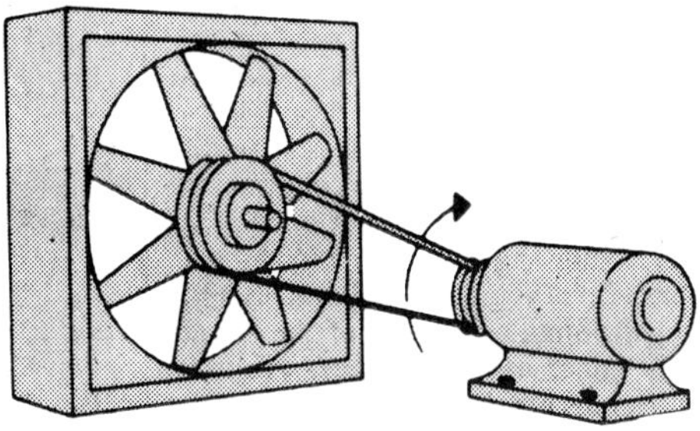

FIG. 8-52 Overrunning clutch lets massive fan rotor coast to a stop without dragging the motor armature along with it.

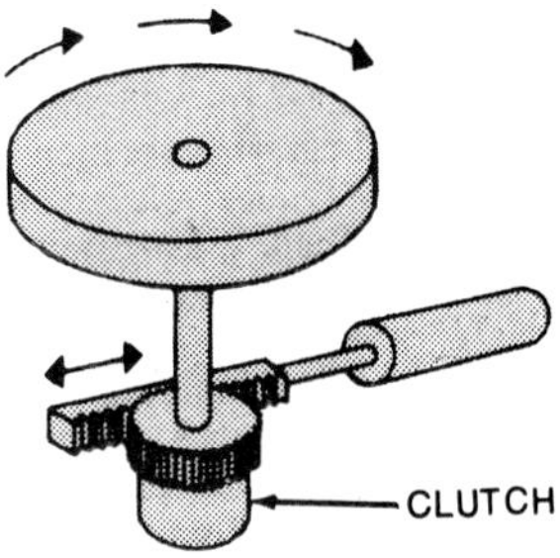

FIG. 8-53 Freewheeling clutch engages on cylinder's return stroke, disengages on extension.

8-42 Indexing

A third application of roller clutches is converting reciprocating motion to one-way rotary impulses. One ring of a clutch connects to a rack and pinion or to a crank linkage, which imparts a series of forward and reverse rotations to the ring.

In one direction, the clutch engages and rotates the other ring. In the other direction, it freewheels. Typical applications include punch presses, food-slicing presses, counters, assembly tables, and packaging machines.

A cylinder actuates a gear rack in Fig. 8-53. The rack drives a gear on the outer ring of a clutch, which transmits rotary impulses to the table shaft. The table rotates a given amount, then remains stationary until the next rotary impulse.

8-43 Ranges

Roller clutches are available in capacities from 15 lb·in to 11,500 lb·ft. Mounting options are numerous.

Cam clutches come in torque capacities to 52,000 lb·ft and maximum speeds to 15,000 r/min.

SPRAG-TYPE OVERRUNNING CLUTCHES[8]

An overrunning clutch transmits torque in one direction only. It releases when input rotation is reversed or when the output overspeeds the input in the drive direction. The sprag-type overrunning clutch generally consists of an inner race, an outer race, a set of sprags, a sprag retainer, energizing springs, and bearings.

Wedging of sprags between races transmits power from one race to the other. The sprags have a greater diagonal dimension across one set of corners than across the other (Fig. 8-54). Wedging occurs when relative rotation of inner and outer races tends to force the sprag to a more upright position where the cross section is greater.

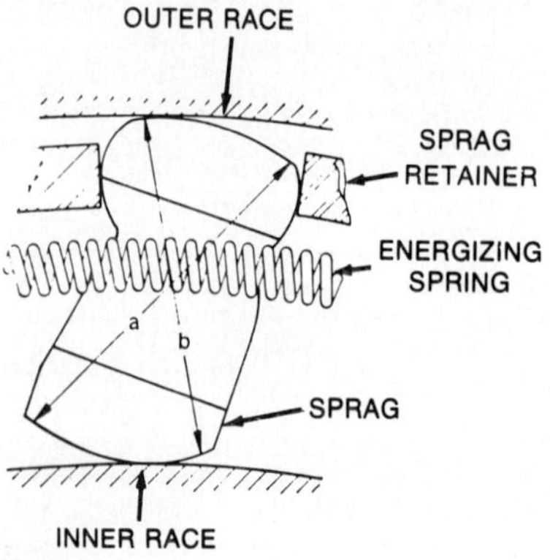

FIG. 8-54 Detail of sprag. Dimension *a* is greater than dimension *b*.

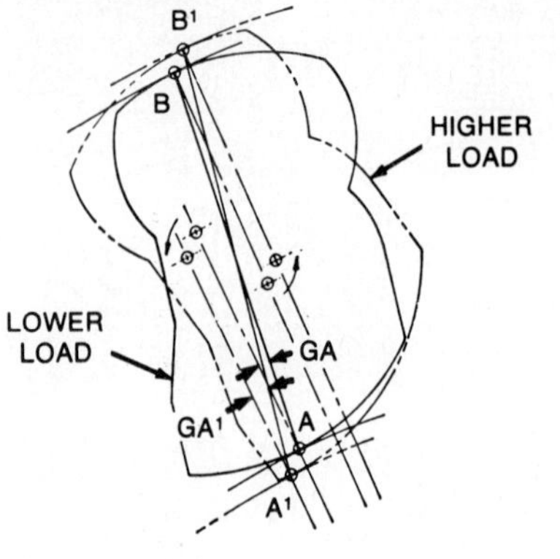

FIG. 8-55 Geometry of sprag, showing gripping angle GA.

8-44 Gripping Angle

Wedging action depends upon the wedging, or gripping, angle of the sprags between the races. The fundamental concept of sprag clutches requires that the coefficient of friction of the sprag, with respect to the inner race at the instant torque is applied in the drive direction, must be greater than the tangent of the gripping angle GA. If the condition is not satisfied, wedging will not occur.

The gripping angle is determined by the construction of Fig. 8-55, where points *A* and *B* are the points of contact of the sprag with the inner and outer races, respectively. Sprags are designed to have a low initial gripping angle to ensure positive initial engagement. As torque increases, the sprags produce radial forces that cause race deflections, which make the sprags roll to new positions. Sprags are usually designed to have an increasing gripping angle as they roll from overrunning position to maximum load-carrying position. A higher gripping angle reduces the radial load imposed by the sprag, thus permitting higher torque to be transmitted within the limits of race stretch and brinelling. Figures 8-56 and 8-57 show this characteristic.

8-45 Torque Capacity

Sprag-clutch torque capacity, for a given race surface and core hardness, is established by three considerations: Surface compressive, or Hertz, stress; race hoop stress; and race deflection. The equations on which these considerations are based follow,

where d_e = outer race external diameter, in
d_o = outer race diameter, in
d_i = inner race diameter, in
d_b = inner race bore diameter, in
d_{si} = effective diameter of sprag inner race cam, in
d_{so} = effective diameter of sprag outer race cam, in

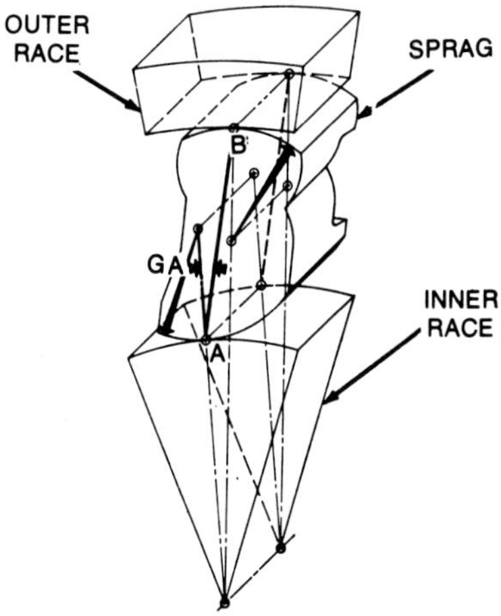

FIG. 8-56 Gripping angle increases as load increases and annular space increases.

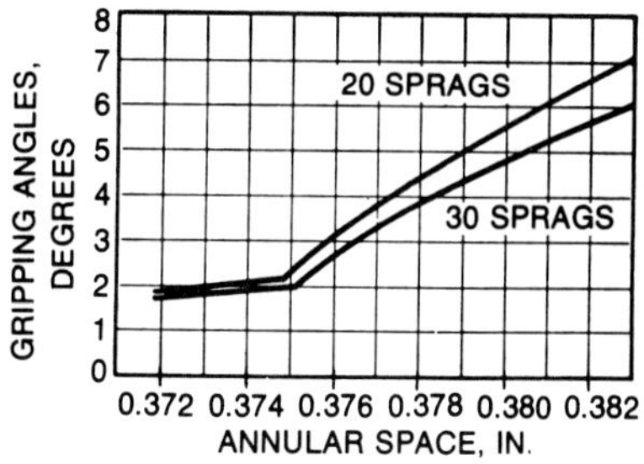

FIG. 8-57 Change of gripping angle with increased annular space.

P_i = normal load of sprag against inner race, lb/in
P_o = normal load of sprag against outer race, lb/in
S_c = Hertz stress, lb/in^2
S_t = hoop stress, lb/in^2
p_i = inner race pressure, lb/in^2 = $p_i N/d_i l$
p_o = outer race pressure, lb/in^2 = $p_i d_i/d_o$
E = modulus of elasticity of race material, lb/in^2
Δ = change in annular space, in
N = number of sprags
l = effective race length, in
r_i = inner race radius, in
L = sprag length, in
GA = gripping angle, degrees

Hertz stress at the inner race is given by

$$S_{c,\max} = 0.591[P_i E(d_i + d_{si})/d_i d_{si}]^{1/2} \tag{8-42}$$

Hertz stress at the outer race is given by

$$S_{c,\max} = 0.591[P_o E(d_o - d_{so})/d_o d_{so}]^{1/2} \tag{8-43}$$

Hoop stresses are calculated by thick-walled equations. For the outer race,

$$S_t = p_o \frac{d_e^2 + d_o^2}{d_e^2 - d_b^2} \tag{8-44}$$

For the inner race,

$$S_t = \frac{-2p_i d_i^2}{d_i^2 - d_b^2} \tag{8-45}$$

Equations for deflections of the inner and outer race are combined to give the change in annular space:

$$\Delta = \frac{p_o d_o}{2E}\left(\frac{d_e^2 + d_o^2}{d_e^2 - d_o^2} + \frac{d_e^2 + d_b^2}{d_e^2 - d_b^2}\right) \tag{8-46}$$

To calculate torque capacity, sum the tangential components of the resulting load, through points A and B of Fig. 8-55, that the applied torque produces. The equation that results is

$$T = P_i r_i LN \tan \text{GA} \tag{8-47}$$

To apply this equation to a given design, consider the specific characteristics of the sprag. Also, you must use an iterative process, because the term P_i is quite elusive. Three items influence it. First, for races of given size, the value of P_i is limited to the maximum allowable working stress for the race materials. Second, it is limited by the allowable Hertz stress to ensure adequate protection from race brinelling. Third, the total race deflections at the race stresses chosen must not exceed a given "cam rise," the difference between dimensions a and b in Fig. 8-54. An appropriate angle can be determined from a curve like that of Fig. 8-57. Note from Fig. 8-57 that a given sprag will have a gripping angle that depends upon the number of sprags. Number of sprags is a function of race diameters.

Current industry standards limit Hertz stress to 450,000 lb/in². Because brinelling occurs at 650,000 lb/in² for steel hardened to R_c 58–62, a safety factor of 2 results. Using sprag-cam radii larger than that defined by the annular space helps minimize Hertz stress (Fig. 8-58).

Should the maximum hoop stress be limited to half the yield strength taken

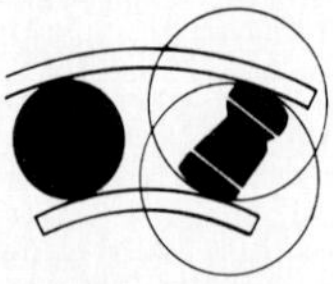

FIG. 8-58 Relative relationship of sprag cam radii to annular space.

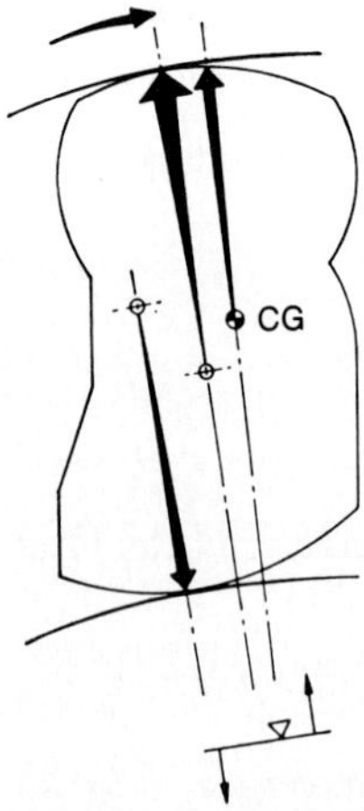

FIG. 8-59 Centrifugal force acting to maintain positive contact.

at the center of the race material, and the total race deflection controlled to half the sprag cam rise available, overall safety factor will be 2.

8-46 Energization

To provide and maintain instantaneous drive engagement, the sprags must engage the two races. The methods of engagement are called methods of *energizing*. One method uses expanding garter springs in a notch at each end of the sprag. This arrangement tends to rotate the sprag at its axis and hold it in contact with both races. Another method has a contracting spring acting on inclined trunnion surfaces, tending to hold the sprags in contact with the two races. A third method uses leaf-type springs mounted on the trunnions of the sprags and held in slots in the trunnions. (This arrangement is most prevalent in large clutches.) The center of gravity of the sprag can be located to provide a moment while the clutch is rotating, thus providing additional energization (Fig. 8-59).

The center of gravity may also be located to deenergize the sprag. As clutch speed increases, the sprags move away from the inner race (see Fig. 8-60). Using this arrangement necessitates reducing speed below throwout speed before the clutch will reengage. This necessity would apply to starter mechanisms (e.g., engine or turbine).

When an expanding garter spring is used for energizing, the amount of spring energizing increases as the speed of clutch rotation increases. The resulting centrifugal force increases the pressure of the sprag contact at the races (Fig. 8-61). The magnitude of the energizing forces is significant to the pressure-velocity relationship, which limits maximum overrunning speed.

For maximum performance, any sprag-type overrunning clutch should have uniform sprag action when loaded and when overrunning. You can get this uniformity in only the independent sprag action of a free-action type sprag

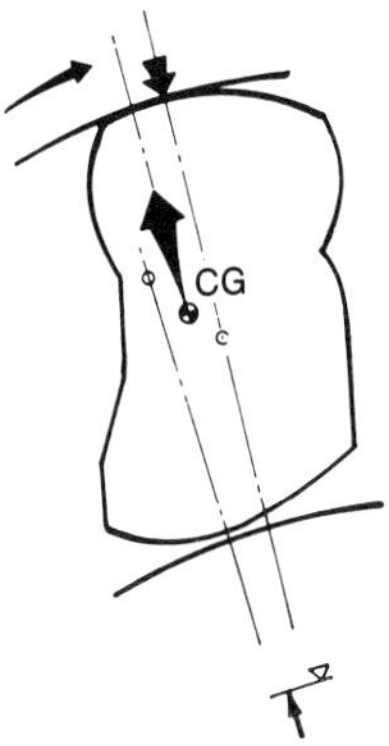

FIG. 8-60 Centrifugal force acting to deenergize the sprag.

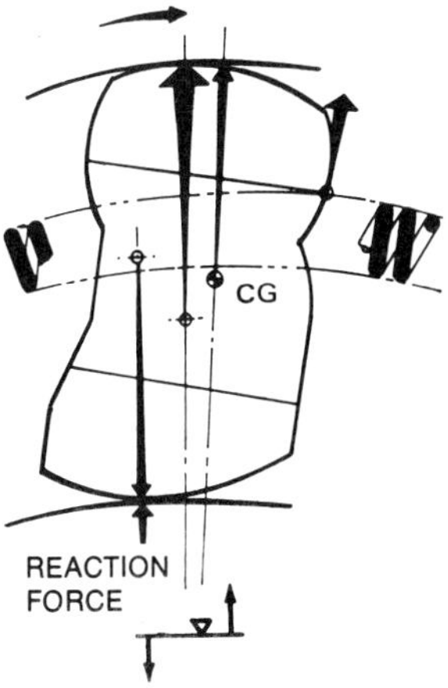

FIG. 8-61 Reaction force caused by centrifugal force and expanding spring.

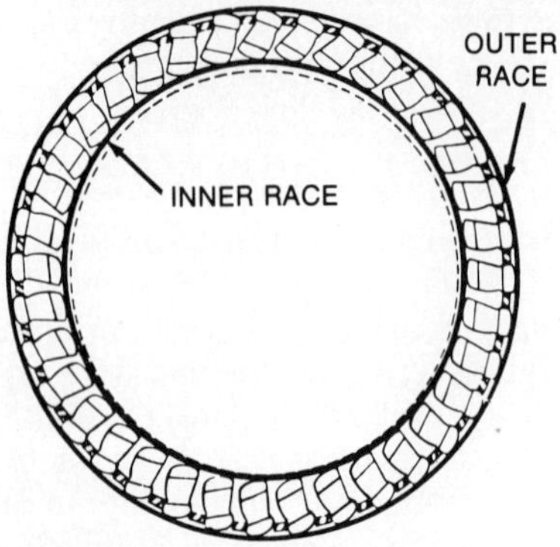

FIG. 8-62 Free-action sprag retainer.

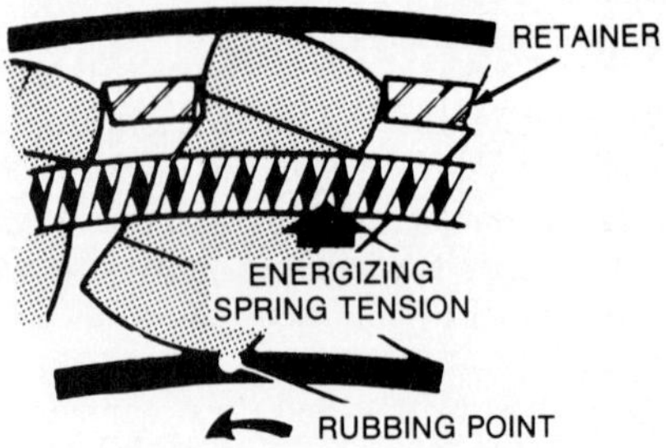

FIG. 8-63 Free-action sprag overrunning.

retainer (Fig. 8-62). During overrunning, free action minimizes wear (Fig. 8-63), because rubbing pressure is required only to overcome the energizing-spring force during inner-race overrunning.

Wear is important in the performance of an overrunning clutch, because the sprag geometry must be preserved if clutch life is to be prolonged with proper performance. An excellent method of increasing resistance to wear is application of a chromallizing process. The surface microstructure is realloyed by diffusing chromium into the surface of the high carbon steel sprag to form chromium carbides, thus producing very hard sprag surfaces practically and economically. This process doubles the surface hardness over that of fully hardened SAE 52100 sprag steel.

Sprag wear in overrunning clutches is determined generally by the relationship of sprag pressure P on the race, rubbing velocity V of the sprag on that race, and lubrication.

The combined centrifugal force of the sprag and energizing spring produces a reaction force between the sprag and the inner race (see Fig. 8-61). If this pressure or bearing load is high and subjected to a high rubbing velocity, the sprag cams may become worn prematurely. Repositioning the center of gravity can increase or diminish the pressure or produce a neutral reaction force at the inner race. Each application requires its own study.

With inner-race overrunning, the reaction force depends on the force that the spring exerts on the sprags, tending to keep them in engagement with the races. With the outer race overrunning, the sprags tend to stay with the outer race and increase the reaction force at the inner race (as shown earlier in Fig. 8-61) because of centrifugal forces.

To control the allowable PV factor with the outer race overrunning, there is a high-speed outer-race overrunning clutch. This is not of the centrifugal throw-out type. The sprags are held to the inner race. There is no reaction force at the inner race. Thus, pressure at the outer race is from spring energizing only.

A development in sprag clutches is the positive continuous-engagement sprag. It was designed to meet the increased demands of aircraft and helicopters. The sprag profile keeps excess torque overload from making sprags roll

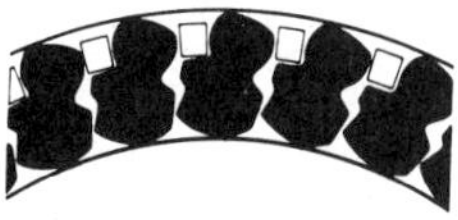

FIG. 8-64 Positive continuous-engagement sprag clutch overrunning.

FIG. 8-65 Positive continuous-engagement sprag clutch driving.

over, yet it does not interfere with normal action and performance in regular clutch engagement or overrunning (Figs. 8-64 to 8-66).

8-47 Lubrication

The best lubricant for sprag clutches is a light-grade straight mineral oil. Slippery additives, such as molybdenum disulfide, graphite, or EP-type lubricants are not recommended. Their properties are incompatible with the basic principle of sprag clutches, which depends on a specific minimum coefficient of friction.

Oil lubrication has some inherent disadvantages. The oil in self-contained clutches is generally lip-sealed, which is a limiting factor on overrunning speed. Oil must be flushed periodically because of sludge accumulation. In many oil-lubricated lip-sealed clutches, maximum seal temperature may limit overrunning speed.

In extremely high-speed applications, such as aircraft engines, forced-feed oil lubrication can accommodate rubbing velocities greater than 10,000 ft/min. Forced-feed oil lubrication coupled with centrifugal disengagement of sprags has been successful in engine-starter applications with overrunning speeds greater than 60,000 r/min.

It is sometimes desirable to use grease and labyrinth seals. Although grease does not lubricate as efficiently as oil, the decrease of friction between a seal lip and spinning race permits significantly higher overrunning speeds. Very often, grease is a preferred lubricant, because it is more convenient for maintenance and relubrication. However, grease lubrication has significant disadvantages at low temperature. It may become tacky and hard, interfering with sprag energizing.

8-48 Application

There are three basic applications for sprag clutches: overrunning, indexing, and backstopping (Fig. 8-67).

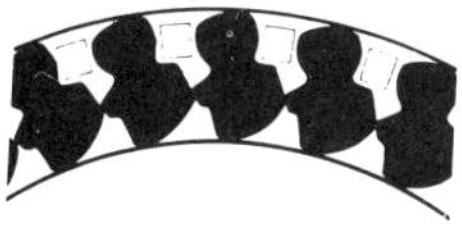

FIG. 8-66 Positive continuous-engagement sprag clutch overloaded. Sprags are in contact and cannot roll farther.

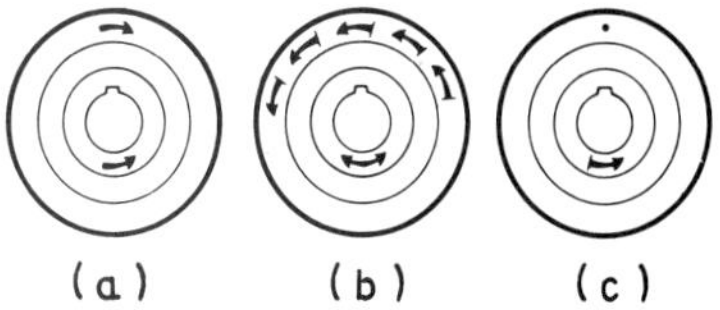

FIG. 8-67 Three basic applications of sprag clutches: (*a*) overrunning, (*b*) indexing, (*c*) backstopping.

8-49 Overrunning

Standby and compound drives are typical applications for overrunning clutches. For example, overrunning clutches may connect a steam turbine and a standby electric motor to a driven shaft. The turbine, the motor, or both can drive the shaft with no further modification of the installation. The turbine-driven clutch automatically engages when the turbine starts to drive, but automatically overruns when the load is transferred to the electric motor.

If both races rotate during the overrunning cycle, there is relative overrunning. The overrunning will fall into one of three categories: inner race rotating with outer race rotating faster in the same direction; outer race rotating with inner race rotating faster in the same direction; and both races rotating, but in opposite directions.

Figure 8-68 shows typical overrunning relationships based on *PV* limitations. Use the curve to select the inner-race speed permissible for a given outer-race speed for self-contained oil clutches. The area inside the curve covers safe relative overrunning speeds. Outside the curve, the relative speeds are too high. Beyond the zero line, the clutch would drive as noted, and no overrunning would occur.

Select an overrunning clutch tentatively according to three considerations: design torque (including service factors), overrunning speed and member, and shaft size. Before finalizing the selection, consider the total drive for its possible effects upon the choice or application of the overrunning clutch.

- Consider location if the design as a whole permits the choice of location for the clutch. In general, a location calling for the lower overrunning speeds will call for higher torque and larger shaft size, and hence a larger clutch.

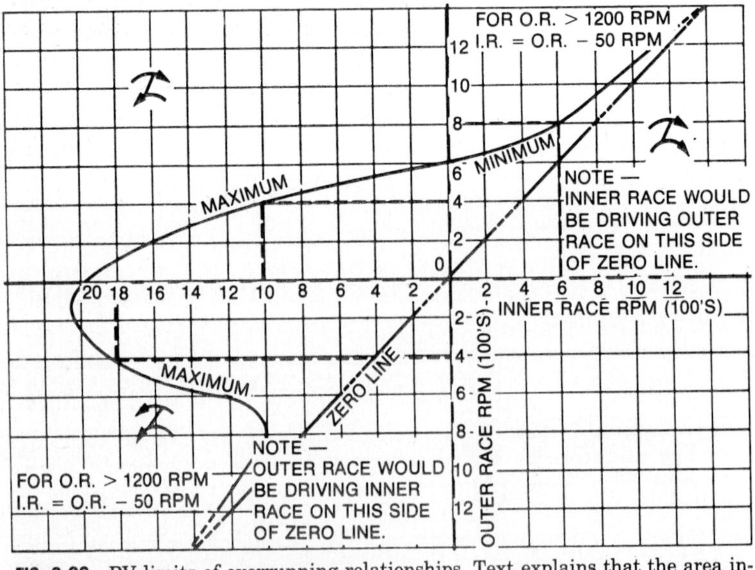

FIG. 8-68 PV limits of overrunning relationships. Text explains that the area inside the curve is the regimen of safe speeds.

You can use a smaller clutch on a lower-torque, smaller-bore application at the expense of a higher overrunning speed requirement.

- Review the entire drive for possible thrust loads. When such loads do exist, review the bearing capacities.
- If the drive as a whole imposes differential or relative overrunning speeds on the clutch, compare these speeds with maximum permissible speeds.
- Mounting considerations, like vertical position or exposed locations, call for special attention. In clutches mounted on vertical shafts, grease lubrication has given better results than oil. Oil tends to flow away from the upper bearing and leave it dry.

Because of its heavier body and tendency to adhere to surfaces, grease remains on the upper bearings and keeps them lubricated. In exposed locations, extra sealing and lubrication provisions may be necessary to protect internal clutch components.

Rated torque capacity for most overrunning clutches is based on a steady-state load, gradually applied and without shock or pulsation. When applying the clutch to overrunning applications, first establish torque on the basis of the torque absorbed by the driven mechanism, if you have this information. If you don't, you can determine the torque from the standard torque equation,

$$T = 5250\, P/N \tag{8-48}$$

This equation gives the torque at the clutch under a steady load at the speed and horsepower used in the equation. Because the equation does not consider the type of load or method of load application, you should apply a service factor to the result to obtain the design torque to be used in selecting the clutch.

In overrunning applications, service factors may vary from 1 to 6 depending on the nature of the application and the type of loading:

Steady load, gradually applied, no shock	1.0 to 1.25
Steady load, applied through chain or gears, minor shock	1.25 to 1.5
Pulsating loads—fans, blowers, pumps, conveyors, etc.	1.5
Critical applications such as hoists, or personnel safety	2.0 to 3.0
Machine tools—arbitrary for long machine-tool life	3.0 to 4.0
High-torque motors, heavy, shock applications such as jogging duty	4.0 to 6.0

Using an internal-combustion engine with an overrunning clutch complicates selection of a service factor. An electric motor or turbine produces steady nonpulsating power flow to the driven mechanism, but an internal combustion engine produces a pulsating input. The fewer the cylinders, the greater the pulsation. For a 4-cylinder engine, service factor is 4.0; for a 6-cylinder engine, 3.0; 8-cylinder, 2.0. Accumulate service factors for an internal-combustion engine to be used with a pulsating load. For example, if a pulsating load like a pump which normally requires a service factor of 1.5 is driven by a 4-cylinder engine which requires a service factor of 4, use a service factor of 6.0 at the clutch ($1.5 \times 4.0 = 6.0$).

A clutch coupling is required when two shafts are coupled end-to-end and overrunning is required in the installation. A clutch-coupling is an overrunning clutch combined with a flexible coupling. The sprag-type overrunning clutch with a grid-type coupling is a popular package. However, the sprag-type overrunning clutch can be coupled with any type of flexible coupling. It is often available as a stock package.

Clutch couplings require higher service factors than clutches alone. The higher factors are needed to protect the flexible coupling elements against fatigue. In general, sprag clutches are not designed to couple two in-line shafts directly. On such applications, the clutches are combined with flexible couplings. The coupling absorbs the motion caused by misalignment and protects the clutch bearings and sprags from excessive load.

For the combination of sprag-type overrunning clutch and grid coupling, service-factor requirements are based on the service factors required by the coupling portion:

Minimum service factor for any sprag clutch with grid coupling	1.5
Pulsating loads like compressors, bucket elevators, and pumps	2.0
Heavy pulsating loads like forced-draft fans	2.5
Heavy variable loads like induced-draft fans and kilns	3.0
4-cylinder engines	6.0
6-cylinder engines	5.0
8-cylinder engines	4.0

Again, accumulate service factors if an internal combustion engine is to be used with a pulsating load. Use a minimum service factor of 1.25 on all combinations of sprag clutch and gear couplings. Here are some typical service factors. Note that they are approximately 20 percent lower than factors for the combination sprag clutch and grid coupling.

Minimum service factor	1.25
Compressors, pumps, etc.	2.0
Forced-draft fans	1.5
Induced-draft fans	2.5
4-cylinder engines	5.0
6-cylinder engines	4.0
8-cylinder engines	3.0

8-50 Indexing

Reciprocating motion applied to the driving race is transformed into intermittent motion in only one direction at the driven race. For example, you can use an eccentric or crank mechanism to convert continuous rotary motion to reciprocating motion through the linkage to the clutch driving race. The clutch will

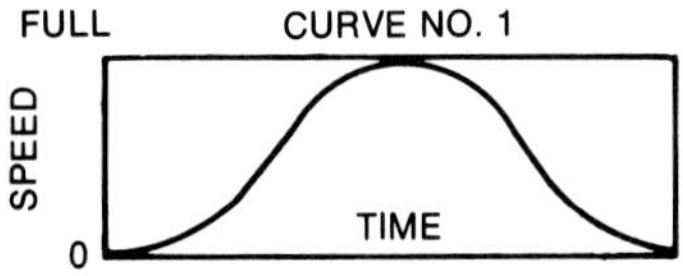

FIG. 8-69 Ideal application and release of a load. Clutches receiving this kind of smooth load variation need smaller service factors than those undergoing abrupt change.

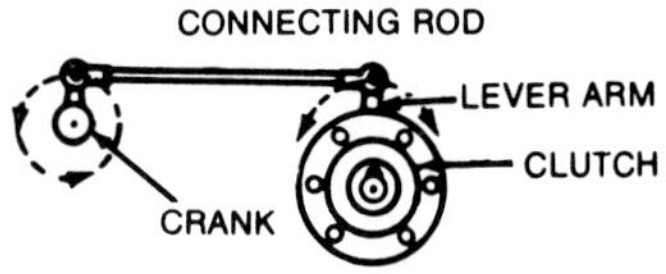

FIG. 8-70 Crank applies the kind of sinusoidal variation most easily tolerated by a clutch.

then advance, or index the work (driven race) on each forward stroke of the link, but the driven race will not return or back up on the return stroke of the link.

If a brake is to be used in the indexing system, the clutch must operate against the brake resistance. The resistance must be calculated in terms of torque combined with the other torques (acceleration torque, etc.) in the mechanism. If, as in a punch-press feed, the indexing mechanism must pull the stock from the coil, you must add the force required to do the pulling. The best way to determine the force is by measurement.

Most sprag clutches are designed to transmit as much torque as the shaft on which the clutch is mounted can carry, subject to the usual safety factors for shaft stress. Therefore, as a general rule, a clutch selected by shaft diameter is adequate for the load.

Generally, the most significant torque load in indexing is the inertia torque of the system and the material being fed. This torque is given by

$$T = \frac{N^2 \theta I}{5225} \tag{8-49}$$

where T = torque, lb·in.
N = number of indexes per minute
θ = angular motion of clutch per index, degrees
I = mass moment of inertia of load, lb·in·s^2

The service factors for indexing applications depend on the nature of the loads. A load which is smoothly applied, uniformly accelerated, and smoothly released (Figs. 8-69 and 8-70) requires the smallest service factor. A load applied abruptly, but cushioned by a dashpot, requires a higher service factor. An abruptly applied load with no dashpot requires the highest service factor (Fig. 8-71).

In indexing, the service factor ranges from 2 to 4. It depends on the rate and magnitude of index and operating loads.

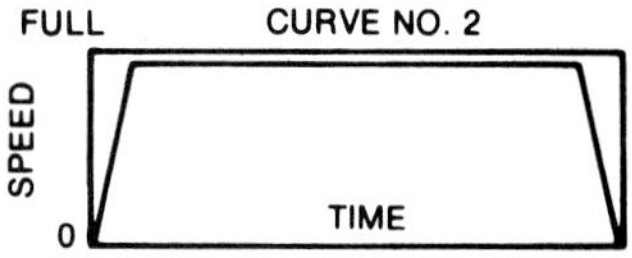

FIG. 8-71 Plot of a suddenly applied or released load. This is the loading pattern imposed by hydraulic cylinders. It calls for a high service factor. The load can be modified by metering valves (trapezoid-shaped curve), and then lower service factors can be used.

For low speeds and short strokes, the smaller, general-purpose clutch is usually appropriate. However, at higher speeds, or where the stroke exceeds 90°, use special high-performance clutches.

The components of the sprag-type overrunning clutch are subject to torsional windup because of the natural elasticity of the parts. The windup is very small and is nearly proportional to the load applied. Therefore, it remains constant for any given load and can be accommodated during setup. It should not be construed as lost motion. Overall accuracy of the complete indexing installation depends on the fits and clearances within the entire train of the indexing mechanism.

Small variations in indexing accuracy, which may range from overfeed to underfeed, are not the result of improper performance within the sprag-type indexing clutch. Small over-and-under variations usually result from looseness or wear of other elements of the indexing system.

8-51 Holdbacks and Backstops

In backstopping or holdback applications, one race is always stationary. The function of a backstop clutch is to permit rotation of the nonstationary race in one direction only, and to prevent any rotation in the reverse direction at any time. *Holdback* is the name given to the backstop clutch when it is mounted on an inclined-conveyor headshaft. This is, therefore, basically an overrunning installation, applied to the one job of holding back. Holdbacks or backstops are used as safety devices on conveyors, gear reducers, and similar equipment.

Service factors for holdbacks vary from 1.0 to 2.5, depending on the nature of the application and the accuracy with which the torque required may be predicted.

Calculate the torque requirements for maximum conditions, not average or typical conditions. Analyze the entire system. Consider both static and dynamic factors as the conveyor is brought to rest. Before final selection, analyze the stiffness of related components and the nature of load sharing in multiple drives.

You can select backstop clutches as you select holdback clutches. However, many backstop clutches are built in when the reducer is built. Because holdback torque cannot be calculated accurately, it is common to select a backstop according to the maximum-horsepower motor which can be used with the reducer.

For holdbacks used on inclined conveyors, the horsepower required to drive the conveyor must be able to overcome friction losses in addition to lifting the load on the conveyor. Because friction losses help the holdback, a holdback selected according to full motor horsepower would be far larger than necessary. For this reason, most torque calculations for holdbacks of inclined belt conveyors are based on the reverse torque generated by the design peak load.

ACKNOWLEDGMENTS

The editor wishes to thank the following authors who originally prepared the material which appeared in *Power Transmission Design Handbook* (1981–1982), from which this section has been developed.

1. Richard C. St. John
   Manager of Engineering
   Mercury Division of Aspro, Inc.
   Canton, Ohio
2. Jim Hanks
   Vice President, Engineering
   Horton Mfg. Co., Inc.
   Minneapolis, Minnesota
3. Don Reiff
   Chief Engineer, R & D
   FMC Corporation
   Power Control Division
   Milwaukee, Wisconsin
4. John Kozdron, Manager
   Application Engineering
   Eaton Corporation
   Industrial Drives Operations
   Airflex Division
   Cleveland, Ohio
5. Gerald R. Eddens
   Vice President—Engineering
   W.J. Industries, Inc.
   Fenton, Missouri
6. Force Control Industries, Inc.
   Hamilton, Ohio
7. H. S. Cummings, Jr.
   President
   Lowell Corporation
   Worcester, Massachusetts
8. Engineering Staff
   Formsprag Company
   Division Dana Corp.
   Warren, Michigan

In addition, the editor wishes to thank the following for their assistance during the preparation of this section.

LaVerne Leonard, Editor
Power Transmission Design
1111 Chester Avenue
Cleveland, Ohio

Richard Shemanske
PT Components, Inc.
Milwaukee, Wisconsin

Ed Brooks, Manager
Engineering Services
Horton Mfg. Company, Inc.
Minneapolis, Minnesota

Jerry Yater, Chief Designer
Force Control Industries, Inc.
Fairfield, Ohio

9 SPRINGS

MEMBERS OF THE STAFF

Spring Manufacturers Institute*/Wheeling, Illinois

* This chapter by the Spring Manufacturers Institute, Inc. (SMI) contains advisory information only. SMI is not a standards-development organization. It does not develop and publish industry standards. It does not provide interpretations of any standards with regard to a manufacturer's specific product. No person has authority in the name of SMI to issue an interpretation of any standard or an interpretation of the technical information and comments published in this chapter that relate to any manufacturer's specific product.

Springs may be produced in an almost infinite variety of sizes, materials, and configurations. Only basic information on the most common types can be presented here in this condensation of the *Handbook of Spring Design,* published by the Spring Manufacturers Institute. Those who are interested in obtaining more information on spring design may wish to obtain this publication or, for more extensive information, *Mechanical Springs* by A. M. Wahl, also available from SMI.

GENERAL INFORMATION

9-1 Basic Considerations

Although various springs require different design techniques, kinds of material, and manufacturing processes, there are some things of a general nature that may be said about all types of mechanical springs.

For intelligent spring design, many factors other than dimensional and load requirements must be considered. One such question concerns the environment in which the spring must operate. How hot or cold will it be? Will the spring be in contact with corrosive media, and, if so, what kind? Another question concerns whether a specific load-deflection characteristic is required, and, if so, over what range of deflection it must be maintained. A third area of interest is the frequency and velocity of load application, and a fourth is the required life. How long must the spring survive under the specific working conditions without breaking or experiencing excessive permanent set? After careful consideration of all the factors which could affect performance, an experienced designer can develop specifications that will yield the greatest possible value to the spring user.

A matter too often neglected is providing sufficient space for functional springs in newly designed machines and equipment. When sufficient space is not allowed, the use of costly, highly stressed, close-tolerance springs may be required, and this increases the risk of early failure. To all equipment and machine designers this admonition is given: Please consider the spring requirements carefully before finalizing the product design.

9-2 Spring Life

The life expectancy of a spring is a subject of growing importance. Engine valves and diesel injector pumps are prime examples of applications in which extreme care is taken to ensure long spring life. There are many factors which contribute to extended spring life, and each one usually increases the cost of the spring. As a result, most applications involve a compromise which assumes an acceptable rate of failure.

Springs, like most components, are designed to last for either the life of the product or some shorter time which is acceptable to the customer. In all applications it is assumed that a certain percentage will fail prematurely, and an attempt should be made to minimize the effect this failure will have on the performance of the product.

No matter how much time, effort, and money are spent to ensure long life, it is practically impossible to guarantee that there will be no failures in a given production lot of springs.

Spring life can be extended by careful design and selection of material, as well as by quality control of both material and manufacturing. When a high spring life is required, it is important to select a manufacturer who understands the problems which may be encountered. These problems may include frequency and velocity of load application, coil stresses, hook stresses, and material capabilities and quality.

Spring breakage is not the only form of failure which must be considered; loss of load, distortion, and other such eventualities may also constitute spring failure. Predicting spring life is not an exact science, and spring life must be carefully considered in all designs in consultation with the spring manufacturer.

9-3 Corrosion

One of the most neglected factors that can adversely affect spring performance is corrosion. Often corrosion of microscopic proportion is the original cause of spring failure, but its presence goes undetected and the cause of failure is erroneously attributed to something else.

Unfortunately, steel, the most commonly used spring material, is the least corrosion-resistant of all. Springs made of uncoated steel must be given some kind of corrosion protection, even during manufacturing, shipping, and storage. The degree of protection required after installation is another matter and depends on the nature of the application. An engine valve spring will be sprayed with oil from other parts of the valve train and needs no other protection. In contrast, a spring in a safety valve could be subjected to live steam that may contain traces of corrosive chemicals, and adequate protection against this harsh environment is necessary.

9-4 Tolerances

Because cost is a major consideration, springs must be produced in the most economical manner. Specified tolerances, therefore, should be generous enough to permit the fabrication of acceptable springs by ordinary production methods.

Also, it is wise to apply tolerances only to functional requirements and dimensions. This practice gives the spring maker an opportunity to make adjustments to compensate for the allowable variations present in the size and mechanical properties of all spring materials.

Another recommendation for product designers: if the standard drawing forms have tolerance boxes for machined dimensions, they are almost sure to be impractical for springs. Delete them and apply realistic tolerances to the mandatory spring requirements.

9-5 Burrs

Burrs are considered undesirable, so some drafters include a note on all drawings reading, "Remove all burrs." This can result in additional cost without adding value to the part. Burrs are produced, to some degree or other, by many of the operations used in manufacturing springs.

The cutting-off operation used in the coiling of helical springs displaces material on spring ends; while it is firmly attached, this material projects beyond the wire diameter. As this type of burr is often harmless, it seems unwise to pay for its removal in such cases. Burrs arising from other operations may sometimes be controlled, within limits, as to size, shape, and location. When the spring maker and the spring user can agree upon such limits, an opportunity for significant savings is created.

9-6 Hydrogen Embrittlement

There is no quick check for hydrogen embrittlement. Possibly the most reliable method is to hold some of the parts under load at their elastic limit for a period of 24 h.

Whenever a carbon steel is pickled in preparation for plating or during some electroplating processes, hydrogen can be absorbed into the material. While cracks can develop in the pickling or plating bath, the usual result is for stress cracks to appear and then failure to occur sometime after plating, perhaps when the plated springs are in service.

The hazard of hydrogen embrittlement becomes more acute where there is (1) high stress concentration, (2) high Rockwell hardness, or (3) high carbon content. Tempered materials are particularly susceptible to hydrogen embrittlement.

To relieve embrittlement, the springs must be baked immediately after plating to drive the hydrogen out of the material. Typical procedure is to heat for 2½ h at 350°F for zinc plating and 3 h at 375°F ± 25°F for cadmium plating.

9-7 Wire Calculations

For initial calculation of wire sizes in the design formulas presented here, multiply the figures in Table 9-1 by the appropriate design stress percentage in Table 9-2.

Once the wire size is chosen, use the actual minimum tensile strength for the wire selected.

TABLE 9-1 Approximate Tensile Strengths for Initial Design Calculations

Steel wire	
Hard-drawn wire	200,000 lb/in^2
Oil-tempered wire	200,000
Chrome vanadium	200,000
Chrome silicon	240,000
Music wire	250,000
Stainless steel wire	
AISI 302	180,000
AISI 316	180,000
17-7 PH	240,000
Nonferrous-alloy wire	
Spring brass	120,000
Phosphorous bronze	135,000
Beryllium copper	170,000

TABLE 9-2 Design Stress Percentages for Static Service

	Design stress, % min. tensile
Steel wire	
Hard-drawn wire	40
Oil-tempered wire	45
Chrome vanadium	45
Chrome silicon	45
Music wire	45
Stainless steel wire	
AISI 302	30–40
AISI 316	40
17-7 PH	45
Nonferrous alloy wire	
Spring brass	40
Phosphorous bronze	40
Beryllium copper	45

COMPRESSION AND EXTENSION SPRINGS

9-8 Design Information

Proper design of compression and extension springs requires knowledge of both the potentials and the limitations of available materials, together with simple formulas. Since spring theory is normally developed on the basis of spring rate (or gradient), the formula for spring rate is the most widely used in spring design. The primary spring characteristics that are useful in designing both compression and extension springs are:

R = rate, lb load per in deflection
P = load, lb
F = deflection, in
D = mean coil diameter, in
d = wire diameter, in
t = side of square wire or thickness of rectangular wire, in
b = width of rectangular wire
G = modulus of rigidity of material, lb/in^2
n = number of active coils
S = torsional stress, lb/in^2
S_k = corrected torsional stress, lb/in^2
OD = outside diameter
ID = inside diameter
C = spring index
N = total number of coils
L = spring length
H = solid height of spring

For compression springs with closed ends, either ground or not ground, the number of active coils n is 2 less than the total number of coils N. For extension springs, all coils are active; body length is wire diameter times the total number of coils plus 1: $d(n + 1)$. The formulas do not apply to extension springs until there has been sufficient deflection to separate the close-wound coils and thus remove all initial tension.

9-9 Design Formulas—Round Wire

Spring Rate

$$R = \frac{Gd^4}{8nD^3} \qquad \text{lb/in} \tag{9-1}$$

Torsional Stress

$$S = \frac{8PD}{\pi d^3} \qquad \text{lb/in}^2 \qquad S_k = KS \qquad \text{lb/in}^2 \tag{9-2}$$

The Wahl curvature-stress correction factor K for compression and extension springs is determined by the following formula:

$$K = \frac{4C - 1}{4C - 4} + \frac{0.615}{C} \qquad \text{where } C = \frac{D}{d} \tag{9-3}$$

and by Fig. 9-1.

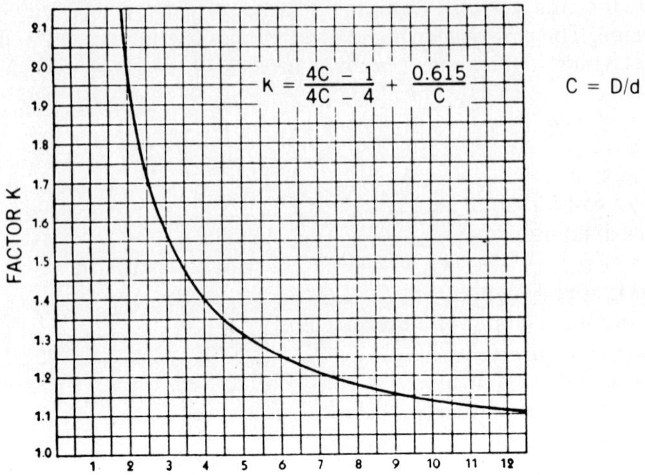

FIG. 9-1 Wahl curvature stress correction.

9-10 Design Formulas—Square and Rectangular Wire

Square

$$R = \frac{Gt^4}{nD^3}(0.180)$$

Uncorrected stress:

$$S = \frac{PD}{0.416t^3}$$

Corrected torsional stress = $K'S$
Correction factor K':

$$K' = 1 + \frac{1.2}{C} + \frac{0.56}{C^2} + \frac{0.5}{C^3}$$

$$C = \frac{D}{t}$$

Rectangular

$$R = \frac{Gbt^3}{nD^3}K^2$$

Uncorrected stress:

$$S = \frac{PD}{K_1 bt^2}$$

Corrected stress:

$$KS = \frac{PD}{bt\sqrt{bt}}\beta$$

Obtain β from Fig. 9-2 and factors for K_1 and K_2 from Table 9-3.

TABLE 9-3 Factors for Square and Rectangular Sections

b/t	1	1.2	1.5	2	2.5	3	5	10	∞
Factor K_1	0.416	0.438	0.462	0.492	0.516	0.534	0.582	0.624	0.666
Factor K_2	0.180	0.212	0.250	0.292	0.317	0.335	0.371	0.398	0.424

9-11 Tolerances

Spring manufacturing, like any other production process, is not exact. It can be expected to produce variations in such spring characteristics as load, mean coil diameter, free length, and relationship of ends or hooks. The very nature of spring forms, materials, and standard manufacturing processes causes inherent variations. The overall quality level for a given spring design, however, can be expected to be superior when it is produced by a spring manufacturer specializing in precision, high-quality components.

Normal or average tolerances on performance and dimensional characteristics may be expected to be different for each spring design. Manufacturing variations in a particular spring depend in large part on variations in spring characteristics, such as spring index, wire diameter, number of coils, free length, deflection, and ratio of deflection to free length.

9-12 Temperature, Stress, and Spring Relaxation

In many applications compression and extension springs are subjected to elevated temperatures at high stresses. This results in relaxation or loss of load, often referred to as "set." In many instances, the set can be predicted and allowances made if the service temperature and requirements are known. When no set is allowed in the application, the spring manufacturer may be able to preset the spring at temperatures and stresses higher than those encountered in operation.

In determining anticipated relaxation, the Wahl corrected stress must be used. For further information regarding relaxation of springs at elevated temperatures, contact your spring maker.

COMPRESSION SPRINGS

9-13 General Data

A compression spring is an open-coil helical spring that offers resistance to a compressive force applied axially. Compression springs are usually coiled as

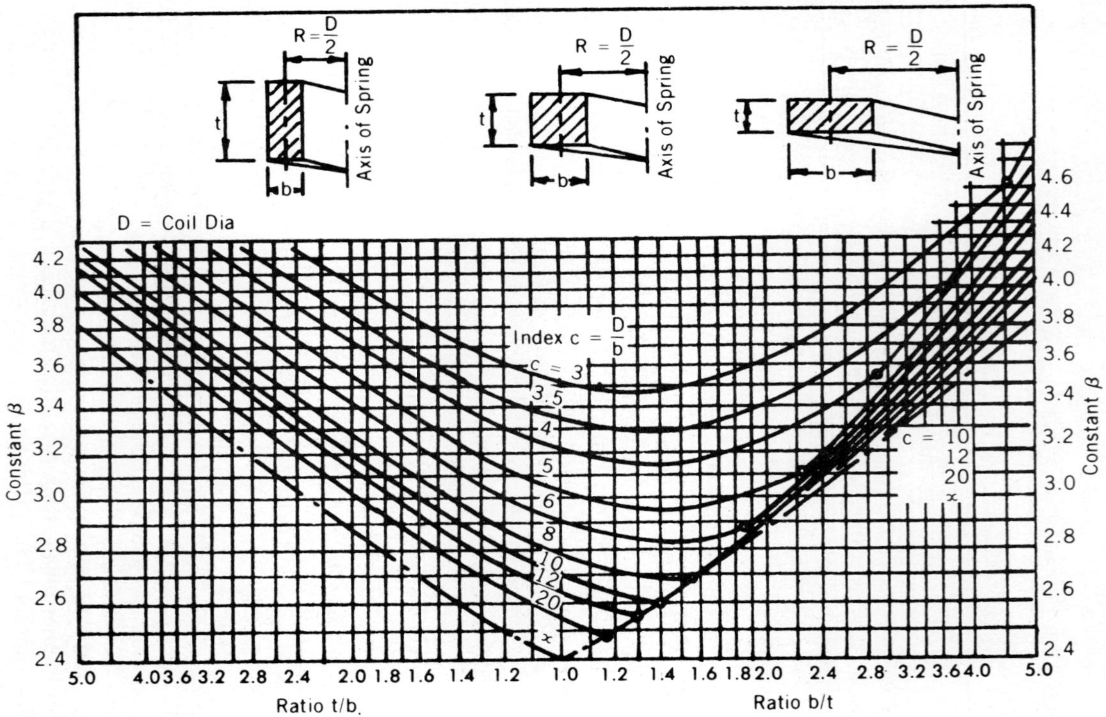

FIG. 9-2 Stress factor β for rectangular wire.

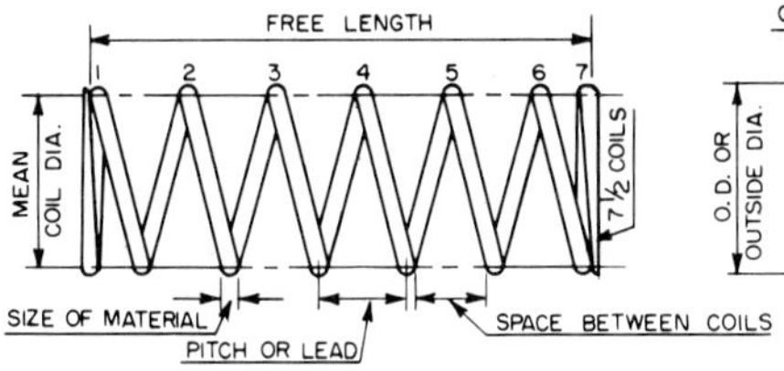

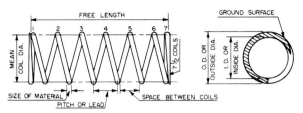

FIG. 9-3 Compression spring characteristics.

constant-diameter cylinders. Other common forms of compression springs—such as conical, tapered, concave, convex, or various combinations of these—are used as required by the application. While square, rectangular, or special-section wire may have to be specified, round wire is predominant in compression springs because it is readily available and adaptable to standard coiler tooling.

The functional design characteristics of the spring should be given as mandatory specifications. These characteristics may include free length, outside diameter, inside diameter, load, rate, maximum solid height, direction of helix, and types of ends (Figs. 9-3 and 9-4). Secondary characteristics, which may be useful for reference, should be identified as advisory data. This practice controls the essential requirements, while giving the spring manufacturer as much design flexibility as possible in meeting these requirements. Secondary characteristics may include wire diameter, mean coil diameter, number of active coils, and total number of coils.

Compression springs should be stress-relieved to remove residual bending stresses produced by the coiling operation. Depending on design and space limitations, compression springs may be categorized according to stress level as follows:

1. Springs which can be compressed solid without permanent set, so that an extra operation for removing set is not needed. These springs are designed with torsional stress levels when compressed solid that do not exceed about 40 percent of the minimum tensile strength of the material.
2. Springs which can be compressed solid without further permanent set after set has initially been removed. These may be preset by the spring manufacturer as an added operation, or they may be preset later by the user prior to

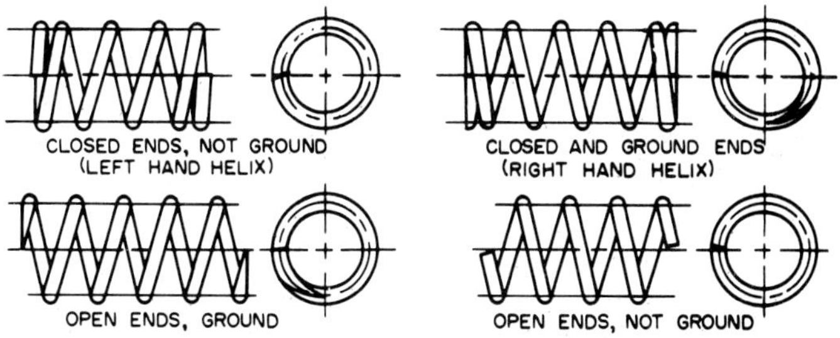

FIG. 9-4 Standard compression spring ends.

or during the assembly operation. These springs are designed with torsional stress levels when compressed solid that do not exceed 60 percent of the minimum tensile strength of the material.

3. Springs which cannot be compressed solid without some further permanent set taking place because set cannot be completely removed in advance. These springs involve torsional stress levels which exceed 60 percent of the minimum tensile strength of the material. The spring manufacturer will usually advise the user of the maximum allowable spring deflection without set whenever springs in this category are specified.

In design of compression springs, the space allotted governs the dimensional limits of a spring with regard to allowable solid height and outside and inside diameters. These dimensional limits, together with the load and deflection requirements, determine the stress level. It is extremely important to carefully consider the space allotted to ensure that the spring will function properly to begin with, thereby avoiding costly design changes.

9-14 Solid Height

The solid height of a compression spring is defined as the length of the spring when under sufficient load to bring all coils into contact with the adjacent coils, with additional load causing no further deflection. The user should specify solid height as a maximum, with the actual number of coils in the spring to be determined by the spring manufacturer.

As square or rectangular wire is coiled, the wire cross section deforms slightly into a keystone or trapezoidal shape, which increases the solid height considerably. This dimensional change is a function of the spring index and the thickness of the material. It may be determined approximately by the following formula:

$$t' = 0.48t\left(\frac{OD}{D} + 1\right)$$

where t' equals new thickness of inner edge (in the axial direction) after coiling and t equals thickness before coiling. When calculating maximum solid height, allowance must be made for all the factors which apply.

9-15 How to Determine Rate

Rate, which is the change in load per unit deflection, may be determined by the following procedure:

1. Deflect spring to approximately 20 percent of available deflection and measure load P_1 and spring length L_1.
2. Deflect spring not more than 80 percent of available deflection and measure

TABLE 9-4 Formulas for Dimensional Characteristics

Spring characteristic	Types of ends: Open	Open, ground	Closed	Closed, ground
Pitch p	$\frac{L-d}{n}$	$\frac{L}{N}$	$\frac{L-3d}{n}$	$\frac{L-2d}{n}$
Solid height H	$d(N+1)$	$d \times N$	$d(N+1)$	$d \times N$
Active coils n	N	$N-1$	$N-2$	$N-2$
Total coils N	$\frac{L-d}{p}$	$\frac{L}{p}$	$\frac{L-3d}{p}+2$	$\frac{L-2d}{p}+2$
Free length L	$(p \times N)+d$	$p \times N$	$(p \times n)+3d$	$(p \times n)+2d$

load P_2 and spring length L_2. Be certain that no coils (other than closed ends) are touching at L_2.

3. Calculate rate R in pounds per inch.

$$R = \frac{P_2 - P_1}{L_1 - L_2}$$

9-16 Spring Ends

There are four basic types of compression spring ends, as shown in Fig. 9-4. The particular types of ends specified affect the pitch, solid height, number of active and total coils, free length, and seating characteristics of the spring.

Table 9-4 gives formulas for calculating dimensional characteristics for various types of ends on compression springs. In applying the given data to solid height, it should be remembered that there are several factors which the formulas do not consider. The actual solid height may not be the same as the calculated value as a result of improper seating of coils, normal variation in wire size, and electroplating, which adds appreciably to the wire size.

9-17 End Coil Effects

A compression spring cannot be closed and ground so consistently that its ends will always be square (in parallel planes at right angles to its axis). In addition, the helix angles adjacent to the end coils will not have uniform configuration and closing tension, and these springs cannot be coiled accurately enough to permit all coils to close out simultaneously under load. As a result of these end coil effects, the spring rate tends to lag over the initial 20 percent of the deflection range, often being considerably less than calculated. As the ends seat during the first stage of deflection, the spring rate rises to the calculated value. In contrast, the spring rate for the final 20 percent of the deflection range tends to increase as coils progressively close out.

The spring rate over the central 60 percent of the deflection range is essentially linear. If possible, critical loads and rates should be specified within this range, which can be increased to about 80 percent of total deflection by special production techniques. However, these techniques add substantially to manufacturing cost and are usually unwarranted.

9-18 Squareness of Ends, Grinding, and Degree of Bearing

The squareness of compression spring ends influences the manner in which the axial force produced by the spring can be transferred to adjacent parts in a mechanism. There are some types of applications in which open ends may be entirely suitable. However, when space permits, closed ends afford a greater degree of squareness and reduce the possibility of tangling with little increase in cost. With closed ends, the degree of squareness depends on the relationship of the wire diameter d and the mean coil diameter D. Unground springs with indexes D/d that are low have less squareness, while unground high-index springs have more squareness. Compression springs with closed ends can often perform well without grinding, particularly in wire sizes smaller than 0.020 in or with spring indexes exceeding 12.

Many applications require grinding of the ends in order to provide greater control over squareness. Among these are applications in which (1) high-duty springs are specified, (2) unusually close tolerances on load or rate are needed, (3) solid height must be minimized, (4) accurate seating and uniform bearing pressures are required, and (5) a tendency toward buckling must be reduced.

Since springs are flexible and external forces tend to tilt the ends, grinding to extreme squareness is difficult. Close tolerances which exceed those available using standard manufacturing methods require special techniques and added operations, which increase manufacturing costs.

A spring may be specified as ground square in the unloaded condition or square under load, but not in both conditions with any degree of accuracy. When squareness at a specific load or height is required, it should be specified.

Well-proportioned, high-quality compression springs which are specified with closed and ground ends should have the spring wire at the ends taper uniformly from the full wire diameter to the tip. A slight gap, which occasionally opens during grinding, between the closed end coil and the adjacent coil is permissible. The bearing surface provided by grinding should extend over a minimum of 240° of the end coils. Results will vary considerably from these nominal attainable values with springs in smaller wire sizes or with higher indexes. In general, it is impractical to adhere to a general rule regarding "degree of bearing," since process capabilities depend so much on the individual configuration of the spring.

9-19 Design Method

Load The design method for helical compression springs is mainly a process of manipulating the two fundamental formulas which were given earlier.

How these formulas are applied depends on what spring characteristics the engineer needs to calculate. These include (1) those spring characteristics which are not specifically fixed by application requirements but must be recorded in order to specify a complete spring, and (2) those characteristics which must be known to determine whether the spring being designed is possible and practical and properly fulfills requirements. The logic of the design method (not the detailed steps involved in reaching a solution) follows. The examples in this section involve the same design logic and can be solved entirely with data given here.

The most common specifications given in designing a compression spring are one or more loads, spring lengths at these loads, dimensions of available space, types of ends, and any factors which govern selection of the spring material. The basic method is to design the spring for maximum economy (of space, weight, and dollars) by working out the wire diameter d corresponding to maximum allowable stress in Eq. (9-2), then using Eq. (9-1) to determine the number of active coils n.

To determine d in Eq. (9-2), it is recommended that, unless mean diameter D is given, the designer make D = OD to get a trial value of d. The trial value of torsional stress S in Eq. (9-2) is determined by multiplying a trial value of minimum tensile strength by the appropriate percentage, as given in Table 9-1. Since this trial value of stress is really the maximum allowable design stress, the P in Eq. (9-2) must be the maximum load that will be applied to the spring, either at solid height $P_s = R(L - H)$ or at maximum deflection F. If no maximum deflection is specified, use P_s in the equation.

If the actual value of minimum tensile strength S_t for the standard wire gage size just larger than the calculated d is fairly close to the trial value, the standard wire gage size may be temporarily assumed as an acceptable value of d. However, if the two values of S_t are far apart, the calculated d is used in turn to determine an approximate value of D in D = OD $- d$; with the approximate D and a value of S corresponding to the first calculated d, Eq. (9-2) is solved again for d. If the second value of d is quite close to the first, it can be assumed to be acceptable, and the next larger standard wire size can be selected. If not, it is necessary to repeat the calculation in Eq. (9-2) once again.

With the acceptable value of d, calculate the number of active coils n in the transposed Eq. (9-1) for rate R. Solid height H is then determined from the appropriate formula given for the total number of coils N and wire diameter d; if necessary, verify that space limitations are being met. Load at solid height $P_s = R(L - H)$.

Stress Stress at solid height (S_s) and stress at load (S_1) are then calculated in Eq. (9-2), using P_s and P_1, respectively.

The Wahl curvature-stress correction factor K is determined in the formula given, with $C = D/d$, and corrected values S_{sK} and S_{1K} of S_s and S_1 are then calculated. These figures are compared to the maximum allowable design stress, which is the product of S_t for the accepted d and the appropriate percentage as given in Table 9-1. Design time may be saved by calculating S_{sK} alone at first, then calculating S_{1K} only if S_{sK} is found to be too high. Comparison of corrected stresses with the maximum allowable indicates one of three conditions:

1. S_{1K} and S_{sK} both below the maximum allowable. Stress is acceptable, and the spring will not set in application. If S_{sK} is quite far below the maximum, however, this is likely to be a somewhat inefficient design. It might then be worthwhile to recalculate, using a smaller coil diameter and/or fewer active coils n.
2. S_{1K} below and S_{sK} above the maximum allowable. The spring will not set at L_1 but will do so at some larger deflection between L_1 and solid height. If the spring is likely to be deflected beyond L_1 in use or assembly, n should be increased (or even D, if space can be made available), with a corresponding increase in wire size to reduce stress. Otherwise, the spring can be preset. If deflection will never exceed L_1, the original design may be acceptable.
3. Both S_{1K} and S_{sK} above the maximum allowable. The spring is impossible with the given specifications; it will set before it reaches the specified load P_1. The basic solution is to increase the amount of material in the spring—increase its energy capacity—by increasing n and D, which will result in a larger d in Eq. (9-1) and then a smaller S in Eq. (9-2). This approach is limited both by maximum available space and by the fact that H cannot be larger than L_1. However, if S_{sK} is less than 60 percent of the minimum tensile strength of the material, the spring may be present by the spring manufacturer or during assembly by the user to minimize load loss in the application.

While the actual step-by-step procedure depends on the particular needs of the design problem, the basic method described here is used in designing most compression springs. The specifications are somewhat different for what are called rate springs, in which the user is interested primarily in load per unit deflection. Rate R and rate tolerance, free length L, solid height H, and any space limitations are usually specified. The only difference in procedure, however, is to calculate the maximum load P_{max} as $P_{max} = R(L - H)$ and use this value in Eq. (9-2).

EXTENSION SPRINGS

9-20 Definition

Extension springs are springs which absorb and store energy by offering resistance to a pulling force. Various types of ends are used to attach the extension spring to the source of the force (Fig. 9-5).

9-21 Initial Tension

Most extension springs are wound with initial tension. This is an internal force that holds the coils tightly together. The measure of the initial tension is the load necessary to overcome the internal force and just start coil separation. Unlike a compression spring, which has zero load at zero deflection, an exten-

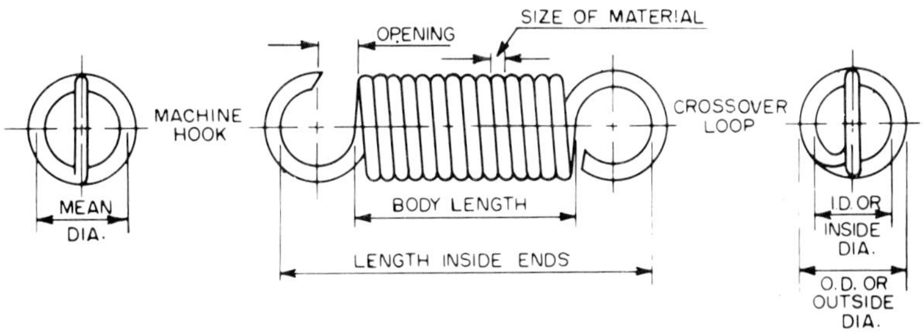

FIG. 9-5 Extension spring characteristics.

sion spring can have a preload at zero deflection. This is graphically illustrated in Fig. 9-6.

This built-in load, called "initial tension," can be varied within limits, decreasing as the spring index increases. Figure 9-7 illustrates this fact. Note that there is a range of stress (and, therefore, force) for any spring index that can be held without problems. If the designer needs an extension spring with no initial tension, the spring should be designed with space between the coils.

9-22 Measuring Rate and Initial Tension

Measuring Rate

1. Extend the spring to a length L_1 such that definite coil separation occurs, and measure the load P_1.
2. Extend the spring further to a second length L_2, and again measure the load P_2.
3. Calculate the rate by dividing the load difference by the length difference:

$$R = \frac{P_2 - P_1}{L_2 - L_1}$$

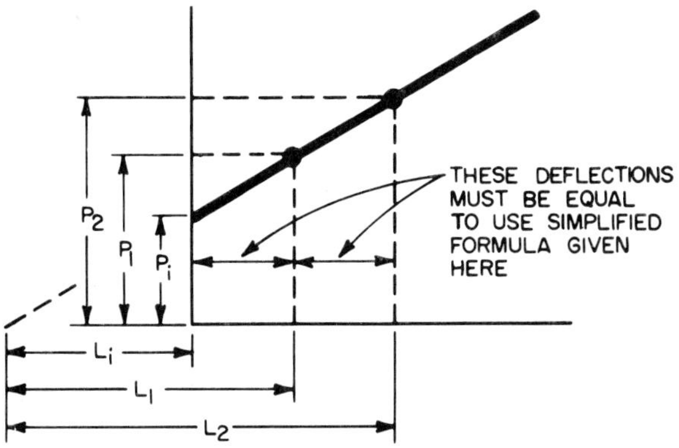

FIG. 9-6 Initial tension.

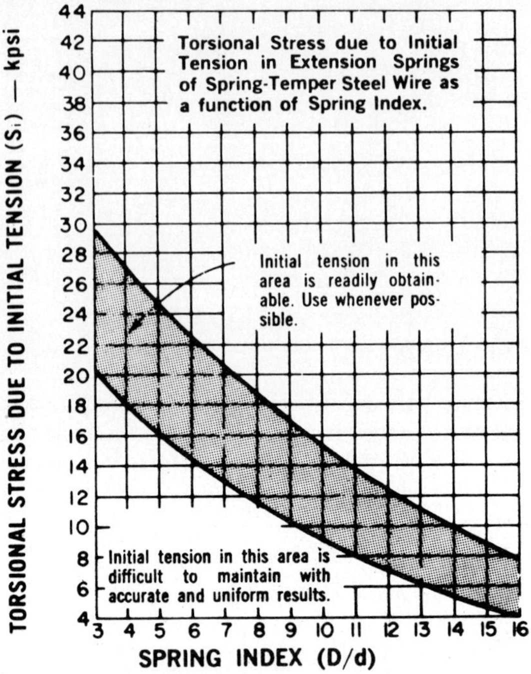

FIG. 9-7 Torsional stress index.

Measuring Initial Tension—Simplified Way

1. Establish the exact initial length L_i of the spring by applying enough load to get slack out but not enough to separate the coils.
2. Extend the spring to a length L_1 sufficient to open the coils and measure the load P_1.
3. Extend the spring to a length L_2 such that the second deflection equals the first deflection and measure the load P_2.
4. Since the two deflections are equal, it can be shown that the initial tension is as follows:

$$P_i = 2P_1 - P_2$$

9-23 Extension Spring Ends

The variety of ends that can be put on extension springs is limited only by the imagination. It may include threaded inserts; reduced and expanded eyes on the side or in the center of the spring; extended loops, hooks, or eyes at varying positions or distances from the body of the spring; and even rectangular or teardrop-shaped ends. (The end is a loop when the opening is less than one wire size; the end is a hook when the opening is greater than one wire size.) By

TABLE 9-5 Standard Extension Spring Ends

LOOP TYPE		RECOMMENDED LENGTH* Min.	RECOMMENDED LENGTH* Max.
Machine	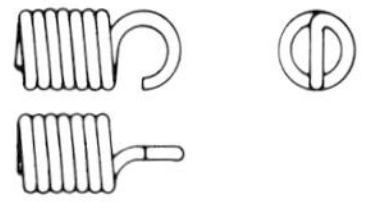	½ I.D.†	1.1 × I.D.
Crossover	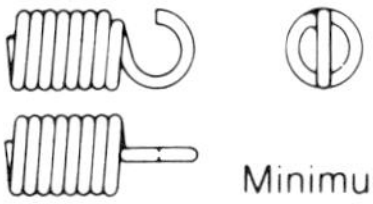	I.D.	I.D.
		Minimum Recommended Index = 7	
Side		I.D.	I.D.
Extended	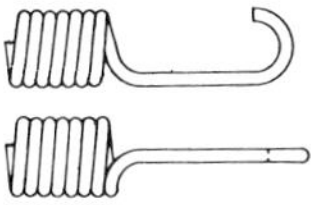	1.1 × I.D.	As Required
Special: as required by design		As Required	

* Length is distance from last body coil to inside of end.
† ID is inside diameter of adjacent coil in spring body.

far the most common, however, are the machine loop and crossover loop, shown in Table 9-5. These ends are made with standard tools in one operation and should be specified whenever possible to minimize cost.

It should be remembered that as the space occupied by the machine loop is shortened, transition radius is reduced and an appreciable stress concentration occurs, contributing greatly to shortened spring life and premature failure.

Most extension spring failures occur in the area of the end. To maximize the life of the spring, the path of the wire should be smooth and gradual as it flows into the end. Tool marks and other stress concentrations should be held to a minimum. A minimum bend radius of 1½ times the wire diameter is recommended.

In the past, many ends were made as a secondary operation. Today, with modern mechanical and computer-controlled machines, many ends can be made as part of the coiling operation. Because of the large variety of machines available for coiling and looping in one operation, it is recommended that the spring manufacturer be consulted before a design is concluded.

9-24 Solving for Initial Tension

Figure 9-7 relates the torsional stress resulting from load due to initial tension to the spring index. For any spring index there is a range of stress (load) that is easily obtainable.

1. Calculate torsional stress due to initial tension S_i:

$$S_i = \frac{8DP_i}{\pi d^3}$$

2. For the calculated value of S_i and the known spring index D/d, determine whether S_i appears in the preferred (shaded) area of the graph.
3. If S_i falls in the shaded area, it is safe to assume that the spring can be readily produced. If S_i is above the shaded area, reduce it by increasing the wire size. If S_i is below the shaded area, select a smaller wire size. In either of the latter two cases, recalculation of stress, number of coils, axial space, and initial tension is necessary.

9-25 Design Method

The fundamental formulas for load-deflection (rate) and stress for compression springs also apply to helical extension springs. The only unique property is that of solving for and including initial tension in the concept and method. Given a certain volume of space in which the spring will act and a certain maximum load P, the basic design approach is to find a wire diameter d based on trial values of mean diameter D, assumed on the basis of the available space, and a reasonable stress S. Remember that an extension spring is not normally preset and must be designed within the torsional proportional limit of the material. This value will be about 40 percent of the tensile strength of the material.

After a wire size has been determined, establish the load-deflection relationship and find out whether the wire size picked will allow the spring to fit in the volume of space available. Involved in this decision are the rate, number of coils, and initial tension, all of which must be solved for. The rate is found from the load-deflection relationship. Using this rate in Eq. (9-1), solve for the number of coils. This number of coils plus the room necessary for end loops takes up a definite amount of space. The final step, then, is to determine whether the available initial tension P_i plus the load added by deflecting to L_1 will add up to the first load required P_i.

TORSION SPRINGS

9-26 General Data

Torsion springs, whose ends are rotated in angular deflection, offer resistance to externally applied torque. The wire itself is subjected to bending stresses rather than torsional stresses, as might be expected from the name. Springs of this type are usually close wound, and reduce in coil diameter and increase in body length as they are deflected. The designer must consider the effects of friction and arm deflection on the torque.

Special types of torsion springs include double torsion springs and springs

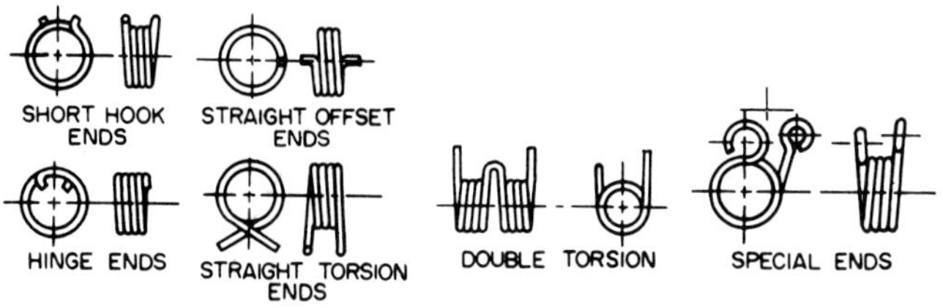

FIG. 9-8 Standard torsion spring ends.

that have a space between the coils to minimize friction. Double torsion springs consist of one right-hand and one left-hand coil section connected together and working in parallel. The sections are designed separately, with the total torque exerted being the sum of the two.

9-27 Types of Ends

The types of ends on torsion springs should be carefully considered. While there is a good deal of flexibility in specifying special ends and end forming (see Fig. 9-8), the cost may be increased and a tool charge incurred.

9-28 Specifications (Fig. 9-9)

In addition to supplying mandatory specifications and advisory data, it is important that a drawing be provided detailing the end configurations. Mandatory specifications may include diameter of shaft to be worked over, outside diameter of spring, inside diameter of spring, torque, length of space available, maximum wound position, length of moment arm, direction of helix, and type of end. Advisory data may include wire diameter, mean coil diameter, number of coils, rate, and free angle reference. Useful special information includes type of material, finish, operating temperature, and end use of application.

9-29 Design Formulas

The basic formulas for torque or moment M and bending stress S used in designing torsion springs are:

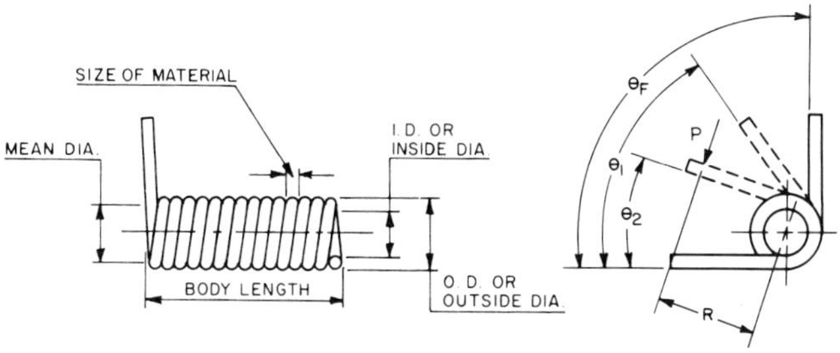

FIG. 9-9 Torsion spring characteristics.

Round Wire

$$M = \frac{Ed^4T}{10.8ND} \tag{9-4}$$

$$S = \frac{32M}{\pi d^3} \tag{9-5}$$

Rectangular Wire

$$M = \frac{Ebt^3T}{6.6ND} \tag{9-6}$$

$$S = \frac{6M}{bt^2} \tag{9-7}$$

Square Wire

$$M = \frac{Et^4T}{6.6ND} \tag{9-8}$$

$$S = \frac{6M}{t^3} \tag{9-9}$$

where D = mean coil diameter, in
d = diameter of round wire, in
N = number of coils
E = modulus of elasticity, lb/in^2
T = number of turns or revolutions of spring
S = bending stress, lb/in^2
M = moment or torque, in·lb
b = width or axial dimension, in
t = thickness or radial dimension, in

In these formulas the constants 10.8 and 6.6, while not strictly theoretical, give results closer to the actual values obtained.

9-30 Design Method

The basic design approach is to begin by calculating the wire diameter d in Eq. (9-5) using the specified maximum torque M and a trial value of maximum design stress S. This value is assumed to be 75 percent of the minimum tensile strength for all materials. If this value of S, taken as 75 percent minimum tensile strength of d just calculated, does not agree with the first trial value for S, calculate a new value for d using Eq. (9-5) and substitute for S 75 percent of the minimum tensile strength of d from the first trial.

When a standard wire diameter just larger than the adjusted value of d is selected, calculate the design stress again, using Eq. (9-5), for the adjusted value of d and compare this value with the maximum allowable design stress. Correction for curvature stress effect in torsion springs is generally negligible.

In planning this design, the engineer should carefully consider the preloading conditions and the change in spring dimensions with deflection, so that adequate clearance is provided (diameter decreases, length increases).

The longer and more extensively formed the spring arms, the higher the cost for tooling and secondary operations. Therefore, relatively short, straight arms should be specified wherever possible.

SPIRAL TORSION SPRINGS

9-31 General Data

Spiral torsion springs, which are usually made of rectangular-section material, are wound flat, generally with an increasing space between the coils. The torque delivered per revolution is linear for the first 360°. At greater angular rotations, the coils begin to close on the arbor, and the torque per turn increases rapidly. For this reason, springs of this type are usually used in applications requiring less than 360° of rotation.

9-32 Design Formula (See Fig. 9-10)

The formula for torque delivered by a spiral torsion spring is

$$M = \frac{\pi E b t^3 \theta}{6L} \quad \text{in·lb} \tag{9-10}$$

where E = modulus of elasticity, lb/in^2
θ = angular deflection, r
L = length of active material, in
M = moment or torque, in·lb
b = material width, in
t = material thickness, in

The stresses imposed on a spiral torsion spring are in bending, and the deflecting beam formula for stress may be used:

$$S = \frac{6M}{bt^2} \quad \text{lb/in}^2 \tag{9-11}$$

Spiral torsion springs for general use can be stressed to 175,000 to 200,000 lb/in^2, depending on material hardness. In applications in which higher stresses and material fatigue are involved, it is suggested that a spring manufacturer be consulted.

The arbor diameter A and outside diameter in the free condition OD_F do not appear in the formulas for torque or stress. But the space occupied by the spring must be considered in design. A spring which is too small may wind up tight on the arbor before the desired deflection is reached. If the outside diameter is too large, the spring will not fit the space available.

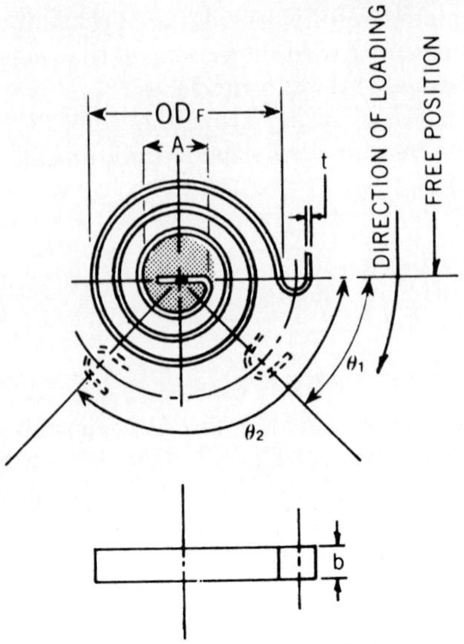

FIG. 9-10 Spiral torsion spring design.

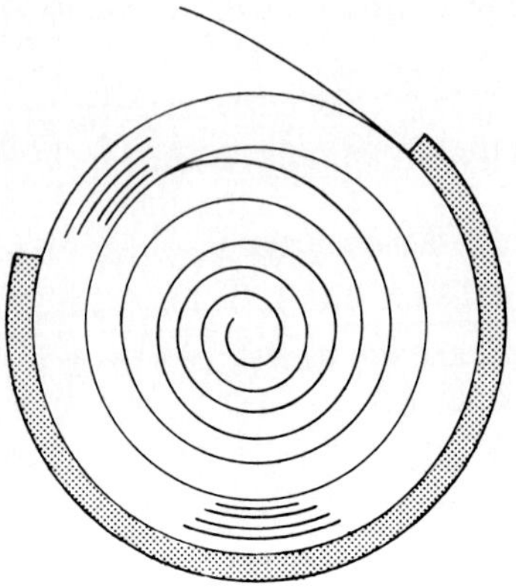

FIG. 9-11 Power spring.

The following formula, based on concentric circles with a uniform space between the coils, gives a close approximation of the minimum OD_F:

$$OD_F = \frac{2L}{\pi[(\sqrt{A^2 + 1.27Lt} - A)/2t - \theta]} - A$$

POWER SPRINGS

9-33 General Data

Power springs (Fig. 9-11) are made of rectangular-section material, are wound flat, and have either special retaining holes or bends on both ends. As the length-to-thickness ratio L/t increases, the spiral space between coils increases rapidly. Therefore, the spring must be retained in some type of housing or case in application. The coils other than the transition coils and the coil which is attached to the arbor are solid against the case. As the arbor rotates, solid material will become active as it pulls away from the case and is wound on the arbor. The amount of active material is constantly changing, making it difficult to develop a formula for the torque per turn delivered by the spring.

The torque-deflection characteristic of a power spring is nonlinear. This condition is caused by the friction between the coils, the varying amount of material that becomes active as the spring is deflected, and normal hysteresis effect.

9-34 Design Method

Design of power springs involves the torque-deflection curve, the number of turns the spring will produce, and the maximum torque. Figure 9-12 shows the maximum torque for various material thicknesses, based on a strip 1 in wide and a modulus of elasticity of 30×10^6 lb/in^2. The torque for any other width is directly proportional to the torque for the 1-in-wide strip (e.g., half as much for ½-in-wide strip).

The formulas for the number of revolutions θ that the spring will deliver and the active length L of material are as follows:

$$\theta = \frac{\sqrt{2(D_h^2 + A^2)} - (D_h + A)}{2.55t}$$

$$L = \frac{D_h^2 - A^2}{2.55t}$$

where D_h = inside diameter of housing or case
A = outside diameter of arbor
t = material thickness

These formulas are based on the optimum condition, in which the active material occupies one-half of the available space between the arbor diameter A and the case inside diameter D_h. This condition gives the maximum number of arbor revolutions; either a longer or a shorter length of active material will decrease the revolutions of the arbor. It is also assumed in the formulas that the arbor and case diameters are known. If the case size is unknown, the graph in Fig. 9-13 gives the case diameter D_h for the material thickness t and the number of revolutions θ desired.

CONSTANT-FORCE SPRINGS

9-35 General Data

A constant-force spring is a strip of flat spring material which has been wound to a given curvature so that in its relaxed condition it is in the form of a tightly wound coil or spiral. A constant force is obtained when the outer end of the spring is extended tangent to the coiled body of the spring. A constant torque is obtained when the outer end of the spring is attached to another spool and caused to wind in either the reverse direction from that in which it was originally wound, or the same direction. Torque is obtained only from the reverse-wound or larger spool.

Unlike conventional springs, constant-force springs produce a constant force or torque regardless of their extension or number of turns. Because there are actually slight variations in the force exerted, these springs are not really "constant-force" in the precise meaning of the term. However, "constant" is the best word available to describe their force-deflection characteristics, especially when they are compared with conventional springs.

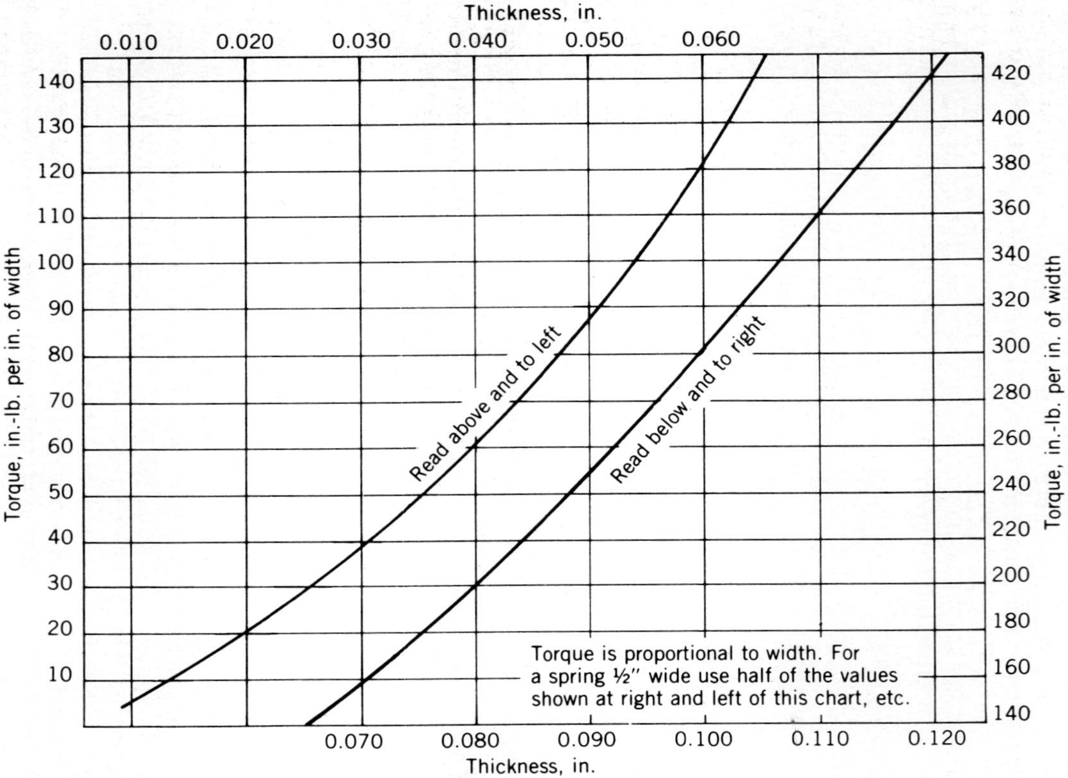

FIG. 9-12 Maximum torque per inch of spring width for power springs. (*Courtesy of Associated Spring, Barnes Group, Inc.*)

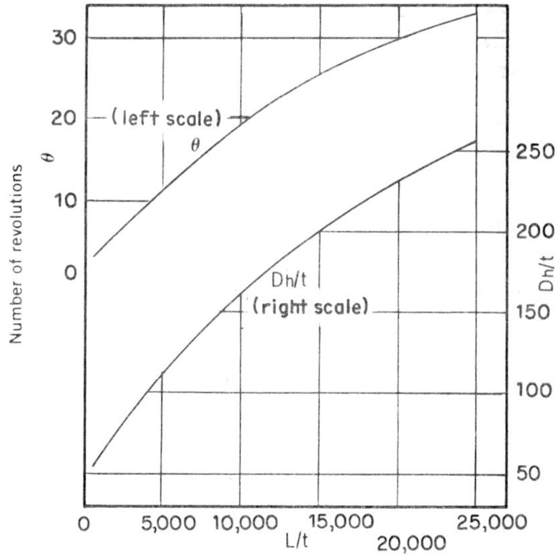

FIG. 9-13 Relationship between number of turns, case diameter, strip length, and thickness for power springs. *(Courtesy of Associated Spring, Barnes Group, Inc.)*

Most constant-force springs are made from AISI 301 stainless steel, and for most applications the spring is "stress-relieved" after rolling at about 500°F. Some of the many applications of constant-force springs are in furniture, toys, electric motors, appliances, space vehicles, and automobiles.

One of the greatest limitations of a constant-force spring is a relatively short operating life. While a life of 1 million cycles is possible, the material is used so inefficiently that such applications become impractical. The most efficient use of the material occurs at a life of 3000 cycles, with most applications for the spring requiring a life of between 3000 and 20,000 cycles. A failure usually begins with a crack that starts in the center of the strip and runs across its width. Unlike conventional springs, which fail suddenly, a constant-force spring usually runs for several hundred cycles between the time the crack first appears and the time complete failure occurs.

Like every spring that is made, whether it be compression, extension, torsion, or power, the constant-force spring has many unique characteristics, which are given here for the three basic forms. While the design formulas are accurate, it is good practice to consult a spring manufacturer before finalizing a design.

9-36 Symbols

P = load, lb ± 10 percent

M = output torque, in·lb ± 10 percent

E = modulus of elasticity, lb/in^2; for stainless steel $E = 28 \times 10^6$ and for high-carbon steel $E = 30 \times 10^6$

S = stress, lb/in^2
b = material width, in
t = material thickness, in
R_n = natural radius, in ± 10 percent
R_2 = storage drum radius, in
R_3 = output drum radius, in
D_2 = storage drum
D_3 = output drum
S_f = stress factor

Note the following:

1. The load formulas give good results for moderately stressed springs. Extremely high-stress springs will not correlate with the formulas, and experience factors must be used.
2. The factor 26.4 is used to compensate for cross curvature occurring in the extension form. Cross curvature is not a factor in the motor forms.
3. The spring obeys Hooke's law as an increment is deflected from curved to straight.
4. Diameter build-up must be compensated for on long springs.
5. The spring is usually not attached to storage drum D_2.
6. At full deflection, 1.5 turns should remain on storage drum D_2.
7. Since the constant-force spring is a highly stressed element, it has an inherently limited fatigue life and should be used cautiously where its failure can result in harm to equipment or personnel.

9-37 Extension Form (See Fig. 9-14)

Design Formulas The formulas for load and stress are

$$P = \frac{Ebt^3}{26.4R_n^2} \quad \text{lb} \qquad (9\text{-}12)$$

$$S = \frac{Et}{2R_n} \quad \text{lb/in}^2 \qquad (9\text{-}13)$$

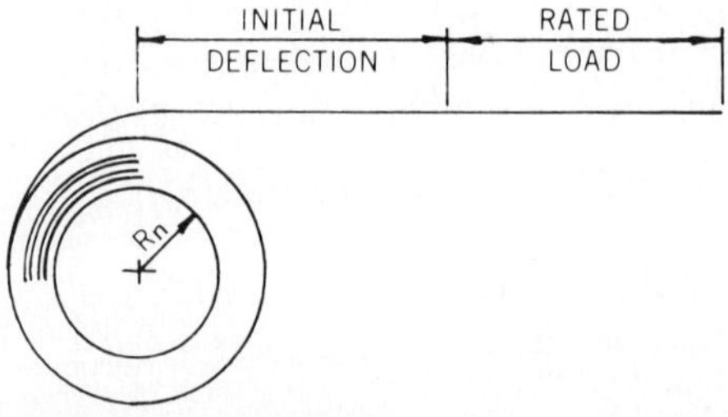

FIG. 9-14 Extension form.

$$R_2 = 1.15R_n \tag{9-14}$$

$$S_f = \frac{t}{R_n} \tag{9-15}$$

Characteristics

1. Constant retracting force (zero gradient)
2. Rated load reached after an initial deflection of 1.25 times the outside diameter
3. No intercoil friction (smooth operation)
4. Low life
   a. Maximum force obtained at a life of 3000 cycles
   b. Life requirements of more than 50,000 cycles use the material too inefficiently to be practical
5. *a.* $S_f = 0.025$ produces 5000 cycles life
   b. $S_f = 0.010$ produces 50,000 cycles life
6. Long deflection possible
7. Unstable when not supported

9-38 Motor Form A (See Fig. 9-15)

Design Formulas

$$M = \frac{Ebt^3R_3}{24}\left(\frac{1}{R_n} - \frac{1}{R_3}\right)^2 \quad \text{in·lb} \tag{9-16}$$

$$S = \frac{Et}{2R_n} \quad \text{lb/in}^2 \tag{9-17}$$

$$R_3 > 2R_n \tag{9-18}$$

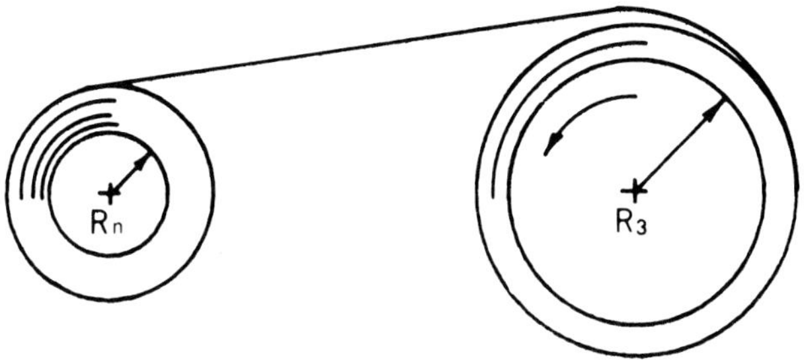

FIG. 9-15 Motor form A.

Characteristics

1. Constant output torque (zero gradient)
2. Low output torque (inefficient use of material)

3. No intercoil friction (smooth operation, low hysteresis losses)
4. Low life
   a. Maximum torque obtained at a life of 2500 cycles
   b. Life increases as R_3 increases
5. Large number of output turns
6. Spring can operate as clutch on D_3

9-39 Motor Form B (See Fig. 9-16)

Design Formulas

$$M = \frac{Ebt^3R_3}{24}\left(\frac{1}{R_n} + \frac{1}{R_3}\right)^2 \quad \text{in}\cdot\text{lb} \tag{9-19}$$

$$S = \frac{Et}{2}\left(\frac{1}{R_n} + \frac{1}{R_3}\right) \quad \text{lb/in}^2 \tag{9-20}$$

$$R_3 = 1.6R_n \quad \text{(recommended)} \tag{9-21}$$

$$S_f = t\left(\frac{1}{R_n} + \frac{1}{R_3}\right) \tag{9-22}$$

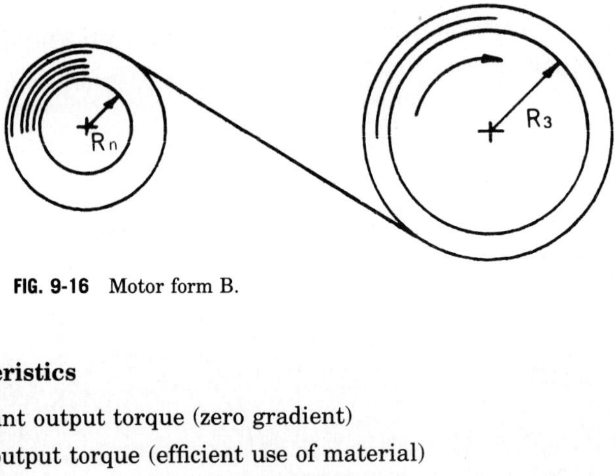

FIG. 9-16 Motor form B.

Characteristics

1. Constant output torque (zero gradient)
2. High output torque (efficient use of material)
3. No intercoil friction (smooth operation, low hysteresis losses)
4. Low life
   a. Maximum torque obtained at a life of 2500 cycles
   b. Maximum practical life = 35,000 cycles
5. *a.* S_f = 0.025 produces 5000 cycles life
   b. S_f = 0.010 produces 50,000 cycles life
6. High number of output turns (over 50 turns possible)

FLAT SPRINGS

9-40 General Data

The term "flat springs" covers a wide range of springs or stampings fabricated from flat strip material which, on being deflected by an external load, will store and then release energy. Only a small portion of a complex-shaped stamping may actually be functioning as a spring. For design purposes, that portion which acts as a spring may often be considered as an independent simple spring form, while the rest of the part is temporarily ignored.

9-41 Cantilever and Beam Springs

In comparing the load and stress equations for both types of springs (Figs. 9-17 and 9-18), it can be seen that an equal active length L of identical cross section b and t produces 16 times the load in a simple beam spring as in a cantilever spring. However, the stress in the simple beam spring is 4 times that of the cantilever spring for a given deflection.

The differences between the two types of springs cancel out when they are compared in terms of volume of active spring material. A simple beam spring can be designed with greater length and decreased thickness so that it will have the same load, same deflection, same stress, and same volume of material as a given cantilever spring.

Because load P varies as the third power of the thickness t, the flat-spring material should have minimum variation in thickness. Load also varies as the third power of length.

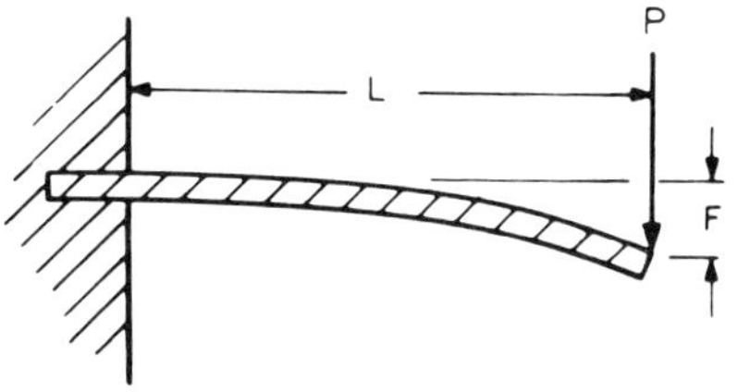

FIG. 9-17 Cantilever spring.

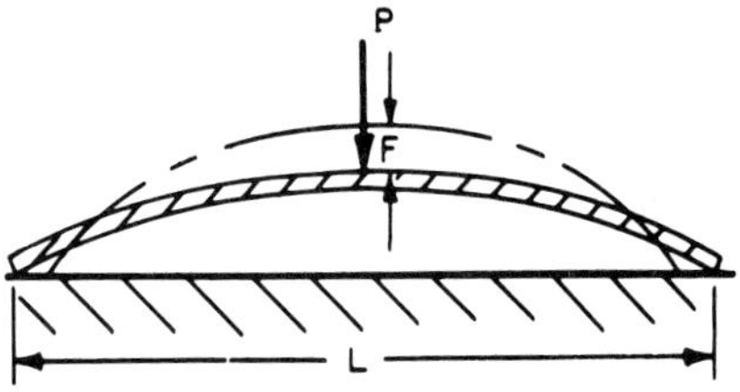

FIG. 9-18 Simple beam spring.

9-42 Design Formulas

The formulas for load P and bending stress S used in cantilever and simple beam springs are as follows:

Cantilever Spring

$$P = \frac{EFbt^3}{4L^3} \quad \text{lb} \tag{9-23}$$

$$S = \frac{3EFt}{2L^2} = \frac{6PL}{bt^2} \quad \text{lb/in}^2 \tag{9-24}$$

Simple Beam Spring

$$P = \frac{4EFbt^3}{L^3} \quad \text{lb} \tag{9-25}$$

$$S = \frac{6EFt}{L^2} = \frac{3PL}{2bt^2} \quad \text{lb/in}^2 \tag{9-26}$$

where P = load, lb
E = modulus of elasticity, lb/in^2
F = deflection, in
t = thickness of material, in*
L = active spring length, in
b = width of material, in
S = design bending stress, lb/in^2

9-43 Design Method

The design of the many special types of flat springs is covered thoroughly in many standard texts and technical articles and will not be covered in this handbook.

The design method for the two basic types of flat springs—cantilever and simple beam springs—is quite easy compared with the design method for helical compression, extension, and torsion springs. While flat springs may appear to be extremely complex in shape, the engineer need consider only those parts which are active in operation. It is likely, then, that the active section is no more than an ordinary cantilever or simple beam spring.

9-44 Flat-Spring Materials

The most commonly used flat-spring material is spring steel, usually the AISI 1050, AISI 1074, and AISI 1095 grades. Type 301, 302, or 17-7 PH stainless is often specified where corrosion resistance is needed, and phosphor bronze and beryllium copper alloys are the most common for high electrical conductivity.

Carbon steels may be fabricated as either pretempered or annealed material. Pretempered steel is usually specified where the bending requirements are not severe (thin sections, large-radius bends). Pretempered flat springs are almost always stress-relieved after forming. Annealed material is used for heavier sections or where forming requirements would not allow a pretempered material. After forming, the parts are hardened and tempered to the required spring properties. Since a finished spring can easily be distorted during hardening, pretempered materials are preferred to annealed materials. In addition, thin, close-tolerance broad flat springs usually require either additional setting operations or jig hardening, both of which add to cost. Where pretempered materials cannot be used, it is often possible to substitute type

* In some manuals h instead of t is used for thickness.

301 or 17-7 PH stainless steel to obtain spring properties similar to those of a pretempered material without the added cost of hardening or setting operations.

As with all other types of springs stressed in bending, the tensile or tensile yield strength of a selected material is an important factor in properly designing a flat spring.

SPRING WASHERS

9-45 General Data

Spring washers are becoming more widely used because of trends toward miniaturization and greater compactness of design. They afford space and weight advantages over conventional wire springs, and, when compared on an installed cost basis, they are often more economical to use. Their applications include keeping fasteners secure, distributing loads, absorbing vibrations, compensating for temperature changes, eliminating side and end play, and controlling end pressure.

Design formulas have been developed for determining the spring characteristics of the three basic types: curved washers, wave washers, and Belleville washers. Together they cover a wide range of loading conditions.

9-46 Curved Washers (Fig. 9-19)

Characteristics

1. Light loads.
2. Spring rate nearly linear.
3. Deflection up to one-third of OD, but not to exceed $0.8h$ when loads are specified.
4. Most expansion under load.
5. Two-point contact.
6. Load tolerances should not be less than ± 20 percent.

Design Formulas The formulas for load P and bending stress S are

$$P = \frac{4Eft^3(\text{OD} - \text{ID})}{(\text{OD})^3} \quad \text{lb} \tag{9-27}$$

$$S = \frac{1.5P(\text{OD})}{t^2(\text{OD} - \text{ID})} \quad \text{lb/in}^2 \tag{9-28}$$

where P = load, lb
E = modulus of elasticity, lb/in^2
f = deflection, in
t = material thickness, in

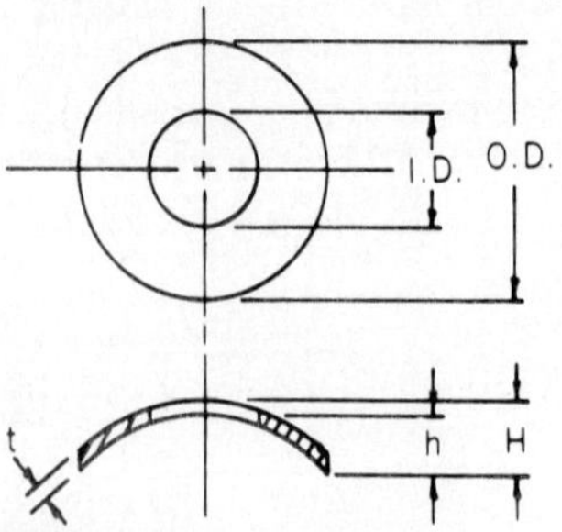

FIG. 9-19 Curved washer.

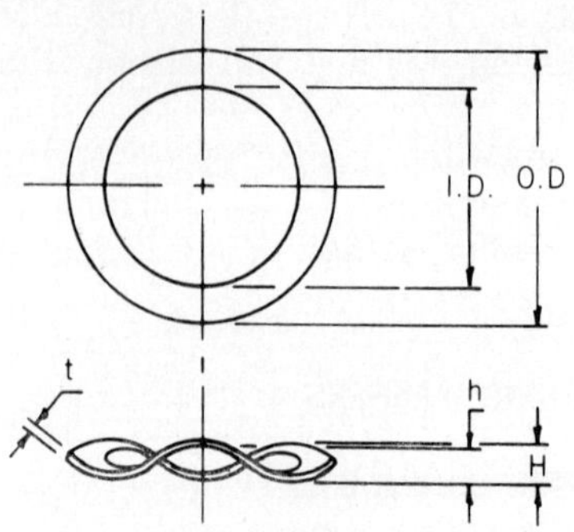

FIG. 9-20 Wave washer.

h = free height minus t, in
H = free overall height, in
S = bending stress, lb/in^2

These formulas give approximate results because of the variables involved.

9-47 Wave Washers (Fig. 9-20)

Characteristics

1. Light to medium loads.
2. Spring rate nearly linear.
3. Deflection up to one-fourth of OD, but not to exceed $0.75h$ when loads are specified.
4. Less expansion than curved washer.
5. Contact points on one side equal number of waves.
6. Load tolerances should not be less than ±20 percent.

Design Formulas The formulas for load P and bending stress S are

$$P = \frac{Efbt^3N^4}{2.4D^3}\left(\frac{\text{OD}}{\text{ID}}\right) \quad \text{lb} \tag{9-29}$$

$$S = \frac{3\pi PD}{4bt^2N^2} \quad \text{lb/in}^2 \tag{9-30}$$

These formulas give approximate results because of the variables involved. For deflections between $0.25h$ and $0.75h$, the OD increases and the new mean diameter D' can be calculated by

$$D' = \sqrt{D^2 + 0.458h^2N^2} \quad \text{in} \tag{9-31}$$

where P = load, lb
E = modulus of elasticity, lb/in^2

f = deflection, in
b = radial width of material, in = (OD − ID)/2
t = material thickness, in
N = number of waves
D = mean diameter, in = (OD + ID)/2
h = free height minus t, in
H = free overall height, in
S = bending stress, lb/in^2

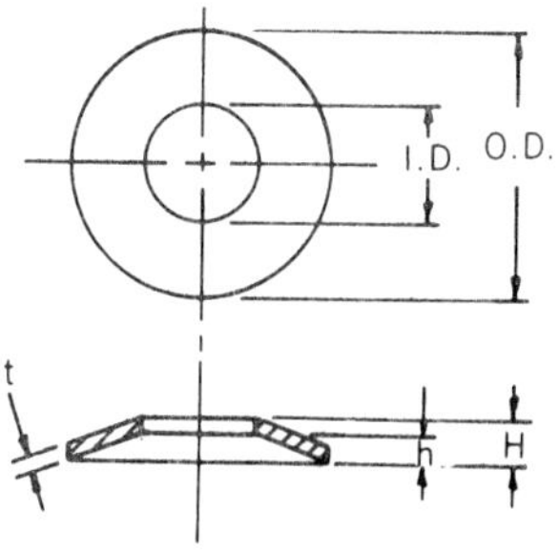

FIG. 9-21 Belleville washer.

9-48 Belleville Washers

Belleville spring washers may be used where small deflections and relatively high loads are required (see Fig. 9-21). It is especially interesting that by varying the ratio of h/t the load-deflection characteristics may be changed, as shown in Fig. 9-22.

Characteristics

1. Medium to heavy loads.
2. Constant spring rate for $h/t < 0.4$.

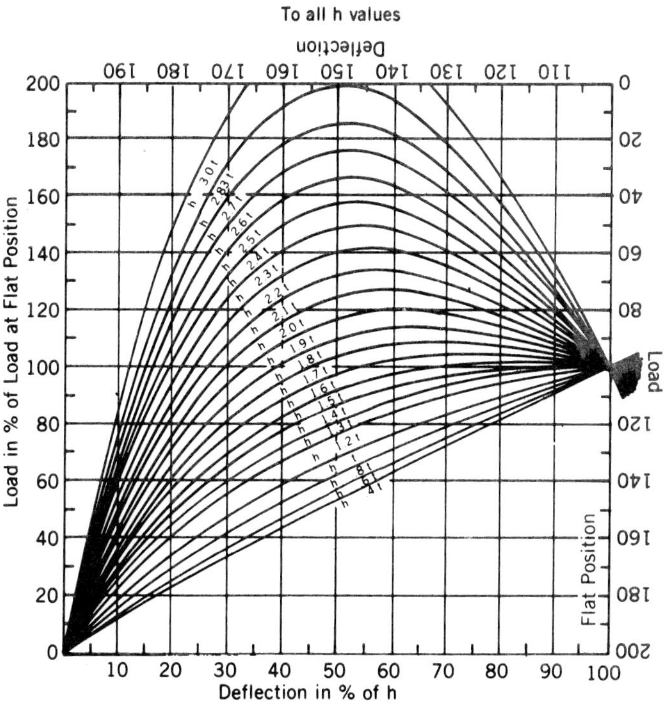

FIG. 9-22 Belleville washer load/deflection curves for various h/t ratios.

3. Nearly constant rate for $h/t > 0.4 < 0.8$.
4. Positive decreasing rate for $0.8 < h/t > 1.41$.
5. Zero rate over part of deflection for $h/t = 1.5$.
6. For $1.4 < h/t > 2.83$, positive decreasing rate which becomes zero, then negative, and increasing before bottoming out.
7. For $h/t > 2.83$, similar to 6, except that the washer will snap past flat position and remain stable. It must be loaded in the opposite direction to restore the original working position.
8. Deflection up to one-tenth of rim width, but not to exceed $0.9h$ when loads are specified.
9. Least expansion under load.
10. Contact all around circumference.
11. Normal load tolerance is ±20 percent.
12. May be stacked in series to increase overall deflection, in parallel to increase load, or a combination of these. Belleville washers with h/t ratios greater than 1.2 should not be used in series.

Design Formulas The formulas for load P and stress S at the convex inner edge are given here. The constants M, C_1, and C_2 may be obtained from Fig. 9-23.

$$P = \frac{4Ef}{M(1-\mu^2)(\text{OD})^2}\left[\left(h-\frac{f}{2}\right)(h-f)t + t^3\right] \quad \text{lb} \qquad (9\text{-}32)$$

$$S = \frac{4Ef}{M(1-\mu^2)(\text{OD})^2}\left[C_1\left(h-\frac{f}{2}\right) + C_2 t\right] \quad \text{lb/in}^2 \qquad (9\text{-}33)$$

where a = OD/ID
μ = Poisson's ratio (0.3 for steel)
ln = natural logarithm

SPRING MATERIALS

9-49 Wire-Spring Materials

Selection of the proper wire material is a major step in designing all types of springs. Four major metallurgical classifications to select from are carbon steels, alloy steels, stainless steels, and nonferrous spring materials.

For proper design, the designer must consider, in order of importance, such factors as temperature and corrosion, conductivity, physical properties, and types of materials available. When there are a large number of stress cycles, and therefore a long fatigue life is required, the quality of the material must be tightly controlled. The wire material must have a uniform internal structure, and its surface must be free of pits, seams, scratches, and any other flaws that can impair its fatigue life. The most common long-service materials are

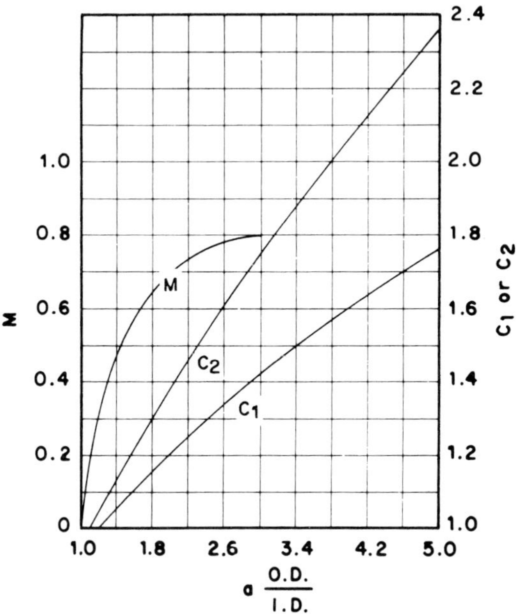

FIG. 9-23 Stress constants for Belleville spring washer calculations.

carbon steels and alloy steels, since both can be free from seams and of aircraft quality, as is specified for valve springs.

9-50 Flat-Spring Materials

The most commonly used flat-spring materials are the cold-rolled spring steels. AISI 301, 302 (18-8), and 17-7 PH (ASTM A693) stainless are often specified where corrosion resistance is needed, and phosphor bronze and beryllium copper alloys are the most common for high electrical conductivity.

Carbon steel may be fabricated from either pretempered or annealed material. Pretempered steel is usually specified when the bending requirements are not severe (thin sections, large-radius bends). Pretempered flat springs are almost always stress-relieved after forming. Annealed material is used for heavier sections or where forming requirements would not allow a pretempered material.

After forming, the parts are hardened and tempered to the required spring properties. Since a finished spring can easily be distorted during hardening, pretempered materials are preferred to annealed materials. In addition, thin, wide flat springs specified to close tolerances usually require either additional setting operations or jig hardening, both of which add to cost. Where pretempered materials cannot be used, it is often possible to substitute AISI 301 or 17-7PH (ASTM A693) stainless steel to obtain spring properties similar to those of a pretempered material without the added cost of hardening or setting operations.

10
THREADED FASTENERS

HARRY S. BRENNER, P. E.
President/Almay Research & Testing Corp., Los Angeles, California

INTRODUCTION

The art—and the necessity—of structural joining of different components undoubtedly dates to the beginning of human history. The first of our ancestors who constructed a hunting device by binding a sharp rock to a length of wood in effect was fashioning a tool using the techniques of fastening and joining. As societies developed, later generations explored the use of malleable metals, with pins and rivets finding wide application as simple fastening devices. Today, a highly developed system of fastening devices is in place on a worldwide basis. This is undoubtedly one of the major cornerstones of current technology, permitting fabrication and use of equipment in such diverse industries as automotive, electronics, aerospace, transportation, construction, energy, nuclear, and appliance.

10-1 Joining Methods

In any general discussion of "fastening," it should be recognized that there are several prominent methods of joining components. Among these are mechanical fasteners, welding, brazing, soldering, and various adhesive systems. The salient feature of the last four joining techniques (welding, brazing, soldering, and adhesives) is that they result in a permanently bonded joint and normally require special equipment and controls to effect a proper structural joint. Mechanical fasteners, while often intended as a permanent system, offer certain distinct advantages not found in bonded joints. Threaded fasteners, in particular, can be disassembled to permit maintenance or repair of equipment, as required. Production, as well as standard manual, tools can be used for assembly and installation, even under restricted conditions. And significantly, the multitude of available sizes, materials, finishes, and strength levels of mechanical fasteners provides a strong design support system for optimum structural compatibility, even under wide environmental extremes.

10-2 Prime Concern

This section will be limited to discussion of threaded-type fasteners. While such fastening devices are often considered to be simple joining elements, in reality the threaded fastener assembly represents a complex mechanical design system. Understanding the critical components of the system, their intended functions, and criteria for successful use is of paramount importance to the designer and user alike. This is especially true in light of the major emphasis on developing longer life integrity of new structures and equipment, as well as on minimizing product liability exposure resulting from premature failure.

THREADED FASTENER SYSTEM

The basic threaded fastener system includes a male or externally threaded element (bolt, screw, or stud), a mating female or internally threaded element (nut, insert, or tapped hole), and very often a washer (plain washer, lockwasher). The production techniques for each of the major elements are unique and different, but obviously careful control of dimensions is required so that all elements in the system will mate and engage properly, irrespective of the manufacturing source. Therefore, standards have been developed for the majority of the commonly used bolts, nuts, washers, etc. Economy and availability dictate reliance on standard products wherever practicable. However, there are still joining applications or individual industry requirements where special fasteners are designed to fill a particular need. The field of threaded fasteners therefore comprises the full gamut of standards and specials. While available configurations, sizes, materials, and styles may vary, all contributing components should be evaluated as part of an integral balanced "fastening system." The ability to see individual fasteners as part of a compatible system is sometimes overlooked, but unless the overall system is properly considered, assuring efficient joint design could be a problem.

10-3 Externally Threaded Fasteners

Loosely defined, bolts and screws are fasteners with a formed head on one end and an external thread on the other end. Studs are fasteners that incorporate external threads at each end. Some examples of representative bolts and screws are illustrated in Fig. 10-1, while various stud samples are shown in Fig. 10-2.

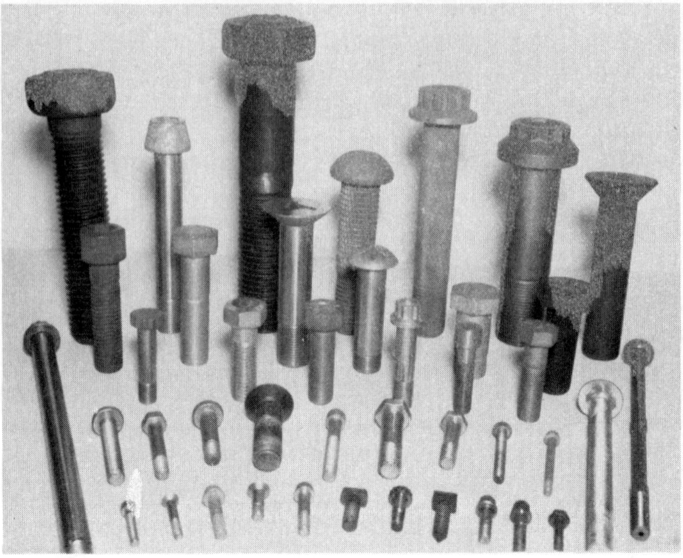

FIG. 10-1 Representative examples of bolts and screws, illustrating versatility of design and configurations.

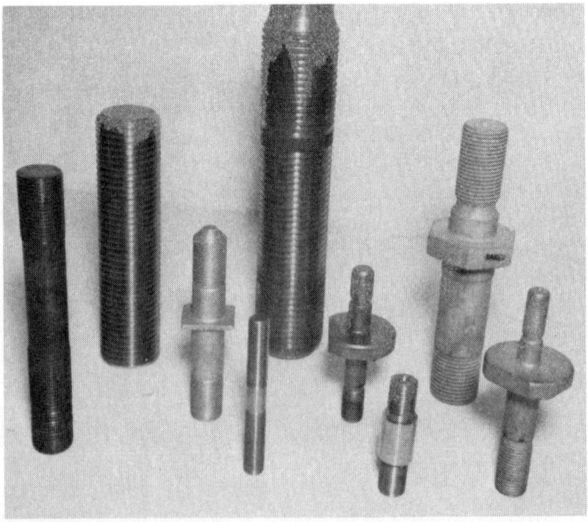

FIG. 10-2 Representative examples of threaded studs.

Early commercial development of fasteners was centered around applications and needs peculiar to specific industries. In turn, terminology describing the fasteners used within an industry was established. As modern technology expanded the application of both old and new types of fasteners, the sometimes arbitrary terminology presented conflicts, especially the distinction between "bolts" and "screws." Currently, the following is the generally accepted criterion for distinguishing between bolts and screws:

1. *Bolts* are externally threaded fasteners intended to be mated with a nut and tightened by turning the nut.
2. *Screws* are externally threaded fasteners intended to be installed in a preformed mating internal thread and tightened or removed by turning the screw head. It should be noted that certain classes of screws are designed to form or cut mating threads during the installation and tightening operation.

A typical example of look-alike fasteners illustrating these definitions is shown in Fig. 10-3. There are still some exceptions to the rule, as in the case where bolt heads are torqued during installation when they are assembled with attached plate nuts or barrel nuts.

Structural bolts and screws are installed through prepared holes in the material to be joined. Since the thickness of the material to be fastened can vary, the lengths of bolts and screws are designed to vary to accommodate both the grip length of the joint and the additional length of thread needed for proper engagement with the nut or with the internal tapped thread. These fasteners are subjected to tensile, shear, bending, and fatigue loads sensed by the joint. They also respond to the environment imposed on the joint, which may include temperature extremes or exposure to various corrosive conditions.

Studs represent a special class of externally threaded fasteners, and a number of configurations have found wide use in design. Double-end studs have standard threads at both ends; they can be mated with two nuts, or one end of

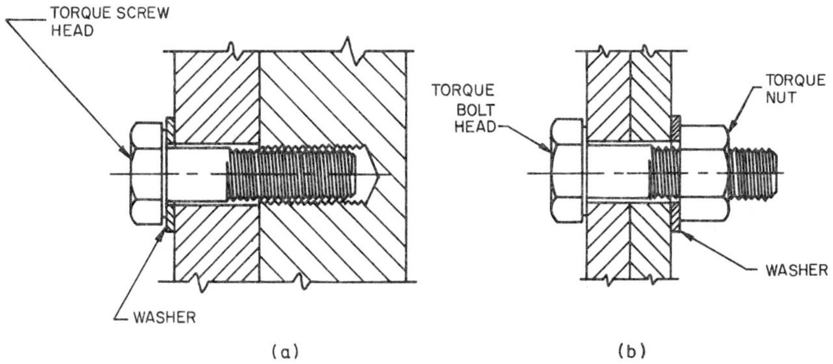

FIG. 10-3 Distinction between "bolt" and "screw" based on assembly torque method.

the stud can be installed in a tapped hole and a nut employed on the other end to tighten or secure the joint. Tap-end studs normally have an interference-fit thread on one end, which essentially anchors or locks the stud when it is installed in a tapped hole. The other end of the stud includes a standard thread for nut engagement. Variations on this concept are locked-in studs and pressed-in studs. Another configuration is the continuous-threaded stud, which has been utilized both in general-purpose joining and in special classes for high-temperature and/or high-pressure bolting.

While not as extensively used as bolts and screws, studs nevertheless do offer certain benefits in selected applications. Installed studs can serve as guides or pilots when assembling cap plates or similar cover sections. They provide additional latitude in nut adjustment, alignment, and hole clearances, especially when such cap plates are relatively thick. And many smaller-size studs have been particularly adapted for automated installation in light structural applications.

10-4 Internally Threaded Fasteners

Specific fastening devices which incorporate internal (female) threads include nuts and inserts. They are intended to engage with the external threads of bolts, screws, and studs, and should be compatible to develop the full rated strength of the external thread.

As previously indicated, structural nuts are normally torqued to tighten the fastener assembly to prevent loosening under possible vibration and fatigue loading conditions or as a result of environmental temperature extremes. In addition, the act of torquing introduces a clamp load in the fastener assembly which minimizes the potential effect of fatigue loading sensed by the system.

One aspect of threaded fastener performance that is of concern is the fastener's susceptibility to loosening as a result of severe vibration or dynamic loadings acting on the joint. Assessment of vibration exposure and prevention of loosening that could possibly lead to loss of the fastener system are basic objectives of sound design practice. Since vibratory stresses cannot be totally eliminated, several methods have been developed which have proven effective in maintaining fastener integrity. The more prominent techniques include the use of self-locking nuts, chemical (adhesive) thread locking, cotter pins, and safety wiring.

- *Self-locking nuts* are integral fasteners which incorporate in the nut element a controlled high-torque feature which is designed to prevent rotation off the external threads, even if the initial tightening torque is completely relaxed. The same principle of an inherent self-locking feature is often extended to screws used in tapped holes.
- *Chemical thread-locking systems* include anerobic and epoxy adhesives which are applied to the fastener threads before installation, and which cure and effect a permanent bond after assembly.

- *Cotter pins* are a supplemental locking device used with a slotted or castellated nut. They are installed through a drilled hole in the bolt threads to prevent rotation or movement of the nut after installation.
- *Safety lock wiring* is also a supplemental locking system. Usually two or more fasteners in series must be wired to prevent rotation in the "off" direction. The nut end can be wired using slotted or castellated nuts and bolts with drilled holes. However, the predominant use of safety wiring is to secure screw heads which have been drilled to accommodate the wire where the screws have been installed in tapped or blind holes.

As with bolts and screws, there are numerous variations of both standard and self-locking nuts. Common configurations for wrenching include square, hexagon, and 12-point (double hex) designs. For ultra-high-strength bolts and nuts designed for aerospace, a new spline drive wrenching design has been introduced to assure strength capability to sustain the higher torque levels required for assembly of these particular fasteners. Nonwrenchable nuts include pressed-in–type nuts and plate nuts, which may be welded or riveted to substructure to provide a permanent in-place nut.

Some typical nuts illustrating the variety of types and sizes available are shown in Fig. 10-4.

The threaded insert incorporates an internal (female) thread similar to that of a nut, but it is considered to be a separate and distinct fastener category. Basically, inserts duplicate the function of a tapped hole. At the same time, they add certain performance features in selected applications not normally associated with a tapped thread. Inserts of various designs can either be

FIG. 10-4 Examples of nuts of various configurations and sizes.

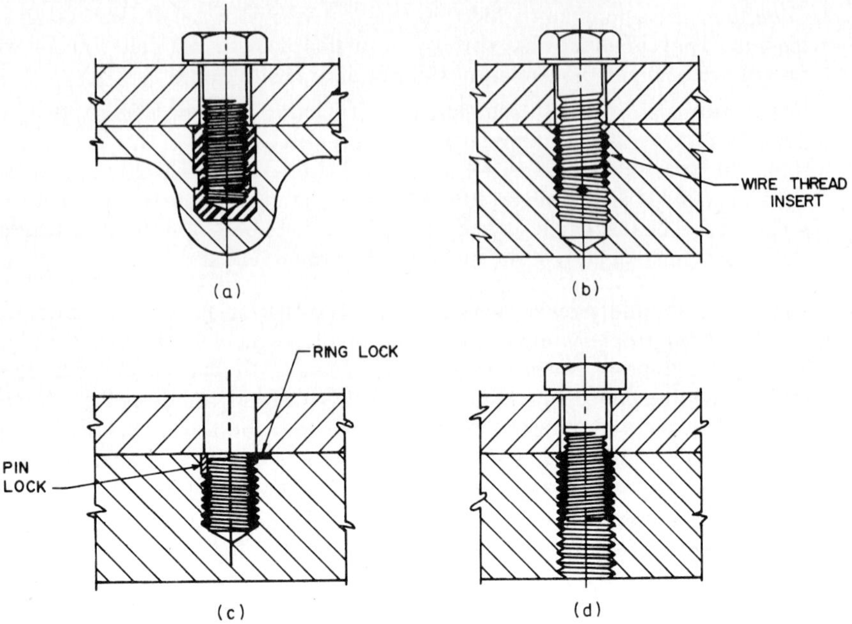

FIG. 10-5 Typical types of threaded insert applications. (*a*) Molded-in threaded insert. (*b*) Wire form threaded insert (plain and self-locking). (*c*) Bushing insert ring locked or pin-locked. (*d*) Solid bushing threaded insert (plain and self-locking).

molded in cast metal or plastic or be threaded into position, as illustrated in Fig. 10-5. In softer parent materials, inserts provide increased strength and improve wear resistance of the threads, particularly where frequent disassembly is encountered. They also are valuable as a repair fastener for original tapped threads which may have been damaged, or as a replacement for threads worn in service. Representative examples are illustrated in Fig. 10-6.

10-5 Washers

The significance of the washer as part of the structural fastener system is occasionally overlooked. Yet, when it is used, it does contribute to the integrity of the joint, and its importance should not be minimized.

As noted earlier, a single washer is normally applied under the fastener element being torqued, i.e., the nut or the screw. However, flat or plain washers provide additional bearing area for distribution of stresses, notably in softer structural materials, and they may be used under bolt heads as well. Washers also permit bridging of oversize or large-clearance holes. They can also be used to compensate for excessive grip length of bolts or screws in assembly. Washers can often be effective as a barrier to prevent or minimize corrosion resulting from contact between dissimilar metals in a joint.

FIG. 10-6 Examples of typical threaded inserts. (*Courtesy of Rexnord/Specialty Fastener Division.*)

When using structural washers, the key is to make sure that the right washer is specified. In particular, the washer should be specifically compatible with the strength level of the threaded fasteners, since too soft a washer may actually extrude and result in subsequent fastener relaxation or loosening with service exposure.

Another class of washer, known as a "lockwasher," is primarily used as a spring tension device. Helical-spring lockwashers, tooth lockwashers, and conical washers are representative of this type of fastener. When the complete fastener assembly is torqued, the washer is compressed flat. Should the bolt or screw tend to relax or loosen, the inherent spring tension of the lockwasher exerts force on the threaded fastener to maintain some degree of tension to inhibit loosening. Lockwashers are generally regarded as applicable for structural use in lightly to medium-loaded joints.

THREADS

Common to threaded fasteners are the external and internal threads, which obviously must properly mate and disengage, and must also be capable of developing the full rated strength of the assembly. Underlying the successful and interchangeable use of bolts, screws, nuts, and other such components is the system of thread standards and tolerances which has evolved through extensive practical experience. Currently, there are a number of apparently similar, yet different, thread forms in use throughout the world. For the most part, the major popular thread forms were first developed during the nineteenth century in the various industrial countries and later adopted as respective national standards. While this was effective for the sponsoring country, detailed comparison of the different thread standards indicated that they were not functionally interchangeable.

Perhaps the first significant effort to achieve international standardization was the adoption of the "Unified Screw Thread" through the Declaration of Accord, signed in 1948 by Canada, Great Britain, and the United States. More recently, international attention and effort have been directed toward the development of metric thread standards under the auspices of the International Organization for Standardization (ISO).

At present, threads and their related standards fall into two major categories: (1) the inch series and (2) the metric series. Threads represent the heart of the threaded fastener system, and knowledge of the standards is invaluable in their proper application.

10-6 Inch Threads

Inch-series thread forms were developed and used by countries employing the English system of measurement. In the United States, standards have been predicated on a 60° screw thread angle, originally standardized as the American National thread. Since 1948, the accepted standard has been the Unified thread form, which incidentally also specifies a 60° angle. Details defining critical thread criteria, tolerances, gaging requirements, etc., are outlined in National Bureau of Standards Handbook H-28, *Screw Thread Standards for Federal Services,* and in American National Standards Institute Specification B1.1, "Unified Inch Screw Threads." Among other things, screw-thread standards reflect essential requirements for diameter-pitch combinations, class fit, and allowances for platings or coatings where needed. For reference, the basic design form and definitions applicable to the Unified thread are illustrated in Fig. 10-7.

The relationship between nominal diameter and the number of threads per inch is referred to as the "diameter-pitch combination." There are several prominent thread forms representing different diameter-pitch series which cover the majority of standards intended for general engineering use. The standards of primary interest are identified as follows:

- *Coarse-thread series.* This is perhaps the most widely used series of commercial and industrial fasteners. The thread form is particularly advantageous for applications requiring rapid assembly or disassembly, or for threading into lower-strength materials, such as castings, soft metals, and plastics.
- *Fine-thread series.* For the same nominal diameter, this series incorporates more threads per inch. The result is a larger tensile stress area than that of the same size coarse thread, contributing to the greater strength capability of fine-thread fasteners. These fasteners are normally used where the length of thread engagement is short, or where a smaller lead angle is desired. Fine-thread-series fasteners are used extensively in aerospace and in applications where coarse threads would not be suitable.

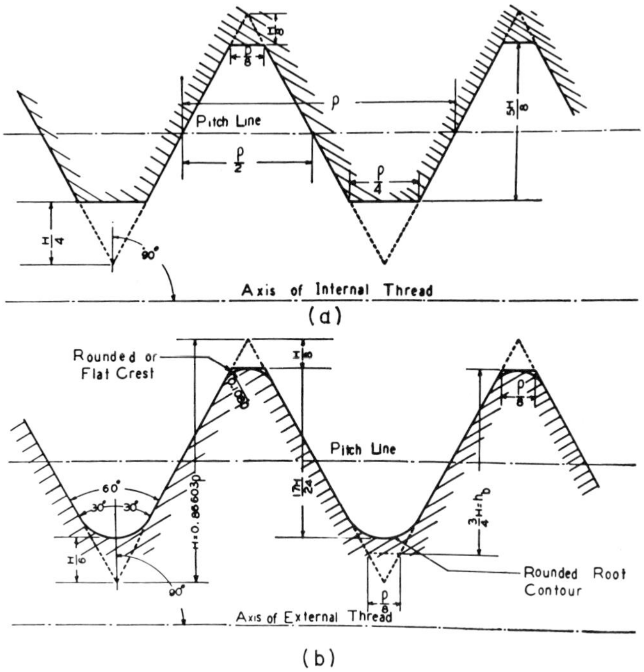

FIG. 10-7 Unified internal and external screw-thread design forms (maximum metal condition). (*Source: Unified Screw Thread Standards; National Bureau of Standards Circular 479.*)

- *Extra-fine-thread series.* Use of this series is normally limited to applications in which depth of thread is critical, such as in threading of thin-wall materials, or in which a maximum practicable number of threads is required within a given thread length. This thread form is occasionally used for adjustment control.
- *8-thread series.* This is one of several uniform-pitch-series thread forms, in which the number of threads per inch is the same, irrespective of the fastener diameter. Although originally intended for high-pressure-joint bolts and nuts, it is now widely used as a substitute for the coarse-thread series for diameters larger than 1 in.
- *12-thread series.* This is a constant-pitch-series thread form that incorporates medium-pitch threads. It is widely used in machine construction for thin nuts on shafts and sleeves. The 12-thread series provides an extension of the fine-thread series for diameters larger than 1½ in.
- *16-thread series.* This is a uniform-pitch-series thread form that incorporates fine-pitch threads. It is considered as an extension of the extra-fine-thread series for large fasteners, usually above 2 in diameter.

A comparison of fastener nominal diameters and threads per inch for the standard thread series noted above is outlined in Table 10-1.

TABLE 10-1 Comparison of Unified Thread Standard Series

Nominal size, in	Threads per inch					
	Coarse (UNC)	Fine (UNF)	Extra fine (UNEF)	8-thread (8UN)	12-thread (12UN)	16-thread (16UN)
1/4	20	28	32			
5/16	18	24	32			
3/8	16	24	32			
7/16	14	20	28			
1/2	13	20	28			
9/16	12	18	24			
5/8	11	18	24			
3/4	10	16	20			
7/8	9	14	20			
1	8	12	20	8		
1 1/8	7	12	18	8	12	16
1 1/4	7	12	18	8	12	16
1 3/8	6	12	18	8	12	16
1 1/2	6	12	18	8	12	16
1 3/4	5		16	8	12	16
2	4 1/2		16	8	12	16
2 1/4	4 1/2			8	12	16
2 1/2	4			8	12	16
2 3/4	4			8	12	16
3	4			8	12	16
3 1/4	4			8	12	16
3 1/2	4			8	12	16
3 3/4	4			8	12	16
4	4			8	12	16

In addition to establishing diameter-pitch combinations, the Unified screw-thread system also defines the distinct profile and identification requirements for several screw-thread forms. The basic thread for the bulk of commercial and industrial fasteners is the Unified form, identified as "UNC" for the coarse-thread series and "UNF" for the fine-thread series. The external thread permits the option of a flat or radius root provided that the flat is not less than $p/8$ ($0.125p$), as shown in Fig. 10-7.

The "UNR" thread form for external threads is similar to the Unified thread, with the exception that a rounded or radius root is mandatory. The advantage of the radius root configuration is increased fatigue life performance, especially when the threads are formed by rolling.

The "UNJ" thread form for external threads also has a mandatory root radius, which is somewhat larger than the radius on the UNR thread. The increased radius results in a larger minor diameter for this thread. The UNJ thread form is used extensively in the aerospace industry and similar critical applications in which optimum strength and fatigue performance are required (for an example, see MIL-S-8879).

Internal threads conforming to the UNC and UNF series are intended to mate with external threads in the UNC, UNF, and UNR series. The internal thread form does not incorporate a root radius, and minor diameters are designed to clear and be compatible with the external threads in these series. However, as indicated earlier, since the UNJ external thread has a larger minor diameter, the internal thread in the UNJ series also has an increased minor diameter in order to be functional. When using the UNJ thread, therefore, it is vital to specify both external and internal threads.

One other criterion of importance relating to threads is the "class of thread," which is a means of defining the permissible amount of allowance and tolerance. To distinguish between external and internal threads, the Unified system codes the external thread as "A," while the internal thread is noted as "B." The overwhelming majority of threads are designed to permit free assembly; these fall into three distinct classes.

10-7 Class 1A and Class 1B

This class has maximum allowance and clearance fit when threads are mated. These threads are normally used where quick and easy assembly is necessary, and where a liberal allowance is required to permit ready assembly, even with slightly bruised or dirty threads. They are often used in ordnance and similar applications.

10-8 Class 2A and Class 2B

Class 2A and 2B threads are the threads most commonly used for general applications, including production of bolts, screws, nuts, and similar threaded fasteners. The allowance is specified on Class 2A (external) threads, which can minimize galling and seizing in high-cycle wrench assembly or can be used to accommodate plated finishes or other coatings. The minimum diameters of Class 2B (internal) threads, whether or not plated or coated, are basic, affording no allowance or clearance in assembly at maximum metal limits.

10-9 Class 3A and Class 3B

These threads are used for applications in which closeness of fit and accuracy of lead and angle of thread are important. These classes are highest-level precision standards, and, as such, require both quality production and gaging. The maximum diameters of Class 3A (external) threads and the minimum diameters of Class 3B (internal) threads, whether or not plated or coated, are basic, affording no allowance or clearance for assembly of maximum material components.

The disposition of tolerances and allowances for Unified Class 1 and Class 2 threads is illustrated in Fig. 10-8, while the disposition of tolerances for the Unified Class 3 thread is illustrated in Fig. 10-9.

Separate note is made of the Class 5 thread, which is a special interference-fit thread. As the identification implies, there is an actual interference of

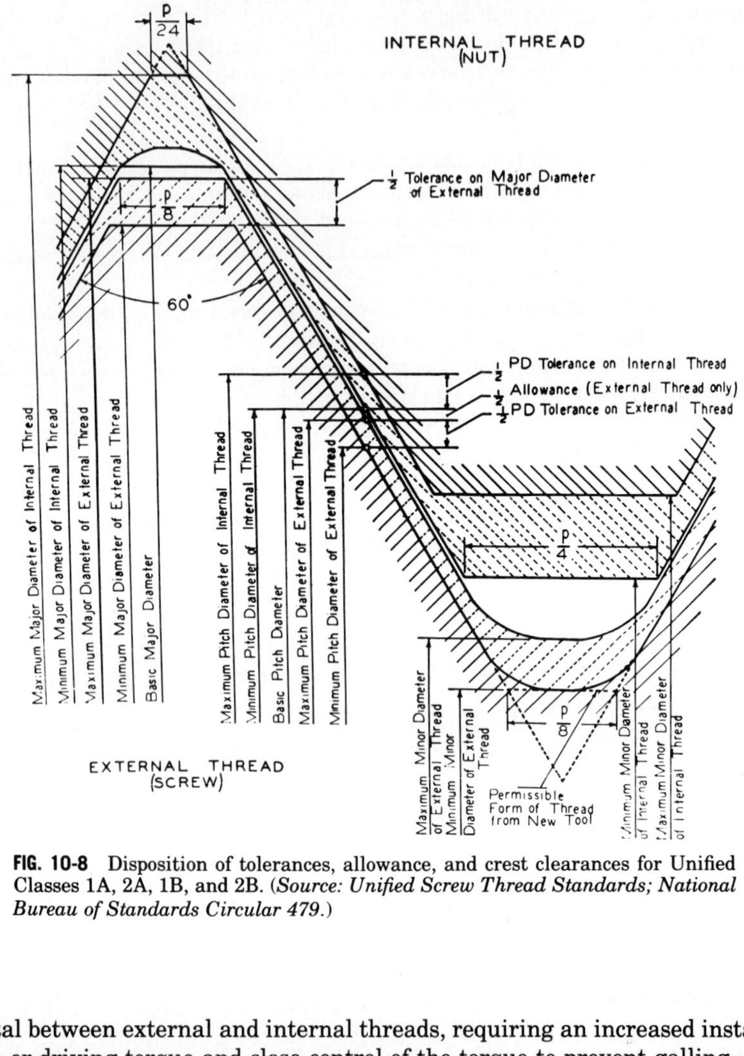

FIG. 10-8 Disposition of tolerances, allowance, and crest clearances for Unified Classes 1A, 2A, 1B, and 2B. (*Source: Unified Screw Thread Standards; National Bureau of Standards Circular 479.*)

metal between external and internal threads, requiring an increased installation or driving torque and close control of the torque to prevent galling, seizing, and/or thread stripping. In the past the Class 5 wrench-fit threads have been used primarily for studs and tapped holes which are to be assembled permanently. Designers interested in this thread class should refer to NBS Handbook H-28, or ANSI Standard B1.12.

The standard method of designating a Unified screw thread is by specifying in sequence the nominal size, number of threads per inch, thread-series symbol, and the thread class symbol, supplemented optionally by pitch diameter and its tolerance or pitch-diameter limits of size. The following is a typical example:

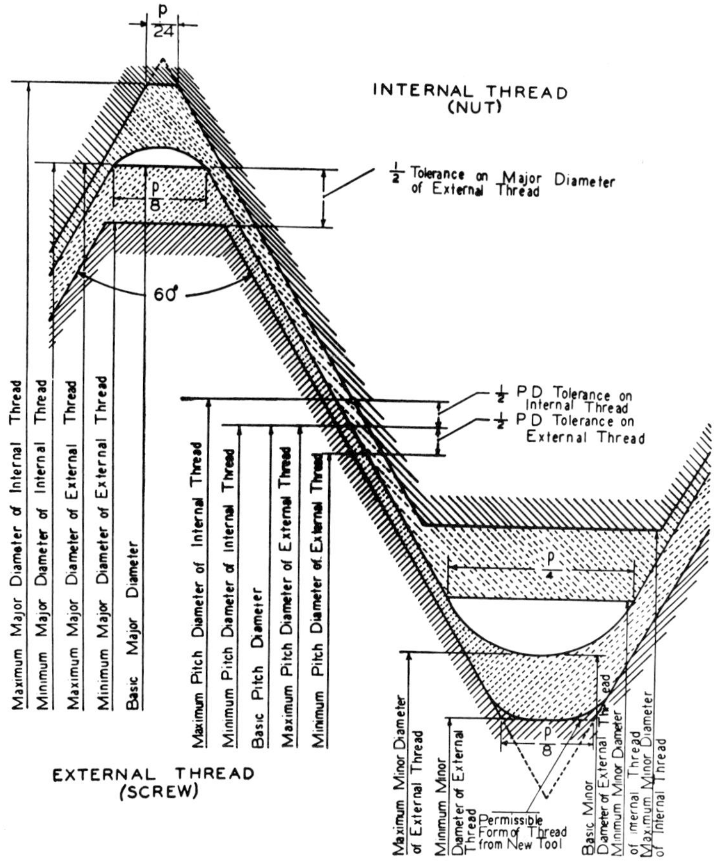

FIG. 10-9 Disposition of tolerances and crest clearances for Unified Classes 3A and 3B. (*Source: Unified Screw Thread Standards; National Bureau of Standards Circular 479.*)

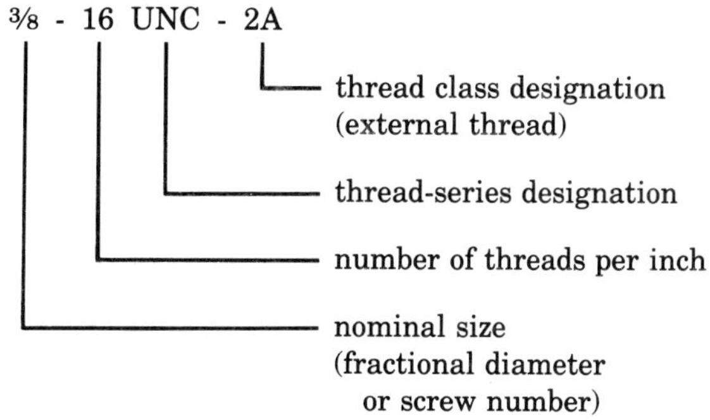

Unless otherwise specified, threads are right-hand. A left-hand thread shall be designated as "LH," as follows:

$^3/_8$ - 16 UNC - 2A - LH

10-10 Metric Threads

Increasing attention is being given to the development of metric-series threads under the auspices of the ISO. Actually, metric fasteners and metric thread forms have long been standard in most countries of the world that rely on the metric system of measurement. As with the inch system, there have been variations in both metric configurations and individual national standards. However, the impact of expanding worldwide trade has emphasized the value of uniform acceptable international standards in this area.

Currently, the applicable engineering standards are being developed through the ISO. In turn, member countries that accept a developed standard normally incorporate its requirements into a national standard for general use within the particular country. It is significant to note that several standards for metric threads predicated on ISO recommendations have been published in the United States, in anticipation of a substantial increase in the production of metric fasteners in the years ahead.

There are two major versions of metric threads covered by existing standards. The "M" series threads are designated for general-purpose industrial and commercial applications, while the "MJ" series threads have been directed specifically toward use in aerospace structures and equipment.

Metric "M" Series Applicable data and details for general-purpose metric screw threads are provided in American Society of Mechanical Engineers (ASME) Interpretive Document B1.13. This document consolidates the following separate ISO recommendations and draft proposals: ISO 68, "ISO General Purpose Screw Threads—Basic Profile"; ISO 261, "ISO General Purpose Metric Screw Threads—General Plan"; and ISO 262, "ISO General Purpose Metric Screw Threads—Selected Sizes for Screws, Bolts, and Nuts."

The ISO metric screw-thread basic profile described in ASME Document B1.13 is shown in Fig. 10-10. This is the profile or thread form to which the deviations that define the limits of the external and internal threads are applied.

The design profiles (forms) for both the internal and external threads at the maximum-material condition with and without an allowance (minimum clearance on flanks) on the external thread are illustrated in Fig. 10-11.

Metric thread series are groups of diameter-pitch combinations distinguished from one another by the pitch applied to specific diameters. Although a number of diameter-pitch combinations have been established under the ISO metric screw-thread standard series (General Plan), the standard series threads to be utilized for the production of commercial screws, bolts, and nuts are listed in Table 10-2. It is essentially intended that each industry should select from the threads shown in Table 10-2 those diameter-pitch combinations which are applicable to its own particular requirements.

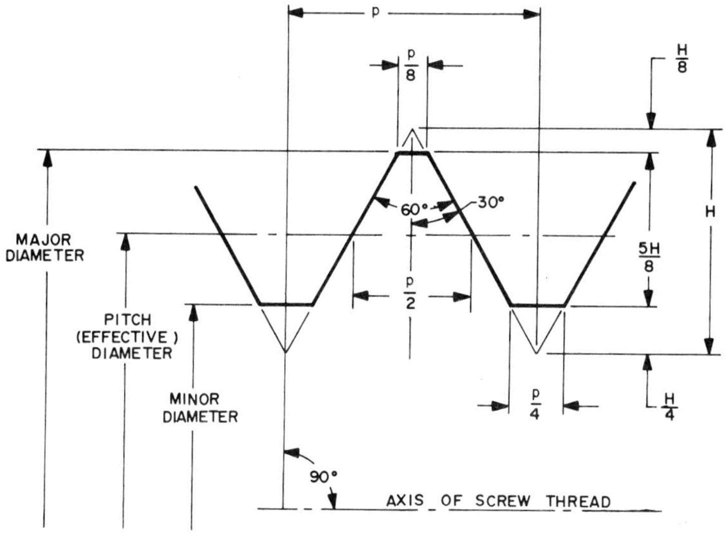

FIG. 10-10 ISO basic thread profile (metric).

The ISO metric screw-thread tolerance system provides for allowances and tolerances defined by tolerance grades, tolerance positions, and tolerance classes. In this system, tolerance grades are designated by a number which reflects the size of the tolerance on the pitch diameter and the crest diameter. For example, grade 4 tolerances are smaller than grade 6 tolerances, and grade 8 tolerances are larger than grade 6 tolerances. Further, tolerance positions define the maximum-material limits of the pitch and crest diameters of the external and internal threads and indicate their relationship to the basic profile. The respective tolerance positions are noted by letters, with lower-case letters used for external threads and capital letters used for internal threads. In conformance with current coating (or plating) thickness requirements and the demands for ease of assembly, a series of tolerance positions reflecting the application of varying amounts of allowance have been established as follows:

For external threads:
- Tolerance position "e" (large allowance)
- Tolerance position "g" (small allowance)
- Tolerance position "h" (no allowance)

For internal threads:
- Tolerance position "G" (small allowance)
- Tolerance position "H" (no allowance)

A specific tolerance class is accordingly defined by a combination of tolerance grade and tolerance position, as for example, 6H. Under the ISO tolerancing system, there are a number of combinations of tolerance grade and tolerance position to cover the possible contingencies of actual design. For

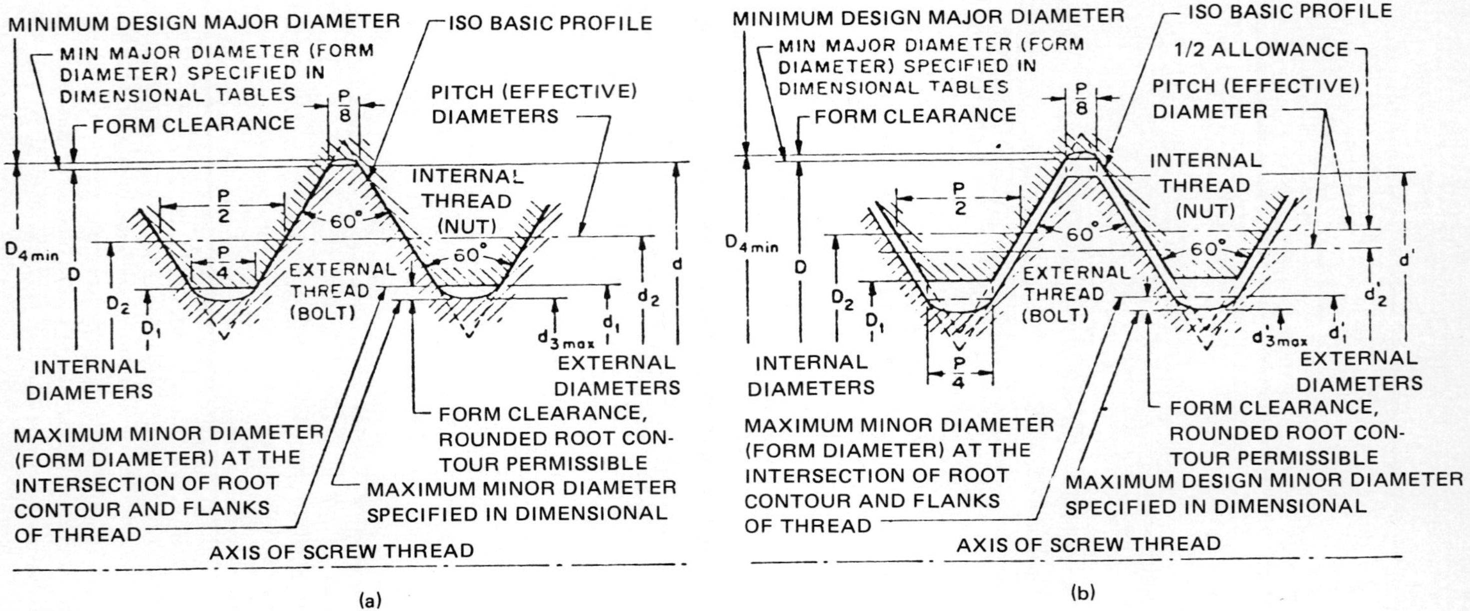

FIG. 10-11 ISO metric internal and external thread design profiles at maximum material condition.

TABLE 10-2 Metric Threads for Commercial Screws, Bolts, and Nuts

Nominal size, mm		Pitch P, mm	
Column*		Coarse thread	Fine thread
1	2		
0.25		0.075	—
0.3		0.08	—
	0.35	0.09	—
0.4		0.1	—
	0.45	0.1	—
			—
0.5		0.125	—
	0.55	0.125	—
0.6		0.15	—
	0.7	0.175	—
0.8		0.2	—
	0.9	0.225	—
1		0.25	—
	1.1	0.25	—
1.2		0.25	—
	1.4	0.3	—
1.6		0.35	—
	1.8	0.35	—
2		0.4	—
	2.2	0.45	—
2.5		0.45	—
3		0.5	—
	3.5	0.6	—
4		0.7	—
	4.5	0.75	—
5		0.8	—
6		1	—
	7	1	—
8		1.25	1
	10	1.5	1.25
12		1.75	1.25
	14	2	1.5
16		2	1.5
	18	2.5	1.5
20		2.5	1.5
	22	2.5	1.5
24		3	2
	27	3	2
30		3.5	2
	33	3.5	2
36		4	3
	39	4	3

* Thread diameter should be selected from column 1 or 2, with preference being given in that order.

threaded fasteners, the following tolerance classes generally approximate the thread classes associated with the more familiar Unified (inch) thread forms reviewed earlier:

- *Fine quality (5H/4h)*. This applies to precision threads where little variation in fit character is permissible. It requires high-quality production and inspection control.
- *Medium quality (6H/6g)*. This applies to general-purpose threads intended for most engineering applications.
- *Coarse quality (7H/8g)*. This class is normally intended for use where quick and easy assembly is necessary, even with slightly bruised or dirty threads.

Metric screw threads are designated by the letter "M" followed by the nominal size (basic major diameter in millimeters) and the pitch in millimeters, separated by the sign "×". For coarse-series threads, the pitch indication is omitted, as illustrated in the following examples:

Coarse-series threads: M6

Other threads: M8 × 1

A complete designation for an ISO metric screw thread includes, in addition to the basic designation, an identification for the tolerance class. The tolerance class designation is separated from the basic designation by a dash and includes the symbol for the pitch diameter tolerance followed immediately by the symbol for the crest diameter tolerance. Each of these symbols, in turn, consists of a numeral indicating the tolerance grade followed by a letter indicating the tolerance position (a capital letter for internal threads and a lower-case letter for external threads), as in the following example:

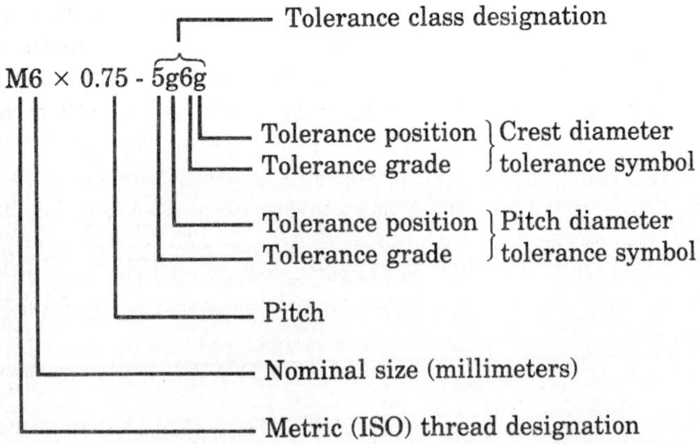

If the pitch and crest diameter tolerance symbols are identical, the symbol need only be given once and not repeated, as in the example:

M20 × 2 - 6H

└── Pitch and crest diameter tolerance symbol (tolerance equal) } Tolerance class designation

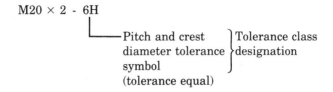

Metric "MJ" Series The MJ series of metric threads differs from the standard M series by incorporating controlled radii at the root of the external thread. Also, the minor diameters of the external and internal threads are increased above the ISO 68 thread form to accommodate the external-thread maximum root radius. The MJ screw thread represents a hard metric version of the UNJ (inch) controlled root radius thread, discussed previously, and is intended for metric-threaded aerospace parts and for other highly stressed applications requiring high fatigue strength or no allowance.

The diameter-pitch combinations for standard series aerospace screws, bolts, and nuts are listed in Table 10-3. It should be noted that one class of threads, tolerance class 4h6h, is specified for all sizes of external threads after all processing, including coating or plating, has been completed.

For internal threads, the following classes of threads are specified after all processing, including coating or plating, has been completed:

Tolerance class 4H6H for sizes 1 through 5 mm

Tolerance class 4H5H for sizes 6 mm and larger

TABLE 10-3 Standard Series Metric Threads for Aerospace Screws, Bolts, and Nuts

Nominal size, mm	Pitch, mm	Nominal size, mm	Pitch, mm
1.6	0.35	14	1.5
2	0.4	16	1.5
2.5	0.45	18	1.5
3	0.5	20	1.5
3.5	0.6	22	1.5
4	0.7	24	2
5	0.8	27	2
6	1	30	2
7	1	33	2
8	1	36	2
10	1.25	39	2
12	1.25		

Note: For threads smaller than 1.6 mm nominal size, use ISO/R 1501 miniature screw threads.

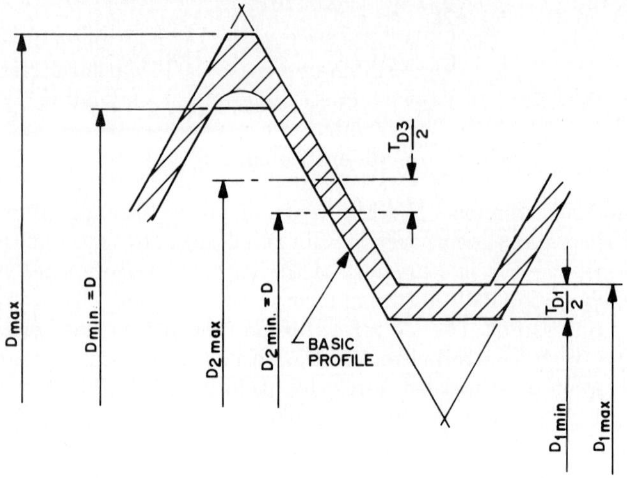

FIG. 10-12 MJ internal thread tolerance. (*Source: SAE AS 1370 and ANSI B1.21 M.*)

Internal-thread tolerances for the MJ ISO metric thread are shown graphically in Fig. 10-12, while external-thread tolerances are depicted in Fig. 10-13.

The aerospace metric screw thread is designated by the letters "MJ" to identify the metric J thread form, with all other identification criteria similar to those specified for the standard ISO metric threads, as in the following examples:

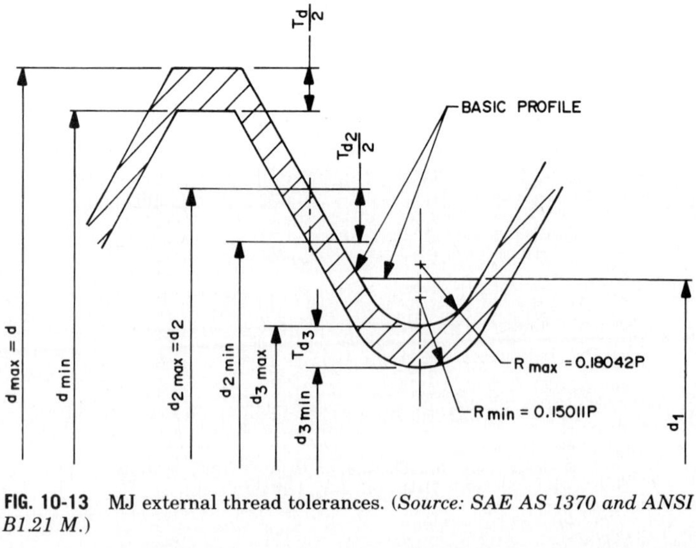

FIG. 10-13 MJ external thread tolerances. (*Source: SAE AS 1370 and ANSI B1.21 M.*)

MJ6 × 1 - 4h6h (external thread)
MJ6 × 1 - 4H5H (internal thread)

Where left-hand threads are used, the letters "LH" are added after the tolerance class, as follows:

MJ6 × 1 - 4h6h LH

Complete dimensional and engineering data for the MJ screw thread form are outlined in Society of Automotive Engineers, Inc., Aerospace Standard AS 1370, and ANSI Standard B1.21M.

STANDARDIZATION

Standardization inherently implies interchangeability irrespective of manufacturing source in terms of critical factors such as form, fit, and function. Because they are so widely used and are produced in so many configurations and materials, threaded fasteners have long been prime candidates for standards programs by major industry groups. Presently, there are a multitude of existing industry-wide standards that have already been developed and that have proved to be of value to the engineering community at large. The opportunity to utilize such standard parts obviously results in significant benefits in terms of cost savings and availability, although it is recognized that individual company standards or "specials" may be just as important to satisfy unique design objectives.

Standards programs for threaded fasteners often involve drawing standards and/or complementing specifications. Understanding the function of this overall support system is an important step in assuring the integrity of the final fastened joint.

10-11 Drawing Standards

Drawing standards essentially define the basic profile, dimensions, and tolerances of the finished fastener. In addition, many standards specify codings or markings to be included on the fastener to assist in identifying the material, strength level, and/or manufacturer. Some standards further include a part number designation to permit control of call-out, procurement, and replacement on a uniform basis.

Besides establishing the dimensional envelope, some drawing standards are complete in themselves in that all support particulars relating to the fastener, such as material, hardness or strength level, finish, and possible special performance requirements, are specified on the standard. However, for the great majority of standards, these applicable details are normally outlined in a companion specification.

Representative drawing standards for typical fasteners are included in published technical literature and standards (see following pages and references at end of section).

10-12 Specifications

The product specification is an adjunct to the drawing standard, and the two documents together constitute a system for assuring both dimensional uniformity and performance capability of the "as manufactured" threaded fastener. The specification permits specific control of materials, finish, and method of manufacture, when these are considered critical to the intended service use of the finished fastener. Just as important, most specifications include provisions for testing and quality assurance inspection sampling to confirm that the stated performance objectives for the fastener have in fact been achieved. Since annual production of threaded fasteners runs in the billions, and since common fasteners are manufactured throughout the world, adherence to the performance and quality assurance provisions of the specification system is assuming increasing importance in light of the attention being directed to minimizing product liability exposure.

The wide range of different fastener applications, mostly by industry groupings, has resulted in the development of various series of specifications by a number of major national technical organizations. For reference, the following are sources for key nationally recognized specifications and related standards covering fasteners of general interest to American industry.

Type of specification	*Source*
Federal specifications	Naval Publications and Forms Center 5801 Tabor Avenue Philadelphia, PA 19120
Military specifications	Naval Publications and Forms Center 5801 Tabor Avenue Philadelphia, PA 19120
Handbook H28 National Bureau of Standards	Superintendent of Documents Government Printing Office Washington, DC 20025
ANSI Standards	American National Standards Institute, Inc. 1430 Broadway New York, NY 10018
ISO Standards	American National Standards Institute, Inc. 1430 Broadway New York, NY 10018
ASTM Specifications	American Society for Testing and Materials 1916 Race Street Philadelphia, PA 19103
SAE Specifications	Society of Automotive Engineers 400 Commonwealth Drive Warrendale, PA 15096

Type of specification	***Source***
ASME Specifications	American Society of Mechanical Engineers United Engineering Center 345 E. 47th Street New York, NY 10017
IFI Documents	Industrial Fasteners Institute 1505 E. Ohio Building Cleveland, OH 44114
NAS Specifications (National Aerospace) Standards Committee)	National Standards Association, Inc. 5161 River Road Bethesda, MD 20816
AMS Specifications (Aeronautical Materials Specifications)	Society of Automotive Engineers 400 Commonwealth Drive Warrendale, PA 15096
AAR Specifications	Association of American Railroads 59 E. Van Buren Street Chicago, IL 60605
EEI Specifications	Edison Electric Institute 750 Third Avenue New York, NY 10017
RCRBSJ Specifications (Research Council on Riveted and Bolted Structural Joints)	American Institute of Steel Construction 101 Park Avenue New York, NY 10017 and Industrial Fasteners Institute 1505 E. Ohio Building Cleveland, OH 44114

FASTENER PERFORMANCE AND MANUFACTURING CONSIDERATIONS

The technique for manufacture of threaded fasteners has undergone dramatic change in the last 40 to 60 years. The improvements in method of manufacture in turn have influenced and affected the mechanical properties of the fastener. Historically, early production bolts and screws were turned and machined from bar stock. The introduction of the header made it possible to forge heads, and the development of the thread roller permitted the forming of threads by the rolling process in lieu of cutting or grinding. These manufacturing methods allowed high-speed production, as well as reducing substantially the amount of raw material needed to make the fastener, as shown in the simplified comparison in Fig. 10-14.

In addition to faster and cheaper production, other benefits have accrued. Test data have indicated, for instance, that externally threaded fasteners with

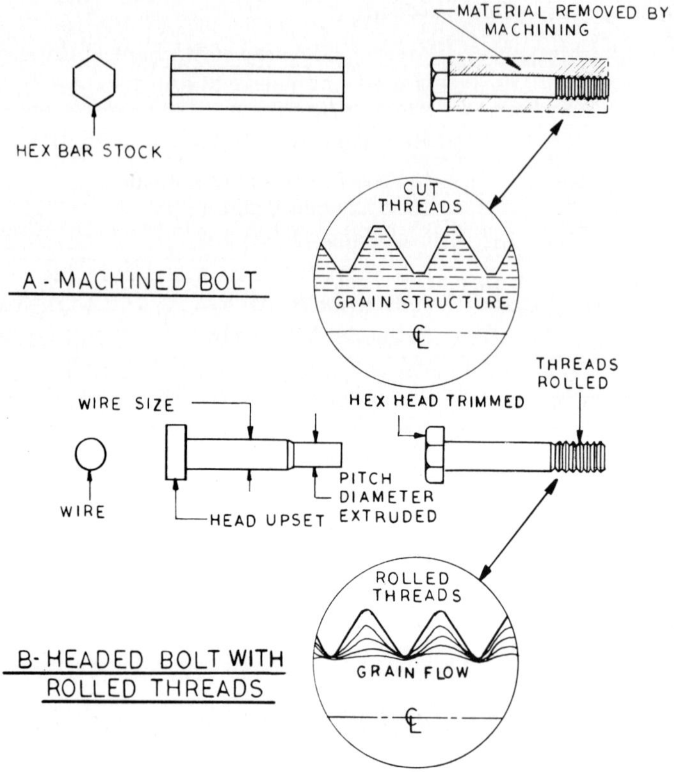

FIG. 10-14 Simplified comparison between machined and headed bolt. (*a*) Machine bolt. (*b*) Headed bolt with rolled threads.

rolled threads develop higher fatigue life than equivalent threads which have been cut or machined. Further, threads rolled after heat treatment exhibit better fatigue life than threads rolled before heat treatment. Other similar technological advances, such as controlled-atmosphere heat-treatment furnaces, better heading materials, and measurement and gaging systems, have also led to improvement in the overall quality of production fasteners.

Coordinating the performance objectives for a type of threaded fastener with the appropriate method of manufacture to assure uniform quality is the function of the product specification, as discussed previously. The product specification, then, is essentially the contract between the user and the manufacturer, guaranteeing that finished fasteners meet strength requirements, are free from injurious defects, and conform dimensionally to the specified standard. Because types of fasteners vary in function and purpose, and since the very nature of the manufacturing method can influence the fastener, actual specifications will vary in extent of detail consistent with defining and supporting requirements for the fastener.

Some of the major criteria relating to performance requirements associated with threaded fasteners are discussed in the following sections.

10-13 Dimensional

Adherence to specified tolerance limits is the basis for dimensional interchangeability. Although in-process inspection during the course of manufacture is routine, final acceptance should be predicated on the dimensions of the finished fastener. Occasionally, special fasteners may require 100 percent inspection. However, in view of the large quantities of fasteners involved in production, statistical sampling is often permitted in accordance with acceptable quality levels (AQLs) established for critical and noncritical dimensional characteristics. A widely used source for such sampling plans is MIL-STD-105.

The gaging of screw threads presents separate and distinct problems. In production, the manufacturer may be concerned with controlling individual characteristics, such as major, minor, and pitch diameter tolerance; lead error; angle error; and others. The user, on the other hand, is concerned that the finished threads are functional within the established class limits, and that the possible cumulative variations conform to the acceptable tolerance range. Details for thread gaging and proper use of gages are summarized in NBS Handbook H28, ANSI standards B1.2 and B1.16, and ISO standards ISO 965 and R1502.

10-14 Tensile Strength

Tensile strength is probably the foremost mechanical property associated with standard threaded fasteners. It is the basis for structural application, and it often dictates the type of material to be employed in manufacture. Certain materials, such as carbon and alloy steels, martensitic stainless steels, and aluminum alloys, may require heat treatment to achieve mechanical properties. Other materials, such as austenitic stainless steels, respond to cold working or drawing to develop specified strength levels.

Tensile strength is reliably confirmed by actual test, and is determined from the following formula:

$$S_t = \frac{P}{A_s} \tag{10-1}$$

where S_t = tensile strength, lb/in^2
P = tensile load, lb
A_s = tensile stress area, in^2

For a specified tensile strength level, the minimum tensile load requirement for a particular fastener can be calculated in terms of $P = S_t A_s$. In this basic formula, a significant factor is the tensile stress area A_s specified for the fastener thread. Since the screw thread is formed on a helix angle, the stress area will vary along the length of the thread. Several stress areas are used to approximate the true area for calculating minimum tensile load values to correspond to specified tensile strength ratings. For inch series threads, the commonly accepted formula for definition of tensile stress area is given in NBS Handbook H28 as

$$A_s = 3.1416 \left(\frac{E}{2} - \frac{3H}{16}\right)^2 \tag{10-2}$$

or

$$A_s = 0.7854 \left(D - \frac{0.9743}{n}\right)^2$$

where E = basic pitch diameter
D = basic major diameter
n = threads per inch

Values for $3H/16$ are given in handbook H28.

The aerospace industry, generally using higher-strength materials, has predicated tensile stress area on the pitch diameter of the thread, as referenced in NAS 1348. For other mechanical properties, such as stress rupture and fatigue strength, the sectional area at the minor diameter of the thread is

TABLE 10-4 Representative Tensile Stress Areas for Inch-Series Threads

Coarse thread series			Fine thread series			
Nominal thread size	Tensile stress area, in²*	Sectional area at minor diameter, in²†	Nominal thread size	Tensile stress area, in²*	Sectional area at minor diameter, in²†	NAS 1348 pitch diameter stress area, in²‡
¼-20	0.0318	0.0269	¼-28	0.0364	0.0326	0.0404
5⁄16-18	0.0524	0.0454	5⁄16-24	0.0580	0.0524	0.0640
⅜-16	0.0775	0.0678	⅜-24	0.0878	0.0809	0.0951
7⁄16-14	0.1063	0.0933	7⁄16-20	0.1187	0.1090	0.1288
½-13	0.1419	0.1257	½-20	0.1599	0.1486	0.1717
9⁄16-12	0.182	0.162	9⁄16-18	0.203	0.189	0.2176
⅝-11	0.226	0.202	⅝-18	0.256	0.240	0.2724
¾-10	0.334	0.302	¾-16	0.373	0.351	0.3953
⅞-9	0.462	0.419	⅞-14	0.509	0.480	0.5392
1-8	0.606	0.551	1-12	0.663	0.625	0.7027

* Tensile stress area from NBS Handbook H28, where

$$A_s = 3.1416 \left(\frac{E}{2} - \frac{3H}{16}\right)^2$$

† Minor-diameter area from NBS Handbook H28 at $D - 2h_b$.

‡ Area of 3A thread at maximum pitch diameter from National Aerospace Standards Committee Standard NAS 1348.

often employed. A comparison of the thread tensile stress areas for inch series threads for the three conditions reviewed is given in Table 10-4. Tensile stress areas for metric series threads are given in Table 10-5.

Several additional considerations are important in understanding and evaluating the tensile strength properties of the finished fastener. Standard bolts and screws are designed to fail in the threads under ultimate tensile test load because of the smaller sectional area of the threads. Test failures of the bolt head may be indicative of a poor manufacturing process or metallurgical problems, and may be cause for rejection under some specification provisions. Also, test experience has shown that the amount of thread engagement, the hardness of the mating components, and the position of the mating nut have an influence on the tensile strength of the bolt. For the same externally threaded fastener, data have indicated a higher tensile strength when the nut was located at a position two threads from the thread run-out, as compared with the nut positioned at six threads from the run-out.

A separate problem is associated with large-diameter bolts. Specifications often permit machining of reduced-gage-section coupons from the bolts in order to establish tensile strength properties, as illustrated in Fig. 10-15. However, caution should be exercised in evaluating the test results from machined coupons, since their mechanical properties do not always correlate with the results obtained from full-scale bolt testing. Beneficial contributions of work effect in rolled threads, and of upsetting or forging in the head, are often lost in the standard machined specimen. This is particularly noticeable in corrosion-

TABLE 10-5 Representative Tensile Stress Areas for Metric Threads

M threads		MJ threads			
Basic major diameter and thread pitch, mm	Tensile stress area, mm²	Basic major diameter and thread pitch, mm	Tensile stress area, mm²	Pitch diameter stress area, mm²	Area at minimum minor diameter, mm²
M3 × 0.5	5.03	MJ 3 × 0.5	5.03	5.6	4.31
M3.5 × 0.6	6.78	MJ 3.5 × 0.6	6.78	7.6	5.80
M4 × 0.7	8.78	MJ 4 × 0.7	8.78	9.9	7.50
M5 × 0.8	14.20	MJ 5 × 0.8	14.20	15.8	12.40
M6 × 1	20.10	MJ 6 × 1	20.10	22.5	17.40
		MJ 7 × 1	28.86	31.7	25.60
M8 × 1.25	36.6	MJ 8 × 1	39.20	42.4	35.40
M10 × 1.5	58.0	MJ 10 × 1.25	61.20	66.3	55.50
M12 × 1.75	84.3	MJ 12 × 1.25	92	98.3	85
M14 × 2	115	MJ 14 × 1.5	125	133.3	115
M16 × 2	157	MJ 16 × 1.5	167	177.3	156
		MJ 18 × 1.5	216	227.7	203
M20 × 2.5	245	MJ 20 × 1.5	272	284.3	257
M22 × 2.5	303	MJ 22 × 1.5	333	347.2	317
M24 × 3	353	MJ 24 × 2.0	384	404.7	362

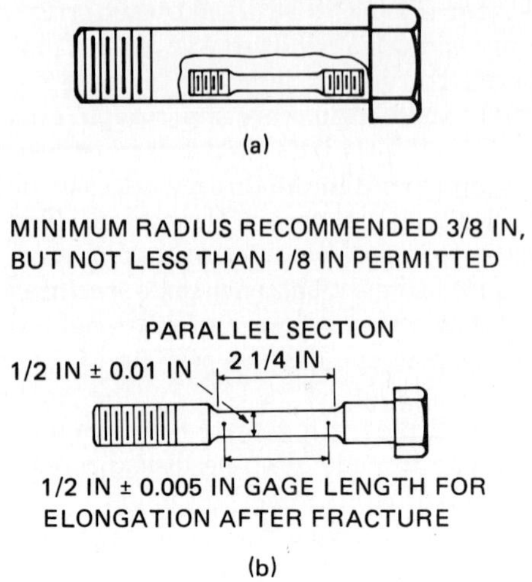

FIG. 10-15 Machined specimen from bolt for evaluating mechanical properties. (*Source: Standard Handbook of Fastening and Joining, McGraw-Hill, New York, 1977.*)

resistant steel and similar materials where fastener mechanical properties have been enhanced by cold work. For this reason, prime emphasis should be directed toward actual testing of full-scale finished bolts wherever possible to assure confirmation of tensile properties of the fasteners which will be used in service.

10-15 Proof Load

The tensile strength test normally implies testing to destruction to develop and confirm the full rated tensile strength of the fastener. Another criterion noted in some specifications is the proof load test, which may be equally as important. This property, in a sense, represents the usable strength range of the fastener for many design functions. When subjected to proof load exposure, the fastener should not exhibit permanent set, as determined by length measurement prior to and after application of tensile load. As with other critical fastener characteristics, evaluation of proof load properties, when specified, should be performed on the actual finished fasteners.

10-16 Wedge Tensile Strength

The wedge tensile strength test is limited to externally threaded fasteners, and is used as a measure of ductility and head integrity. In conducting a wedge tensile strength test, a wedge is positioned under the bolt head, as shown in Fig. 10-16. Wedge angles vary with both the size and the type of bolt, and the applicable specification should be consulted for the correct angle. The use of

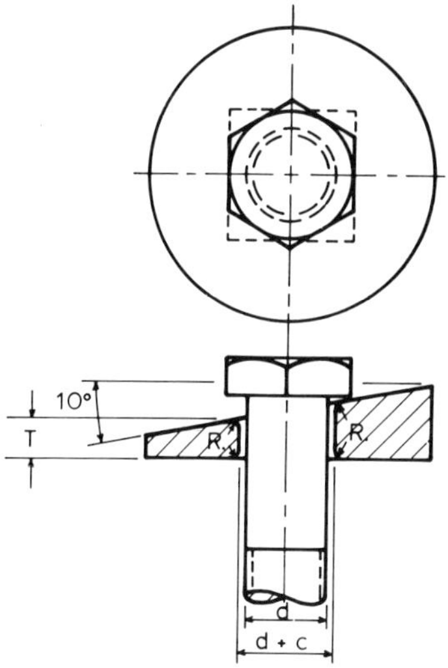

FIG. 10-16 Set-up for wedge tensile test. (*Source: Standard Handbook of Fastening and Joining, McGraw-Hill, New York, 1977.*)

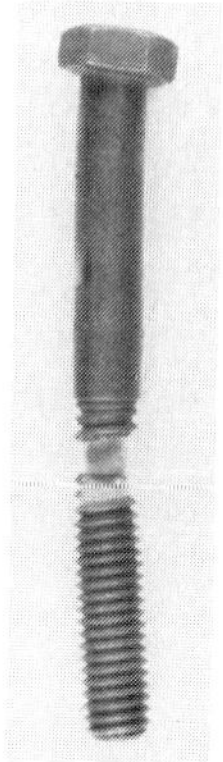

FIG. 10-17 Example of bolt after wedge tensile test. Note fracture in thread section. (*Courtesy of Almay Research & Testing Corp., Los Angeles, Calif.*)

the wedge introduces artificial bending in the bolt, forcing stress concentrations at the bolt head and in the threads. When such requirements are specified, failures under ultimate tensile load are limited to the bolt threads or bolt shank section, as shown in the test sample in Fig. 10-17.

10-17 Cone Proof Load

The cone proof load test is intended for nuts, and is designed to subject the nut to both dilation and stripping forces. In this test, a hardened cone with an included angle of 120° is positioned under the nut prior to the application of the tensile test load. The presence of surface discontinuities will affect the integrity of the nut under this type of stress and its ability to sustain design loads.

10-18 Shear Strength

Shear strength properties of externally threaded fasteners are often as important as tensile load capability, since many designs attempt to load fasteners in shear. The aerospace industry, in particular, has rigid requirements for shear strength as well as tensile strength. Now, commercial specifications are including similar performance objectives where such fasteners may be sensitive to shear loadings in service.

10-19 Hardness

Brinell or Rockwell hardness testing is a popular technique for estimating the tensile strength properties of steel fasteners. It is recognized as a valuable tool for generally indicating tensile properties, but should not be construed as a direct measure of actual tensile strength. It is emphasized that true tensile strength can be determined only from full-scale fastener tests. There are instances, though, in which production bolts are too short to permit a tensile strength test; in such cases, lot acceptance may be predicated on the results of hardness tests.

While hardness testing is of value, sole reliance on hardness does not give a true picture of the overall quality of the fastener. Fasteners may contain metallurgical defects, for instance, which will go undetected on the basis of hardness evaluation alone.

10-20 Microhardness

Microhardness testing is normally used in conjunction with metallurgical evaluation, primarily to determine evidence of surface case hardening, decarburization, carburization, and so on. While visual metallurgical examination can detect obvious decarburization, specification criteria for partial decarburization are normally based on a differential of three Rockwell points between carburization zone and uniform parent material. The ability of microhardness test instruments to measure hardness at intervals as close as 0.001 in is an important support tool in determining depth of case hardening or confirming decarburization and similar properties.

10-21 Metallurgical

Metallurgical characteristics of finished fasteners are often affected by the method of manufacture and heat treatment or processing operations, as well as by the quality of the initial material used in fabrication. Of critical significance is the importance of metallurgical integrity to the performance of structural fasteners, especially under extended service conditions.

Generally, metallurgical requirements fall into two distinct categories. One main area of examination covers evidence of correct manufacturing method and uniformity of heat treatment, while the second area relates to examination for indications of defects, such as laps, cracks, bursts, seams, discontinuities, and decarburization.

The technique of heading (or forging) bolt heads and rolling threads produces characteristic grain flow patterns that can readily be distinguished from those of comparable bolts which have been machined. Representative examples of grain flow patterns are illustrated in Figs. 10-18 and 10-19. Proper examination can confirm that patterns are acceptable, as well as detect marginal or poor practice which can introduce undesirable stress concentrations, possibly resulting in premature failure.

Metallurgical study is also valuable in inspecting for possible defects that are not apparent by normal visual examination. Recognizing the nature of

FIG. 10-18 Illustration of grain flow pattern in 12-point head bolt. (*Courtesy of Almay Research & Testing Corp., Los Angeles, Calif.*)

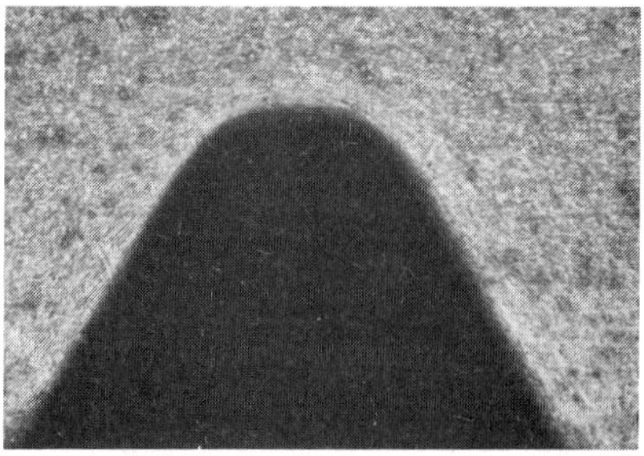

FIG. 10-19 Illustration of work effect and grain flow pattern at root section by virtue of thread rolling process. (*Courtesy of Almay Research & Testing Corp., Los Angeles, Calif.*)

mass production methods for fasteners, some specifications permit certain noncritical discontinuities within maximum defined limits, but usually defects such as laps in the threads at or below the pitch diameter, laps at the thread roots, and laps or cracks at the head fillet are considered to be cause for rejection. The thread crest lap illustrated in Fig. 10-20 just conformed to the permissible limit for depth. However, thread laps detected at the pitch diameter, illustrated in Fig. 10-21, were not acceptable. Metallurgical examination also detected obvious evidence of decarburization, illustrated in Fig. 10-22.

A special aspect of metallurgical sectioning is the opportunity to undertake difficult profile examinations, particularly of internal threads. The defective nut threads illustrated in Fig. 10-23 supported prior rejection by standard thread gaging practice.

10-22 Surface Inspection

Seams occasionally appear in the basic material used in the fabrication of fasteners. Under the pressure of forging, seams may have a tendency to open

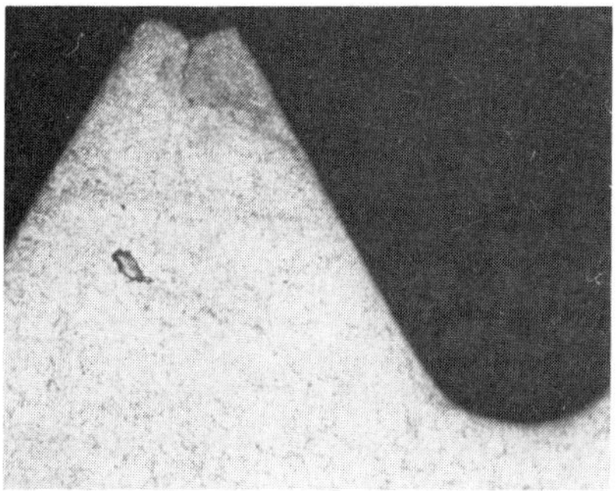

FIG. 10-20 Illustration of thread crest lap. (*Courtesy of Almay Research & Testing Corp., Los Angeles, Calif.*)

FIG. 10-21 Thread laps at pitch diameter are basis for metallurgical rejection. (*Courtesy of Almay Research & Testing Corp., Los Angeles, Calif.*)

FIG. 10-22 Decarburization detected by means of metallurgical examination. (*Courtesy of Almay Research & Testing Corp., Los Angeles, Calif.*)

or burst, causing an obvious defect. Sometimes, though, the seam is carried through to the manufactured fastener. Also, improper heat treatment can result in a quench crack, as illustrated in Fig. 10-24. Accordingly, surface inspection of finished fasteners is designed to detect the presence of such surface indications as seams, quench cracks, pits, and so on. Magnetic particle inspection is the technique used for ferrous materials, while fluorescent particle inspection is employed for nonferrous fasteners.

Surface indications noted as a result of inspection are not necessar-

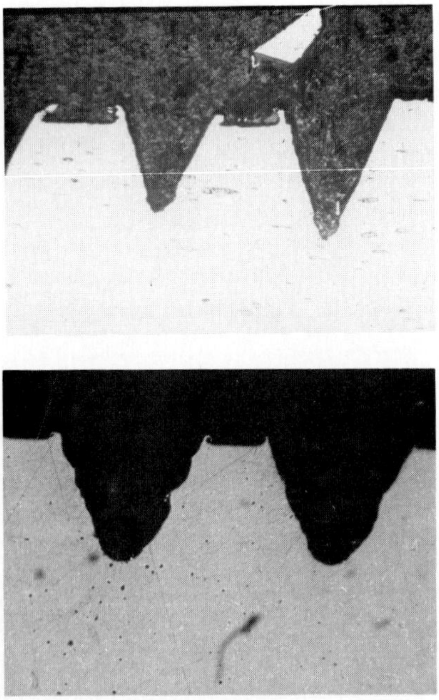

FIG. 10-23 Defective nut threads confirmed by metallurgical examination. (*Courtesy of Almay Research & Testing Corp., Los Angeles, Calif.*)

FIG. 10-24 Quench crack as a result of heat-treating operation. (*Courtesy of Almay Research & Testing Corp., Los Angeles, Calif.*)

ily cause for rejection, since it is not normally possible to visually ascertain the nature, depth, or severity of the observation. If such indications, such as the example illustrated in Fig. 10-25, are present, subsequent metallurgical examination to define the type and seriousness of the indication before final acceptance or rejection is usually warranted.

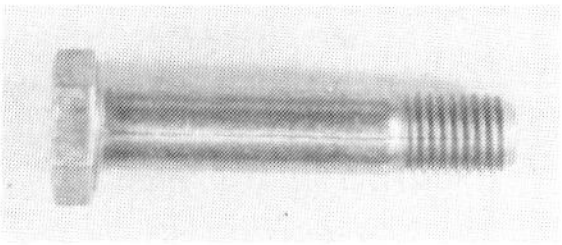

FIG. 10-25 Longitudinal surface indication in bolt detected by magnetic particle inspection. (*Courtesy of Almay Research & Testing Corp., Los Angeles, Calif.*)

10-23 Hydrogen Embrittlement

The phenomenon of hydrogen embrittlement in threaded fasteners has been mostly associated with high-strength (over 160 ksi) steel parts which have been furnished with either zinc or cadmium electroplating. During the plating process, atomic hydrogen can be trapped in or under the plating. Other sources of hydrogen can be traced to material pickling or alkaline or acid cleaning. Unless the free hydrogen is removed, when the fastener is used and stressed, as in a structural application, the hydrogen can attack the grain boundaries. The result is rapid crack propagation and often catastrophic failure of the steel fastener.

To eliminate potential surface trapped hydrogen in electroplated fasteners, it is common procedure to specify baking at 375°F immediately after plating.

Up until the present, visual and/or metallurgical examination for possible hydrogen contamination has not been effective. The technique used for evaluating or screening fasteners is to simulate typical stress conditions for anywhere from 24 to 200 hours. If hydrogen is in fact present, brittle fracture will generally be encountered within the specified test period at points of high stress concentration, such as head-to-shank junction, or in the threads. Test loadings to screen for the presence of hydrogen will vary from 75 to 90 percent of rated ultimate tensile load.

10-24 Fatigue Strength

The fatigue strength capability of threaded fasteners is particularly important in applications subjected to repetitive or cycling structural service loads. Although fatigue strength is mainly identified with aerospace-type fasteners, other classes of commercial and industrial threaded fasteners are similarly exposed in critical "fatigue-rated" installations. The importance of fatigue strength properties is associated with the fact that when failure is encountered, it is invariably catastrophic in nature, and often occurs without warning.

In discussing fatigue properties, perhaps the most essential consideration is the need to build fatigue performance into the fastener during the production cycle. As was noted at the outset of this section, research has established that rolled threads exhibit higher fatigue life than machined or ground threads. Further, threads rolled after heat treatment show better fatigue performance than comparable threads rolled before heat treatment. In addition, factors such as proper bolt head design, cold work of the head-to-shank fillet, quality control of the basic material used, and minimization of possible metallurgical defects all contribute significantly to improved fatigue life.

With respect to fatigue performance, it has been observed that failures normally develop at stress levels well below the static strength of the fastener. While most current commercial and industrial fastener standards have not specifically been designed for fatigue-type applications, they have found use at acceptable, if not reduced, strength levels. Because of the seriousness of fatigue as a service environment, more detailed attention to both fastener design and manufacturing method is warranted to take advantage of optimum higher-strength performance, and improved long life reliability.

10-25 Environmental Temperature Strength

Modern structures and equipment are no longer limited to service at ambient temperatures. Both high-temperature and low-temperature (cryogenic) service exposures are experienced in practice in nuclear systems, aerospace, electronics, transportation, energy systems, construction, and similar applications. The ability to maintain structural integrity in these actual operating environments is a prime requisite for mechanical fasteners. Part of the solution is the initial choice and selection of specialty materials developed for particular environmental temperature service, since most of the traditional structural materials suffer appreciable degradation of strength or performance at or after temperature exposure.

Characteristically, materials used at cryogenic temperatures will show an increase in tensile strength, but may sacrifice ductility. Conversely, at elevated temperatures, tensile strength properties are usually reduced, and above critical service temperature limits they may drop off dramatically.

Where high-temperature materials and requirements are referred to in fastener product specifications, a distinction is normally made between "short-time" tensile properties and "long-time" or stress rupture strength at temperature. Rupture strength, as its name implies, is the stress the fastener can sustain at a particular elevated temperature for a specified period of time without failure; it is usually lower than the short-time tensile strength for the same temperature exposure. Although actual mechanically fastened structures may need to be reliable for thousands of hours at the specified temperature, product specifications often give stress rupture strength test requirements of from 16 to 100 hours. This approach takes into account prior correlation of design long-time strength properties, and provides a basis for verifying the performance integrity of as-manufactured fasteners. Again, manufacturing control and correct processing of the special materials intended for use at service temperatures is essential to ensure long-life reliability.

DESIGN AND APPLICATION CONSIDERATIONS

The end use of threaded fasteners is as fastening and joining components designed to carry structural loads and to provide long-life joint integrity. Obviously, it is vital that the individual fasteners be of sound quality and free from defects which could impair their service function. Earlier sections have touched on the importance of proper manufacture of the fasteners and its influence on performance properties. Additionally, there are equal concerns related to the performance of the installed fastener system in the joint.

The design function dictates where and how mechanical fasteners are to be used. Considerations may include type of material being joined, required safety factors, type of loading (i.e., static or dynamic), environmental exposure, expected service life, installation clearance for maintenance or replacement needs, necessity for special assembly tooling, and so forth. Also, basic requirements and experience vary among industry groups, and as a result, the approach to joint design is often different and independent. If there is no one standard for uniform joint design, there are guidelines and design principles which have evolved by experience, and which have been valuable in contributing to structural integrity. An understanding of some of the fundamentals of the threaded fastening system and its function therefore becomes an important asset to designers and users alike.

10-26 Torquing and Preload

Aside from ensuring the inherent quality of the fastener, there is probably no function as important to the efficiency and success of a structural joint as the installation torque. For the common range of structural bolt sizes, the action of torquing a bolt-nut combination elongates and introduces a clamp load (or preload) in the bolt. It is the clamp load which effectively does the work of maintaining joint tightness and minimizing fatigue as long as the clamp load exceeds the forces acting on the joint.

For every fastener system, there is an optimum torque range to develop the design clamp load. This is normally referred to as the "torque-tension relationship." Overtorquing can result in excessive bolt yielding and possible subsequent relaxation, or even thread stripping and failure on installation. Too low an initial torque can contribute to potential fatigue and/or joint loosening with extended service life.

The approach to defining design clamp load varies from approximately 80 percent of yield strength to 70 to 75 percent of ultimate tensile strength, depending on the bolt type and material. Where feasible, using a higher-strength bolt increases the usable clamp load available in the bolt; this constitutes one of the major advantages of utilizing the technology of high-strength bolting.

There are a number of factors which affect and influence the nominal torque-tension relationship, including condition of the threads, condition and squareness of the joint, method and equipment for torquing, installation from the nut or bolt head end, and lubrication. Possibly the most influential factor is the lubrication (plating and/or supplemental lubricant) on the fastener system, since the effective coefficient of friction can alter the installation torque requirements by as much as 50 to 100 percent.

Occasionally, because of the variables identified with threaded fasteners, it is necessary to develop torque requirements empirically. However, the key objective of developing required design clamp loads is to observe specified installation torque values. Manual torque wrenches are the most common torque control tools, although new advances in automatic preload control instrumentation, strain gage systems, and special load-indicating devices are being used in critical joints which warrant the support cost. Notably in the construction industry, and mainly with larger-size fasteners, some reliance has been placed on the turn-of-the-nut method, which requires the nut to be subjected to an additional amount of rotation after first reaching a "snug" condition in the joint.

The essential fact is that care taken to assure correct installation is as crucial to joint integrity as any other effort expended in the design process.

10-27 Bolt and Nut Compatibility

In design, major attention is invariably focused on the selection or replacement of the bolt for the intended application. Particularly where high-strength bolts are used, the same critical attention must be given to specifying the correct mating nut. Inadvertent specification of a lower-strength (grade) nut invites the possibility of nut thread stripping under high tensile loading. But more significantly, a weaker nut will not adequately develop the full clamp load capability of a high-strength bolt when subjected to the necessary installation torque, due in part to "dilation" of the nut at the bearing surface.

Standards for various grades of bolts normally include comparable standards for matching nuts, making it possible to call out the right nut for each bolt. However, when special or nonstandard bolts are used, similar care should be exercised in specifying nuts which are designed to develop the full mechanical properties of the bolt. And, as discussed earlier, compatibility must also take into account the mating washers intended for the structural joint.

As a rule of thumb, the thickness (height) of a nut should approximate the diameter of the equivalent mating bolt to develop the full tensile strength properties of the bolt. In the case of screws engaged in threads tapped in softer parent materials or in threaded inserts, the design criterion is different. For these applications, the shear area of the external and internal threads is used, and the correct length of thread engagement to assure that the full strength of the screw is being achieved depends on the strength of the parent material.

Essentially, an application of threaded fasteners should be thought of as a "system," with each of the individual components making a major contribution to the success of the overall system.

10-28 Fatigue and Vibration

Fatigue and vibration are two separate and distinct environments encountered in actual service. Although they are initially unrelated, the mechanism of these failure modes is such that vibration loosening can, in fact, result in fatigue failure. Both environments can be especially damaging to the threaded fastener system and the integrity of the structure.

Fatigue is the condition in which repetitive or cyclic loading is sensed by the structure or equipment. The two main types of joint fatigue loading are shear fatigue and tension fatigue. For shear-loaded joints, fatigue failure normally occurs in the plate or sheet material. The applied fatigue or dynamic stresses, hole preparation, hole clearance, amount of induced bending, and fastener preload are some of the factors which influence shear joint fatigue life.

The choice of threaded fasteners for joints subjected to tension fatigue loading is more critical. Foremost is the necessity for proper and adequate preloading of the fastener to meet or exceed anticipated dynamic loads imposed on the joint. A correctly tightened and preloaded bolt will realize only minimal external loads as a result of repetitive tensile loadings. And, as was pointed out previously, bolts specifically designed and manufactured for high fatigue performance should be used in fatigue-sensitive joints. The fact that a good portion of the service failures observed are fatigue-related suggests the importance of good joint design and proper installation.

Whereas fatigue loading is presumed to be relatively high with respect to the strength of the threaded fastener or the joint, vibration loads are relatively low, but may be associated with various ranges of cyclic frequencies. Critical combinations of frequency, loading, and amplitude can force a structure into resonance, often with catastrophic results. While the overwhelming majority of operating structures are not subjected to conditions of resonance, the vibration forces present (including random and steady-state vibration, shock, and impact) are sometimes serious enough to drastically affect the threaded fastener system.

Under repeated or extensive vibration, there is a tendency for the nut to rotate or loosen off the bolt threads. Continued vibration can actually result in the nut completely disengaging from the bolt, with subsequent loss of the bolt from the joint. Not as severe, but just as important, vibration loosening can reduce or completely relax the original preload in the bolt, causing the bolt to sense increased fatigue loads with continued exposure. What may have first started as vibration loosening may actually end as a fatigue failure because of the complex mechanism involved.

Where vibration, shock, or impact loadings are recognized to be present, additional requirements for positive safety of the threaded fastener system are recommended. Approaches to securing and safety to maintain preload and prevent potential loss of the fasteners include self-locking nuts, self-locking screws or bolts, cotter pins, safety wire, or adhesive locking techniques.

10-29 Joint Design

Experience has indicated that there is a wide range of versatility in concepts of structural joining. No single "standard" joint design has been accepted. Rather, there has been an evolution of optimum designs characteristic of typical structures within various industries which have proved to be successful for their intended purpose. For these structures, this evolution has included the development of particular fastener standards designed to support the structural design needs of the industry.

Typically, most joints can be classified as representing tensile, shear, or bending (combined tensile and shear) loadings. Each condition requires detailed analysis and consideration by the designer when specifying threaded fasteners, taking into account special needs for permanent or removable connections. For example, joints designed for routine maintenance or removal, especially under field conditions, may require adequate wrench clearance, fastener spacing, and accessibility not necessarily needed for permanent fasteners installed under factory conditions.

Characteristically, for rigid tensile-type joints, the proof load or yield strength rating of the bolt may be the significant design criterion. The higher the basic tensile strength of the fastener system, the higher the yield strength, with a correspondingly higher clamp load capability. For nonrigid or flexible joints, as in the case where gaskets are used, a different design analysis is necessary. For the gasket to be effective, it is important that all fasteners in the joint be uniformly loaded to the design stress value. Any increase in tensile loading in a gasketed joint adds to the bolt preload, usually requiring either a higher-strength or larger-diameter fastener than would be required for the same clamp load objective in a rigid joint.

Shear-type joints also fall into two separate categories. The first is the friction connection, which depends on high clamp load induced in the threaded fasteners. As long as the mating faying surfaces are clean and free of dirt, lubricant, paint, and other contaminants, the high clamp load contributes to

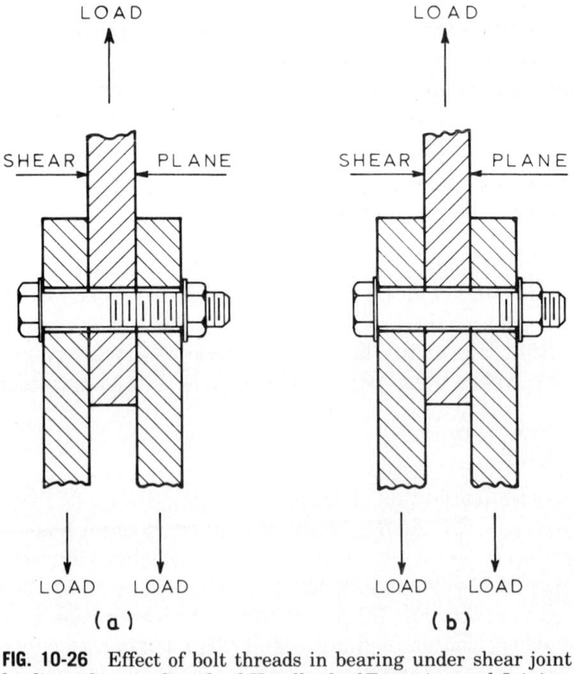

FIG. 10-26 Effect of bolt threads in bearing under shear joint loading. (*Source: Standard Handbook of Fastening and Joining, McGraw-Hill, New York, 1977.*)

high frictional forces in the joint. The frictional forces developed are normally well within the allowable design shear stress ratings, particularly in the construction industry. Should joint shear load exceed the frictional stress, joint slipping would occur to the full tolerance of the clearance hole, resulting in shear forces acting on the bolt and bearing forces acting on the joint material. The second type of shear joint, therefore, is based on actual bolt shear strength and plate bearing strength properties.

One special caution is noted with bolts employed in shear-type joints. For bearing-type joints, care should be taken to specify bolts which are long enough to permit the full bolt shank to act through the shear planes. Short bolts or improper bolts with a portion of the bolt threads resisting shear loading, as illustrated in Fig. 10-26, can reduce bolt shear strength by as much as 20 to 35 percent because of the smaller net sectional area in the screw threads. To take full advantage of shear strength properties, the preferred design requires the full shank body to be positioned in the shear planes, as also shown in Fig. 10-26.

REFERENCES

1. Parmley, R. O.: *Standard Handbook of Fastening and Joining,* McGraw-Hill, New York, 1977.
2. Fisher, J. W., and J. H. A. Struik: *Guide to Design Criteria for Bolted and Riveted Joints,* Wiley, New York, 1974.
3. Brenner, H. S.: "Development of Technology for Installation of Mechanical Fasteners", NASA report no. CR-103179, 1971.
4. *Metals Handbook,* 9th ed., vol. 1, American Society for Metals, 1978.
5. *Metric Fastener Standards,* 2d ed., Industrial Fasteners Institute, 1983.
6. "Fastener Test Methods," MIL-STD-1312, U.S. Department of Defense.
7. "Standard Method for Conducting Tests to Determine the Mechanical Properties of Externally and Internally Threaded Fasteners, Washers, and Rivets", Specification F 606, vol. 15.08, American Society for Testing and Materials, 1983.

11
Retaining Rings

DR. EDMUND KILLIAN, P.E.
Chief Engineer/Waldes Kohinoor, Inc./Long Island City, New York

DESIGN AND APPLICATION

11-1 Design Variety

Retaining rings are precision-engineered fasteners designed to provide an accurately located shoulder for positioning and securing components on shafts and in bores and housings. More than 50 functionally different types of rings for a variety of assembly needs are fabricated in some 1200 standard inch and metric sizes for shafts and bores with diameters of 0.040 in (1 mm) to 10 in (250 mm). Rings with diameters as large as 40 in (1 m) have been made for special applications. Representative types are shown in Fig. 11-1.

The most popular rings, the *basic series,* are shaped somewhat like a horse-

FIG. 11-1 Representative collection of retaining rings. Fasteners are available in more than 50 functionally different types, some 1200 standard inch and metric sizes for shafts and bores 0.040 in (1 mm) to 10 in (250 mm) diameter. (*Courtesy of Truarc® Retaining Rings Division, Waldes Kohinoor, Inc., Long Island City, New York.*)

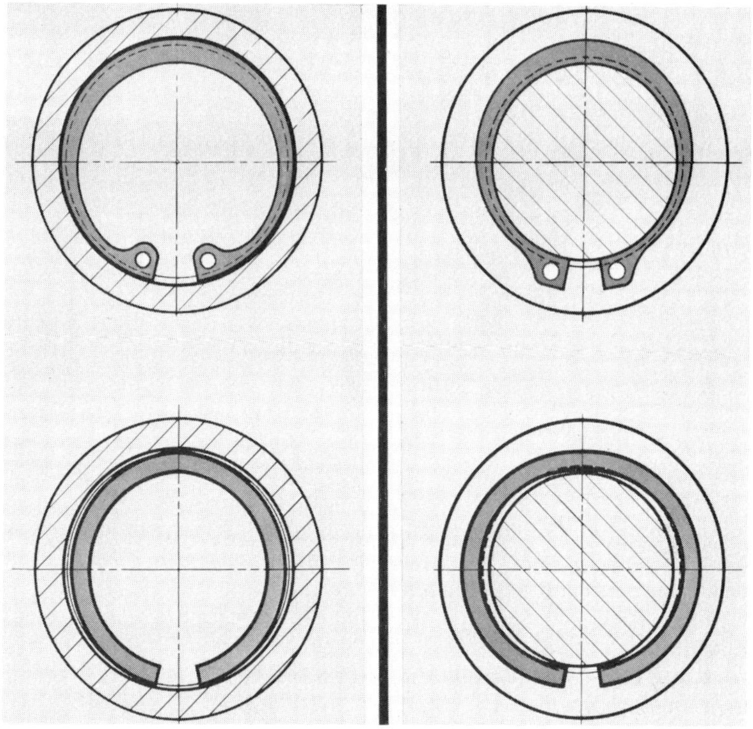

FIG. 11-2 Tapered-section retaining rings (top) remain *circular* after assembly in a bore or on a shaft to assure maximum contact surface with groove. Uniform-section rings undergo *oval deformation* which limits groove contact and reduces thrust-load capacity.

shoe and are stamped from relatively thin carbon spring steel or other materials with good spring properties. Unlike wire-formed rings, which have a uniform section height, stamped rings are characterized by a tapered radial width which decreases symmetrically from the center section to the free ends. In contrast to uniform-section rings, which undergo *oval deformation* when they are expanded for assembly over a shaft or compressed for insertion into a bore, tapered-section rings remain *circular* after installation (Fig. 11-2). This constant circularity assures maximum contact surface with the groove bottom and is an important factor in achieving high static and dynamic thrust load capacities and reducing high stress concentration, which can lead to ring failure. Holes for pliers (Fig. 11-3), another characteristic of tapered-section rings, are located in the lugs at the free ends and greatly facilitate ring assembly and disassembly during both initial factory installation and field service.

11-2 Advantages over Other Fastening Methods

Retaining rings have been designed into a great many different consumer, industrial, and military products because they offer several advantages over other fastening methods. These include the following.

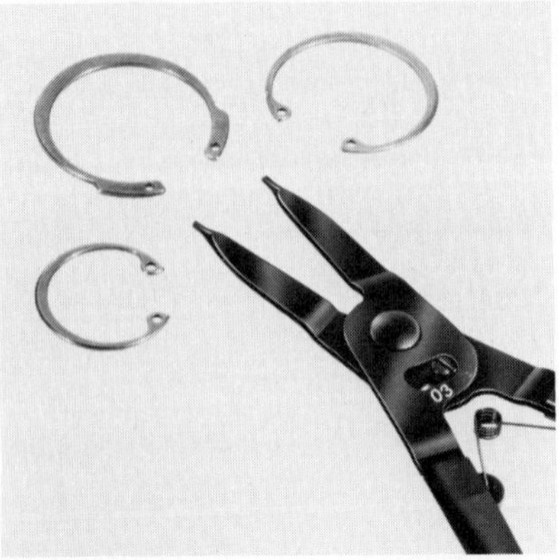

FIG. 11-3 Lugs with holes for pliers facilitate installation and removal of axially assembled rings.

Providing a Simpler Design Rings may be used to replace machined shoulders, collars and set screws, cotter pins and washers, threaded plugs and sleeves, and many other bulkier and more expensive devices. By reducing the number and complexity of parts in an assembly, they often make possible substantial savings in size, weight, and energy consumption.

Lowering Costs for Materials and Labor Retaining rings may be used with smaller-diameter shafts and thinner housings than other fasteners and so may reduce shaft extensions needed for threaded parts. They eliminate expensive drilling, tapping, threading, facing, and other machining. For most product applications, ring grooves can be cut inexpensively, often simultaneously with other production operations, such as shaft cut-off and chamfering, bore drilling, or facing. Self-locking rings, which are also popular, do not require a groove or other preparatory machining.

Providing Faster Assembly and Disassembly Rings can be assembled on the production line quickly and inexpensively, even by unskilled labor. A variety of hand tools, semiautomatic magazine-fed tools, and automatic assembly equipment is available for specific manufacturing operations and volume requirements. Retaining rings can be removed easily for product service or adjustment, and most rings are reusable after disassembly.

AVAILABLE RING TYPES

Retaining rings generally are classified into two major groups: *internal rings,* for bores and housings, and *external rings,* for securing components on shafts, studs, and similar parts. Within these two broad categories, the fasteners may be separated into four subdivisions based on ring function and assembly characteristics: *axially assembled rings, radially assembled rings, rings for end-play take-up,* and *self-locking rings,* which, as we indicated earlier, are used without grooves.

11-3 Axially Assembled Internal and External Rings

Basic Internal and External Types The first rings used in American industry were the *basic internal and external types* (Fig. 11-4), which were developed for aircraft in 1942 and are still among the most popular types.

Both rings are installed in an axial direction, the internal type after com-

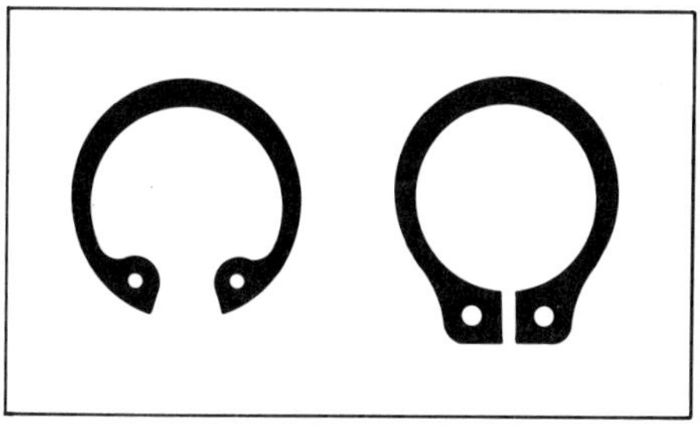

FIG. 11-4 Basic internal (left) and external rings.

pression for insertion into a bore, the external type following expansion to clear a shaft. The tapered section described earlier provides maximum flexibility with a minimum of permanent set and permits the rings to be installed in relatively deep grooves.

Both types maintain constant pressure against the bottom of the groove in which they are assembled and are secure against heavy thrust loads. External rings, often used in assemblies subjected to strong centrifugal forces, also are secure against high rotational speeds. High rotational speeds are of no consequence with the internal types, since centrifugal forces merely cause the rings to seat more tightly against the bottom of the groove.

Inverted Rings These rings (Fig. 11-5) derive their name from the location of the lugs, which has been changed from that of the basic types so that the lugs abut the bottom of the groove. In addition, the section height of the rings has been increased.

Because of the lugs, inverted rings provide less contact surface with the groove wall and, as a result, provide somewhat lower thrust load capacity than the basic types. The rings offer several advantages, however, over the basic types:

Better clearance results from the positioning of the lugs in the groove and the smaller maximum overall diameter, compared with the basic types.

A higher, uniformly circular shoulder, better suited to positioning and se-

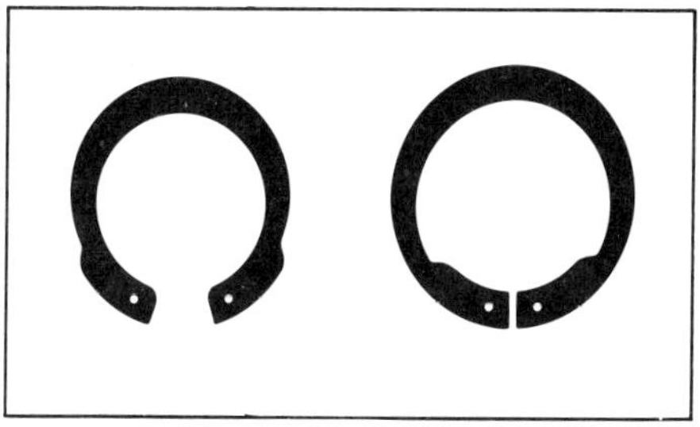

FIG. 11-5 Inverted internal (left) and external rings.

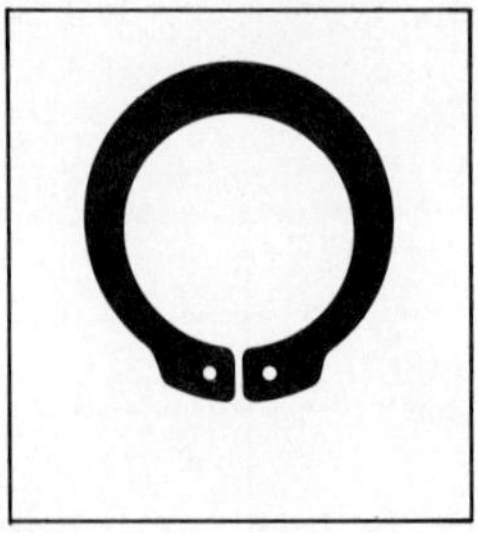

FIG. 11-6 Heavy-duty external ring.

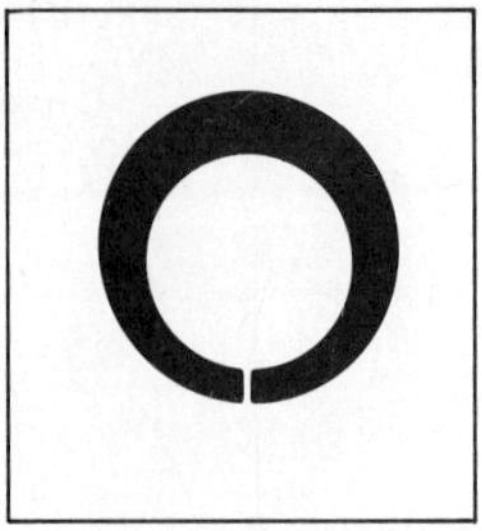

FIG. 11-7 Miniature high-strength ring.

curing seals, lenses, and other components with curved abutting surfaces is formed. The inverted rings' high shoulder also makes it possible for the fasteners to accommodate ball and roller bearings and other parts with large corner radii or chamfers.

These rings have a better appearance for applications in which they will be left exposed. The uniform, circular shoulder has a tendency to "blend in" with other components and, as a result, offers a more pleasing appearance than the basic types.

Heavy-Duty External Rings These rings (Fig. 11-6) are designed for assemblies in which the fasteners will be subjected to unusually high loads or where components to be secured have extra-large corner radii or chamfers. The heavy-duty rings are substantially thicker than the basic external type and have a greatly increased section height. Wider lugs which extend into the body of the ring provide more uniform load distribution and compensate for the ring's added section height, providing better dynamic balancing under high rotational speeds.

Miniature High-Strength Rings These rings (Fig. 11-7) have a tapered section without lugs or holes for pliers. They are designed to provide a strong, tamperproof shoulder on small-diameter shafts, with maximum assembly clearance. The rings are thicker than conventional retaining rings of the same size and have a higher section—features which, combined with the maximum groove contact made possible by the tapered design, assure high thrust and impact resistance. Used on straight shafts, the high-strength rings are assembled with a tapered pin and sleeve which expands the fastener until it reaches its groove; if the pin on which the ring is to be installed is already tapered, only the sleeve is needed for assembly.

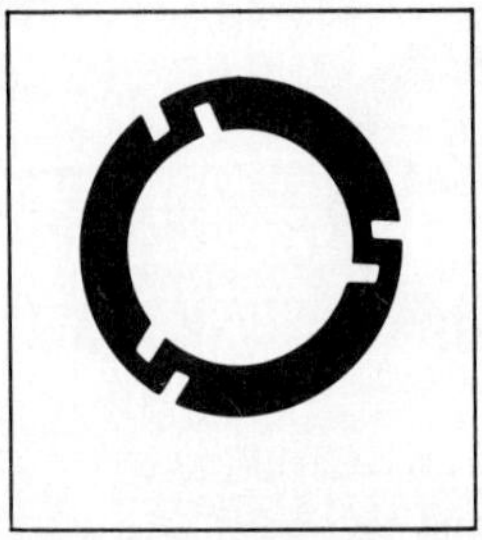

FIG. 11-8 Permanent-shoulder ring.

Permanent-Shoulder Rings These rings (Fig. 11-8) are designed to provide nonremovable shoulders for positioning and securing parts on small-diameter shafts, studs, and similar parts. When the ring is compressed in a V-shaped groove, the notches deform into small triangles, closing the gaps in the ring and simultaneously reducing the inside and outside diameters. This causes the ring to grip the shaft tightly and form a 360° shoulder.

Typical product applications of axially assembled internal and external rings are illustrated in Figs. 11-9 through 11-14. Dimensions, load capacities, and other engineering data are given in Tables 11-1 and 11-2.

11-4 Radially Assembled Rings

In many product designs, it is not feasible to install retaining rings in an axial direction. Assembly clearance, in relation to other components, or the need for easy accessibility during field service or maintenance dictates the use of radially installed fasteners. It was for these reasons that radial rings were developed. Most radial rings are designed to be assembled radially on a shaft, directly in the plane of the groove; the self-locking type, which does not need a groove, is installed in the same direction, adjacent to the part to be retained.

Unlike the axial rings, which have lugs for pliers, the radial rings have a large gap between the free ends which permits the rings to be pushed over a shaft or stud. Since they do not provide as much contact with the shaft as the axial external rings, most radial rings offer somewhat lower thrust load capacity. The heavy-duty rings are an exception to this; they were designed for both

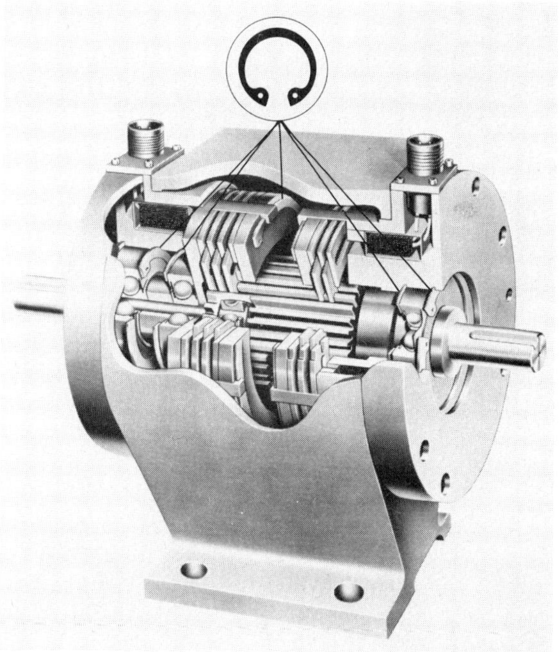

FIG. 11-9 Electromagnetic clutch brake. Seven basic internal rings position and secure bearings, transmission disks, and other parts. The rings eliminate machined shoulders and counterbores and permit fast assembly with precise tolerances.

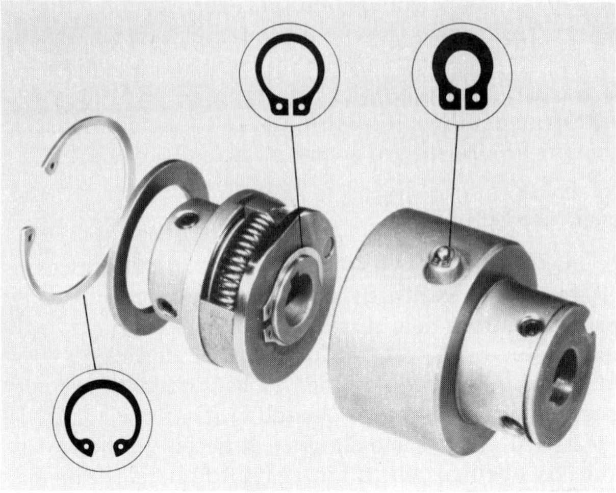

FIG. 11-10 Overload protector coupling. Basic internal ring (left) locks drive shaft assembly in housing. Basic external ring (center) holds slides and stops on drive shaft. Axial clamp ring (right) secures actuating pin. The rings replace threaded fasteners and save machining of shaft and housing.

FIG. 11-11 Turret nozzle for pressure sprayer. Inverted internal ring holds disk in housing, replacing expensive threaded collar. The ring provides compact design and speeds assembly and field service.

FIG. 11-12 Plastic-strapping tool. Axial clamp rings (left inset) hold pivot pins. Because no grooves are needed, rings automatically compensate for tolerances in pins and bracket. Heavy-duty ring (right) positions and secures feed wheel and assures required thrust and impact resistance.

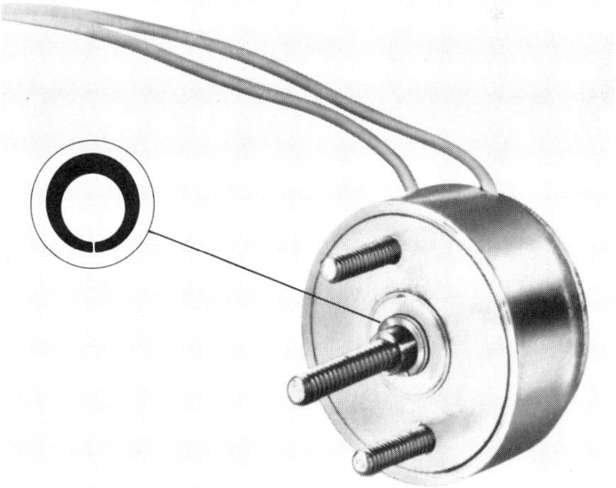

FIG. 11-13 Push-pull solenoid. Miniature high-strength ring secures plunger in stator. Rigidity and resistance to return impact force prevent deformation of plunger and help maintain stroke for 100 million actuations.

FIG. 11-14 Wet-plunger ac/dc solenoid. Basic external ring (top) holds tube assembly in housing. Internal self-locking ring (bottom left) retains O-ring seal; axial clamp ring (right) holds spring on override button. Rings simplify design, lower costs for parts and labor.

ease of radial assembly and the greater gripping strength needed for many products.

Special applicators, dispensers, and other assembly tools are available for radial rings. These make possible fast, economical installation by unskilled labor.

Crescent-Shaped Rings These rings (Fig. 11-15) have tapered sections like the basic axial types. As a consequence, they remain circular after they are installed and maintain a tight fit against the bottom of the groove. Because the taper is located on the inner circumference of the ring, the fastener provides a uniform shoulder concentric to the shaft. This feature, combined with the narrow section, makes the crescent type especially suitable for assemblies in which clearance is critical. The rings withstand moderate thrust and impact loads, have greater gripping strength than hardened C washers, and eliminate

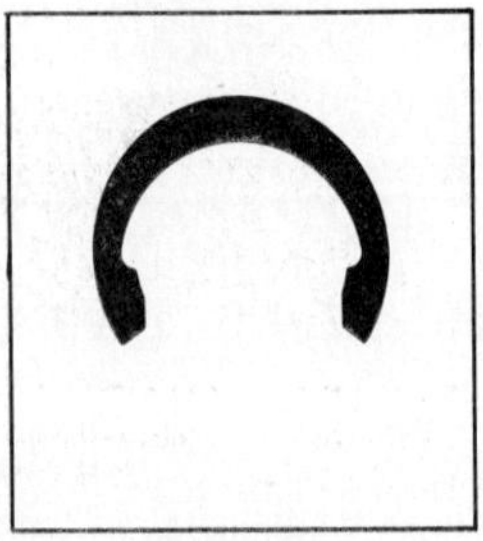

FIG. 11-15 Crescent-shaped ring.

FIG. 11-16 E ring.

FIG. 11-17 Reinforced E ring.

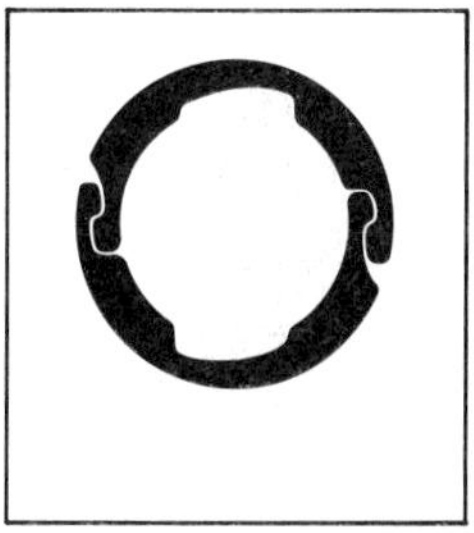

FIG. 11-18 Two-part interlocking ring.

problems that sometimes arise when soft C washers have variations in forming.

E Rings These rings (Fig. 11-16) function as large shoulders on small-diameter shafts. Contact surface with the groove, which is relatively deep, is provided by the three heavy prongs, spaced approximately 120° apart. Use of a deep groove—made possible because the radius of the bending arms is substantially greater than the radius of the shaft—increases the fastener's thrust-load capacity.

Reinforced E Rings These rings (Fig. 11-17) were designed for product applications in which conventional E rings do not provide adequate holding power. They differ from regular E rings in that the bending arms have a tapered construction which gives the rings approximately 5 times greater gripping strength and resistance to radial push-out forces. The reinforced E rings also are secure against high rotational speeds and relative rotation between the retained parts.

Two-Part Interlocking Rings These rings (Fig. 11-18) have identical semicircular halves with prongs which engage at the free ends. When the ring is assembled in a groove, it forms a high 360° shoulder that is uniformly concentric with the shaft. The ring is secure against high rotational speeds, relative rotation between the parts being retained, and heavy thrust loads. The fastener's attractive appearance makes it particularly suitable for exposed installations.

The interlocking ring may be assembled easily with snub-nosed pliers, clamps, or vises, which squeeze the two halves together so that the prongs become engaged. The parts can be disassembled with a screwdriver.

High-Strength Radially Assembled Rings These rings (Fig. 11-19) provide substantially greater gripping strength and load capacity than conventional E rings and other radial types. They are thicker and have ta-

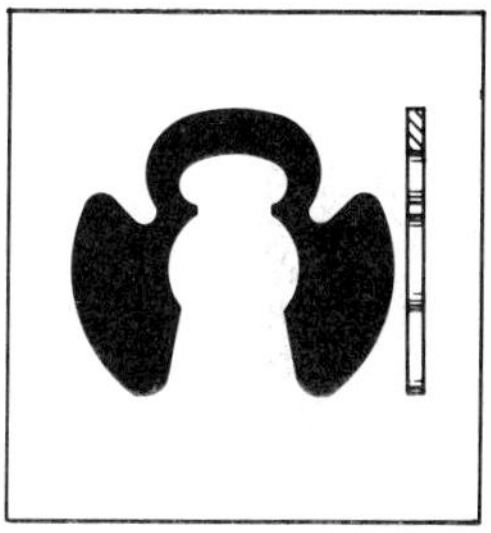

FIG. 11-19 High-strength radially assembled ring.

TABLE 11-1 Axially Assembled Tapered-Section Internal Rings

Ring type	Housing diameter, dec. equiv. in.		Nominal ring thickness, in	Allowable static thrust load, lb* when rings abut parts with sharp corners				Maximum allowable corner radii or chamfers of retained parts, in†				Allowable thrust load, lb, when rings abut parts with listed corner radii or chamfers
				Groove material having minimum tensile yield strength of 150,000 lb/in²		Groove material having minimum tensile yield strength of 45,000 lb/in²		Radii		Chamfers		
	From	Through		From	Through	From	Through	From	Through	From	Through	P_r'
Basic internal,	0.250	0.312	0.015	420	530	190	240	0.011	0.016	0.0085	0.013	190
Bowed internal,	0.375	0.453	0.025	1,050	1,280	350	460	0.023	0.027	0.018	0.021	530
Beveled internal	0.500	0.750	0.035	1,980	3,000	510	1,460	0.027	0.032	0.021	0.025	1,100
in grooved	0.777	1.023	0.042	4,550	6,050	1,580	3,000	0.035	0.042	0.028	0.034	1,650
housings	1.062	1.500	0.050	7,450	10,550	3,050	6,000	0.044	0.048	0.035	0.038	2,400
	1.562	2.000	0.062	13,700	17,500	6,350	10,300	0.064	0.064	0.050	0.050	3,900
	2.047	2.531	0.078	22,750	27,600	10,850	15,650	0.076	0.078	0.061	0.062	6,200
	2.562	3.000	0.093	33,700	39,500	16,500	23,150	0.088	0.092	0.070	0.074	9,000
	3.062	5.000	0.109	47,100	77,000	24,100	55,000	0.097	0.158	0.078	0.126	12,000

	5.250	6.000	0.125	92,700	105,900	60,000	68,600	0.168	0.168	0.134	0.134	15,000
	6.250	7.000	0.156	137,700	154,300	74,100	93,100	0.177	0.196	0.142	0.142	23,000
	7.250	10.000	0.187	191,500	264,200	99,600	190,700	0.202	0.270	0.162	0.162	34,000
Double beveled internal	1.562	1.688	0.053	11,550	12,450	3,600	4,300	0.064	0.064	0.050	0.050	2,850
	1.750	2.000	0.052	12,850	14,700	4,700	6,100	0.064	0.064	0.050	0.050	2,750
	2.062	2.531	0.068	19,850	23,900	6,500	9,600	0.078	0.078	0.062	0.062	4,700
	2.562	2.812	0.082	29,700	10,200	10,200	12,200	0.088	0.088	0.070	0.070	7,000
Inverted internal in grooved housings	0.750	—	0.035	1,650	—	600	—	0.050	—	0.031	—	850
	0.812	1.000	0.042	2,600	3,300	700	1,150	0.054	0.064	0.034	0.040	1,250
	1.062	1.500	0.050	4,150	5,850	1,250	2,500	0.069	0.081	0.043	0.051	1,800
	1.562	2.000	0.062	7,600	9,750	2,650	4,300	0.088	0.118	0.055	0.074	2,900
	2.062	2.500	0.078	12,650	15,300	4,500	6,500	0.125	0.144	0.078	0.090	4,600
	2.625	3.000	0.093	19,200	21,900	7,200	9,600	0.150	0.169	0.094	0.106	6,700
	3.156	4.000	0.109	27,000	34,200	10,600	16,900	0.174	0.174	0.109	0.109	9,000

* Where rings are of intermediate size—or groove materials have intermediate tensile yield strengths—loads may be obtained by interpolation.

† Approximate corner radii and chamfers limits for parts with intermediate diameters can be determined by interpolation. Corner radii and chamfers smaller than those listed will increase the thrust load proportionately, approaching but not exceeding allowable static thrust loads of rings abutting parts with sharp corners.

TABLE 11-2 Axially Assembled Tapered-Section External Rings

Ring type	Shaft diameter, dec. equiv. in.		Nominal ring thickness, in	Allowable static thrust load, lb* when rings abut parts with sharp corners: Groove material having minimum tensile yield strength of 150,000 lb/in^2		Groove material having minimum tensile yield strength of 45,000 lb/in^2		Maximum allowable corner radii or chamfers of retained parts, in†: Radii‡		Chamfers‡		Allowable thrust load, lb, when rings abut parts with listed corner radii or chamfers
	From	Through		From	Through	From	Through	From	Through	From	Through	P'_r
Basic external,	§0.125	§0.156	0.010	110	130	35	55	0.010	0.015	0.006	0.009	45
Bowed external,	§0.188	§0.236	0.015	240	310	80	120	0.014	0.0165	0.0085	0.010	105
Beveled external	0.250	0.469	0.025	590	1,100	175	450	0.018	0.031	0.011	0.018	470
on grooved	0.500	0.672	0.035	1,650	2,200	550	950	0.034	0.040	0.020	0.024	910
shafts	0.688	1.023	0.042	3,400	5,050	1,000	2,250	0.042	0.058	0.025	0.035	1,340
	1.062	1.500	0.050	6,200	8,800	2,400	5,000	0.060	0.079	0.036	0.047	1,950
	1.562	2.000	0.062	11,400	14,600	5,200	8,050	0.082	0.096	0.049	0.057	3,000
	2.062	2.688	0.078	18,950	24,700	8,450	13,850	0.098	0.1115	0.059	0.067	5,000
	2.750	3.438	0.093	30,100	37,700	14,400	21,900	0.112	0.129	0.067	0.077	7,350
	3.500	5.000	0.109	44,900	64,200	22,800	37,100	0.1295	0.165	0.078	0.099	10,500
	5.250	6.000	0.125	77,300	88,300	40,800	53,800	0.169	0.184	0.101	0.110	13,500
	6.250	7.000	0.156	114,800	128,600	58,300	72,700	0.187	0.208	0.112	0.125	21,000
	7.500	10.000	0.188	165,200	220,200	84,800	149,800	0.220	0.294	0.132	0.176	30,000
Inverted external	0.500	0.672	0.035	1,100	1,450	280	470	0.051	0.065	0.032	0.041	680
on grooved	0.688	1.000	0.042	2,300	3,300	500	1,050	0.066	0.091	0.042	0.057	1,000
shafts	1.062	1.500	0.050	4,150	5,850	1,200	2,500	0.092	0.100	0.058	0.063	1,460
	1.562	2.000	0.062	7,600	9,750	2,600	4,000	0.104	0.127	0.066	0.080	2,250
	2.125	2.625	0.078	13,000	16,100	4,550	6,650	0.133	0.159	0.084	0.099	3,750
	2.750	3.346	0.093	20,100	24,500	7,200	10,500	0.165	0.194	0.103	0.121	5,500
	3.500	4.000	0.109	29,900	34,300	11,500	14,000	0.202	0.213	0.127	0.133	7,850

Heavy-duty external on grooved shafts	0.394	—	0.035	2,000	—	700	—	0.047	—	0.039	—	450
	0.473	—	0.042	3,000	—	1,000	—	0.070	—	0.058	—	550
	0.500	0.669	0.050	3,900	5,200	1,100	1,900	0.070	0.077	0.058	0.064	650–900
	0.750	1.000	0.078	9,000	11,500	2,400	4,000	0.089	0.100	0.074	0.083	2,500
	1.062	1.378	0.093	15,000	19,500	4,800	8,200	0.106	0.128	0.088	0.107	4,000
	1.500	1.772	0.109	24,500	29,000	10,000	12,400	0.128	0.128	0.107	0.107	5,000
	1.938	2.000	0.125	37,000	38,000	15,300	17,000	0.153	0.153	0.128	0.128	6,000
Miniature high-strength external on grooved shafts	0.101	0.134	0.020	250	330	60	90	0.013	0.014	0.010	0.011	200
	0.156	0.203	0.025	490	650	130	200	0.021	0.023	0.017	0.018	320
	0.219	0.328	0.035	1,200	1,800	220	460	0.028	0.038	0.022	0.030	¶600
Permanent shoulder on grooved shafts	Average sizes shaft diameter, in			CRS SAE 1010 on soft steel shaft	Cabra 353 brass	Cabra 110 copper	Type 3003 aluminum	Not applicable				
	0.375		0.050	900	750	600	300					
	0.500		0.062	1,200	1,200	900	450					
	0.625		0.062	1,900	1,600	1,100	650					

* Where rings are of intermediate size—or groove materials have intermediate tensile yield strengths—loads may be obtained by interpolation.

† Approximate corner radii and chamfers limits for parts with intermediate diameters can be determined by interpolation. Corner radii and chamfers smaller than those listed will increase the thrust load of rings proportionately, approaching but not exceeding allowable static thrust loads of rings abutting parts with sharp corners.

‡ Exceptions: for shafts 0.551, 3.062, 3.500, 3.543, 3.625, 4.000, 4.500, 4.750, 6.000, and 6.250 in diameter, refer to manufacturer's specifications for data.

§ Rings for shafts 0.125 through 0.236 diameter are made of beryllium copper only.

¶ Note: $P_r' = 700$ lb for ring used with 0.260-in-diameter shaft.

pered-section bending arms which exert very strong spring force when the ring is installed. The design of the bridge between the two "ear-muff" arms gives the ring unusual flexibility and permits the fastener to be seated in grooves with large diameter tolerances. Because of its thickness, section height, and tapered-section construction, the ring has high thrust-load and impact capacity and resistance to axial and radial push-out forces.

Thinner-Gage High-Strength Radial Rings These rings have the same configuration as the high-strength rings just described, but with reduced thickness. They are designed to be installed in the same width grooves as E rings where higher thrust capacity is needed and where the rings must accommodate large tolerances in groove diameter. The rings are especially suitable for use on die-cast shafts and studs and on injection-molded parts with preformed grooves.

Both heavy-duty and thinner-gage rings may be assembled with special applicators and dispensers, or for field repair can be installed with a pair of pliers and removed easily with a screwdriver.

Typical product applications of radially assembled rings are illustrated in Figs. 11-20 through 11-25. Dimensions, load capacities, and other engineering data are given in Table 11-3.

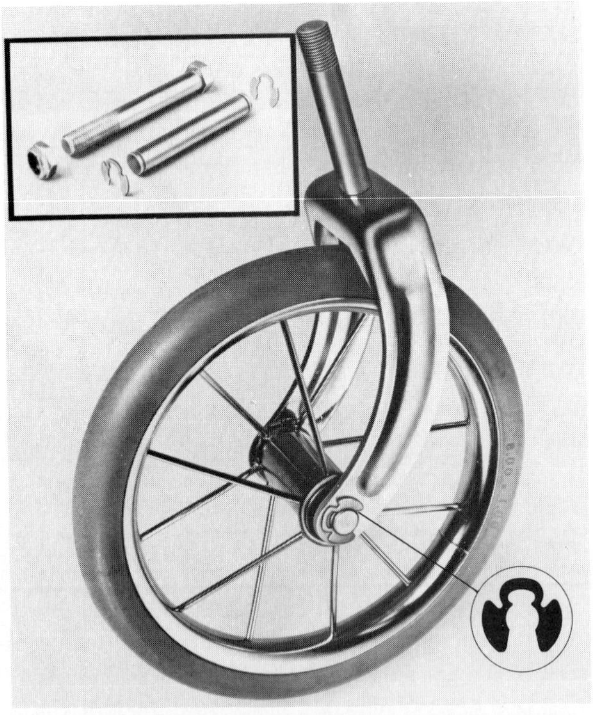

FIG. 11-20 Wheelchair axle. High-strength radially assembled rings, installed on chamfered rod, replace bolt and nut in front caster wheel. The rings speed assembly and eliminate objectionable protrusions which could scratch patients and damage furniture.

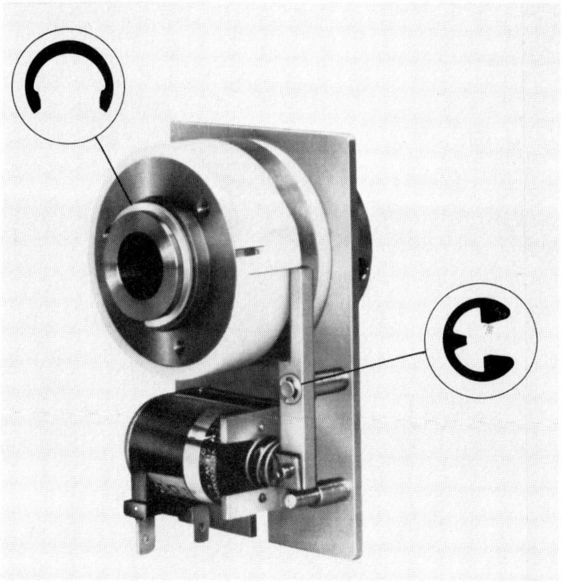

FIG. 11-21 Solenoid-operated clutch. Crescent-shaped ring (top) retains flange and collar on sleeve; it permits use of shorter sleeve and eliminates threading for panel nut. E ring secures latching mechanism, replaces washer and nut, and shortens shaft.

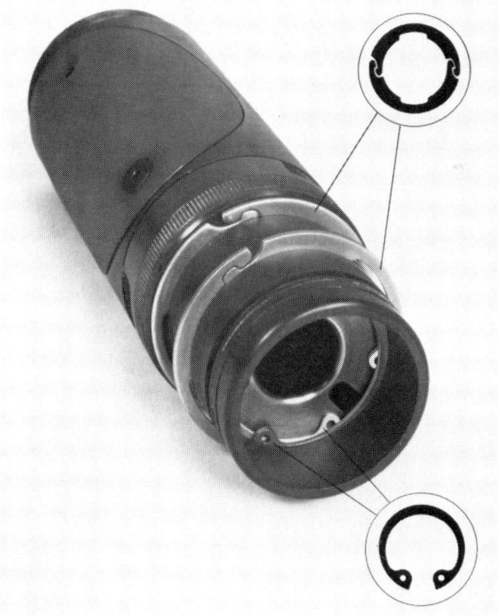

FIG. 11-22 Underwater cable connector. Two-part interlocking rings, made of stainless steel, lock split collars in position, form back-up shoulders for external couplings. Two basic internal rings secure housing in end fitting. Rings withstand high tensile loads when cable is in use, facilitate on-board replacement of damaged cables.

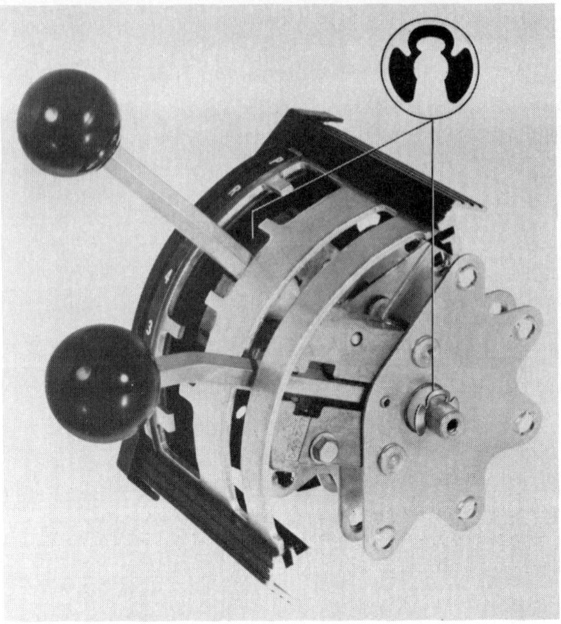

FIG. 11-23 Transmission control. Heavy-duty radially assembled rings lock shift lever shaft, spacers, and other components. Rings, seated in accurately located grooves at each end of shaft, assure precise positioning of parts. Fasteners are secure against vibration, shaft rotation, and other loads.

11-5 Rings for Taking up End Play

In many designs, accumulated tolerances or wear in the parts being retained can cause objectionable end play in the assembly. A number of axially and radially installed rings have been developed to eliminate this problem.

Bowed Internal and External Rings These rings (Fig. 11-26) are similar in appearance to the basic axial types, but are bowed around an axis normal to the diameter bisecting the ring gap.

The bowed construction permits the rings to function as springs as well as fasteners, taking up end play resiliently and dampening vibrations and oscillations. The rings are intended for relatively small assemblies involving shaft, bore, or housing diameters no larger than 1.5 in. In addition to providing resilient take-up in an axial direction, the bowed rings maintain a tight grip radially against the groove bottom. For optimum ring performance, the internal ring should be installed with the *convex* surface against the retained part; external rings should be assembled with the *concave* surface abutting the part.

Bowed Radial Rings These rings are available in two types: an E ring (Fig. 11-27) and a locking-prong ring (Fig. 11-28).

The bowed E ring is similar in shape to its flat counterpart and is bowed cylindrically in the same manner as the axial types. It provides resilient end-play take-up in an axial direction and maintains a tight grip against the

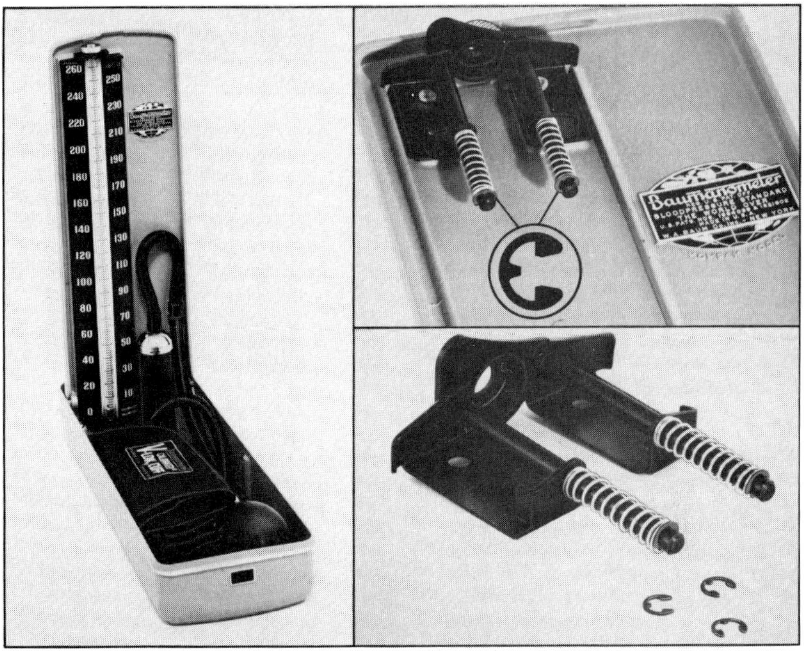

FIG. 11-24 Sphygmomanometer. Tiny E rings, installed on 0.140-in-diameter shafts, secure springs on slide assembly. Radially assembled rings replace small pins inserted into holes drilled through shafts. They reduce costs for parts and machining and speed assembly of unit.

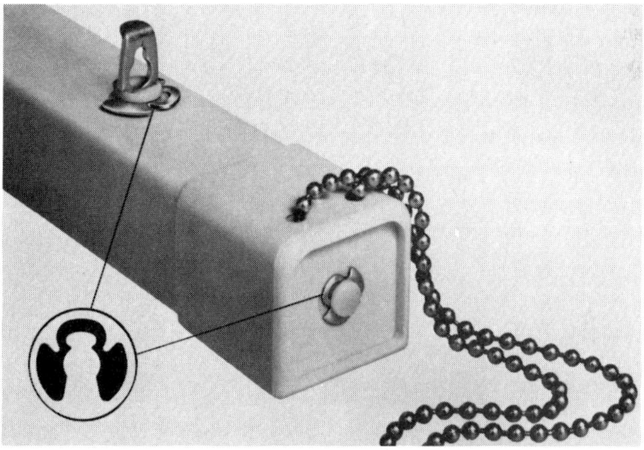

FIG. 11-25 Vertical blind track. Thinner-gage heavy-duty radially assembled rings hold plastic drive shaft, other components. Rings provide large bearing surface with gripping power to accommodate tolerances in premolded grooves of plastic parts.

TABLE 11-3 Radially Assembled External Rings

Ring type	Shaft diameter, dec. equiv. in.		Nominal ring thickness, in	Allowable static thrust load, lb* when rings abut parts with sharp corners: Groove material having minimum tensile yield strength of 150,000 lb/in²		Allowable static thrust load, lb* when rings abut parts with sharp corners: Groove material having minimum tensile yield strength of 45,000 lb/in²		Maximum allowable corner radii or chamfers of retained parts, in†: Radii‡		Maximum allowable corner radii or chamfers of retained parts, in†: Chamfers‡		Allowable thrust load, lb, when rings abut parts with listed corner radii or chamfers P'_r
	From	Through		From	Through	From	Through	From	Through	From	Through	
Crescent-shaped on grooved shafts	0.125	0.188	0.015	85	130	45	70	0.014	0.021	0.011	0.016	
	0.219	0.438	0.025	260	520	100	350	0.021	0.029	0.016	0.022	Same values as sharp corner abutment
	0.500	0.625	0.035	830	1,030	450	700	0.030	0.033	0.023	0.025	
	0.688	1.000	0.042	1,700	2,480	800	1,800	0.034	0.046	0.026	0.035	
	1.125	1.500	0.050	3,320	4,420	2,200	4,000	0.052	0.069	0.040	0.053	
	1.750	2.000	0.062	6,430	7,300	5,300	7,000	0.081	0.091	0.062	0.070	
Two-part interlocking on grooved shafts	0.469	0.625	0.035	2,000	2,650	620	830	0.052	0.052	0.040	0.040	610
	0.669	0.875	0.042	3,350	4,400	1,250	1,600	0.065	0.065	0.050	0.050	880
	0.984	1.500	0.050	5,850	8,950	2,900	4,450	0.086	0.086	0.066	0.066	1,250
	1.562	1.875	0.062	11,750	14,100	5,650	6,800	0.100	0.100	0.077	0.077	1,900
	1.969	2.625	0.078	18,250	24,300	9,000	12,000	0.114	0.114	0.088	0.088	3,050
	2.750	3.250	0.093	30,200	35,750	15,000	17,800	0.143	0.143	0.110	0.110	4,300
	3.375	—	0.109	43,500	—	20,600	—	0.182	0.182	0.140	—	5,950
Bowed E ring on grooved shafts	0.125	—	0.010	43	—	45	—	0.040	—	0.030	—	
	0.140	0.219	0.015	75	115	60	75	0.060	0.060	0.045	0.045	Same values as sharp corner abutment
	0.250	0.312	0.025	255	325	115	225	0.060	0.060	0.045	0.045	
	0.375	0.438	0.035	690	830	315	480	0.065	0.065	0.050	0.050	
	0.500	0.625	0.042	1,110	1,420	600	1,050	0.080	0.080	0.060	0.060	
	0.750	1.000	0.050	2,000	2,650	1,500	1,900	0.085	0.077	0.065	0.057	
	1.188	1.375	0.062	3,450	4,100	1,500	2,350	0.090	0.090	0.070	0.070	

E ring on grooved shafts	§0.040	0.062	0.010	13	20	6	7	0.015	0.030	0.010	0.020	Same values as sharp corner abutment
	0.094	0.140	0.015	45	75	20	45	0.040	0.040	0.030	0.030	
	0.140	0.312	0.025	170	325	60	225	0.060	0.060	0.045	0.045	
	0.375	0.438	0.035	690	830	315	480	0.065	0.065	0.050	0.050	
	0.500	0.625	0.042	1,110	1,420	600	1,050	0.080	0.080	0.060	0.060	
	0.750	1.000	0.050	2,000	2,650	1,500	1,900	0.085	0.077	0.065	0.057	
	1.188	1.375	0.062	3,450	4,100	1,500	2,350	0.090	0.090	0.070	0.070	
Reinforced E ring on grooved shafts	0.094	0.125	0.015	50	75	13	25	0.045	0.045	0.033	0.033	Same values as sharp corner abutment
	0.156	0.250	0.025	150	250	40	75	0.065	0.065	0.050	0.050	
	0.312	0.438	0.035	420	600	135	285	0.070	0.070	0.055	0.055	
	0.500	0.562	0.042	820	930	460	480	0.080	0.080	0.060	0.060	
Locking-prong ring on grooved shafts	0.092	0.156	0.010	80	120	35	100	Not applicable				
	0.188	0.312	0.015	200	350	140	300					
	0.375	0.438	0.020	550	700	450	600					
High-strength on grooved shafts	0.188	0.250	0.035	600	900	130	200	0.050	0.050	0.040	0.040	250
	0.312	0.375	0.042	1,300	1,550	250	300	0.065	0.065	0.050	0.050	350
	0.438	0.625	0.050	2,200	3,000	400	1,100	0.080	0.080	0.060	0.060	600
	0.750	—	0.062	4,600	—	1,600	—	0.085	—	0.065	—	1,000
	1.000	—	0.078	7,500	—	2,600	—	0.090	—	0.070	—	1,800
Thinner-gage high-strength on grooved shafts	0.188	0.312	0.025	430	780	130	250	0.050	0.050	0.040	0.040	150
	0.375	0.438	0.035	1,300	1,850	300	400	0.065	0.065	0.050	0.050	300
	0.500	0.625	0.042	2,100	2,500	400	600	0.080	0.080	0.060	0.060	400
	0.750	1.000	0.050	3,700	4,800	1,600	2,600	0.090	0.090	0.070	0.070	1,000

* Where rings are of intermediate size—or groove materials have intermediate tensile yield strengths—loads may be obtained by interpolation.

† Approximate corner radii and chamfers limits for parts with intermediate diameters can be determined by interpolation. Corner radii and chamfers smaller than those listed will increase the thrust load of rings proportionately, approaching but not exceeding allowable static thrust loads of rings abutting parts with sharp corners.

‡ Exceptions: For shafts 1.000, 1.772, 2.156 and 3.156 in diameter refer to manufacturer's specifications for data. Bowed E rings for shafts 0.110, 0.140, 0.438, 0.744, and 0.750 in diameter available in other sizes varying from standard design. (Refer to manufacturer's specifications for complete data.) E rings for shafts 0.062, 0.094, 0.110, 0.140, 0.438, 0.744, 0.750 and 1.000 in diameter are available in one or more sizes varying from standard design. Refer to manufacturer's specifications for complete data.

§ Ring for shaft 0.040 in diameter is made of beryllium copper only.

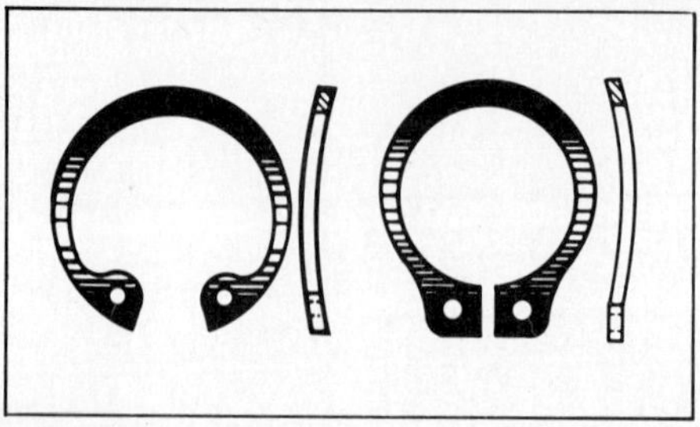

FIG. 11-26 Bowed internal (left) and external (right) rings.

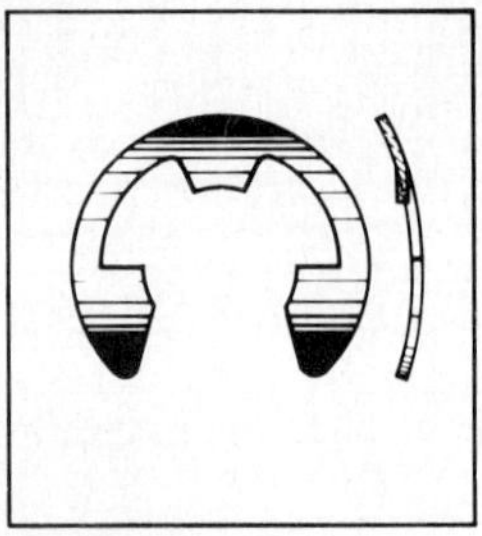

FIG. 11-27 Bowed E ring.

bottom of the groove. It should be installed with the *concave* surface abutting the retained part. The ring is not recommended as a shoulder against rotating parts, however, since rotation may force the ring from the groove. For such applications, the locking-prong ring should be considered.

The locking-prong ring derives its name from two prongs extending from the inner circumference to the open end. The ring must be compressed with an applicating tool so that the bowed ends with the prongs can enter the groove. The ring is then pushed forward in the groove until the prongs pass the outer circumference of the shaft, at which time the ring springs back to its bowed position and the prongs lock around the shaft. The fastener has good thrust-load capacity and is especially suitable as a retainer for rotating parts. Because it functions as a spring and shoulder, it eliminates the need for separate springs, bowed washers, and other accessories often used in assemblies with rotating parts.

Beveled Rings These rings (Fig. 11-29) are designed to provide rigid end-play take-up in an assembly. The fasteners are similar in design to the basic internal and external axially assembled rings, except for the groove-engaging edge, which is beveled to a 15° angle. The bevel is located on the *outer* circumference of internal rings, and around the *inner* circumference of external rings. (Internal rings are available in two types, one with a single beveled edge, the other a newer design with a double-beveled construction. The double-beveled

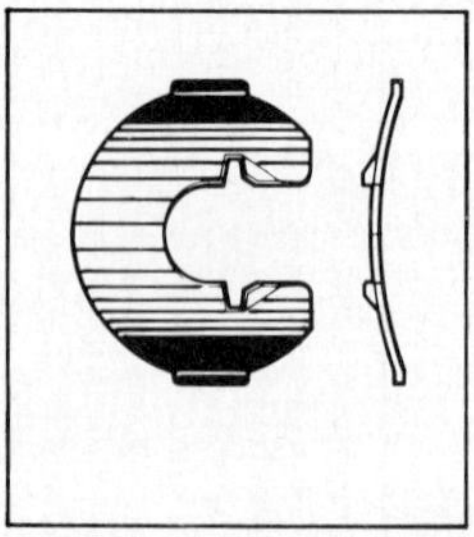

FIG. 11-28 Locking-prong ring.

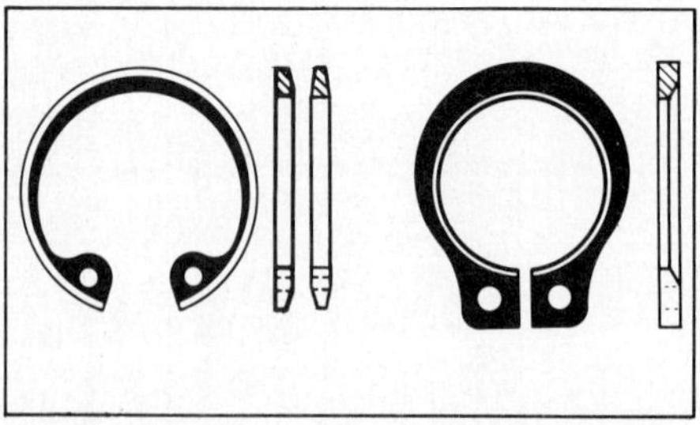

FIG. 11-29 Beveled and double-beveled internal rings (left) and beveled external rings (right).

rings eliminate the need for orienting rings prior to assembly and assure correct installation during field service.)

The beveled rings are installed in grooves that have a corresponding 15° bevel on the load-bearing wall. When the ring is seated in the groove, it functions as a wedge between the retained part and the outer groove wall. When there is end play between the ring and the abutting face of the part, the ring's spring action causes the fastener to seat more deeply in the groove, compensating for the end play. The ring also exerts an axial force against the retained part.

Typical product applications of rings used for taking up end play are illustrated in Figs. 11-30 through 11-34. Dimensions, load capacities, and other data are given in Tables 11-1 to 11-3.

11-6 Self-Locking Rings

All the axial and radial ring types described previously are designed to be assembled in a groove on a shaft or inside a bore or housing. In many product designs, however, cutting a groove is neither practical nor necessary. This is especially true for small appliances, instruments, plastic products, and other applications in which the ring will serve merely as a positioning and locking device against small impacts and vibrations and need not absorb any sizable thrusts.

In addition to eliminating grooves, threads, and other preparatory machining, the self-locking rings provide another advantage: Because they may be positioned at any point on a shaft or inside a bore or housing, they automatically compensate for accumulated tolerances in the retained parts. The rings

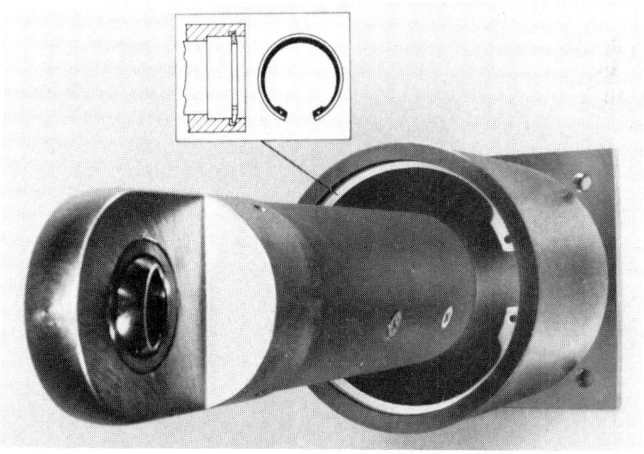

FIG. 11-30 Mechanical shock arrestor. Beveled internal ring secures support cylinder in housing. Ring has 15° bevel on outer circumference and is seated in groove with comparable bevel on load-bearing wall. Wedge action of ring in groove (inset drawing) provides rigid end-play take-up in assembly.

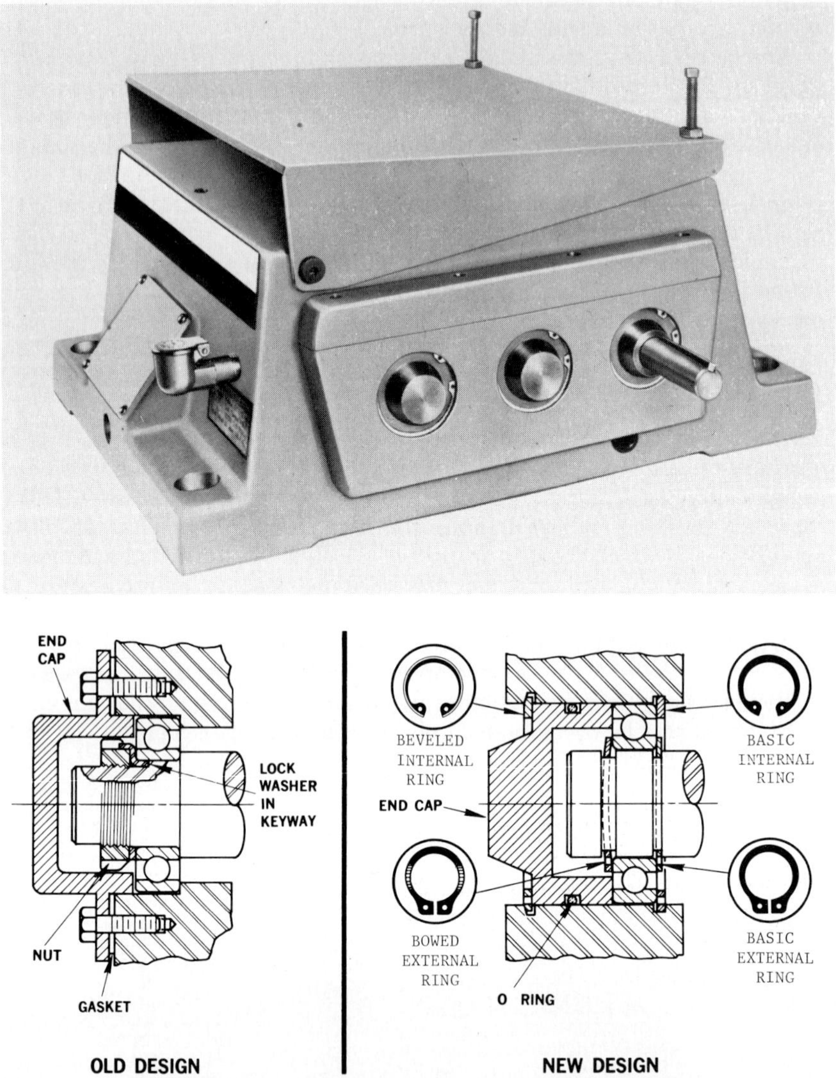

FIG. 11-31 Power transmission drive. In original design (left) components were positioned and secured by bolted end caps, machined shoulders, lock washers, and threaded ring nuts. Redesign (right) utilizes beveled and basic internal rings, bowed and basic external rings to simplify assembly and lower costs.

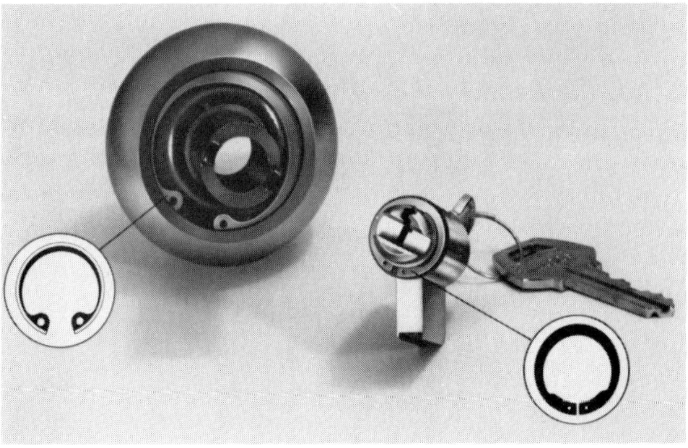

FIG. 11-32 Cylindrical lockset. Beveled internal ring (left) secures knob and shank assembly, eliminating need for drilling and tapping parts, then fastening with screws. Inverted external ring (right) couples cylinder plug and housing, replacing drilling and pinning operations.

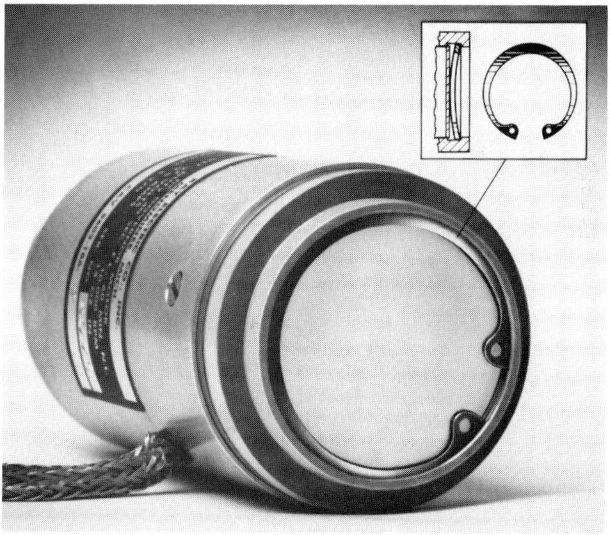

FIG. 11-33 Aerial camera motor. Bowed internal ring, used in conjunction with aluminum cover plate, eliminates bolted end cover. Ring's bowed construction provides resilient end-play take-up, assures tight fit, and automatically compensates for tolerances in parts.

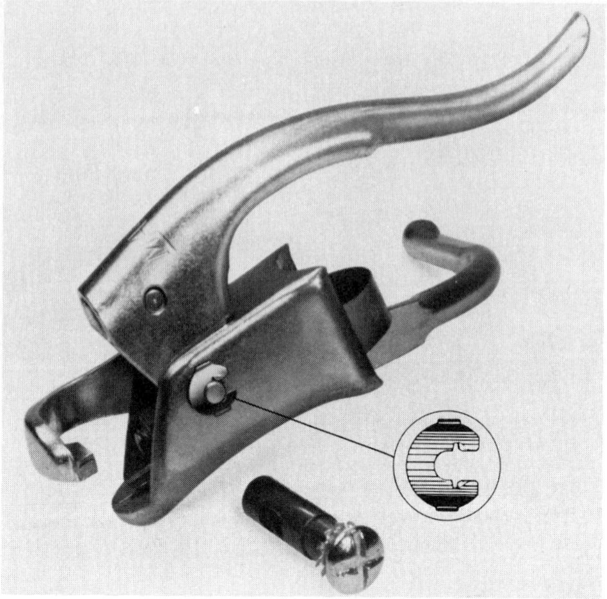

FIG. 11-34 Bicycle hand brake. Bowed locking-prong ring holds hinge pin in frame under required spring pressure and automatically compensates for tolerances in formed frame. Ring replaces screw and lock washer and eliminates costly drilling and tapping operations.

can be installed rapidly with simple hand tools, ensuring economical assembly even by unskilled labor.

Axial Clamp Rings These rings (Fig. 11-35) utilize the same tapered-section design as the basic external ring. The radial section height is higher, however, so that when the fastener is expanded and then allowed to snap back and grip a shaft, it forms a high shoulder against the retained parts and exerts a strong frictional hold, secure against displacement by thrust loads in either axial direction. Like basic rings, the fastener has lugs with holes for assembly pliers. Because it is reusable after disassembly, it may be adjusted to any position on a shaft to accommodate wear or tolerances in the parts.

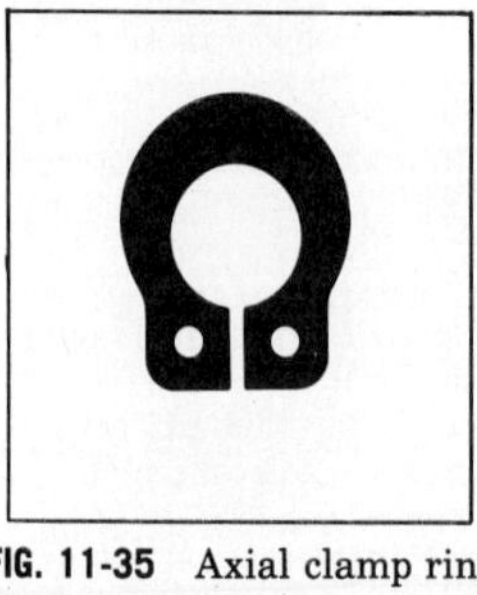

FIG. 11-35 Axial clamp ring.

FIG. 11-36 Radial clamp ring.

FIG. 11-37 Internal self-locking ring.

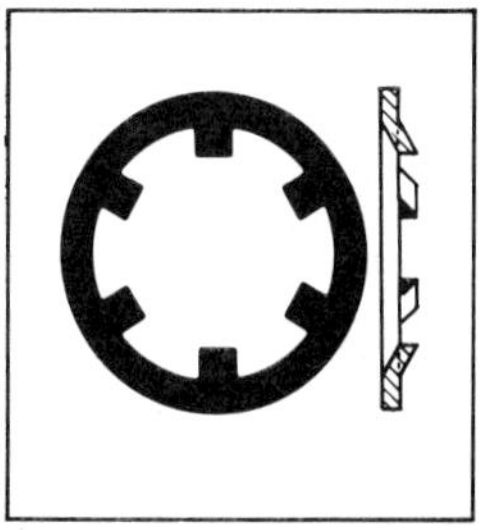

FIG. 11-38 External self-locking ring.

Radially Applied Self-Locking Clamp Rings These rings (Fig. 11-36) are of a new design which also has a high tapered section. The ring differs from the axial type, however, in that it has a wide gap between the "legs" which permits the fastener to be pushed over a shaft in a radial direction. Once assembled on the shaft, the ring exerts a strong frictional hold against push-out forces in either axial direction. Radial clamp rings can be pushed onto a shaft with simple hand tools; cut-outs in the legs permit the rings to be adjusted in position on a shaft or removed with pliers or other tools. The fasteners are reusable after disassembly.

Circular Self-Locking Rings These rings are push-on fasteners available in internal (Fig. 11-37), external (Fig. 11-38), and reinforced external (Fig. 11-39) axial types.

The internal rings have a flat rim surrounded by inclined prongs, which spread when the fastener is pushed into a bore or housing. When load is applied from the opposite direction, the prongs "bite into" the wall of the bore and lock the ring in position. The external rings have the same design, except that the prongs are located inside the rim so that when the ring is pushed over a shaft, the prongs can spread in the opposite direction and lock the fastener against the retained part.

Both the internal and external rings are secure against moderate thrust loads and small impacts and vibrations. They can be installed with simple plunger-type tools, but, since the prongs must be pried loose for removal, cannot be reused after disassembly.

FIG. 11-39 Reinforced external self-locking ring.

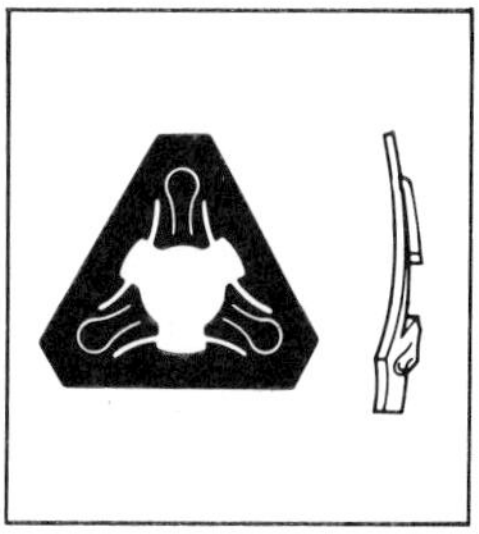

FIG. 11-40 Triangular self-locking ring.

The reinforced external ring derives its name from the arched rim, which provides extra strength and resistance to bending, twisting, or buckling. It also has longer locking prongs than the flat push-ons. The combination of rim design and elongated prongs gives the reinforced ring higher thrust-load capacity and the ability to compensate for larger shaft tolerances.

Triangular Self-Locking Rings These rings (Fig. 11-40) are bowed and have three slightly inclined prongs which, in smaller sizes, are ribbed for extra stiffness and gripping power. When the fastener is pushed over a shaft, the prongs spread apart and the body flattens against the retained part. When pressure on the fastener is released, the body springs back to its bowed shape,

TABLE 11-4 Self-Locking Rings

Ring type	Housing diameter or shaft diameter, dec. equiv. in.		Nominal ring thickness, in	Allowable static thrust load, lb* when rings abut parts with sharp corners: Groove material having minimum tensile yield strength of 150,000 lb/in²		Groove material having minimum tensile yield strength of 45,000 lb/in²	
	From	Through		From	Through	From	Through
Reinforced self-locking external on shafts, no grooves	0.094	0.375	0.010	—	—	27	65
	0.094	0.375	0.015	—	—	45	120
	0.438	1.000	0.015	—	—	120	140
Self-locking external on shafts, no grooves	0.094	0.375	0.010†	—	—	13	45
	0.438	1.000	0.015	—	—	50	65
Self-locking internal in housings, no grooves	0.312	0.625	0.010	—	—	80	45
	0.750	2.000‡	0.015	—	—	75	55
Triangular retainer on shafts, no grooves	0.062	—	0.010	—	—	25	—
	0.062	—	0.015	—	—	40	—
	0.094	0.156	0.010	—	—	60	75
	0.094	0.156	0.015	—	—	80	120
	0.188	0.312	0.015	—	—	140	200
	0.375	—	0.020	—	—	250	—
	0.437	§	0.025	—	—	270	—
Triangular nut on threaded parts	6/32	10/32	0.015	140	170	140	145
	6/32	10/32	0.020	200	220	180	190
	1/4–20	1/4–28	0.020	220	—	220	—
	1/4–20	1/4–28	0.025	220	—	220	—

TABLE 11-4 Self-Locking Rings (*continued*)

Tapered-section self-locking clamp ring on shafts with or without grooves	Shaft diameter, dec. equiv. in.		Nominal ring thickness, in	Allowable static thrust load, lb*			
				Shaft without groove		Shaft (45,000 lb/in²) with groove¶	
	From	Through		From	Through	From	Through
Inch type	0.094	0.156	0.025	10	22	—	—
	0.187	0.250	0.035	25	35	—	90
	0.312	0.375	0.042	45	60	110	180
	0.437	0.500	0.050	60	65	290	390
	0.625	0.750	0.062	85	90	570	850
Millimeter type	0.079	0.118	0.024	10	15	—	—
	0.197	—	0.032	30	—	40	—
	0.236	0.276	0.039	35	40	70	100
	0.354	0.394	0.047	50	55	130	170
	0.533	0.590	0.059	75	80	340	370
Radially applied self-locking clamp rings on shafts without grooves							
Inch type	0.093	0.156	0.025	8	13	—	—
	0.187	0.250	0.035	18	22	—	—
	0.312	0.375	0.042	32	42	—	—
Millimeter type	0.078	0.156	0.024	7	12	—	—
	0.197	0.276	0.035	19	23	—	—
	0.312	0.393	0.043	33	49	—	—

* Where rings are of intermediate size—or groove materials have intermediate tensile yield strengths—loads may be obtained by interpolation.

† Ring for shaft 0.240 in diameter is available only in 0.015-in thickness; allowable thrust load = 40 lb.

‡ Ring for housing 1.375 in diameter is available only as reinforced ring having an allowable thrust load = 150 lb.

§ Round and hexagonal shafts

¶ Grooved shafts are recommended *only* for rings used on shafts 0.197 in (5.0 mm) or larger.

causing the prongs to lock against the shaft and remain secure under spring tension. The triangular ring provides a larger shoulder and higher thrust-load capacity than the flat circular push-on types and is especially useful where the shape complements other components in the assembly.

Typical product applications of self-locking retaining rings are illustrated in Figs. 11-41 through 11-47. Dimensions, load capacities, and other data are given in Table 11-4.

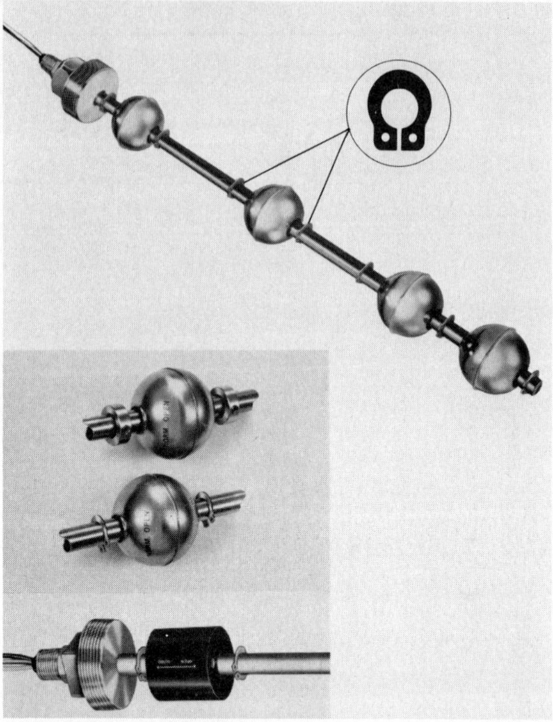

FIG. 11-41 Liquid level switch. Axial clamp rings replace costly brass or stainless steel collars (inset, top) as float stops. Rings exert frictional hold against displacement and do not require any groove or other preparatory machining. Lower costs for parts and assembly.

MATERIALS AND FINISHES

11-7 Spring Properties

To function properly, retaining rings must be able to deform elastically during assembly and disassembly. For this reason, the materials from which they are fabricated must have good spring properties. These properties include high tensile and yield strengths, in addition to a ratio of ultimate tensile strength to modulus of elasticity which permits the required deformation without excessive permanent set. For most tapered-section designs, a ratio of 1 : 100 is satisfactory.

Among the most popular materials for retaining rings are carbon spring steel (SAE 1060-1090), aluminum (Alclad 7075-T6), beryllium copper (Alloy #25, CDA #17200), and precipitation-hardening stainless steel (PH 15-7 Mo or equivalent; AISI 632-AMS 5520). Rings made of carbon spring steel, stainless steel, and beryllium copper should be hardened to develop the spring properties needed for satisfactory ring performance.

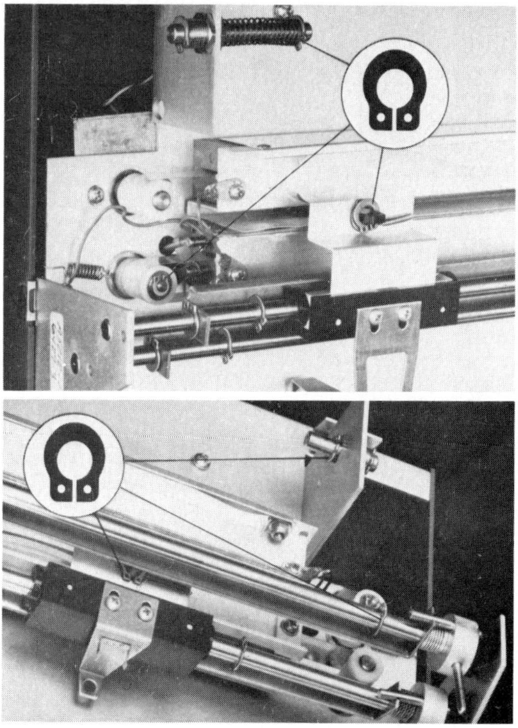

FIG. 11-42 Strip chart recorder. Extensive use of axial clamp rings—used to secure many different moving parts—reduces costs for both parts and assembly. Rings provide reliable means for accurate location of components required for faultless recorder performance.

FIG. 11-43 Retrofit electronic ignition. External self-locking ring fastens wheel and sensor. Permits "do-it-yourself" consumer to install conversion kit without threading or other machining of distributor cam and sleeve.

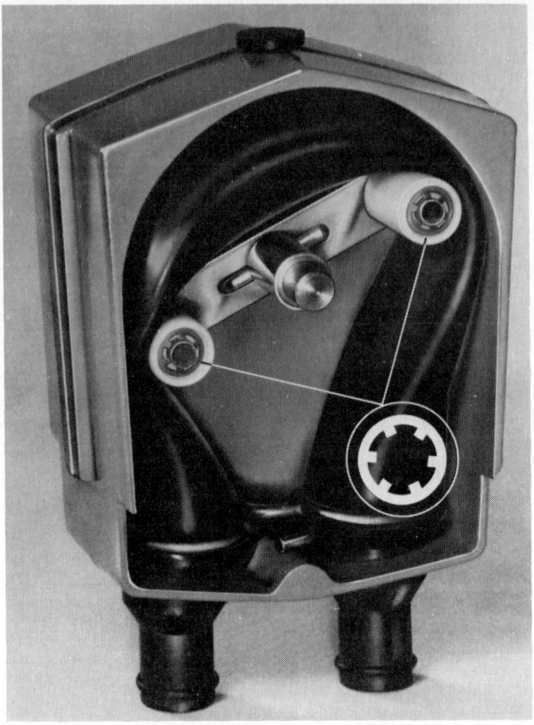

FIG. 11-44 Industrial pump. External self-locking rings hold plastic rollers on shafts for simple, economical fastening system. In original design of roller assembly, shafts were drilled and tapped and rollers secured by screws and washers. Stainless steel rings do not require any groove or other machining.

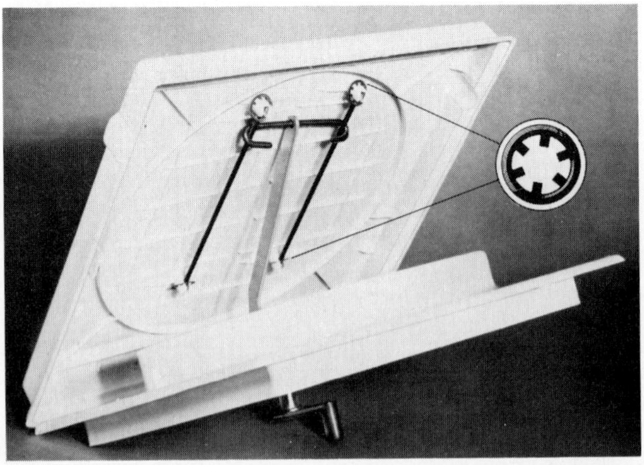

FIG. 11-45 Recreational vehicle roof vent. Reinforced self-locking rings hold formed wire linkages for cranking mechanism to molded plastic studs. Rings' inclined prongs "bite" into studs to assure necessary resistance to push-out forces when crank is operated; reinforced, arched rim prevents fasteners from twisting or buckling.

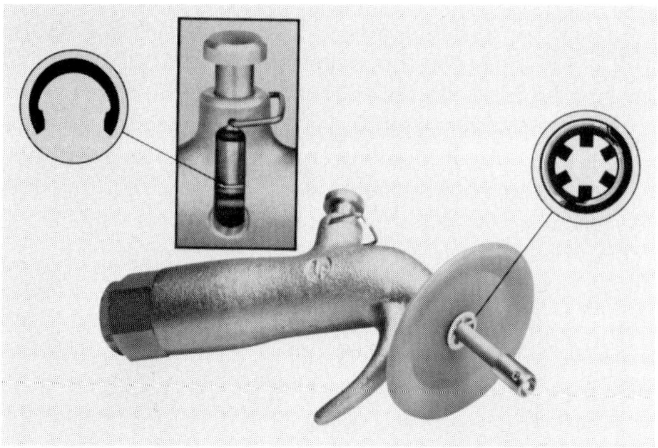

FIG. 11-46 Air gun. Reinforced self-locking ring (right) positions and secures plastic chip guard behind nozzle. Because no groove is needed, the ring can be installed flush against guard for tight fit, compensating for dimensional variations in plastic part. Crescent-shaped ring (left) secures push-button assembly which activates air valve.

FIG. 11-47 Knife sharpener. Triangular self-locking rings hold grinding wheel between aluminum "roving cones." Wheel and cones are installed on keyed aluminum shaft which permits parts to spin freely as single unit when plastic housing is assembled around them. Rings replace nuts, eliminate costly threading operation.

11-8 Protective Coatings

Carbon spring steel rings are often supplied with a black, corrosion-retardant phosphate coating which provides adequate protection for most product applications; some ring manufacturers rely on an oil-dip finish for protection. Zinc plating with a dichromate dip and other special finishes are also available for extra corrosion resistance. Aluminum and beryllium copper rings generally are furnished without additional finish; stainless steel rings are passivated.

11-9 Heat Treating and Plating

Heat treating and plating procedures used during the manufacture of retaining rings can have a direct effect on ring performance because the fasteners are subjected to extremely high stresses during installation and removal. (Carbon spring steel rings, for example, sometimes undergo stresses of as high as 250,000 lb/in during assembly.) The rings also must withstand a certain amount of stress when they are seated in an assembly. Improper heat treating, which produces brittleness, or poor plating techniques, which can cause hydrogen embrittlement, can cause rings to fail suddenly during assembly or disassembly, or even when the rings are subjected to relatively light loading conditions. Special heat-treating procedures such as austempering and plating processes such as mechanical plating, which avoids hydrogen embrittlement, are used to prevent this kind of ring failure.

CALCULATING LOAD CAPACITIES

The static thrust load capacity of a retaining ring assembly is dependent on several factors: the strength of the wall of the groove in which the ring is installed, the ring's shear strength, and the configuration of the part abutting the ring. Ideally, the ring should be seated against a square-cornered part; if the abutting surface of the retained part has a corner radius or chamfer, the ring may dish under load or may even be forced out of the groove. The smallest of the three values—groove wall strength, ring shear strength, and capacity with a corner radius or chamfer—determines the thrust-load capacity of the ring assembly.

11-10 Groove Wall Strength

To calculate this quantity, use the formula

$$P_g = \frac{C_F S d \pi s_y}{Fq} \tag{11-1}$$

where P_g = allowable static thrust load on the groove wall
C_F = conversion factor (see Table 11-5)
S = shaft or housing diameter
d = groove depth
s_y = tensile yield strength of groove material, lb/in^2 (see Table 11-6)

TABLE 11-5 Conversion Factor C_F for Calculating P_r and P_g

Ring type	Conversion factor C_F Ring: P_r	Conversion factor C_F Groove: P_g
Basic, bowed internal	1.2	1.2
Beveled internal	1.2	1.2
Double-beveled internal		Use $d/2$ instead of d
Inverted internal, external	⅔	½
Basic, bowed external	1	1
Beveled external	1	1 Use $d/2$ instead of d
Crescent-shaped	½	½
Two-part interlocking	¾	¾
E ring, bowed E ring	⅓	⅓
Reinforced E ring	¼	¼
Locking-prong ring	See manufacturer's specifications	½
Heavy-duty external	1.3	2
High-strength radial	½	½
Miniature high-strength	See manufacturer's specifications	
Thinner-gage high-strength radial	½	½

F = safety factor (a factor of 2 is satisfactory for most conditions)
q = reduction factor, taken from curve illustrated in Fig. 11-48 (dimension Z is the distance of the outer groove wall from the end of the shaft or bore, as shown in Fig. 11-49). If $Z/d \geqq 3$, use $q = 1$)

11-11 Ring Shear Strength

The strength of a retaining ring in shear is calculated from the formula

$$P_r = \frac{C_F S t \pi s_s}{F} \qquad (11\text{-}2)$$

TABLE 11-6 Tensile Yield Strength of Groove Material

Groove material	Tensile yield strength, lb/in^2
Cold-rolled steel	45,000
Hardened steel (Rockwell C40)	150,000
Hardened steel (Rockwell C50)	200,000
Aluminum (2024-T4)	40,000
Brass (naval)	30,000

where P_r = allowable thrust load of the ring, lb
C_F = conversion factor (see Table 11-5)
S = housing or shaft diameter, in
t = ring thickness, in
s_s = shear strength of ring material, lb/in^2 (see Table 11-7)
F = safety factor

11-12 Capacity with Corner Radii or Chamfers

The maximum allowable corner radius or chamfer of a retained part for various types of rings is given in Tables 11-1 through 11-4 (condensed load tables). In no case should a corner radius or chamfer exceed the given maximum values of radius and chamfer. The load capacity of the ring must be selected from the tables, listed under the symbol P_r'. If the actual corner radius or chamfer is less than the listed maximum value, the load capacity of the ring may be increased according to the following formulas:

$$P_r'' = P_r' \frac{R_{\max}}{R} \qquad \text{(for radius)} \qquad (11\text{-}3)$$

$$P_r'' = P_r' \frac{Ch_{\max}}{Ch} \qquad \text{(for chamfer)} \qquad (11\text{-}4)$$

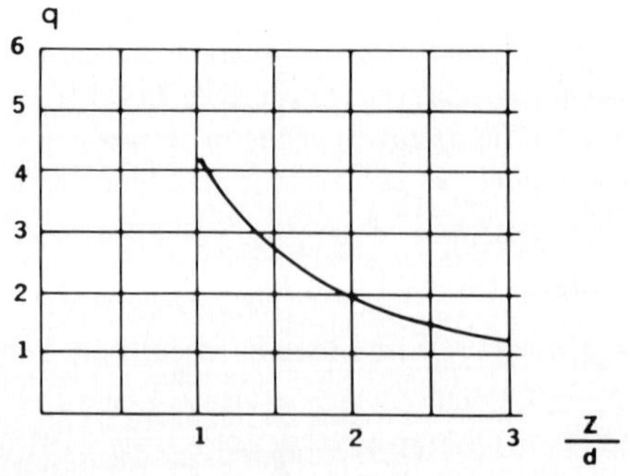

FIG. 11-48 Reduction curve.

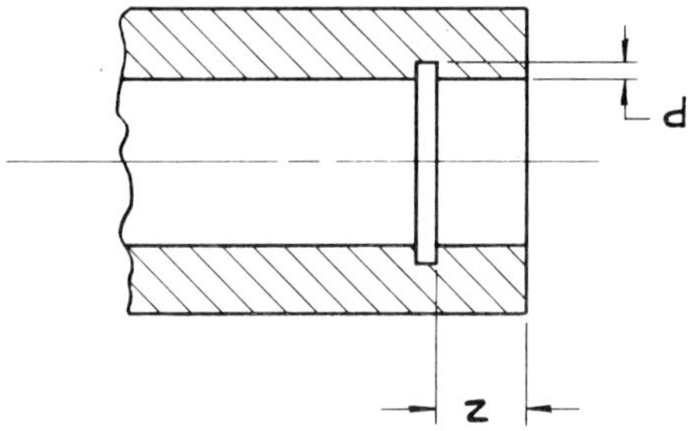

FIG. 11-49 Edge margin.

TABLE 11-7 Shear Strength of Ring Material

Ring material	Ring type	Ring thickness, in	Shear strength, lb/in²
Carbon spring steel (SAE 1060–1090)	Basic, bowed, beveled, inverted: internal and external rings and crescent-shaped. Double-beveled internal rings	Up to and including 0.035	120,000
		0.042 and over	150,000
	Heavy-duty external	0.035 and over	150,000
	Miniature high-strength	0.020 and 0.025	120,000
		0.035 and over	150,000
	Two-part interlocking, reinforced E ring, high-strength radial	All available	150,000
	Thinner high-strength radial	All available	150,000
	E ring, bowed E ring	0.010 and 0.015	100,000
		0.025	120,000
		0.035 and over	150,000
	Locking-prong	All available	130,000
Beryllium copper (CDA 17200)	Basic external	0.010 and 0.015 (sizes 12 through 23)	110,000
	Bowed external	0.015 (sizes 18 through 23)	110,000
	E ring	0.010 (size X4 only)	95,000

where P''_r = allowable assembly load when corner radius or chamfer is *less* than listed maximum (in no case can P''_r exceed P'_r)
P'_r = listed allowable assembly load with maximum corner radius or chamfer
R_{max} = listed maximum allowable corner radius
R = actual corner radius
Ch_{max} = listed maximum allowable chamfer
Ch = actual chamfer

Where static loading conditions exist, these formulas are generally sufficient to determine the allowable strength of a retaining ring assembly.

11-13 Relative Rotation

When a part which is exerting a static load against the ring also rotates against the ring, the allowable load that may be exerted by the part on the ring may be determined by the following formula:

$$P_{rr} \leqq \frac{stE^2}{\mu 18S} \tag{11-5}$$

where P_{rr} = allowable thrust load exerted by adjacent part, lb
s = maximum working stress of ring during expansion or contraction (see Table 11-8)
t = ring thickness, in
E = largest section of ring, in
μ = coefficient of friction between ring and retained part, whichever is higher (consult appropriate reference)
S = shaft or housing diameter, in

11-14 Dynamic Loading

Where repeated loading of the ring may lead to conditions of fatigue, actual tests of the assembly must be made. The following six formulas may be used

TABLE 11-8 Maximum Working Stress of Ring During Expansion or Contraction

Ring material	Maximum allowable working stress, lb/in^2
Carbon spring steel (SAE 1075)	250,000
Stainless steel (PH 15-7 Mo)	250,000
Beryllium copper (CDA 17200)	200,000
Aluminum (Alclad 7075-T6)	70,000

for calculating the capacities of a retaining ring assembly under various conditions of dynamic loading. (Definitions of terms for all formulas follow the last formula.)

No Play between Ring and Parts For tight assemblies with no play between the ring and retained parts, four formulas are used.

Allowable sudden load on ring:

$$P_{SR} \leqq 0.5P_r \tag{11-6}$$

Allowable sudden load on groove:

$$P_{SG} \leqq 0.5P_g \tag{11-7}$$

Allowable vibration loading on ring:

$$wa \leqq 540P_r{}^* \tag{11-8}$$

Allowable vibration loading on groove:

$$wa \leqq 400P_g{}^* \tag{11-9}$$

Play between Ring and Part When there is play between the ring and the retained part and the part strikes the ring, the load must be calculated as impact. Formulas are:

Allowable impact loading on ring:

$$I_r = \frac{P_r t}{2} \tag{11-10}$$

Allowable impact loading on groove:

$$I_g = \frac{P_r d}{2} \tag{11-11}$$

Definitions of Formula Terms

P_{SR} = allowable sudden load on ring, lb
P_{SG} = allowable sudden load on groove, lb
I_g = allowable impact load on groove, in·lb
I_r = allowable impact load on ring, in·lb
P_g = allowable static thrust load on groove, lb
P_r = allowable static thrust load on ring, lb
w = weight of retained parts, lb
a = acceleration of retained parts, in/s^2

* *Note:* Actual tests should be made because of repeated or cyclic conditions.

For harmonic oscillation,

$$a \approx 40\rho f^2 \tag{11-12}$$

where ρ = amplitude of vibration, in
f = frequency of vibration, cycles/s

ASSEMBLY TOOLS

There are a variety of assembly tools available to assure fast, economical ring installation, even by unskilled labor. These range from simple hand tools such as pliers and applicators to more sophisticated magazine-fed tools and both semiautomatic and automatic assembly equipment for high-speed mass-production lines.

11-15 Pliers

Pliers for Internal Rings These pliers (Fig. 11-50) are designed to compress the rings for installation in or removal from a bore or housing. The pliers have formed handles, sheathed with red plastic sleeves for extra comfort and control. An adjustable positive stop may be preset to align the tips with the lug holes in the rings for faster ring pickup. Springs connected to the handles return the tips to the free position after ring assembly or removal.

External-Ring Pliers These pliers are similar in construction to the internal type, except that the tips expand the rings so they may be slipped over

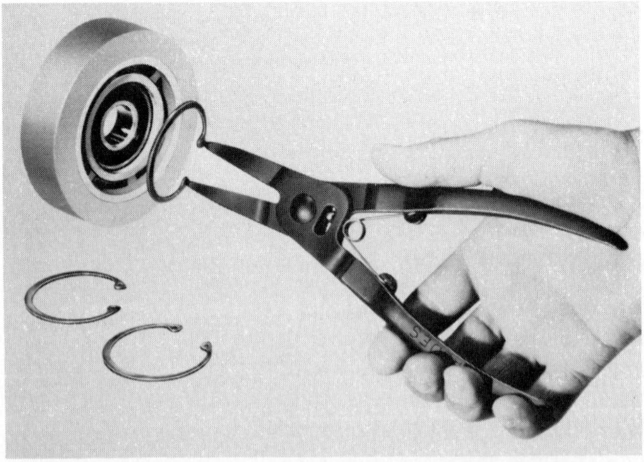

FIG. 11-50 Pliers for internal rings compress fasteners for insertion into bore or housing. Adjustable stop below pivot pin permits tips to be preset for alignment with lug holes for faster ring pickup. External ring pliers expand rings for assembly over shaft.

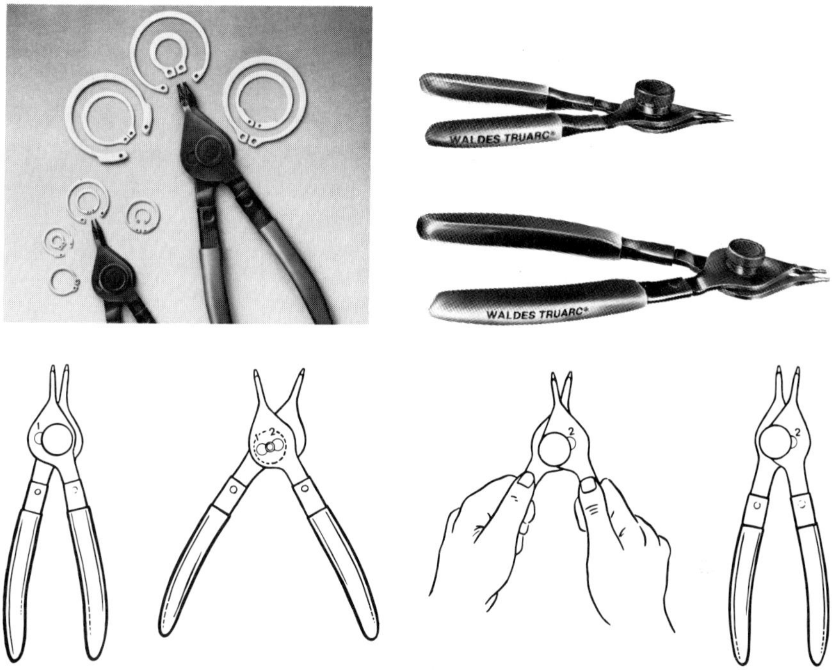

FIG. 11-51 Convertible pliers may be used with both internal and external rings simply by changing setting as shown in line drawings. They are intended primarily for low-volume production or maintenance and repair operations.

the end of a shaft or similar part. On the external-ring pliers, the adjustable stop is used to limit the spreading of the tips to avoid accidental overspreading of the rings.

Convertible Pliers These pliers (Fig. 11-51) may be used with either internal or external rings by simply changing the pivot setting. They are intended primarily for maintenance and repair operations or for production where rings are installed in small quantities.

Internal, external, and convertible pliers normally are supplied with straight tips. For assembly operations in which clearance is limited, tips bent to a 45° or 90° angle can also be supplied.

Ratchet Pliers These pliers (Fig. 11-52) are designed to facilitate the installation and removal of large rings, which exert considerable spring pressure during expansion or contraction. The pliers have a unique double ratchet which serves two purposes: It reduces the effort needed to spread or compress a ring, and it locks the pliers at any desired point of expansion or contraction. The locking action is an important safety feature because it eliminates the need for the operator to maintain pressure on the handles during ring assembly and disassembly and prevents the ring from springing loose accidentally.

FIG. 11-52 Ratchet pliers for large rings. Double ratchet reduces effort needed to compress or expand rings and locks pliers at any given point so operator need not maintain pressure on handles. Ratchet also serves as safety device to prevent ring from springing loose accidentally.

11-16 Applicators and Dispensers

These tools (Fig. 11-53) are widely used with radially assembled rings, supplied in convenient tape-wrapped cartridges. The rings are slipped over a spring rail on the dispenser and allowed to fall to the base, after which the tape is peeled away. The operator then picks up the rings from the base with an applicator which has a split, fork-like blade set into the handle. When the applicator is withdrawn from the base, the first ring in the stick is held in a recess on the underside of the blade and the next ring in the stack falls into position automatically. To install the ring, the operator simply pushes the ring into a groove on the shaft. Because the gripping power of the ring on the shaft is greater than the tension of the tool jaws, the ring remains seated in the groove when the applicator is withdrawn. Applicators are usually furnished with straight blades, but angled (45° or 90°) or offset blades are also available for assemblies where clearance is limited.

11-17 Magazine-Fed Tools

These tools (Fig. 11-54) combine an applicator and dispenser in a single portable tool which is loaded with the same type of tape-wrapped ring cartridges. The operator squeezes the trigger to activate a flip-up applicator blade which grasps the first ring in the magazine. When pressure on the handle is released, the blade returns to its forward position, the ring is pushed into the groove, and the tool is withdrawn. The applicator blade is offset to grasp the first ring in the magazine and to facilitate ring assembly.

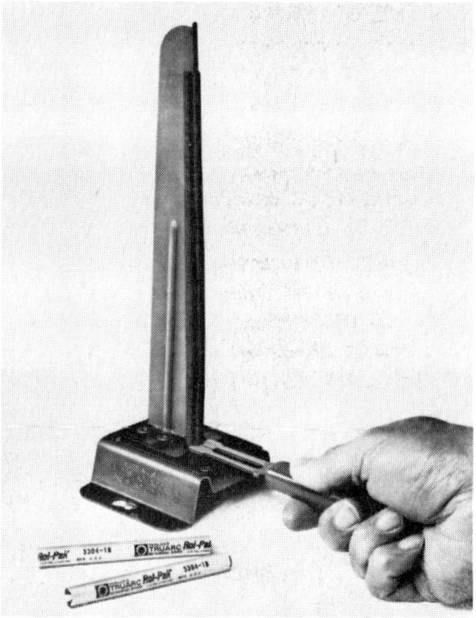

FIG. 11-53 Applicator and dispenser for radially assembled rings. Dispenser is loaded with tape-wrapped ring cartridges shown in foreground; applicator is used to grasp bottom ring in stack and assemble it on work part.

11-18 Air-Driven Tools

These tools (Fig. 11-55) are designed for high-speed, repetitive ring assembly. The unit illustrated works some-

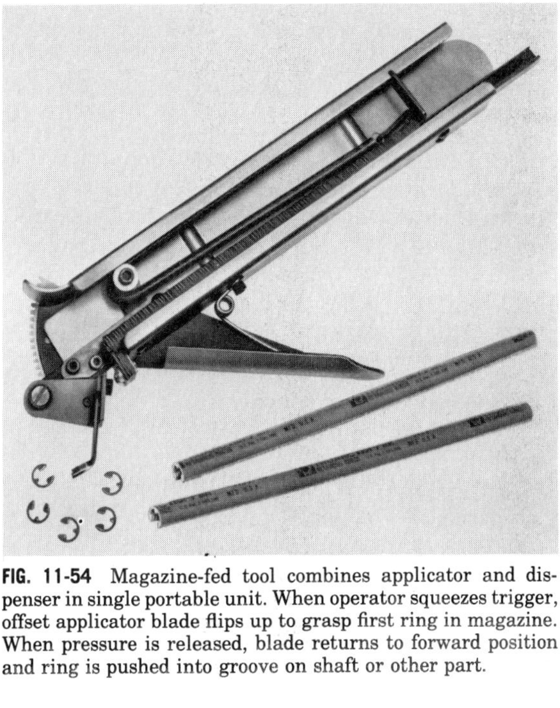

FIG. 11-54 Magazine-fed tool combines applicator and dispenser in single portable unit. When operator squeezes trigger, offset applicator blade flips up to grasp first ring in magazine. When pressure is released, blade returns to forward position and ring is pushed into groove on shaft or other part.

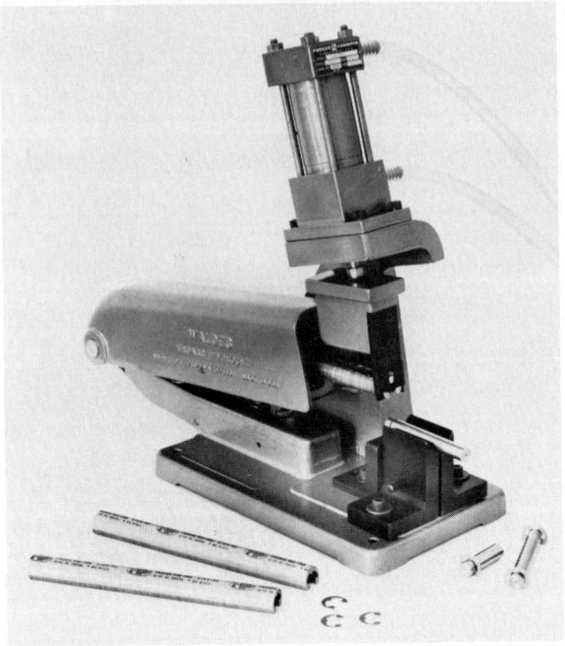

FIG. 11-55 Air-driven tool for high-speed, repetitive ring assembly. To install ring, operator merely places part in nest at front and activates hand or foot pedal. Air cylinder forces applicator blade down to grasp first ring in magazine, then push it into groove on shaft. Unit is loaded with same taped cartridges used for manually operated tools.

what like a paper stapler: The work part is placed in a nest at the front of the unit, and the tool is activated with a hand or foot control. The air cylinder pushes an applicator blade downward to grasp the first ring in the magazine and push it into the groove on the work part. When the ring is seated in the groove, the applicator is retracted automatically. Because nesting fixtures must be custom designed for specific work parts, they are not supplied with the tools.

11-19 Portable Air-Operated Tools

These tools (Fig. 11-56) are intended for high-volume application of basic and heavy-duty axially assembled external rings. They are available as standard tools designed to replace more expensive custom-designed assembly equipment.

The tools can be suspended over a fixed work station or a mechanized conveyer line wherever an 80-lb/in^2 (gage) air supply is available. A spring-loaded or counterbalanced cable is suggested to facilitate handling of the tool, which weighs approximately 9 lb.

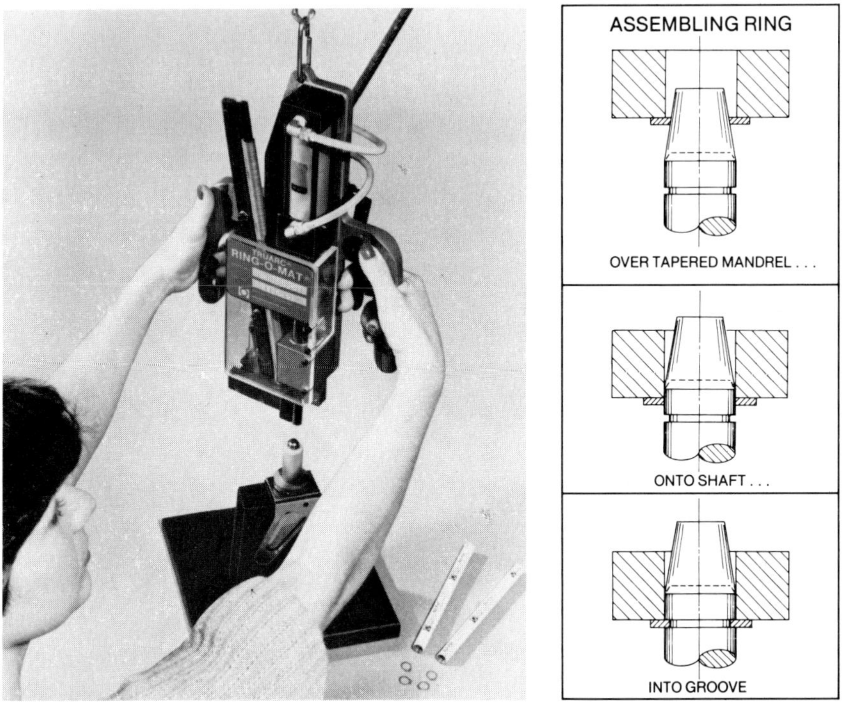

FIG. 11-56 Portable air-operated tool can be hung over fixed work station or mechanized conveyor. To install ring, operator brings tool over work part and squeezes trigger. Air-driven sleeve pushes ring over tapered mandrel, onto shaft, and into groove (inset drawings). When trigger is released, sleeve is withdrawn automatically and next ring is positioned for assembly. Tool is especially suitable for high-volume, mass-production ring installation.

After the tool is loaded with taped ring cartridges, the operator merely places the work part in a nest, brings the tool down over the work, and squeezes the trigger. An air-driven sleeve pushes the ring over a tapered mandrel, spreading the fastener so that it can ride over the work part and into the groove. When the ring is seated, the operator releases the trigger, the sleeve is retracted automatically, and the next ring in the stack is positioned for assembly. Speed of installation is limited only by the speed of the operator.

Many different custom-designed air-driven tools also can be developed for both axial and radial rings. The automatic equipment for axially assembled external rings uses the basic sleeve and tapered mandrel used in the portable tool previously described; internal axial rings require a tapered sleeve and plunger. Various equipment designs for radially assembled rings use applicators and dispensers to assemble rings at high speeds.

12 LOCKING COMPONENTS

ROBERT O. PARMLEY, P.E.
President, Morgan & Parmley, Ltd./Ladysmith, Wisconsin

Generally, mechanical locking is accomplished by the use of one or more standard components assembled in such a way as to secure several parts of a device or structure. This segment of the handbook will discuss those locking components not covered in other sections of this work.

Retaining rings, threaded fasteners, fittings, and related components are discussed in separate sections. Refer to the Contents or the Index for their exact location.

STANDARD PINS

This portion of Sec. 12 will describe the major types and designs of standard pins in present-day industrial use.

The pin, as a fastener, has taken on many shapes and forms during its evolution from a solid, straight cylindrical component.

A pin is a machine component or fastener which secures the position of two or more parts of a device, mechanism, or assembly relative to each other. The following discussion illustrates major pin designs.

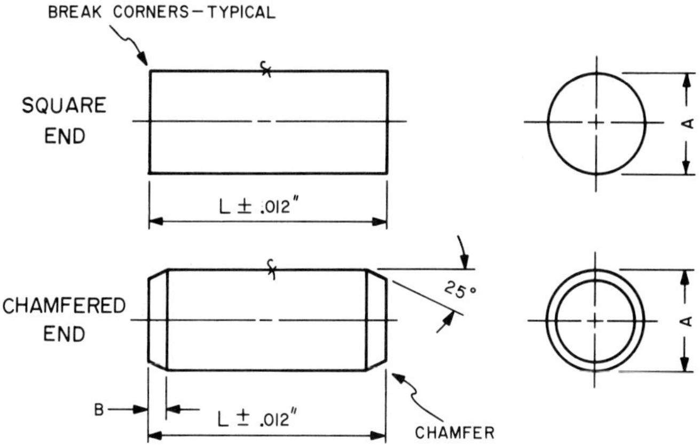

FIG. 12-1 Straight cylindrical pins.

12-1 Straight Pins

The original straight or solid cylindrical pin is illustrated in Fig. 12-1. This pin is usually cut from bar stock. Sizes range from 0.062 in diameter (nominal) to 0.500 in diameter (nominal), with those of ¼ in diameter or less having a tolerance of ±0.010 in and those from ¼ to ½ in diameter having a tolerance of ±0.015 in.

12-2 Dowel Pins

Dowel pins are a refinement of straight cylindrical pins and are generally used in jigs and fixtures. They are fabricated from commercial bar stock.

Dowel pins may be hardened or soft. Hardened dowel pins have a bullet nose on the lead end, while soft dowel pins generally are chamfered on both ends. See Fig. 12-2.

12-3 Tapered Pins

Standard tapered pins are usually employed to transmit light torques or to accurately position components. They are designed and sized to be installed by drive fit.

Standard tapered pin diameters at the "large" end range from 7/0 (0.0625 in) to 10 (0.7060 in), with sizes from 11 (0.8600 in) to 14 (1.523 in) being special. To find the diameter of the "small" end of a taper pin, multiply the pin length in inches by 0.02083 and subtract the result from the large-end diameter. The taper is ¼ in per foot. See Fig. 12-3. Refer to ANS B5.20-1958 for complete tables.

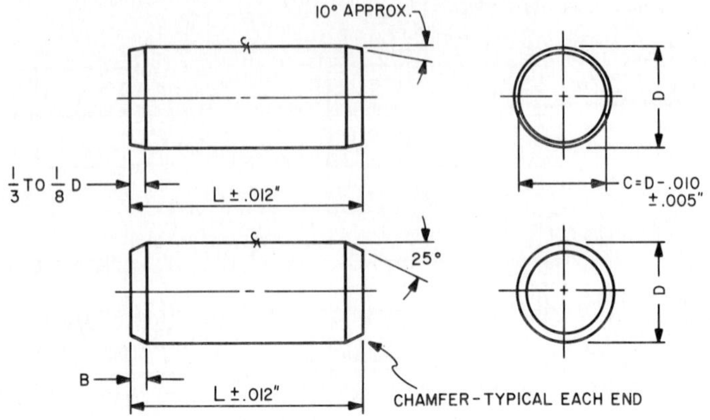

FIG. 12-2 Dowel pins (hardened and soft).

12-4 Clevis Pins

Clevis pins use cotter pins to secure their position and lock the total assembly. Standard sizes are listed in ANS B18.8.1-1972, as published by the American Society of Mechanical Engineers. Figure 12-4 illustrates this pin's basic design. Standard sizes range from 3⁄16 to 1 in nominal diameter, with lengths of 0.58 to 2.41 in, respectively.

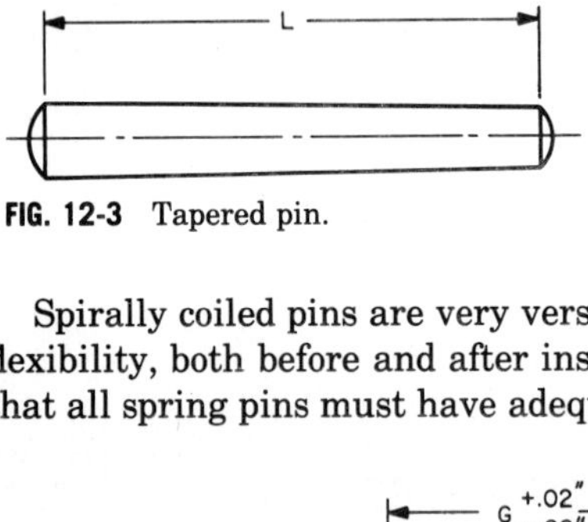

FIG. 12-3 Tapered pin.

12-5 Spirally Coiled Pins

The spirally coiled or wrapped pin, illustrated in Fig. 12-5, has a cross section which is very similar to the geometrical configuration of the Spiral of Archimedes.

Spirally coiled pins are very versatile locking components because of their flexibility, both before and after insertion. It is well known among designers that all spring pins must have adequate flexibility to allow them to be driven

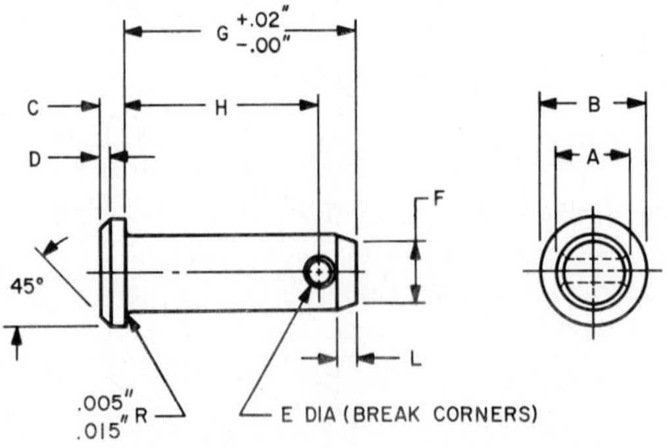

FIG. 12-4 Clevis pin.

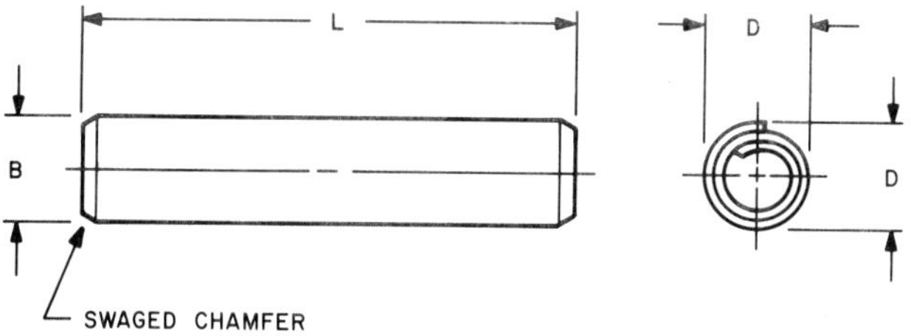

FIG. 12-5 Spirally coiled pin.

or forced into a hole that has a diameter less than their expanded pin diameter. Spirally coiled pins are fabricated not only to be flexible during insertion, but also to remain flexible once they are positioned in a hole. This flexibility allows them to absorb shock or dynamic loads and reduces the possibility of damage to the pin hole.

Additionally, this characteristic has a damping effect on transmission of vibration from one element to another, which protects the total mechanism.

Another feature of these pins is the fact that they can effectively replace other pins, and on special occasions even substitute for rivets, bolts, retainers, stops, and locators.

Standard sizes for spirally coiled pins range from 1⁄32 to 3⁄4 in diameter. There is a wide range of materials and finishes. The user is advised to confer with the manufacturer, prior to final selection, so that the proper material is specified.

The basic design of headed spirally coiled pins is identical, except that they are flanged on their driven end. This flange provides an insertion stop and cleans up the mechanism. They can be successfully removed by withdrawal or reversal from the entry direction.

12-6 Grooved Pins

The grooved pin is an evolution of a straight pin. Three equally spaced, parallel grooves are pressed lengthwise on the exterior surface, as illustrated in Fig. 12-6.

During the fabrication process, pin grooves are formed by a specially designed tool which penetrates below the pin's surface and thereby displaces a predetermined quantity of metal stock. No metal is cut or removed from the basic stock. Metal is displaced to each side of the groove, resulting in a raised crest or flute constituting a slightly expanded diameter, shown as Dx in Fig. 12-6.

As the grooved pin is forced into a drilled hole slightly larger than its nominal diameter, the flutes are significantly recoiled into the preformed grooves. The resistance of the metal forced back into the grooves produces a radial force exerting pressure against the hole wall, thus securing the pin in a positive position. All six raised flutes provide a balanced, controlled radial force, since the three grooves are equally spaced at 120°.

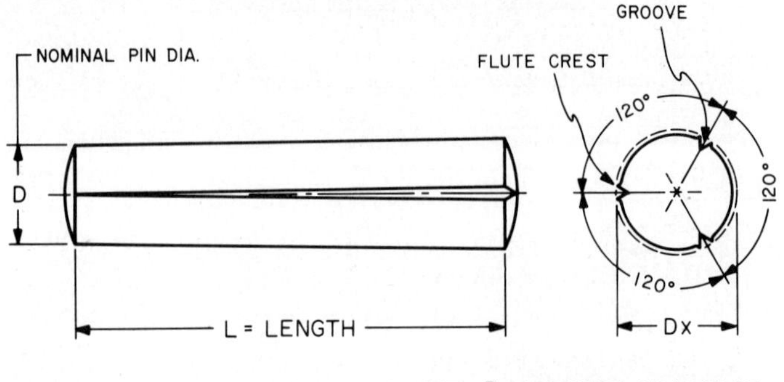

FIG. 12-6 Typical grooved pin. *Dx* (expanded diameter) can be determined accurately only with ring gages.

Pin material is generally cold-drawn low-carbon steel. However, many other materials (such as alloy steel, brass, silicon bronze, and stainless steel) are used, generally for special purposes.

Finishes usually are zinc electroplate, with a thickness of 0.00015 in. Heavier deposits, chromate treatment, and special finishes of nickel, cadmium, brass, and black oxide are sometimes specified.

Drilling tolerances and pin sizing tolerances are very critical, and depend on the application and type of stock. Obtaining tolerance data from the manufacturer is recommended.

Figure 12-7 shows a variety of types of grooved pins which have evolved from the original concept. Refer to ANS B5.20-1958 for a complete tabulation of sizes and design data.

Grooved studs are a recent expansion of the basic grooved pin. They replace conventional screws, rivets, peened brackets, and springs. Consult the manu-

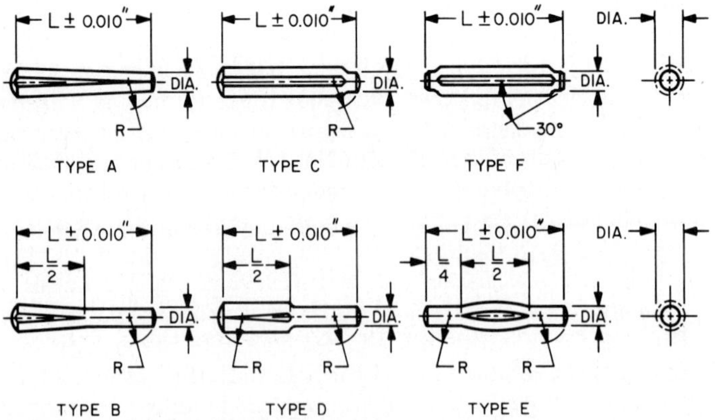

FIG. 12-7 Types of grooved pins.

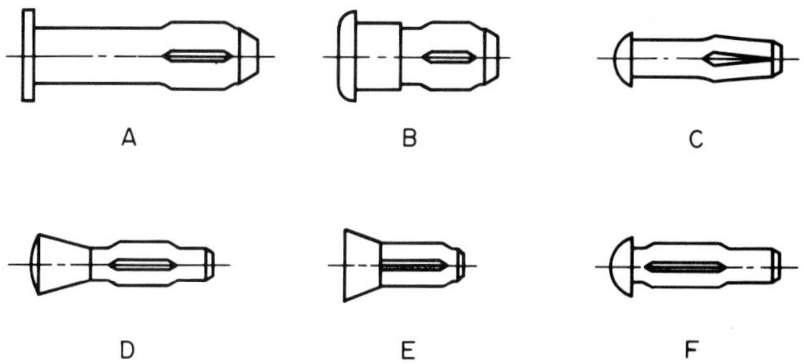

FIG. 12-8 Stud pins. A, flat head; B, flat head with shoulder; C, round head reverse taper; D, stud with conical head; E, countersunk head; F, T head.

facturer for sizing and related design data. Figure 12-8 illustrates six current styles or types of grooved studs.

12-7 Slotted Tubular Pins

Slotted tubular pins or spring pins are readily used as fasteners in a wide range of industrial applications. They are a stock item and generally available. Typical nominal sizes range from 0.062 to 0.500 in diameter. Lengths range from 3⁄4 to 3 in, respectively.

These pins are chamfered, they are heat-treated for optimum toughness, and they have a good resilience and shear strength. Additionally, the slotted tubular pin is self-locking and reusable. It exerts continuous spring pressure against the sides of the hole, providing a positive force that retards loosening or unwanted removal.

Holes drilled to normal production tolerances without reaming are satisfactory to provide a tight fit.

As with most pins, the spring pin is usually made from carbon steel. Of course, other material is often used for special designs to resist attacks from corrosion and provide superior electrical, antimagnetic, and nonsparking characteristics.

Refer to Fig. 12-9 for the basic design of the slotted tubular (spring) pin.

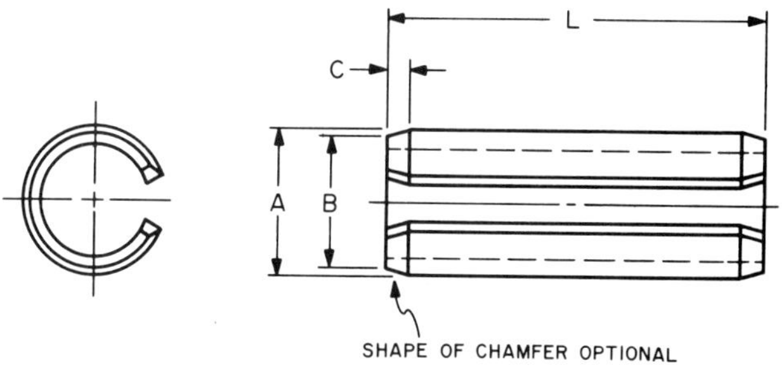

FIG. 12-9 Slotted spring pin.

12-8 Knurled Pins

Die castings and plastic assemblies frequently employ knurled pins. They are a superior fastener for assembly of thin sections. The knurled areas provide a biting contact. These pins adapt to nonuniform configurations. See Fig. 12-10 for some typical designs.

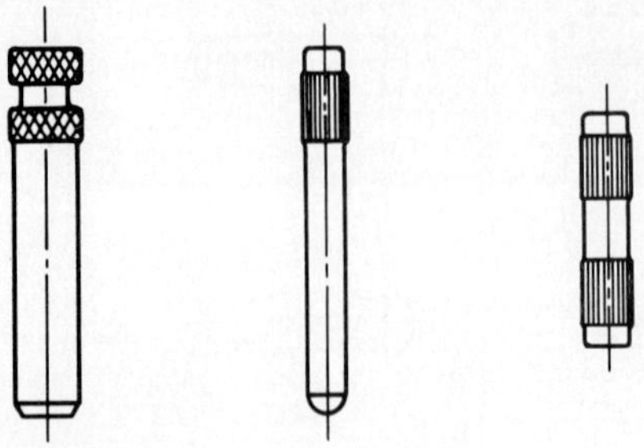

FIG. 12-10 Knurled pin designs.

12-9 Quick-Release Pins

There are many designs for quick-release pins, but all are basically similar in concept. See Fig. 12-11 for a standard design.

The need, in modern machinery, for a safe and quick connect/release feature led to the creation of this mechanical component. Its *push* insert and speedy *pull* release action is the key element. Either a conventional handle or, as shown in the illustration, a ring handle is employed for easy use. Two lock balls are internally spring-loaded for light contact and speedy locking.

12-10 Cotter Pins

Perhaps the most common pin known today is the cotter pin. It is universally used to lock or secure clevises, nuts, and similar mechanical parts. Generally this pin is driven into the shaft hole; the eyelet stops penetration and stabilizes the components, and one leg is spread 90° to prevent removal. Sizes are almost limitless. Refer to Fig. 12-12 for the basic design and a pictorial review of alternative point styles.

Section 14 includes some innovative uses of the cotter pin.

12-11 Wire Pins

Generally this style of pin is custom made from either carbon or stainless steel. It is usually light weight and used where reuse is mandatory. A typical shape is pictured in Fig. 12-13.

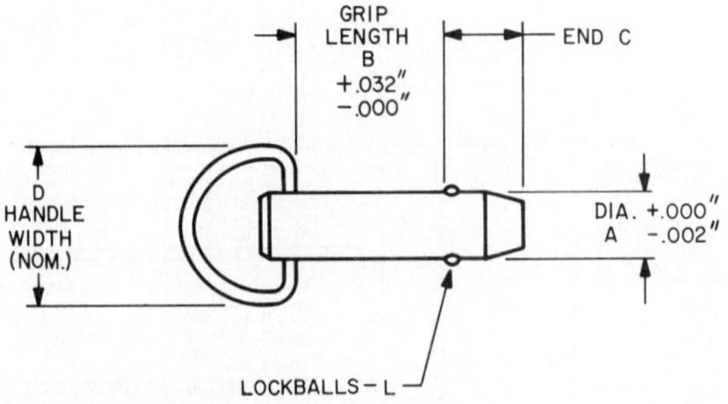

FIG. 12-11 Typical quick-release pin.

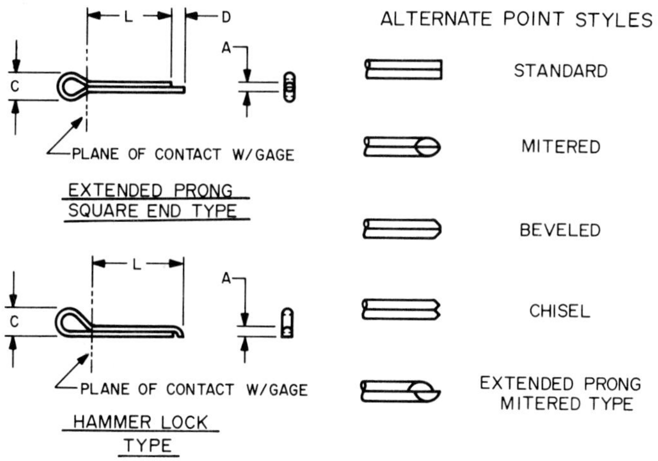

FIG. 12-12 Basic cotter pin design.

KEYS AND KEYSEATS

A key is a removable mechanical component which, when positioned in a machine or assembly, seats into a keyway or keyseat to provide a secure lock between hub and shaft, thus providing torque transfer.

Keys are made from bar stock and key stock, depending on desired tolerances, and must be properly sized relative to shaft diameter. Figure 12-14 reveals accepted keyway proportions. Refer to ANS B17.1-1967 (Re-1973) for a tabulation of nominal shaft diameters versus nominal key and keyway sizes. Square keys are recommended through 6½-in-diameter shafts, and rectangular keys for larger shafts.

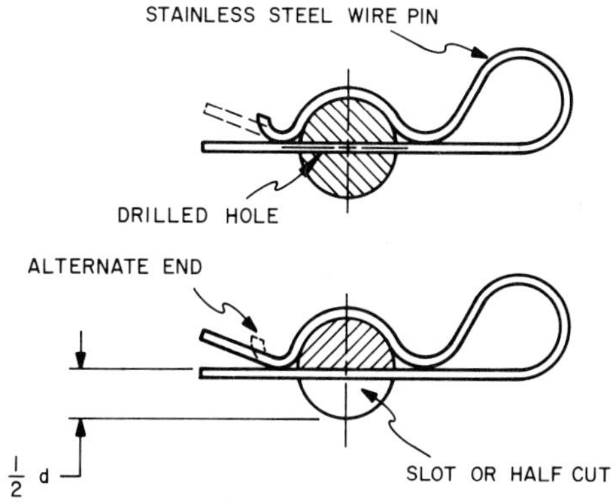

FIG. 12-13 Wire pin.

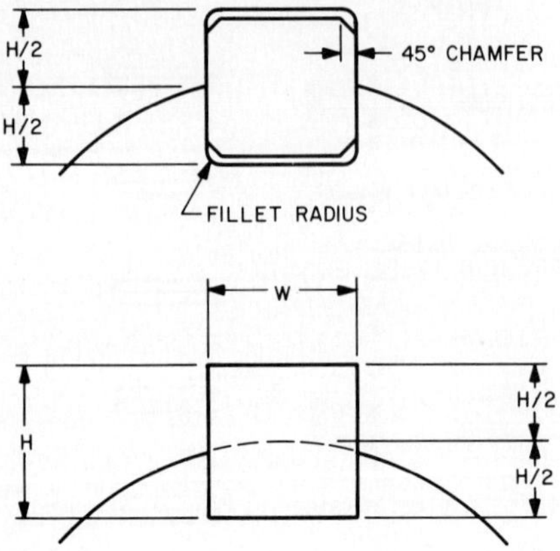

FIG. 12-14 Key size vs. shaft keyway.

12-12 Keyways and Keyseats

Figure 12-15 graphically reveals the proportional formulas of keyways for both hubs and shafts. Typical designs of standard keyseats for hubs and shafts are shown in Fig. 12-16.

12-13 Parallel and Plain Taper Keys

A parallel key is simply a square key cut from bar stock. The plain taper key is also cut from bar stock, but has a taper, either partial or full, as shown in Fig. 12-17.

12-14 Gib-Head Taper Keys

This key is used in applications where disassembly may be difficult. Figure 12-17 also pictures this key's basic configuration.

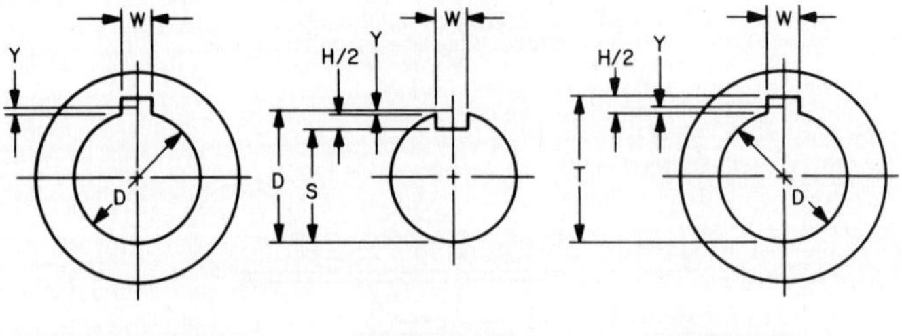

FIG. 12-15 Depth-control formulas for keyways.

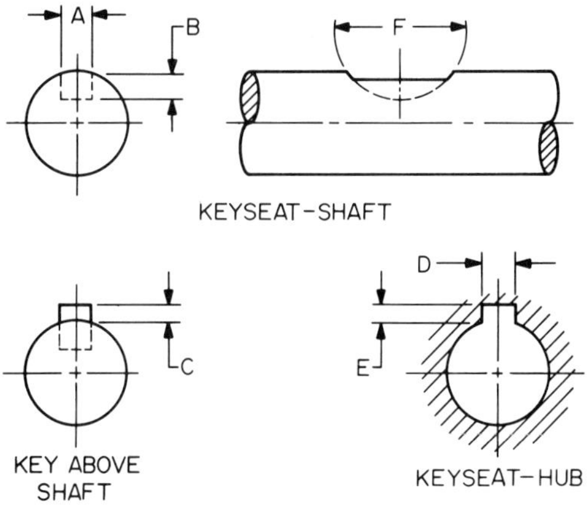

FIG. 12-16 Typical design of keyseats.

12-15 Square Machine Keys

This key is most valuable as a locking component for machines. This square key with chamfered ends and without a head or taper has proven its value as an important mechanical component. Lengths vary from ½ to 3 in in ¼-in steps and from 3 to 4½ in in steps of ½ in. Available widths are ⅛, 3⁄16, ¼, 5⁄16, ⅜, 7⁄16, and ½ in; the keys are, of course, square. They are manufactured plain or plated, or fabricated from stainless steel. See Fig. 12-18 for the typical design.

12-16 Woodruff Keys

These keys are widely used in machine assembly and in the automotive industry. Construction equipment and agricultural machinery also employ woodruff

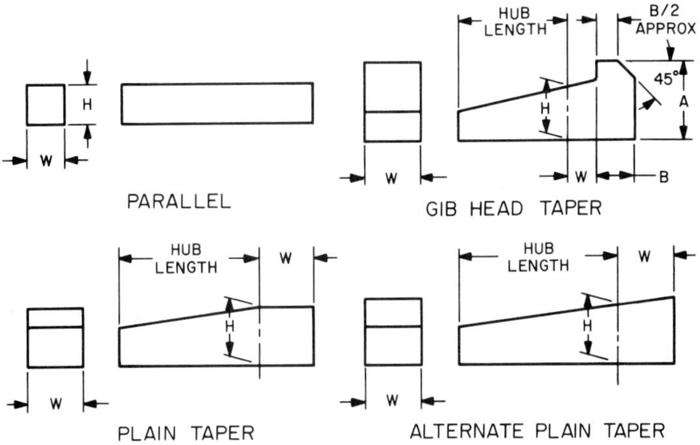

FIG. 12-17 Design of plain and gib-head taper keys.

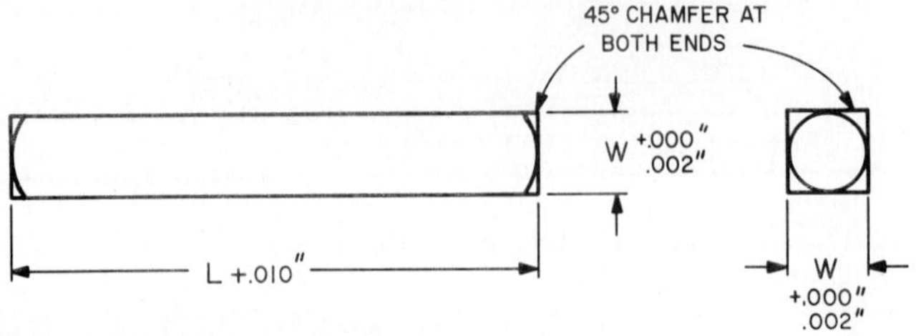

FIG. 12-18 Square machine key.

keys. Standard designs of full radius and flat bottom types are detailed in Fig. 12-19. Nominal standard sizes range from 1⁄16 × 1⁄4 in to 3⁄4 in to 3½ in, as tabulated in ANS B17.2-1967 (Re-1972).

12-17 Splines

When it becomes necessary to avoid an extra component, splines are used. Spline shafts are very efficient for transmitting torque.

Two types of spline styles are standardized by the SAE square and involute. Refer to Fig. 12-20 for basic spline layouts: 4, 6, 10, and 16 parallel side splines.

Since splines are strictly speaking not components, but rather a configuration of a shaft and/or mating member, we will not go into a detailed discussion. However, their torque capacity can be calculated from the following formula:

$$T = 1000NrHL \tag{12-1}$$

where N = number of splines

r = mean radial distance from center of hole to center of spline, in

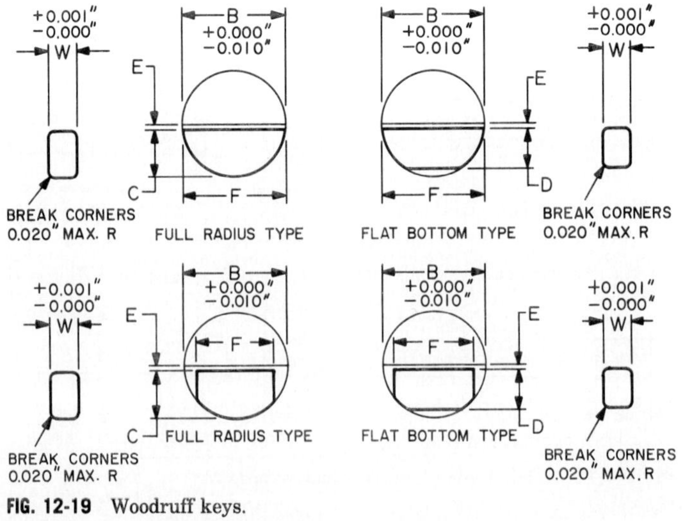

FIG. 12-19 Woodruff keys.

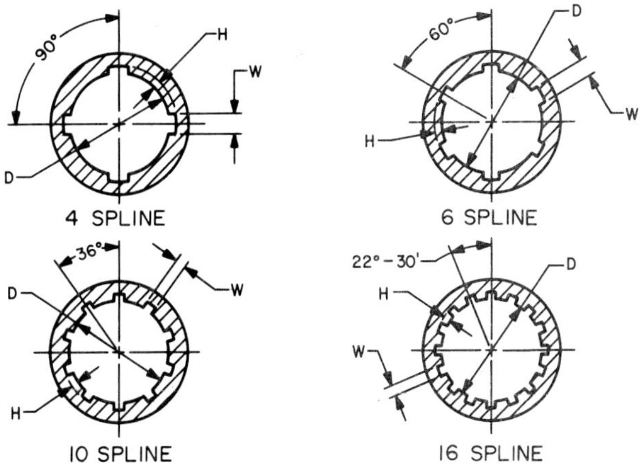

FIG. 12-20 Parallel side splines.

H = depth of spline
L = length of spline bearing surface, in

The formula is based on 1000 lb/in^2 pressure on the spline side.

LOCK WASHERS

Lock washers provide greater bolt tension per unit of applied force (or torque), protection from component losses resulting from vibration, and a uniform load distribution.

Lock washers are made from a variety of materials, including carbon steel, corrosion-resistant steel, aluminum-zinc alloy, phosphor bronze, silicon bronze, and K-Monel.

12-18 Standard Lock Washers

There is a wide range of lock washer designs which have become standard in the industry. Major types are discussed in the following text.

12-19 Helical-Spring Lock Washers

This is certainly the most widely used type of lock washer and one that is easily recognized by the public.

Standard sizes range from no. 2 (0.086 in A or ID) to 3 in (A or ID). These are nominal dimensions, they are tabulated in ANS B-18.21.1-1972. See Fig. 12-21.

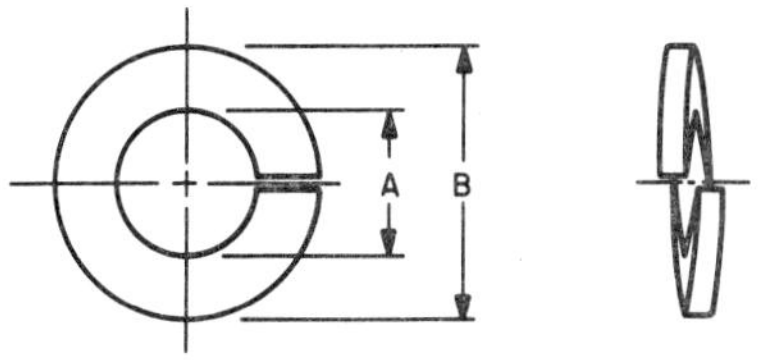

FIG. 12-21 Helical-spring lock washer.

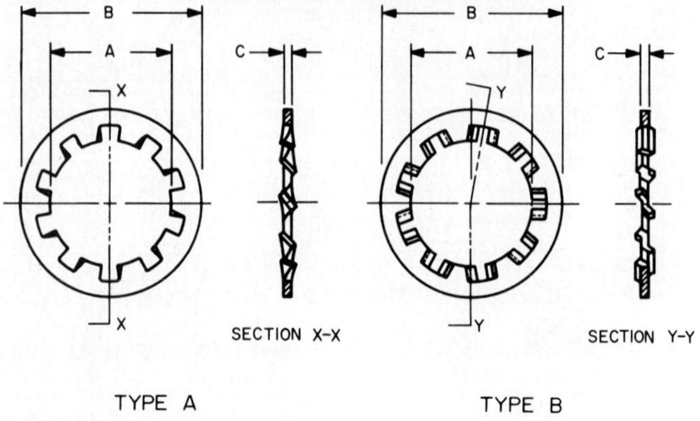

FIG. 12-22 Internal-tooth lock washers.

12-20 Internal-Tooth Lock Washers

Figure 12-22 illustrates the design of both type A and type B internal-tooth lock washers. Nominal ID standard dimensions commence with no. 2 (0.086 in) size and increase in steps to 1¼-in ID size. Heavy-duty washers begin at ¼ in ID and top out at ⅞ in ID. Note in Fig. 12-22 that ID is depicted as dimension *A*.

12-21 External-Tooth Lock Washers

These lock washers have teeth on the outside circumference of the disk. In assembly, these washers lock the mating elements together, forming a mechanical lock. Figure 12-23 shows two types, A and B.

ANS B18.21-1-1972 tabulates basic dimensions. Nominal washer sizes commence with no. 3 (0.099 in, dimension *A*) and range up to 1 in ID (dimension *A*).

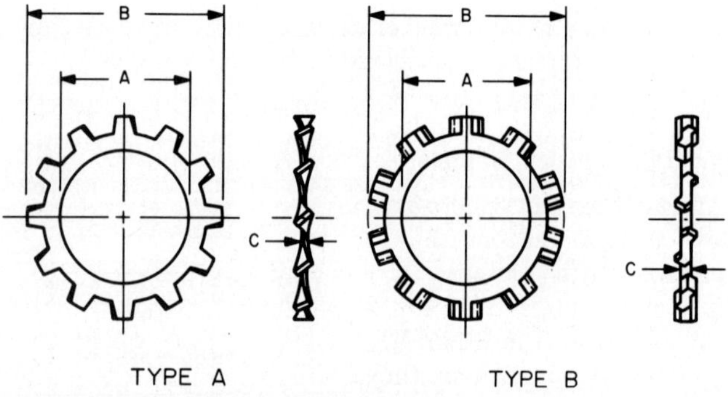

FIG. 12-23 External-tooth lock washers.

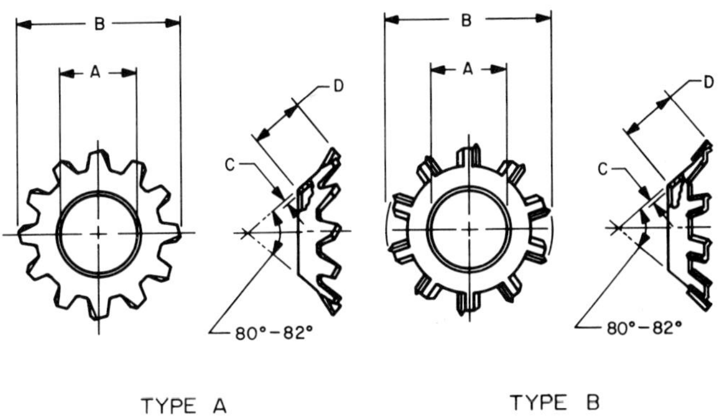

FIG. 12-24 Countersunk external-tooth lock washers.

12-22 Countersunk External-Tooth Lock Washers

The dished or countersunk configuration results in increased locking strength in final assembly. Figure 12-24 pictures standardized design, given in ANS B18.21.1-1972. Inside diameter (*A*) ranges from no. 4 (0.112 in) size through ½ in.

12-23 Internal-External-Tooth Lock Washers

This washer combines two features into one design, providing an exceptional double holding power. Figure 12-25 depicts types A and B. Nominal sizes (*A* dimension) range from no. 4 (0.112 in) through ⅝ in.

12-24 Beveled Washers

Beveled or tapered washers are generally used in the assembly of structural steel to compensate for beam and channel flange slopes.

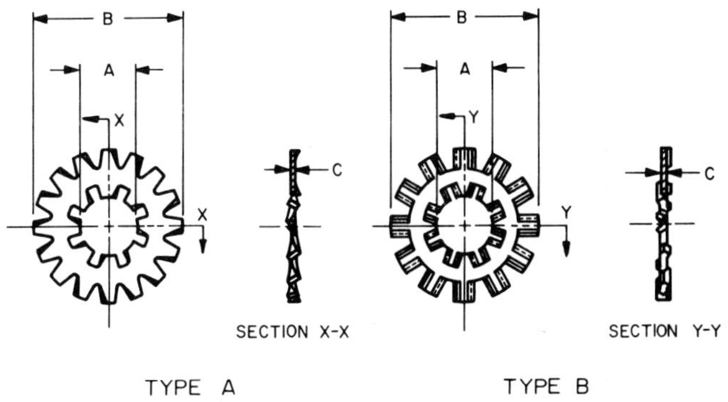

FIG. 12-25 Internal-external-tooth lock washers.

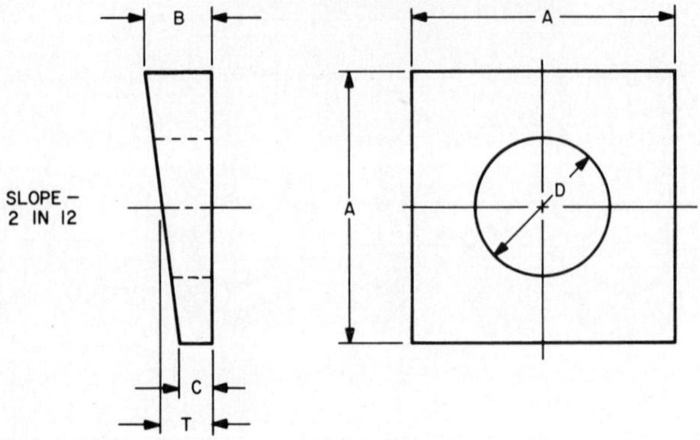

FIG. 12-26 Design of beveled washer.

There are two basic types: A (malleable iron) and B (steel). ANS B27.4-1967 tabulates the standard sizes. See Fig. 12-26.

12-25 Belleville Washers

These dished components are also known as conical-spring washers. They are manufactured in a wide range of sizes and materials. Refer to Fig. 12-27.

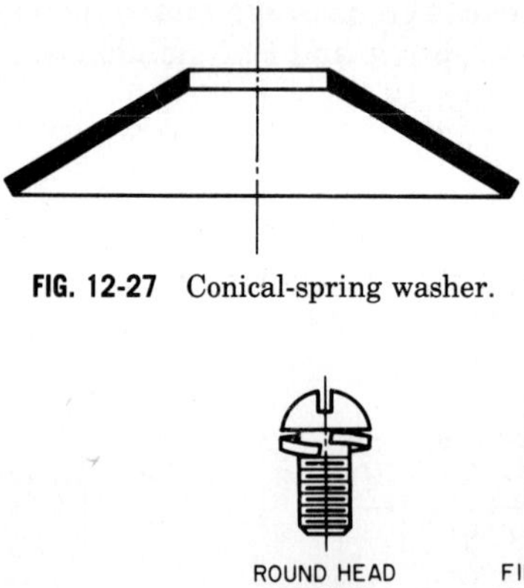

FIG. 12-27 Conical-spring washer.

SEMS

Sems are assemblies of standard screws and their captive lock washers. These combinations are numerous, employing a wide range of screws and the gamut of lock washers. Figure 12-28 illustrates six combinations using a helical-spring lock washer.

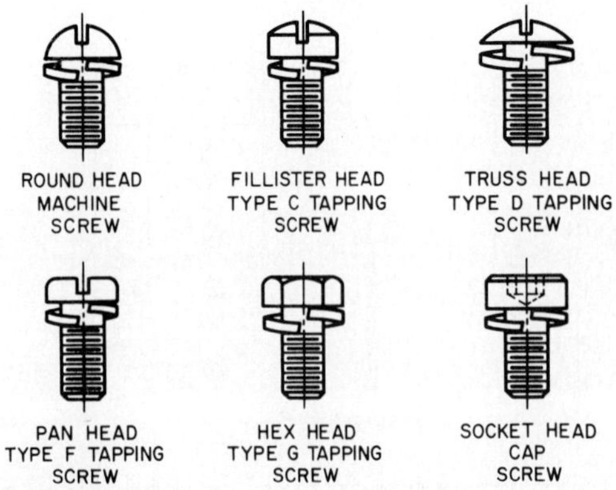

FIG. 12-28 Sems.

12-26 Screw and Washer Assemblies

As previously stated, sems are varied. Refer to ANS B18.13-1965 for a pictorial review of representative examples. Occasionally flat washers are used, but generally lock washers or conical-spring washers are employed.

RIVETS

Riveting is one of the most widely used methods of fastening and joining. Its growth is significantly due to the modern refinement of high-speed riveting machines and techniques. Economics also plays a part in the expanding use of rivets, especially in manufacturing. Rivets are themselves much lower in cost than comparable fastening components. Additionally, unit labor costs for installing rivets are low when automatic-feeding high-speed machines are used.

Since space is limited, only a very brief discussion will be presented. Refer to Sec. 13 of *Standard Handbook of Fastening and Joining* for an in-depth presentation of riveting technology.

12-27 Small Rivets

Small rivets are classified into many categories, but the two most important are solid and tubular types. Both solid and tubular rivets are employed when two or more parts are to be fastened together. In each case, the rivet head holds a position on one side, and the opposite end is held in place by the final formed shape of the shank, as illustrated in Fig. 12-29. This figure shows a cross-sectional comparison of both solid and tubular rivets before and after setting.

12-28 Compression Rivets

Compression rivets are formed of two members, a solid or blank (male) rivet and a deep-drilled tubular (female) component. Figure 12-30 depicts the typical form of male and female compression rivets.

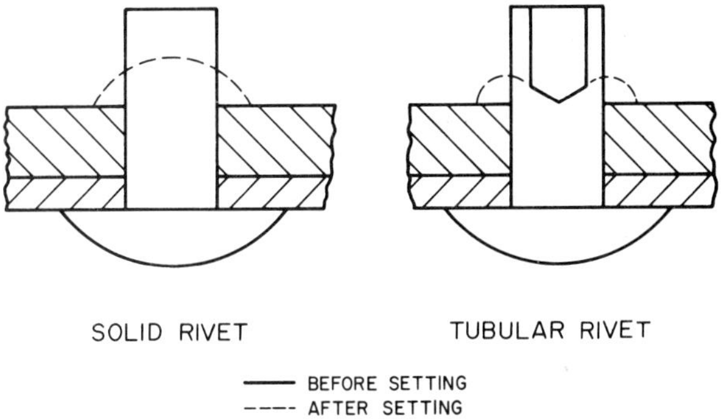

FIG. 12-29 Comparison of cross-sectional shapes of solid and tubular rivets before and after setting.

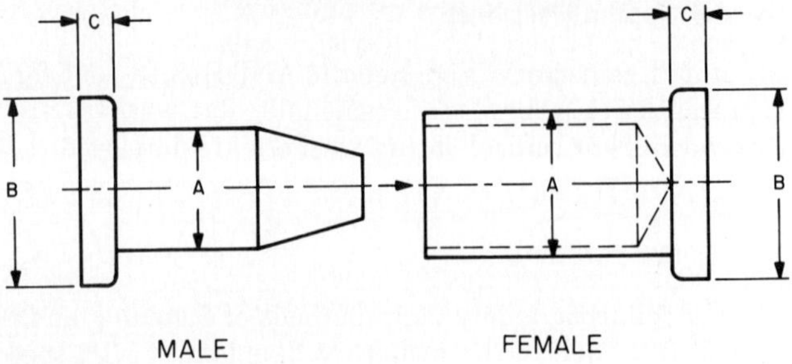

FIG. 12-30 Typical compression rivets.

The diameters of the solid shank (A) and the drilled hole are designed so that a compression or press fit will be produced when the parts mate during assembly.

Full tubular (Fig. 12-31) and bifurcated (Fig. 12-32) rivets are often used in combination with a cap, which acts as a washer to prevent the clinch area from snagging soft materials.

12-29 Cold-Headed Parts

Cold-headed parts can be either tubular or solid and made in a variety of special shapes. These special designs include fluted or knurled shanks, ornamental heads, and shapes used for secondary functions in addition to their primary function as a fastener.

12-30 Standard Sizes

Standard rivets are available from stock. The user should specify a standard size to avoid delay and expense. Slight alterations in dimensions or tolerances seem insignificant to the uninitiated, but they often require completely new dies, resulting in a more expensive rivet.

Shank diameters of standard rivets can be maintained within ±0.001 in on rivets to 1⁄16-in diameter and ±0.003 in on rivets 1⁄4 in and larger. Tolerances

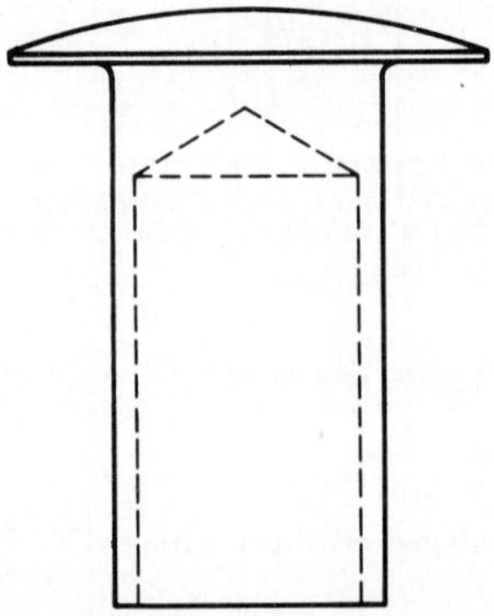

FIG. 12-31 Full tubular rivet.

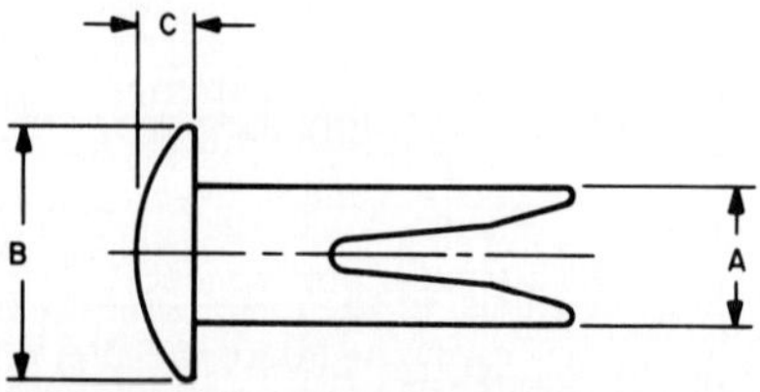

FIG. 12-32 Bifurcated (split) rivet.

of ±0.005 in on head diameter and shank length generally are recommended on most rivet sizes.

Standard sizes of semitubular, bifurcated, and compression rivets are available from major rivet manufacturers.

12-31 Solid Rivet System

The solid rivet, when installed, provides a permanent fastener. The requirement that there be access and adequate clearance on *both* sides of the assembly is extremely critical. Driving tools and installation equipment must be properly positioned to upset the rivet and obtain a secure connection.

Pneumatic-hammer riveting requires two-person teams, one on each side of the assembly. On large-size assemblies, squeeze or pressure riveting, employing large production-type machines, may be used. Automatic riveting machines are extensively used. These automatic machines are capable of being programmed to automatically drill the assembly, insert the rivet, drive the rivet, and shave the rivet (if required), then complete the operation by upsetting the head.

High-speed automatic solid riveting systems ensure top-level uniformity at less cost.

12-32 Blind Rivet System

Blind rivets can be inserted and completely installed in a joint from only one side of an assembly. The back, or blind, side of this type of rivet is mechanically expanded to form an upset head, producing a permanently installed fastener.

The ability of the blind rivet to be installed from one side has led to its use in a wide range of fastening applications. Blind fastening is used in industry in situations where crude and costly methods would otherwise have to be used to achieve a permanent connection.

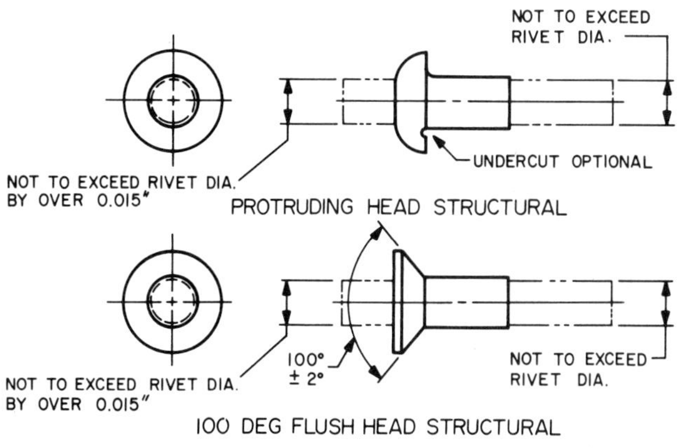

FIG. 12-33 Self-plugging pull-mandrel blind rivets.

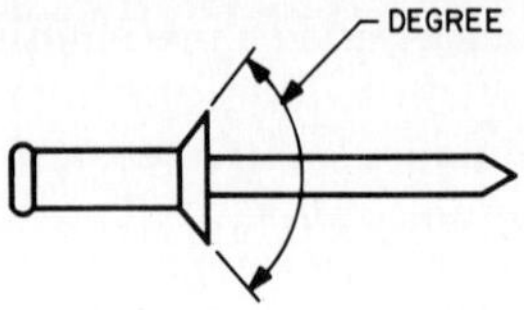

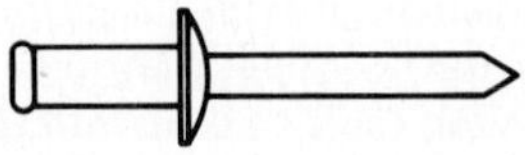

FIG. 12-34 Break-mandrel blind rivets.

Modern designs often require the use of thin structural sheets, which has posed problems. Damage and coining results if solid riveting is used to join these thin components. Using blind rivets often reduces this condition.

12-33 Types of Blind Rivets

Mechanically expanded blind rivets are the type most commonly used by industry. The pull-type blind rivet and the drive-pin-type blind rivet are the basic designs we will briefly discuss.

In the pull-type design, the rivet body normally uses a through hole. The mating mandrel is positioned in the rivet body, which includes a preformed head with the mandrel extending above the rivet head. A mating tool is used to drive the blind rivet. This tool grips and pulls the mandrel while reacting, to hold the rivet in the assembly. This action of drawing the mandrel through the rivet's body expands the rivet, which fills the joint hole, and expands the portion of the rivet sleeve extending on the back or blind side of the assembly, thus forming an upset or blind head.

The pull-type blind rivet is available in a self-plugging design and a pull-through type. In the first, a segment of the mandrel is retained permanently in the rivet body, while the pull-through blind rivet's mandrel is completely drawn through the rivet's body following expansion.

Figure 12-33 illustrates two styles of self-plugging pull mandrel blind rivets, and Fig. 12-34 pictures two styles of break mandrel blind rivets.

Drive-pin rivet design includes a partial hole in the rivet's body and a mating protruding pin which is positioned in the hole. The rivet's body is slotted, and the pin is driven into the rivet, resulting in a flaring of the rivet and the blind-side upset.

REFERENCE

Parmley, R. O.: *Standard Handbook of Fastening and Joining,* McGraw-Hill, New York, 1977.

13
METRICS AND CONVERSION DATA

ROBERT O. PARMLEY, P.E.
President, Morgan & Parmley, Ltd./Ladysmith, Wisconsin

13-1 History

Records of measurement date back to approximately 5000 B.C. to ancient civilizations located on the Chaldean plains and areas adjacent to the Nile River. These fledgling attempts at standardization of dimensions involved length units and were directly associated with the human body, specifically the thumb, hand, foot, forearm, and pace. Commencing about 2500 B.C., balances were used to weigh gold, as shown on records buried in ancient Egyptian tombs.

The "cubit," a measurement often mentioned in the Bible, is the distance from the outstretched tip of a person's middle finger to the point of the elbow. It had become the principal length unit by 4000 B.C. and was somewhat standardized at what is now approximately 460 mm.

The "span" is approximately one-half a cubit; it is measured from the tips or points of the outstretched thumb and little finger.

As civilization progressed, each country introduced its own standard for weights and measures. English history tells us that by A.D. 1500, the English used the following set of conversions:

1 inch = 3 barley corns
1 foot = 12 inches
1 yard = 3 feet
1 span = 9 inches
1 ell = 5 spans
1 pace = 5 feet
1 furlong = 125 paces
1 rod = 5½ yards
1 statute mile = 8 furlongs
1 league = 12 furlongs

13-2 Metric System

As world trade became more common, there was an ever-increasing need for a universal standard of measurement based on a logical, scientific system.

TABLE 13-1 Basic SI Units*

Quantity	Unit
Length	meter (m)
Mass	kilogram (kg)
Time	second (s)
Electric current	ampere (A)
Temperature (thermodynamic)	kelvin (K)
Amount of substance	mole (mol)
Luminous intensity	candela (cd)

* From *Metrication Manual* by Tyler G. Hicks. © 1972 McGraw-Hill Book Company. Used with permission of the publisher.

TABLE 13-2 Prefixes for SI Units*

Multiple and submultiple	Prefix	Symbol
1,000,000,000,000 = 10^{12}	tera	T
1,000,000,000 = 10^{9}	giga	G
1,000,000 = 10^{6}	mega	M
1,000 = 10^{3}	kilo	k
100 = 10^{2}	hecto	h
10 = 10	deka	da
0.1 = 10^{-1}	deci	d
0.01 = 10^{-2}	centi	c
0.001 = 10^{-3}	milli	m
0.000 001 = 10^{-6}	micro	μ
0.000 000 001 = 10^{-9}	nano	n
0.000 000 000 001 = 10^{-12}	pico	p
0.000 000 000 000 001 = 10^{-15}	femto	f
0.000 000 000 000 000 001 = 10^{-18}	atto	a

* From *Metrication Manual* by Tyler G. Hicks. © 1972 McGraw-Hill Book Company. Used with permission of the publisher.

A commission of French scientists developed the metric system, and France adopted it as their legal system of weights and measures in the year A.D. 1799. Several revisions have occurred since that time. In 1960, the present form, known as the "Système International d'Unités" (International System of Units), or SI, was approved.

SI metric has seven basic units of measure and two supplemental units. All other units of measure can be derived from these basic units. (See Tables 13-1 and 13-2.)

13-3 Simpler Language

The metric system is a much simpler language; however, conversion from the common U.S. Customary System is very expensive and time-consuming for American engineering disciplines.

The seven basic SI units previously mentioned are relatively easy to use and convert to, or from, common English units. However, in the case of derived units, those which are derived from one or more basic SI units, there can be some difficulty. Table 13-3 includes the most common derived units used in SI.

13-4 Conversion Factors

Table 13-4 gives the definitions of various units of measure that are exact numerical multiples of coherent SI units, and provides multiplication factors for converting numbers and miscellaneous units to corresponding new numbers and SI units.

The first two digits of each numerical entry represent a power of 10. An

TABLE 13-3 Derived Units of the International System*

Quantity	Name of unit	Unit symbol or abbreviation, where differing from basic form	Unit expressed in terms of basic or supplementary units†
Area	square meter		m^2
Volume	cubic meter		m^3
Frequency	hertz, cycle per second‡	Hz	s^{-1}
Density	kilogram per cubic meter		kg/m^3
Velocity	meter per second		m/s
Angular velocity	radian per second		rad/s
Acceleration	meter per second squared		m/s^2
Angular acceleration	radian per second squared		rad/s^2
Volumetric flow rate	cubic meter per second		m^3/s
Force	newton	N	$kg \cdot m/s^2$
Surface tension	newton per meter, joule per square meter	N/m, J/m^2	kg/s^2
Pressure	newton per square meter, pascal‡	N/m^2, Pa‡	$kg/m \cdot s^2$
Viscosity, dynamic	newton-second per square meter, poiseuille‡	$N \cdot s/m^2$, Pl‡	$kg/m \cdot s$
Viscosity, kinematic	meter squared per second		m^2/s
Work, torque, energy, quantity of heat	joule, newton-meter, watt-second	J, N·m, W·s	$kg \cdot m^2/s^2$

Power, heat flux	watt, joule per second	W, J/s	$kg\cdot m^2/s^3$
Heat flux density	watt per square meter	W/m^2	kg/s^3
Volumetric heat release rate	watt per cubic meter	W/m^3	$kg/m\cdot s^3$
Heat transfer coefficient	watt per square meter-degree	$W/m^2\cdot deg$	$kg/s^3\cdot deg$
Heat capacity (specific)	joule per kilogram-degree	$J/kg\cdot deg$	$m^2/s^2\cdot deg$
Capacity rate	watt per degree	W/deg	$kg\cdot m^2/s^3\cdot deg$
Thermal conductivity	watt per meter-degree	$W/m\cdot deg$, $J\cdot m/s\cdot m^2\cdot deg$	$kg\cdot m/s^3\cdot deg$
Quantity of electricity	coulomb	C	$A\cdot s$
Electromotive force	volt	V,W/A	$kg\cdot m^2/A\cdot s^3$
Electric field strength	volt per meter		V/m
Electric resistance	ohm	Ω, V/A	$kg\cdot m^2/A^2\cdot s^3$
Electric conductivity	ampere per volt-meter	$A/V\cdot m$	$A^2s^3/kg\cdot m^3$
Electric capacitance	farad	$F, A\cdot s/V$	$A^3s^1/kg\cdot m^2$
Magnetic flux	weber	$Wb, V\cdot s$	$kg\cdot m^2/A\cdot s^2$
Inductance	henry	$H, V\cdot s/A$	$kg\cdot m^2/A^2s^2$
Magnetic permeability	henry per meter	H/m	$kg\cdot m/A^2s^2$
Magnetic flux density	tesla, weber per square meter	T, Wb/m^2	$kg/A\cdot s^2$
Magnetic field strength	ampere per meter		A/m
Magnetomotive force	ampere		A
Luminous flux	lumen	lm	cd sr
Luminance	candela per square meter		cd/m^2
Illumination	lux, lumen per square meter	lx, lm/m^2	$cd\cdot sr/m^2$

* From *Metrication Manual* by Tyler G. Hicks. © 1972 McGraw-Hill Book Company. Used with permission of the publisher.

† Supplementary units are plane angle, radian (rad), solid angle, steradian (sr).

‡ Not used in all countries.

TABLE 13-4 Conversion Factors as Extracted Multiples of SI Units†

To convert from	To	Multiply by
abampere	ampere	+01 1.00*
abcoulomb	coulomb	+01 1.00*
abfarad	farad	+09 1.00*
abhenry	henry	−09 1.00*
abmho	mho	+09 1.00*
abohm	ohm	−09 1.00*
abvolt	volt	−08 1.00*
acre	$meter^2$	+03 4.046 856 422 4*
ampere (international of 1948)	ampere	−01 9.998 35
angstrom	meter	−10 1.00*
are	$meter^2$	+02 1.00*
astronomical unit	meter	+11 1.495 978 9
atmosphere	$newton/meter^2$	+05 1.013 25*
bar	$newton/meter^2$	+05 1.00*
barn	$meter^2$	−28 1.00*
barrel (petroleum, 42 gallons)	$meter^3$	−01 1.589 873
barye	$newton/meter^2$	−01 1.00*
British thermal unit (ISO/TC 12)	joule	+03 1.055 06
British thermal unit (International Steam Table)	joule	+03 1.055 04
British thermal unit (mean)	joule	+03 1.055 87
British thermal unit (thermochemical)	joule	+03 1.054 350 264 488
British thermal unit (39°F)	joule	+03 1.059 67
British thermal unit (60°F)	joule	+03 1.054 68
bushel (U.S.)	$meter^3$	−02 3.523 907 016 688*
cable	meter	+02 2.194 56*
caliber	meter	−04 2.54*
calorie (International Steam Table)	joule	+00 4.1868
calorie (mean)	joule	+00 4.190 02
calorie (thermochemical)	joule	+00 4.184*
calorie (15°C)	joule	+00 4.185 80
calorie (20°C)	joule	+00 4.181 90
calorie (kilogram, International Steam Table)	joule	+03 4.1868
calorie (kilogram, mean)	joule	+03 4.190 02
calorie (kilogram, thermochemical)	joule	+03 4.184*
carat (metric)	kilogram	−04 2.00*
Celsius (temperature)	kelvin	$t_K = t_C = 273.15$
centimeter of mercury (0°C)	$newton/meter^2$	+03 1.333 22
centimeter of water (4°C)	$newton/meter^2$	+01 9.806 38
chain (engineer or ramden)	meter	+01 3.048*
chain (surveyor or gunter)	meter	+01 2.011 68*
circular mil	$meter^2$	−10 5.067 074 8
cord	$meter^3$	+00 3.624 556 3
coulomb (international of 1948)	coulomb	−01 9.998 35
cubit	meter	−01 4.572*

† From *Metrication Manual* by Tyler G. Hicks. © 1972 McGraw-Hill Book Company. Used with permission of the publisher.

TABLE 13-4 Conversion Factors as Extracted Multiples of SI Units† (*continued*)

To convert from	To	Multiply by
cup	meter^3	−04 2.365 882 365*
curie	disintegration/second	+10 3.70*
day (mean solar)	second (mean solar)	+04 8.64*
day (sidereal)	second (mean solar)	+04 8.616 409 0
degree (angle)	radian	−02 1.745 329 251 994 3
denier (international)	kilogram/meter	−07 1.00*
dram (avoirdupois)	kilogram	−03 1.771 845 195 312 5*
dram (troy or apothecary)	kilogram	−03 3.887 934 6*
dram (U.S. fluid)	meter^3	−06 3.696 691 195 312 5*
dyne	newton	−05 1.00*
electron-volt	joule	−19 1.602 10
erg	joule	−07 1.00*
Fahrenheit (temperature)	kelvin	$t_K = (5/9)(t_F + 459.67)$
Fahrenheit (temperature)	Celsius	$t_C = (5/9)(t_F - 32)$
farad (international of 1948)	farad	−01 9.995 05
faraday (based on carbon 12)	coulomb	+04 9.648 70
faraday (chemical)	coulomb	+04 9.649 57
faraday (physical)	coulomb	+04 9.652 19
fathom	meter	+00 1.828 8*
fermi (femtometer)	meter	−15 1.00*
fluid ounce (U.S.)	meter^3	−05 2.957 352 956 25*
foot	meter	−01 3.048*
foot (U.S. survey)	meter	+00 1200/3937*
foot (U.S. survey)	meter	−01 3.048 006 096
foot of water (39.2°F)	newton/meter^2	+03 2.988 98
foot-candle	lumen/meter^2	+01 1.076 391 0
foot-lambert	candela/meter^2	+00 3.426 259
furlong	meter	+02 2.011 68*
gal (galileo)	meter/second^2	−02 1.00*
gallon (U.K. liquid)	meter^3	−03 4.546 087
gallon (U.S. dry)	meter^3	−03 4.404 883 770 86*
gallon (U.S. liquid)	meter^3	−03 3.785 411 784*
gamma	tesla	−09 1.00*
gauss	tesla	−04 1.00*
gilbert	ampere-turn	−01 7.957 747 2
gill (U.K.)	meter^3	−04 1.420 652
gill (U.S.)	meter^3	−04 1.182 941 2
grad	degree (angular)	−01 9.00*
grad	radian	−02 1.570 796 3
grain	kilogram	−05 6.479 891*
gram	kilogram	−03 1.00*
hand	meter	−01 1.016*
hectare	meter^2	+04 1.00*
henry (international of 1948)	henry	+00 1.000 495
hogshead (U.S.)	meter^3	−01 2.384 809 423 92*
horsepower (550 foot lbf/second)	watt	+02 7.456 998 7

TABLE 13-4 Conversion Factors as Extracted Multiples of SI Units† (*continued*)

To convert from	To	Multiply by
horsepower (boiler)	watt	+03 9.809 50
horsepower (electric)	watt	+02 7.46*
horsepower (metric)	watt	+02 7.354 99
horsepower (U.K.)	watt	+02 7.457
horsepower (water)	watt	+02 7.460 43
hour (mean solar)	second (mean solar)	+03 3.60*
hour (sidereal)	second (mean solar)	+03 3.590 170 4
hundredweight (long)	kilogram	+01 5.080 234 544*
hundredweight (short)	kilogram	+01 4.535 923 7*
inch	meter	−02 2.54*
inch of mercury (32°F)	newton/meter^2	+03 3.386 389
inch of mercury (60°F)	newton/meter^2	+03 3.376 85
inch of water (39.2°F)	newton/meter^2	+02 2.490 82
inch of water (60°F)	newton/meter^2	+02 2.4884
joule (international of 1948)	joule	+00 1.000 165
kayser	1/meter	+02 1.00*
kilocalorie (International Steam Table)	joule	+03 4.186 74
kilocalorie (mean)	joule	+03 4.190 02
kilocalorie (thermochemical)	joule	+03 4.184*
kilogram mass	kilogram	+00 1.00*
kilogram force (kgf)	newton	+00 9.806 65*
kilopond force	newton	+00 9.806 65*
kip	newton	+03 4.448 221 615 260 5*
knot (international)	meter/second	−01 5.144 444 444
lambert	candela/meter^2	+04 $1/\pi$*
lambert	candela/meter^2	+03 3.183 098 8
langley	joule/meter^2	+04 4.184*
lbf (pound force, avoirdupois)	newton	+00 4.448 221 615 260 5*
lbm (pound mass, avoirdupois)	kilogram	−01 4.535 923 7*
league (British nautical)	meter	+03 5.559 552*
league (international nautical)	meter	+03 5.556*
league (statute)	meter	+03 4.828 032*
light year	meter	+15 9.460 55
link (engineer or ramden)	meter	−01 3.048*
link (surveyor or gunter)	meter	−01 2.011 68*
liter	meter^3	−03 1.00*
lux	lumen/meter^2	+00 1.00*
maxwell	weber	−08 1.00*
meter	wavelengths Kr 86	+06 1.650 763 73*
micron	meter	−06 1.00*
mil	meter	−05 2.54*
mile (U.S. statute)	meter	+03 1.609 344*

TABLE 13-4 Conversion Factors as Extracted Multiples of SI Units† (*continued*)

To convert from	To	Multiply by
mile (U.K. nautical)	meter	+03 1.853 184*
mile (international nautical)	meter	+03 1.852*
mile (U.S. nautical)	meter	+03 1.852*
millibar	newton/meter^2	+02 1.00*
millimeter of mercury (0°C)	newton/meter^2	+02 1.333 224
minute (angle)	radian	−04 2.908 882 086 66
minute (mean solar)	second (mean solar)	+01 6.00*
minute (sidereal)	second (mean solar)	+01 5.983 617 4
month (mean calendar)	second (mean solar)	+06 2.628*
nautical mile (international)	meter	+03 1.852*
nautical mile (U.S.)	meter	+03 1.852*
nautical mile (U.K.)	meter	+03 1.853 184*
oersted	ampere/meter	+01 7.957 747 2
ohm (international of 1948)	ohm	+00 1.000 495
ounce force (avoirdupois)	newton	−01 2.780 138 5
ounce mass (avoirdupois)	kilogram	−02 2.834 952 312 5*
ounce mass (troy or apothecary)	kilogram	−02 3.110 347 68*
ounce (U.S. fluid)	meter^3	−05 2.957 352 956 25*
pace	meter	−01 7.62*
parsec	meter	+16 3.083 74
pascal	newton/meter^2	+00 1.00*
peck (U.S.)	meter^3	−03 8.809 767 541 72*
pennyweight	kilogram	−03 1.555 173 84*
perch	meter	+00 5.0292*
phot	lumen/meter^2	+04 1.00
pica (printers)	meter	−03 4.217 517 6*
pint (U.S. dry)	meter^3	−04 5.506 104 713 575*
pint (U.S. liquid)	meter^3	−04 4.731 764 73*
point (printers)	meter	−04 3.514 598*
poise	newton-second/meter^2	−01 1.00*
pole	meter	+00 5.0292*
pound force (lbf avoirdupois)	newton	+00 4.448 221 615 260 5*
pound mass (lbm avoirdupois)	kilogram	−01 4.535 923 7*
pound mass (troy or apothecary)	kilogram	−01 3.732 417 216*
poundal	newton	−01 1.382 549 543 76*
quart (U.S. dry)	meter^3	−03 1.101 220 942 715*
quart (U.S. liquid)	meter^3	−04 9.463 529 5
rad (radiation dose absorbed)	joule/kilogram	−02 1.00*
Rankine (temperature)	kelvin	$t_K = (5/9)t_R$
rayleigh (rate of photon emission)	1/second-meter^2	+10 1.00*
rhe	meter^2/newton-second	+01 1.00*
rod	meter	+00 5.0292*
roentgen	coulomb/kilogram	−04 2.579 76*
rutherford	disintegration/second	+06 1.00*

TABLE 13-4 Conversion Factors as Extracted Multiples of SI Units† (*continued*)

To convert from	To	Multiply by
second (angle)	radian	−06 4.848 136 811
second (ephemeris)	second	+00 1.000 000 000
second (mean solar)	second (ephemeris)	Consult American Ephemeris and Nautical Almanac
second (sidereal)	second (mean solar)	−01 9.972 695 7
section	$meter^2$	+06 2.589 988 110 336*
scruple (apothecary)	kilogram	−03 1.295 978 2*
shake	second	−08 1.00
skein	meter	+02 1.097 28*
slug	kilogram	+01 1.459 390 29
span	meter	−01 2.286*
statampere	ampere	−10 3.335 640
statcoulomb	coulomb	−10 3.335 640
statfarad	farad	−12 1.112 650
stathenry	henry	+11 8.987 554
statmho	mho	−12 1.112 650
statohm	ohm	+11 8.987 554
statute mile (U.S.)	meter	+03 1.609 344*
statvolt	volt	+02 2.997 925
stere	$meter^3$	+00 1.00*
stilb	$candela/meter^2$	+04 1.00
stoke	$meter^2/second$	−04 1.00*
tablespoon	$meter^3$	−05 1.478 676 478 125*
teaspoon	$meter^3$	−06 4.928 921 593 75*
ton (assay)	kilogram	−02 2.916 666 6
ton (long)	kilogram	+03 1.016 046 908 8*
ton (metric)	kilogram	+03 1.00*
ton (nuclear equivalent of TNT)	joule	+09 4.20
ton (register)	$meter^3$	+00 2.831 684 659 2*
ton (short, 2000 pound)	kilogram	+02 9.071 847 4*
tonne	kilogram	+03 1.00*
torr (0°C)	$newton/meter^2$	+02 1.333 22
township	$meter^2$	+07 9.323 957 2
unit pole	weber	−07 1.256 637
volt (international of 1948)	volt	+00 1.000 330
watt (international of 1948)	watt	+00 1.000 165
yard	meter	−01 9.144*
year (calendar)	second (mean solar)	+07 3.1536*
year (sidereal)	second (mean solar)	+07 3.155 815 0
year (tropical)	second (mean solar)	+07 3.155 692 6
year 1900, tropical, Jan., day 0, hour 12	second (ephemeris)	+07 3.155 692 597 47*
year 1900, tropical, Jan., day 0, hour 12	second	+07 3.155 692 597 47

TABLE 13-4 Conversion Factors as Extracted Multiples of SI Units† (*continued*)

To convert from	To	Multiply by
LISTING BY PHYSICAL QUANTITY		
	Acceleration	
foot/second^2	meter/second^2	−01 3.048*
free fall, standard	meter/second^2	+00 9.806 65*
gal (galileo)	meter/second^2	−02 1.00*
inch/second^2	meter/second^2	−02 2.54*
	Area	
acre	meter^2	+03 4.046 856 422 4*
are	meter^2	+02 1.00*
barn	meter^2	−28 1.00*
circular mil	meter^2	−10 5.067 074 8
foot^2	meter^2	−02 9.290 304*
hectare	meter^2	+04 1.00*
inch^2	meter^2	−04 6.4516*
mile^2 (U.S. statute)	meter^2	+06 2.589 988 110 336*
section	meter^2	+06 2.589 988 110 336*
township	meter^2	+07 9.323 957 2
yard^2	meter^2	−01 8.361 273 6*
	Density	
gram/centimeter^3	kilogram/meter^3	+03 1.00*
lbm/inch^3	kilogram/meter^3	+04 2.767 990 5
lbm/foot^3	kilogram/meter^3	+01 1.601 846 3
slug/foot^3	kilogram/meter^3	+02 5.153 79
	Energy	
British thermal unit (ISO/TC 12)	joule	+03 1.055 06
British thermal unit (International Steam Table)	joule	+03 1.055 04
British thermal unit (mean)	joule	+03 1.055 87
British thermal unit (thermochemical)	joule	+03 1.054 350 264 488
British thermal unit (39°F)	joule	+03 1.059 67
British thermal unit (60°F)	joule	+03 1.054 68
calorie (International Steam Table)	joule	+00 4.1868
calorie (mean)	joule	+00 4.190 02
calorie (thermochemical)	joule	+00 4.184*
calorie (15°C)	joule	+00 4.185 80
calorie (20°C)	joule	+00 4.181 90
calorie (kilogram, International Steam Table)	joule	+03 4.1868
calorie (kilogram, mean)	joule	+03 4.190 02
calorie (kilogram, thermochemical)	joule	+03 4.184*

TABLE 13-4 Conversion Factors as Extracted Multiples of SI Units† (*continued*)

To convert from	To	Multiply by
electron-volt	joule	−19 1.602 10
erg	joule	−07 1.00*
foot-lbf	joule	+00 1.355 817 9
foot-poundal	joule	−02 4.214 011 0
joule (international of 1948)	joule	+00 1.000 165
kilocalorie (International Steam Table)	joule	+03 4.1868
kilocalorie (mean)	joule	+03 4.190 02
kilocalorie (thermochemical)	joule	+03 4.184*
kilowatt-hour	joule	+06 3.60*
kilowatt-hour (international of 1948)	joule	+06 3.600 59
ton (nuclear equivalent of TNT)	joule	+09 4.20
watt-hour	joule	+03 3.60*
	Energy/area time	
Btu (thermochemical)/foot²-second	watt/meter²	+04 1.134 893 1
Btu (thermochemical)/foot²-minute	watt/meter²	+02 1.891 488 5
Btu (thermochemical)/foot²-hour	watt/meter²	+00 3.152 480 8
Btu (thermochemical)/inch²-second	watt/meter²	+06 1.634 246 2
calorie (thermochemical)/centimeter²-minute	watt/meter²	+02 6.973 333 3
erg/centimeter²-second	watt/meter²	−03 1.00*
watt/centimeter²	watt/meter²	+04 1.00*
	Force	
dyne	newton	−05 1.00*
kilogram force (kgf)	newton	+00 9.806 65*
kilopond force	newton	+00 9.806 65*
kip	newton	+03 4.448 221 615 260 5*
lbf (pound force, avoirdupois)	newton	+00 4.448 221 615 260 5*
ounce force (avoirdupois)	newton	−01 2.780 138 5
pound force, lbf (avoirdupois)	newton	+00 4.448 221 615 260 5*
poundal	newton	−01 1.382 549 543 76*
	Length	
angstrom	meter	−10 1.00*
astronomical unit	meter	+11 1.495 978 9
cable	meter	+02 2.194 56*
caliber	meter	−04 2.54*
chain (surveyor or gunter)	meter	+01 2.011 68*
chain (engineer or ramden)	meter	+01 3.048*
cubit	meter	−01 4.572*
fathom	meter	+00 1.8288*
fermi (femtometer)	meter	−15 1.00*

TABLE 13-4 Conversion Factors as Extracted Multiples of SI Units† (*continued*)

To convert from	To	Multiply by
foot	meter	−01 3.048*
foot (U.S. survey)	meter	+00 1200/3937*
foot (U.S. survey)	meter	−01 3.048 006 096
furlong	meter	+02 2.011 68*
hand	meter	−01 1.016*
inch	meter	−02 2.54*
league (U.K. nautical)	meter	+03 5.559 552*
league (international nautical)	meter	+03 5.556*
league (statute)	meter	+03 4.828 032*
light year	meter	+15 9.460 55
link (engineer or ramden)	meter	−01 3.048*
link (surveyor or gunter)	meter	−01 2.011 68*
meter	wavelengths Kr 86	+06 1.650 763 73*
micron	meter	−06 1.00*
mil	meter	−05 2.54*
mile (U.S. statute)	meter	+03 1.609 344*
mile (U.K. nautical)	meter	+03 1.853 184*
mile (international nautical)	meter	+03 1.852*
mile (U.S. nautical)	meter	+03 1.852*
nautical mile (U.K.)	meter	+03 1.853 184*
nautical mile (international)	meter	+03 1.852*
nautical mile (U.S.)	meter	+03 1.852*
pace	meter	−01 7.62*
parsec	meter	+16 3.083 74
perch	meter	+00 5.0292*
pica (printers)	meter	−03 4.217 517 6*
point (printers)	meter	−04 3.514 598*
pole	meter	+00 5.0292*
rod	meter	+00 5.0292*
skein	meter	+02 1.097 28*
span	meter	−01 2.286*
statute mile (U.S.)	meter	+03 1.609 344*
yard	meter	−01 9.144*
	Mass	
carat (metric)	kilogram	−04 2.00*
dram (avoirdupois)	kilogram	−03 1.771 845 195 312 5*
dram (troy or apothecary)	kilogram	−03 3.887 934 6*
grain	kilogram	−05 6.479 891*
gram	kilogram	−03 1.00*
hundredweight (long)	kilogram	+01 5.080 234 544*
hundredweight (short)	kilogram	+01 4.535 923 7*
kgf-second^2-meter (mass)	kilogram	+00 9.806 65*
kilogram mass	kilogram	+00 1.00*
lbm (pound mass, avoirdupois)	kilogram	−01 4.535 923 7*
ounce mass (avoirdupois)	kilogram	−02 2.834 952 312 5*
ounce mass (troy or apothecary)	kilogram	−02 3.110 347 68*

TABLE 13-4 Conversion Factors as Extracted Multiples of SI Units† (*continued*)

To convert from	To	Multiply by
pennyweight	kilogram	−03 1.555 173 84*
pound mass, lbm (avoirdupois)	kilogram	−01 4.535 923 7*
pound mass (troy or apothecary)	kilogram	−01 3.732 417 216*
scruple (apothecary)	kilogram	−03 1.295 978 2*
slug	kilogram	+01 1.459 390 29
ton (assay)	kilogram	−02 2.916 666 6
ton (long)	kilogram	+03 1.016 046 908 8*
ton (metric)	kilogram	+03 1.00*
ton (short, 2000 pound)	kilogram	+02 9.071 847 4*
tonne	kilogram	+03 1.00*
	Power	
Btu (thermochemical)/second	watt	+03 1.054 350 264 488
Btu (thermochemical)/minute	watt	+01 1.757 250 4
calorie (thermochemical)/second	watt	+00 4.184*
calorie (thermochemical)/minute	watt	−02 6.973 333 3
foot-lbf/hour	watt	−04 3.766 161 0
foot-lbf/minute	watt	−02 2.259 696 6
foot-lbf/second	watt	+00 1.355 817 9
horsepower (550 foot lbf/second)	watt	+02 7.456 998 7
horsepower (boiler)	watt	+03 9.809 50
horsepower (electric)	watt	+02 7.46*
horsepower (metric)	watt	+02 7.354 99
horsepower (U.K.)	watt	+02 7.457
horsepower (water)	watt	+02 7.460 43
kilocalorie (thermochemical)/minute	watt	+01 6.973 333 3
kilocalorie (thermochemical)/second	watt	+03 4.184*
watt (international of 1948)	watt	+00 1.000 165
	Pressure	
atmosphere	newton/meter^2	+05 1.013 25*
bar	newton/meter^2	+05 1.00*
barye	newton/meter^2	−01 1.00*
centimeter of mercury (0°C)	newton/meter^2	+03 1.333 22
centimeter of water (4°C)	newton/meter^2	+01 9.806 38
dyne/centimeter^2	newton/meter^2	−01 1.00*
foot of water (39.2°F)	newton/meter^2	+03 2.988 98
inch of mercury (32°F)	newton/meter^2	+03 3.386 389
inch of mercury (60°F)	newton/meter^2	+03 3.376 85
inch of water (39.2°F)	newton/meter^2	+02 2.490 82
inch of water (60°F)	newton/meter^2	+02 2.4884
kgf centimeter^2	newton/meter^2	+04 9.806 65*
kgf/meter^2	newton/meter^2	+00 9.806 65*
lbf/foot^2	newton/meter^2	+01 4.788 025 8
lbf/inch^2(psi)	newton/meter^2	+03 6.894 757 2
millibar	newton/meter^2	+02 1.00*

TABLE 13-4 Conversion Factors as Extracted Multiples of SI Units† (*continued*)

To convert from	To	Multiply by
millimeter of mercury (0°C)	newton/meter^2	+02 1.333 224
pascal	newton/meter^2	+00 1.00*
psi (lbf/inch^2)	newton/meter^2	+03 6.894 757 2
torr (0°C)	newton/meter^2	+02 1.333 22
	Speed	
foot/hour	meter/second	−05 8.466 666 6
foot/minute	meter/second	−03 5.08*
foot/second	meter/second	−01 3.048*
inch/second	meter/second	−02 2.54*
kilometer/hour	meter/second	−01 2.777 777 8
knot (international)	meter/second	−01 5.144 444 444
mile hour (U.S. statute)	meter/second	−01 4.4704*
mile/minute (U.S. statute)	meter/second	+01 2.682 24*
mile/second (U.S. statute)	meter/second	+03 1.609 344*
	Temperature	
Celsius	kelvin	$t_K = t_C + 273.15$
Fahrenheit	kelvin	$t_K = (5/9)(t_F + 459.67)$
Fahrenheit	Celsius	$t_C = (5/9)(t_F - 32)$
Rankine	kelvin	$t_K = (5/9)t_R$
	Time	
day (mean solar)	second (mean solar)	+04 8.64*
day (sidereal)	second (mean solar)	+04 8.616 409 0
hour (mean solar)	second (mean solar)	+03 3.60*
hour (sidereal)	second (mean solar)	+03 3.590 170 4
minute (mean solar)	second (mean solar)	+01 6.00*
minute (sidereal)	second (mean solar)	+01 5.983 617 4
month (mean calendar)	second (mean solar)	+06 2.628*
second (ephemeris)	second	+00 1.000 000 000
second (mean solar)	second (ephemeris)	Consult American Ephemeris and Nautical Almanac
second (sidereal)	second (mean solar)	−01 9.972 695 7
year (calendar)	second (mean solar)	+07 3.1536*
year (sidereal)	second (mean solar)	+07 3.155 815 0
year (tropical)	second (mean solar)	+07 3.155 692 6
year 1900, tropical, Jan., day 0 hour 12	second (ephemeris)	+07 3.155 692 597 47*
year 1900, tropical, Jan., day 0, hour 12	second	+07 3.155 692 597 47
	Viscosity	
centistoke	meter^2/second	−06 1.00*
stoke	meter^2/second	−04 1.00*

TABLE 13-4 Conversion Factors as Extracted Multiples of SI Units† (*continued*)

To convert from	To	Multiply by
foot²/second	meter²/second	−02 9.290 304*
centipoise	newton-second/meter²	−03 1.00*
lbm/foot-second	newton-second/meter²	+00 1.488 163 9
lbf-second/foot²	newton-second/meter²	+01 4.788 025 8
poise	newton-second/meter²	−01 1.00*
poundal-second/foot²	newton-second/meter²	+00 1.488 163 9
slug/foot-second	newton-second/meter²	+01 4.788 025 8
rhe	meter²/newton-second	+01 1.00*
	Volume	
acre-foot	meter³	+03 1.233 481 9
barrel (petroleum, 42 gallons)	meter³	−01 1.589 873
board foot	meter³	−03 2.359 737 216*
bushel (U.S.)	meter³	−02 3.523 907 016 688*
cord	meter³	+00 3.624 556 3
cup	meter³	−04 2.365 882 365*
dram (U.S. fluid)	meter³	−06 3.696 691 195 312 5*
fluid ounce (U.S.)	meter³	−05 2.957 352 956 25*
foot³	meter³	−02 2.831 684 659 2*
gallon (U.K. liquid)	meter³	−03 4.546 087
gallon (U.S. dry)	meter³	−03 4.404 883 770 86*
gallon (U.S. liquid)	meter³	−03 3.785 411 784*
gill (U.K.)	meter³	−04 1.420 652
gill (U.S.)	meter³	−04 1.182 941 2
hogshead (U.S.)	meter³	−01 2.384 809 423 92*
inch³	meter³	−05 1.638 706 4*
liter	meter³	−03 1.00*
ounce (U.S. fluid)	meter³	−05 2.957 352 956 25*
peck (U.S.)	meter³	−03 8.809 767 541 72*
pint (U.S. dry)	meter³	−04 5.506 104 713 575*
pint (U.S. liquid)	meter³	−04 4.731 764 73*
quart (U.S. dry)	meter³	−03 1.101 220 942 715*
quart (U.S. liquid)	meter³	−04 9.463 529 5
stere	meter³	+00 1.00*
tablespoon	meter³	−05 1.478 676 478 125*
teaspoon	meter³	−06 4.928 921 593 75*
ton (register)	meter³	+00 2.831 684 659 2*
yard³	meter³	−01 7.645 548 579 84*

asterisk follows each number that expresses an exact definition. For example, the entry "−02 2.54*" expresses the fact that 1 inch = 2.54 × 10^{-2} meter, exactly, by definition. Most of the definitions are extracted from National Bureau of Standards documents. Numbers not followed by an asterisk are only approximate or are the results of physical measurements. The conversion factors are listed alphabetically and by physical quantity.

The listing by physical quantity includes only relationships which are frequently encountered and deliberately omits the many combinations of units which are used for more specialized purposes. Conversion factors for combinations of units are easily generated from numbers given in the alphabetical listing by the technique of direct substitution or by other well-known rules for manipulating units. These units are adequately discussed in many science and engineering textbooks and are not repeated here.

14
Innovative Design

ROBERT O. PARMLEY, P.E.
President, Morgan & Parmley, Ltd./Ladysmith, Wisconsin

This section has been included for the single purpose of illustrating that individual standard components have wide potential uses, far beyond their original concept. All component designers must be aware of this fact so that they can use stock items to ensure cost-effective products.

The following illustrated examples depict a broad range of basic components, each with a multitude of uses.

Most of the following pages originally appeared in *Product Engineering*. The remaining articles were extracted from *Assembly Engineering* and *Machine Design*. Written permission has been granted for their use, and proper credit is given each publication.

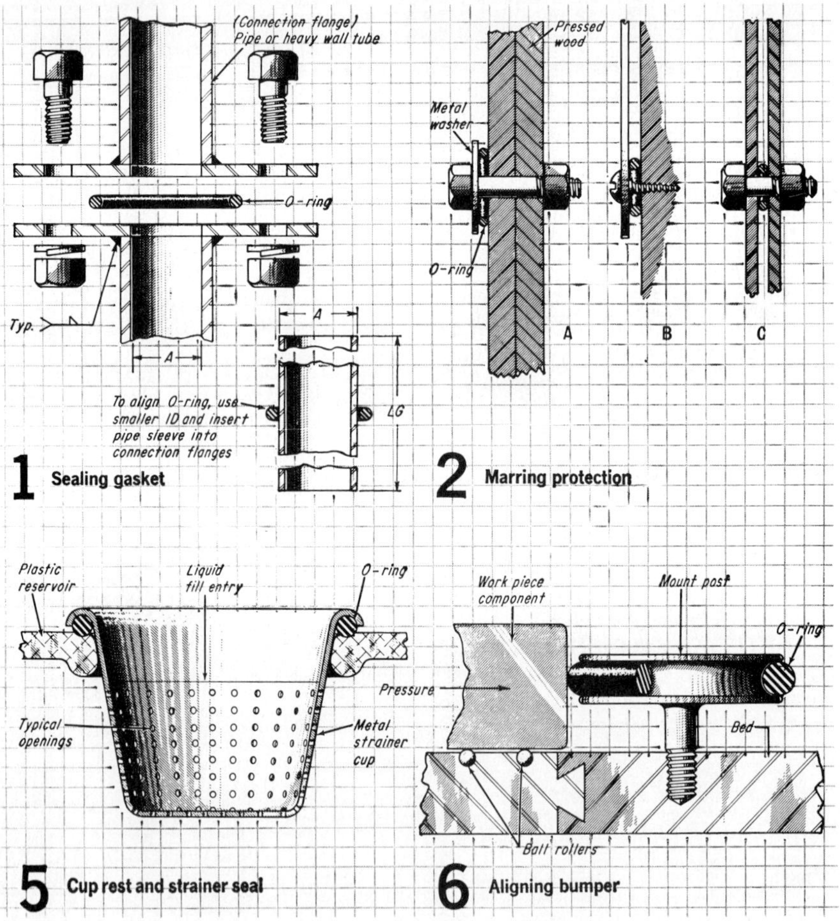

FIG. 14-1 A different look at O rings. Although they are primarily seals, O rings can do a variety of other jobs as well as more sophisticated pieces of hardware. (*From R. O. Parmley, "Look At O-Rings Differently," in Product Engineering, Aug. 16, 1965. Reprinted by permission, Morgan-Grampian Publishing Co.*)

14-1 O Rings

Rubber O rings are excellent seals, but they may be used in a wide variety of other applications, as shown in Figs. 14-1 to 14-5.

Playing many different roles, O rings can perform a variety of functions never dreamed of by their originators.

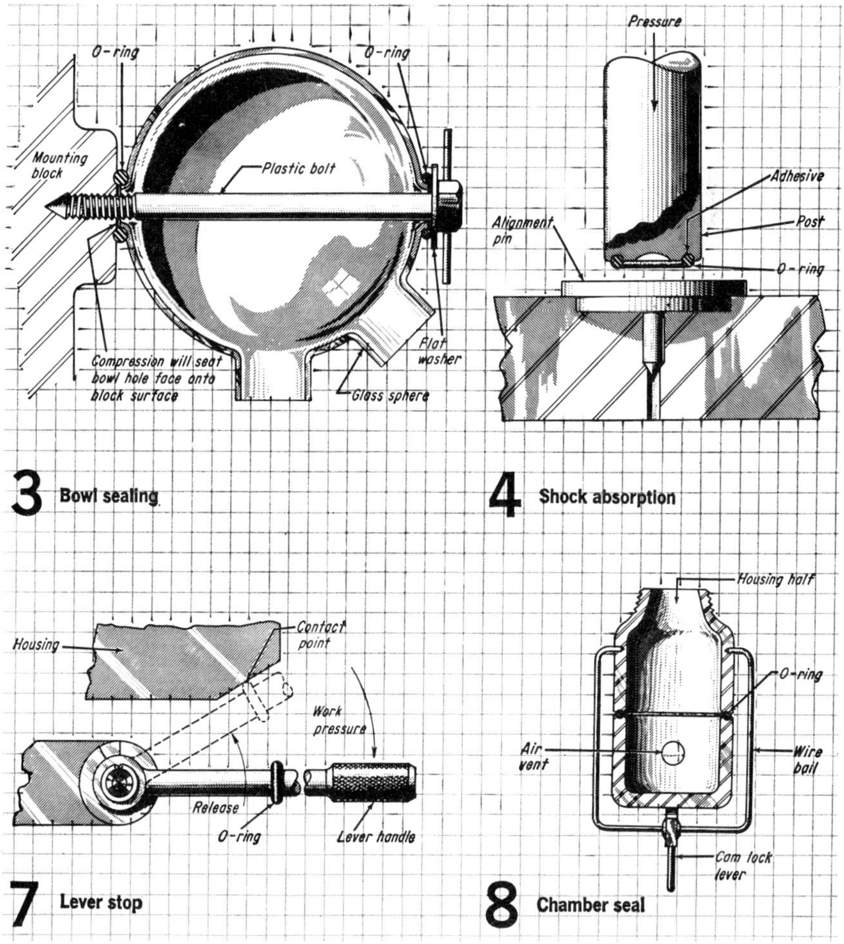

FIG. 14-1 *(continued)*

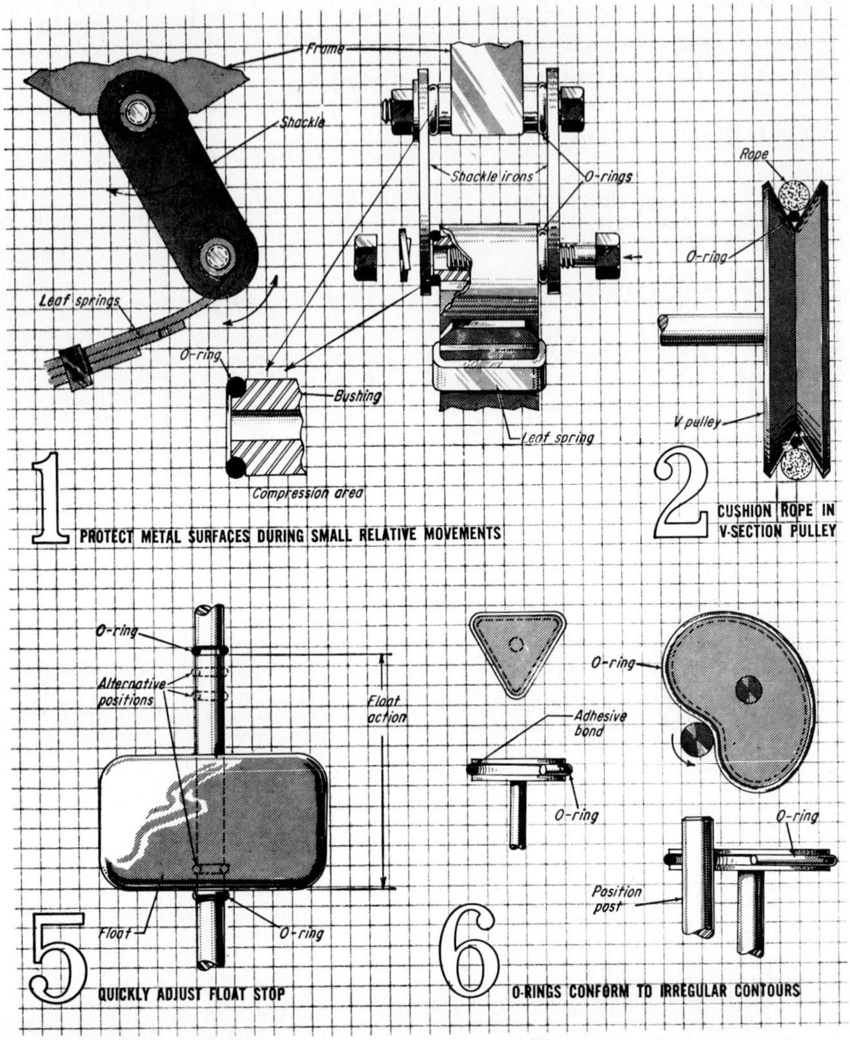

FIG. 14-2 Unusual O ring applications. Playing many different roles, O rings can perform as protective devices, hole liners, float stops, and other key design components. (*From R. O. Parmley, "8 Unusual Applications for O-Rings," in Product Engineering, Nov. 25, 1963. Reprinted by permission, Morgan-Grampian Publishing Co.*)

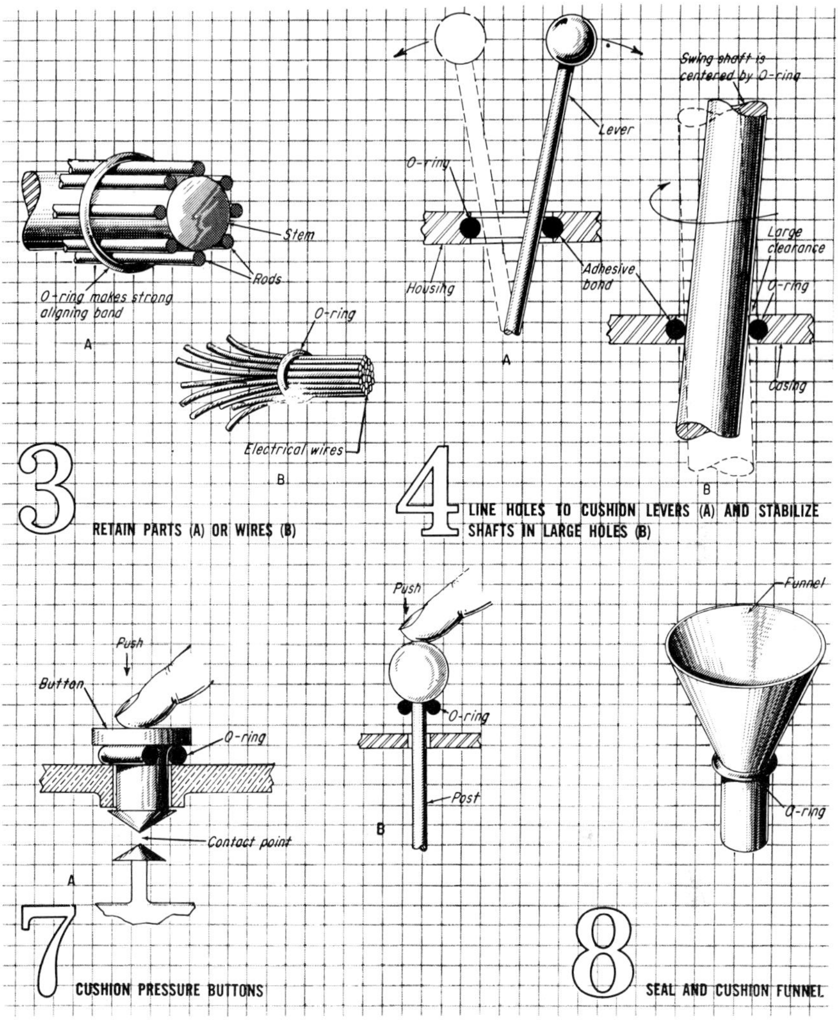

FIG. 14-2 *(continued)*

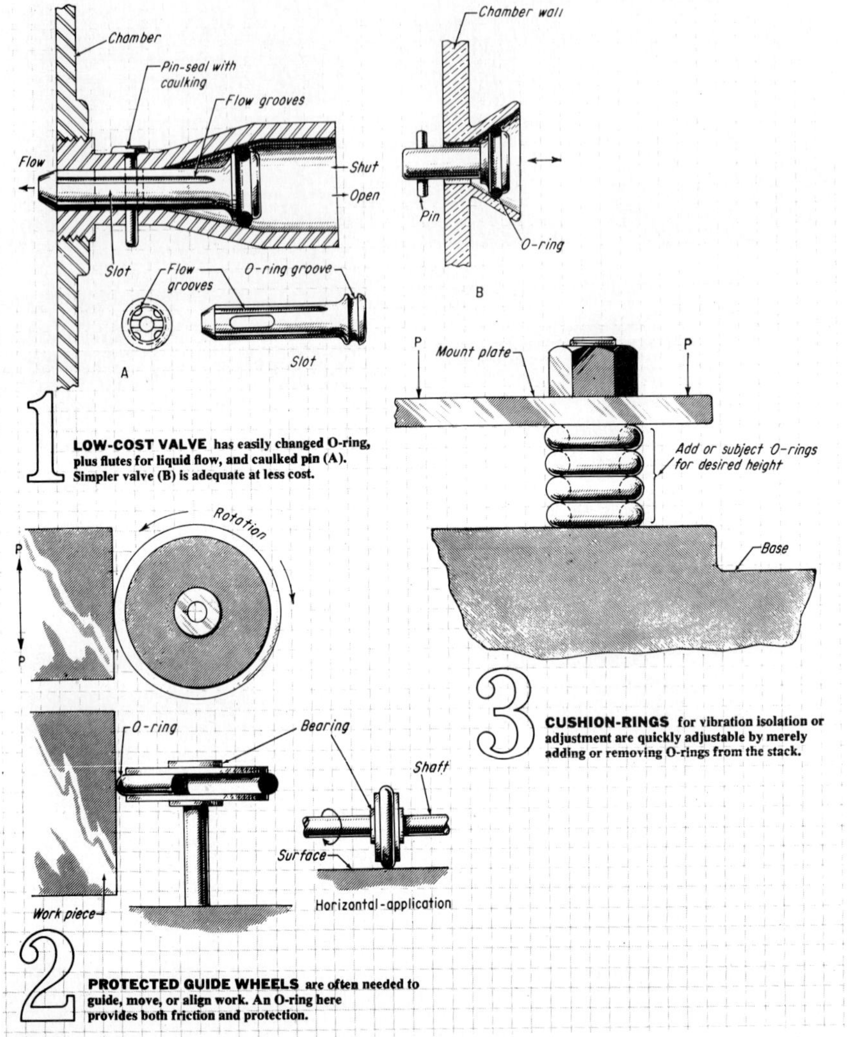

FIG. 14-3 More ways to use O rings. O-Rings are shown here performing in valves, on guide wheels, and as cushioning components. (*From R. O. Parmley, "7 More Applications for O-Rings," in Product Engineering, Dec. 9, 1963. Reprinted by permission, Morgan-Grampian Publishing Co.*)

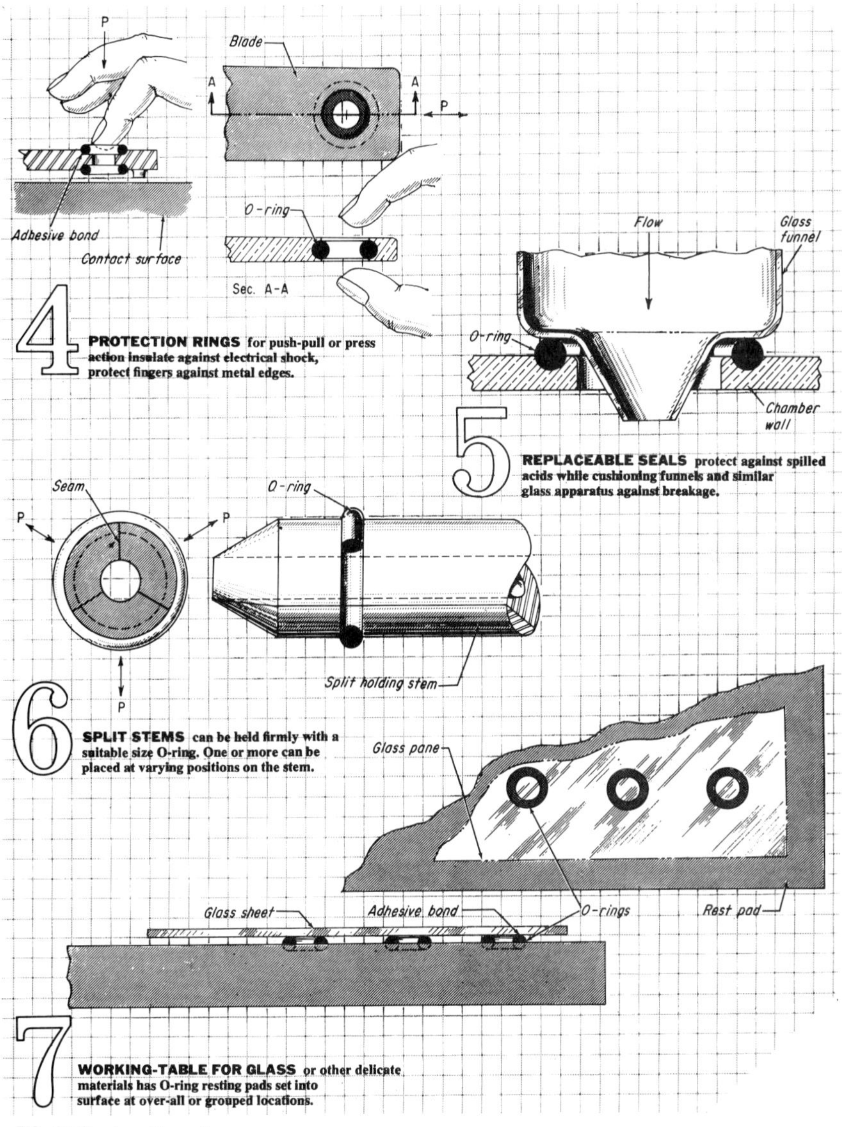

FIG. 14-3 *(continued)*

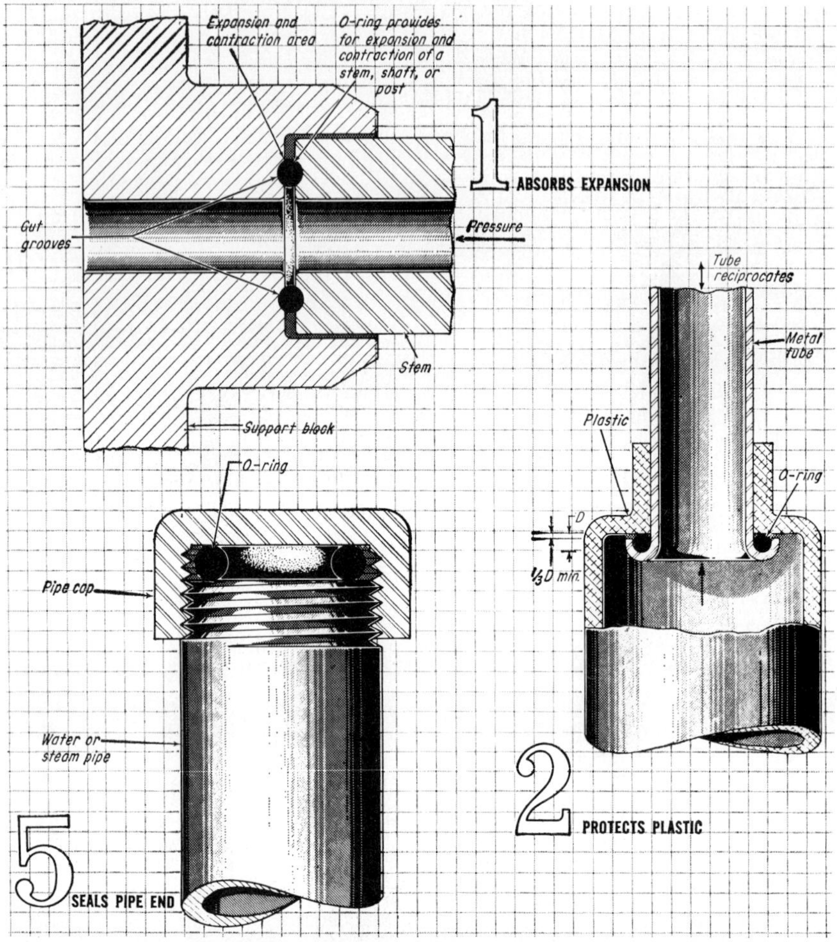

FIG. 14-4 Design problems solved with O rings. Rubber O rings provide thermal expansion, protect surfaces, seal pipe ends and connections, and prevent slipping. (*From R. O. Parmley, "O-Rings Solve Design Problems," in Product Engineering, May 11, 1964. Reprinted by permission, Morgan-Grampian Publishing Co.*)

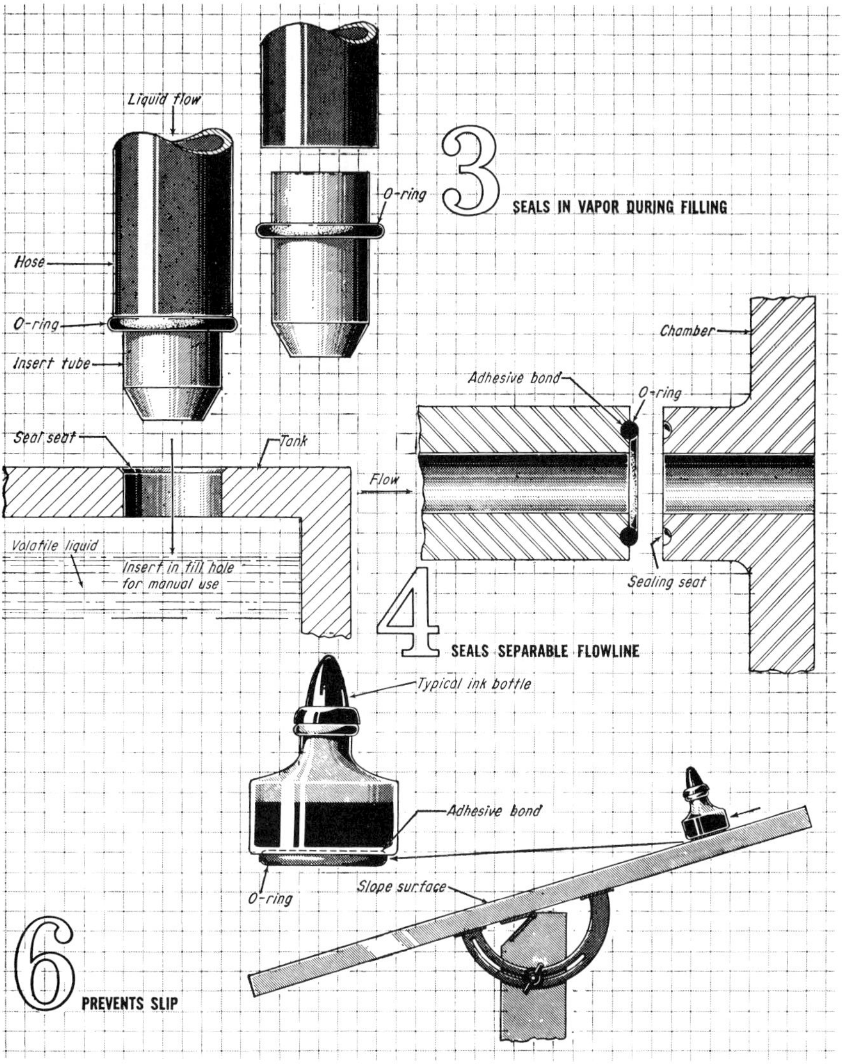

FIG. 14-4 *(continued)*

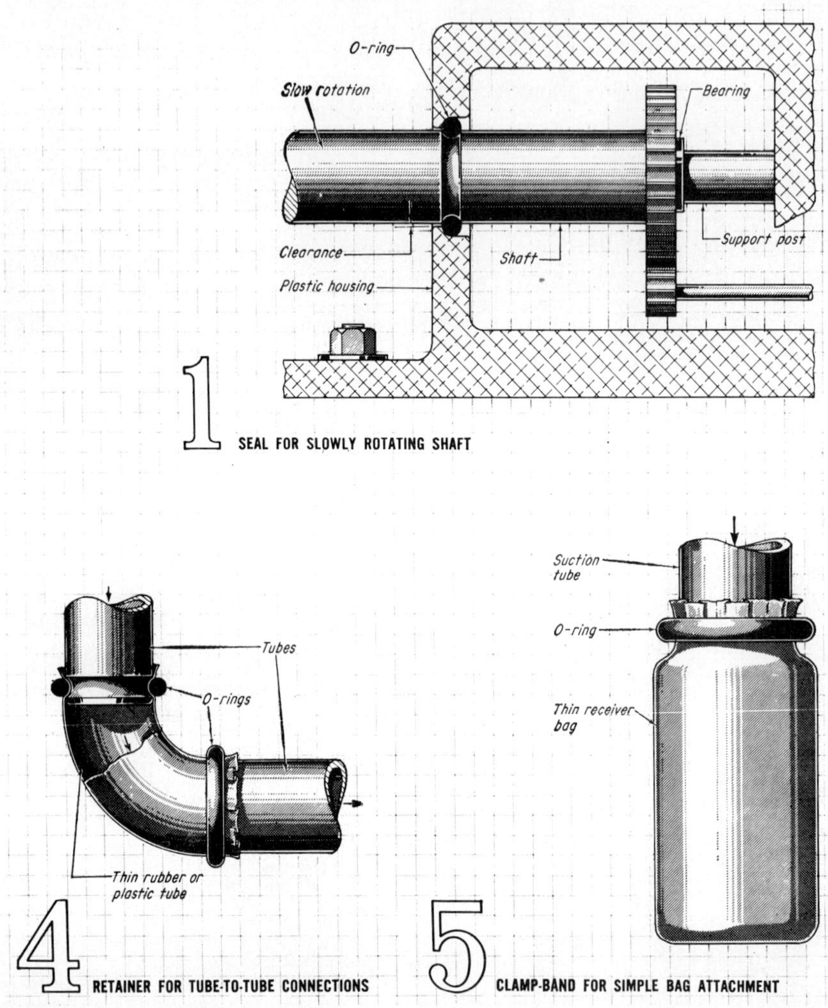

FIG. 14-5 Solve design problems with O rings. More examples of how rubber O rings provide seals for shafts, lids, nozzles, and elbows, as well as protect corners and cushion metal surfaces. (*From R. O. Parmley, "O-Rings Solve Design Problems II," in Product Engineering, May 25, 1964. Reprinted by permission, Morgan-Grampian Publishing Co.*)

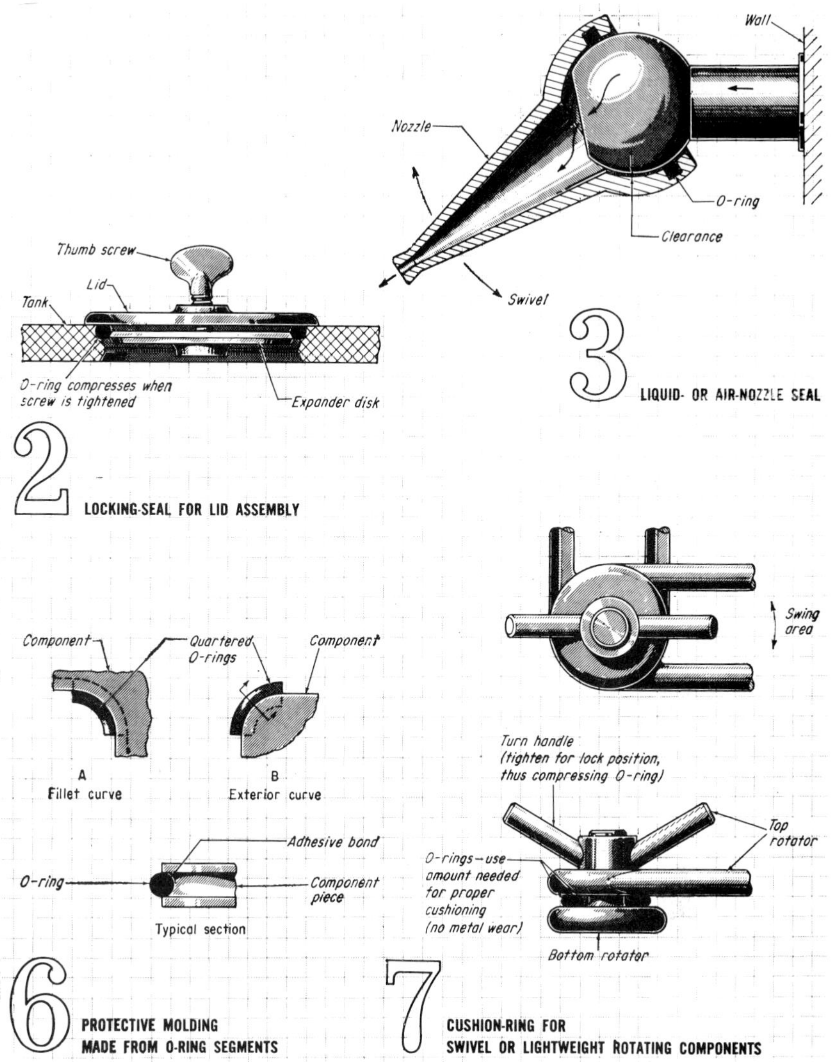

FIG. 14-5 *(continued)*

14-2 Rubber Balls

Rubber or plastic balls, whether solid or hollow, can have a variety of important applications in many designs. Figures 14-6 and 14-7 picture many potential uses.

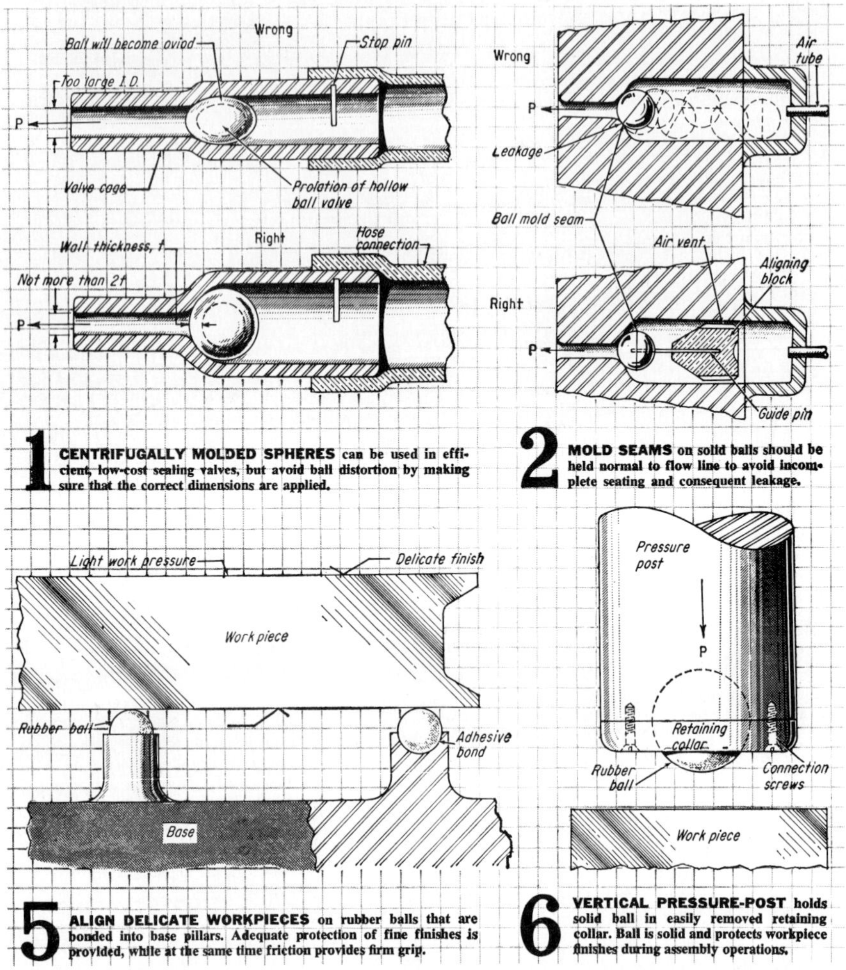

FIG. 14-6 Many jobs for rubber balls. Plastic and rubber balls, whether solid or hollow, can find a variety of important applications in many designs. (*From R. O. Parmley, "Rubber Balls Find Many Jobs," in Product Engineering, December 23, 1963. Reprinted by permission, Morgan-Grampian Publishing Co.*)

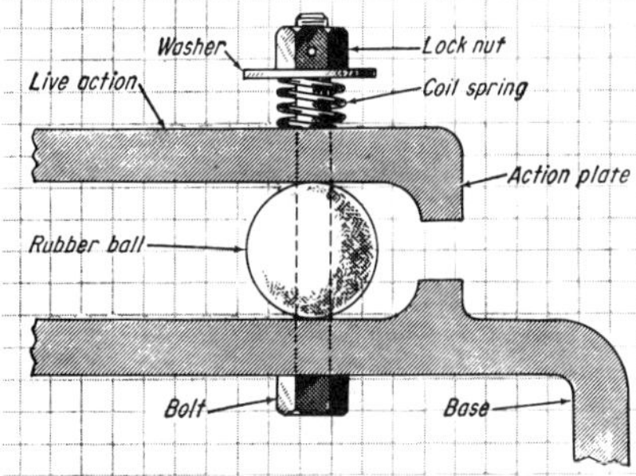

3 **CUSHION PAD AND SHOCK ABSORBER** is easily made for mounting-plates that must either carry dynamic loads or absorb shock forces.

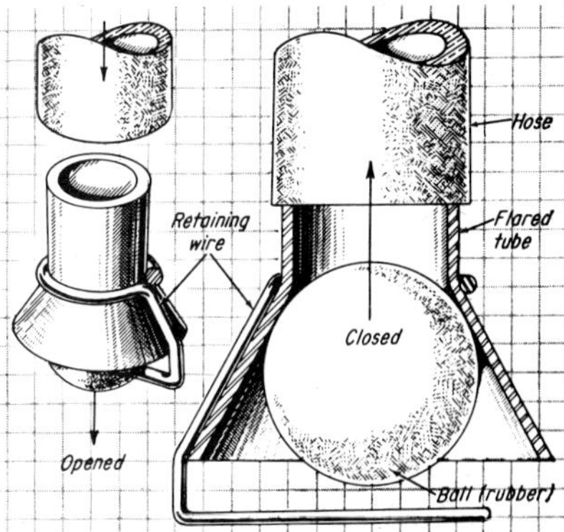

4 **DISCHARGE VALVE** as shown here is an effective way of controlling liquid displacement at the end of hoses where suction would be unwanted or harmful.

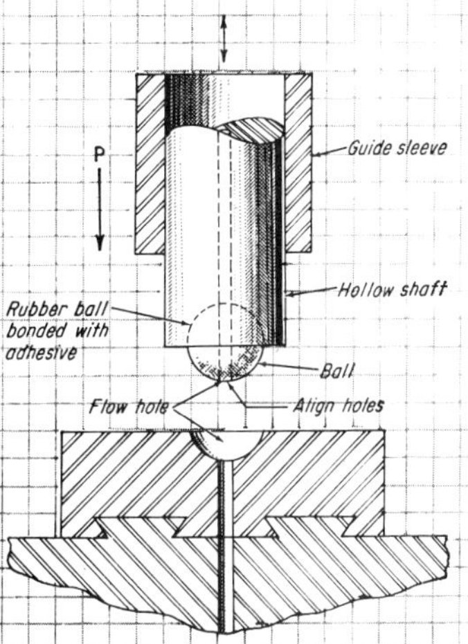

7 **HOLLOW SHAFT-SEAL** embodies adhesive-bonded rubber ball with flow hole. Quick connection of leakproof joint for lubricant or other liquid is gained.

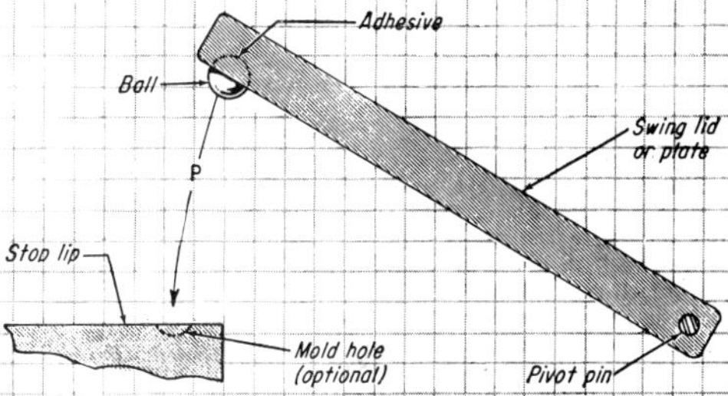

8 **BUMPER STOP** is another example of the simple but effective way a rubber ball can be employed to protect surfaces or parts.

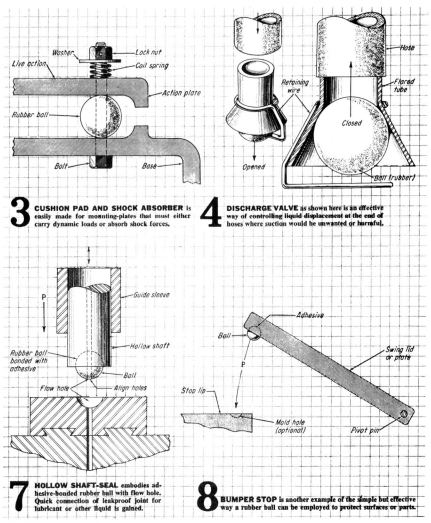

FIG. 14-6 *(continued)*

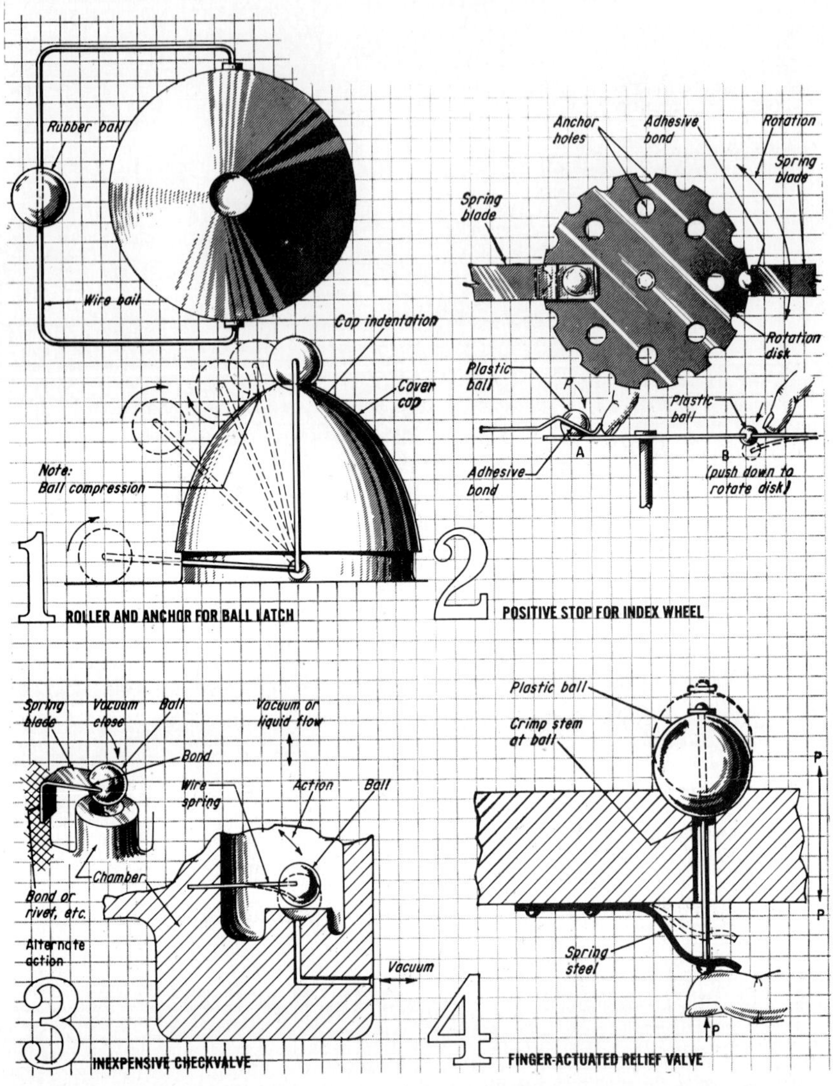

FIG. 14-7 Soft balls simplify design. Balls made from flexible material can perform as latches, stops for index disks, inexpensive valves, and buffers for compression springs. (*From R. O. Parmley, "How Soft Balls Can Simplify Design," in Product Engineering, June 22, 1964. Reprinted by permission, Morgan-Grampian Publishing Co.*)

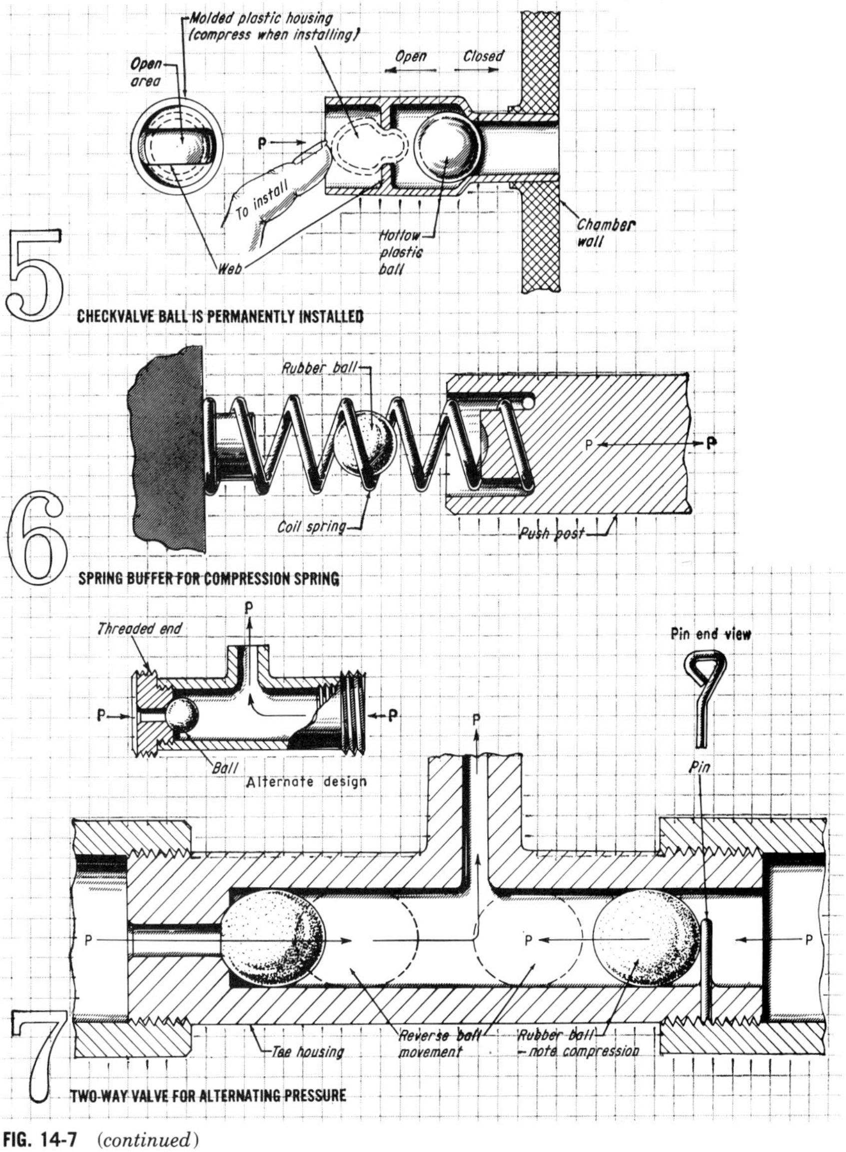

FIG. 14-7 *(continued)*

14-3 Washers

Washers have many designs, the most common being flat and plain. Figures 14-8 to 14-10 illustrate examples of uses for flat steel and flat rubber washers.

The serrated flat washer is also a stock item with a large scope for potential usage. See Fig. 14-11.

Cupped washers are not as common, but have as much potential as their more common cousins. Figure 14–12 pictures some thought-provoking designs.

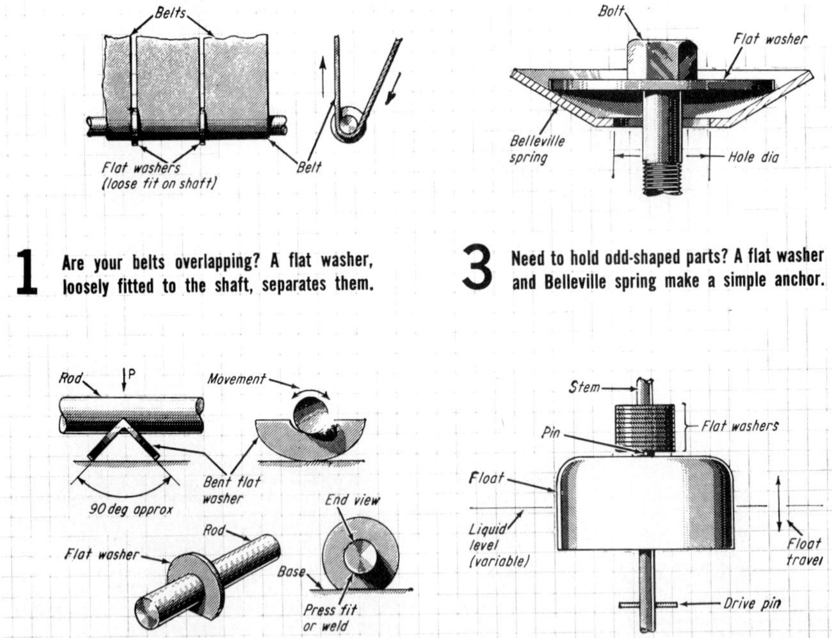
Belts
Flat washers (loose fit on shaft)
Belt
1 Are your belts overlapping? A flat washer, loosely fitted to the shaft, separates them.
Bolt
Flat washer
Belleville spring
Hole dia
3 Need to hold odd-shaped parts? A flat washer and Belleville spring make a simple anchor.
Rod
P
Movement
Bent flat washer
90 deg approx
End view
Rod
Flat washer
Base
Press fit or weld
2 How about a rod support? A bent washer permits some rocking; if welded, the support is stable.
Stem
Pin
Flat washers
Float
Liquid level (variable)
Float travel
Drive pin
4 Got a weight problem? Adding or subtracting flat washers can easily control float action.

FIG. 14-8 Ideas for flat washers. You can do more with flat washers than you may think. Here are 10 ideas that may save the day next time you need a simple, quick, inexpensive design. *(From R. O. Parmley, "10 Ideas for Flat Washers," in Product Engineering, June 21, 1965. Reprinted by permission, Morgan-Grampian Publishing Co.)*

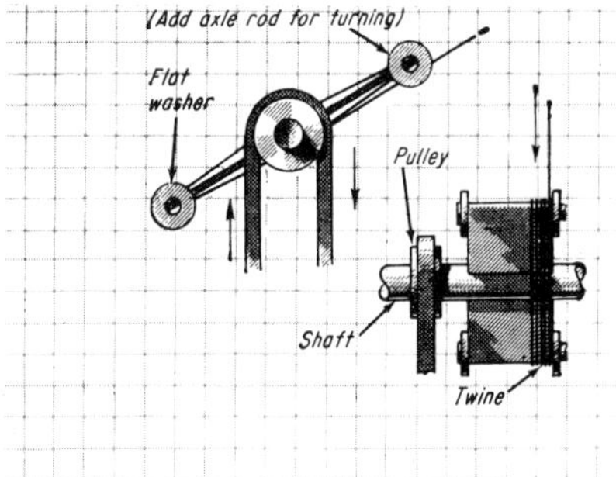

5 How about some simple flanges? Here the washers guide the twine and keep it under control.

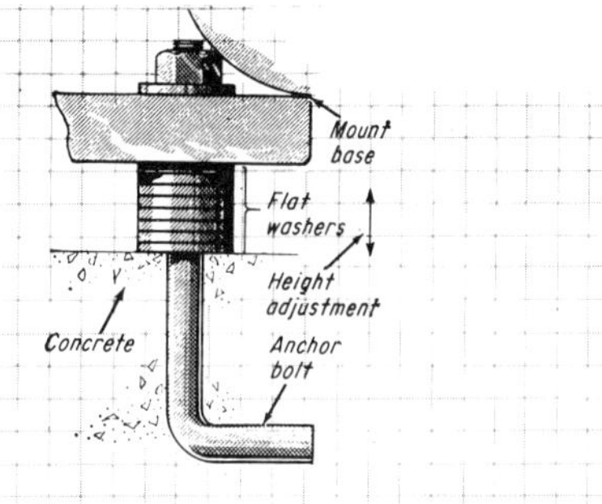

8 Does your floor tilt? Stacked washers can level machines, or give a stable height adjustment.

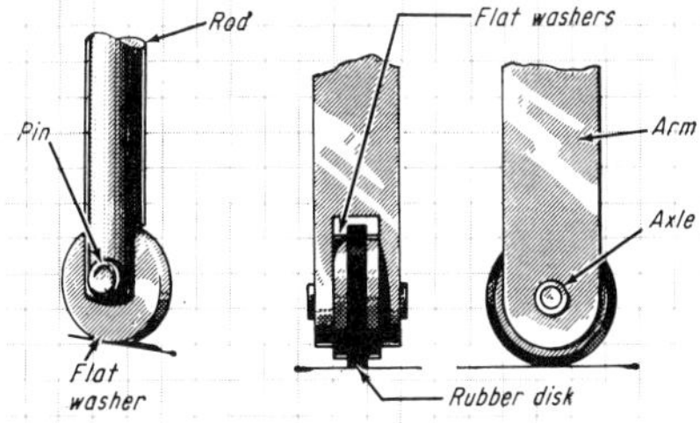

6 Need some wheels? Here flat washer is the wheel. A rubber disk quiets the assembly.

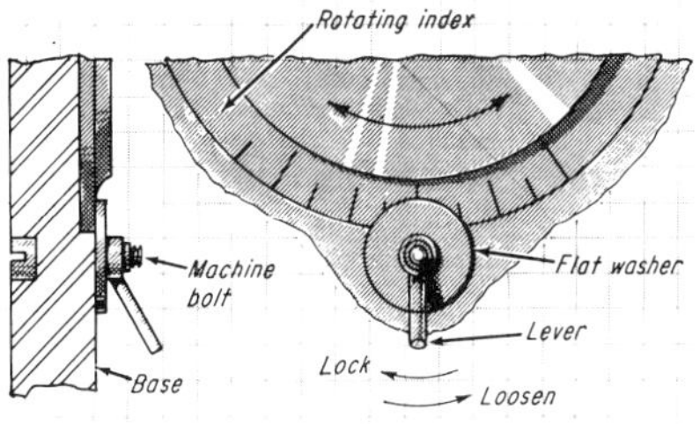

9 Here's a simple lock. A machine bolt, a washer, and a wing or lever nut make a strong clamp.

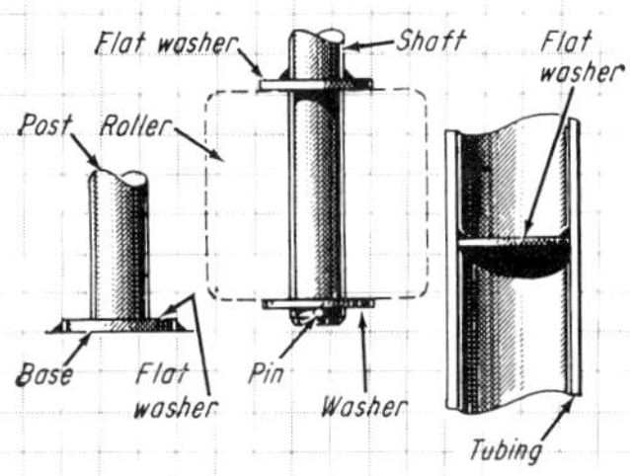

7 Want to avoid machining? Washers can make anchors, stop shoulders, even reduce tubing ID.

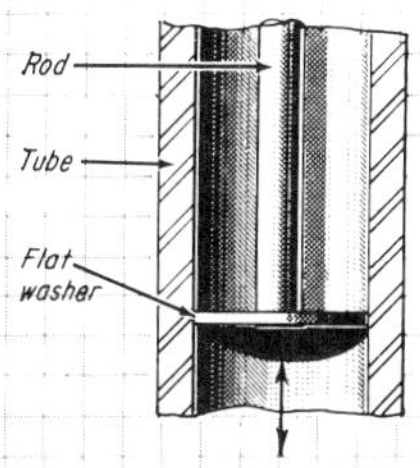

10 Need a piston in a hurry? For light service, tube, rod, and washer will be adequate.

FIG. 14-8 *(continued)*

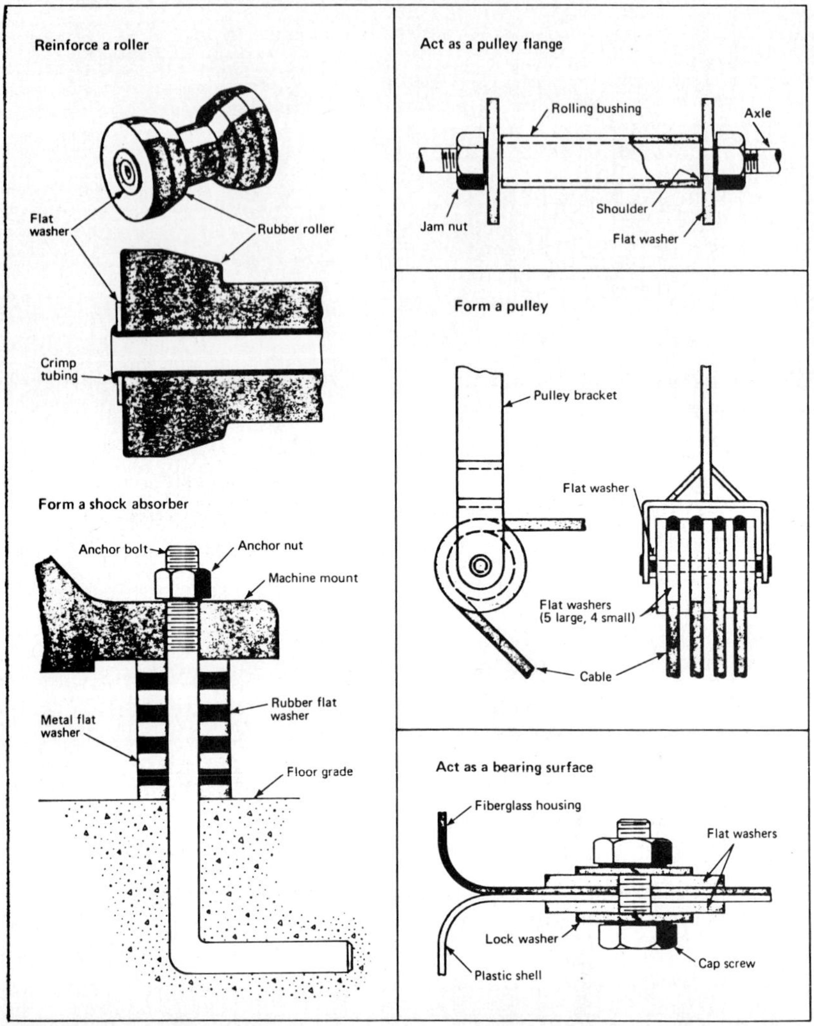

FIG. 14-9 Versatile flat washers. Washers are usually thought of as bearing surfaces placed under bolt heads. However, they can be used in a variety of ways that could simplify design or be an immediate fix until a designed part is available. (*From R. O. Parmley, "Versatile Flat Washers," in Product Engineering, January 1973. Reprinted by permission, Morgan-Grampian Publishing Co.*)

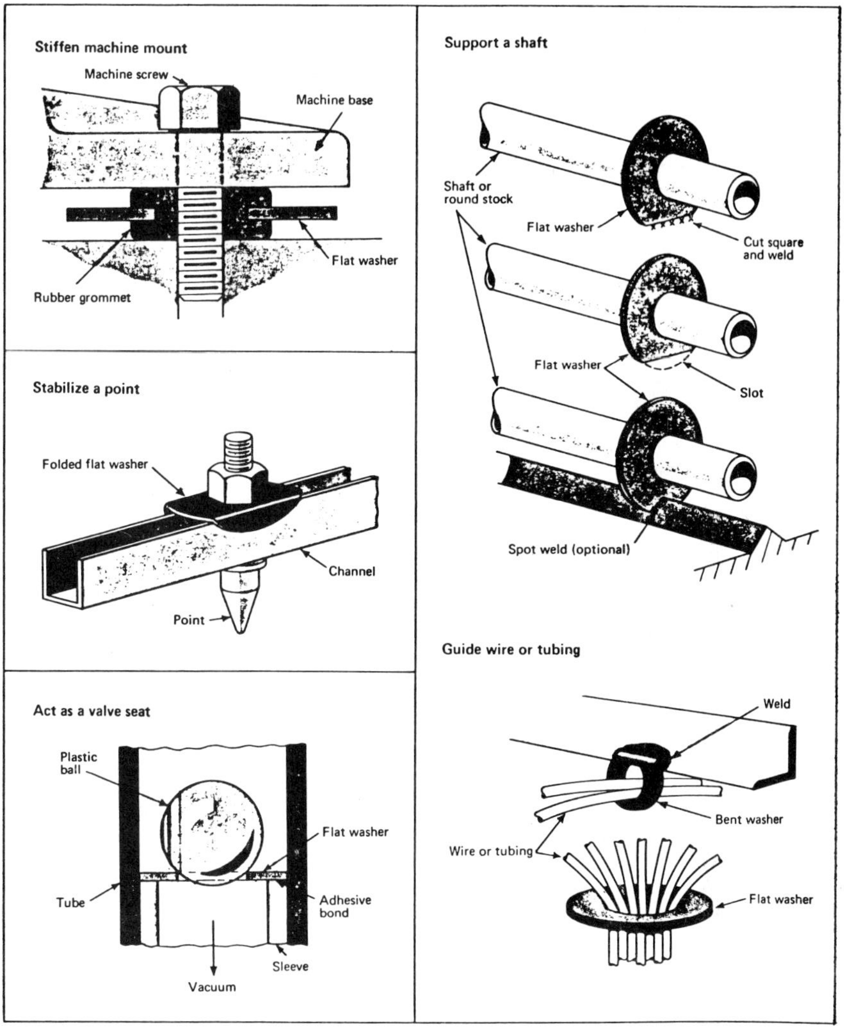

FIG. 14-9 *(continued)*

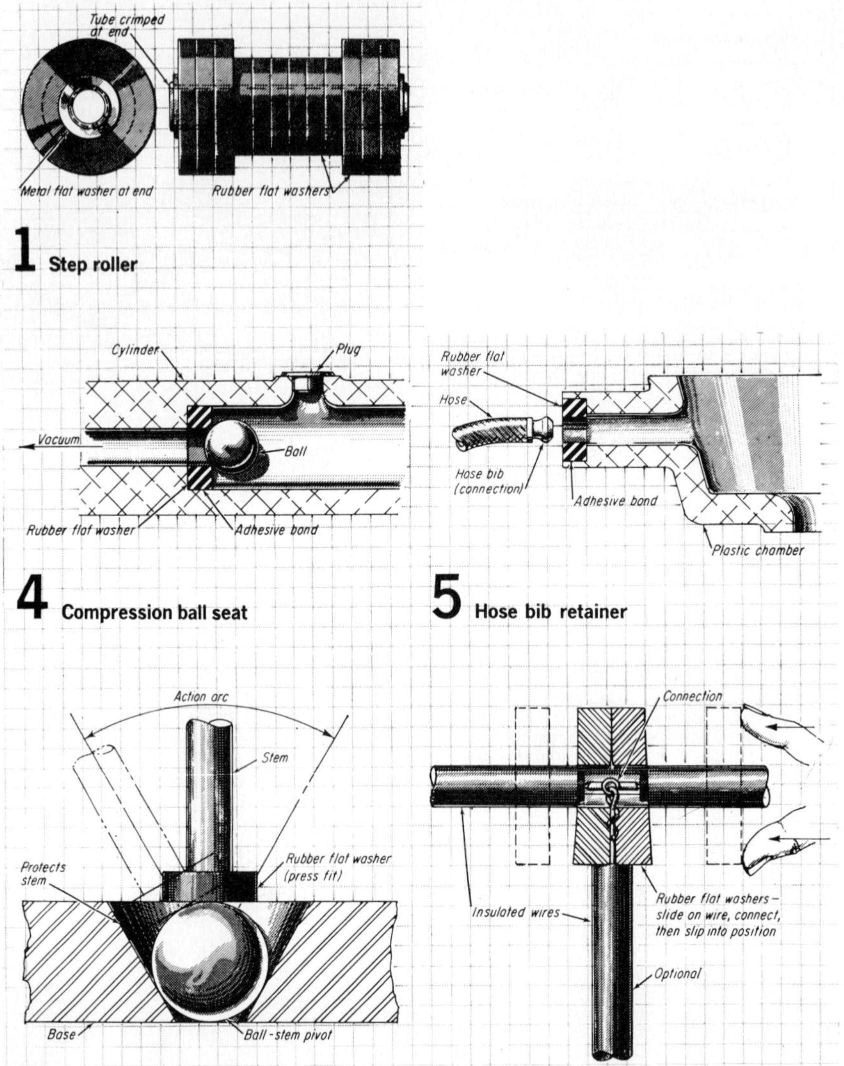

FIG. 14-10 Jobs for flat rubber washers. Rubber washers are far more versatile than you realize. Here are some odd jobs they can do that may make your next design job easier. (*From R. O. Parmley, "Jobs for Flat Washers," in Product Engineering, Nov. 22, 1965. Reprinted by permission, Morgan-Grampian Publishing Co.*)

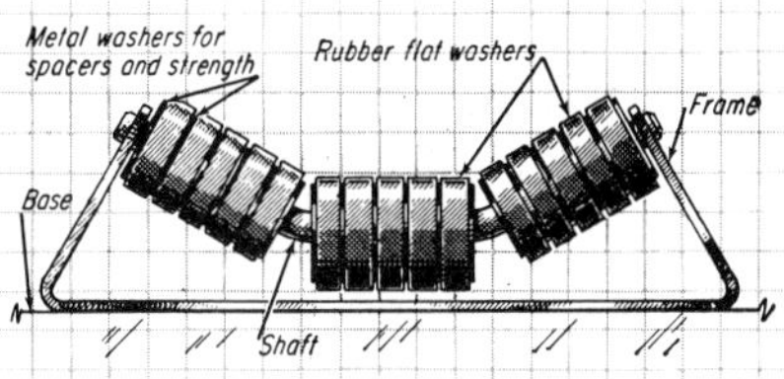

2 **Impact absorption idler roller**

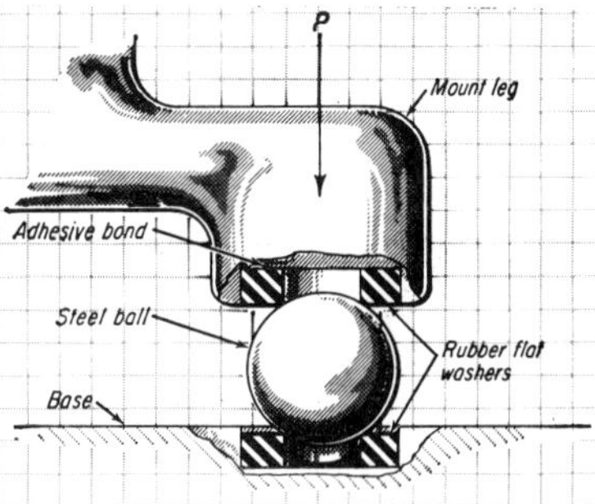

3 **Compression mount**

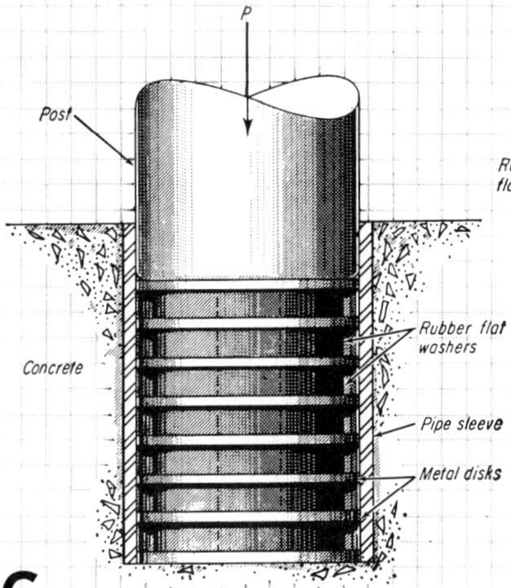

6 **Heavy shock load support**

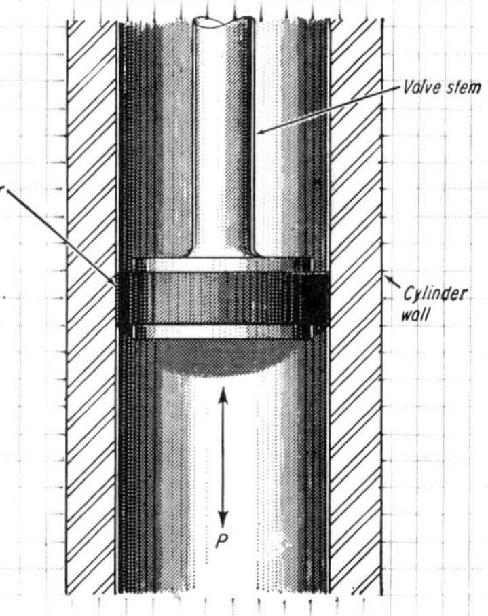

7 **Cylinder valve**

FIG. 14-10 *(continued)*

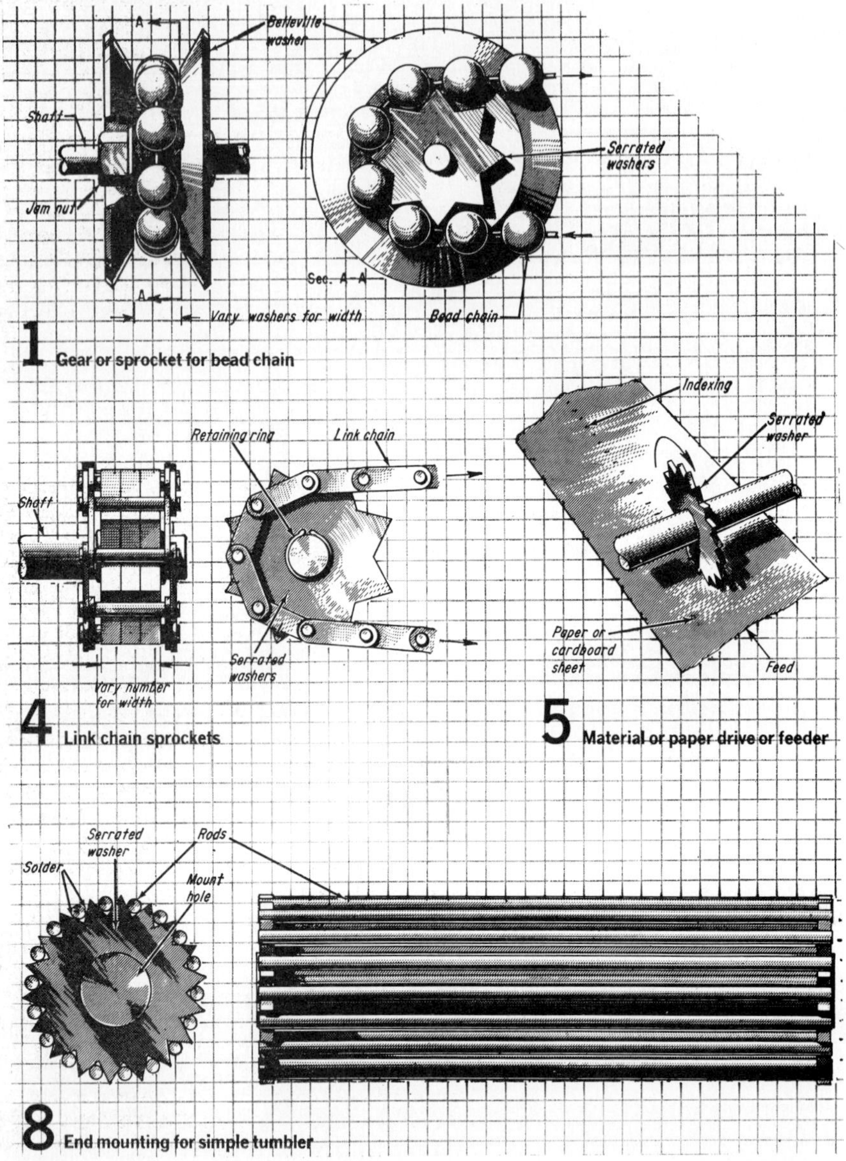

FIG. 14-11 Serrated washers are a stock item and come in a wide variety of sizes. With little thought they can do a variety of jobs. Here are just eight. (*From R. O. Parmley, "Take Another Look at Serrated Washers," in Product Engineering, Sept. 27, 1965. Reprinted by permission, Morgan-Grampian Publishing Co.*)

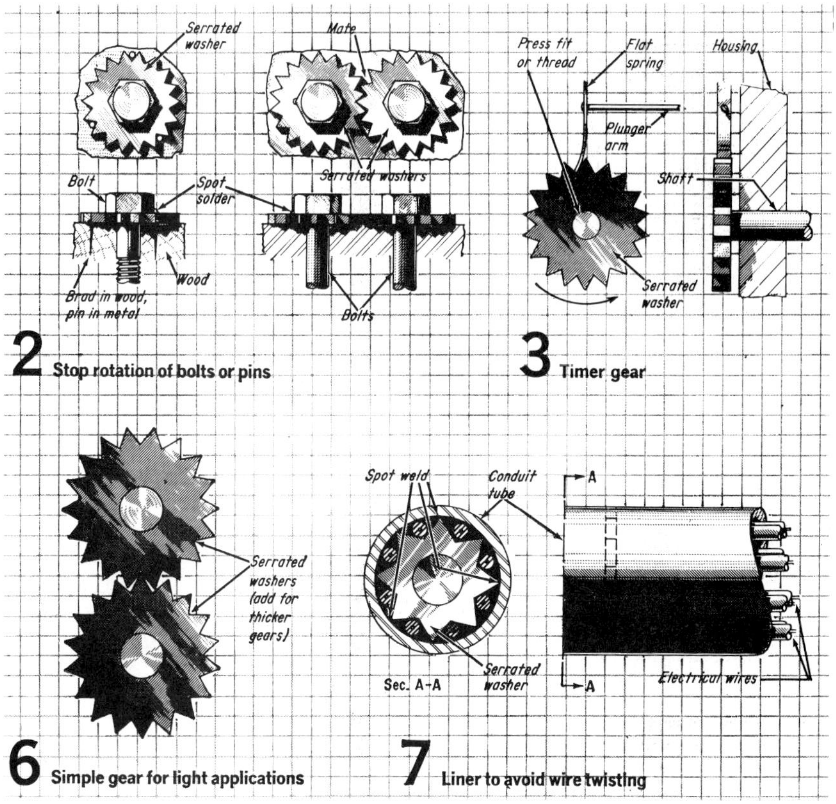

FIG. 14-11 *(continued)*

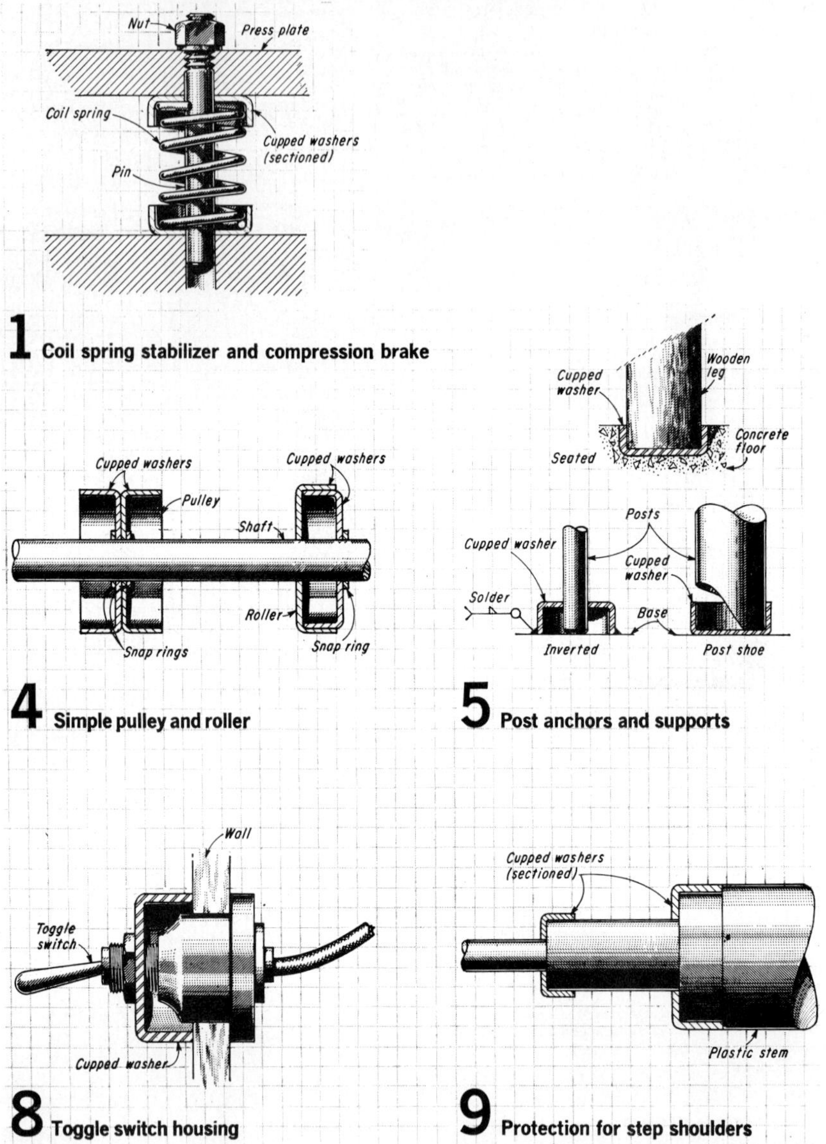

FIG. 14-12 Creative ideas for cupped washers. A standard "off-the-shelf" item with more uses than many ever considered. (*From R. O. Parmley, "Creative Ideas for Cupped Washers," in Product Engineering, Dec. 20, 1965. Reprinted by permission, Morgan-Grampian Publishing Co.*)

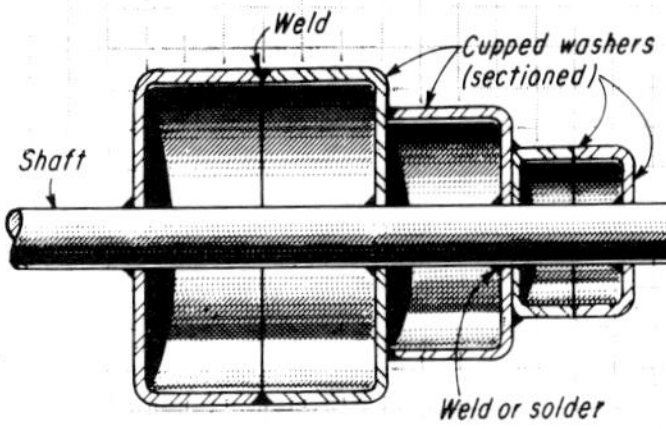

2 Simple step pulley

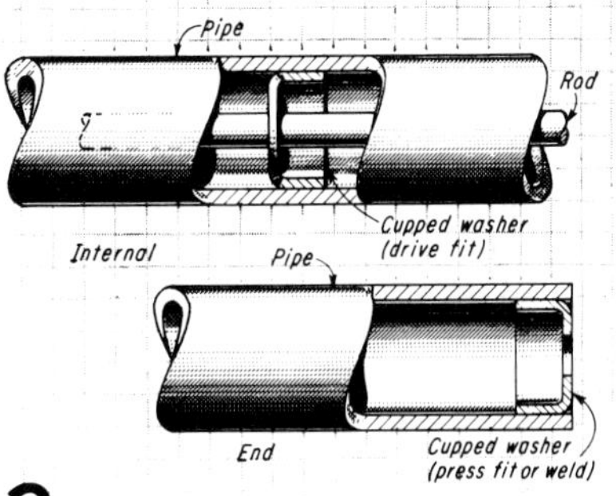

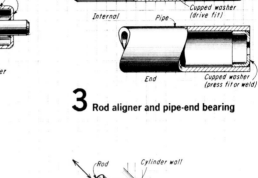

3 Rod aligner and pipe-end bearing

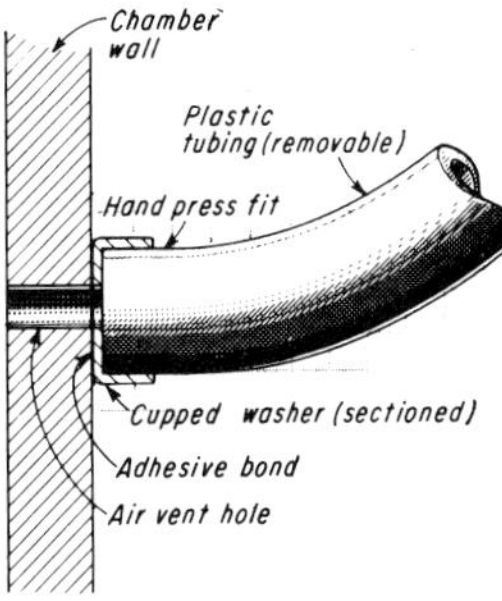

6 Tubing connector

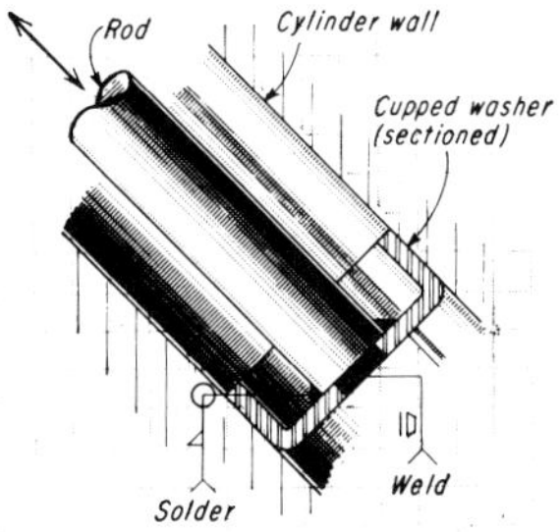

7 Simple piston for cylinder

FIG. 14-12 *(continued)*

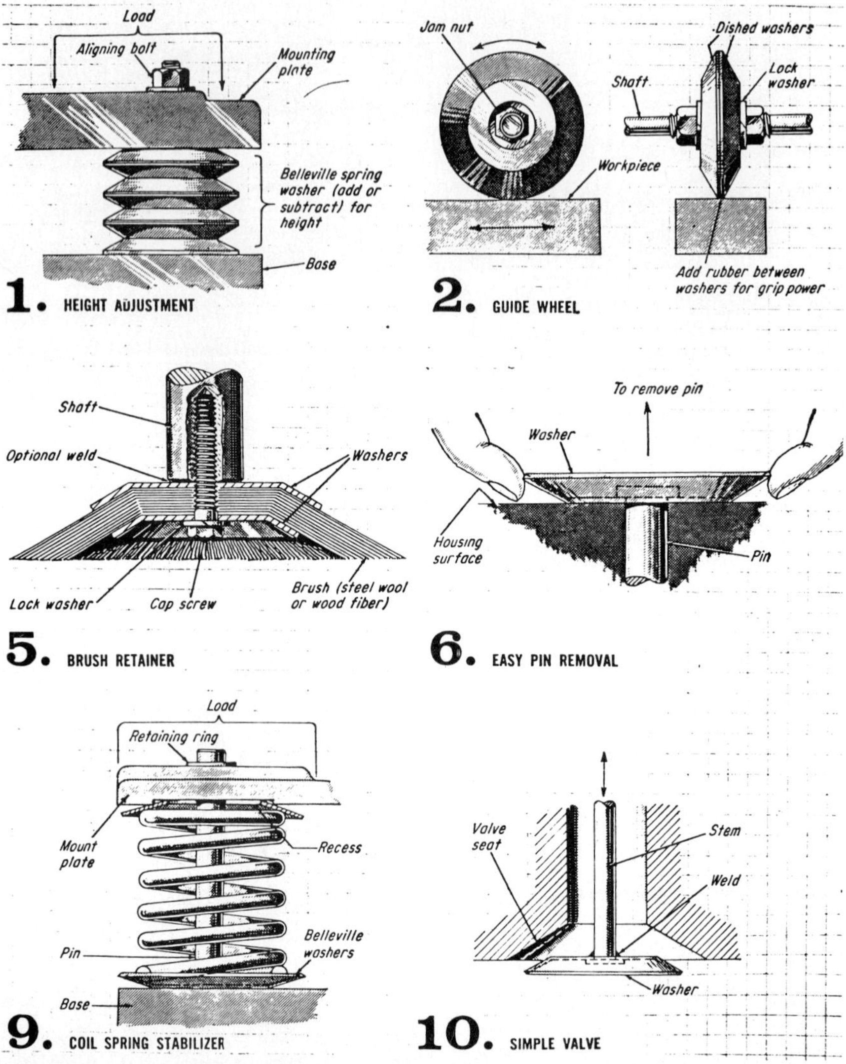

FIG. 14-13 Dished washer designs. Let these ideas spur your own design creativity. Sometimes Belleville washers will suit; otherwise you can easily dish your own. (*From R. O. Parmley, "Dished Washers Get Busy," in Product Engineering, Dec. 7, 1964. Reprinted by permission, Morgan-Grampian Publishing Co.*)

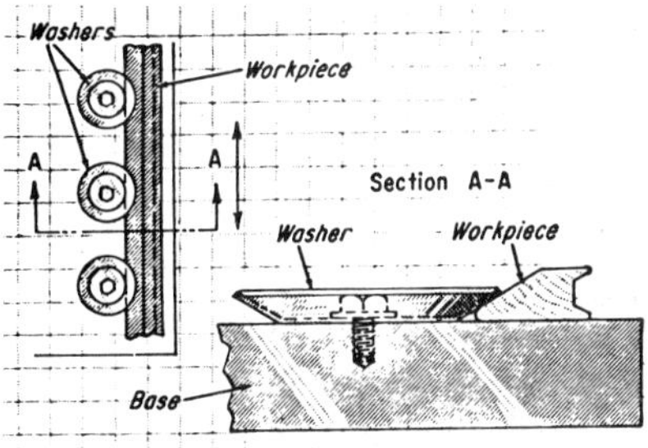

3. ALIGNING BUTTONS
will rotate if holding screw is shouldered

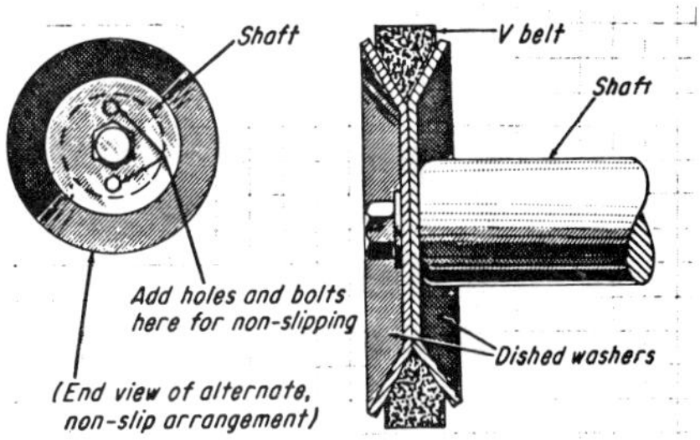

4. V-BELT PULLEY

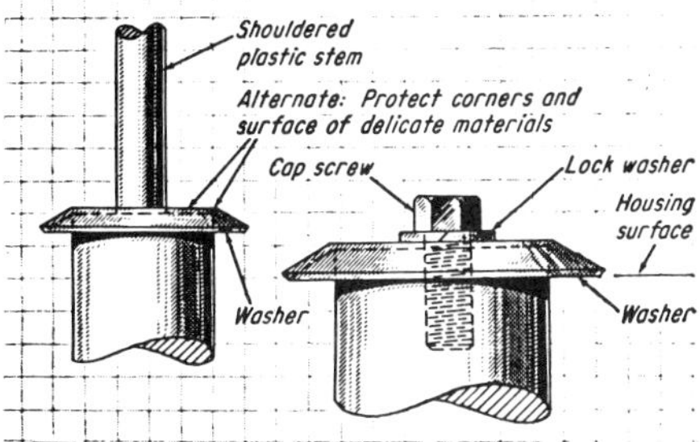

7. END AND CORNER PROTECTION

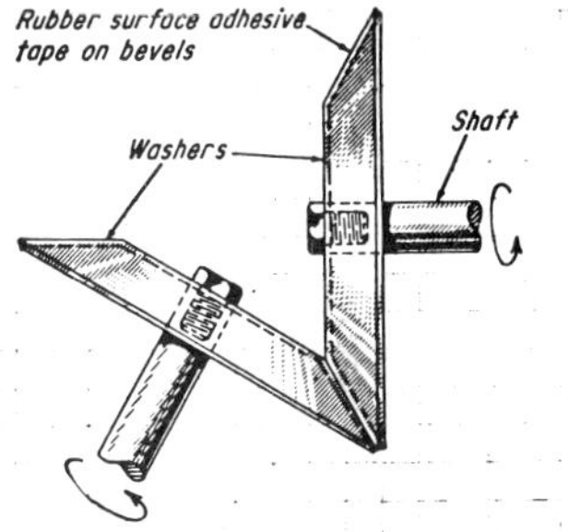

8. SIMPLE BEVEL DRIVE

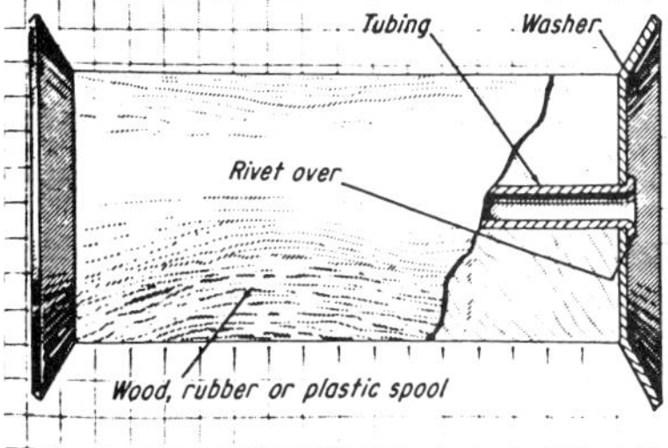

11. FLARED SPOOL-FLANGES

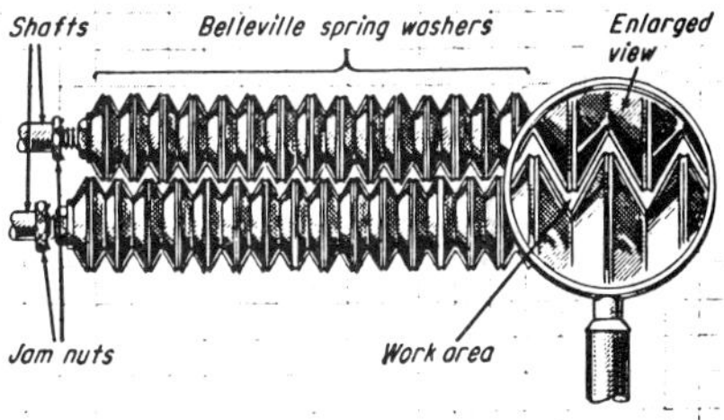

12. CORRUGATING ROLLERS FOR PAPER OR CARDBOARD

FIG. 14-13 *(continued)*

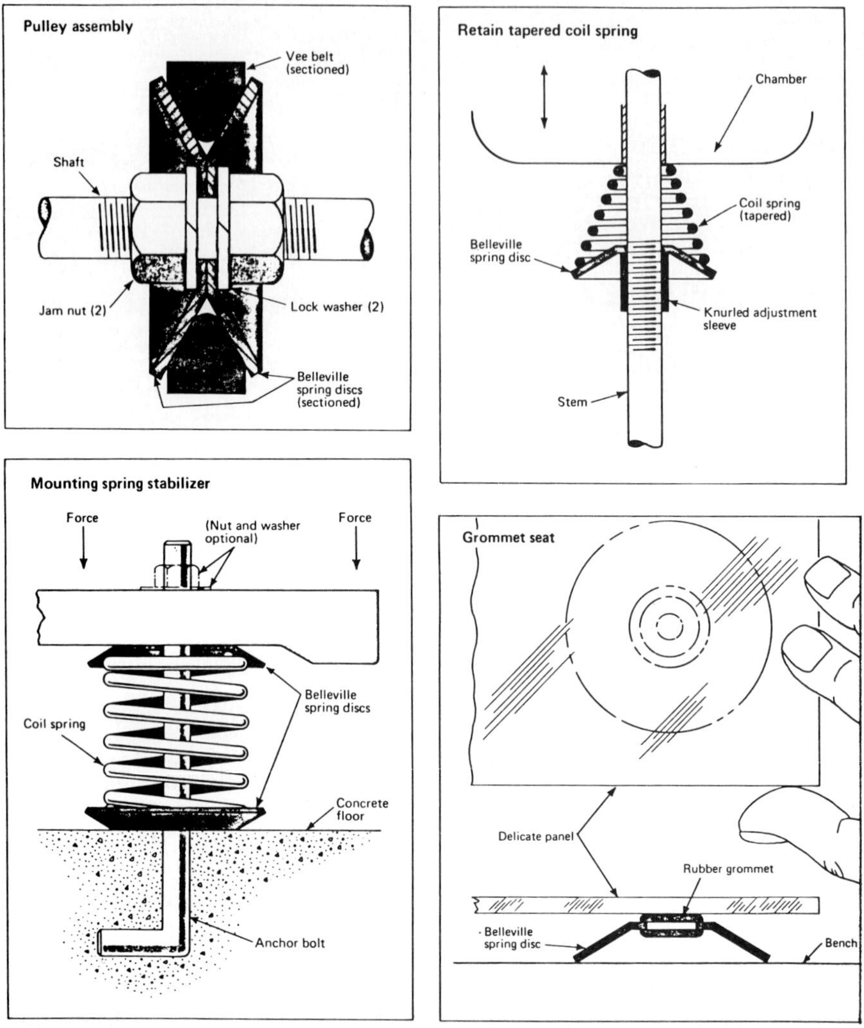

FIG. 14-14 Disk springs solve design problems. Disk springs or Belleville springs are a versatile component that offers a wide range of applications. There are many places where the component can be used and its availability as a stock item should be considered when confronted with a design problem that requires a fast solution. *(From R. O. Parmley, "Disk Springs Are an Easy Way to Solve Some Design Problems," in Product Engineering, February 1973. Reprinted by permission, Morgan-Grampian Publishing Co.)*

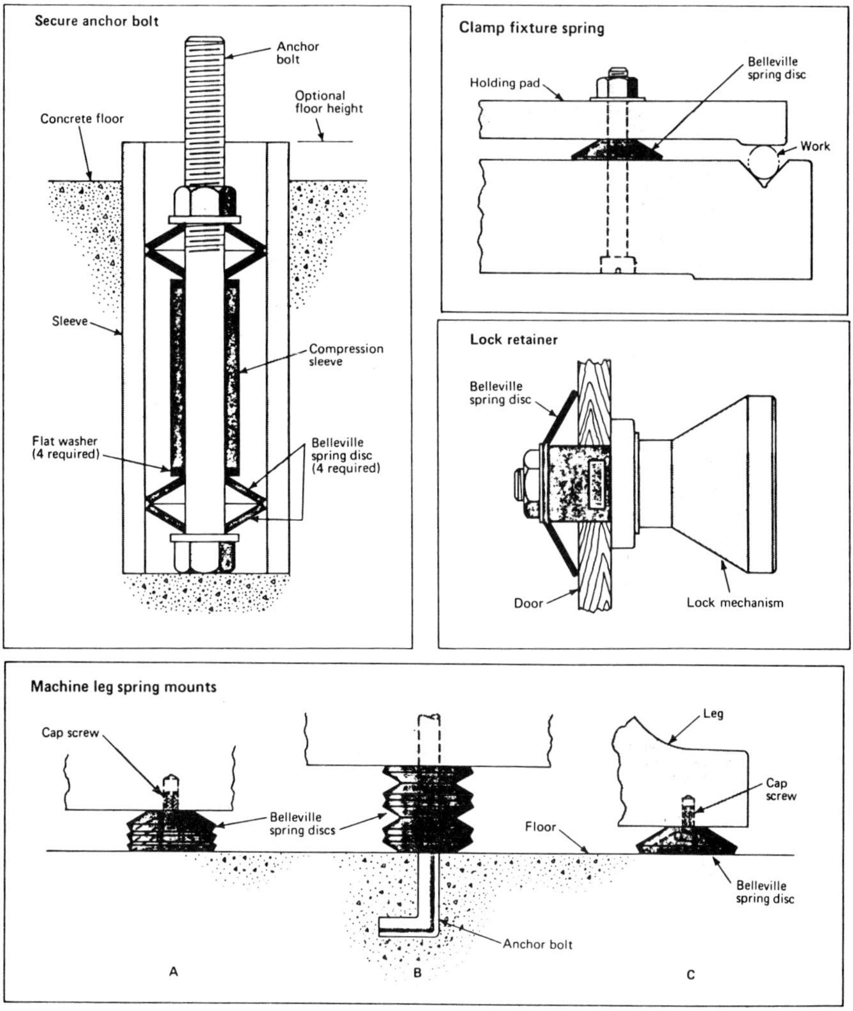

FIG. 14-14 *(continued)*

14-4 Springs

Coil springs have many uses, most of which are well known. However, to illustrate their ever-expanding application, Fig. 14-15 reveals eight examples of how to stiffen or strengthen rubber bellows.

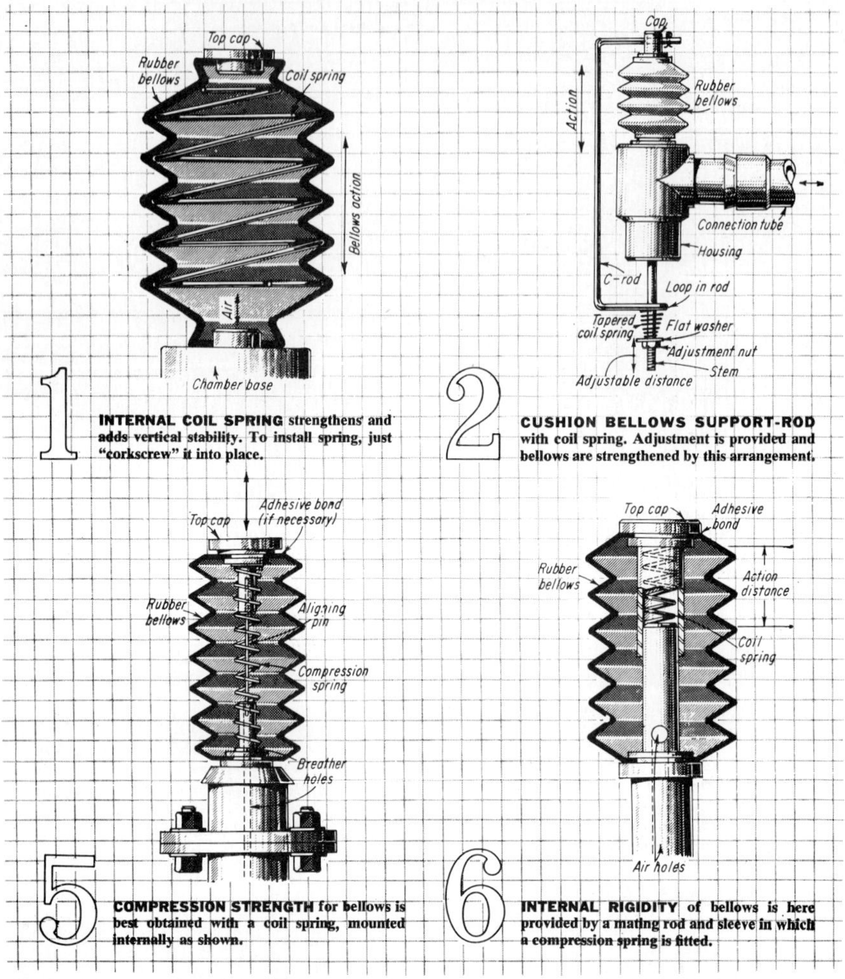

FIG. 14-15 Stiffen bellows with springs. Rubber bellows are an essential part of many products. Here are eight ways to strengthen, cushion, and stabilize bellows with springs. (*From R. O. Parmley, "How To Stiffen Bellows With Springs," in Product Engineering, Oct. 26, 1964. Reprinted by permission, Morgan-Grampian Publishing Co.*)

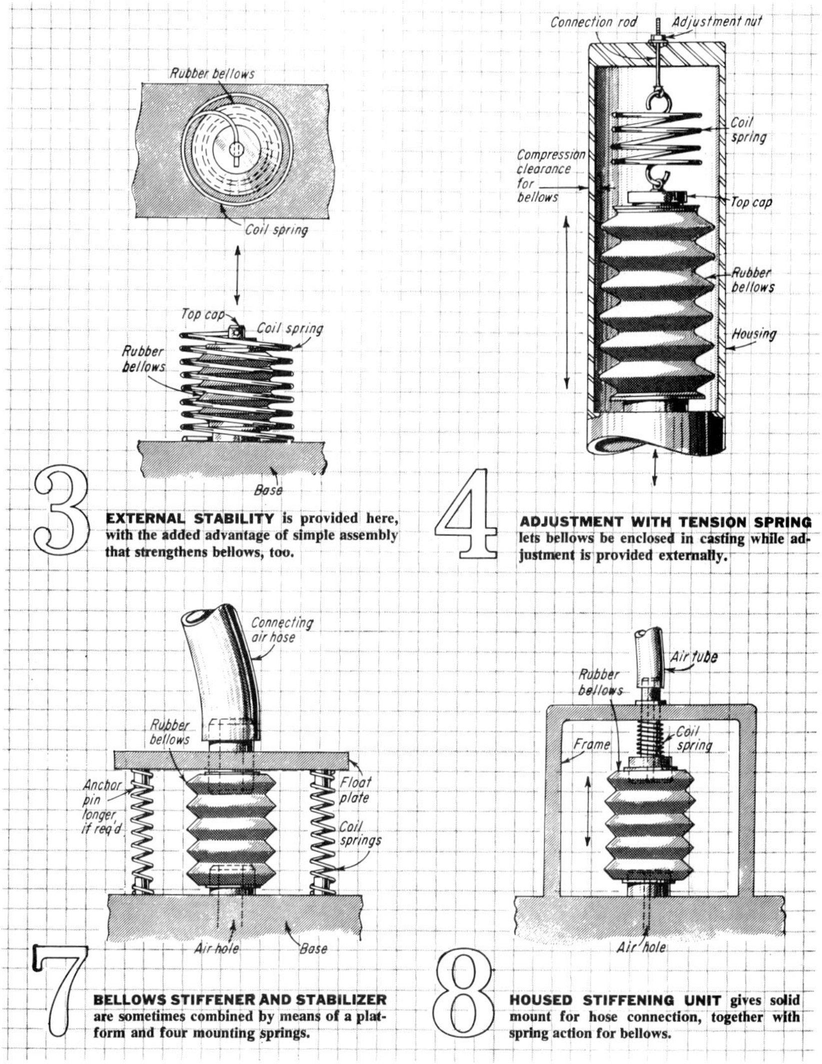

EXTERNAL STABILITY is provided here, with the added advantage of simple assembly that strengthens bellows, too.

ADJUSTMENT WITH TENSION SPRING lets bellows be enclosed in casting while adjustment is provided externally.

BELLOWS STIFFENER AND STABILIZER are sometimes combined by means of a platform and four mounting springs.

HOUSED STIFFENING UNIT gives solid mount for hose connection, together with spring action for bellows.

FIG. 14-15 *(continued)*

14-5 Flanged Bushings

Steel flanged bushings (journal bushings) and flanged rubber bushings are widely known and used. Figures 14-16 and 14-17 depict some common and unusual uses.

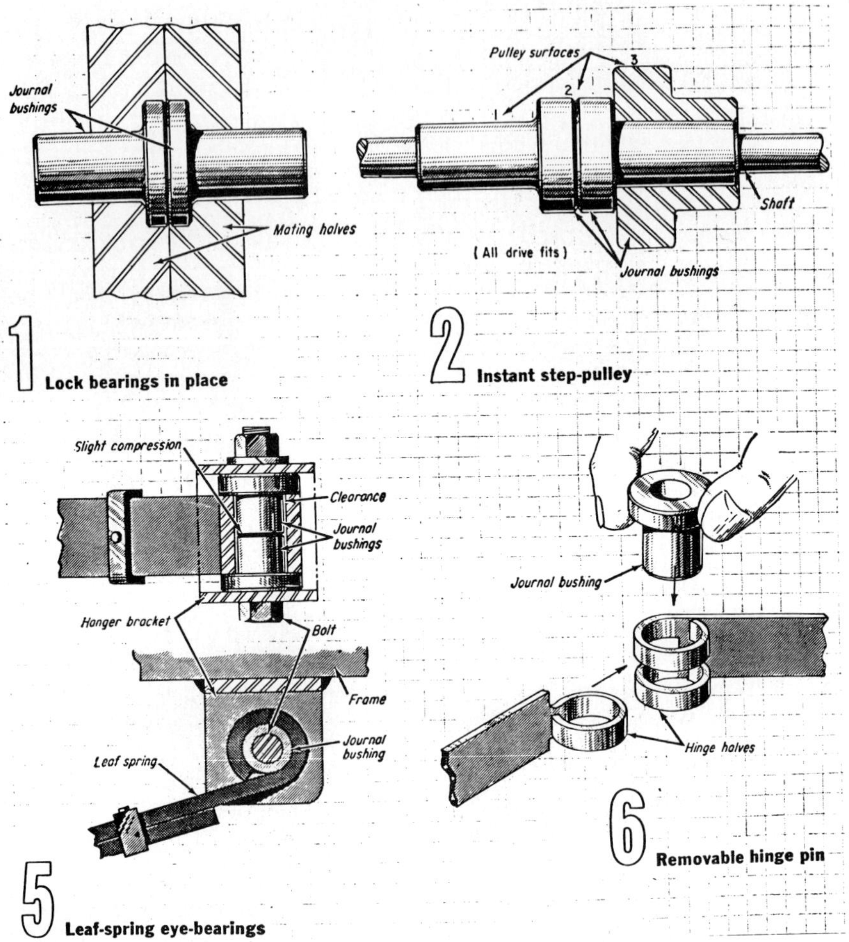

FIG. 14-16 Creative usage of flanged bushings. These sintered bushings find a variety of jobs and are available in 88 sizes, from 1/8 to 1⅝ in internal diameter. (*From R. O. Parmley, "Go Creative With Flanged Bushings," in Product Engineering, March 15, 1965. Reprinted by permission, Morgan-Grampian Publishing Co.*)

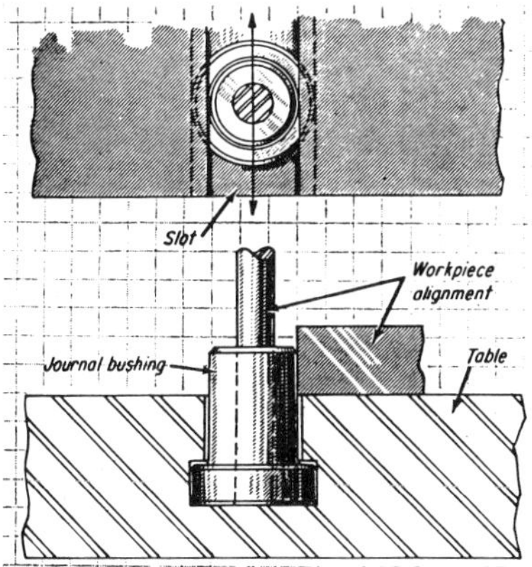

3 **Post or location-pin holder**

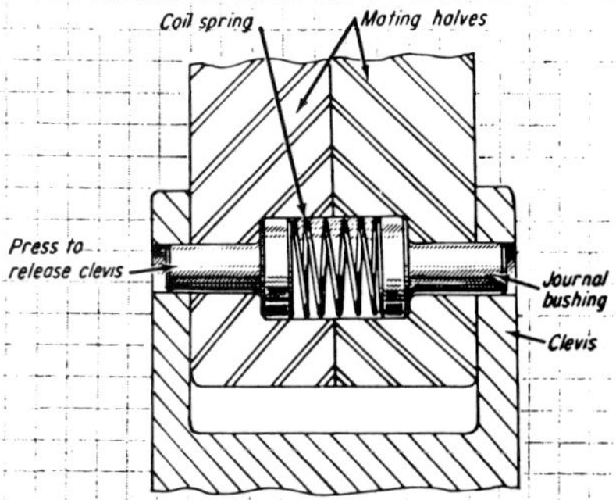

4 **Spring-loaded pins**

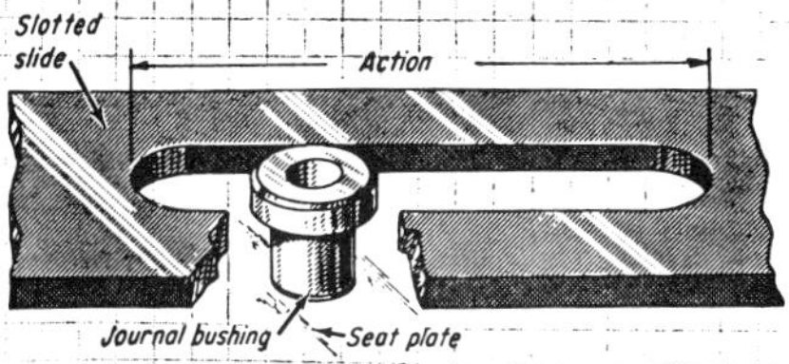

7 **Slider pin is self-lubricating**

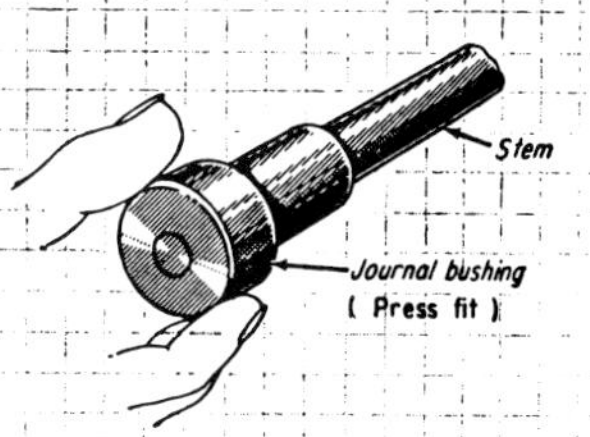

8 **Handle or knob**

FIG. 14-16 *(continued)*

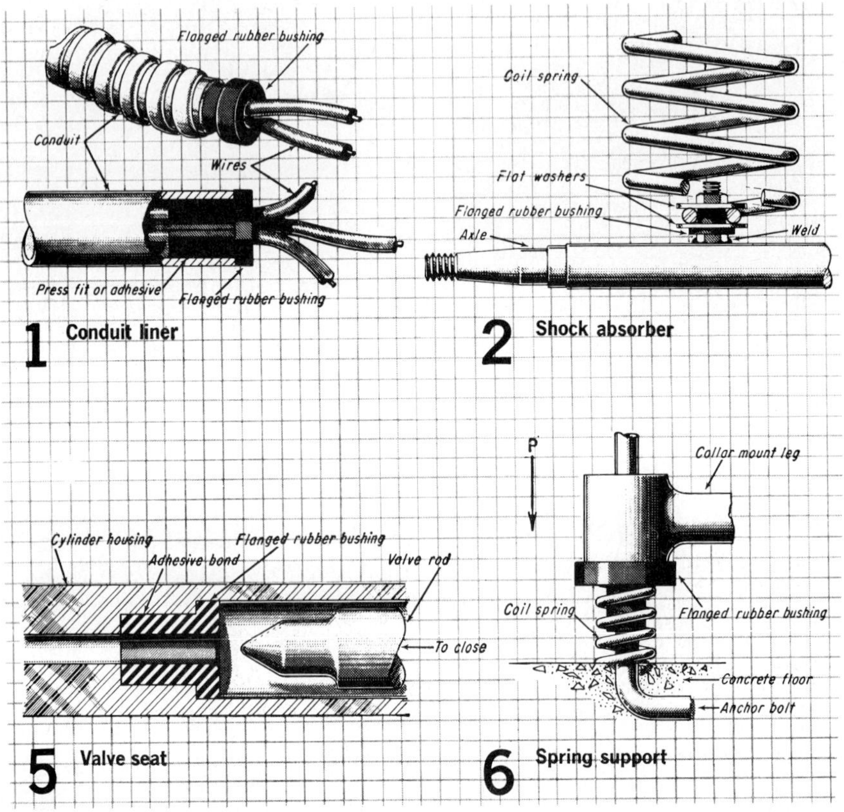

FIG. 14-17 Ideas for flanged rubber bushings are simple, inexpensive, and often overlooked. Check your design for places where rubber bushings may be a solution to a design problem. (*From R. O. Parmley, "Seven Creative Ideas for Flanged Rubber Bushings," Product Engineering, Feb. 14, 1966. Reprinted by permission, Morgan-Grampian Publishing Co.*)

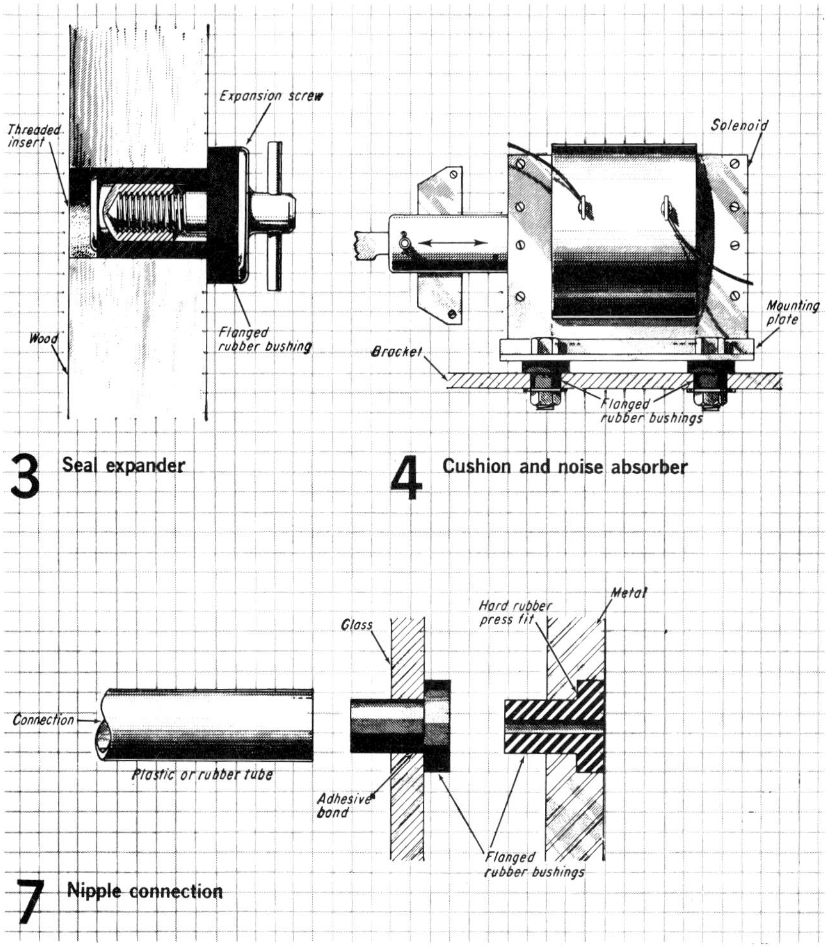

FIG. 14-17 *(continued)*

14-6 Grommets and Bumpers

Rubber grommets and bumpers have a wide range of uses as key components.

Figure 14-18 reveals eight unusual applications of rubber grommets, ranging from seals to shock absorbers.

Rubber mushroom bumpers also serve a broad range of applications, as shown in Fig. 14-19.

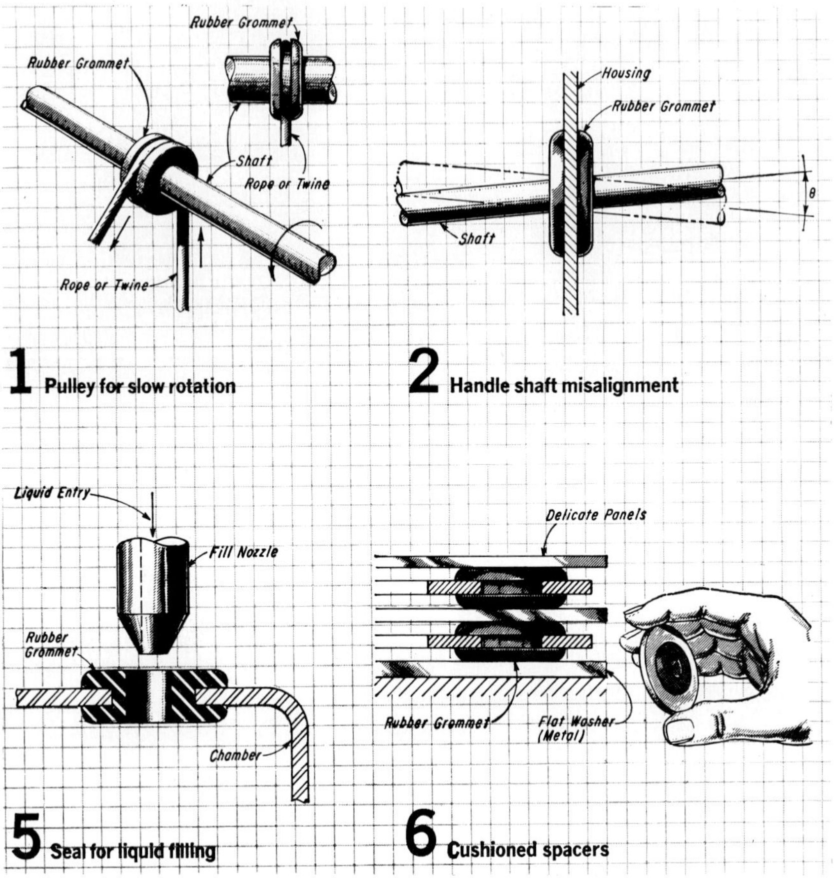

FIG. 14-18 A fresh look at rubber grommets. A small component that is often neglected in the details of a design is shown here in eight unusual applications. (*From R. O. Parmley, "A Fresh Look at Rubber Grommets," Product Engineering, Jan. 3, 1966. Reprinted by permission, Morgan-Grampian Publishing Co.*)

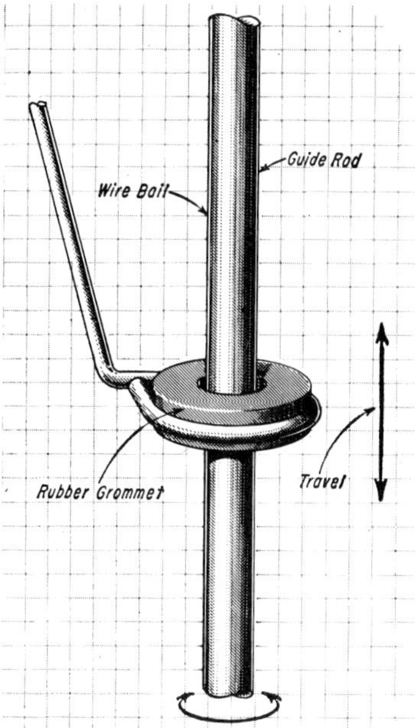

3 **Guide liner**

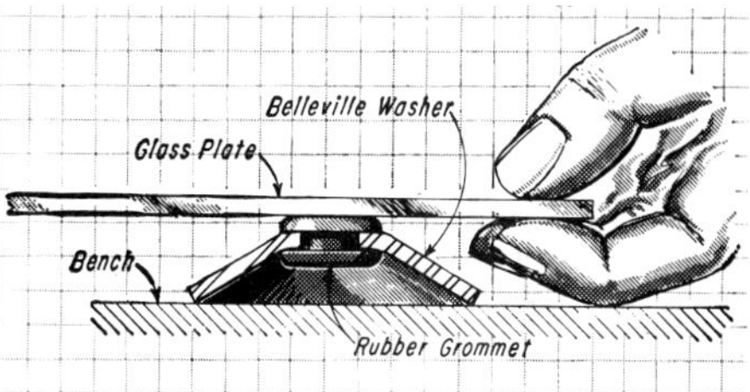

4 **Support delicate work plates**

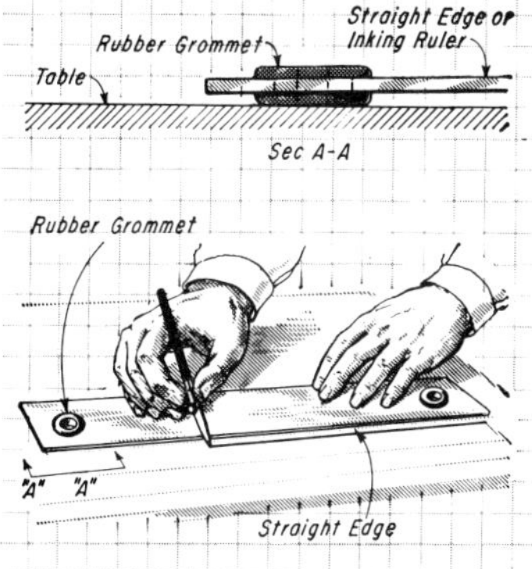

7 **Straightedge pads**

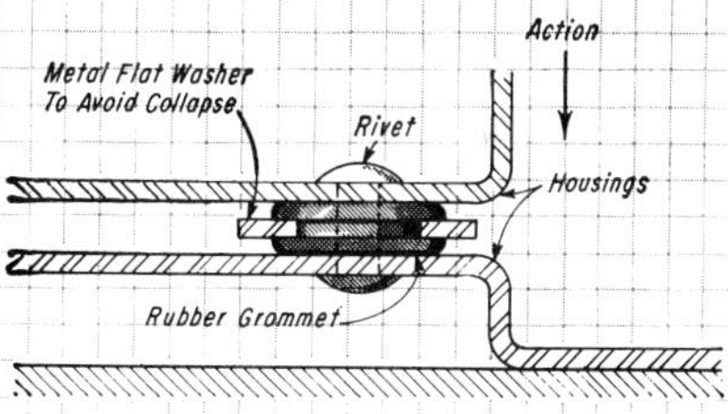

8 **Shock absorber**

FIG. 14-18 *(continued)*

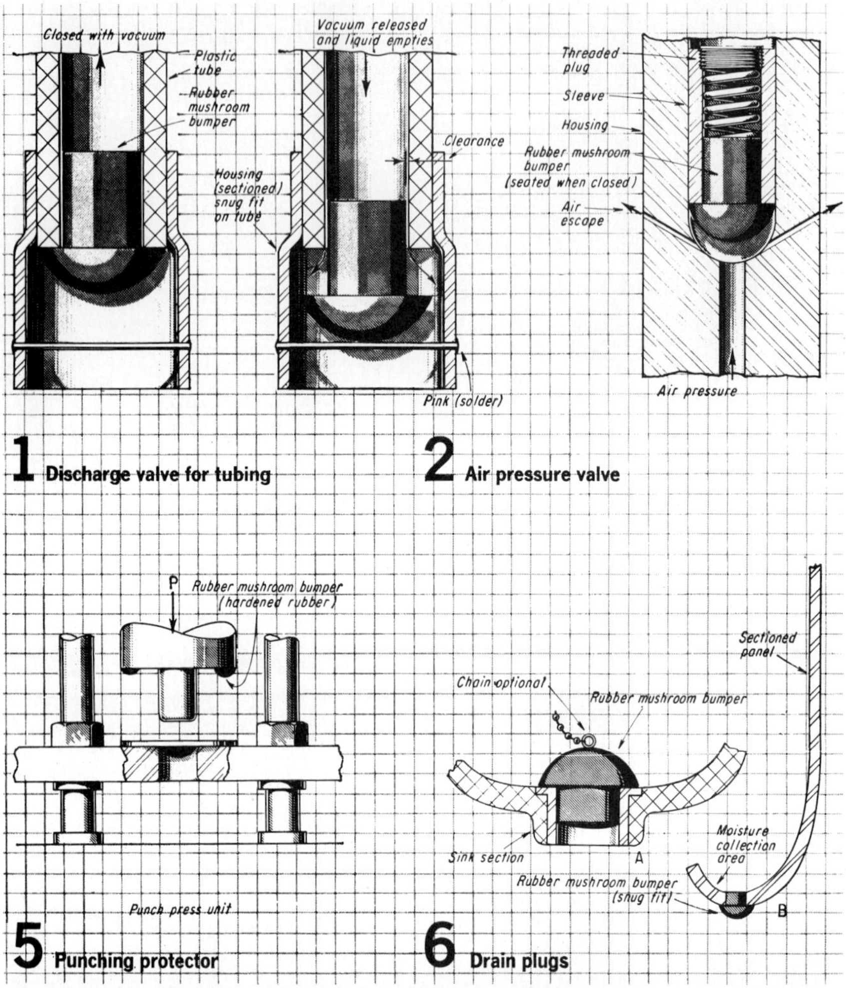

FIG. 14-19 Rubber mushroom bumpers. High energy absorption at low cost is the way mushroom bumpers are usually billed. But they have other uses; here are seven that are rather unconventional. (*From R. O. Parmley, "Odd Jobs for Rubber Mushroom Bumpers," Product Engineering, Jan. 31, 1966. Reprinted by permission, Morgan-Grampian Publishing Co.*)

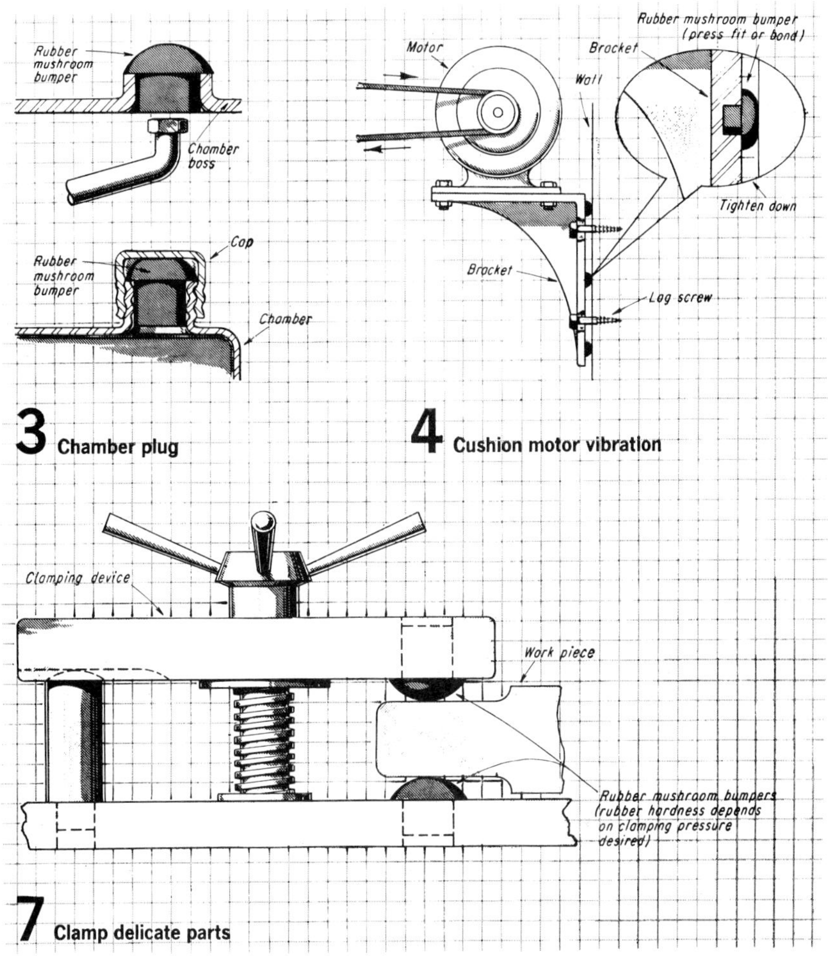

FIG. 14-19 *(continued)*

14-7 Chains

Chains, of course, have long been used, mainly as transmitters of power. For light power transfer, bead chain is excellent. Figure 14-20 illustrates some basic and unusual applications.

Figure 14-21 shows some unusual designs of joints for conveyor chain systems which employ some basic stock components.

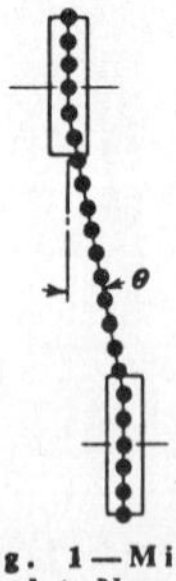

Fig. 1—Misaligned sprockets. Nonparallel planes usually occur when alignment is too expensive to maintain. Bead chain can operate at angles up to θ=20 degrees.

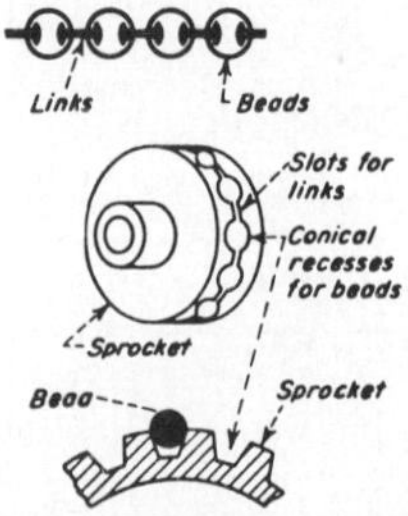

Fig. 2—Details of bead chain and sprocket. Beads of chain seat themselves firmly in conical recesses in the face of sprocket. Links ride freely in slots between recesses in sprocket.

Fig. 3—Skewed shafts normally acquire two sets of spiral gears to bridge space between shafts. Angle misalignment does not interfere with qualified bead chain operation on sprockets.

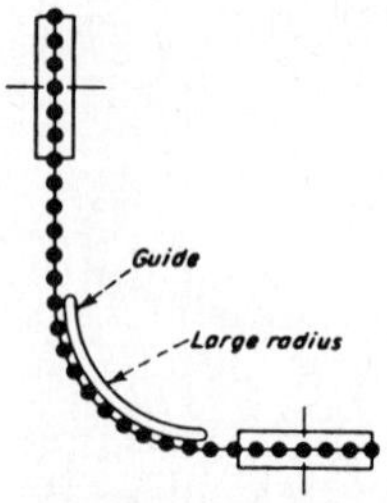

Fig. 4—Right angle drive does not require idler sprockets to go around corner. Suitable only for very low torque application because of friction drag of bead chain against guide.

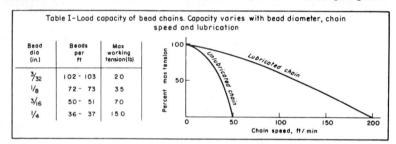

Table I-Load capacity of bead chains. Capacity varies with bead diameter, chain speed and lubrication

Bead dia (in.)	Beads per ft	Max working tension(lb)
3/32	102 - 103	20
1/8	72 - 73	35
3/16	50 - 51	70
1/4	36 - 37	150

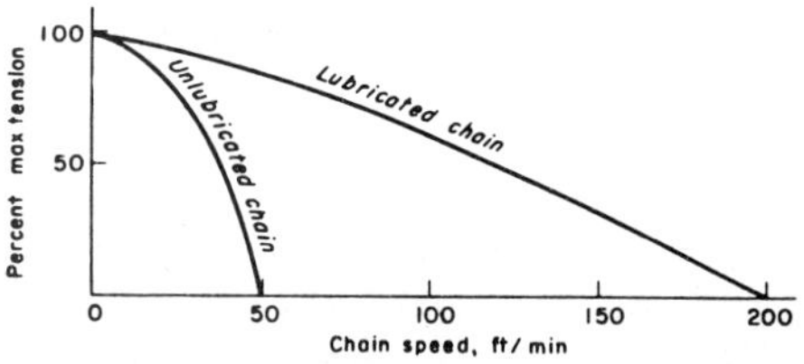

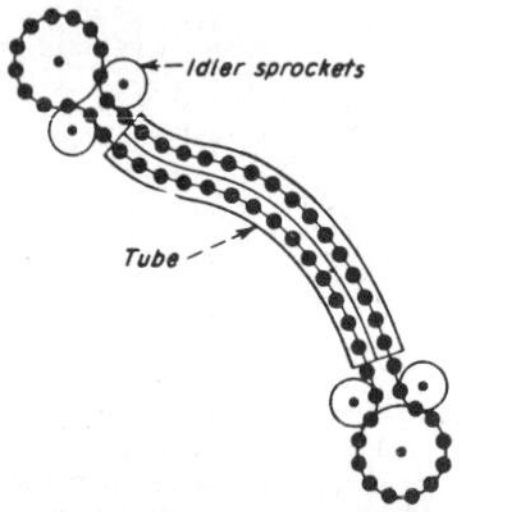

Fig. 5—Remote control through rigid or flexible tube has almost no backlash and can keep input and output shafts synchronized.

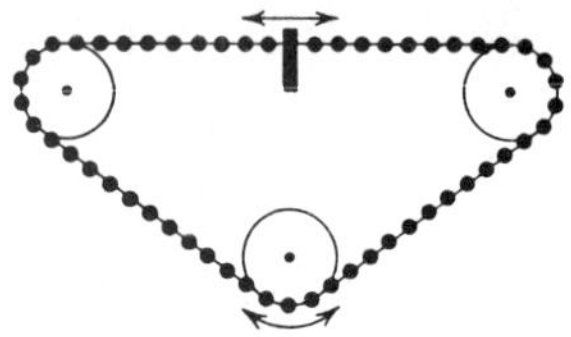

Fig. 6—Linear output from rotary input. Beads prevent slippage and maintain accurate ratio between the input and output displacemonts.

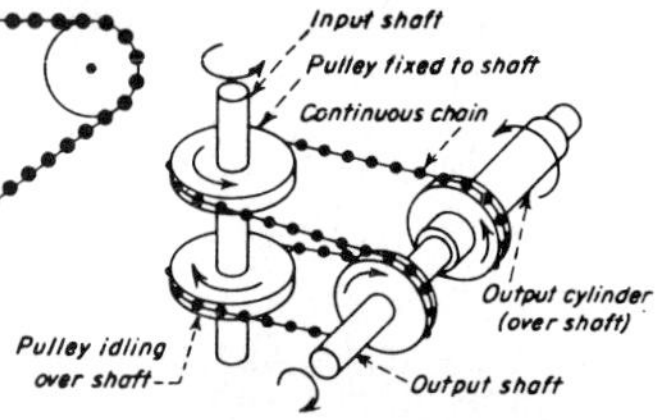

Fig. 7—Counter-rotating shafts. Input shaft drives two counter-rotating outputs (shaft and cylinder) through a continuous chain.

FIG. 14-20 Bead chains for light service. Where torque requirements and operating speeds are low, qualified bead chains offer a quick and economical way to couple misaligned shafts, convert from one type of motion to another, counterrotate shafts, obtain high-ratio drives and overload protection, control switches, and serve as mechanical counters. (*From Bernard Wasko, "Bead Chains for Light Service," Product Engineering and Product Engineering Design Manual, edited by D. C. Greenwood, 1959. Reprinted by permission, Morgan-Grampian Publishing Co.*)

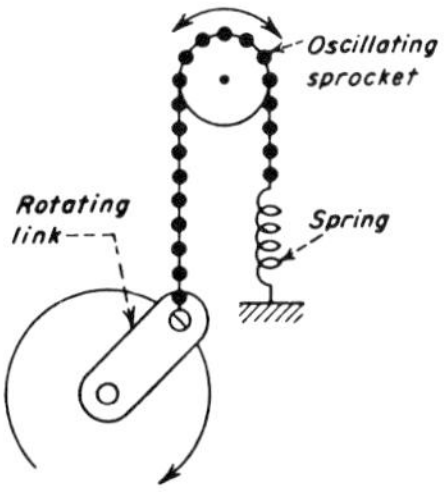

Fig. 8—Angular oscillations from rotary input. Link makes complete revolutions causing sprocket to oscillate. Spring maintains chain tension.

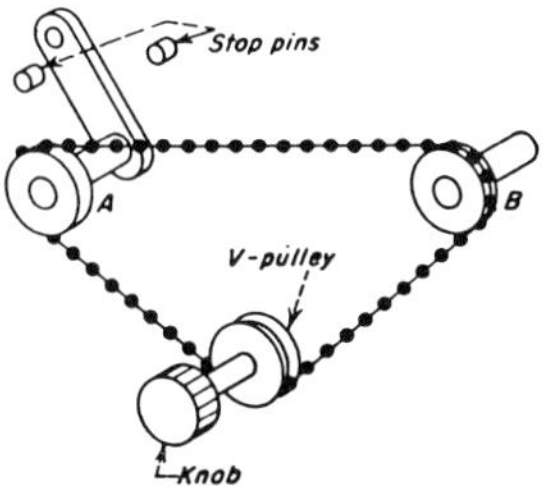

Fig. 9—Restricted angular motion. Pulley, rotated by knob, slips when limit stop is reached; shafts A and B remain stationary and synchronous.

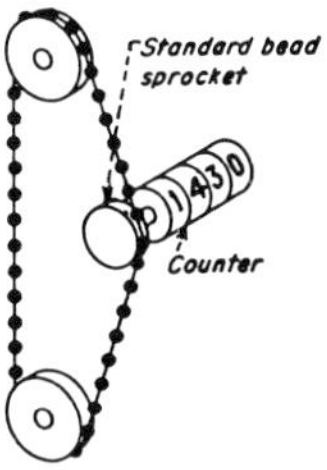

Fig. 10—Remote control of counter. For applications where counter cannot be coupled directly to shaft, bead chain and sprockets can be used.

Fig. 11—High-ratio drive less expensive than gear trains. Qualified bead chains and sprockets will transmit power without slippage.

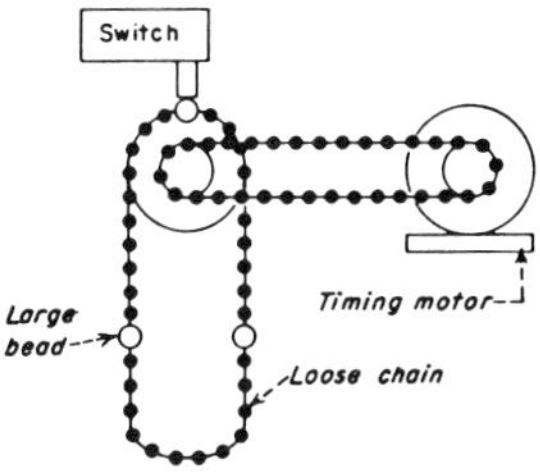

Fig. 12—Timing chain containing large beads at desired intervals operates microswitch. Chain can be lengthened to contain thousands of intervals for complex timing.

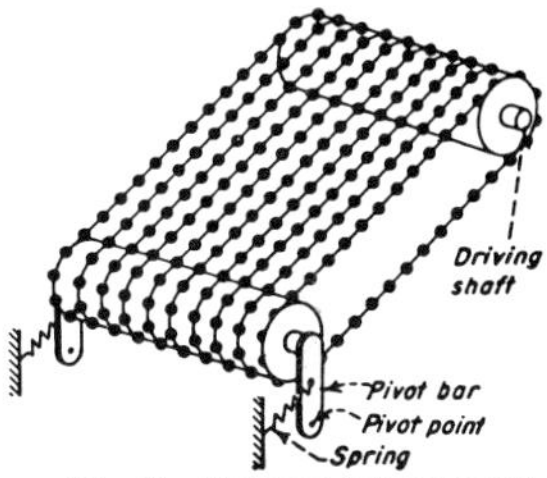

Fig. 13—Conveyor belt composed of multiple chains and sprockets. Tension maintained by pivot bar and spring. Width of belt easily changed.

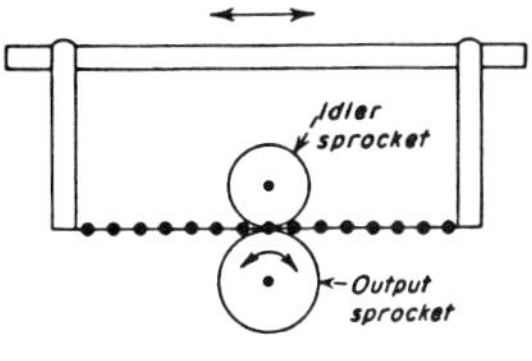

Fig. 14—Gear and rack duplicated by chain and two sprockets. Converts linear motion into rotary motion.

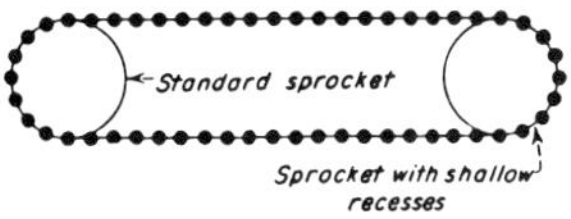

Fig. 15 — Overload protection. Shallow sprocket gives positive drive for low loads; slips one bead at a time when overloaded.

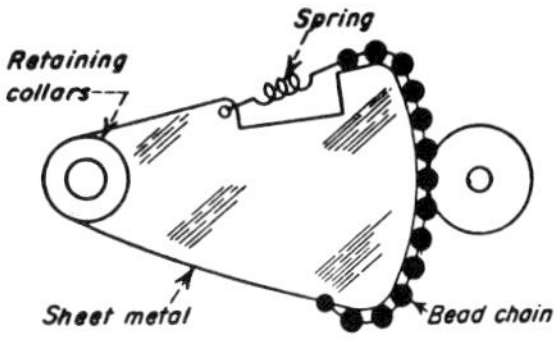

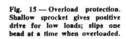

Fig. 16—Gear segment inexpensively made with bead chain and spring wrapped around edge of sheet metal. Retaining collars keep sheet metal sector from twisting on the shaft.

FIG. 14-20 *(continued)*

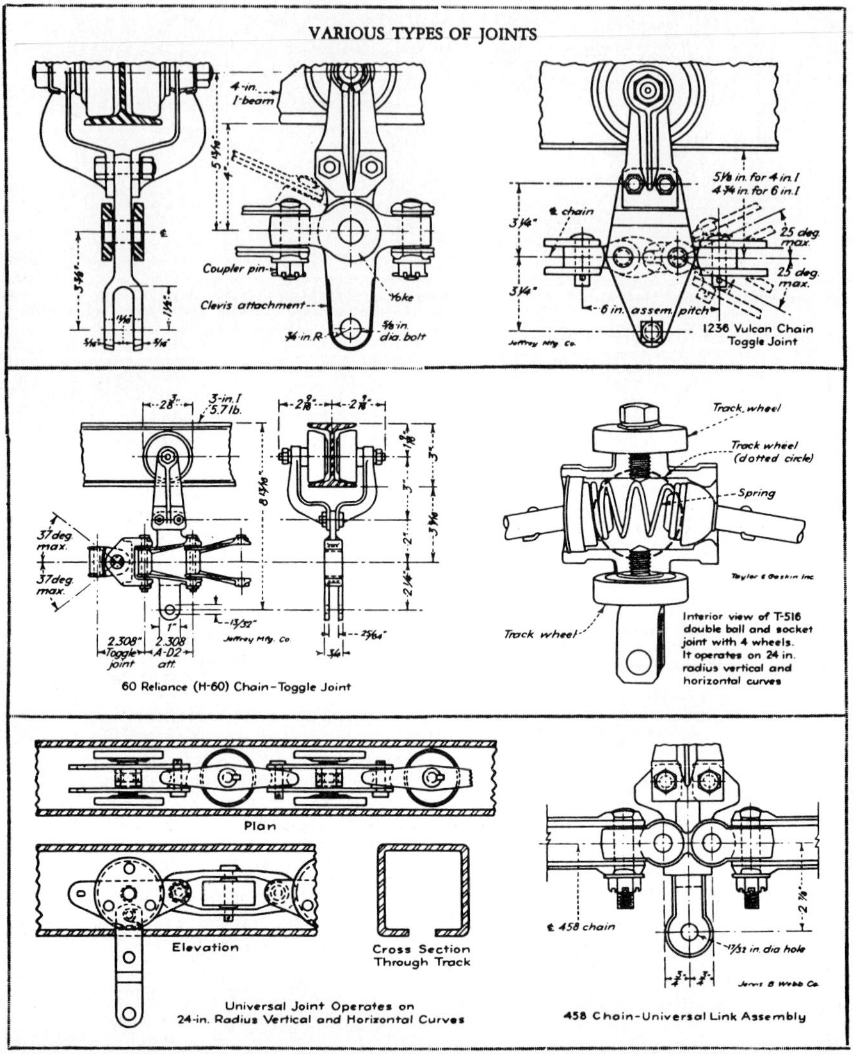

FIG. 14-21 Trolley conveyor chain links. Joints used in conveyor chain systems are designed for specific applications. The type of joint specified for a certain installation depends entirely on the layout of the supporting track. Joints can be designed for use with bends in the vertical or horizontal planes and for combination service. The success of the overhead trolley conveyor is largely the result of the development and use of drop-forged, rivetless, keystone chain. The dimensions of several sizes of keystone chain links are shown below with two examples of pin-jointed chain. Standard keystone chain parts are shown in three views. (*From Sidney Reibel, "Types of Trolley Conveyor Chain Links and Joints," Product Engineering and Mechanical Details for Product Design, edited by D. C. Greenwood, 1964. Reprinted by permission, Morgan-Grampian Publishing Co.*)

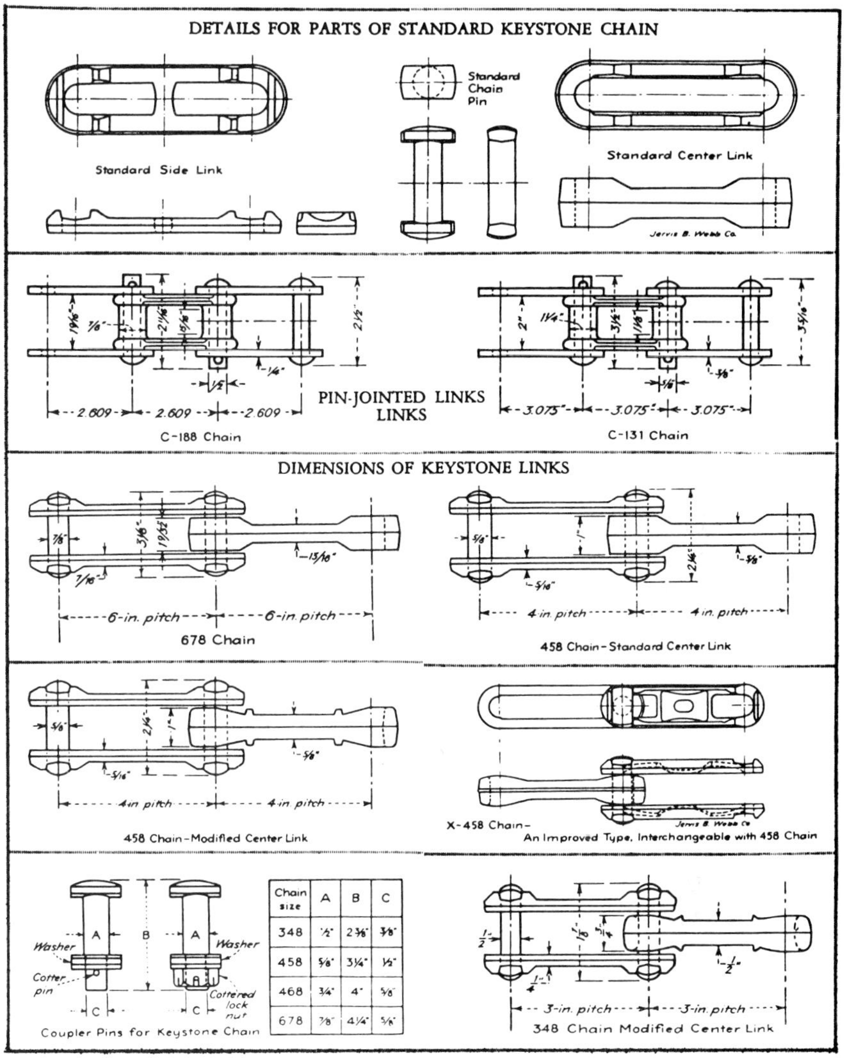

Chain size	A	B	C
348	½"	2⅜"	⅜"
458	⅝"	3¼"	½"
468	¾"	4"	⅝"
678	⅞"	4¼"	⅝"

FIG. 14-21 *(continued)*

14-8 Backlash Prevention

Since backlash prevention is an ever-present problem and one that sometimes requires unusual solutions, we have included some examples in which stock components are used to effectively prevent backlash. See Figs. 14-22, 14-23, and 14-24.

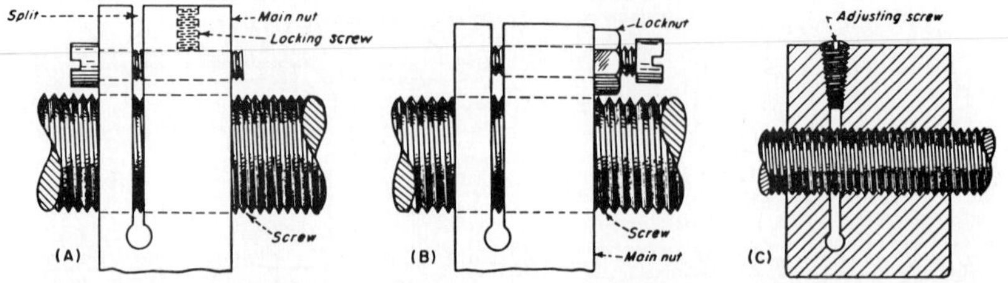

THREE METHODS of using slotted nuts. In *(A)*, nut sections are brought closer together to force left-hand nut flanks to bear on right-hand flanks of screw thread and vice versa. In *(B)*, and *(C)* nut sections are forced apart for same purpose.

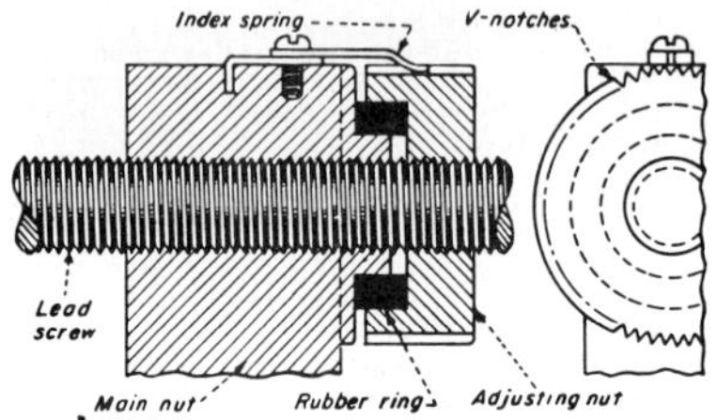

AROUND THE PERIPHERY of the backlash-adjusting nut are "v" notches of small pitch which engage the index spring. To eliminate play in the lead screw, adjusting nut is turned clockwise. Spring and adjusting nut can be calibrated for precise use.

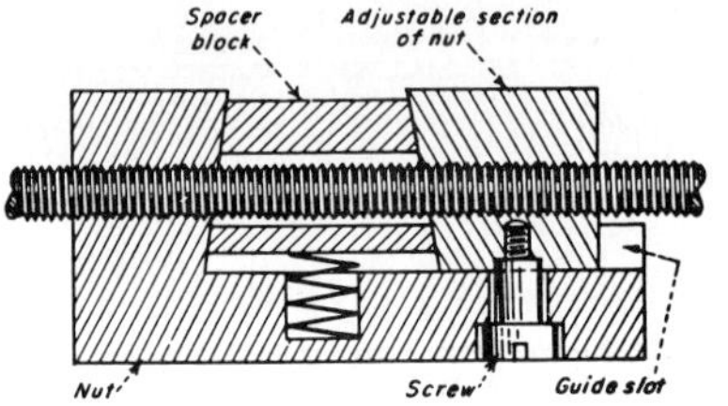

SELF-COMPENSATING MEANS of removing backlash. Slot is milled in nut for an adjustable section which is locked by a screw. Spring presses the tapered spacer block upwards, forcing the nut elements apart, thereby taking up backlash.

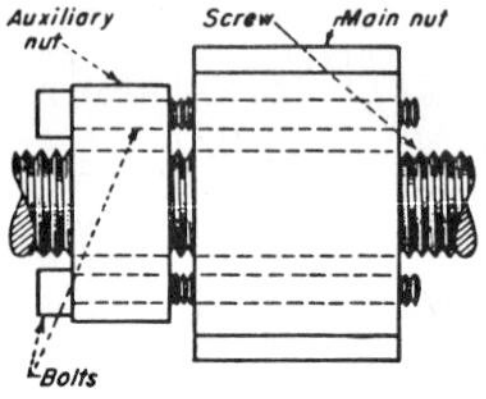

MAIN NUT is integral with base attached to part moved by screw. Auxiliary nut is positioned one or two pitches from main nut. The two are brought closer together by bolts which pass freely through the auxiliary nut.

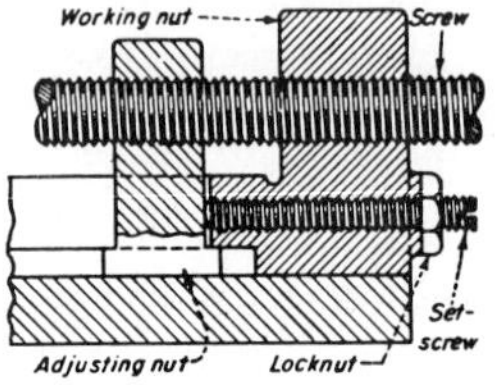

ANOTHER WAY to use an auxiliary or adjusting nut for axial adjustment of backlash. Relative movement between the working and adjusting nuts is obtained manually by the set screw which can be locked in place as shown.

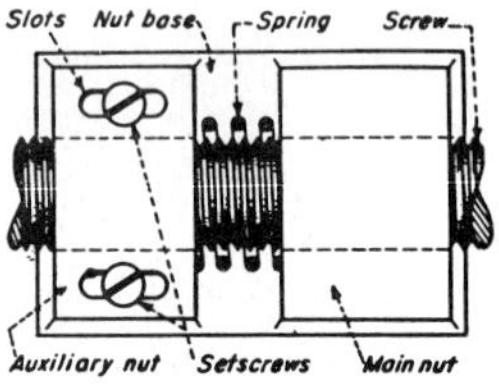

COMPRESSION SPRING placed between main and auxiliary nuts exerts force tending to separate them and thus take up slack. Set screws engage nut base and prevent rotation of auxiliary nut after adjustment is made.

FIG. 14-22 Provide for backlash in threaded parts. These illustrations are based on two general methods of providing for lost motion or backlash. One allows for relative movement of the nut and screw in the plane parallel to the thread axis; the other method involves a radial adjustment to compensate for clearance between sloping faces of the threads on each element. *(From Clifford T. Bower, "How To Provide For Backlash in Threaded Parts," Product Engineering and Product Engineering Design Manual, edited by D. C. Greenwood, 1959. Reprinted by permission, Morgan-Grampian Publishing Co.)*

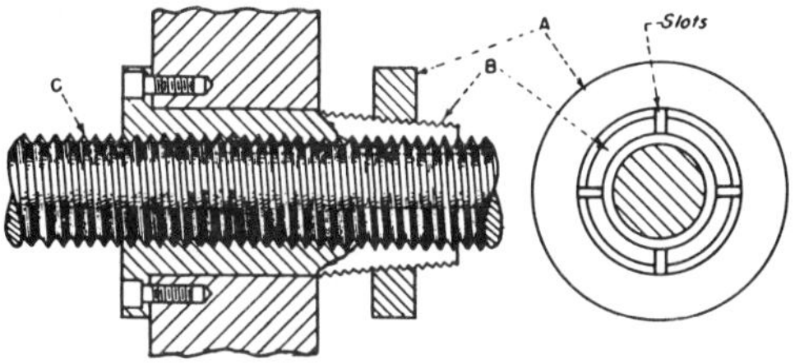

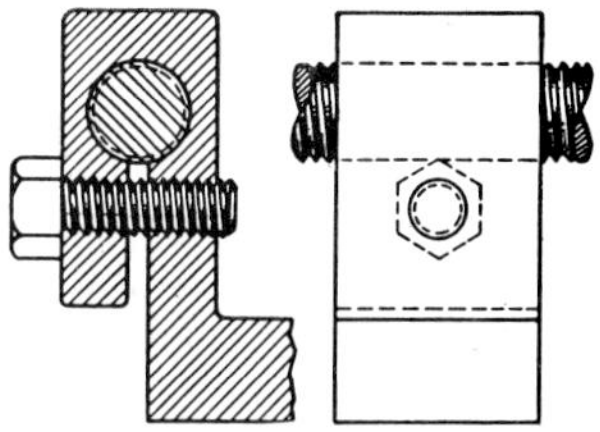

NUT *A* IS SCREWED along the tapered round nut, *B*, to eliminate backlash or wear between *B* and *C*, the main screw, by means of the four slots shown.

ANOTHER METHOD of clamping a nut around a screw to reduce radial clearance.

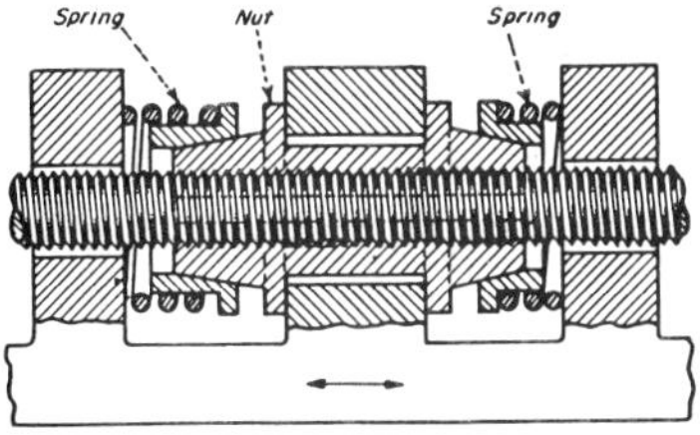

AUTOMATIC ADJUSTMENT for backlash. Nut is flanged on each end, has a square outer section between flanges and slots cut in the tapered sections. Spring forces have components which push slotted sections radially inward.

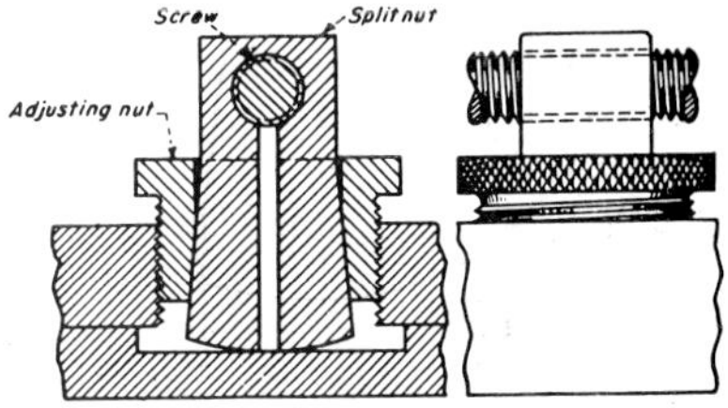

SPLIT NUT is tapered and has a rounded bottom to maintain as near as possible a fixed distance between its seat and the center line of the screw. When the adjusting nut is tightened, the split nut springs inward slightly.

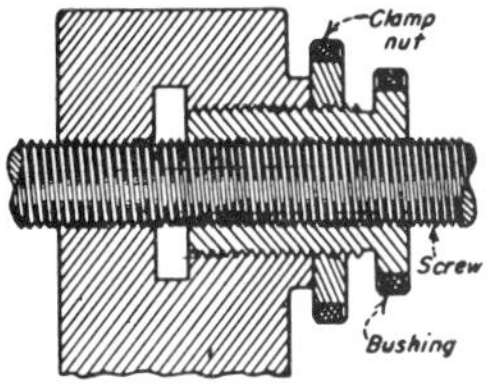

CLAMP NUT holds adjusting bushing rigidly. Bushing must have different pitch on outside thread than on inside thread. If outer thread is the coarser one, a relatively small amount of rotation will take up backlash.

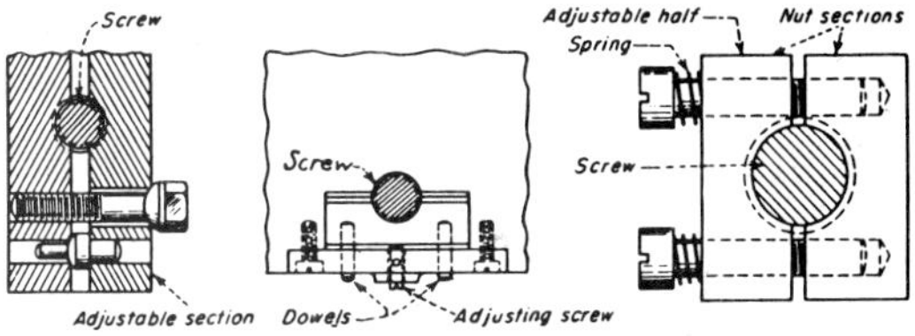

TYPICAL CONSTRUCTIONS based on the half nut principle. In each case, the nut bearing width is equal to the width of the adjustable or inserted slide piece. In the sketch at the extreme left, the cap screw with the spherical seat provides for adjustments. In the center sketch, the adjusting screw bears on the movable nut section. Two dowels insure proper alignment. The third illustration is similar to the first except that two adjusting screws are used instead of only one.

FIG. 14-22 *(continued)*

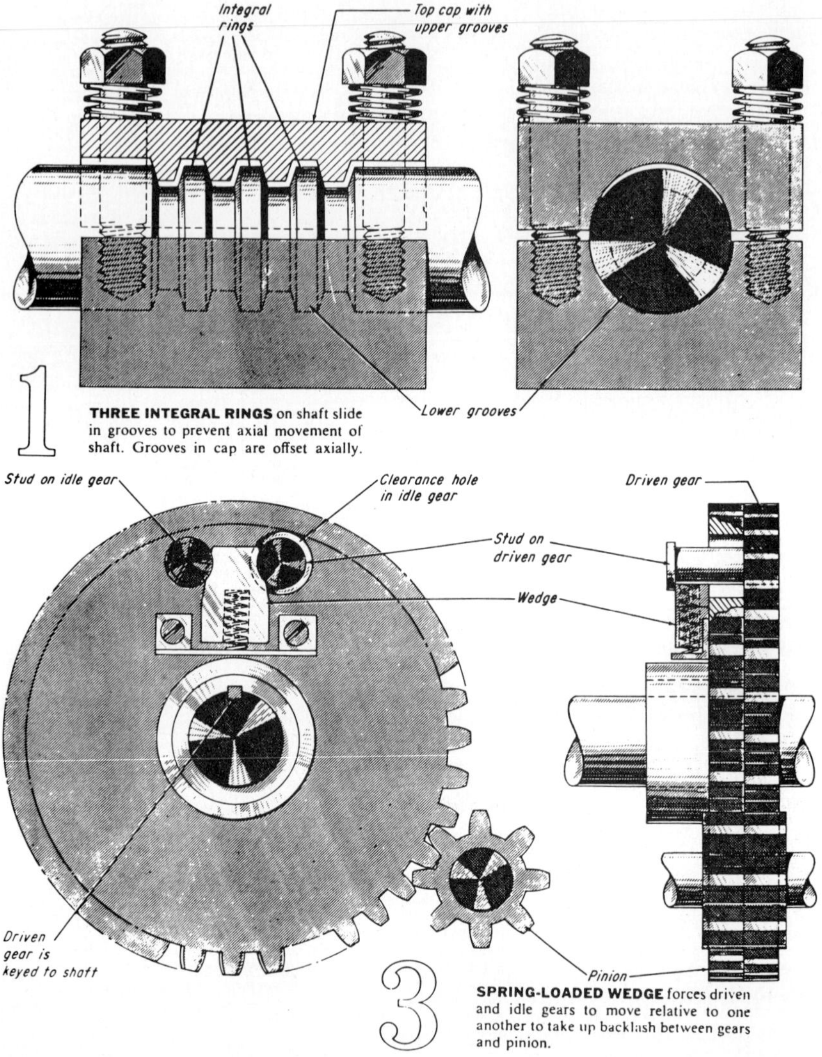

THREE INTEGRAL RINGS on shaft slide in grooves to prevent axial movement of shaft. Grooves in cap are offset axially.

SPRING-LOADED WEDGE forces driven and idle gears to move relative to one another to take up backlash between gears and pinion.

FIG. 14-23 Eliminate backlash. Wedges take up freedom in threads and gears and hold shafts snug against bearings. (*From L. Kasper, "4 Ways to Eliminate Backlash," Product Engineering and Mechanical Details for Product Design, edited by D. C. Greenwood, 1964. Reprinted by permission, Morgan-Grampian Publishing Co.*)

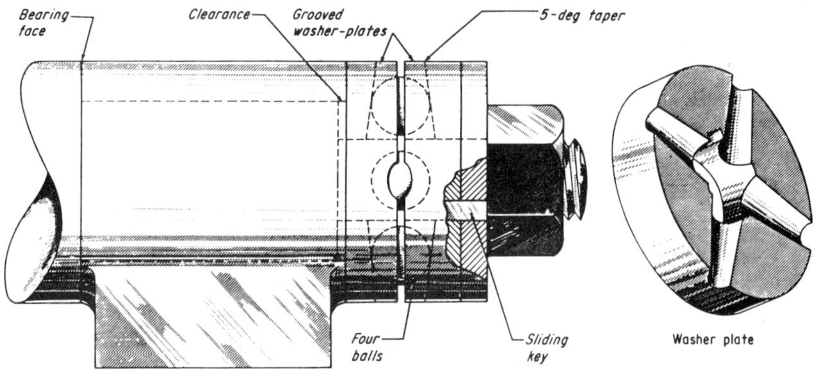

2

CENTRIFUGAL FORCE causes balls to exert force on grooved washer-plates when shaft rotates, pulling it against bearing face.

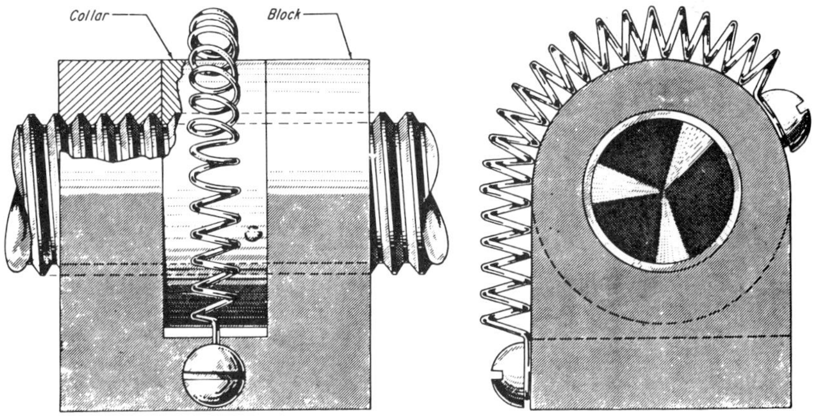

4

COLLAR AND BLOCK have continuous V-thread. When wear takes place in lead screw, the collar always maintains pressure on threads.

FIG. 14-23 *(continued)*

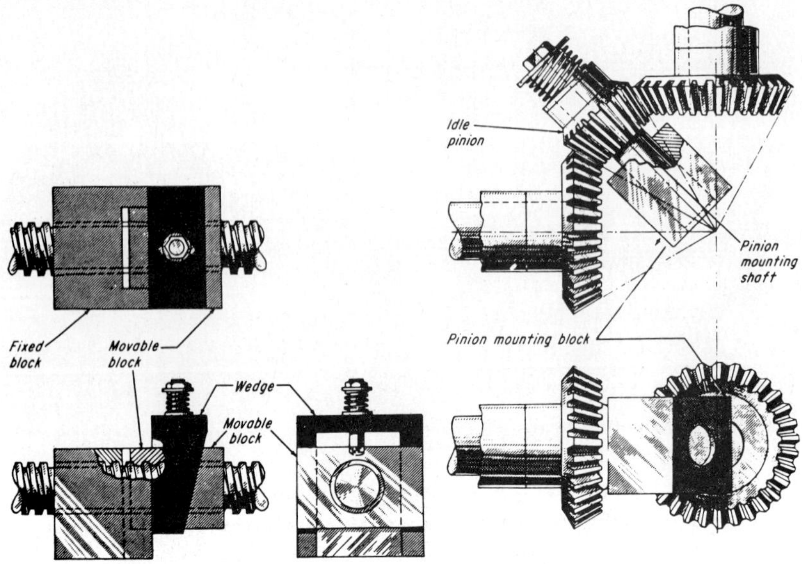

1 **MOVABLE BLOCK** is forced away from fixed block by spring-loaded wedge. Pressure is applied to both sides of lead screw, thus ensuring snug fit.

2 **SPRING-LOADED PINION** is mounted on a shaft located so that the spring forces pinion teeth into gear teeth to take up lost motion or backlash.

FIG. 14-24 More ways to prevent backlash. Springs combine with wedging action to ensure that threads, gears, and toggles respond smoothly. (*From L. Kasper, "4 More Ways to Prevent Backlash," Product Engineering and Mechanical Details for Product Design, edited by D. C. Greenwood, 1964. Reprinted by permission, Morgan-Grampian Publishing Co.*)

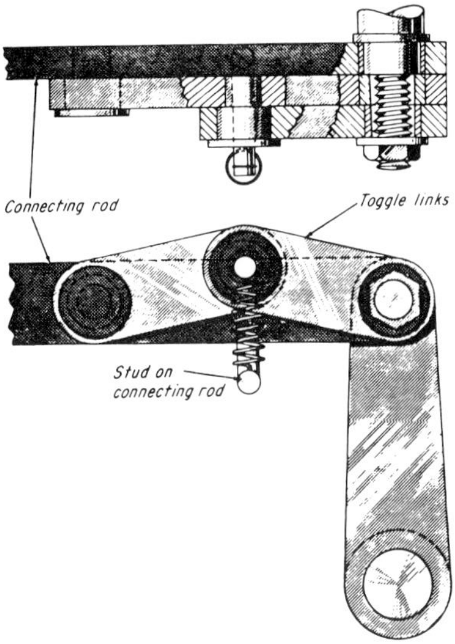

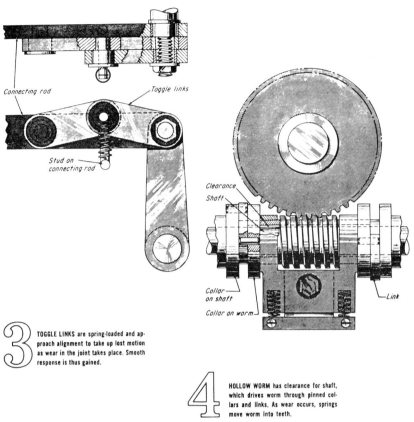

3 **TOGGLE LINKS** are spring-loaded and approach alignment to take up lost motion as wear in the joint takes place. Smooth response is thus gained.

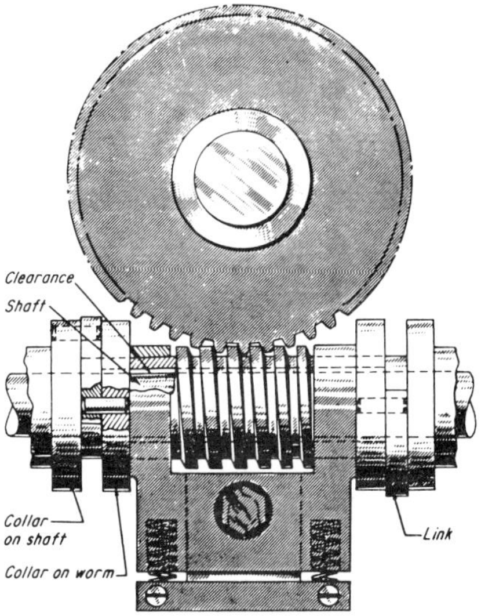

4 **HOLLOW WORM** has clearance for shaft, which drives worm through pinned collars and links. As wear occurs, springs move worm into teeth.

FIG. 14-24 *(continued)*

14-9 Drives and Components

Some unusual cable drives are pictured in Fig. 14-25. In Fig. 14-26, a variety of components for trolley conveyor systems are shown.

Feeders, take-ups, drives, and idlers are shown in Fig. 14-27. These items contain stock components that blend to form the unit.

1 **CHANGE OF DRIVE PLANE** from vertical to horizontal is no problem with a cable drive. One or more idlers may be needed—and watch friction.

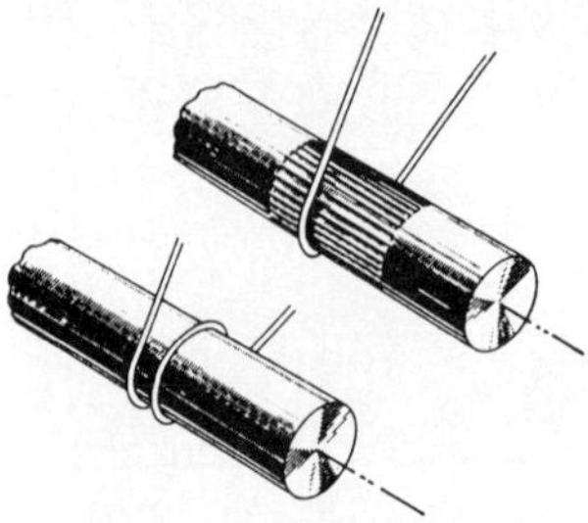

2 **FINE STRAIGHT-KURL** or double-wrapped smooth shaft provides adequate non-slip friction for most drives. Slip occurs before drive is damaged.

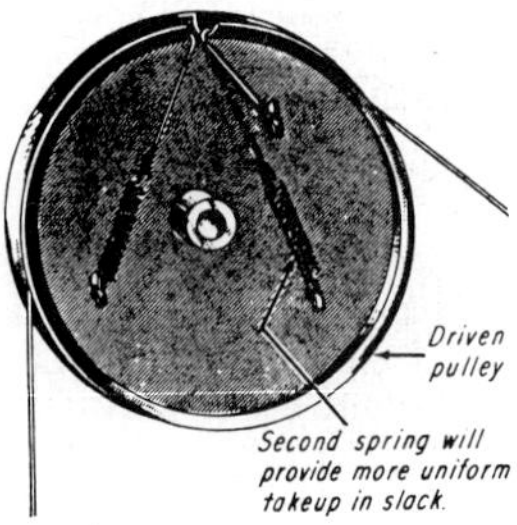

5 **"INSIDE"** location for spring is right on the driven pulley itself. While one spring will often suffice, a second one provides uniformity.

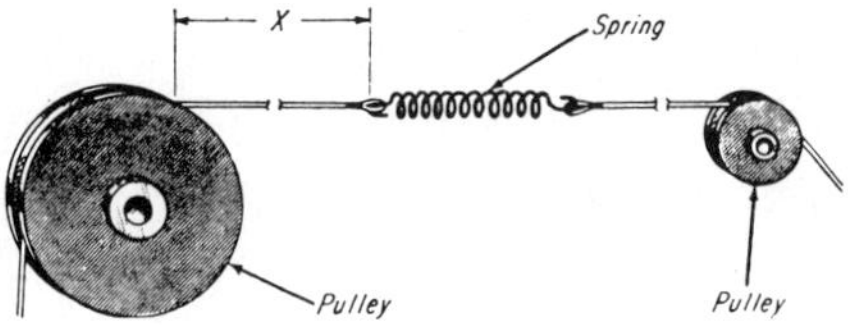

6 **SPRINGS** can be located along any part of the cable, so long as distance X is sufficient to prevent the spring from engaging pulley.

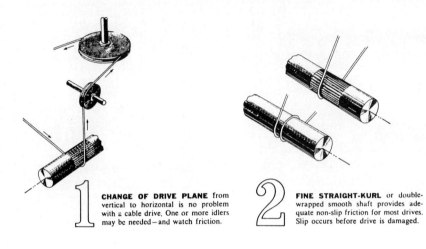

FIG. 14-25 Cable drives. When gears are too expensive, try a cable drive; low cost, durability, and reliability are features. (*From Frank W. Wood, Jr., "For Cable Drives—These Design Hints," Product Engineering and Mechanical Details for Product Design, edited by D. C. Greenwood, 1964. Reprinted by permission, Morgan-Grampian Publishing Co.*)

3 **BACKLASH** in a drive system can be eliminated by using an extension spring preloaded so it won't stretch under normal drive forces.

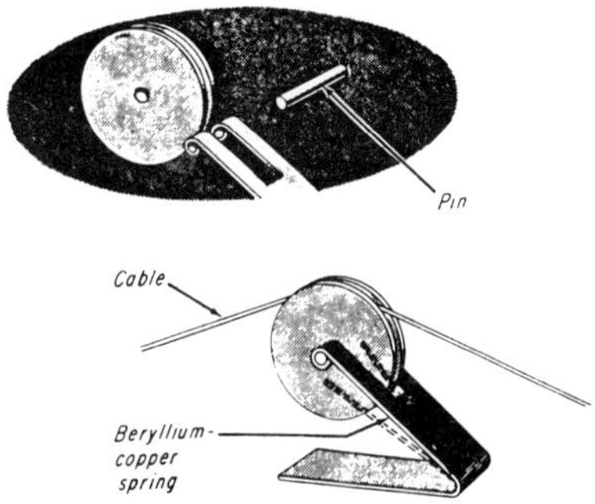

4 **"OUTSIDE"** location for takeup spring often allows more freedom of design. Beryllium-copper springs need no protective plating.

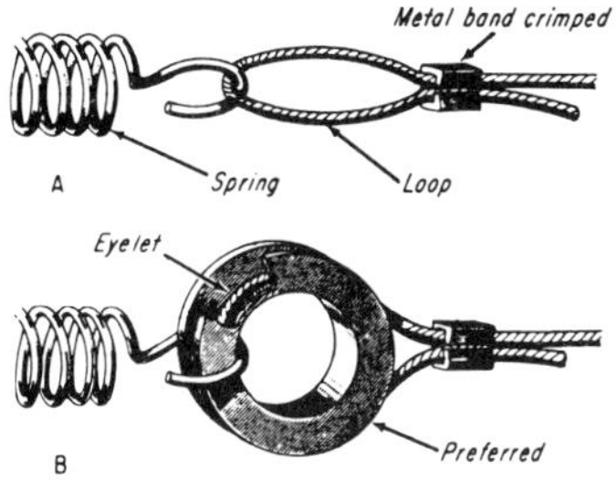

7 **SPRING ATTACHMENT** to the drive cable should be as secure as possible. Cable loop (A) is good; eyelet (B) causes less cable wear.

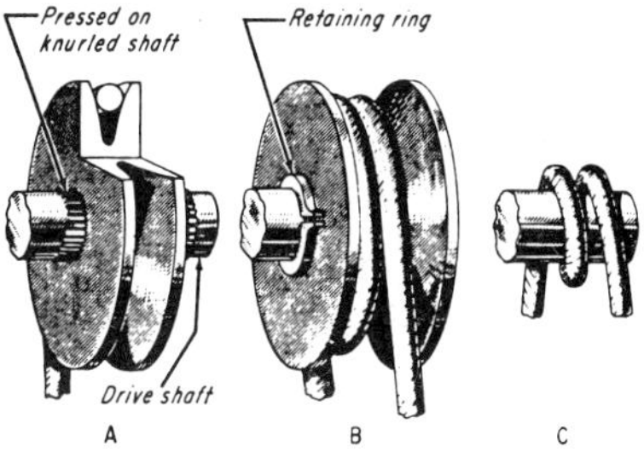

8 **PULLEYS** of nylon (A) give needed grip, yet slip at unsafe loads. Phenolic pulley (B) is free fit on shaft. Fixed polished shaft (C) adds friction.

FIG. 14-25 *(continued)*

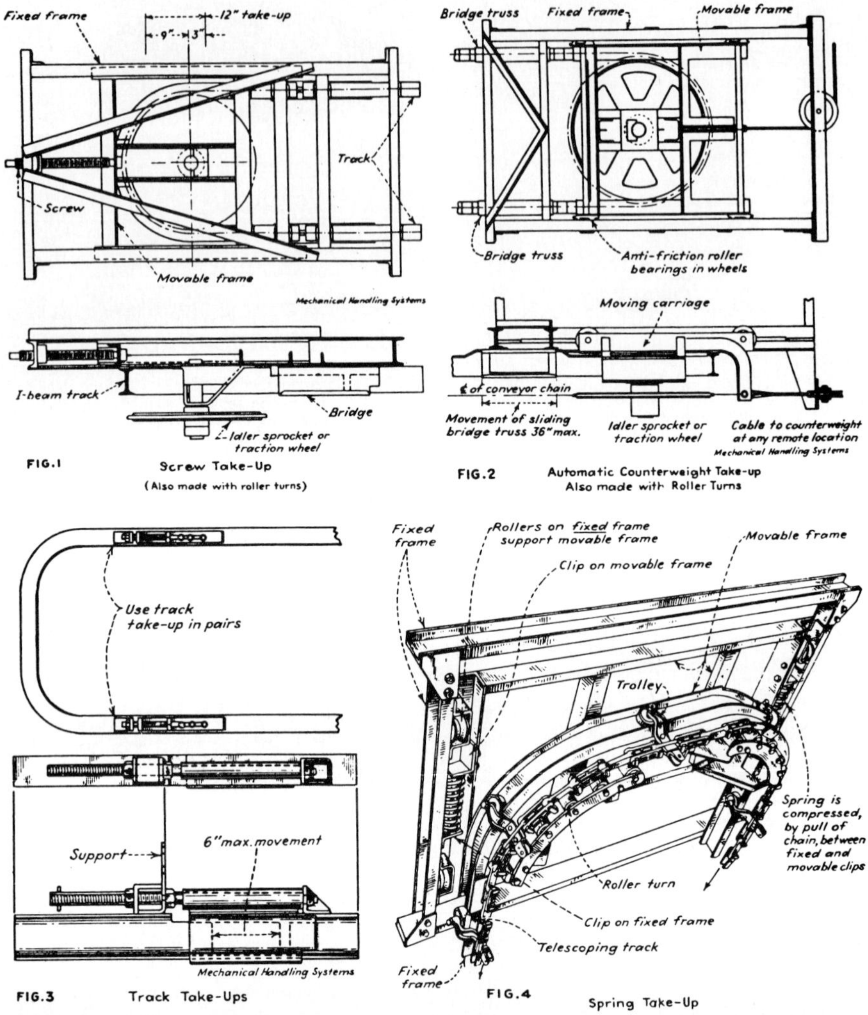

FIG. 14-26 Trolley conveyor system components. Take-ups: Take-ups are not necessary when there is a definite down run following a drive unit, but should be used in level conveyor systems. It is often necessary to put an extra loop in the system so that the take-up can use a 180° turn. Conveyors operating through ovens should have automatic take-ups. Short systems use spring type, and long conveyors are built with counterweighted take-ups for larger differential movement.

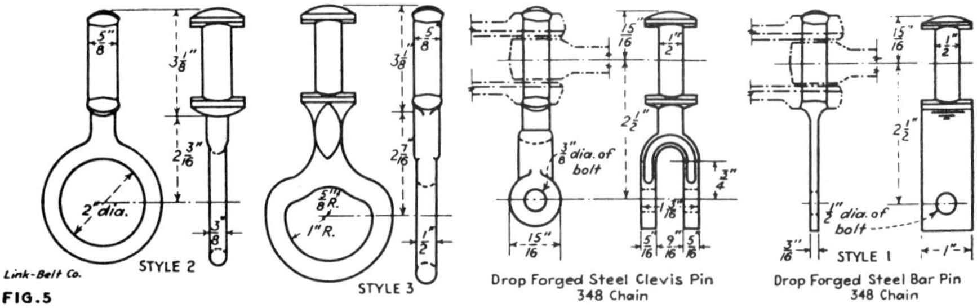

Link-Belt Co.

FIG.5

Drop Forged Steel Clevis Pin
348 Chain

Drop Forged Steel Bar Pin
348 Chain

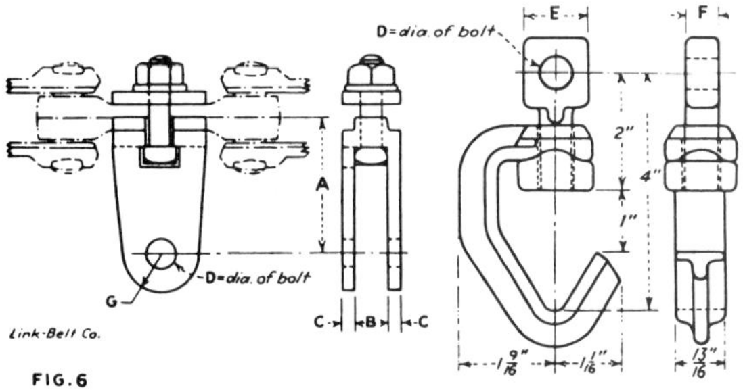

Link-Belt Co.

FIG.6

Malleable Iron
Intermediate Clevis Attachment

Malleable Iron
Swivel Hook

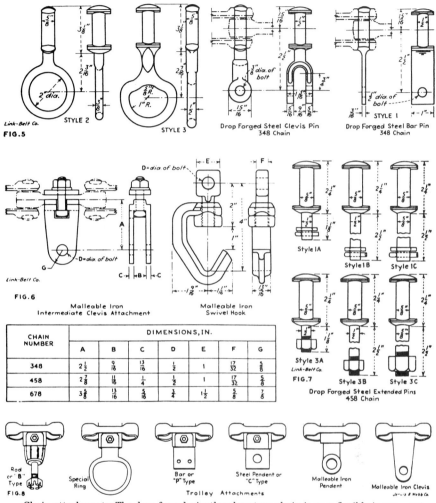

CHAIN NUMBER	DIMENSIONS, IN.						
	A	B	C	D	E	F	G
348	$2\frac{1}{2}$	$\frac{9}{16}$	$\frac{13}{16}$	$\frac{1}{2}$	1	$\frac{17}{32}$	$\frac{5}{8}$
458	$2\frac{7}{8}$	$\frac{11}{16}$	$\frac{1}{4}$	$\frac{1}{2}$	1	$\frac{17}{32}$	$\frac{5}{8}$
678	$3\frac{5}{8}$	$\frac{13}{16}$	$\frac{5}{16}$	$\frac{3}{4}$	$1\frac{1}{2}$	$\frac{5}{8}$	$\frac{7}{8}$

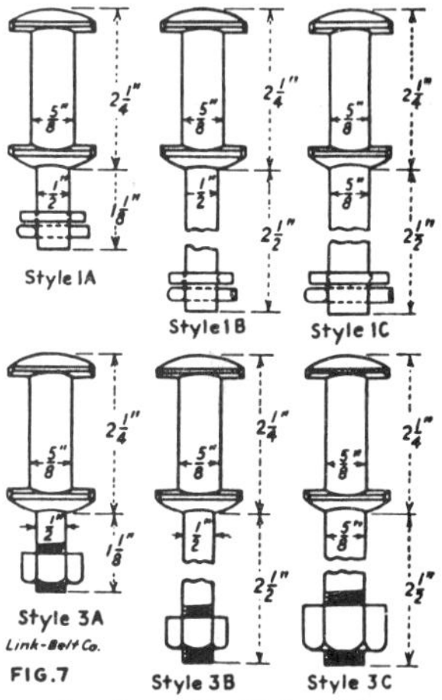

Link-Belt Co.

FIG.7

Drop Forged Steel Extended Pins
458 Chain

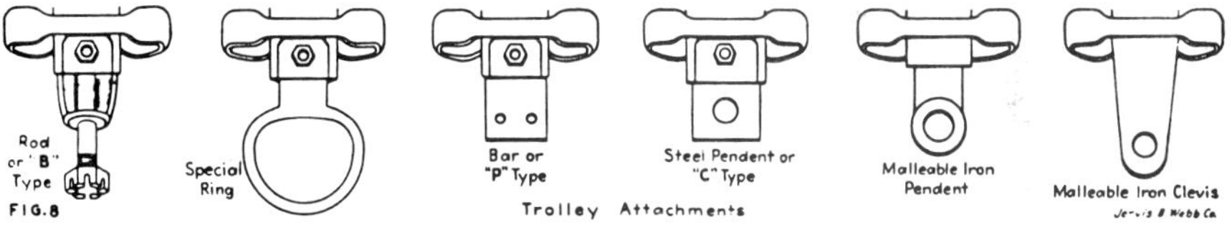

FIG.8

Trolley Attachments

Jervis B Webb Co.

Chain attachments: The drop-forged, rivetless keystone chain is very flexible in use—not only because of its design, but also because of its application possibilities. Light objects can be carried by pin attachments and heavy objects can be transported on trolley attachments. Typical pin and trolley attachments are shown below. Attachments for special applications can be designed for each installation. (*From Sidney Reibel, "Typical Trolley Conveyor System Components," Product Engineering and Product Design, edited by D. C. Greenwood, 1964. Reprinted by permission, Morgan-Grampian Publishing Co.*)

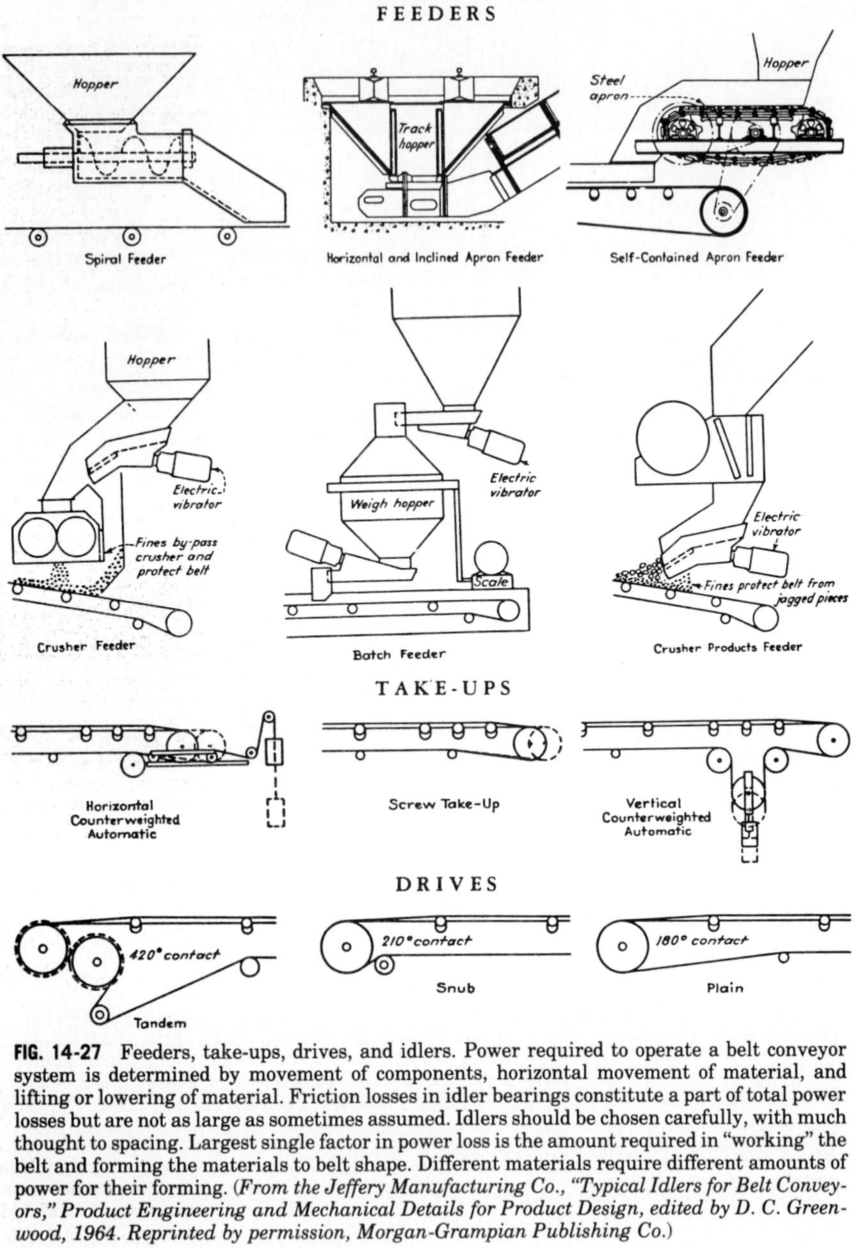

FIG. 14-27 Feeders, take-ups, drives, and idlers. Power required to operate a belt conveyor system is determined by movement of components, horizontal movement of material, and lifting or lowering of material. Friction losses in idler bearings constitute a part of total power losses but are not as large as sometimes assumed. Idlers should be chosen carefully, with much thought to spacing. Largest single factor in power loss is the amount required in "working" the belt and forming the materials to belt shape. Different materials require different amounts of power for their forming. (*From the Jeffery Manufacturing Co., "Typical Idlers for Belt Conveyors," Product Engineering and Mechanical Details for Product Design, edited by D. C. Greenwood, 1964. Reprinted by permission, Morgan-Grampian Publishing Co.*)

IDLERS

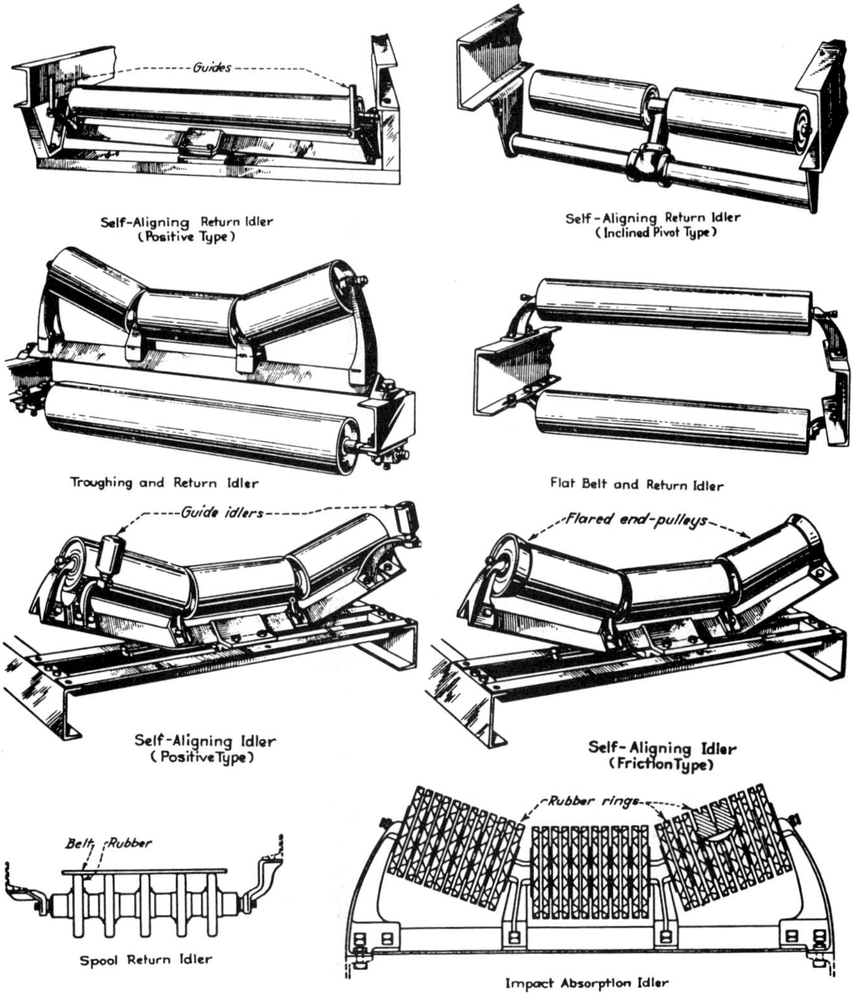

FIG. 14-27 *(continued)*

14-10 Pins

Space does not allow full coverage of all of today's uses of stock pins. However, to show a small segment of the versatility of pins in general, two have been selected.

Slotted or split-spring pins are widely used today. Figures 14-28, 14-29, and 14-30 present a clear picture of their universal potential in machine design.

Spiral-wrapped, coiled, or rolled pins come in a wide range of sizes, and their use is almost limitless, as illustrated in Fig. 14-31.

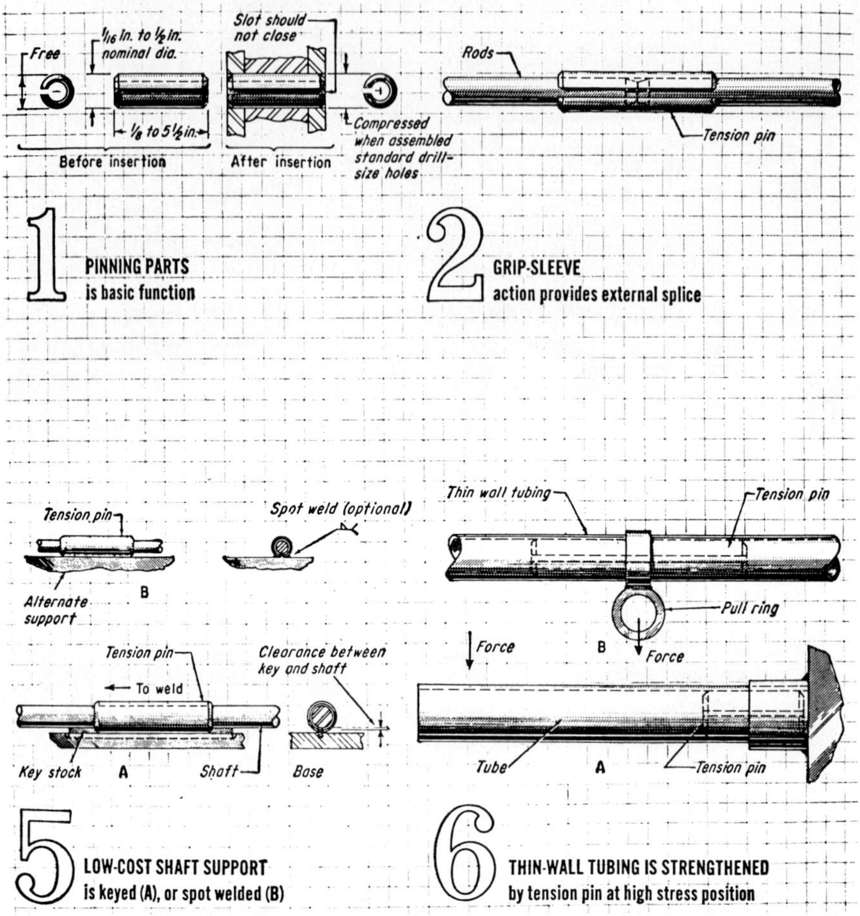

FIG. 14-28 Slotted spring pins. Assembled under pressure, these fasteners provide powerful gripping action to locate parts and hold them together. (*From R. O. Parmley, "Slotted Spring Pins Find Many Jobs," in Product Engineering, Aug. 31, 1964. Reprinted by permission, Morgan-Grampian Publishing Co.*)

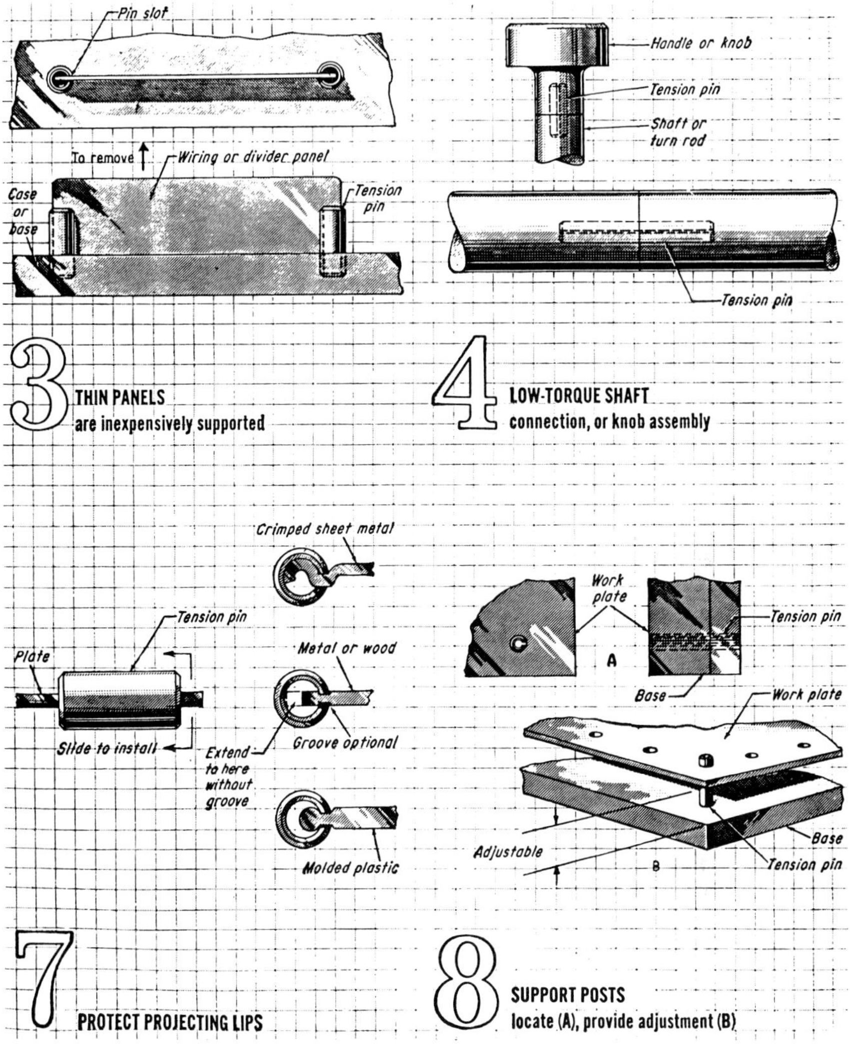

FIG. 14-28 *(continued)*

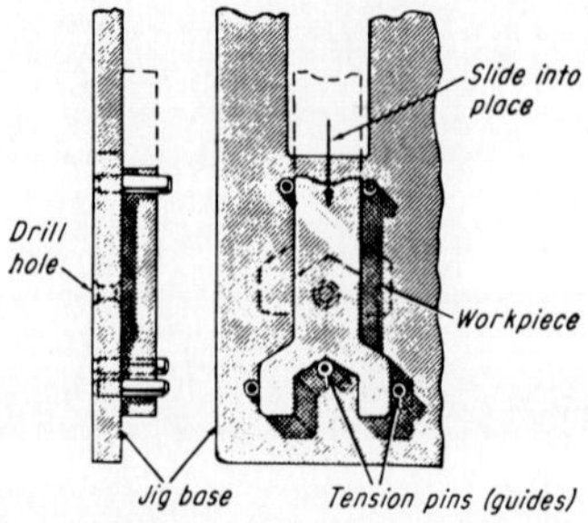

1 **LOCATE WORKPIECES** with this inexpensive jig

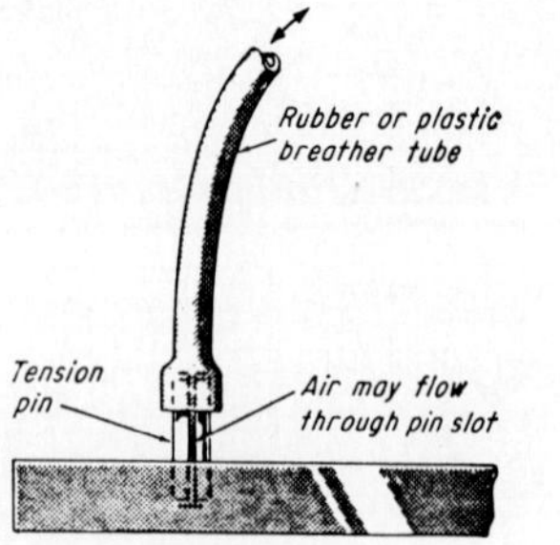

2 **FLEXIBLE BREATHER TUBE** is enclosed to prevent swinging and to keep it away from moving machine parts

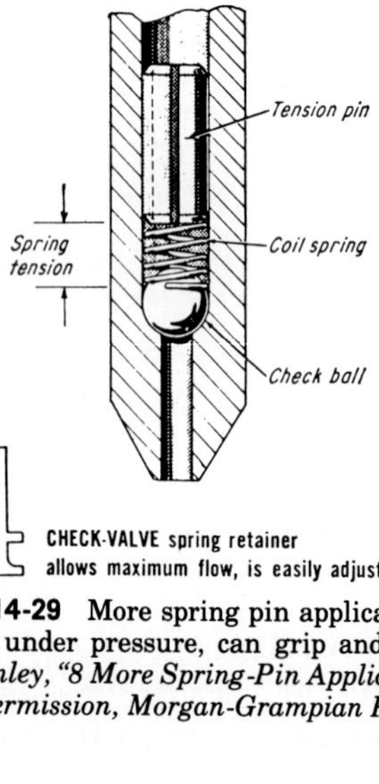

4 **CHECK-VALVE** spring retainer allows maximum flow, is easily adjustable

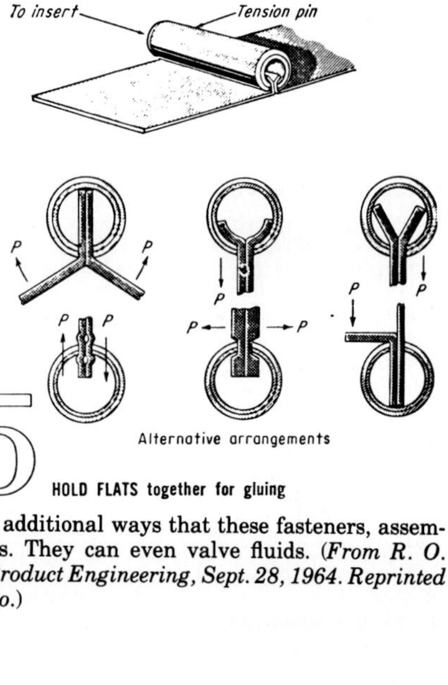

5 **HOLD FLATS** together for gluing

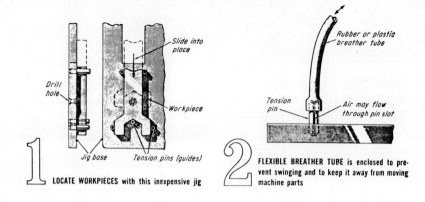

FIG. 14-29 More spring pin applications. Some additional ways that these fasteners, assembled under pressure, can grip and locate parts. They can even valve fluids. (*From R. O. Parmley, "8 More Spring-Pin Applications," in Product Engineering, Sept. 28, 1964. Reprinted by permission, Morgan-Grampian Publishing Co.*)

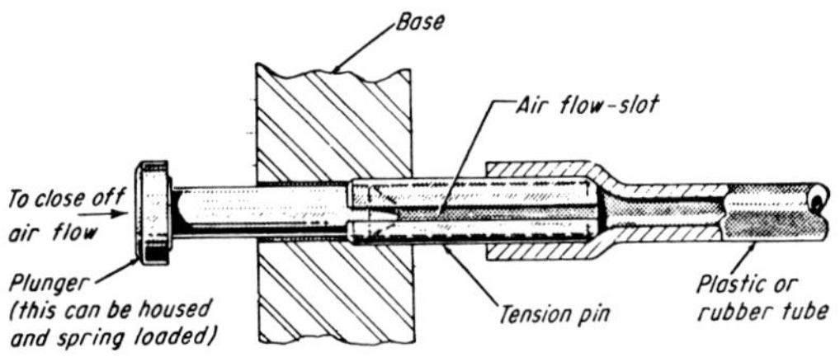

3 **AIR VALVE** is simple yet effective

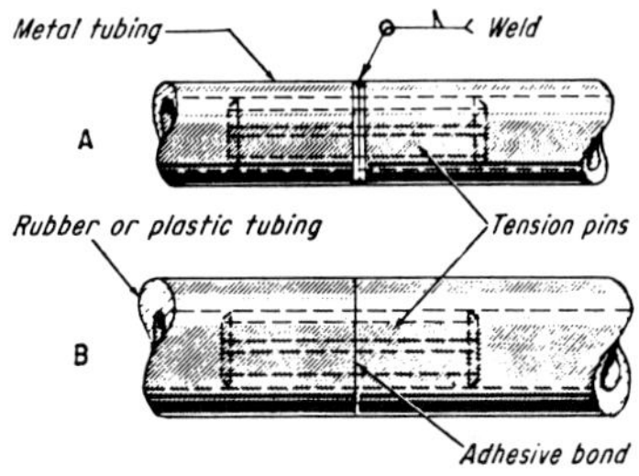

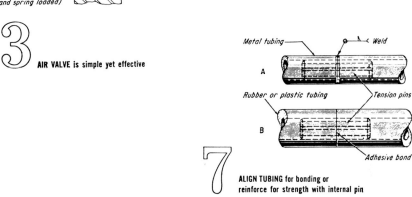

7 **ALIGN TUBING** for bonding or reinforce for strength with internal pin

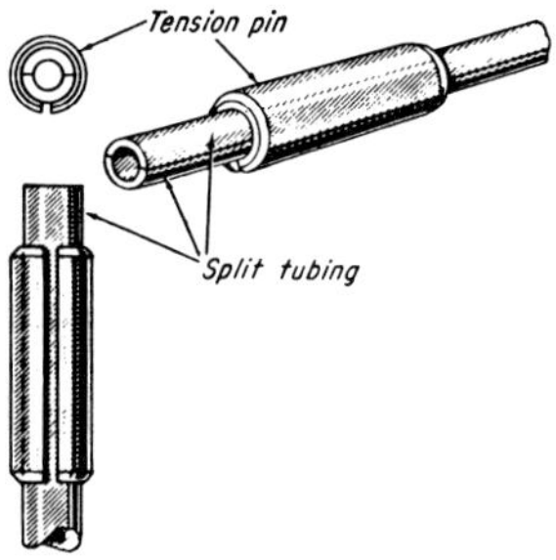

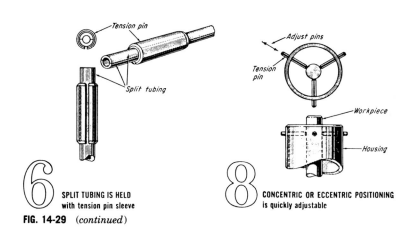

6 **SPLIT TUBING IS HELD** with tension pin sleeve

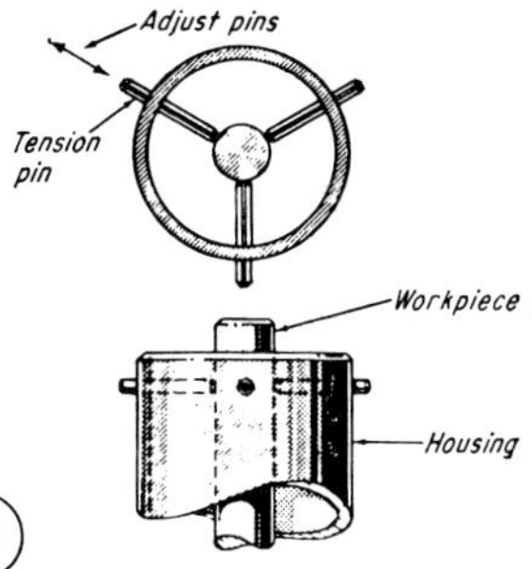

8 **CONCENTRIC OR ECCENTRIC POSITIONING** is quickly adjustable

FIG. 14-29 (*continued*)

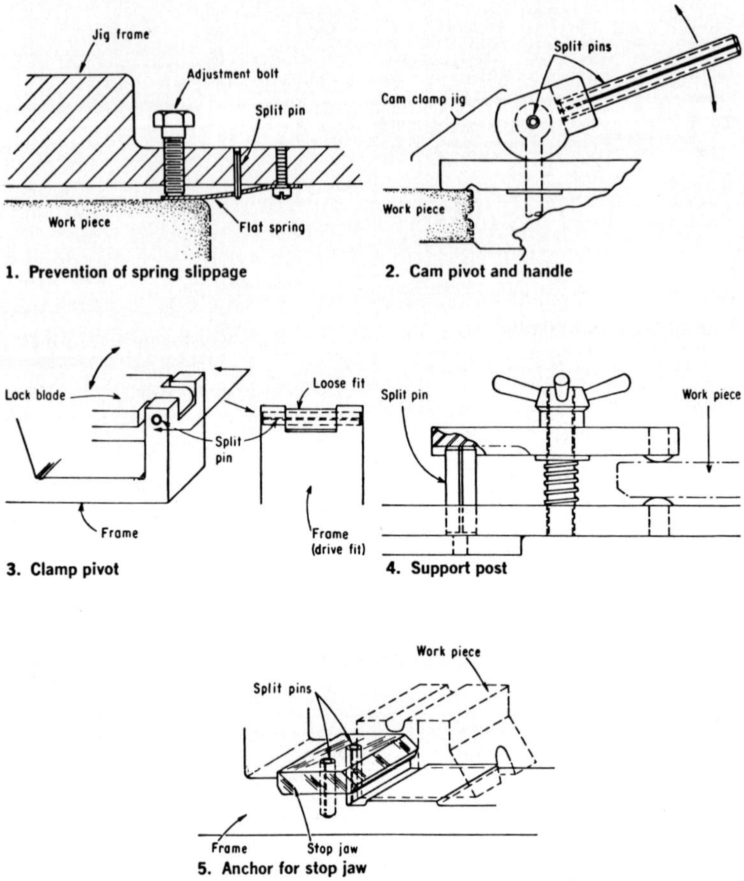

FIG. 14-30 Uses of split pins. Ten examples show how these pins simplify assembly of jigs and fixtures. The pins are easily removed. (*From R. O. Parmley, "Uses of Split Pins, 1," in American Machinist, March 25, 1968. Reprinted by permission, American Machinist.*)

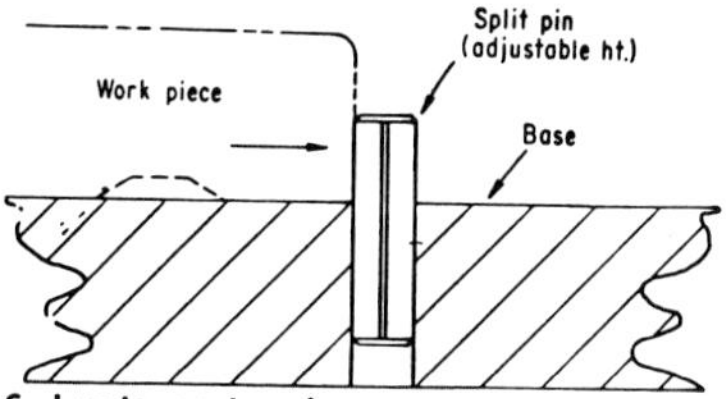

6. Locator or stop pin

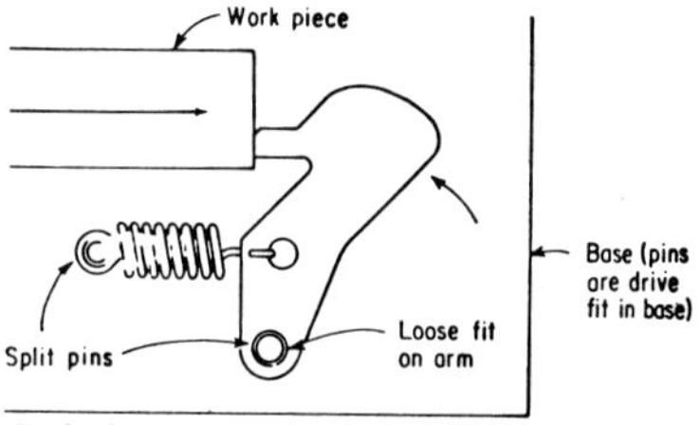

7. Spring anchor and arm pivot

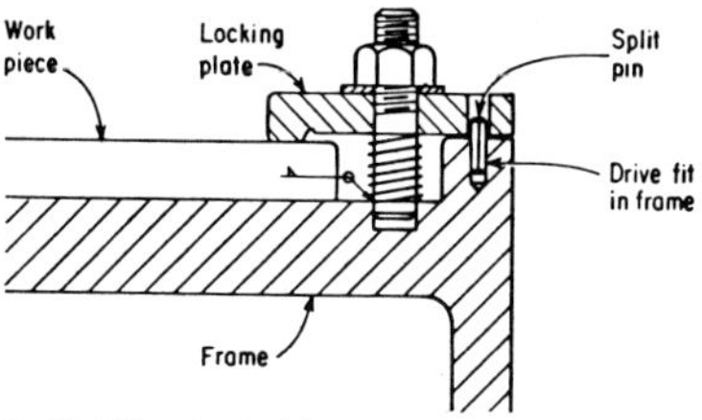

8. Stabilizer for locking plate

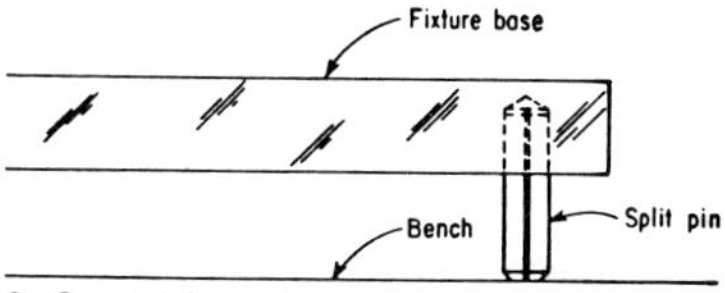

9. Support for post leg

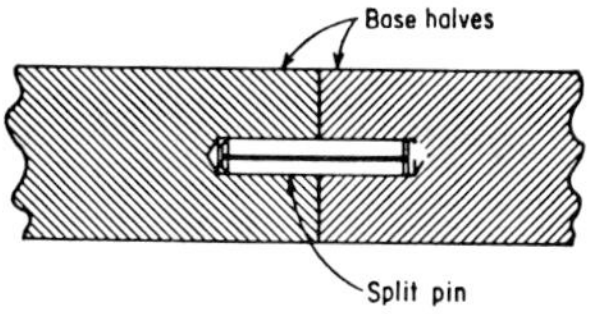

10. Dowels for fixture base

Slotted tubular pins are intended to be forced into their locations; free diameter should be larger than hole diameter so the pin exerts radial force all along its mounting hole to resist axial motion when properly mounted. Maximum compression is controlled by the amount of gap when the pin is free. When it acts as a pivot, the hole through the pivoting member should be a free fit (see figures 7 and 8) so the pin will not be worked loose from its anchor hole. These pins may be made of heat-treated carbon steel, corrosion-resistant steel, or beryllium-copper.

FIG. 14-30 *(continued)*

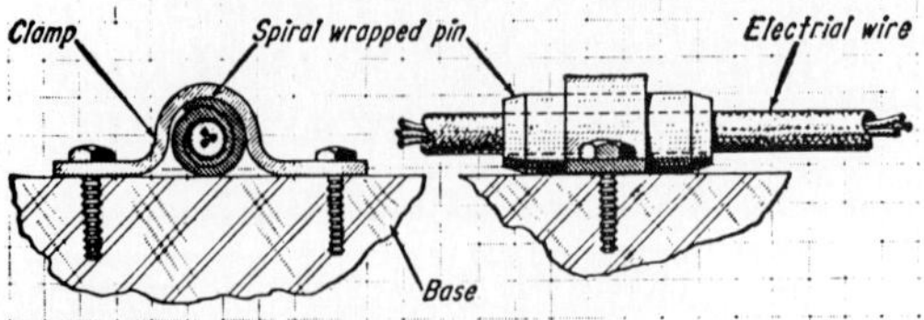

1 **Wire gripper. As clamp is tightened in place, pin coils, and secures the wire**

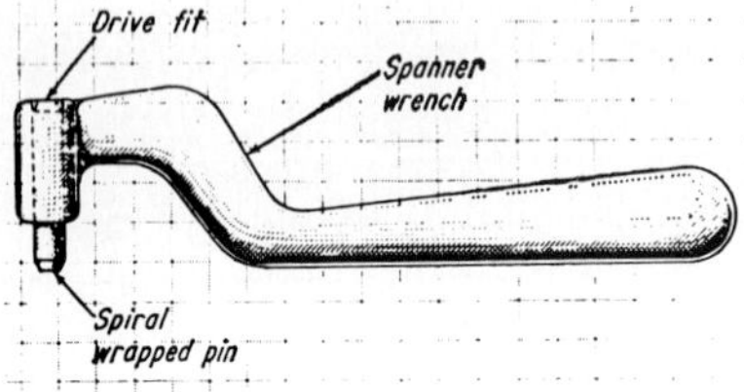

4 **Wrench pin. Coil construction permits pin to fit holes with large tolerances**

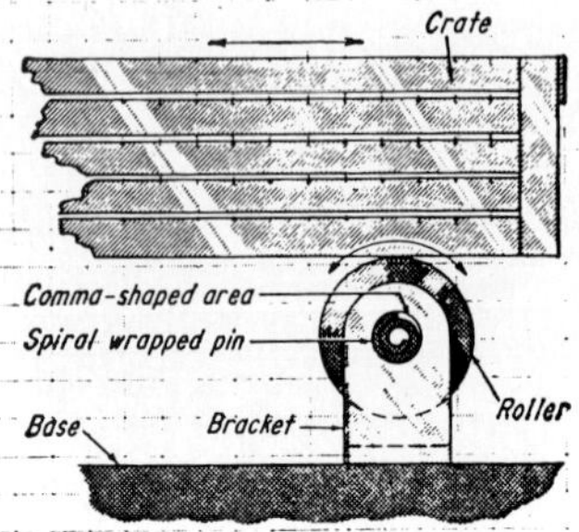

5 **Lubricated shaft for work roller. Comma-shaped area of the spiral wrapped pin forms an oil reservoir for the roller**

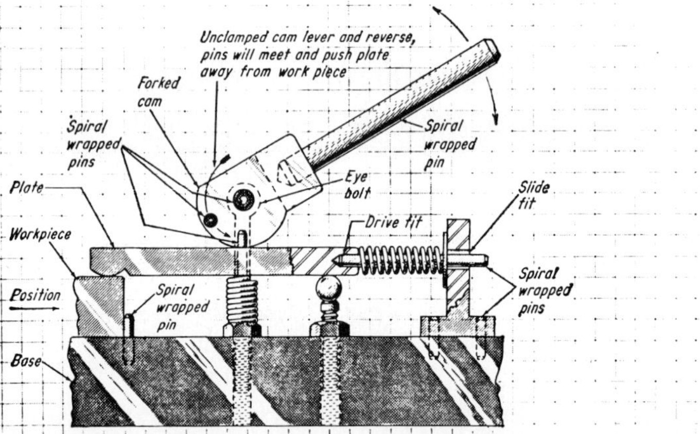

8 **Pivot, stop, locator, handle, and anchor are typical applications in the design of a clamp. The pins are drive or slide fits, and can be removed and reused if the clamp position must be changed**

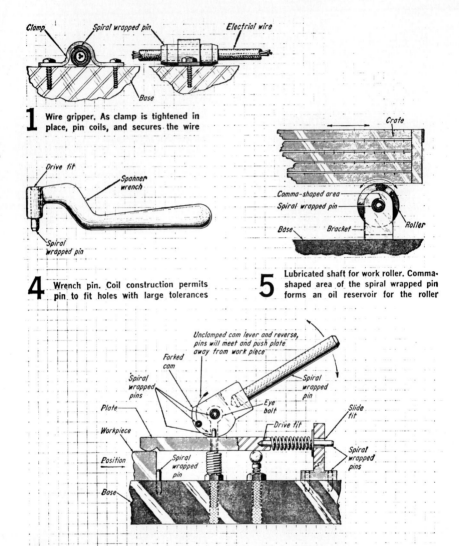

FIG. 14-31 Spiral-wrapped pins. Coil, rolled, or spiral-wrapped pins come in a wide range of lengths and diameters. Their applications are limitless; here are eight. (*From R. O. Parmley, "Design around Spiral-Wrapped Pins," in Product Engineering, Oct. 25, 1965. Reprinted by permission, Morgan-Grampian Publishing Co.*)

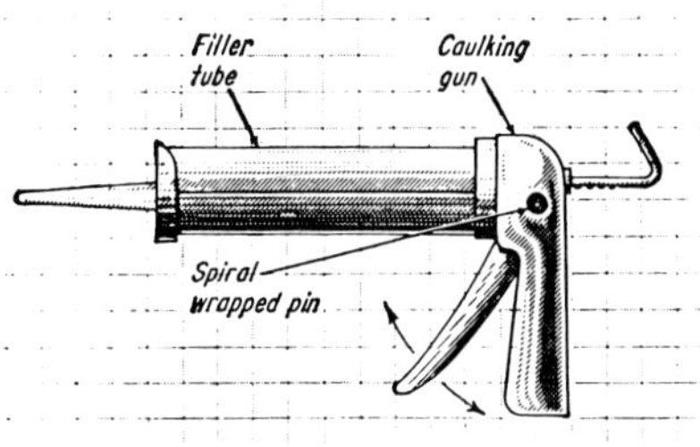

2 Pivot. Pin is a drive fit in the handle housing, and acts as a pivot for trigger

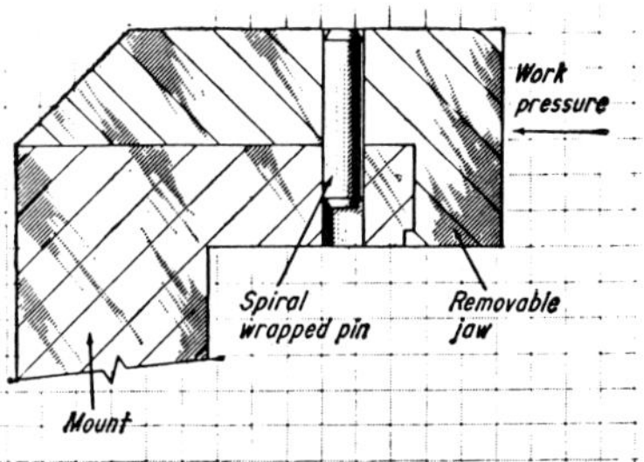

3 Dowel. Here, the dowel acts as a locator anchor pin that can be removed and reused

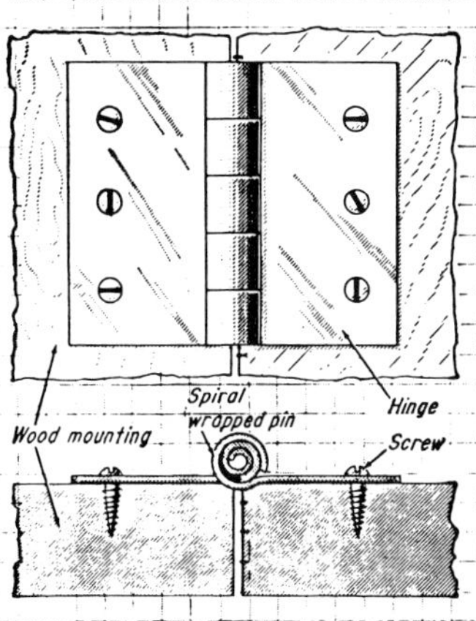

6 Hinge pins. If hole size in both members is different hinge will be free moving; if the hole size is the same in both members a friction hinge is the result

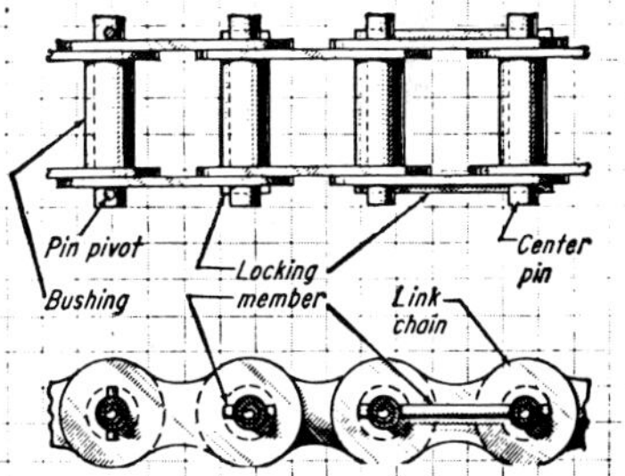

7 Link chain connection. Pins are used as pivot or locking members. An advantage: Both types are removable and reusable

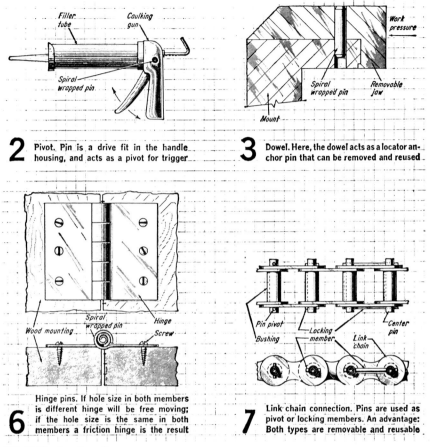

FIG. 14-31 *(continued)*

14-11 Retaining Rings

The retaining ring comes in a variety of sizes, shapes, and designs. This versatile fastener functions both as a shoulder and as a locking device, thus reducing machine and assembly complexity.

Figures 14-32 to 14-36 illustrate the versatility of this mechanical component, not only as a fastener, but also as a key machine component.

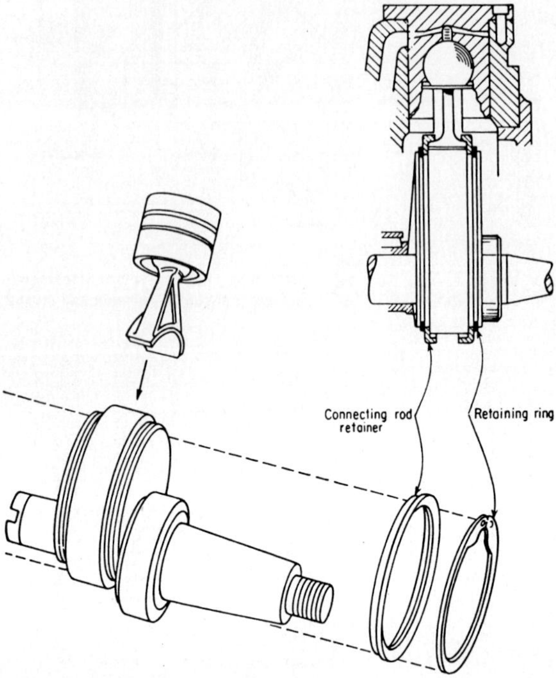

Slow-moving piston of hydraulic motor is assembled to the crank throw by two retainers. These are held in place by two retaining rings that fit into grooves in the crankthrow

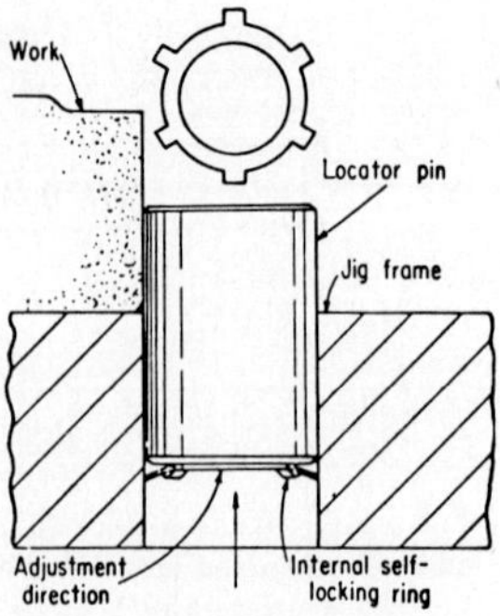

Internal self-locking ring supports a locator. Elevation of the pin may be altered in the entry direction only; the pin won't push down into the frame

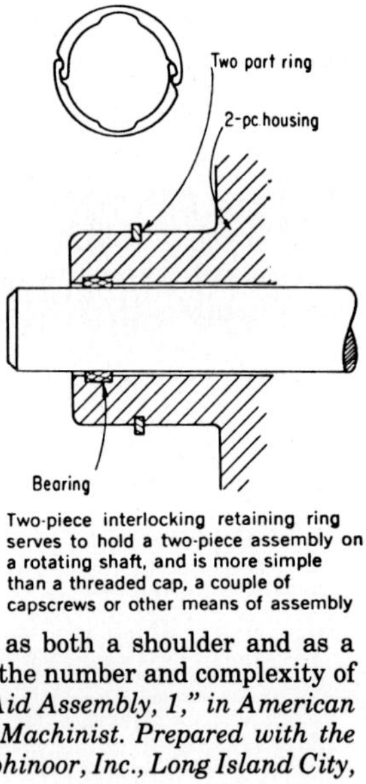

Two-piece interlocking retaining ring serves to hold a two-piece assembly on a rotating shaft, and is more simple than a threaded cap, a couple of capscrews or other means of assembly

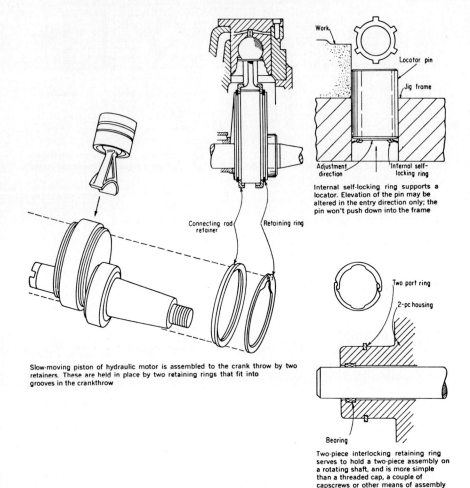

FIG. 14-32 Retaining rings aid assembly—1. By functioning as both a shoulder and as a locking device, these versatile fasteners reduce machining and the number and complexity of parts in an assembly. (*From R. O. Parmley, "Retaining Rings Aid Assembly, 1," in American Machinist, June 3, 1968. Reprinted by permission, American Machinist. Prepared with the cooperation of the Truarc Retaining Rings Division of Waldes Kohinoor, Inc., Long Island City, N.Y.*)

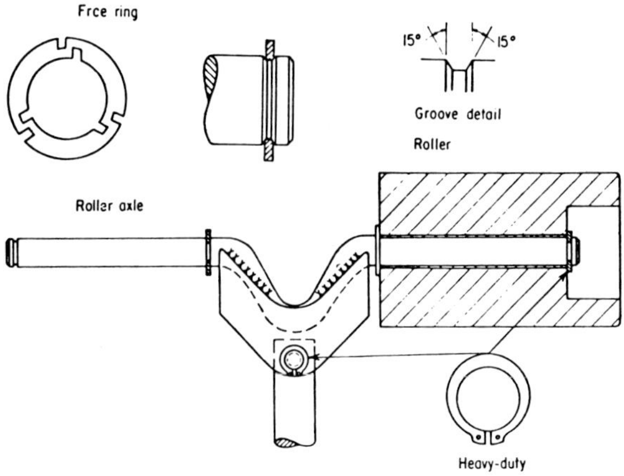

Two types of rings may be used on one assembly. Here permanent-shoulder rings provide a uniform axle step for each roller, without spotwelding or the like. Heavy-duty rings keep the rollers in place

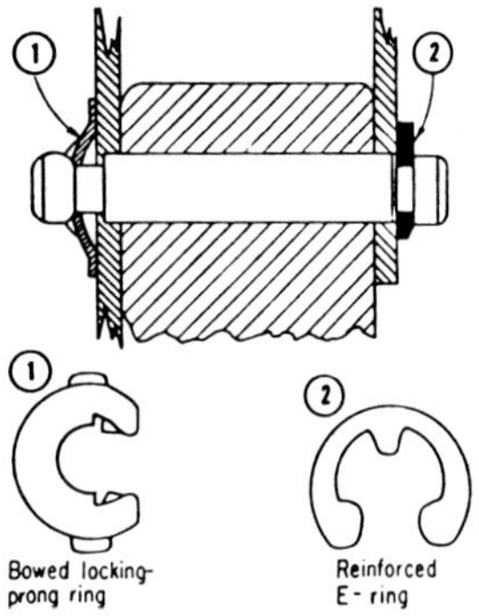

Snug assembly of side members to a casting with cored hole is secured with two rings: 1—spring-like ring has high thrust capacity, eliminates springs, bow washers, etc; 2—reinforced E-ring acts as a retaining shoulder or head. Each ring can be dismantled with a screwdriver

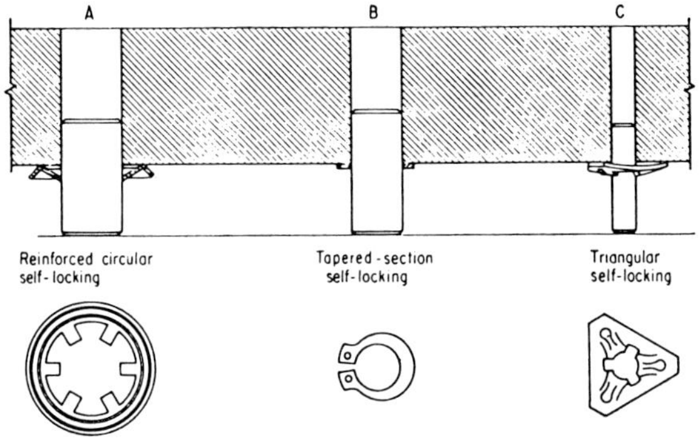

These three examples show self-locking retaining rings used as adjustable stops on support members (pins made to commercial tolerances): A—external ring provides positive grip, and arched rim adds strength; B—ring is adjustable in both directions, but frictional resistance is considerable, and C—triangular ring with dished body and three prongs will resist extreme thrust. Both A and C have one-direction adjustment only

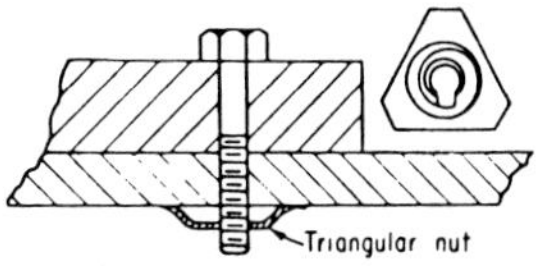

Triangular retaining nut eliminates the need for tapping mounting holes and using a large nut and washer. Secure mounting of small motors and devices can be obtained in this manner

FIG. 14-32 *(continued)*

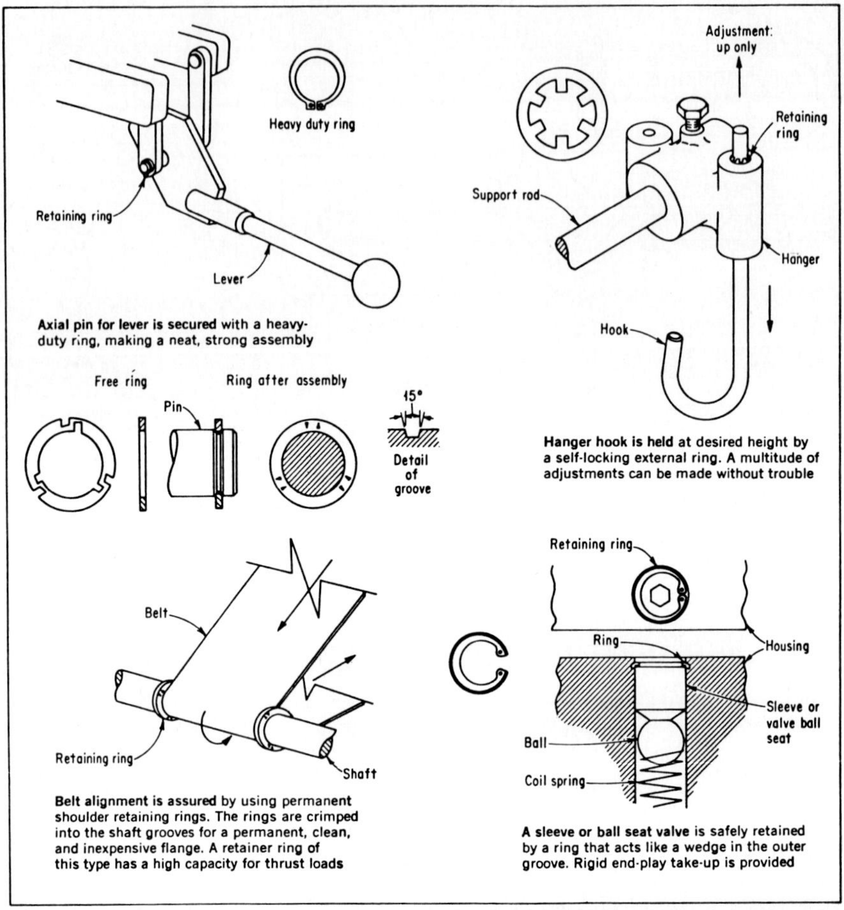

FIG. 14-33 Retaining rings aid assembly—2. Here are eight more thought-provoking uses for retaining rings, to be added to the previous assortment of ideas. (*From R. O. Parmley, "Retaining Rings Aid Assembly," in American Machinist, Feb. 22, 1971. Reprinted by permission, American Machinist.*)

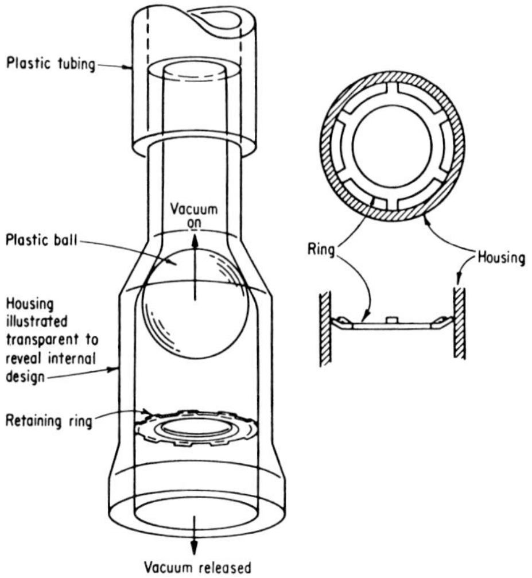

Internal self-locking ring supports the plastic ball valve when the vacuum is released, thus providing a support during the "off" cycle. Air or liquid is released when ball is at rest and exits through the areas between the grip points of the ring, which is adjustable at entry position

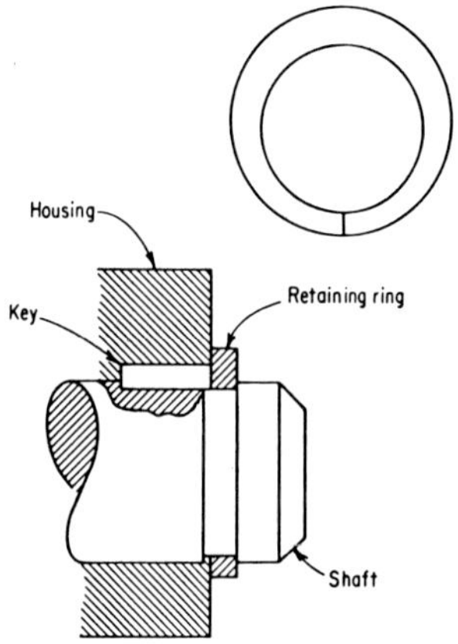

Tamper-proof lock for a shaft in a housing provides location of the shaft and at the same time retains the key. Heavy axial loading and permanent retention of the key are double values in this application

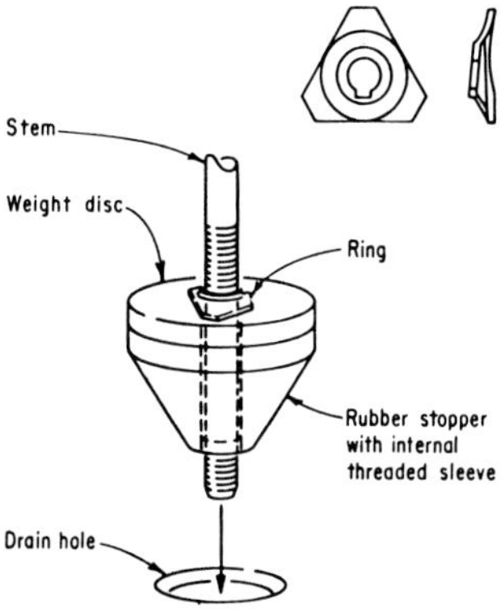

Triangular retainer nut positions and unifies components of the tank drain assembly. The triangular nut eliminates the need for a large standard nut and lockwasher or spring-type component and simplifies the design

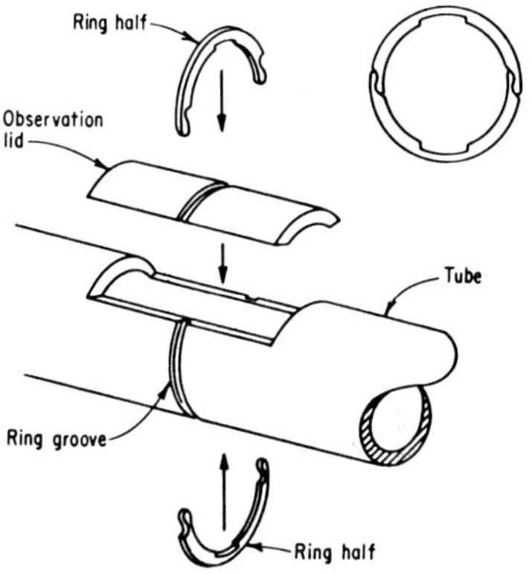

Observation lid on tubing makes it possible to inspect wiring at will. The two-part balanced retainer ring has identical semicircular halves, which are held together by the interlocking prongs at the free ends

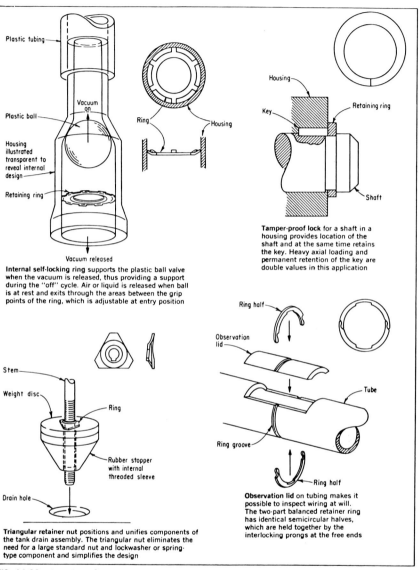

FIG. 14-33 *(continued)*

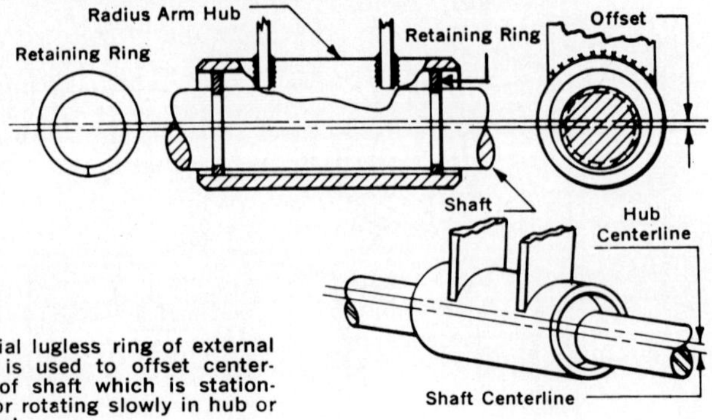

1

Special lugless ring of external type is used to offset centerline of shaft which is stationary or rotating slowly in hub or journal.

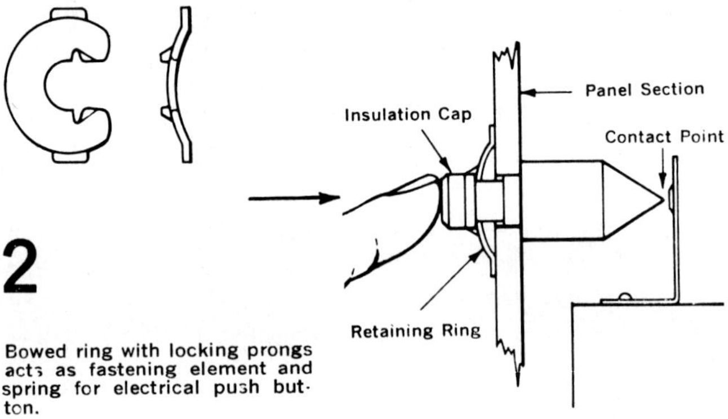

2

Bowed ring with locking prongs acts as fastening element and spring for electrical push button.

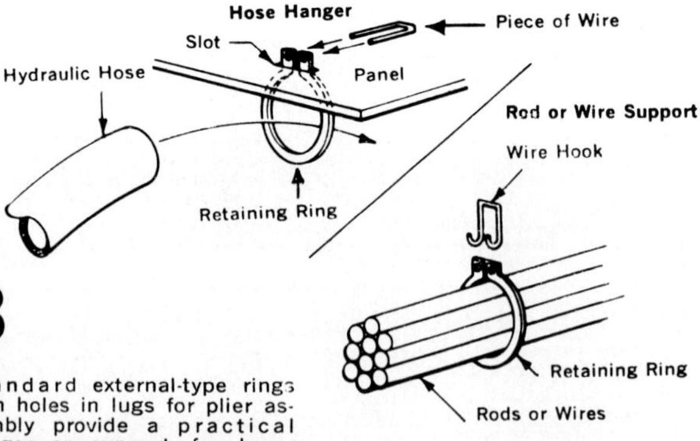

3

Standard external-type rings with holes in lugs for plier assembly provide a practical hanger or support for hose, rods, and wires.

FIG. 14-34 Multiple-purpose retaining ring. A roundup of 10 unusual ways for putting retaining rings to work in assembly jobs. Examples shown are based on stamped ring designs produced by Truarc Retaining Rings Division of Waldes Kohinoor, Inc. (*From R. O. Parmley, "The Multiple Purpose Retaining Ring," in Assembly Engineering, July 1966, vol. 9, no. 7. Reprinted by permission, Assembly Engineering.*)

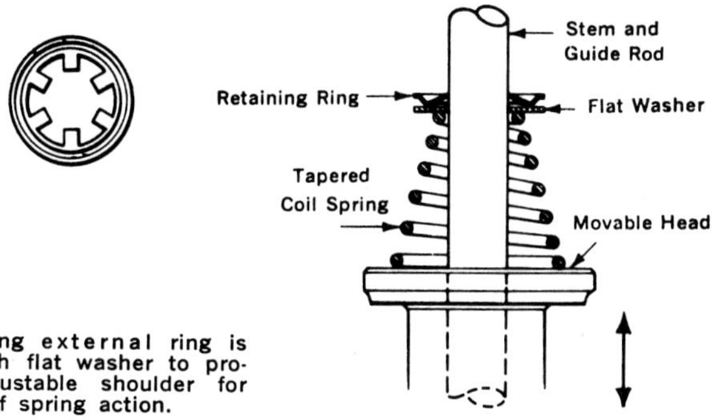

4

Self-locking external ring is used with flat washer to provide adjustable shoulder for control of spring action.

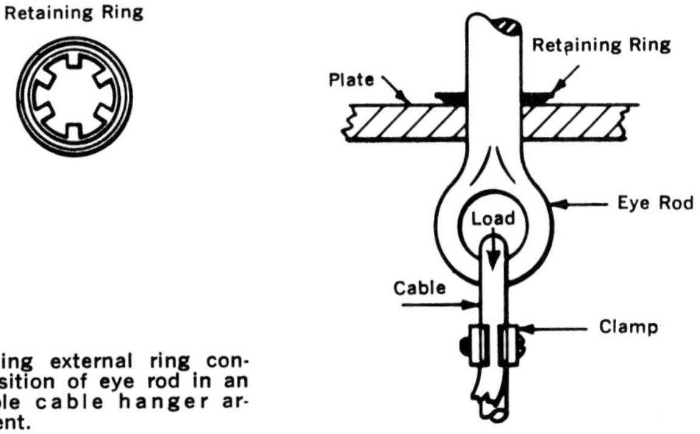

5

Self-locking external ring controls position of eye rod in an adjustable cable hanger arrangement.

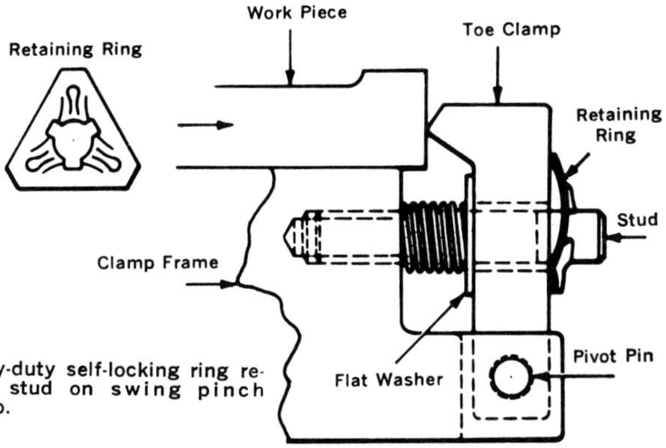

6

Heavy-duty self-locking ring retains stud on swing pinch clamp.

FIG. 14-34 *(continued)*

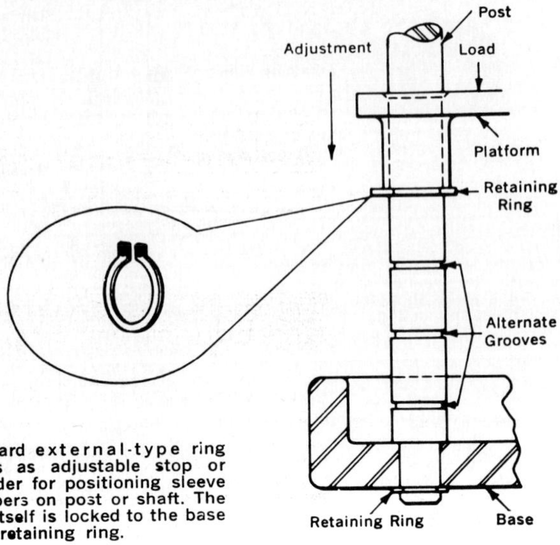

7

Standard external-type ring serves as adjustable stop or shoulder for positioning sleeve members on post or shaft. The post itself is locked to the base by a retaining ring.

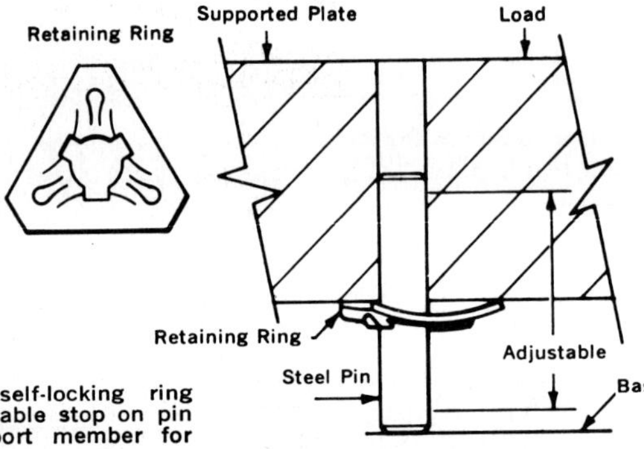

8

Heavy duty self-locking ring acts as adjustable stop on pin used as support member for plate section.

FIG. 14-34 *(continued)*

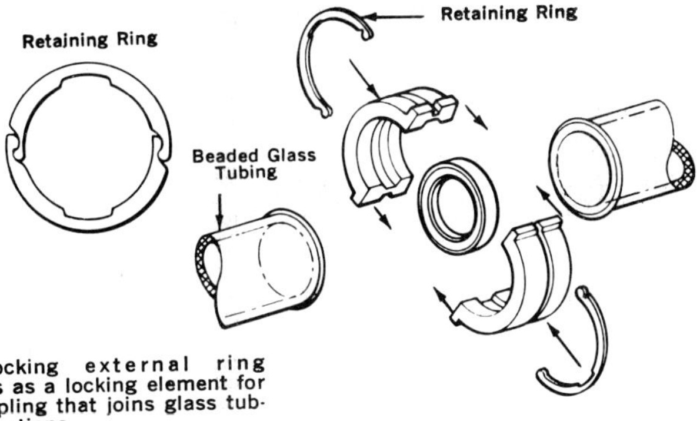

9

Interlocking external ring serves as a locking element for a coupling that joins glass tubing sections.

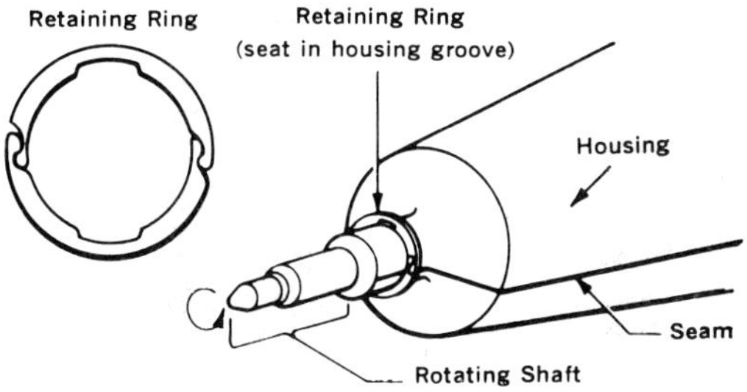

10

Interlocking external ring locks two-piece housing that fits around a rotating shaft.

FIG. 14-34 *(continued)*

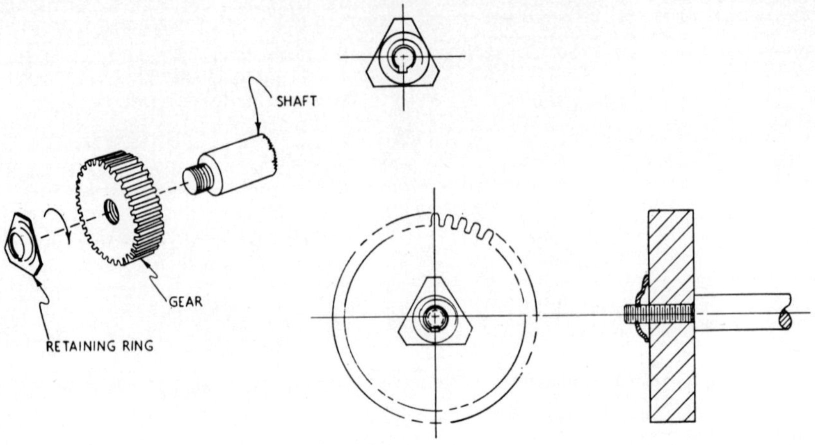

(a)

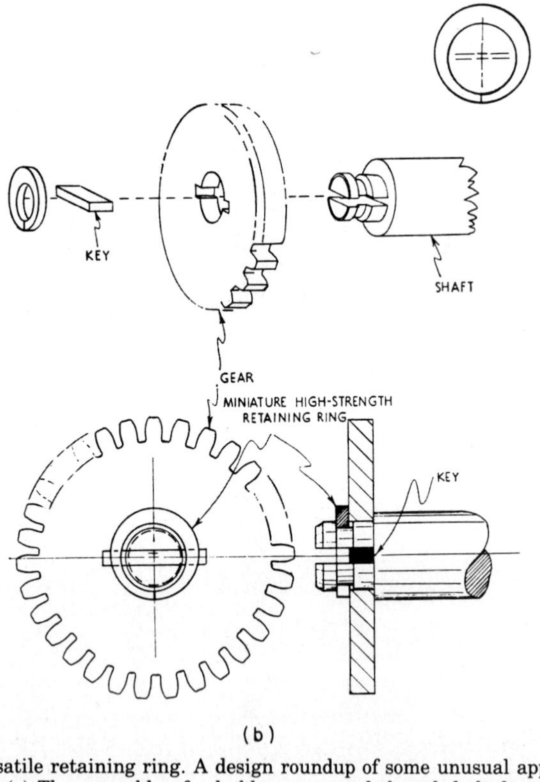

(b)

FIG. 14-35 Versatile retaining ring. A design roundup of some unusual applications for retaining rings. (*a*) The assembly of a hubless gear and threaded shaft may be accomplished by using a triangular nut retaining component which eliminates the need for a large standard nut and lock washer or other spring-type part. The dished body of the triangular nut flattens under torque to lock the gear to the shaft. (*b*) This heavy-duty hubless gear and shaft are designed for high torque and end thrusts. The retaining ring seated in a square groove and the key in a slot provide a tamper-proof lock. This design is recommended for permanent assemblies in which the ring may be subjected to heavy loads from either or both axial directions. An angled groove can be provided which has one wall cut at a 40° angle to the shaft axis. This will permit the ring to be removed without

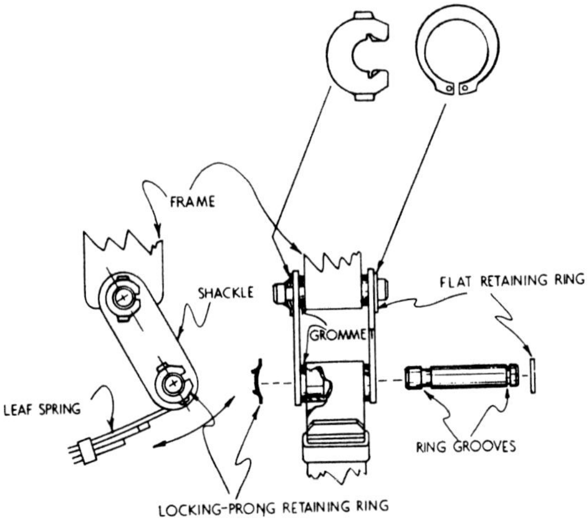

(c)

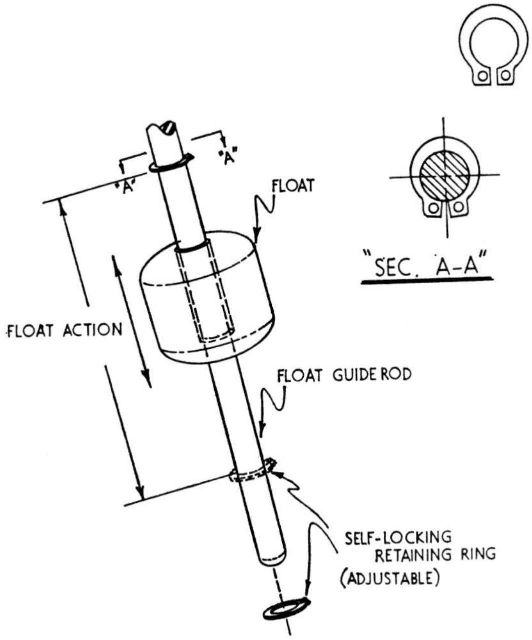

(d)

damage. (*c*) Two different types of retaining rings are used in this application involving a leaf spring and shackle assembly. A locking-prong retaining ring is bowed for tension while the prongs act as fastening elements to secure the pivot bolt. A flat or standard external ring is used as a flange or bolt head. (*d*) The self-locking retaining rings used in this application provide stops for a float. The rings are adjustable on the guide rod and yet the friction force produced by the heavy spring pressure makes axial displacement from the lightweight hollow float impossible. (*e*) Retaining rings provide a uniform circular shoulder for small-diameter parts such as the pipe nipple shown here. In this case the retaining ring shoulder is used as a stop for the plastic tube. The wall thickness of the nipple should be at least 3 times as thick as the depth of the groove. When assembling the ring in the groove, the nipple should be supported by inserting a mandrel of rod. (*f*) This

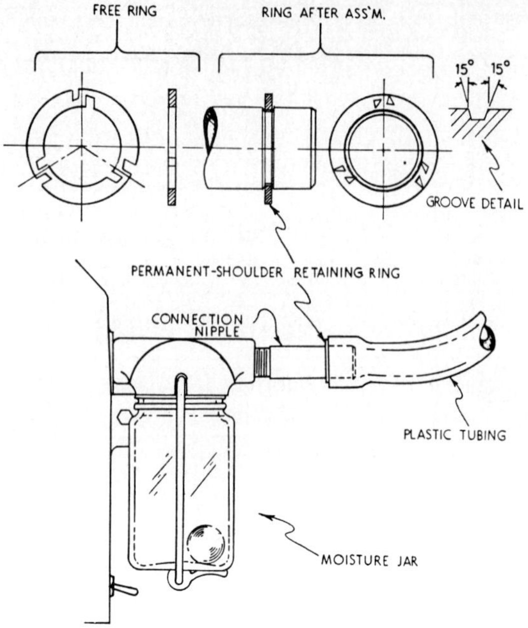

(e)

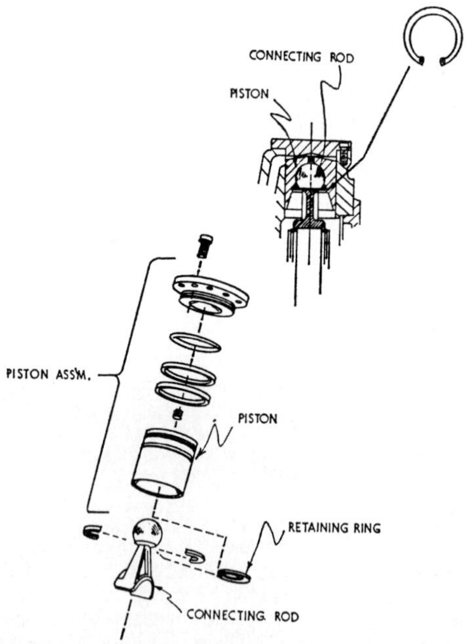

(f)

(**FIG. 14-35,** *continued*) internal retaining ring is a key part in the assembly of a connecting rod and piston for a hydraulic motor. The ring's lug holes make rapid assembly and disassembly possible when the proper pliers are used. The piston assembly in this case is slow moving and is not subject to heavy cycle loading. (*g*) Internal self-locking rings can act as a support carrier when the ID of a sleeve or housing cylinder is too large to center and stabilize small rods or conduit. The rings are adjustable in the entry direction only,

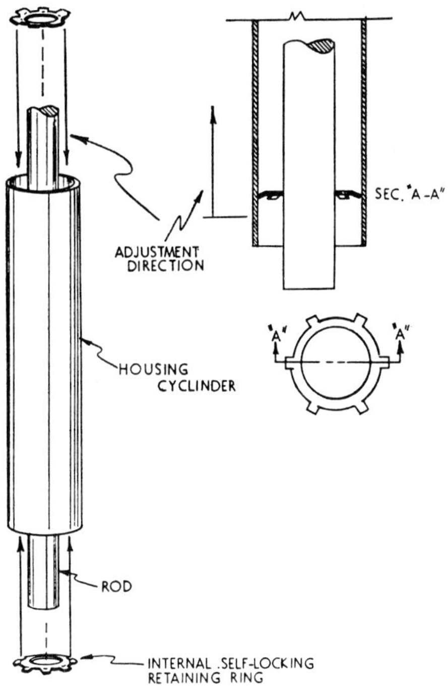

(g)

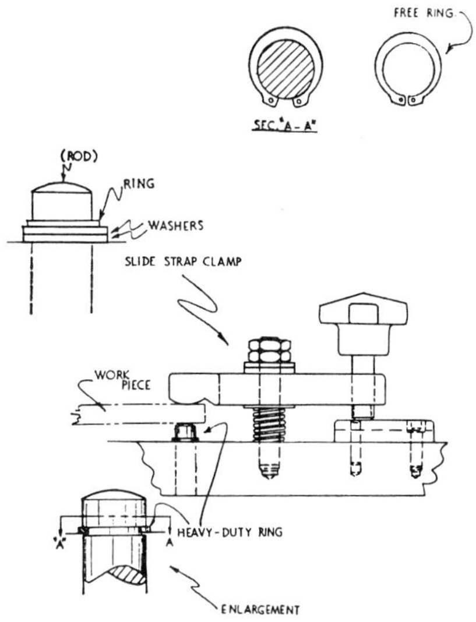

(h)

however, and a sufficient number should be used to secure the rod. (*h*) The heavy-duty external retaining ring shown here controls the elevation or position of a support post in a holding clamp. This type of ring is ideal for heavy-duty applications where extreme loading conditions are encountered. By adding washers under the ring the elevation of the support post can be adjusted as required. (*From R. O. Parmley, "The Versatile Retaining Ring," in Assembly Engineering, February 1968, vol. 2, no. 2. Reprinted by permission, Assembly Engineering.*)

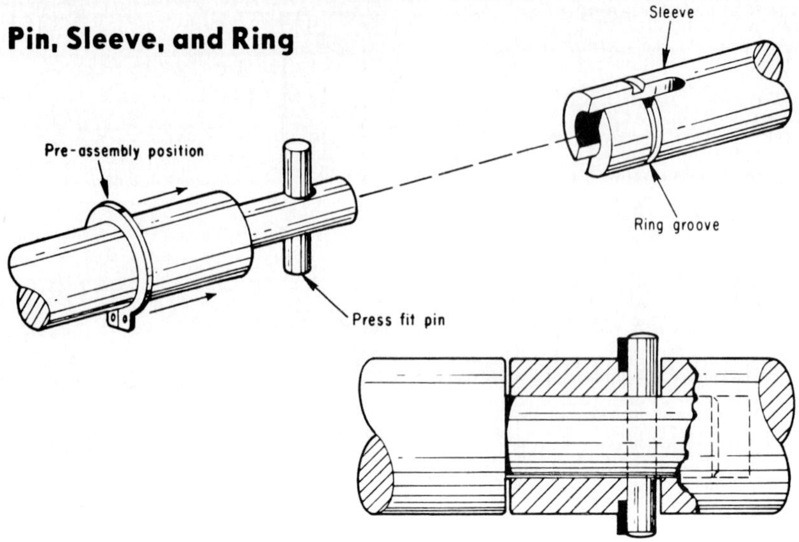

(a)

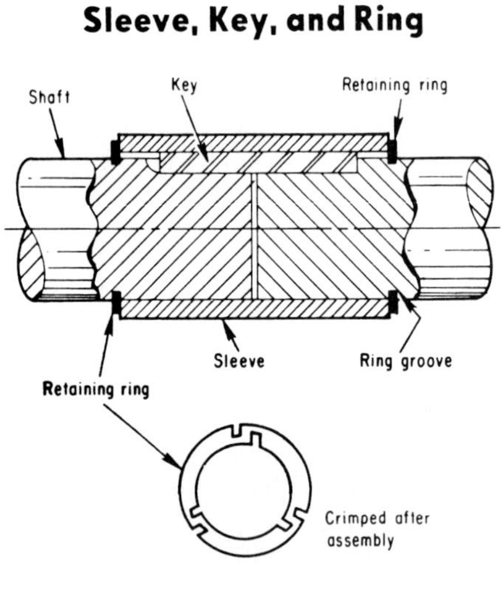

(b)

FIG. 14-36 Coupling shafts with retaining rings. These simple fasteners can provide an original way around certain design snags. For example, here are eight ways they are used to solve shaft-coupling problems. (*a*) This inexpensive connection is for light torques and moderate loads where accurate positioning is not required. A heavy-duty ring is used to resist high-impact and thrust loads. (*b*) Crimping the retaining ring into the groove produces a permanent, simple, and clean connection. This method is used to avoid machining shoulders in expensive materials, and to permit use of smaller-diameter shafts. When the ring is compressed into a V-shaped groove on the shaft, the notches permanently deform into small triangles, causing a reduction of the inner and outer diameters of the ring. Thus, the fastener tightly grips the groove, and provides a 360° shoulder around the shaft. Good torsional strength and high thrust-load capacity are provided by this connection. (*c*) A balanced two-

Two-Shaft Splice

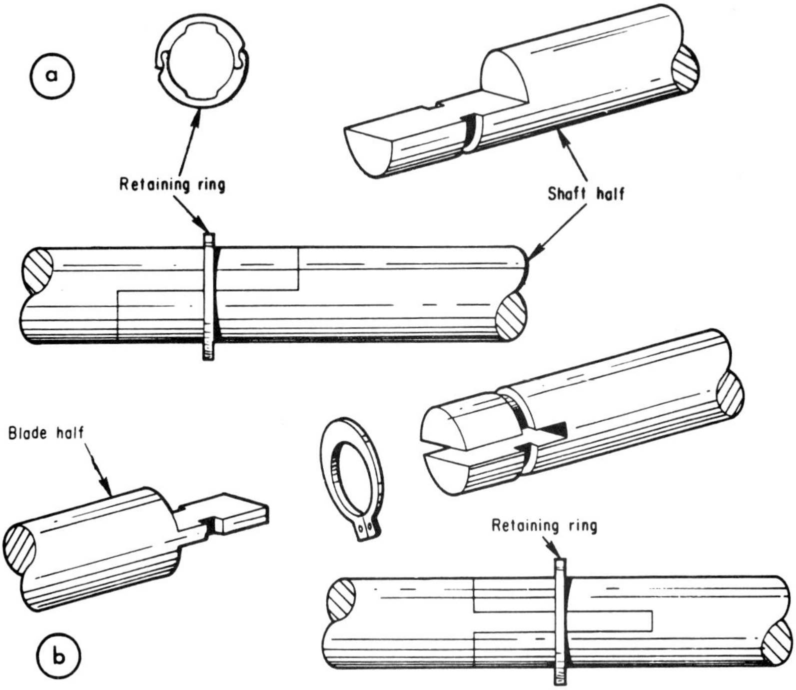

(c)

End-Flange Connection

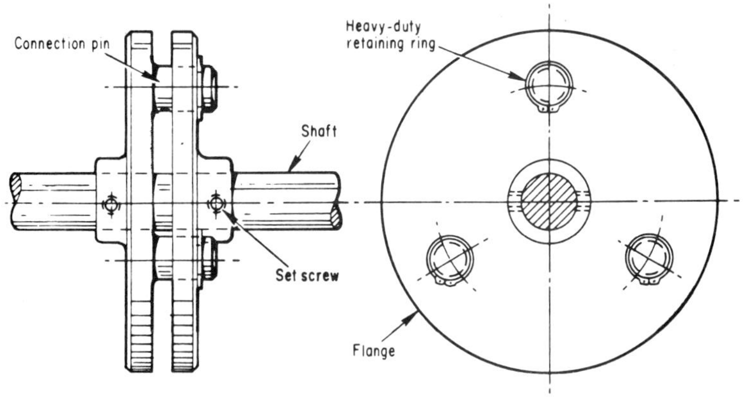

(d)

part ring provides an attractive appearance in addition to withstanding high rotational speeds and heavy thrust loads, 1. The one-piece ring, 2, secures the shafts in a high-torque capacity design. (*d*) This assembly for heavy-duty service requires minimum machining. Ring thickness should be substantial, and extra ring-section height is desirable. (*e*) For a connection that requires axial shaft adjustment, the self-locking ring requires no groove ⓐ An alternative solution ⓑ uses an inverted-lug ring seated in an internal groove. Extra ring-section height

Collar, Rings, and Threaded Shaft

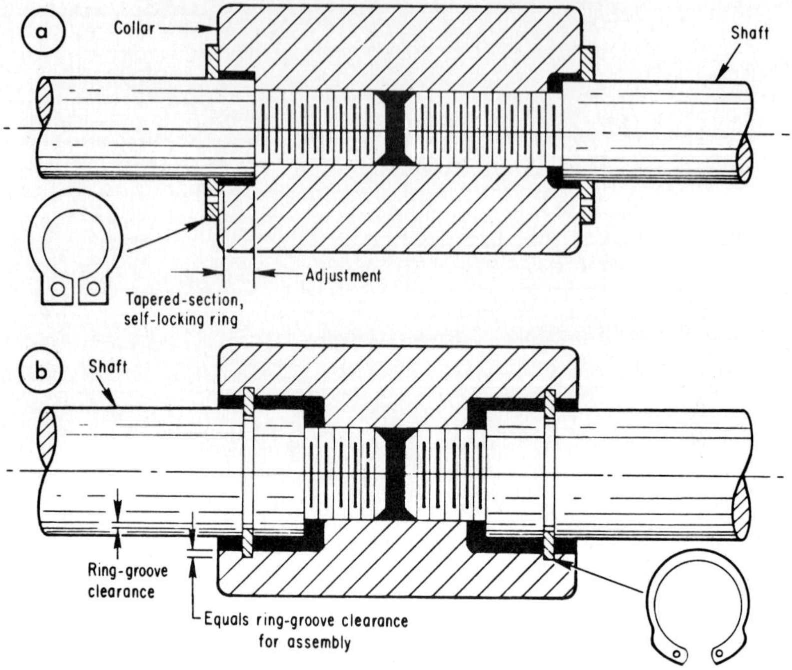

(e)

Coupler and Ring

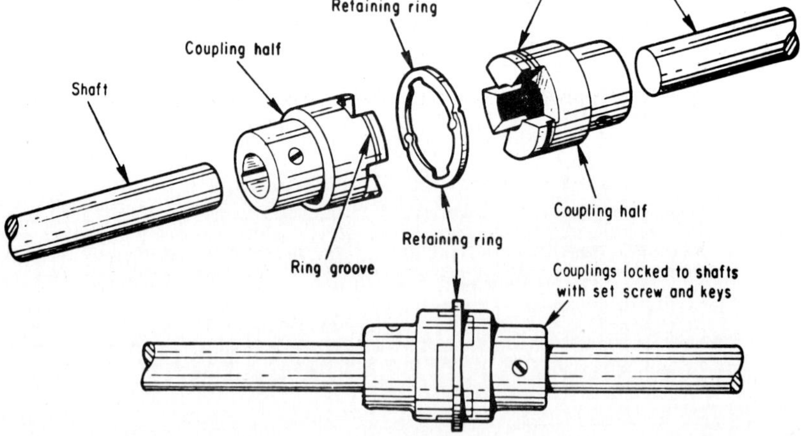

(f)

(**FIG. 14-36,** *continued*) provides a good shoulder. The ring is uniformly concentric with housing and shaft. (*f*) Where attractive appearance is desired in a dependable locking device, this connector and ring can be used. (*g*) A slotted sleeve with tapered threads connects shafts which cannot be machined. Prongs on the retaining ring provide positive shaft gripping to stop collar movement. The arched rim adds extra strength. (*h*) An alternative solution for coupling

Slotted Sleeve With Tapered Threads

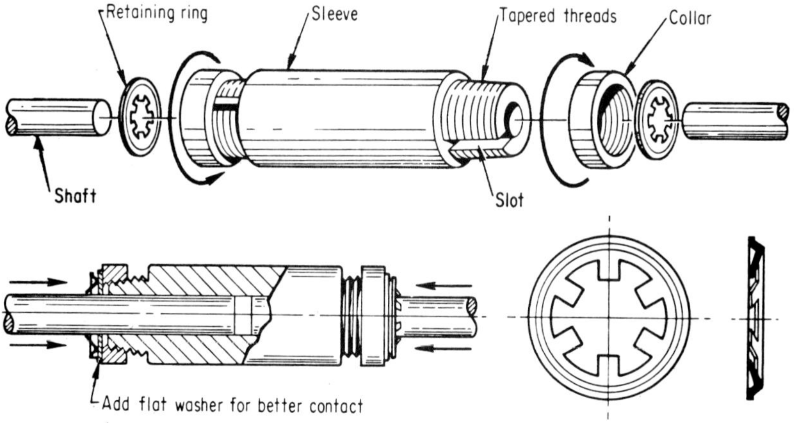

(g)

Bossed Coupling and Rings

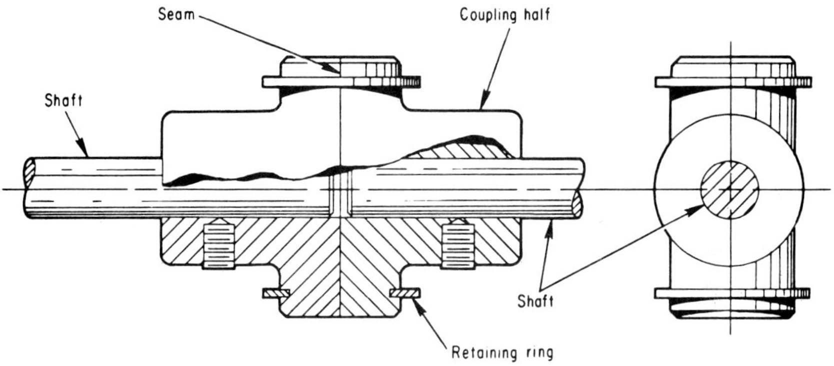

(h)

unmachined shafts uses bossed coupling halves with locking retaining rings. (*From R. O. Parmley, "Coupling Shafts With Retaining Rings," in Machine Design, Jan. 19, 1967. Reprinted by permission, Machine Design.*)

14-12 Splined Connections

Use of standard components with or without modification can result in economical splined connections. Ten examples are shown in Fig. 14-37.

CYLINDRICAL TYPES

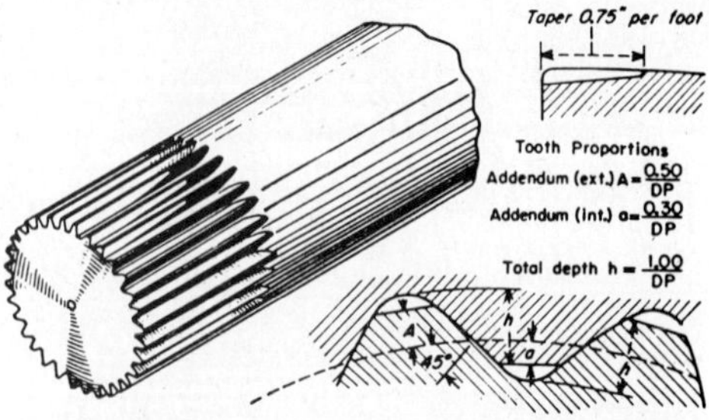

1 SQUARE SPLINES make a simple connection and are used mainly for applications of light loads, where accurate positioning is not important. This type is commonly used on machine tools; a cap screw is necessary to hold the enveloping member.

2 SERRATIONS of small size are used mostly for applications of light loads. Forcing this shaft into a hole of softer material makes an inexpensive connection. Originally straight-sided and limited to small pitches, 45 deg serrations have been standardized (SAE) with large pitches up to 10 in. dia. For tight fits, serrations are tapered.

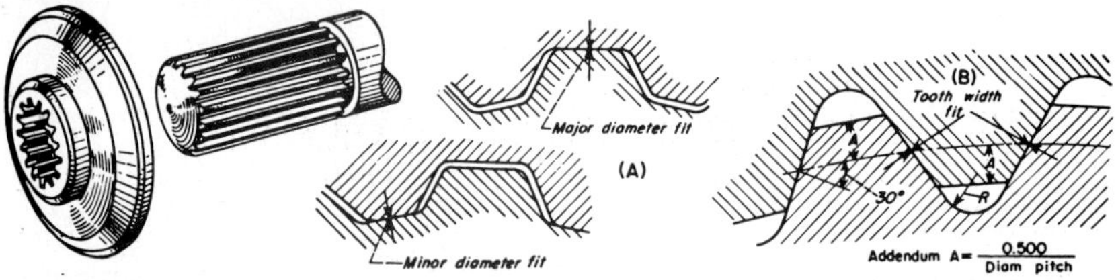

5 INVOLUTE-FORM splines are used where high loads are to be transmitted. Tooth proportions are based on a 30 deg stub tooth form. (A) Splined members may be positioned either by close fitting major or minor diameters. (B) Use of the tooth width or side positioning has the advantage of a full fillet radius at the roots. Splines may be parallel or helical. Contact stresses of 4,000 psi are used for accurate, hardened splines. Diametral pitch above is the ratio of teeth to the pitch diameter.

FACE TYPES

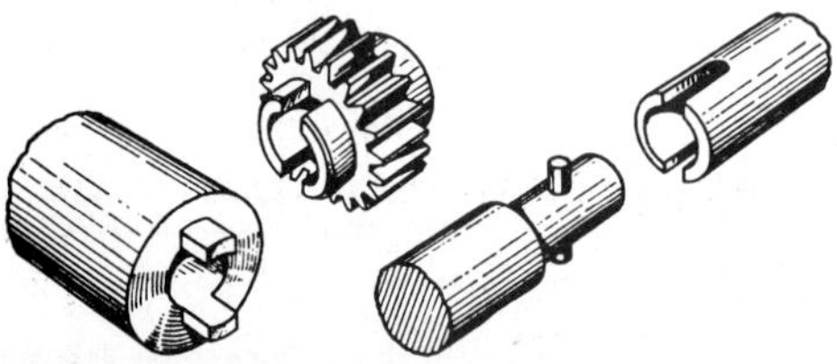

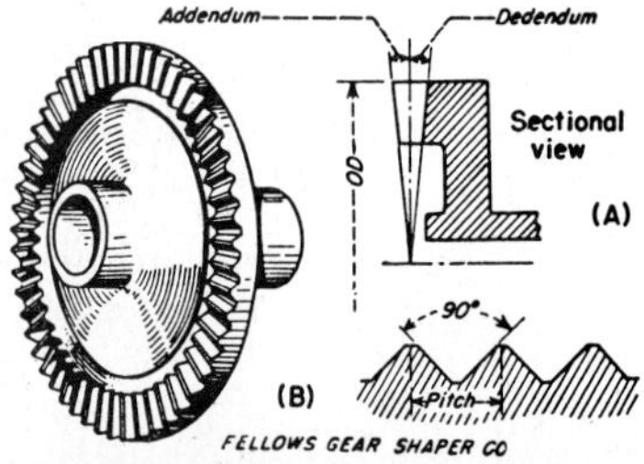

8 MILLED SLOTS in hubs or shafts make an inexpensive connection. This type is limited to moderate loads and requires a locking device to maintain positive engagement. Pin and sleeve method is used for light torques and where accurate positioning is not required.

9 RADIAL SERRATIONS by milling or shaping the teeth make a simple connection. (A) Tooth proportions decrease radially. (B) Teeth may be straight-sided (castellated) or inclined; a 90 deg angle is common.

FIG. 14-37 Ten types of splined connections. *(From W. W. Heath, "Ten Different Types of Splined Connections," in Product Engineering and Product Engineering Design Manual, edited by D. C. Greenwood, 1959. Reprinted by permission, Morgan-Grampian Publishing Co.)*

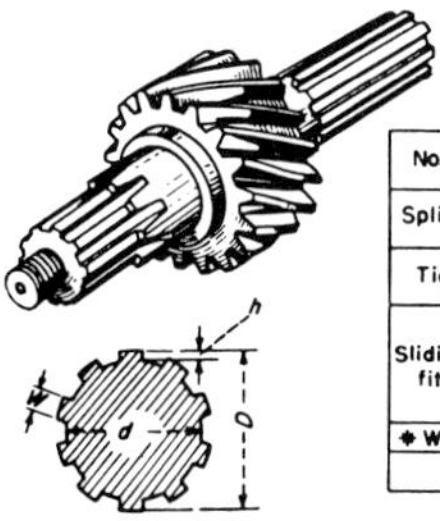

SAE STANDARD
SPLINE PROPORTIONS

No. of splines			4	6	10 or 16
Spline width		W	0.241 D	0.250 D	0.156 D*
Tight fit		h	0.075 D	0.050 D	0.045 D
Sliding fit	Unloaded	h	0.125 D	0.075 D	0.070 D
	loaded	h	——	0.100 D	0.095 D
* Width of 16-tooth spline is 0.098 D					
Root dia, d = D − 2h					

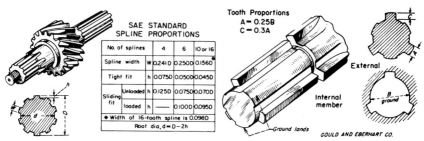

3 STRAIGHT-SIDED splines have been widely used in the automotive field. Such splines are often used for sliding members. The sharp corner at the root limits the torque capacity to pressures of approximately 1,000 psi on the spline projected area. For different applications, tooth height is altered as shown in the table above.

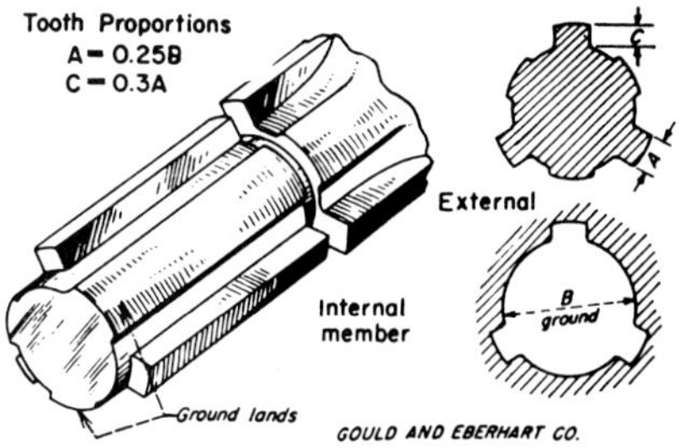

4 MACHINE-TOOL spline has a wide gap between splines to permit accurate cylindrical grinding of the lands—for precise positioning. Internal parts can be ground readily so that they will fit closely with the lands of the external member.

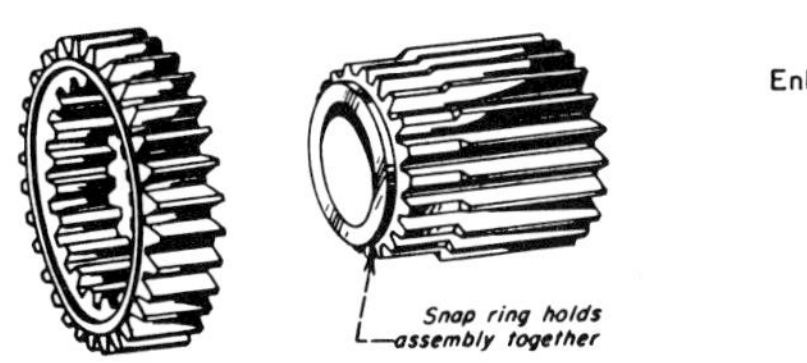

6 SPECIAL INVOLUTE splines are made by using gear tooth proportions. With full depth teeth, greater contact area is possible. A compound pinion is shown made by cropping the smaller pinion teeth and internally splining the larger pinion.

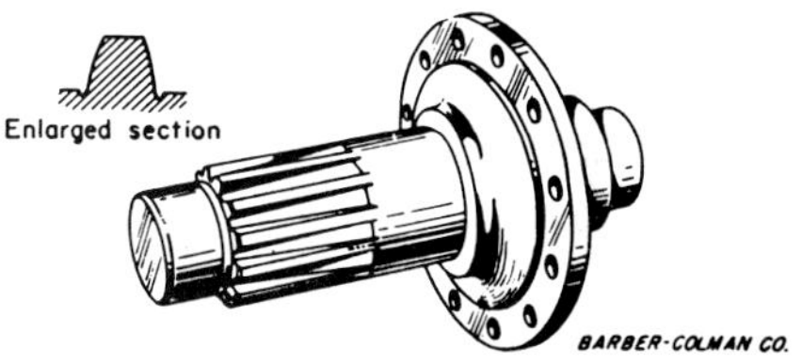

7 TAPER-ROOT splines are for drives which require positive positioning. This method holds mating parts securely. With a 30 deg involute stub tooth, this type is stronger than parallel root splines and can be hobbed with a range of tapers.

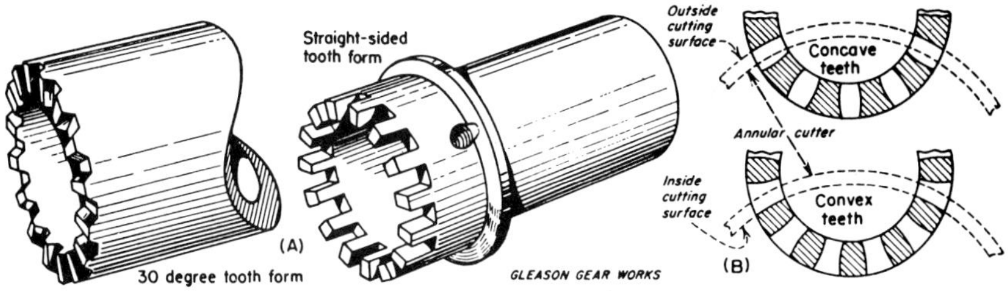

10 CURVIC COUPLING teeth are machined by a face-mill type of cutter. When hardened parts are used which require accurate positioning, the teeth can be ground. (A) This process produces teeth with uniform depth and can be cut at any pressure angle, although 30 deg is most common. **(B)** Due to the cutting action, the shape of the teeth will be concave (hour-glass) on one member and convex on the other—the member with which it will be assembled.

FIG. 14-37 *(continued)*

14-13 Modified Geneva Drives

The sketches shown in Fig. 14-38 were selected as practical examples of often used mechanisms. Most of them add a velocity component to the conventional Geneva motion.

Fig. 1—(Below) In the conventional external Geneva drive, a constant-velocity input produces an output consisting of a varying velocity period plus a dwell. In this modified Geneva, the motion period has a constant-velocity interval which can be varied within limits. When spring-loaded driving roller *a* enters the fixed cam *b*, the output-shaft velocity is zero. As the roller travels along the cam path, the output velocity rises to some constant value, which is less than the maximum output of an unmodified Geneva with the same number of slots; the duration of constant-velocity output is arbitrary within limits. When the roller leaves the cam, the output velocity is zero; then the output shaft dwells until the roller re-enters the cam. The spring produces a variable radial distance of the driving roller from the input shaft which accounts for the described motions. The locus of the roller's path during the constant-velocity output is based on the velocity-ratio desired.

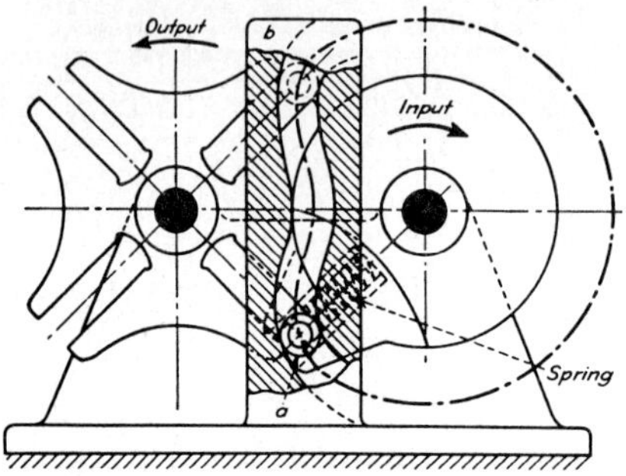

Saxonian Carton Machine Co., Dresden, Germany

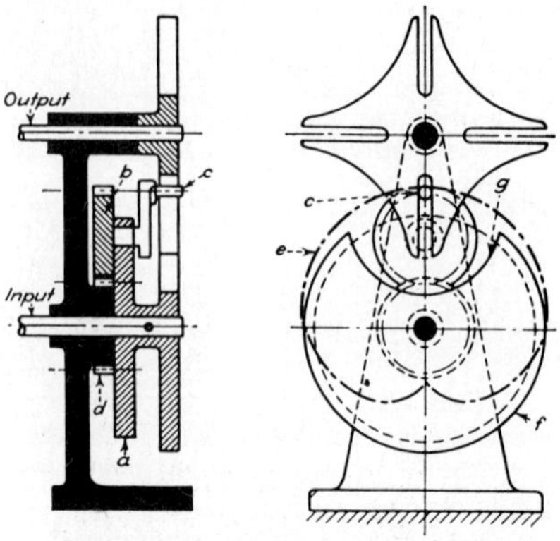

Fig. 2—(Above) This design incorporates a planet gear in the drive mechanism. The motion period of the output shaft is decreased and the maximum angular velocity is increased over that of an unmodified Geneva with the same number of slots. Crank wheel *a* drives the unit composed of plant gear *b* and driving roller *c*. The axis of the driving roller coincides with a point on the pitch circle of the planet gear; since the planet gear rolls around the fixed sun gear *d*, the axis of roller *c* describes a cardioid *e*. To prevent the roller from interfering with the locking disk *f*, the clearance arc *g* must be larger than required for unmodified Genevas.

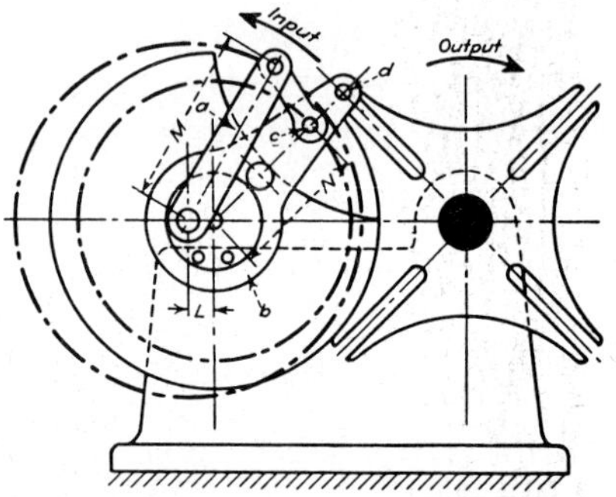

Fig. 3—A motion curve similar to that of Fig. 2 can be derived by driving a Geneva wheel by means of a two-crank linkage. Input crank *a* drives crank *b* through link *c*. The variable angular velocity of driving roller *d*, mounted on *b*, depends on the center distance *L*, and on the radii *M* and *N* of the crank arms. This velocity is about equivalent to what would be produced if the input shaft were driven by elliptical gears.

FIG. 14-38 Modified Geneva drives and special mechanisms. These sketches were selected as practical examples of uncommon, but often useful mechanisms. Most of them serve to add a varying velocity component to the conventional Geneva motion. The data were based in part on material and figures in AWF and VDMA Getriebeblaetter, published by Ausschuss für Getriebe beim Ausschuss für wirtschaftiche Fertigung, Leipzig, Germany. (*From Sigmund Rappaport, Product Engineering and Product Engineering Design Manual, edited by D. C. Greenwood, 1959. Reprinted by permission, Morgan-Grampian Publishing Co.*)

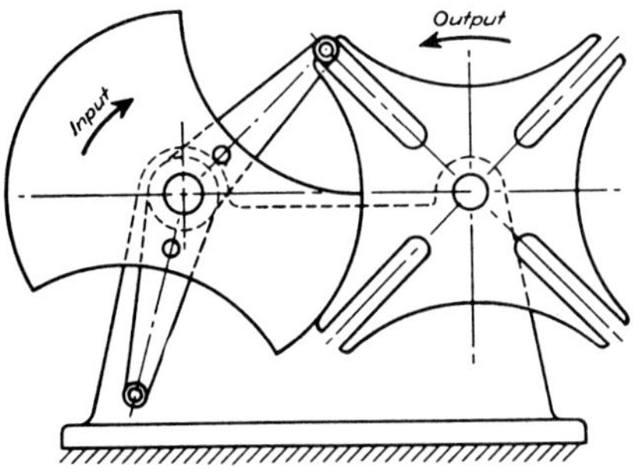

Fig. 4—(Left) The duration of the dwell periods is changed by arranging the driving rollers unsymmetrically around the input shaft. This does not affect the duration of the motion periods. If unequal motion periods are desired as well as unequal dwell periods, then the roller crank-arms must be unequal in length and the star must be suitably modified; such a mechanism is called an "irregular Geneva drive."

Fig. 5—(Below) In this intermittent drive, the two rollers drive the output shaft as well as lock it during dwell periods. For each revolution of the input shaft the output shaft has two motion periods. The output displacement ϕ is determined by the number of teeth; the driving angle, ψ, may be chosen within limits. Gear *a* is driven intermittently by two driving rollers mounted on input wheel *b*, which is bearing-mounted on frame *c*. During the dwell period the rollers circle around the top of a tooth. During the motion period, a roller's path *d* relative to the driven gear is a straight line inclined towards the output shaft. The tooth profile is a curve parallel to path *d*. The top land of a tooth becomes the arc of a circle of radius *R*, the arc approximating part of the path of a roller.

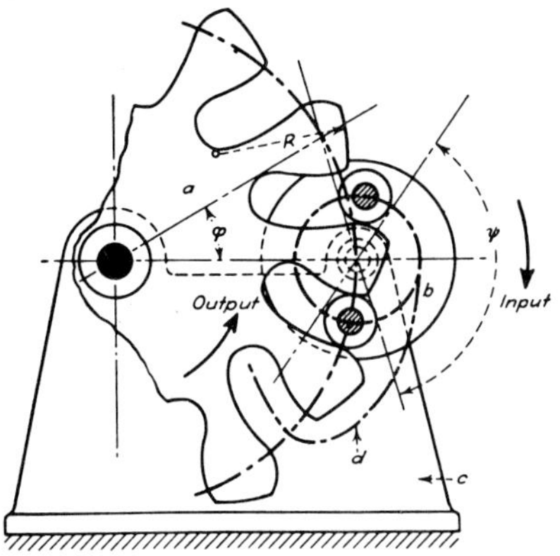

FIG. 14-38 *(continued)*

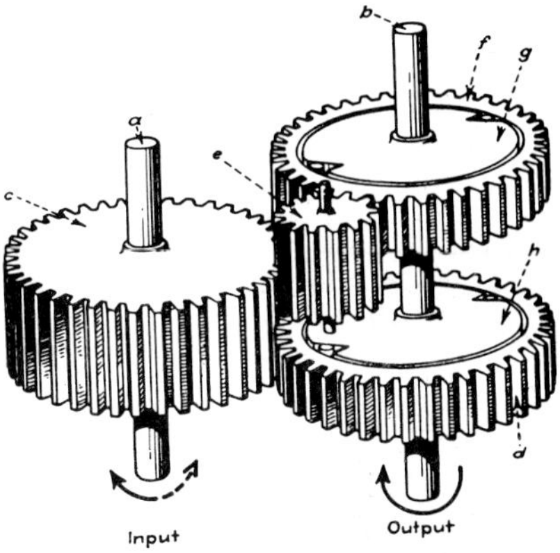

Fig. 6—This uni-directional drive was developed by the author and to his knowledge is novel. The output shaft rotates in the same direction at all times, without regard to the direction of the rotation of the input shaft; angular velocity of the output shaft is directly proportional to the angular velocity of the input shaft. Input shaft *a* carries spur gear *c*, which has approximately twice the face width of spur gears *f* and *d* mounted on output shaft *b*. Spur gear *c* meshes with idler *e* and with spur gear *d*. Idler *e* meshes with spur gears *c* and *f*. The output shaft *b* carries two free-wheel disks *g* and *h*, which are oriented uni-directionally.

When the input shaft rotates clockwise (bold arrow), spur gear *d* rotates counter-clockwise and idles around free-wheel disk *h*. At the same time idler *e*, which is also rotating counter-clockwise, causes spur gear *f* to turn clockwise and engage the rollers on free-wheel disk *g*; thus, shaft *b* is made to rotate clockwise. On the other hand, if the input shaft turns counter-clockwise (dotted arrow), then spur gear *f* will idle while spur gear *d* engages free-wheel disk *h*, again causing shaft *b* to rotate clockwise.

INDEX

ABOUT THE EDITOR IN CHIEF

Robert O. Parmley, P.E., CMfgE, is President and Consulting Engineer of Morgan & Parmley, Ltd., Professional Consulting Engineers, Ladysmith, Wisconsin. Mr. Parmley is also a member of the National Society of Professional Engineers, the American Society of Mechanical Engineers, the Constructions Specifications Institute, the American Institute for Design & Drafting, the Society of Manufacturing Engineers, and is listed in the AAES *Who's Who in Engineering.* He is a registered professional engineer in Wisconsin, California, and Canada and a certified manufacturing engineer under SME's national certification program. In a career covering more than 25 years, Mr. Parmley has worked on the design and construction supervision of a wide variety of structures, systems, and machines—from dams and bridges to pollution control systems and municipal projects. The author of over 40 technical articles, he is also the Editor in Chief of the *Standard Handbook of Fastening and Joining* and the *Field Engineer's Manual,* both published by McGraw-Hill.